CYLINDRIC ALGEBRAS

TO THE MEMORY OF

ANATOLIĬ IVANOVIČ MAL′CEV

WHO TAUGHT US THE WAY

FROM LOGIC TO ALGEBRA

STUDIES IN LOGIC

AND

THE FOUNDATIONS OF MATHEMATICS

VOLUME 64

Editors

A. HEYTING, *Amsterdam*
H. J. KEISLER, *Madison*
A. MOSTOWSKI, *Warszawa*
A. ROBINSON, *New Haven*
P. SUPPES, *Stanford*

Advisory Editorial Board

Y. BAR-HILLEL, *Jerusalem*
K. L. DE BOUVÈRE, *Santa Clara*
H. HERMES, *Freiburg i. Br.*
J. HINTIKKA, *Helsinki*
J. C. SHEPHERDSON, *Bristol*
E. P. SPECKER, *Zürich*

NORTH-HOLLAND PUBLISHING COMPANY
AMSTERDAM · LONDON

CYLINDRIC ALGEBRAS

PART I

WITH AN INTRODUCTORY CHAPTER:

GENERAL THEORY OF ALGEBRAS

LEON HENKIN

Professor of Mathematics, University of California, Berkeley

J. DONALD MONK

Professor of Mathematics, University of Colorado

ALFRED TARSKI

Professor of Mathematics, University of California, Berkeley

1971

NORTH-HOLLAND PUBLISHING COMPANY

AMSTERDAM · LONDON

Library of Congress Catalog Card Number 77–126903

ISBN: 0 7204 2043 1

Publishers:

NORTH-HOLLAND PUBLISHING COMPANY - AMSTERDAM

NORTH-HOLLAND PUBLISHING COMPANY, LTD. - LONDON

Transferred to digital printing 2005

CONTENTS

FOREWORD

In the middle of the nineteenth century George Boole initiated the investigation of a class of algebraic structures which were subsequently called Boolean algebras. The theory of these algebras is directly related to the development of the most elementary part of mathematical logic — sentential calculus. As is well known, however, the theory can be developed in a purely algebraic fashion, it has at present numerous connections with several branches of mathematics — set theory, topology, and analysis — and hence it can be understood and appreciated by mathematicians unfamiliar with the logical problems to which it owes its birth.

The work of Boole was the starting point for a continuous flow of inquiries into the algebraization of logic which, through various intermediate stages, led to (but did not end with) the foundation of the theory of cylindric algebras. This theory, which is the main subject matter of our work, was originally designed to provide an apparatus for an algebraic study of first-order predicate logic, a portion of mathematical logic which plays a central role in contemporary logical research. Indeed, the theory does for predicate logic what has been done for sentential calculus by the theory of Boolean algebras. In its subsequent development, however, this theory, just as that of Boolean algebras, has found interesting realizations and applications outside of logic. Also, the theory can be presented, and will be presented here, in a manner primarily algebraic in character, which will make the main bulk of this work fully accessible to mathematicians who do not have a detailed knowledge of logical concepts and methods.

Our first task is to give the reader some intuitive appreciation as to what cylindric algebras are. To this end we begin with consideration of special algebraic structures referred to as cylindric set algebras. These structures are defined in terms of general set theory, and their construction generalizes the familiar construction of Boolean set algebras.

In describing cylindric set algebras it is convenient to use a terminology borrowed from analytic geometry. A cylindric set algebra may be regarded as a kind of multi-dimensional Boolean set algebra. Indeed, associated with each cylindric set algebra is an ordinal number α, finite or transfinite, which indicates the dimensionality of the algebra. Whereas the elements of a Boolean set algebra may be subsets of a quite arbitrary set V, the elements of a cylindric set algebra of dimension α are subsets of the Cartesian power ${}^{\alpha}U$ of some set U. The set ${}^{\alpha}U$ is called the *α-dimensional Cartesian space with the base U*. Its elements, i.e., the sequences $x = \langle x_0, x_1, \ldots, x_\kappa, \ldots\rangle$ of length α in which all terms belong to U, are referred to as *points* with the *coordinates* $x_0, x_1, \ldots, x_\kappa, \ldots$. In a Boolean set algebra the empty set 0 and the unit set V are treated as distinguished elements. In a cylindric set algebra we treat as distinguished elements, not only the empty set 0 and the whole space ${}^{\alpha}U$, but also certain additional point sets $\mathsf{D}_{\kappa\lambda}$ (for any $\kappa, \lambda < \alpha$), called the *diagonal sets*. The set $\mathsf{D}_{\kappa\lambda}$ consists of all those points x of ${}^{\alpha}U$ whose κth coordinate x_κ equals the λth coordinate x_λ. Thus, in case $\kappa \neq \lambda$, the diagonal sets $\mathsf{D}_{\kappa\lambda}$ are seen to be the hyperplanes defined by the equations $x_\kappa = x_\lambda$; if, in particular, $\alpha = 2$, then D_{01} is the main diagonal line of the coordinate system.

In a cylindric set algebra, just as in a Boolean set algebra, we are concerned with the set-theoretical operations of forming unions $X \cup Y$, intersections $X \cap Y$, and complements $\sim X$ of members of the algebra. In the case of a Boolean set algebra these are the only fundamental operations. In a cylindric set algebra, however, we single out certain further fundamental operations called cylindrifications. In fact, with each ordinal $\kappa < \alpha$ we correlate the unary operation C_κ, the *κth cylindrification*, which, when applied to a subset X of the space ${}^{\alpha}U$, produces the cylinder $\mathsf{C}_\kappa X$ swept out by all translations of X parallel to the κth coordinate axis; in other words, a point y of ${}^{\alpha}U$ belongs to the cylinder $\mathsf{C}_\kappa X$ if and only if it can be obtained from some point x of X by changing at most the κth coordinate. Thus, e.g., we have for any $\kappa, \lambda, \mu < \alpha$:

$$\mathsf{C}_\kappa 0 = 0, \quad \mathsf{C}_\kappa {}^{\alpha}U = {}^{\alpha}U;$$

$$\mathsf{C}_\kappa \mathsf{D}_{\kappa\lambda} = \mathsf{C}_\lambda \mathsf{D}_{\kappa\lambda} = {}^{\alpha}U, \quad \mathsf{C}_\mu \mathsf{D}_{\kappa\lambda} = \mathsf{D}_{\kappa\lambda} \text{ in case } \mu \neq \kappa, \lambda.$$

Now let S be any family of subsets of ${}^{\alpha}U$ which contains as elements all the distinguished sets — 0, ${}^{\alpha}U$, and $\mathsf{D}_{\kappa\lambda}$ for $\kappa, \lambda < \alpha$ — and which is closed under all the fundamental operations — $\cup$, $\cap$, $\sim$, and C_κ, for $\kappa < \alpha$. The structure

$$\mathfrak{S} = \langle S, \cup, \cap, \sim, 0, {}^{\alpha}U, \mathsf{C}_\kappa, \mathsf{D}_{\kappa\lambda}\rangle_{\kappa,\lambda<\alpha}$$

is then called a *cylindric set algebra of dimension* α; S is, of course, the universe (the set of all elements) of this algebra.

Many basic properties of cylindric set algebras can be derived directly from the definition. We know, in particular, a great variety of laws which have the form of algebraic equations identically satisfied in all cylindric set algebras. Since, for any given cylindric set algebra

$$\mathfrak{S} = \langle S, \cup, \cap, \sim, 0, {}^{\alpha}U, \mathsf{C}_{\kappa}, \mathsf{D}_{\kappa\lambda} \rangle_{\kappa,\lambda<\alpha},$$

the structure

$$\mathfrak{Bl}\,\mathfrak{S} = \langle S, \cup, \cap, \sim, 0, {}^{\alpha}U \rangle$$

is a Boolean set algebra, called the *Boolean part* of $\mathfrak{S}$, all the familiar Boolean laws are seen to hold in $\mathfrak{S}$. Besides these, there are numerous laws involving the cylindrifications and diagonal sets which hold in every cylindric set algebra. Some such laws will be formulated below in this Foreword, and many other examples will be found in Chapter 1.

The general notion of a Boolean algebra is obtained from that of a Boolean set algebra by a process of abstraction: we consider the algebraic identities which hold in all Boolean set algebras and we select some of them as postulates for the general, abstract theory of Boolean algebras. An analogous procedure can be applied to arrive at the general notion of a cylindric algebra. Changing the notation which we have introduced for cylindric set algebras, we now consider algebraic structures

$$\mathfrak{A} = \langle A, +, \cdot, -, 0, 1, \mathsf{c}_{\kappa}, \mathsf{d}_{\kappa\lambda} \rangle_{\kappa,\lambda<\alpha}$$

where A is an arbitrary set, $+$ and $\cdot$ are binary operations on A, $-$ and c_{κ} (for each $\kappa < \alpha$) are unary operations on A, while 0, 1, and $\mathsf{d}_{\kappa\lambda}$ (for any given $\kappa, \lambda < \alpha$) are elements of A. Here $+$, $\cdot$, and $-$ are called *Boolean addition*, *Boolean multiplication*, and *complementation*; 0 and 1 are the *Boolean zero* and *unit elements*; for c_{κ} and $\mathsf{d}_{\kappa\lambda}$ we preserve the terminology adopted in discussing cylindric set algebras, i.e., c_{κ} is called the *κth cylindrification* and $\mathsf{d}_{\kappa\lambda}$ the *κ, λ-diagonal element*. Next, we select certain equations which are identically satisfied in all cylindric set algebras of a given dimension α, and we take them as postulates for the general theory of cylindric algebras; in other words, we declare cylindric algebras of dimension α to be just those structures in which all the selected postulates hold. The postulates are divided into two groups. Those of the first group involve only Boolean notions, and jointly they characterize the structure

$$\mathfrak{Bl}\,\mathfrak{A} = \langle A, +, \cdot, -, 0, 1 \rangle$$

as a Boolean algebra; systems of equations which are adequate for this purpose can be found in the literature. The postulates of the second group involve, in addition, cylindrifications and diagonal elements, and are thus specific for cylindric algebras. The following simple equations, assumed to be identically satisfied by arbitrary elements x, y of A and arbitrary ordinals κ, λ, μ less than α, are used as the postulates of the second group:

(C_1) $c_\kappa 0 = 0$.

(C_2) $x \cdot c_\kappa x = x$.

(C_3) $c_\kappa(x \cdot c_\kappa y) = c_\kappa x \cdot c_\kappa y$.

(C_4) $c_\kappa c_\lambda x = c_\lambda c_\kappa x$.

(C_5) $c_\kappa d_{\kappa\lambda} = 1$.

(C_6) $c_\mu(d_{\kappa\mu} \cdot d_{\mu\lambda}) = d_{\kappa\lambda}$ provided $\mu \neq \kappa, \lambda$.

(C_7) $c_\kappa(d_{\kappa\lambda} \cdot x) \cdot c_\kappa(d_{\kappa\lambda} \cdot -x) = 0$ provided $\kappa \neq \lambda$.

As an example of an equation which is implied by these postulates and hence is identically satisfied in all cylindric algebras we mention the distributive law for cylindrifications over Boolean addition,

$$c_\kappa(x+y) = c_\kappa x + c_\kappa y.$$

From the above we see that cylindric algebras are Boolean algebras enriched by new distinguished elements and new fundamental operations which are distributive over Boolean addition; they form, therefore, a subclass of a comprehensive class of algebraic structures known in the literature as Boolean algebras with operators.

We have thus described the construction of cylindric set algebras and have introduced the general notion of a cylindric algebra. Several important points, however, remain obscure in this presentation; they concern primarily our motivation in the whole procedure. The reader may wonder why, when constructing algebras of subsets of a Cartesian space, we have selected just the cylindrifications and the diagonal sets as additional notions to supplement the familiar fundamental notions of Boolean set algebras. He may also ask what our motives are in singling out just (C_1)–(C_7), from among the infinitely many identities holding in cylindric set algebras, as the only postulates which, together with Boolean postulates, characterize the general notion of a cylindric algebra. It seems impossible to find a satisfactory clarification of these points within the conceptual framework of cylindric set algebras themselves. We have to recall that the theory of cylindric algebras is rooted in logic and was originally developed as an instrument for the algebraization of predicate logic. Thus, to explain the selection of fundamental notions and postulates for this theory, we have to examine its logical origin more closely. In the first place, we have to discuss another special class of cylindric algebras; the class is described in metalogical terms and is directly related to the main purpose for which the theory of cylindric algebras has been designed.

We consider a system of predicate logic (with identity). The formal language in which this system is developed will be referred to as the language Λ. The vocabulary of Λ (i.e., the set of all symbols occurring in expressions of Λ) consists, as usual, of individual variables, logical constants, and non-logical constants. There are infinitely many different variables in Λ; they are arranged in a simple infinite sequence $\langle \mathbf{v}_0, \ldots, \mathbf{v}_\kappa, \ldots \rangle_{\kappa<\omega}$. The logical constants are: the disjunction symbol $\mathbf{v}$, the conjunction symbol $\boldsymbol{\wedge}$, the negation symbol $\neg$, the existential quantifier $\exists$, the identity symbol $=$, the truth symbol $\boldsymbol{T}$, and the falsehood symbol $\boldsymbol{F}$. The non-logical constants are various predicates, again arranged in a sequence $\langle \boldsymbol{P}_0, \ldots, \boldsymbol{P}_\xi, \ldots \rangle_{\xi<\beta}$ where β is an arbitrary ordinal, finite or infinite. Each predicate has a definite finite rank $\rho = 0, 1, 2, \ldots$; we refer to a predicate with rank ρ as a ρ-place predicate.

The expressions of the language Λ are arbitrary finite sequences of symbols; single symbols are identified with one-termed sequences and hence are also regarded as expressions. An expression formed by a ρ-place predicate $\boldsymbol{P}_\xi$ followed by ρ variables is called an *atomic formula*. In this connection the truth symbol $\boldsymbol{T}$, the falsehood symbol $\boldsymbol{F}$, and the identity symbol $=$ are also treated as predicates (although they are not included in the sequence $\langle \boldsymbol{P}_0, \ldots, \boldsymbol{P}_\xi, \ldots \rangle_{\xi<\beta}$) — in fact, $\boldsymbol{T}$ and $\boldsymbol{F}$ as zero-place predicates and $=$ as a two-place predicate. Thus, $\boldsymbol{T}$ and $\boldsymbol{F}$ by themselves (i.e., without any following variables) are atomic formulas. On the other hand, the sequence of three symbols $=$, $\mathbf{v}_\kappa$, and $\mathbf{v}_\lambda$ is also an atomic formula; we denote it by $\mathbf{v}_\kappa = \mathbf{v}_\lambda$ (thus inverting the order in which the symbols $=$ and $\mathbf{v}_\kappa$ occur in the sequence) and call it an *equation* or, specifically, the *κ,λ-equation.*

With each of the logical constants $\mathbf{v}$, $\boldsymbol{\wedge}$, $\neg$, and $\exists$ there is correlated a certain operation on expressions. Thus, from two given expressions φ and ψ we form their disjunction $\varphi \mathbin{\mathbf{v}} \psi$, or conjunction $\varphi \mathbin{\boldsymbol{\wedge}} \psi$, by combining them by means of the disjunction symbol $\mathbf{v}$, or the conjunction symbol $\boldsymbol{\wedge}$. Notice that "$\mathbf{v}$" and "$\boldsymbol{\wedge}$" (which served originally as metalogical designations of certain symbols in Λ) are now used in a new sense — as designations of certain binary operations on expressions of Λ. Similarly, we obtain the negation $\neg\varphi$, or the κth existential quantification $\exists_{\mathbf{v}_\kappa}\varphi$, of a given expression φ, by prefixing to φ the negation symbol $\neg$, or the expression formed by the existential quantifier $\exists$ and the immediately following variable $\mathbf{v}_\kappa$. Thus, $\neg$ and $\exists_{\mathbf{v}_\kappa}$ (for each natural number κ) are viewed as unary operations on expressions.

A system of predicate logic may be provided with some further logical constants, such as the implication symbol, the biconditional symbol, and the universal quantifier. With these new logical constants are correlated new operations on expressions — formation of implications $\rightarrow$, formation of biconditionals $\leftrightarrow$, and the κth universal quantification $\forall_{v_\kappa}$. We prefer here to obtain these new operations by defining them in terms of old operations; the definitions can be given the following form:

$$\varphi \rightarrow \psi = (\neg\varphi) \vee \psi,$$

$$\varphi \leftrightarrow \psi = (\varphi \rightarrow \psi) \wedge (\psi \rightarrow \varphi),$$

$$\forall_{v_\kappa}\varphi = \neg\exists_{v_\kappa}\neg\varphi.$$

An expression obtained from atomic formulas by applying the operations $\vee$, $\wedge$, $\neg$, and $\exists_{v_\kappa}$ an arbitrary number of times is called a *formula*. We assume known the conditions under which a variable v_κ is said to occur free in a formula φ. Formulas in which no variables occur free are referred to as *sentences*. For instance, T, F, $\exists_{v_0}v_0 = v_0$ and $\neg\exists_{v_0}v_0 = v_0$ are sentences.

The set Φ of all formulas is closed under the operations $\vee$, $\wedge$, $\neg$, and $\exists_{v_\kappa}$ and contains as elements the expressions F, T, and $v_\kappa = v_\lambda$ (for any $\kappa, \lambda < \omega$). Hence the structure

$$\mathfrak{F} = \langle \Phi, \vee, \wedge, \neg, F, T, \exists_{v_\kappa}, v_\kappa = v_\lambda \rangle_{\kappa,\lambda<\omega}$$

is an algebra (in a wide sense of the word). We refer to $\mathfrak{F}$ as the *free algebra of formulas* (in the language Λ). Indeed, $\mathfrak{F}$ is what is called an absolutely free algebra in the general theory of algebraic structures. As a consequence, no algebraic equation formulated in terms of fundamental operations and distinguished elements of $\mathfrak{F}$ is identically satisfied in this algebra unless it is a pure tautology of the form "$x = x$" — so that, for example, the operations $\vee$ and $\wedge$ are neither commutative nor associative. Thus $\mathfrak{F}$ presents but little interest from an algebraic point of view. It should also be emphasized that $\mathfrak{F}$ in itself is by no means a proper tool for an algebraic discussion of predicate logic. This appears clear if only from the fact that the notion of consequence, which is a most basic element in the development of any logical system, has not been involved at all in the construction of $\mathfrak{F}$.

The situation will change essentially when we introduce the notion of consequence and use it to construct from $\mathfrak{F}$ some derivative algebras. To this end we have first to explain how the formal language Λ can be interpreted, i.e., how definite meanings can be ascribed to symbols, formulas, and sentence in Λ. As a base for an interpretation we choose an arbitrary relational structur

$\mathfrak{R} = \langle A, R_0, ..., R_\xi, ... \rangle_{\xi<\beta}$ where A is an arbitrary non-empty set and $R_0, ..., R_\xi, ...$ are finitary relations among elements of A.

The sequence of fundamental relations R_ξ in $\mathfrak{R}$ has the same length as the sequence of predicates $\boldsymbol{P}_\xi$ in Λ; each relation R_ξ is assumed to be of the same rank as the corresponding predicate $\boldsymbol{P}_\xi$, so that, e.g., R_ξ is a binary relation if $\boldsymbol{P}_\xi$ is a two-place predicate. We interpret Λ in $\mathfrak{R}$ by stipulating that variables $\boldsymbol{v}_0, \boldsymbol{v}_1, ...$ represent arbitrary elements of A (in other words, that any elements $a_0, a_1, ...$ of A can be assigned as values to $\boldsymbol{v}_0, \boldsymbol{v}_1, ...$) and that the predicates $\boldsymbol{P}_0, \boldsymbol{P}_1, ...$ denote the corresponding relations $R_0, R_1, ...$. Furthermore, we stipulate the logical constants of Λ to be synonymous with certain familiar words and phrases of common language, e.g., the disjunction symbol $\mathbf{v}$ with the word "or" and the existential quantifier $\exists$ with the phrase "there is a ... such that". To the truth symbol $\boldsymbol{T}$ we ascribe the meaning of some particular sentence in common language which is obviously (logically) true, say the sentence "for every x, x is identical with x"; the falsehood symbol $\boldsymbol{F}$ is assumed to mean the same as the negation of this sentence. Thus the meaning assigned to logical constants is independent of any special properties of the structure $\mathfrak{R}$.

The interpretation of Λ in $\mathfrak{R}$ extends in an obvious way from symbols to formulas and sentences. An essential semantical difference between sentences and those formulas which are not sentences comes to light now. Under our interpretation, every sentence σ acquires the meaning of a definite statement (in the common language) concerning $\mathfrak{R}$; hence we can meaningfully ask the question whether σ *holds* (is true) or *fails* (is false) in $\mathfrak{R}$. If, however, φ is a formula with free variables, we can only ask whether or not, under a given assignment of elements of A to variables of Λ, the elements assigned to the free variables of φ satisfy φ in $\mathfrak{R}$. For instance, given a predicate $\boldsymbol{P}_\xi$ of rank ρ and elements $a_1, ..., a_\rho$ assigned to variables $\boldsymbol{v}_1, ..., \boldsymbol{v}_\rho$, we can ask whether or not $a_1, ..., a_\rho$ satisfy the atomic formula $\boldsymbol{P}_\xi \boldsymbol{v}_1 ... \boldsymbol{v}_\rho$; this amounts to asking whether or not the relation R_ξ holds among the elements $a_1, ..., a_\rho$. It may happen that a formula φ is satisfied in $\mathfrak{R}$ under every assignment of elements to its free variables. In this case we say that φ is universally or identically satisfied in $\mathfrak{R}$; by extending to such formulas the terminology suggested for sentences, we can simply say that φ holds in $\mathfrak{R}$.

Now let Σ be any set of sentences of Λ. By a *model* of Σ we understand any relational structure $\mathfrak{R} = \langle A, R_0, ..., R_\xi, ... \rangle_{\xi<\beta}$ in which all sentences of Σ hold. A sentence — or, more generally, a formula — is said to be a *consequence* of Σ, or to be *implied* by Σ, if it holds in every model of Σ. The set Θ of all sentences implied by some set Σ is called a *theory*. It is sometimes convenient to think of Σ as an axiom system for Θ and refer to Θ as the theory axiomatized or generated by Σ; sentences belonging to Θ are called, in this context, theorems of Θ. If, in particular, we take the empty set for Σ, then Θ coincides with the set of all logically valid sentences, i.e., sentences holding in every relational structure. Given any relational structure $\mathfrak{R}$, or any family F of such structures, the set of all sentences which hold in $\mathfrak{R}$, or in every structure belonging to F, is always a theory; it is referred to as the (first-order) theory of $\mathfrak{R}$, or of F, formalized in the language Λ. Conversely, every theory Θ is the theory of some (possibly empty) family of structures, for instance, of the family of all models of Θ.

Two arbitrary formulas φ and ψ are called equivalent under a set of sentences Σ if the biconditional $\varphi \leftrightarrow \psi$ is a consequence of Σ; we write, symbolically, $\varphi \equiv_\Sigma \psi$. It is easily seen that, for each set Σ, $\equiv_\Sigma$ is an equivalence relation on the set Φ of all formulas, and therefore it partitions Φ into mutually exclusive equivalence classes φ_Σ; for each formula φ of Λ, φ_Σ is the set of all formulas equivalent with φ under Σ. The set of all these equivalence classes may be denoted by Φ_Σ. Moreover, by relating $\equiv_\Sigma$ to the free algebra of formulas $\mathfrak{F}$, we realize that $\equiv_\Sigma$ is a congruence relation of that algebra, in the sense of the general theory of algebras. Hence, by applying a well-known algebraic procedure, we can construct the quotient algebra $\mathfrak{F}/{\equiv_\Sigma}$, which we denote for brevity by $\mathfrak{F}_\Sigma$. The elements of $\mathfrak{F}_\Sigma$ are arbitrary equivalence classes φ_Σ, and hence the universe of $\mathfrak{F}_\Sigma$ is the set Φ_Σ. The fundamental operations and distinguished elements of $\mathfrak{F}_\Sigma$ are those induced in a familiar manner by the fundamental operations and distinguished elements of $\mathfrak{F}$. We denote the induced operations and elements by means of the same symbols which are used in $\mathfrak{F}$, providing them sometimes, for clarity, with the subscript "Σ". We thus arrive at the algebra

$$\mathfrak{F}_\Sigma = \langle \Phi_\Sigma, \vee_\Sigma, \wedge_\Sigma, \neg_\Sigma, \boldsymbol{F}_\Sigma, \boldsymbol{T}_\Sigma, (\exists_{v_\kappa})_\Sigma, (v_\kappa = v_\lambda)_\Sigma \rangle_{\kappa,\lambda<\omega}.$$

It is a trivial matter to check that the algebra $\mathfrak{F}_\Sigma$ satisfies all the postulates for ω-dimensional cylindric algebras. We refer to $\mathfrak{F}_\Sigma$ as a *cylindric algebra of formulas* and, more specifically, as the cylindric algebra of formulas in the language Λ associated with the set of sentences Σ. If Θ is the theory generated by Σ, then the relations $\equiv_\Sigma$ and $\equiv_\Theta$ clearly coincide, and so do the algebras $\mathfrak{F}_\Sigma$ and $\mathfrak{F}_\Theta$.

The significance of the algebras $\mathfrak{F}_\Sigma$ for an algebraic study of predicate logic is rather obvious. Theories are basic entities in metalogical discussion just as algebraic structures are basic entities in algebraic research. By correlating the algebra $\mathfrak{F}_\Theta$ with any theory Θ, we have established a correspondence, which turns out to be one-one, between first-order theories and cylindric algebras of formulas. The basic metalogical problems for a fixed set of sentences Σ, or for a fixed theory Θ, are problems of the type: "Is a given sentence φ implied by the set Σ?" or "Is a given sentence φ a theorem of the theory Θ?" Each such problem clearly reduces to an algebraic problem concerning the associated algebra of formulas: "Does the equation $\varphi_\Sigma = \boldsymbol{T}_\Sigma$, or $\varphi_\Theta = \boldsymbol{T}_\Theta$, hold in the algebra $\mathfrak{F}_\Sigma$, or $\mathfrak{F}_\Theta$?"

The discussion of cylindric algebras of formulas leads in a most natural way to cylindric set algebras. Consider, indeed, a theory Θ in the language Λ; to simplify the discussion, assume that Θ is the first-order theory of a single relational structure, $\langle U, R_0, \ldots, R_\xi, \ldots\rangle_{\xi<\beta}$. Let φ be any formula in Λ and let $\boldsymbol{v}_{\kappa_1}, \ldots, \boldsymbol{v}_{\kappa_\nu}$ be all of its free variables. We correlate with φ a subset X_φ of the Cartesian space ${}^\omega U$, in fact, the set of all those points $x = \langle x_0, \ldots, x_\kappa, \ldots\rangle_{\kappa<\omega}$ such that the coordinates $x_{\kappa_1}, \ldots, x_{\kappa_\nu}$ satisfy the formula φ. Given any other formula ψ, the set X_ψ coincides with X_φ if and only if ψ is equivalent with φ under Θ, i.e., belongs to the equivalence class φ_Θ. We thus obtain a "natural" one-one mapping of the set Φ_Θ of all equivalence classes φ_Θ onto the family S of all sets X_φ. We can now consider those sets of S and those operations on members of S which correspond under this mapping to the distinguished elements and fundamental operations of the algebra $\mathfrak{F}_\Theta$. Obviously the induced notions corresponding to the fundamental Boolean operations $\vee$, $\wedge$, $\neg$ and the distinguished Boolean elements $\boldsymbol{F}$, $\boldsymbol{T}$ of $\mathfrak{F}_\Theta$ are the usual set-theoretical operations $\cup$, $\cap$, $\sim$ and the distinguished sets 0, ${}^\omega U$ from the theory of Boolean set algebras. But it is just as easy to realize that the operation corresponding to the κth existential quantification $\exists_{\boldsymbol{v}_\kappa}$ is just the κth cylindrification C_κ as it was defined for subsets of a Cartesian space in an earlier part of this Foreword; similarly, the set corresponding to the κ,λ-equation $\boldsymbol{v}_\kappa = \boldsymbol{v}_\lambda$ is the κ,λ-diagonal set $\mathsf{D}_{\kappa\lambda}$. Thus we arrive at a cylindric set algebra

$$\mathfrak{S}_\Theta = \langle S, \cup, \cap, \sim, 0, {}^\omega U, \mathsf{C}_\kappa, \mathsf{D}_{\kappa\lambda}\rangle_{\kappa,\lambda<\omega}$$

obtained by a "natural" isomorphic transformation from the cylindric algebra of formulas $\mathfrak{F}_\Theta$. We are confronted here with a simple extension of the Boolean parallelism between logical and set-theoretical notions, which is familiar to every student of Boolean algebras; we believe that this fully clarifies our motivation in selecting cylindrifications and diagonal elements as fundamental notions for the theory of cylindric set algebras.

A thorough study of cylindric algebras of formulas is, of course, of paramount importance for our purposes. To secure the applicability of modern algebraic notions and methods to this study, we enlarge the class of algebras $\mathfrak{F}_\Sigma$ by including in it all the isomorphic images of its members (thus, in particular, the cylindric algebras $\mathfrak{S}_\Theta$ just described). We shall refer here to algebras of the enlarged class as *special cylindric algebras.*

A problem which confronts us immediately in an algebraic discussion of special cylindric algebras is that of providing a purely algebraic characterization of these algebras, i.e., a characterization which refers exclusively to intrinsic properties of algebras and which, in particular, does not use any metalogical notions. The solution of this problem proves to be rather simple. First, we know that all special cylindric algebras are infinite-dimensional and in fact have dimension ω. The infinite dimensionality is an algebraic expression of the fact that the language Λ of predicate logic is provided with infinitely many variables. At the same time, however, the language Λ has a clearly finitary character in the sense that each of its expressions is a finite sequence of symbols. Thus, in particular, every formula φ of Λ has only finitely many free variables. As a consequence, for any given set of sentences Σ there are at most finitely many indices κ such that φ is not equivalent with $\exists_{v_\kappa}\varphi$ under Σ. This in turn implies that every special cylindric algebra

$$\mathfrak{A} = \langle A, +, \cdot, -, 0, 1, \mathsf{c}_\kappa, \mathsf{d}_{\kappa\lambda}\rangle_{\kappa,\lambda<\omega}$$

satisfies the following postulate:

(C_8) For every element x of A there are at most finitely many ordinals κ such that $\mathsf{c}_\kappa x \neq x$.

A cylindric algebra $\mathfrak{A}$, of arbitrary dimension α, which satisfies postulate (C_8) (where κ is assumed to range over all ordinals less than α) is referred to as *locally finite.* All finite-dimensional cylindric algebras are, of course, locally finite. As we have just seen, all special cylindric algebras are ω-dimensional and also locally finite. It turns out that the converse is also true: the class of special cylindric algebras simply coincides with the class of locally finite cylindric algebras of dimension ω.

Historically, cylindric algebras were first defined as algebraic structures $\langle A, +, \cdot, -, 0, 1, \mathsf{c}_\kappa, \mathsf{d}_{\kappa\lambda}\rangle_{\kappa,\lambda<\omega}$ satisfying the postulates for Boolean algebras and the additional postulates (C_1)–(C_8) (with $\alpha = \omega$). Thus, what was originally meant by a cylindric algebra is what we mean here by a special cylindric algebra. The class of these algebras was the sole subject matter of the theory of cylindric algebras in the early stage of its development. Indeed, this class provided an adequate and convenient tool for the algebraization of predicate logic, and no wider class of algebras was needed for this purpose.

Soon, however, it was realized that the class of special cylindric algebras has some serious defects when treated as the sole subject of research in an autonomous algebraic theory. In modern algebraic research one prefers to deal with *equational classes* of algebras, i.e., classes of algebras characterized by postulate systems in which every postulate has the form of an equation (an identity); such classes are also referred to as *varieties.* Classes of algebras which are not varieties are usually introduced in discussions as specialized subclasses of varieties; e.g., one treats fields as a special class of rings. Probably the main reason for this preference is the fact that every variety is closed under certain general operations frequently used to construct new algebras from given ones; we mean here the operations of forming subalgebras, homomorphic images, and direct products. By a well known result of Garrett Birkhoff, the varieties are precisely those classes of algebras which have all three of these closure properties. The class of special cylindric algebras, however, is not a variety. One of the postulates characterizing this class, in fact, (C_8), does not have the form of an identity and cannot be equivalently replaced by any identity or any system of identities. This follows from the simple observation that the direct product of infinitely many locally finite cylindric algebras is not, in general, locally finite.

Also for other, less technical reasons some modifications in the original definition of cylindric algebras seemed desirable. The definition contained certain assumptions which, from an intuitive point of view, have a highly specialized character and considerably restrict the applicability of results implied by this definition. One such assumption is the fixed dimension, ω, of all algebras defined. Another is their local finiteness; thus, postulate (C_8) is again involved here. The restrictive character of these two assumptions becomes obvious when we turn our attention to cylindric set algebras: we find there algebras of all possible dimensions, and we easily construct algebras which are or which are not locally finite. All these structures have, nevertheless, many algebraic properties in common; it would not be purposeful to study these properties within a theory whose results are established only for a narrow class of such structures.

For all these reasons the original conception of a cylindric algebra has been considerably extended: the restriction to dimension ω has been removed, postulate (C_8) has been deleted, and the definition has assumed the form given in an earlier part of this Foreword.

* * *

This work is intended to give a detailed and comprehensive account of the theory of cylindric algebras in its present stage of development. The work is to appear in two parts, of which the first is presented in this volume.

It is hoped that the reader will find this work to be largely self-contained. Inevitably, certain elements of mathematical logic and set theory have been presupposed; but these are of such a nature that they have undoubtedly been absorbed by anyone who has studied some branch of modern mathematics. On the other hand, the work has been based squarely upon the general theory of Boolean algebras, and the reader who is unfamiliar with the basic parts of that theory will find difficulty in following the text.

Immediately following this Foreword will be found a section entitled "Preliminaries" in which we describe our notation for basic set-theoretical and metalogical notions and explain concisely their meanings. For the most part this material is standard and well known.

There follows a long chapter sketching a general theory of algebraic structures; we have designated it "Chapter 0" to emphasize that it is not properly a part of the theory of cylindric algebras. Perhaps some words of explanation are due the reader in this connection.

In the development of the theory of cylindric algebras various concepts are involved which recur in many parts of modern algebra — concepts such as subalgebra, isomorphism, and homomorphism. To give all the definitions and derive all the needed properties of such concepts *ab initio* within the theory of cylindric algebras would have been very tedious for the many readers already acquainted with these ideas from their study of groups, rings, Boolean algebras, etc. Moreover, the generality of the results and the simplicity of their proofs would have been obscured by the specialized context of cylindric algebras. For these reasons it was decided to segregate this material in a separate chapter and present it in a more general form, as a kind of preface to the theory of cylindric algebras. Subsequently, recognizing the increasing attention which the general theory of algebraic structures is now receiving, we decided to extend the material beyond the minimum needed for the purposes of the present work, in order to achieve a unified and self-contained treatment.

Chapter 0 is an embodiment of these decisions and ideas. Essentially, it is an outline of a course, entitled "General theory of algebraic structures", which has been given by Tarski at the University of California, Berkeley, evolving over many years. The principal topics discussed consecutively in the five sections, 0.1–0.5, of the chapter are: algebras and their subalgebras; isomorphisms, homomorphisms, and congruence relations; direct products and some related notions (subdirect products, reduced products); polynomials and free algebras; reducts. (The last of these notions is less known than the remaining ones. A reduct of an algebra $\mathfrak{A}$ is an algebra obtained by deleting some of the fundamental operations of $\mathfrak{A}$, and preserving the universe and the remaining operations unchanged. For instance, the additive group of a ring $\mathfrak{R}$ is a reduct of $\mathfrak{R}$.) Some space in Chapter 0 is devoted to metamathematical and, specifically, model-theoretical topics. Thus, metamathematical aspects of free algebras are briefly discussed. Also, the model-theoretical notions of elementary, universal, and equational classes of algebras are considered, and a concise account is given of known results providing purely algebraic characterizations of these notions.

In Chapter 1 we commence the proper study of the theory of cylindric algebras. The first section of the chapter, 1.1, contains a re-statement of the postulate system for cylindric algebras, as well as formal definitions of two important classes of cylindric algebras previously mentioned in this Foreword — the cylindric set algebras and the cylindric algebras of formulas. In the present volume these two classes are not discussed in detail, but are extensively used as an invaluable source of examples and illustrations. The section is concluded with a preliminary discussion of the main representation problem for cylindric algebras — the problem of representing these algebras isomorphically by means of cylindric set algebras.

The succeeding sections of Chapter 1, 1.2–1.10, are largely devoted to a development of the most elementary and basic consequences of the postulate system for cylindric algebras. The procedure applied throughout these sections can be described as follows: an arbitrary cylindric algebra, about which no special assumptions are made, is regarded as fixed; within this algebra special elements, sets of elements, and operations on elements are defined in terms of the fundamental operations and distinguished elements of the algebra, and by means of these definitions the basic properties of the notions defined are derived from the general postulates for cylindric algebras. The use of set-theoretical tools in this discussion is reduced to a minimum; no variables are employed which represent arbitrary sets of elements, or operations on elements, of the fixed algebra, or arbitrary members of some class of algebras. For these reasons the portion of the work we are discussing may be called the elementary, or arithmetical, part of the theory of cylindric algebras. (In metamathematics the term "elementary", or "arithmetical", is used to denote that part of the theory of a class of mathematical structures which can be formalized within first-order predicate logic; here, however, we use this term in a looser and wider sense.)

Among the operations on elements of a fixed cylindric algebra $\mathfrak{A}$ which are discussed in Chapter 1, the unary substitution operations introduced in Section 1.5 deserve special attention. With each pair of ordinals κ, λ less than the dimension of $\mathfrak{A}$ there is correlated an operation of this kind, denoted by $\mathsf{s}_{\lambda}^{\kappa}$ and called the *κ, λ-substitution*; it is defined by setting, for every element x of the algebra, $\mathsf{s}^{\kappa}_{\lambda}x = \mathsf{c}_{\kappa}(\mathsf{d}_{\kappa\lambda}\cdot x)$ if $\kappa \neq \lambda$, and $\mathsf{s}^{\kappa}_{\lambda}x = x$ if $\kappa = \lambda$. The operation acquires a simple intuitive meaning when applied to cylindric algebras of formulas. Indeed, it is easily seen that in this case $\mathsf{s}^{\kappa}_{\lambda}$ (defined as above) is just the operation on equivalence classes of formulas induced by the familiar substitution operation on the formulas themselves. As is well known, the latter operation consists in substituting the variable $\boldsymbol{v}_{\lambda}$ for all the free occurrences of the variable $\boldsymbol{v}_{\kappa}$ in a given formula (and in changing bound variables of the formula if this is necessary to avoid collisions). The reader is probably aware of the significant role played by this substitution operation in formal logical arguments in which some sentences are shown to be implied by others.

In the last section of Chapter 1, Section 1.11, we study elementary properties of locally finite cylindric algebras. The importance of these algebras for the main purposes of the present work has been emphasized in our earlier discussion. From an abstract algebraic point of view the locally finite algebras form a rather narrow and highly specialized class of cylindric algebras; for this very reason their theory is richer and actually simpler than the general theory. We shall be particularly interested in those properties of substitution operations in locally finite algebras which cannot be extended to arbitrary cylindric algebras. Thus, for instance, we shall show that the operation of multiple substitution (which in its application to formulas, consists in simultaneously substituting new variables for several free variables) can be adequately defined and studied for all locally finite algebras of infinite dimension; we see no way of extending this discussion to the general theory.

Chapter 2, which is the last chapter of this volume, contains what may be called the properly algebraic part of the theory of cylindric algebras, in contrast to the arithmetical part presented in Chapter 1. General set-theoretical concepts are used in Chapter 2 to a much larger extent than in Chapter 1. As a rule, the discussion in Chapter 2 does not refer to a single, fixed cylindric algebra. Instead, the emphasis here is on relations between algebras and on operations — such as formation of subalgebras, quotient algebras, or direct products — which, when performed on given algebras, yield new ones.

In five sections, 2.1 and 2.3–2.6, we deal with the general algebraic notions introduced in Chapter 0. However, our interest here is to find and study special properties acquired by these notions when they are applied either to the class of all cylindric algebras or to comprehensive subclasses of it. For instance, in Section 2.4 there is a series of rather deep theorems that concern direct products of countably complete cylindric algebras (i.e., cylindric algebras in which the Boolean sum and product exist, not only for every pair, but for every countable collection of elements). Section 2.6 contains a comprehensive discussion of reducts of cylindric algebras. The term "reduct" is used here in a restricted and specialized sense. Given a cylindric algebra $\mathfrak{A}$ of dimension α and an ordinal β at most equal to α, by the *β-reduct* of $\mathfrak{A}$ we understand the cylindric algebra $\mathfrak{B}$ (of dimension β) obtained from $\mathfrak{A}$ by deleting the cylindrifications c_κ with $\kappa \geqq \beta$ and the diagonal elements $\mathsf{d}_{\kappa\lambda}$ with $\kappa \geqq \beta$ or $\lambda \geqq \beta$; in particular, the 0-reduct is simply the Boolean part of $\mathfrak{A}$. Subalgebras of reducts are called, for brevity, subreducts; certain special subreducts, the so-called neat subreducts, are singled out. A number of interesting and deep problems concern embedding given cylindric algebras in algebras of higher dimensions, i.e., representing them as subreducts (or possibly neat subreducts) of such algebras; these problems are exhaustively discussed in 2.6.

In the remaining two sections of Chapter 2, 2.2 and 2.7, some more specialized notions are discussed, which do not have a general algebraic character. In Section 2.2 we concern ourselves with structures obtained by relativization of a cylindric algebra to one of its elements. This extends a construction well-known from the theory of Boolean algebras; however, in contrast to the latter, the structures obtained by relativizing cylindric algebras are not, in general, cylindric algebras themselves. The discussion in Section 2.7 is closely related to the basic results on Boolean algebras due to Stone: the embedding theorem by which every Boolean algebra can be embedded as a subalgebra in a complete atomic Boolean algebra, and the representation theorem by which every Boolean algebra can be represented isomorphically as a Boolean set algebra. It is known from the literature that these theorems partially extend to arbitrary Boolean algebras with operators. In 2.7 we reconstruct the relevant results of the theory of Boolean algebras with operators, and by applying them to cylindric algebras we conclude that every cylindric algebra is embeddable in a complete atomic cylindric algebra of the same dimension. With the help of this embedding theorem we then show that every cylindric algebra $\mathfrak{A}$ can be represented isomorphically as a well determined algebra $\mathfrak{A}^*$ in which, just as in cylindric set algebras, the universe consists of subsets of a set and the Boolean notions are the usual set-theoretical notions from the calculus of sets. However, the specific "cylindric" notions of $\mathfrak{A}^*$, i.e., cylindrifications and diagonal elements, are not unambiguously defined in set-theoretical terms and are essentially as "abstract" in character as the corresponding notions in arbitrary cylindric algebras. For this reason the result discussed cannot be regarded as an intuitively satisfactory representation theorem for cylindric algebras. It presents, nevertheless, some intrinsic interest; and, in addition, it serves as an auxiliary device in certain portions of the proper representation theory for cylindric algebras, which will be developed in the second part of our work.

At the end of this volume the reader will find a list of symbols and symbolic expressions, a bibliography, and an index of names and subjects. The list of symbols is quite long; this is largely caused by the fact that in various portions of the volume we deal simultaneously with concepts from several different domains — algebra, set theory, and metalogic. The list is included solely for reference purposes, and probably there will be no need for the reader to consult it until he covers a considerable portion of the volume. The bibliography is intended to contain references to the whole existing literature on cylindric algebras and closely related structures, such as relation and polyadic algebras (not including, however, Boolean algebras and lattices). In addition, the bibliography lists all the papers in other domains which are actually referred to in the text.

The second part of the present work is still under preparation. It will begin with a discussion of cylindric set algebras. In particular, some special cylindric set algebras, with significant applications outside of the theory of cylindric algebras, will be described in some detail. These set algebras are correlated, in a manner described earlier in this Foreword, with certain first-order theories familiar from various parts of mathematics.

A comprehensive account of the representation theory for cylindric algebras will occupy the central position in the second volume. As a starting point for this portion of our discussion we may consider the problem: is every cylindric algebra isomorphic to a cylindric set algebra? We keep in mind, of course, the fact that an analogous problem for Boolean algebras has an affirmative solution and leads thus to a general representation theorem for those algebras. It turns out, however, that the solution of the problem for cylindric algebras is negative. The reason is that the class of cylindric set algebras proves to be highly specialized from a purely algebraic point of view. This is especially easily seen in the case of algebras of finite dimension: a finite-dimensional cylindric set algebra is always simple, in the sense of the general theory of algebras, and hence is never isomorphic to a cylindric algebra which is not simple (such as a direct product of two non-trivial cylindric algebras).

For this reason we re-formulate the problem and, in fact, replace in it the cylindric set algebras by a wider class of algebras, much less specialized in its algebraic properties, namely, the so-called generalized cylindric set algebras. These are algebras which differ from ordinary cylindric set algebras in one respect only: the α-dimensional Cartesian space ${}^{\alpha}U$ is replaced everywhere in their construction by any set which is a union of arbitrarily many pairwise disjoint spaces of the same dimension. The class of generalized cylindric set algebras, just as that of ordinary cylindric set algebras, has many features which make it well qualified for representing other "abstractly" defined classes of algebras. The construction of the algebras in this class is in a sense "concrete" and rather simple; all the fundamental operations and distinguished elements are unambiguously defined in set-theoretical terms, and the definitions are uniform over the whole class; geometrical intuitions underlying the construction give us a good insight into the structure of the algebras. Various properties common to all generalized cylindric set algebras are intuitively evident and hardly require rigorous proof, and many of them automatically extend to all isomorphic algebras.

It is therefore regrettable that, except for algebras of dimension 1, the reformulated problem still turns out to have a negative solution: for every $\alpha \neq 1$ there is a cylindric algebra of dimension α which is not isomorphic to any generalized cylindric set algebra. The proof (disregarding the trivial case $\alpha = 0$) is less simple than that of the corresponding result for ordinary cylindric set algebras. For each $\alpha \geqq 2$ it consists, first, in exhibiting an algebraic equation (essentially of the type given in Postulates (C_1)–(C_7)) which is seen to hold identically in all generalized cylindric set algebras of dimension α and, then, in constructing a cylindric algebra of the same dimension in which this equation fails. A number of such equations are known at present; the simplest of them (which works for every $\alpha \geqq 3$) is

$$\mathsf{c}_2(\mathsf{d}_{20}\cdot\mathsf{c}_0(\mathsf{d}_{01}\cdot\mathsf{c}_1(\mathsf{d}_{12}\cdot\mathsf{c}_2x))) = \mathsf{c}_2(\mathsf{d}_{21}\cdot\mathsf{c}_1(\mathsf{d}_{10}\cdot\mathsf{c}_0(\mathsf{d}_{02}\cdot\mathsf{c}_2x))),$$

i.e.,

$$\mathsf{s}^2_0\mathsf{s}^0_1\mathsf{s}^1_2\mathsf{c}_2x = \mathsf{s}^2_1\mathsf{s}^1_0\mathsf{s}^0_2\mathsf{c}_2x.$$

In this situation it seems worthwhile to undertake a thorough study of those algebras which can be represented isomorphically as generalized cylindric set algebras. We call such algebras representable, for short, and by the representation theory we understand the totality of results concerning the class of representable cylindric algebras. We could refer, more specifically, to the representations discussed here as set-theoretical or geometrical representations, in order to distinguish them from other representations of cylindric algebras which are occasionally mentioned in this Foreword.

The following are among the more important results of the representation theory which will be found in the second volume. Some rather simple and interesting characteristic properties of representable algebras have been established. For instance, it has been shown (rather unexpectedly) that a cylindric algebra of dimension α is representable if and only if, for every ordinal $\beta > \alpha$, it can be embedded as a neat subreduct in some cylindric algebra of dimension β. This result permits us to derive various results of the representation theory as direct corollaries from some theorems established in Section 2.6 of this volume. Notice that the characterization of a representable algebra in terms of neat reducts, just like the characterization used in the definition of such an algebra, is not intrinsic, in the sense that it involves mathematical objects (in fact, certain cylindric algebras) from "outside" of the algebra involved. However, it has been shown that for every $\alpha \neq 0$ the class of representable cylindric algebras is a variety. Hence the representable cylindric algebras can be intrinsically characterized by means of a system of algebraic equations identically satisfied in all such algebras. For each α a suitable system of equations has been constructed in an effective manner. For $\alpha = 0, 1, 2$ the systems turns out to be finite. It has been proved that for $\alpha > 2$ each suitable system must be infinite; no such systems known at present are provided with a simple and perspicuous description. On the other hand, some rather general and simple intrinsic conditions are known which are sufficient (though not necessary) for an algebra to be representable. Thus some comprehensive and simply defined classes of cylindric algebras prove to consist exclusively of representable algebras; such is, e.g., the class of infinite-dimensional locally finite algebras, as well as the wider class of infinite-dimensional algebras which are semi-simple in the general algebraic sense. Certain fairly general methods for constructing non-representable algebras have also been developed.

An outstanding open problem is that of exhibiting a class of cylindric algebras which contains an isomorphic image of every cylindric algebra and hence serves to represent the class of all these algebras, and which is at the same time sufficiently "concrete" and simply constructed to qualify for this purpose from an intuitive point of view. It is by no means certain or even highly plausible that a satisfactory solution of this problem will ever be found.

In a portion of the second volume following the representation theory we shall explore the connections between cylindric algebras and certain other structures of related origin, primarily relation algebras and polyadic algebras.

A characteristic feature of the last portion of the volume will be the essential role played in the discussion by notions and methods from the domain of metalogic and metamathematics. In the first place the reader will find there a detailed study of the relationship between cylindric algebras and predicate logic. Thus, we shall establish the result, mentioned in an earlier part of this Foreword, by which every special cylindric algebra, i.e., every locally finite algebra of dimension ω, can be isomorphically represented as a cylindric algebra of formulas (in some formal language of predicate logic). This result may be referred to as the metalogical representation theorem for special cylindric algebras; it proves to be closely related to the well-known completeness theorem for predicate logic, due to Gödel. We are also concerned with other metalogical results about predicate logic, in particular those which can be conveniently re-formulated in terms of cylindric algebras of formulas and which in this new formulation express some interesting algebraic properties of the algebras involved. By the metalogical representation theorem these results extend at once to all special cylindric algebras. It may be interesting to notice that, although the results thus extended are purely algebraic in both form and content, no proofs of these results are available at present which are not based, at least in part, on arguments of metalogical origin. Examples in the opposite direction can also be given: some metalogical theorems concerning predicate logic can be obtained most naturally as consequences of certain results on special cylindric algebras which are proved by purely algebraic methods.

The applicability of the metalogical representation theorems to the study of special cylindric algebras suggests the idea of extending the notion of predicate logic. In fact, it leads us to investigating new forms of this logic, which are likely to yield analogous metalogical representation for various classes of non-special cylindric algebras and thus to make metalogical methods available for the discussion of these classes. We mention here three classes of cylindric algebras for which the construction of suitable new systems of predicate logic has brought interesting results, or at least has led to interesting problems. These and some related classes, along with the corresponding systems of logic, will be discussed in the last portion of the second volume, though not necessarily in a detailed way.

The first of the three classes is that of locally finite cylindric algebras of arbitrary infinite dimensions — thus, a natural extension of the class of special cylindric algebras. The corresponding systems of predicate logic differ from the ordinary systems at most in having more comprehensive sets of variables; actually, systems are admitted in which the set of variables has any infinite cardinality prescribed in advance. Obviously, the modification is not essential, since the number of variables occurring in each particular formula continues to be finite. As a consequence, the metalogical representation theorem and other results previously mentioned extend with the greatest ease from special cylindric algebras and ordinary systems of predicate logic to all locally finite infinite-dimensional algebras and the modified systems of logic. Actually, in our work all the results in question will be formulated and established from the outset in this more general form.

The second class we have in mind is that of all finite-dimensional cylindric algebras; the corresponding systems of predicate logic differ from the ordinary ones only in that each of them has but finitely many variables (the number of variables serving in effect as a bound on the ranks of the predicates of the system). The third class consists of infinite-dimensional cylindric algebras, possibly with certain restricted completeness properties, i.e., algebras in which the Boolean sum and products exist for every set of elements with cardinality smaller than some infinite cardinal given in advance. The corresponding logical systems are so-called systems of infinitary predicate logic, which promise to become important tools in the model-theoretical discussion of mathematical structures. The most outstanding feature of these systems is the occurrence of formulas with infinitely many symbols. Not only may atomic formulas be infinitely long, but some operations are admitted which always yield infinitely long formulas; such, in fact, are the operations of forming disjunctions and conjunctions of a transfinite sequence of formulas (where the length of the sequence is bounded from above by some fixed infinite cardinal). The relationship between each of the last two classes of cylindric algebras and the corresponding form of predicate logic is more complicated than in the case of special cylindric algebras; the matter has not yet been thoroughly studied, so that many important problems remain open. Naturally, then, our discussion of this subject will be far from exhaustive.

To conclude, we shall mention still another group of problems which will be studied in the last portion of the second volume. These are problems in which various classes of cylindric algebras — or rather, more precisely, formalized theories of such classes — appear, not as tools, but as subjects of metamathematical investigation. Given a class of algebras (e.g., the class of all cylindric algebras of a fixed dimension), the formalized theories which we discuss are the elementary theory and the equational theory; the former is the set of all elementary sentences, i.e., sentences formulated in predicate logic, which hold in every algebra of the given class, while the latter is the (narrower) set of all algebraic equations which hold in every such algebra. The problems with which we are primarily concerned are those of decidability and finite axiomatizability of such theories. It will be seen from our discussion that the study of the most natural and interesting problems of this kind, insofar as they concern theories of cylindric algebras, has nearly been completed.

* * *

With the exception of the last portion of the second volume and some fragments of Chapter 0, metamathematical (and metalogical) notions are used in this work only in informal remarks and in the construction of certain examples and counterexamples, but they do not occur in the statements and proofs of results. Normally a theorem formulated without the help of metamathematical notions is provided with a proof in which these notions are not involved; sometimes we shall return to the theorem in the last portion of the second volume and show how it can be demonstrated by a different, metamathematical method. There are, however, a few exceptional cases where we are confronted with a purely algebraic result in the theory of cylindric algebras, but where the only proof we know at present is a metamathematical one — at least "in spirit", if not in form; in such cases the theorem is stated without proof at an appropriate place in the text, while its proof is given later, in the metamathematical portion of our work.

It will be seen from these remarks that throughout this work we have been at some pains to separate the algebraic from the metamathematical. In doing so we have been motivated partly by the desire to make the bulk of the work accessible to those readers whose knowledge of mathematical logic, and in particular of that portion known as metalogic and metamathematics, may be somewhat skimpy. But the procedure we have adopted has another, more fundamental goal: the theory of cylindric algebras affords a means of investigating the relationship between algebraic structures and logical systems, and, in order to obtain a clear perception of this relationship, it seems quite essential to keep each of these realms in distinct focus.

Nothing stated above implies that in our opinion there is any fundamental difference between metamathematics and mathematics "proper". Quite the contrary: we believe that, from every reasonable point of view, metamathematics is an integral part of mathematics. It is true that metamathematics can be regarded as an autonomous mathematical discipline based upon its own undefined notions and axioms. However, it can also be included, by means of a suitable interpretation of its undefined notions, in a branch of mathematics such as number theory (provided with a suitable set-theoretical basis) or set theory. Metamathematics does not differ in these respects from other mathematical disciplines, e.g., geometry. On the other hand, one could try to distinguish metamathematics from other mathematical disciplines by the peculiar character of its subject matter: while, say, number theory deals with number systems, in particular with the system of integers, and topology deals with certain kinds of spatial configurations, metamathematics deals with logical and mathematical disciplines themselves, or at least with those which have been subjected to the process of formalization. One should not forget, however, that what constitutes the subject of a mathematical discipline is often far from being intuitively clear. For instance, as a result of the high degree of generality and abstraction attained by present-day topology, the claim that certain kinds of spatial configurations form the subject of that discipline will probably be doubted by many of those who work in it. For different reasons, one can wonder whether three-dimensional analytic geometry deals with elements and parts of a space, i.e., with points and sets of points, or with triples of real numbers and sets of such triples characterized by algebraic equations.

Perhaps a more defensible means of distinguishing between mathematical disciplines may be achieved by looking into the kinds of intuitions which underlie their concepts and arguments, rather than by attempting to classify their subject matters. From this point of view, for example, the theory of vector spaces is essentially geometric in character, even though its theorems are literally "about" arbitrary elements and operations on these elements assumed to satisfy certain algebraic identities. Similarly, one can discern specific intuitions underlying metamathematical notions and arguments (although it is not our intention here to analyze these intuitions). Thus, even if we had decided to treat metamathematics consistently as a part of some classical mathematical discipline such as number theory, and if we had translated all metamathematical arguments into conventional number-theoretical terminology (concealing as much as possible their metamathematical origin and content), this would not by itself make our work more accessible to non-logicians. If a number-theorist to whom metamathematical intuitions are foreign were to study such a number-theoretical translation of a metamathematical proof, he would probably be able to say at best that the argument seems to him technically correct, but intuitively incomprehensible. For an analogous reason, although our intention has been to provide purely algebraic proofs for those results not involving metamathematical notions, we have avoided proofs which are algebraic in form, but not "in spirit".

To conclude these remarks on mathematics *versus* metamathematics, we should like to emphasize strongly our view that there is nothing objectionable in principle to the use of metamathematical concepts and methods in establishing algebraic results — just as there is nothing objectionable to the use of algebraic techniques in solving metamathematical problems. Certainly there is an esthetic satisfaction which derives from maintaining a sense of unity throughout the development of a mathematical theory by avoiding any interposition of extraneous concepts or methods. On the other hand, there is also undeniably a feeling of astonishment, of intellectual excitement and challenge, which accompanies the solution of a difficult problem in one scientific domain by means of concepts and methods of another discipline thought to be very remote and unconnected. This kind of cross-over, which is actually well known, seems to be characteristic of contemporary mathematics and is probably one of the principal factors in preserving its unity against breakup by the centrifugal forces of specialization. The phenomenon is abundantly illustrated in the theory of cylindric algebras.

* * *

As is seen from the title page, this work is a joint effort by three authors.

The theory of cylindric algebras was founded by Tarski, in collaboration with his former students Louise H. Chin (Lim) and Frederick B. Thompson, during the period 1948–52. Soon thereafter Henkin became interested in the subject and began to work with Tarski on its further development. In 1961 they published a fairly extensive outline of their research, and the plan was first formulated to prepare a detailed monograph on the subject. Subsequently Monk's substantial contributions to the theory made a joining of efforts desirable, and thus the present team of authors was finally formed.

As regards the authorship of particular results incorporated in the work, whenever these were first announced in an abstract, an article, or a doctoral dissertation, whether by one or more co-authors or by others, the appropriate citations are indicated by bibliographical references. Where a substantial result appears here in print for the first time, we note its authorship by a direct reference.

We wish to express our thanks to Gebhard Fuhrken, Bjarni Jónsson, and Ralph McKenzie, who read parts of an earlier version of the work and made many helpful suggestions. The authors have profited much from the devoted work of several research assistants: Stephen Comer, Daniel Demaree, James S. Johnson, Alan Kostinsky, Don Pigozzi, and Benjamin F. Wells III. We feel impelled particularly to express here our deep gratitude to Pigozzi for contributions far beyond the call of duty. Over a period of years he labored on the technical preparation of the manuscript with devotion and care exceeding all our expectations. In the course of his labors he obtained many specific results which we are pleased to be able to include in this work, giving him, of course, explicit credit at the appropriate points; a number of them, which will appear in the second volume of our work, form an essential part of his doctoral dissertation. Over and beyond these major contributions, he proposed illustrations, simplifications, and other minor improvements of the original text in such abundance that any attempt to list them all would be utterly impractical.

Able secretarial assistance was also an essential ingredient in the production of the manuscript; here we credit Mrs. June Lewin, Mrs. Susan Moss, Mrs. Dale Ogar, and Mrs. Mae Jean Ruehlman.

Support for the research and preparation of this work was provided by the National Science Foundation through grants to the University of California, Berkeley (Grants No. GP-4608, GP-6232X, GP-6232X2, and GP-6232X3), and to the University of Colorado (Grants No. GP-1232, GP-3948, GP-7387, and GP-12209). The authors are pleased to acknowledge their debt.

While the bulk of the work on this book was done at the University of California, Berkeley, and the University of Colorado, some portions of it were prepared at other institutions visited at various times by the several authors. Visits of substantial length took place at Oxford University (All Souls College and the Mathematical Institute), and at the University of California, Los Angeles; the use of the facilities of these institutions is gratefully acknowledged.

Finally, the authors express their appreciation to the North-Holland Publishing Company and, in particular, to its director, Mr. M. D. Frank, for cooperation in meeting extensive typographical demands.

L. Henkin, J. D. Monk, A. Tarski

August 1969
Berkeley, California,
and Boulder, Colorado.

PRELIMINARIES

We briefly discuss here the set-theoretical and metalogical notation which will be applied throughout this work. A reader unfamiliar with the notions introduced may find our discussion inadequate for understanding some of the later material. In such cases he may wish to consult the books Suppes [60*] (see the bibliography at the end of this volume), Bernays-Fraenkel [58*], Monk [69*], and Sierpinski [65*] for further references on set-theoretical matters, and Church [56*] for logical notions; the articles Tarski [54*], [55*], and [65*] may also be of use in connection with the latter.

I. SET-THEORETICAL NOTIONS

As with most mathematical theories, the theory of cylindric algebras can be adapted to any one of several underlying systems of set theory. We shall make no effort to select a particular one and to exhibit in detail the development of our algebraic theory within it. However, for the sake of definiteness the reader may consider the well-known system of Bernays developed in Bernays-Fraenkel [58*] or the one presented in Morse [65*]. Both systems can be modified in various details so as to enhance their usefulness for our purposes. In particular, Morse's system presents certain peculiarities, the most striking of which is the abolishment of all differences in the treatment of sentences and class terms. For our purposes we prefer to deal with a more conventional variant of Morse's system, free of these peculiarities, as it is outlined in the appendix of Kelley [55*] and developed in detail in Monk [69*]. With a few exceptional points (which will be noted explicitly) this variant provides an adequate framework for our developments.

In Morse's (as well as Bernays') system the universe of discourse is restricted to those objects which are classes; we find it convenient to assume that this restriction has been removed. The fundamental relation between objects is that of *membership*. Any object, whether a class or not, which is a member of some class will be referred to as an *element*. A class is called respectively a *set*, or a *proper class*, dependent on whether it is, or is not, an element. (This distinction between two kinds of classes is a basic feature of both Bernays' and Morse's systems.) For instance, the class of all elements is a proper class. Another example of a proper class, to which we shall frequently refer in this work, is the class of Boolean algebras. In general the proper classes are "very large", since any class whose elements are in one-one correspondence with those of a set is itself a set. Sometimes we shall speak of a *family of sets* instead of a class of sets.

The formulas $x, y, \ldots \in A$ and $x, y, \ldots \notin A$ respectively express the facts that $x, y, \ldots$ belong — i.e., are members or elements of A — and do not belong to the class A. In this particular instance we have followed a common notational convention which reserves the use of lower case letters to elements, never to proper classes. However, we shall make no effort to observe this convention rigorously throughout.

Given a formula $\varphi(x)$ we can consider those objects x which *satisfy* $\varphi(x)$; i.e., for which this formula *holds*. We use the symbolic expression $\{x : \varphi(x)\}$ to denote the class of all those elements x for which $\varphi(x)$ holds. Hence we have $y \in \{x : \varphi(x)\}$ iff y is an element and $\varphi(y)$ holds; here, as well as throughout the monograph, we use "iff" as an abbreviation for "if and only if". If $\varphi(x)$ holds for a proper class A, thus for a non-element, we have of course $A \notin \{x : \varphi(x)\}$. In Morse's system (as opposed to Bernays' system) the class $\{x : \varphi(x)\}$ exists for each formula $\varphi(x)$. In a rigorous set-theoretical development it is necessary to investigate for each formula $\varphi(x)$ whether or not the class $\{x : \varphi(x)\}$ introduced in the discussion is a set. In this work, however, we shall make no explicit analysis of this kind; we shall simply treat the classes denoted by expressions $\{x : \varphi(x)\}$ as sets whenever it appears intuitively appropriate on the grounds that the set in question is not "too large". Instead of $\{x : \varphi(x)$ and $\psi(x)\}$ we sometimes write $\{x : \varphi(x), \psi(x)\}$, and similarly in the case of more than two formulas. Also, in place of $\{x : x \in A$ and $\varphi(x)\}$ we sometimes write $\{x \in A : \varphi(x)\}$.

If $\tau(x)$ is a symbolic expression which represents an object for each value of x satisfying $\varphi(x)$, then we let

$$ {}_x\{\tau(x) : \varphi(x)\} = \{y : y = \tau(x) \text{ for some element } x \text{ satisfying } \varphi(x)\}. $$

The prescript x is necessary in principle because the expressions $\tau(x)$ and $\varphi(x)$ may contain variables other than x. However, in these and analogous symbolic expressions we shall omit the prescript whenever there appears to be no danger of confusion. Similarly we let

$$ {}_{x,y}\{\tau(x, y) : \varphi(x, y)\} = \{z : z = \tau(x, y) \text{ for some } x, y \text{ satisfying } \varphi(x, y)\}. $$

Note that in general the three expressions ${}_x\{\tau(x, y): \varphi(x, y)\}$, ${}_y\{\tau(x, y): \varphi(x, y)\}$, and ${}_{x,y}\{\tau(x, y): \varphi(x, y)\}$ have different meanings.

Using this notation we define in particular

$$0 = \{x : x \neq x\} \quad (\text{the } \textit{empty set}),$$

$$\{x\} = \{y : y = x\} \quad (\text{the } \textit{singleton} \text{ of } x),$$

$$\{x, y\} = \{z : z = x \text{ or } z = y\} \quad (\text{the } \textit{unordered pair } x, y),$$

etc. *Inclusion* and *proper inclusion* are represented by $\subseteq$ and $\subset$ respectively, and their negations by $\nsubseteq$ and $\not\subset$; the symbols $A \subseteq B$ and $B \supseteq A$ will be used interchangeably, and the same applies to $A \subset B$ and $B \supset A$, etc. $Sb\,A$ denotes the *class of all subsets* of A. $A \cup B$ and $A \cap B$ denote respectively the *union* and the *intersection* of A and B. More generally, $\bigcup C$ and $\bigcap C$ denote respectively the *union* and the *intersection* of all members of the class C. So, in particular, $\bigcap 0$ is the *universal class*, i.e., the class of all elements; if however in a given context we are dealing exclusively with subsets of a fixed class U, then the notation $\bigcap 0$ will be used to denote U. A still more general notation than $\bigcup A$ and $\bigcap A$ is available. If $\tau(x)$ is an expression which denotes an object for every value of x satisfying the condition $\varphi(x)$, we respectively denote by ${}_x\bigcup_{\varphi(x)} \tau(x)$ and ${}_x\bigcap_{\varphi(x)} \tau(x)$ the *union* and the *intersection* of all those $\tau(x)$ which are classes and for which x satisfies $\varphi(x)$. In this notation $\bigcup A$ becomes $\bigcup_{x \in A} x$. Further we let $A \sim B = \{x : x \in A, x \notin B\}$ (the set-theoretical *difference* of A and B); in case $B \subseteq A$ we refer to $A \sim B$ as the *complement of B with respect to A* and we sometimes use the alternative notation ${}_A\!\sim B$.

The *ordered pair* with *first term* x and *second term* y is the set $\langle x, y\rangle = \{\{x\}, \{x, y\}\}$; the *ordered triple* with *first term* x, *second term* y, and *third term* z is defined to be $\langle x, y, z\rangle = \langle\langle x, y\rangle, z\rangle$; etc. By a (binary) *relation* we shall understand a class of ordered pairs. By the above convention $\{\langle x, y\rangle : \varphi(x, y)\}$ is the class of all ordered pairs $\langle x, y\rangle$ satisfying the condition $\varphi(x, y)$. Thus in particular $\{\langle x, y\rangle : x = y\}$ is the *identity relation*; we shall denote it by Id. Consider further the relation $\{\langle x, y\rangle : x \subseteq y\}$; this is what can be called the *inclusion relation*. We shall not introduce any special symbol to denote this relation, but we shall sometimes use the symbol $\subseteq$ for this purpose (although this symbol was originally introduced only in the context $A \subseteq B$, and not to denote any specific set-theoretical entity). We shall apply the same convention to other formulas of the form $x\varphi y$ (where φ is a single symbol) which will be introduced in our further discussion: we shall use φ as a relation symbol denoting the relation $\{\langle x, y\rangle : x\varphi y\}$, so that, in case x and y are elements (and not proper classes), the formula $x\varphi y$ becomes equivalent to

$\langle x, y\rangle \in \varphi$. Conversely, if R is a relation we shall sometimes use xRy instead of $\langle x, y\rangle \in R$. Instead of "$x\varphi y$ and $y\psi z$" we shall often use for brevity "$x\varphi y\psi z$". For any classes $A, B, C, \ldots$ we write

$$A \times B = \{\langle x, y\rangle : x \in A, y \in B\},$$
$$A \times B \times C = (A \times B) \times C,$$

etc. $A \times B$ is called the *product* (or the *Cartesian product*) of A and B. For any relations R and S we let

$R|S = \{\langle x, z\rangle$: for some y, $xRySz\}$ (the *relative product* of R and S).

For any relation R we let

$R^{-1} = \{\langle x, y\rangle : yRx\}$ (the *converse* of the relation R);

$Do\,R = \{x$: for some y, $xRy\}$ (the *domain* of the relation R);

$Rg\,R = \{y$: for some x, $xRy\}$ (the *range* of the relation R);

$Fd\,R = Do\,R \cup Rg\,R$ (the *field* of the relation R).

Given any relation R and class A we let

$A 1 R = \{\langle x, y\rangle : x \in A, xRy\}$ (R *domain-restricted* to A);

$R^*A = \{y$: for some x, $x \in A$ and $xRy\}$ (the R-*image* of A);

we also put $R^\star x = R^*\{x\}$.

A *function* is a relation f such that for every $x \in Do f$ there is exactly one y with $\langle x, y\rangle \in f$. This unique y is referred to as the xth *value* of f and is denoted usually by fx; alternative notations are $f(x)$, f_x, f^x, $f^{(x)}$, etc. If $\tau(x)$ is an expression which denotes an element for each element x satisfying $\varphi(x)$, then $\{\langle x, \tau(x)\rangle : \varphi(x)\}$ is clearly a function; as an abbreviated notation for this function we shall use ${}_x\langle \tau(x) : \varphi(x)\rangle$, or sometimes ${}_x\langle \tau(x)\rangle_{\varphi(x)}$. (The prescript x will be omitted from these notations whenever there appears to be no danger of confusion, in particular when x is the only variable appearing in the expressions $\tau(x)$ and $\varphi(x)$.) Moreover, we adopt a convention which allows us in some cases to obtain an even shorter designation for this function. This convention is analogous to the one adopted above for introducing relation symbols. Instead of formulating this convention in general terms, we shall explain it by examples. The symbol *Sb* was originally introduced, not to denote a set-theoretical entity, but as a part of symbolic expressions such as $Sb\,A$ denoting the class of all subsets of the class A. As is known from set theory, $Sb\,x$ is always a set when x is an element. Hence $\langle Sb\,x : x$ is an element$\rangle$ is a function whose domain is the universal class. According to our convention, we may use the same symbol Sb to denote this function, so that $Sb = \langle Sbx : x$ is an element$\rangle$. Similarly, in the expression $R^{\bigstar}x$ (or $R^{*}A$) neither $\bigstar$ nor $R^{\bigstar}$ were supposed heretofore to denote any set-theoretical entity. Our convention, however, permits us to use $R^{\bigstar}$ as a function symbol denoting $\langle R^{\bigstar}x : R^{\bigstar}x$ is an element$\rangle$. For each relation R this is a function whose domain consists of all those elements x for which $R^{\bigstar}x$ is also an element. Thus, if D is this domain, we have $R^{\bigstar} = \langle R^{\bigstar}x : x \in D\rangle$. Of course, if f is introduced from the outset as a function symbol, we have analogously $f = \langle fx : x \in Dof\rangle$. It is important to note that in connection with the symbolism $\langle \tau(i) : i \in I\rangle$ mathematicians frequently use a different terminology (and we shall follow here this custom). They speak of $S = \langle \tau(i) : i \in I\rangle$, not as a function with domain I, but as a *system of elements* $\tau(i)$ *indexed by* I (or *by arbitrary elements* i *of* I); the element $\tau(i)$, i.e., the ith value of S, is referred to as the ith *term* of S and is denoted much more frequently by S_i than by Si. The difference is, of course, purely verbal.

If $Dof = A$, we say that f is a function *on* A. When dealing with functions we use instead of relative multiplication the dual operation $\circ$ called *composition;* thus, if f and g are functions, then $f \circ g$ is also a function, $f \circ g = g | f$, and $(f \circ g)x = f(gx) = fgx$ for any $x \in Do(f \circ g)$. If both f and f^{-1} are functions we say that f is *one-one*, and we refer to f^{-1} as the *inverse* of f. If $Dof = A$ and $Rgf = B$, or $Rgf \subseteq B$, we say that f is a function *from* A *onto* B, or *into* B, respectively; we also say that f maps A *onto* B, or *into* B.

Further, we let

$$^{A}B = \{f : f \text{ maps } A \text{ into } B\}.$$

^{A}B is referred to as the *(Cartesian) power of* the set B *with the exponent* A or, more simply, the Ath *(Cartesian) power of* B. If f is a one-one function from A onto B, we also say that f establishes a *one-one correspondence* between the elements of A and the elements of B. If A and B are sets we can express this symbolically by the conjunction of the formulas $f \in {}^{A}B$ and $f^{-1} \in {}^{B}A$. If in particular $f \in {}^{A}A$ and $f^{-1} \in {}^{A}A$, f is called a *permutation* of A.

By the *axiom of choice* we mean the statement that for every class A of non-empty sets there is a function F such that $Fx \in x$ for each $x \in A$. Throughout the work we shall apply the axiom of choice whenever needed.

Given a function F we define $\mathsf{P}F$, the *(Cartesian) product of the* function F, by the formula

$$\mathsf{P}F = \{f : Dof = DoF, \quad f_a \in F_a \text{ for all } a \in Dof\}.$$

If in particular $F = \langle \tau(x) : \varphi(x)\rangle$, we use the notation $\mathsf{P}F = {}_x\mathsf{P}_{\varphi(x)}\tau(x)$, and we refer to $\mathsf{P}F$ as the *(Cartesian) product of all the sets $\tau(x)$ with x satisfying $\varphi(x)$*. If A and B are sets and F is the constant function with $DoF = A$ and $RgF = \{B\}$, then $\mathsf{P}F$ coincides with ${}^AB$, the *A^{th} Cartesian power* of B. If F is a function whose domain is a set and whose range consists exclusively of non-empty sets, then $\mathsf{P}F \neq 0$ by the axiom of choice.

For any given $g \in \mathsf{P}F$ the set P of all $f \in \mathsf{P}F$ such that $f_a \neq g_a$ only for finitely many $a \in Dof$ is called a *weak (Cartesian) product* of F or, specifically, the *weak (Cartesian) product of F relative to g*. Again, if F is the constant function with $DoF = A$ and $RgF = \{B\}$, then P is called the *A^{th} weak (Cartesian) power of B relative to g*. These notions are rather rarely used, and we do not introduce any special symbolic notation for them.

With every element a we correlate a function pj_a called the *a^{th} projection* and defined by the formula

$$\mathsf{pj}_a = {}_f\langle f_a : f \text{ is a function, } a \in Dof\rangle;$$

thus $Do\,\mathsf{pj}_a$ is the class of all functions f such that $a \in Dof$, and for every such function f we have $\mathsf{pj}_a f = f_a$. By this stipulation $(\mathsf{P}_{i \in I} A_i) \upharpoonright \mathsf{pj}_i$ is a function from $\mathsf{P}_{i \in I} A_i$ onto A_i, provided $A_j \neq 0$ for all $j \in I \sim \{i\}$.

An *equivalence relation* is a relation R which is *transitive* and *symmetric* (i.e., $R|R \subseteq R$ and $R^{-1} = R$). An equivalence relation R with field A is always *reflexive over A* (i.e., $A \upharpoonright Id \subseteq R$). In connection with equivalence relations R with field A we often use special notations. Namely, we write x/R instead of $R^{\bigstar}x$ for $x \in A$, X/R instead of $(R^{\bigstar})^*X$ for $X \subseteq A$, and in particular A/R instead of $(R^{\bigstar})^*A$; x/R is called the *equivalence class under R* (the *R-equivalence class*) containing x, or correlated with x. Similarly we let $S/R = \{\langle x/R, y/R\rangle : xSy\}$ for any relation $S \subseteq A \times A$. The notion of an equivalence relation can be characterized in terms of $R^{\bigstar}$ by the following condition: for all $x, y \in FdR$, $R^{\bigstar}x = R^{\bigstar}y$ iff xRy. Thus if $R^{\bigstar}x$ is a set for every $x \in FdR$, then $R^{\bigstar}$ is a function whose domain includes FdR and which, loosely speaking, "identifies" any two elements of FdR between which R holds. However, even in case $R^{\bigstar}x$ is a proper class for some x's in FdR, we can still correlate with R a definite "identifying" function, namely a function τ_R whose domain is FdR and which satisfies the condition: $\tau_R x = \tau_R y$ iff xRy, for all $x, y \in FdR$ (for a possible definition of such a function see Montague-Scott-Tarski [72*]). For $x \in FdR$ the set $\tau_R x$ is called the *R-type* of x.

Two classes X, Y are called *disjoint* if $X \cap Y = 0$. A *partition* of a class A is a family P of pairwise disjoint non-empty subsets of A such that $\bigcup P = A$; thus the only partition of the empty set is empty. If P is a partition of A and $R = \{\langle x, y\rangle : x, y \in M \text{ for some } M \in P\}$, then R is an equivalence relation with field A. Under the assumption that A is a set, the converse also holds: if R is an equivalence relation with field A, then A/R is a partition of A, and a one-one correspondence between partitions of A and equivalence relations over A has been established.

A relation R which is reflexive over its field, *anti-symmetric*, and transitive (i.e., $(FdR)1Id \subseteq R$, $R \cap R^{-1} \subseteq Id$, and $R|R \subseteq R$) is called a *partially ordering relation* or a *partial ordering*. We frequently use the symbol $\leqq$ (or a symbol of related shape) to represent a partial ordering R and the symbol $\geqq$ to represent its converse R^{-1} (which is also a partial ordering). The most typical example of a partial ordering is the inclusion relation $\subseteq$. An extensive terminology has been introduced in the discussion of partially ordering relations. Thus given any such relation R we say that an element z is an *upper bound* (strictly speaking an *R-upper bound*), or a *lower bound*, of the elements x and y if, respectively, xRz and yRz, or zRx and zRy. More generally we say that z is an *upper*, or *lower*, *bound* of a class A if xRz, or zRx, for every $x \in A$. An upper bound z is called the *least upper bound* if zRu for any other upper bound u. If the least upper bound of a class A belongs to A, it is referred to as the *greatest element* of A. A *maximal element* of a class A is an element $x \in A$ such that $x = y$ whenever $xRy \in A$. Analogously we define the notions of *greatest lower bound*, *least element*, and *minimal element*. A partial ordering R is said to be *upper directing*, or simply *directing*, if any two elements $x, y \in FdR$ have an upper bound; it is said to be a *lattice ordering*, or a *complete lattice ordering*, if, respectively, any pair of elements $x, y \in FdR$, or every class $A \subseteq FdR$, has both a least upper bound and a greatest lower bound. If R is a complete lattice ordering, then clearly FdR has least and greatest elements.

A partial ordering R is called a *simple ordering* if it is *connected* (i.e., $(FdR) \times (FdR) \subseteq R \cup R^{-1}$); it is called a *well ordering* if every non-empty class $A \subseteq FdR$ has a least element.

R being a partial ordering, every class $A \subseteq FdR$ is said to be *partially ordered* by R. In an analogous sense we say that A is *directed* by R, A is *simply ordered* by R, etc.

We shall sometimes use a set-theoretical principle known as *Zorn's lemma*, which is known to be equivalent to the axiom of choice. By Zorn's lemma, if A is a non-empty set partially ordered by a relation R, and if every subset of A simply ordered by R has an upper bound in A, then A has a maximal element.

Ordinal numbers, sometimes briefly called just *ordinals*, are denoted by lower case Greek letters α, β, γ, ..., and sets of ordinals by upper case Greek letters Γ, Δ, Θ, ... We assume that the notion of an ordinal has been defined in such a way that all ordinals are sets, every element of an ordinal is an ordinal, and the class of all ordinals is well-ordered by the membership relation $\in$. $\xi < \eta$ ($\eta > \xi$) means by definition that ξ and η are ordinals and $\xi \in \eta$, and $\xi \leqq \eta$ ($\eta \geqq \xi$) means that $\xi < \eta$ or else $\xi = \eta$ and ξ is an ordinal. Consequently every ordinal coincides with the set of all smaller ordinals; $\xi \cup \eta$ proves to be the larger and $\xi \cap \eta$ the smaller of two ordinals ξ, η. The *sum* of two ordinals ξ and η is denoted by $\xi + \eta$. The least ordinals are of course 0, $1 = \{0\}$, $2 = \{0, 1\}$, etc. The least ordinal which is different from 0 and for which no immediately smaller ordinal exists is denoted by ω.

If Γ is a set of ordinals, then $\bigcup\Gamma$ is the least upper bound of all ordinals in Γ with respect to the relation $\leqq$. Since every ordinal γ is also a set of ordinals, we can consider in particular $\bigcup\gamma$. It is easily seen that $\bigcup\gamma$ coincides with the immediate predecessor of γ if such an immediate predecessor exists, and otherwise with γ itself; thus the formula $\bigcup\gamma = \gamma$ expresses the fact that γ is a limit ordinal (not excluding the possibility that $\gamma = 0$). If Γ is any non-empty class of ordinals, then $\bigcap\Gamma$ proves to coincide with the least ordinal in Γ.

It is known that every set A can be mapped in a one-one way onto an ordinal; the least such ordinal is called the *cardinality*, or *power*, or the *number of elements* of A, and is denoted by $|A|$. An ordinal α is said to be a *cardinal number*, or simply a *cardinal*, if $\alpha = |\alpha|$. A set A is called *finite*, or *infinite*, dependent on whether $|A| < \omega$, or $|A| \geqq \omega$; it is called *denumerable* if $|A| = \omega$, and *countable* if $|A| \leqq \omega$. Finite ordinals coincide with finite cardinals; we identify them with *natural numbers* (non-negative integers). Thus ω is the set of all natural numbers $0, 1, 2, \ldots$. With every ordinal ξ we correlate an infinite cardinal ω_ξ by the following stipulations: $\omega_0 = \omega$; for $\xi \neq 0$, ω_ξ is the least cardinal greater than all ω_η with $\eta < \xi$. It turns out that every infinite cardinal can be represented in the form ω_ξ for some ξ.

The notions of the ordinal sum $\alpha+\beta$, product $\alpha\cdot\beta$, and power α^β of two ordinals α and β are assumed to be known, and so are the corresponding cardinal operations (restricted to the case when α and β are cardinals). We do not need special symbols for the cardinal sum and product of cardinals α and β since they coincide respectively with the ordinal sum and product in case both α and β are finite or at least one of them is 0, and in the remaining cases both the cardinal sum and product equal $\alpha \cup \beta$; in case $\alpha = 0$ or $\beta = 0$ we can use, of course, $\alpha+\beta$ or $\alpha \cup \beta$ interchangeably to represent the cardinal sum of α and β. On the other hand, we shall use the symbol α^β ambiguously, denoting by it either the ordinal or the cardinal power (the latter, of course, only in case both α and β are cardinals); if it is not quite clear from the context in which sense the symbol is used, we shall expel possible doubts by an explicit remark. The successor of a cardinal α, i.e., the least cardinal $>\alpha$, is denoted by α^+; thus if α is finite, then $\alpha^+ = \alpha+1$, and if $\alpha = \omega_\xi$, then $\alpha^+ = \omega_{\xi+1}$. By *continuum hypothesis* we understand the statement that $2^\omega = \omega_1$; by the *generalized continuum hypothesis*, abbreviated $\mathscr{GCH}$, the statement that $2^\alpha = \alpha^+$ for every infinite cardinal α. As is now known (see P. J. Cohen [63*], [64*]), neither hypothesis can be confirmed or refuted on the basis of the familiar axiom systems of set theory.

By an *α-termed sequence* or a *sequence of length* α we understand any function f such that $Dof = \alpha$. If a sequence (as a function) is one-one, then we refer to it as a sequence *without repeating terms* or *without repetitions.* When using this terminology we usually write f_ξ instead of $f\xi$ and we call f_ξ the *ξ^{th} term* of the sequence f; i.e., we treat a sequence of length α as a system indexed by α. The set of all α-termed sequences f with $Rgf \subseteq A$ coincides of course with ${}^\alpha A$. For any α-termed sequence x we have $x = \langle x_\xi : \xi < \alpha \rangle = \langle x_\xi \rangle_{\xi<\alpha}$. We also sometimes use the notation $x = \langle x_0, \ldots, x_\xi, \ldots \rangle_{\xi<\alpha}$, as well as $x = \langle x_0, \ldots, x_{\alpha-1} \rangle$ in case $0 < \alpha < \omega$. Analogously the range of the sequence x is expressed by $\{x_0, \ldots, x_\xi, \ldots\}_{\xi<\alpha}$, or $\{x_0, \ldots, x_{\alpha-1}\}$. By $\langle a \rangle$, $\langle a, b \rangle$, etc. we respectively denote the 1-termed sequence x with $x_0 = a$, the 2-termed sequence x with $x_0 = a$ and $x_1 = b$, etc. In many situations we can identify a 1-termed sequence $\langle a \rangle$ with the element a without causing any confusion. The expression $\langle a, b \rangle$ is of course ambiguous and can be interpreted as denoting either an ordered pair or a 2-termed sequence. For most purposes it is irrelevant which interpretation is meant, because the principal property needed for the notation is that, for any x, y, u, and v, $\langle x, y \rangle = \langle u, v \rangle$ iff $x = u$ and $y = v$, and this property holds both for ordered pairs and for 2-termed sequences. Notice that under the second interpretation the Cartesian product $B \times C$ of two sets B and C falls as a particular case under the general notion of the Cartesian product $\mathsf{P}A$ of a system A of sets; in fact we have $\mathsf{P}\langle B, C \rangle = B \times C$. If f is a function with domain $B \times C$, and $x \in B \times C$ with $x_0 = y$ and $x_1 = z$, then we use interchangeably the expressions fx, $f\langle y, z \rangle$, f_{yz}, and $f(y, z)$. If x is an α-termed and y a β-termed sequence, then by the *concatenation* $x^\frown y$ of x and y we understand the $(\alpha+\beta)$-termed sequence z such that $z_\xi = x_\xi$ for $\xi < \alpha$ and $z_{\alpha+\xi} = y_\xi$ for $\xi < \beta$. There is a consequence of the axiom of choice known as the *principle of dependent choices* which it is convenient to formulate in terms of sequences. According to this principle, if R is a binary relation, A is a non-empty set, and if for every $x \in A$ there is a $y \in A$ such that xRy, then corresponding to each $a \in A$ there is an ω-termed sequence $\langle x_\kappa \rangle_{\kappa<\omega}$ of elements of A such that $x_0 = a$ and $x_\kappa R x_{\kappa+1}$ for every $\kappa < \omega$. It is sometimes convenient to apply this principle rather than the axiom of choice itself; see, e.g., the proof of 2.4.25. For certain purposes it proves convenient to have an extended notion of an α-termed sequence and, more generally, of a system indexed by a class I, in such a way that not only elements but also proper classes may appear as terms of a system. To define these new notions let us agree to denote by $\overline{X}$ either X itself in case X is a proper class, or else $\{X\}$ in case X is an element. A relation S will now be called a *system indexed by I (in the new sense)* if $Do\,S = I$ and, for every $i \in I$, $S^\star i = \overline{X}$ for some X; given $i \in I$, X is called the *i^{th} term* of S if $S^\star i = \overline{X}$. In case I is an ordinal α we refer to S as an *α-termed sequence*; in case $\alpha = 2, 3, \ldots$ we may call S an *ordered couple, triple, ...* and may even use the notation $S = \langle A, B \rangle$, $S = \langle A, B, C \rangle, \ldots$, where A is the 0th term of S, B the 1st term, etc. (assuming it is clear from the context that this notation is not used in the old sense). Obviously there is a natural one-one correspondence between systems in the old sense and those systems in the new sense in which no term is a proper class. The use of the new notion in our discussion will be very restricted. Compare an analogous construction in R. M. Robinson [45*], applied to systems of set theory in which no individuals are admitted.

Any class of α-termed sequences is referred to as an *α-ary relation.* Note that, since 0 is the only 0-termed sequence, 0 and 1 are the only 0-ary relations. By the *field* of an α-ary relation R we understand the set $\bigcup_{x \in R} Rg\,x$; this is the least class A such that $R \subseteq {}^\alpha A$. A relation whose field is included in A is called a *relation on A*. By the *characteristic function of an α-ary relation R on a class A* we understand the function H defined by the formula $H = (R \times \{1\}) \cup (({}^\alpha A \sim R) \times \{0\})$, i.e., the function H with $Do\,H = {}^\alpha A$ such that $Hx = 1$ if $x \in R$ and $Hx = 0$ if $x \in {}^\alpha A \sim R$. If R is an α-ary relation, then α is referred to as a *rank* of R. A non-empty relation R uniquely determines its rank, so that we can speak of *the* rank of R. Unfortunately, the empty relation has every ordinal as a rank. This is a cause of some inconveniences. Restricting ourselves to relations on an arbitrary non-empty class A fixed in advance, we can avoid these inconveniences by means of the following observations: We notice that, if H is the characteristic function of some R which, for some α, is an α-ary relation on A, then α as well as R is uniquely determined by H; thus α can be referred to as *the rank* of H. We also notice that for every α there is a one-one correspondence between α-ary relations on A and their characteristic functions. Thus we can redefine the notion of an α-ary relation on A simply by identifying them with characteristic functions, and in this way every α-ary relation on A will have a definite rank; see, for instance, Fraïssé [54*]. The only essential difference between the old and the new notion of an α-ary relation on A is that under the new conception we have many copies of the old empty relation, one copy in each rank. Whenever in our further discussion we shall speak of the rank of a relation on A without assuming that $R \neq 0$, the reader should bear in mind the possibility of re-defining the notion of a relation on A in the way just indicated.

Given an ordinal α and a non-empty class A, by an *α-ary operation* on A we understand any function O from ${}^{\alpha}A$ into A, i.e., in case A is a set, any member of ${}^{{}^{\alpha}A}A$. For an α-termed sequence $x = \langle x_0, ..., x_\xi, ...\rangle_{\xi<\alpha}$ we frequently write $O(x_0, ..., x_\xi, ...)_{\xi<\alpha}$ instead of Ox; the notations $O(x_\xi : \xi < \alpha)$ and $O(x_0, ..., x_{\alpha-1})$ (in case $0 < \alpha < \omega$) are similarly understood. If O is an α-ary operation on A for some α and A, then α is uniquely determined by O; we call α the *rank* of O, in symbols $\rho O = \alpha$. We say that a class B is *closed* under O if $O^*({}^{\alpha}B) \subseteq B \subseteq A$. For $\alpha = 1, 2, 3, ...$ we speak of *unary, binary, ternary, ...* relations and operations. In general an α-ary relation or operation is called *finitary* if $\alpha < \omega$. The term "binary relation" is of course ambiguous, but from what was pointed out before this ambiguity can hardly lead to any confusion. As examples of binary operations on the class of all sets we may mention the operation $\cup$ of forming union and the operation $\cap$ of forming intersection. (Thus we extend to binary operations the conventions for introducing relation and function symbols which were previously stated.)

With every α-ary operation O on a class A we can correlate an $(\alpha+1)$-ary relation R with field $\subseteq A$ by letting $R = \{x^\frown\langle Ox\rangle : x \in {}^{\alpha}A\}$. (If $\alpha \neq 0$, the field of R is actually equal to A; if $\alpha = 0$ and $A \neq 0$, R is of the form $\{\langle a\rangle\}$ and so its field is in general not equal to A.) This correlation is of course one-one; in certain situations it proves convenient to identify α-ary operations with the correlated $(\alpha+1)$-ary relations.

Every binary operation O on a class A can be extended in a natural way to an operation on finite sequences $x = \langle x_0, ..., x_{\kappa-1}\rangle$ of positive length with all terms x_λ in A. The result of this extended operation is denoted by $x_0 O ... O x_{\kappa-1}$ or sometimes by $x_0 O x_1 O ... O x_{\kappa-1}$; we define recursively

$$x_0 O ... O x_{\kappa-1} = x_0 \text{ for } \kappa = 1,$$

$$x_0 O ... O x_{\kappa-1} = (x_0 O ... O x_{\kappa-2}) O x_{\kappa-1} \text{ for } \kappa > 1.$$

(We may say that this extension is made by association to the left; in general there are also other, equally natural but not equivalent, ways of extending the operation O.) Thus starting with the binary operation $\times$ we arrive at the notion of the Cartesian product $A_0 \times ... \times A_{\kappa-1}$ of sets $A_0, ..., A_{\kappa-1}$. In a similar way, starting with $|$ and $\circ$ we arrive at the notions of the relative product $R_0 | ... | R_{\kappa-1}$ of relations $R_0, ..., R_{\kappa-1}$ and the composition $f_0 \circ ... \circ f_{\kappa-1}$ of functions $f_0, ..., f_{\kappa-1}$. For any $x \in Do(f_0 \circ ... \circ f_{\kappa-1})$ we can write $f_0 ... f_{\kappa-1} x$

instead of $(f_0 \circ \ldots \circ f_{\kappa-1})x$. In case $R_\lambda = S$, or $f_\lambda = g$, for all $\lambda < \kappa$ we let $R_0 | \ldots | R_{\kappa-1} = S^{[\kappa]}$ (the κth *relative power* of S), or $f_0 \cdots f_{\kappa-1} = g^{[\kappa]}$ (the κth *iteration* of g).

In informal discourse we shall often use the term "operation" in a much looser sense than in the above discussion. Thus, for example, we shall apply this term to any function whose domain consists of arbitrary systems $\langle x_i : i \in I \rangle$ with $\{x_i : i \in I\} \subseteq A$, or of sets $X \subseteq A$, and whose range is a subset of A or sometimes even a subset of $Sb\,A$.

An element of ${}^A A$ is called a *transformation* of A. By a *finite transformation* of a set A we understand a function $f \in {}^A A$ such that the set $B = \{x : x \in A, fx \neq x\}$ is finite. Since $(A \sim B) 1 f = (A \sim B) 1 Id$, f is uniquely determined by its restriction $B 1 f$. If B is represented as the range of a finite sequence without repeating terms, $B = \{x_\kappa : \kappa < \nu\}$, and if $fx_\kappa = y_\kappa$ for every $\kappa < \nu$, then the finite transformation f is denoted by $[x_\kappa / y_\kappa : \kappa < \nu]_A$ or $[x_0/y_0, \ldots, x_{\nu-1}/y_{\nu-1}]_A$. (We usually omit the subscript $_A$ in the expression $[x_0/y_0, \ldots, x_{\nu-1}/y_{\nu-1}]_A$.) As particular cases of the general notation we use in the obvious sense the expressions $[x/y]$, $[x/y, x'/y']$ where $x \neq x'$, etc. The transformation $[x/y]$ is called a *replacement on* A or, more specifically, the *replacement of* x *by* y *in* A; $[x/y, y/x]$ (with $x \neq y$) is called a *transposition on* A or the *transposition of* x *and* y *in* A. It is easily shown that every finite transformation of A can be expressed as the composition of a finite sequence of replacements and transpositions. In certain portions of this work we shall deal with finite transformations of an ordinal or a set of ordinals. In such a case we shall assume that B has been represented as the range of a *strictly increasing* sequence of ordinals; i.e., $B = \{\xi_\kappa : \kappa < \nu\}$ where $\xi_0 < \xi_1 < \ldots < \xi_{\nu-1}$. Under this assumption the representation $f = [\xi_0/\eta_0, \ldots, \xi_{\nu-1}/\eta_{\nu-1}]$ (where $\eta_\kappa = f\xi_\kappa$ for $\kappa < \nu$) is uniquely determined by the finite transformation f and we can refer to it as the *canonical representation* of f.

By a *relational structure* we understand an ordered triple $\mathfrak{A} = \langle A, R, O \rangle$ where A is a non-empty set, called the *universe* of $\mathfrak{A}$, $R = \langle R_i : i \in I \rangle$ is a system of finitary relations on A, and $O = \langle O_j : j \in J \rangle$ is a system of finitary operations on A. Instead of $\langle A, R, O \rangle$ we shall sometimes write $\langle A, R_i, O_j \rangle_{i \in I, j \in J}$. We do not exclude the possibility that I or J is empty, in which case we write $\langle A, O \rangle$ or $\langle A, R \rangle$. Since ν-ary operations on A can be identified with special $(\nu+1)$-ary relations on A, there would be no loss of generality if we restricted ourselves to relational structures of the form $\langle A, R \rangle$; however, the use of the notation $\langle A, R, O \rangle$ proves to be more convenient in many situations. If R equals $\langle S \rangle$, or $\langle S, T \rangle$, etc., and J is empty, we write $\langle A, S \rangle$, or $\langle A, S, T \rangle$, etc. instead of $\langle A, R, O \rangle$, and similarly in other analogous cases. (The reader may have noticed that our symbolism here, as well as in some other places, is not unequivocal; since, for instance, we do not have any fixed stipulations concerning the use of the variables $R, O, S, T, \ldots$, the expression $\langle A, S, T\rangle$ may represent either an arbitrary relational structure just as $\langle A, R, O\rangle$, or a special structure $\langle A, R, O\rangle$ in which $R = \langle S, T\rangle$ and $O = 0$. We hope, however, that in each individual case the meaning of our symbolic expressions will be unambiguous.) When discussing a finitary relation $S \neq 0$, it is for the most part irrelevant whether we consider S itself or the relational structure $\langle A, S\rangle$ in which $A = FdS$, e.g., instead of discussing partial orderings, or simple orderings, or well-orderings, S we can discuss the corresponding structures $\langle A, S\rangle$ called respectively *partial ordering*, or *simple ordering*, or *well-ordering structures*. Two relational structures $\langle A, R_i, O_j\rangle_{i\in I, j\in J}$ and $\langle A', R'_i, O'_j\rangle_{i\in I', j\in J'}$, are called *similar* if $I = I'$, $J = J'$, any two relations R_i and R'_i with $i \in I$ are of the same rank, and so are any two operations O_j and O'_j with $j \in J$.

A relational structure $\langle A, O_j\rangle_{j\in J}$ in which all the operations O_j are of positive rank is called an *algebraic structure*, or an *algebra*. (The exclusion of operations of rank 0 is not essential, but proves convenient for our purposes.) A general discussion of algebraic structures will be found in Chapter 0 of this work. The subsequent chapters will be devoted to a detailed study of a special kind of algebraic structures, namely cylindric algebras. It should be noticed that various terminological conventions, notions, and results in Chapter 0 can easily be extended to arbitrary relational structures; see, e.g., the remarks in 0.1.3, the notions of isomorphism in 0.2.1 and isomorphism type in 0.2.12, and the discussion of direct products in Section 0.3.

With every well-ordering structure $\langle A, S\rangle$ we can correlate an ordinal α by the following condition: there is a sequence x of length α such that $Rgx = A$ and such that, for any ordinals $\xi, \eta < \alpha$ the formulas $\xi \leqq \eta$ and $x_\xi S x_\eta$ are equivalent. The ordinal uniquely determined by this condition is called the *order type* or, simply, the *type* of the structure $\langle A, S\rangle$. We often refer to α as the (order) type of the set A under the relation S (where an explicit reference to S is occasionally omitted). The notion of an order type is usually introduced in a more general way so that it applies to arbitrary simple ordering structures; it is then defined as a particular case of the notion of isomorphism type.

Besides the relational structures just considered, which should properly be called *first-order relational structures*, we can also consider *second-order relational structures*, such as pairs $\langle A, R\rangle$ where A is a non-empty set and R is a system of finitary relations with fields included in SbA, triples $\langle A, R, O\rangle$ where $\langle A, R\rangle$ is as before and O is a system of finitary operations on SbA, and others; analogously we can consider third and higher-order relational structures.

Two important examples of second-order structures are closed-set structures and closure structures. A *closed-set structure* is a pair $\langle A, F\rangle$ satisfying the conditions: $F \subseteq SbA$, and $\bigcap L \in F$ whenever $L \subseteq F$ (so that, in particular, $A = \bigcap 0 \in F$). F is called the *family of closed sets* of the closed-set structure $\langle A, F\rangle$. A *closure structure* is a pair $\langle A, C\rangle$ which satisfies the conditions: C is a unary operation on SbA; $X \subseteq CX = CCX$; whenever $X \subseteq Y \subseteq A$ we have also $CX \subseteq CY$. C is called the *closure operation* of the closure structure $\langle A, C\rangle$. There is a close connection between these two kinds of structures. In fact, with every closed-set structure $\mathfrak{A} = \langle A, F\rangle$ we can correlate a closure structure $\mathfrak{A}^* = \langle A, C\rangle$ by letting $CX = \bigcap\{Y : X \subseteq Y \in F\}$ for every $X \subseteq A$. This correlation proves to be one-one, and given a closure structure $\mathfrak{A}^* = \langle A, C\rangle$ we can recover the closed-set structure $\mathfrak{A} = \langle A, F\rangle$ with which $\mathfrak{A}^*$ is correlated by letting $F = \{X : CX = X \subseteq A\}$. A closed-set structure $\mathfrak{A} = \langle A, F\rangle$ is called *inductive* if $\bigcup L \in F$ for every non-empty subfamily L of F directed by inclusion. Analogously, a closure structure $\mathfrak{A} = \langle A, C\rangle$ is called *inductive* if $C(\bigcup L) = \bigcup\{CX : X \in L\}$ for every non-empty subfamily L of SbA directed by inclusion. Under the correlation described above inductive closed-set structures go into inductive closure structures and conversely. A closed-set structure $\mathfrak{A} = \langle A, F\rangle$ is called *complete* if $\bigcup L \in F$ for *every* non-empty subfamily L of F; such structures are correlated with *complete closure structures*, i.e., closure structures $\langle A, C\rangle$ in which $C(\bigcup L) = \bigcup\{CX : X \in L\}$ for every non-empty $L \subseteq SbA$. Obviously, complete closed-set structures are inductive, and similarly for complete closure structures. As we shall see in Chapter 0, several important instances of closed-set structures and closure structures are involved in the general theory of algebraic structures.[1)]

To finish this survey of set-theoretical notions we discuss the notion of a filter. A *filter* on a set A is a family $F \subseteq SbA$ such that: (i) $A \in F$, (ii) $B \cap C \in F$ whenever $B, C \in F$, and (iii) $C \in F$ whenever $B \in F$ and $B \subseteq C \subseteq A$. F is called *principal* if it has the form $\{X : B \subseteq X \subseteq A\}$ for some $B \subseteq A$. F is *proper* if $F \subset SbA$. An *ultrafilter* on A is a proper filter F on A such that either $B \in F$ or ${}_A{\sim}\, B \in F$ for every $B \subseteq A$. Every proper filter is included in an ultrafilter, and consequently non-principal ultrafilters exist on every infinite set. A family $F \subseteq SbA$ has the *finite intersection property* provided that

1) Inductive closure structures were first considered in Tarski [30*], where the additional assumption $|A| \leqq \omega$ occurs which is, however, of no consequence for most of the notions and results there. See also J. Schmidt [52*], [53*].

$\bigcap L \neq 0$ for every finite subset L of F. Every family with the finite intersection property is included in a proper filter and hence in an ultrafilter.

II. METALOGICAL NOTIONS

Various metalogical notions will be involved in the discussions of this work; all of them concern the well-known first-order predicate logic. The role of these notions in our development is twofold. On the one hand, as mentioned in the Foreword, cylindric algebras have been specifically designed as an algebraic apparatus for studying predicate logic. On the other hand, when discussing cylindric algebras, or for that matter any other kind of algebraic structures, we usually base the discussion on a broad set-theoretical basis without restricting ourselves to any specific formal language; frequently however we are interested in the most fundamental and elementary part of the theory of these algebras, which can be developed without set-theoretical apparatus and can therefore be formalized within predicate logic.

We begin with the description of the language of predicate logic or, as it is frequently called, the *first-order predicate language*. Actually we will consider not one language of predicate logic but a whole variety of such languages with essentially the same structure. The basic components of each of these languages (as of any other formal language) are *symbols* and (symbolic) *expressions*. Expressions are treated as finite sequences of symbols, and symbols are systematically identified with expressions of length 1. The fundamental operation by which we form compound expressions from simpler ones is that of concatenation.

The set of symbols may be called the *vocabulary* of the language. It is divided into four disjoint sets: the *variables*, the *logical constants*, the *relation symbols* (also called *predicates*), and the *operation symbols*; symbols of the last two sets are sometimes referred to jointly as *non-logical constants*. We may assume that the form of the symbol determines to which of the four sets it belongs. The variables are thought of as forming the set of all terms of a sequence $\upsilon = \langle \upsilon_0, \ldots, \upsilon_\xi, \ldots \rangle_{\xi<\alpha}$ of an arbitrary length α. Analogously, the relation symbols are represented as the terms of a system $\Gamma = \langle \Gamma_i : i \in I \rangle$ and the operation symbols as the terms of a system $\Delta = \langle \Delta_j : j \in J \rangle$ with arbitrary index sets I and J. The set Σ of logical constants, the sequence υ, and the systems Γ and Δ actually determine a given language Λ of predicate logic; we can denote this language by $\Lambda_{\Sigma,\upsilon,\Gamma,\Delta}$. Conversely, a language Λ uniquely determines the set of logical constants, the sequence of variables, and the systems of relation symbols and operation symbols from which its expressions are formed. (If we wish, we can identify Λ with the ordered quadruple $\langle \Sigma, \upsilon, \Gamma, \Delta \rangle$, thus interpreting the term "language" in a rather unusual way.)

It seems convenient to assume that essentially the same symbols are used as variables in all languages. More precisely, if $\langle \upsilon_0, \ldots, \upsilon_\xi, \ldots \rangle_{\xi<\alpha}$ and $\langle \upsilon'_0, \ldots, \upsilon'_\xi, \ldots \rangle_{\xi<\alpha'}$ are the sequences of variables in languages Λ and Λ', then $\upsilon_\xi = \upsilon'_\xi$ for every ξ which is less than both α and α'. Thus, with every ordinal ξ a definite symbol is correlated which we shall denote by $\boldsymbol{v}_\xi$, and in every language Λ the sequence of variables is $\langle \boldsymbol{v}_0, \ldots, \boldsymbol{v}_\xi, \ldots \rangle_{\xi<\alpha}$ for some α; thus only α varies from one language to another. For most purposes, however, we can stipulate that $\alpha = \boldsymbol{\omega}$; in our further discussion, unless specified to the contrary, this stipulation will always be tacitly assumed, and all references to α will be omitted. For the first terms of the sequence $\boldsymbol{v}$ we introduce a typographically more convenient and somewhat shorter notation: $\boldsymbol{x} = \boldsymbol{v}_0$, $\boldsymbol{y} = \boldsymbol{v}_1$, etc.

We assume that Σ consists of seven symbols: three sentential connectives — the *disjunction symbol* $\boldsymbol{A}$, the *conjunction symbol* $\boldsymbol{K}$, and the *negation symbol* $\boldsymbol{N}$—, the *existential quantifier* $\boldsymbol{Q}$, the *equality* (*identity*) *symbol* $\boldsymbol{E}$, the *truth symbol* $\boldsymbol{T}$, and the *falsehood symbol* $\boldsymbol{F}$. With the first five of these symbols we correlate operations on expressions. In fact, for any two expressions φ and ψ we put

$$\varphi \vee \psi = \boldsymbol{A}^\frown\varphi^\frown\psi \quad (\textit{the disjunction of } \varphi \textit{ and } \psi);$$

$$\varphi \wedge \psi = \boldsymbol{K}^\frown\varphi^\frown\psi \quad (\textit{the conjunction of } \varphi \textit{ and } \psi);$$

furthermore given any variable υ and any expression φ we let

$$\neg\varphi = \boldsymbol{N}^\frown\varphi \qquad (\textit{the negation of } \varphi);$$

$$\exists_\upsilon\varphi = \boldsymbol{Q}^\frown\upsilon^\frown\varphi \quad (\textit{the existential quantification of } \varphi \textit{ with respect to } \upsilon).$$

Thus, for every variable υ, $\exists_\upsilon$ is a unary operation on the set of expressions. The operation correlated with the equality symbol will be discussed later. Actually in the whole subsequent discussion we shall use not the metalogical designations $\boldsymbol{A}$, $\boldsymbol{K}$, ... of logical constants but the designations of the correlated operations $\vee, \wedge, \ldots$. (Note that in a disjunction $\varphi \vee \psi$ or a conjunction $\varphi \wedge \psi$ the sentential connective is assumed to precede the expressions φ and ψ. Due to this assumption we avoid the need for introducing parentheses in our formal language; see Łukasiewicz [63*].)

In terms of these operations we define several related operations on expressions. For any given expressions φ, ψ and any variable υ we let

$\varphi \rightarrow \psi = (\neg\varphi) \vee \psi$ (*the implication with hypothesis* φ *and conclusion* ψ);

$\varphi \leftrightarrow \psi = (\varphi \rightarrow \psi) \wedge (\psi \rightarrow \varphi)$ (*the equivalence between* φ *and* ψ);

$\forall_{\upsilon}\varphi = \neg\exists_{\upsilon}\neg\varphi$ (*the universal quantification of* φ *with respect to* υ).

The composition of finite sequences of existential quantifications $\exists_{\upsilon_0}, \ldots, \exists_{\upsilon_{\kappa-1}}$, or universal quantifications $\forall_{\upsilon_0}, \ldots, \forall_{\upsilon_{\kappa-1}}$, will be denoted respectively by $\exists_{\upsilon_0 \ldots \upsilon_{\kappa-1}}$, or $\forall_{\upsilon_0 \ldots \upsilon_{\kappa-1}}$.

It should be emphasized that we can simultaneously modify in various ways the set of logical constants for all languages of predicate logic if such a modification proves convenient for certain purposes. For instance, we can omit one of the sentential connectives A and K, say K, and define the operation $\wedge$ in terms of the operations $\neg$ and $\vee$: $\varphi \wedge \psi = \neg(\neg\varphi \vee \neg\psi)$. Or we can add a new sentential connective, say the *implication symbol* C, and define the operation $\rightarrow$ in its terms: $\varphi \rightarrow \psi = C^\frown\varphi^\frown\psi$. Or, finally, we can either replace the existential quantifier by the universal quantifier or include both quantifiers in the set Σ.

Since we have agreed to use the same symbols for variables and logical constants in all languages, a language Λ turns out to be determined entirely by the systems Γ and Δ of relation and operation symbols, and possibly by the length α of the sequence of variables. We can thus use a simplified notation: $\Lambda = \Lambda_{\alpha,\Gamma,\Delta}$, and even $\Lambda = \Lambda_{\Gamma,\Delta}$ in case $\alpha = \omega$.

As regards the operation and relation symbols we stipulate that with each of them some finite ordinal is correlated which is called the *rank* (*place-number*) of the symbol; we may assume that the rank of the symbol can be decoded from its form. Operation symbols of rank 0 are called *individual constants*. With every operation symbol Δ_j of rank $\kappa \neq 0$ we correlate a κ-ary operation $\hat{\Delta}_j$ on the set of all expressions, defined by the formula

$$\hat{\Delta}_j(\sigma_0, \ldots, \sigma_{\kappa-1}) = \Delta_j{}^\frown\sigma_0{}^\frown\ldots{}^\frown\sigma_{\kappa-1}$$

for any sequence $\langle\sigma_0, \ldots, \sigma_{\kappa-1}\rangle$ of expressions. If Δ_j is of rank two we shall most often write $\sigma_0\hat{\Delta}_j\sigma_1$ instead of $\hat{\Delta}_j(\sigma_0, \sigma_1)$. The sign over Δ_j will frequently be omitted if this omission does not seem to cause any confusion. The intersection of all sets of expressions which contain all the variables and individual constants as elements and are closed under the operations $\hat{\Delta}_j$ correlated with all the operation symbols Δ_j of positive rank occurring in the language Λ is called the set of *terms* and is denoted by $T\mu^{(\Lambda)}$ or simply $T\mu$. In some situations it is convenient to make evident the dependence of the set $T\mu^{(\Lambda)}$, not on the whole language Λ, but only on the number α of variables occurring in Λ. In such cases we shall denote this set by $T\mu^{(\alpha)}$ and we shall refer to its elements as *α-terms*. A similar notation will be applied when needed to other symbolic expressions introduced below, such as $\Phi\mu^{(\Lambda)}$.

Relation symbols of rank 0 are called *sentential constants.* With every relation symbol Γ_i of rank $\kappa \neq 0$ we again correlate a κ-ary operation on expressions which is denoted by $\hat{\Gamma}_i$, often simplified to Γ_i, and is defined analogously to $\hat{\Delta}_j$. Again in case $\kappa = 2$ we usually write $\sigma_0\hat{\Gamma}_i\sigma_1$ instead of $\hat{\Gamma}_i(\sigma_0, \sigma_1)$. The logical constants $\boldsymbol{T}$, $\boldsymbol{F}$, and $\boldsymbol{E}$, though not included in the system $\langle\Gamma_i : i \in I\rangle$, are treated as relation symbols — $\boldsymbol{T}$ and $\boldsymbol{F}$ as sentential constants, and $\boldsymbol{E}$ as a symbol of rank 2. The binary operation correlated with $\boldsymbol{E}$ is denoted by $=$; thus $\sigma \neq \tau$ coincides with $\boldsymbol{E}^\frown\sigma^\frown\tau$. For $\neg(\sigma = \tau)$ we shall write $\sigma = \tau$.

By an *atomic formula* in the language Λ we understand either a sentential constant of Λ, or an expression $\hat{\Gamma}_i(\sigma_0, \ldots, \sigma_{\kappa-1})$ where Γ_i is any relation symbol of Λ of rank $\kappa \neq 0$ and $\sigma_0, \ldots, \sigma_{\kappa-1} \in T\mu$, or finally an expression $\sigma = \tau$ where σ and τ are terms; the atomic formulas of the last kind are called *equations.* The intersection of all sets of expressions containing every atomic formula of Λ as an element and closed under the operations $\vee$, $\wedge$, $\neg$, and $\exists_\upsilon$ for every variable υ is called the set of *formulas* in Λ and is denoted by $\Phi\mu^{(\Lambda)}$ or simply $\Phi\mu$.

We assume known what it means that a variable υ *occurs free at the* κ^{th} *place* in a formula φ, and also what it means that a formula ψ has been obtained from a formula φ by *simultaneously substituting* the terms $\sigma_0, \sigma_1, \ldots$ for all free occurrences of variables $\upsilon_0, \upsilon_1, \ldots$. We say that a variable occurs *free* or *bound* in a given formula if it occurs so at some place; a formula in which no variable occurs free is called a *sentence*, and the set of all sentences in Λ is denoted by $\Sigma\nu^{(\Lambda)}$ or simply $\Sigma\nu$. If φ is a formula and $\langle\upsilon_0, \ldots, \upsilon_{\lambda-1}\rangle$ is the sequence of all variables occurring free in φ with $\upsilon_0 = \boldsymbol{v}_{\kappa_0}, \ldots, \upsilon_{\lambda-1} = \boldsymbol{v}_{\kappa_{\lambda-1}}$, and $\kappa_0 < \ldots < \kappa_{\lambda-1}$, then the *closure* of φ, in symbols $[\varphi]$, is the sentence $\forall_{\upsilon_0\ldots\upsilon_{\lambda-1}}\varphi$. A sentence which is the closure of a quantifier-free formula is called a *universal sentence*; the closure of an equation is called an *identity*.

We now turn to semantical (model-theoretical) notions which are relevant for our discussion. The most fundamental among them is the notion of a model. We want to explain roughly under what conditions a relational structure is a model of a set of sentences in a language Λ.

Let $\mathfrak{A} = \langle A, R, O\rangle$ be a relational structure, with $R = \langle R_i : i \in I\rangle$ and $O = \langle O_j : j \in J\rangle$, and let $\Lambda = \Lambda_{\alpha,\Gamma,\Delta}$ be a language of predicate logic, with $\Gamma = \langle\Gamma_i : i \in I'\rangle$ and $\Delta = \langle\Delta_j : j \in J'\rangle$. We say that $\mathfrak{A}$ is a *realization structure* (or a *possible realization*, or a *possible model*) *for* Λ in case $I = I'$, $J = J'$, the relation R_i and the corresponding relation symbol Γ_i are of the same rank for every $i \in I$, and so are the operation O_j and the operation symbol Δ_j; under the same conditions we say that Λ is a *discourse language* for $\mathfrak{A}$. Obviously, if K is any class of similar relational structures, and one of them is a realization structure for a language Λ, then so is any other structure in K; in this case K is called a *realization class* for Λ, and Λ a *discourse language* for K. An analogous remark can be made in the opposite direction: if $\Lambda = \Lambda_{\alpha,\Gamma,\Delta}$ is a discourse language for a given structure $\mathfrak{A}$ or a given class K of similar structures, the same applies to any other language $\Lambda' = \Lambda_{\alpha',\Gamma',\Delta'}$ such that the index sets of Γ and Γ', as well as of Δ and Δ', coincide, and two corresponding relation symbols Γ_i and Γ'_i, as well as two corresponding operation symbols Δ_j and Δ'_j, always have the same rank. Except for the possible difference between α and α' two such languages Λ and Λ' differ in a very inessential way, in fact, merely in the form of their relation and operation symbols. In by far the most important case, when the ordinals α and α' are infinite, the possible difference between them has very little influence on the content of the notions we shall introduce. We shall talk as if a well-determined discourse language Λ were correlated with any relational structure $\mathfrak{A}$ or any class K of similar structures; we may sometimes refer to Λ simply as the language of $\mathfrak{A}$ or K. All of the notions introduced in the next few pages should be relativized to Λ, but we do not indicate this in our notation.

With any given realization structure $\mathfrak{A} = \langle A, R_i, O_j \rangle_{i \in I, j \in J}$ for Λ and any given term $\tau \in T\mu^{(\Lambda)}$ we now correlate a definite α-ary operation on A, denoted by $\tilde{\tau}^{(\mathfrak{A},\alpha)}$ or $\tilde{\tau}^{(\mathfrak{A})}$, and called the *operation in* $\mathfrak{A}$ *represented by* τ. A precise definition of $\tilde{\tau}^{(\mathfrak{A})}$ involves a recursion on terms; we shall not formulate it explicitly and shall restrict ourselves to an example. Suppose that O_j is a binary operation in $\mathfrak{A}$ and Δ_j is the corresponding binary operation symbol in Λ. Take $\mathbf{v}_0 \Delta_j(\mathbf{v}_1 \Delta_j \mathbf{v}_0)$ for τ; then $\tilde{\tau}^{(\mathfrak{A})}$ is defined by the formula

$$\tilde{\tau}^{(\mathfrak{A})} = \langle x_0 O_j(x_1 O_j x_0) : x \in {}^{\alpha}A \rangle.$$

More generally we can speak of a *β-ary operation in* $\mathfrak{A}$ *represented by* τ, in symbols $\tilde{\tau}_\beta^{(\mathfrak{A},\alpha)}$ or $\tilde{\tau}_\beta^{(\mathfrak{A})}$, for any $\beta \leqq \alpha$. This operation is defined only for those terms τ in which no variable $\mathbf{v}_\xi$ with $\xi \geqq \beta$ occurs; since the value of $\tilde{\tau}^{(\mathfrak{A})}x$, $x \in {}^{\alpha}A$, does not depend on values of x_ξ with $\xi \geqq \beta$, $\tilde{\tau}_\beta^{(\mathfrak{A})}$ is fully determined by the stipulation

$$\tilde{\tau}_\beta^{(\mathfrak{A})}(\beta 1 x) = \tilde{\tau}^{(\mathfrak{A})} x$$

for every $x \in {}^{\alpha}A$.

We next correlate with the structure $\mathfrak{A}$ and an arbitrary formula φ in Λ an α-ary relation on A denoted by $\tilde{\varphi}^{(\mathfrak{A},\alpha)}$ or $\tilde{\varphi}^{(\mathfrak{A})}$ and called the *relation in* $\mathfrak{A}$ *defined by* φ. This is done by recursion on formulas. As an example, suppose

O_j and Δ_j are as before, and take $\exists_{v_1} v_0 \Delta_j v_1 = v_2$ for φ; then

$$\tilde{\varphi}^{(\mathfrak{A})} = \{x : x \in {}^{\alpha}A \text{ and } x_0 O_j y = x_2 \text{ for some } y \in A\}.$$

If $\boldsymbol{T}$, or $\boldsymbol{F}$, is taken for φ, we let $\tilde{\varphi}^{(\mathfrak{A})} = {}^{\alpha}A$, or $\tilde{\varphi}^{(\mathfrak{A})} = 0$, respectively. Again we can speak of the *β-ary relation* $\tilde{\varphi}_{\beta}^{(\mathfrak{A},\alpha)}$ or $\tilde{\varphi}_{\beta}^{(\mathfrak{A})}$ *in* $\mathfrak{A}$ *defined by* φ for any $\beta \leqq \alpha$; it is assumed that no variable v_{ξ}, $\xi \geqq \beta$, occurs free in φ, and the relation is determined by the stipulation

$$\tilde{\varphi}_{\beta}^{(\mathfrak{A})} = \{\beta 1 x : x \in \tilde{\varphi}^{(\mathfrak{A})}\}.$$

A β-ary operation Q on A, $\beta < \alpha$, is said to be *defined by* φ if the corresponding $(\beta+1)$-ary relation on A is defined by φ in the sense just indicated.

Instead of saying that a sequence $x \in {}^{\alpha}A$ belongs to $\tilde{\varphi}^{(\mathfrak{A})}$, we say that x *satisfies* φ in $\mathfrak{A}$, in symbols $\mathfrak{A} \models \varphi[x]$. We now stipulate that a formula φ is *valid* or *holds in a structure* $\mathfrak{A}$ or that $\mathfrak{A}$ *is a model of* φ, in symbols $\mathfrak{A} \models \varphi$, if x satisfies φ for every $x \in {}^{\alpha}A$; this terminology is applied mostly in case the formula φ is a sentence. More generally, φ is said to *hold in a class* K of realization structures for Λ, in symbols $\mathsf{K} \models \varphi$, if $\mathfrak{A} \models \varphi$ for every $\mathfrak{A} \in \mathsf{K}$. By the *elementary* (*first-order*) *theory* of a structure $\mathfrak{A}$, or a class K of structures, in symbols $\Theta\rho\,\mathfrak{A}$, or $\Theta\rho\,\mathsf{K}$, we understand the set of all sentences of the language Λ which hold in $\mathfrak{A}$, or K, respectively. Two structures $\mathfrak{A}$ and $\mathfrak{B}$ are called *elementarily equivalent* if $\Theta\rho\,\mathfrak{A} = \Theta\rho\,\mathfrak{B}$. If K is the class of all realization structures for Λ, then $\Theta\rho\,\mathsf{K}$ is the class of all *logically valid sentences*; we could call $\Theta\rho\,\mathsf{K}$ the *logic* of the language Λ. A structure $\mathfrak{A}$ is called a *model* of a set Φ of formulas (sentences) if $\mathfrak{A}$ is a model of every $\varphi \in \Phi$. The class of all models of a formula φ, or a set Φ, is denoted by $\mathsf{Md}\,\varphi$, or $\mathsf{Md}\,\Phi$. By saying that a formula φ, or a set Φ, *characterizes* a class K of structures we simply mean that $\mathsf{K} = \mathsf{Md}\,\varphi$, or $\mathsf{K} = \mathsf{Md}\,\Phi$.

A class K of structures is called an *elementary class in the narrower sense*, for brevity an $\mathscr{EC}$, if K can be characterized by a single sentence or, what amounts to the same, by a finite set of sentences in the discourse language Λ. It is called an *elementary class in the wider sense* or simply an *elementary class*, for brevity an $\mathscr{EC}_{\Delta}$, if there is any set of sentences in Λ which characterizes it, or, equivalently, if $\mathsf{K} = \mathsf{Md}\,\Theta\rho\,\mathsf{K}$. K is said to be respectively a *universal class in the narrower sense*, or a *universal class* (*in the wider sense*), — for brevity, a $\mathscr{UC}$, or a $\mathscr{UC}_{\Delta}$, — if there is a universal sentence in Λ, or a set of universal sentences in Λ, which characterizes K. Finally we say that K is respectively an *equational class in the narrower sense*, an $\mathscr{EQC}$, or an *equational class* (*in the wider sense*), an $\mathscr{EQC}_{\Delta}$, if there is a finite, or an arbitrary, set of identities which characterizes K. An $\mathscr{EQC}_{\Delta}$ is also called a *primitive class* or a *variety*.

Given a formula φ and a set Σ of sentences we say that φ is a *consequence* of Σ, in symbols $\Sigma \vdash \varphi$, if $\mathsf{Md}\,\Sigma \subseteq \mathsf{Md}\,\varphi$. Closely related to the notion of consequence is that of derivability. To define this latter notion we first single out a certain set $A\xi$ of sentences (all of which are logically valid), called *logical axioms*, and we then say that a formula φ in Λ is *derivable* from Σ if its closure $[\varphi]$ is in every set Θ of sentences in Λ which includes $A\xi \cup \Sigma$ and is such that ψ is in Θ whenever χ and $\chi \to \psi$ are in Θ; see, e.g., Quine [55*] or Tarski [65*]. In case the length α of the sequence of variables is infinite, it is well known how the set $A\xi$ may be chosen, and under such a choice of $A\xi$ the notions of consequence and of derivability prove to be coextensive; this is the content of the famous *completeness theorem*. In case α is finite, the situation is more involved; we shall encounter this problem in the second part of our work.

Two terms σ and τ in a language Λ are called *equivalent* with respect to a set Σ of sentences, in symbols $\sigma \equiv_\Sigma \tau$, if $\Sigma \vdash \sigma = \tau$. Similarly, two formulas φ and ψ are called *equivalent* with respect to Σ, in symbols $\varphi \equiv_\Sigma \psi$, if $\Sigma \vdash \varphi \leftrightarrow \psi$. Strictly speaking we should use the symbol $\equiv_\Sigma^{(\Lambda)}$; as usual, however, the reference to Λ is omitted. Still in certain situations it proves convenient to make evident the length α of the sequence of variables; we write then $\equiv_\Sigma^{(\alpha)}$. In case $\Sigma = 0$, we use the simpler symbol $\equiv$ or $\equiv^{(\alpha)}$. (Notice that for two terms σ and τ the formulas $\sigma \equiv \tau$ and $\sigma = \tau$ are equivalent, i.e., σ and τ are equivalent with respect to the empty set of formulas iff they are identical; this is not true, however, in the case of formulas.) In case $\Sigma = \Theta\rho\,\mathsf{K}$, where K is a given class of structures, we write $\equiv_{\mathsf{K}}$ instead of $\equiv_\Sigma$.

An important corollary of the completeness theorem is the *compactness theorem* for predicate logic; by this theorem, whenever φ is a consequence of a set Σ of sentences, it is also a consequence of some finite subset of Σ. This result, originally established for α infinite, easily extends to finite α. Hence, if we denote by Σ the set of all sentences in Λ and by $C\Theta$ the set of all consequences of a set $\Theta \subseteq \Sigma$, then the pair $\langle \Sigma, C\rangle$ proves to be an inductive closure structure. A set Θ of sentences is called a *theory* or a *deductively closed set* if $C\Theta = \Theta \subseteq \Sigma$; instead of saying that a sentence belongs to a theory Θ, we sometimes say that it is *valid* in Θ. More generally, we say that a formula φ is *valid* in Θ if its closure $[\varphi] \in \Theta$. Similarly, a formula φ is said to be *logically valid* if its closure is a logically valid sentence in the sense previously explained. Thus, if T denotes the set of all theories in the language Λ, then the pair $\langle \Sigma, T\rangle$ is the inductive closed-set structure correlated with $\langle \Sigma, C\rangle$. It should be pointed out that, for any class K of realization structures, the theory $\Theta\rho\,\mathsf{K}$ is a theory in the sense just established. Conversely, for any theory Θ there is a class K of structures such that $\Theta = \Theta\rho\,\mathsf{K}$, namely $\mathsf{K} = \mathsf{Md}\,\Theta$.

Every theory uniquely determines the language Λ in which the sentences of this theory are formulated; we can thus speak of the *language of a given theory*. Of course the language of $\Theta\rho\mathsf{K}$ is the same as the discourse language of K. If Θ and Θ' are two theories in the same language and $\Theta \subseteq \Theta'$, we refer to Θ as a *subtheory* of Θ', and to Θ' as an *extension* of Θ.

In a portion of the second part of this work and occasionally in other places, we shall be concerned with various special classes of theories such as complete and consistent as well as axiomatizable, finitely axiomatizable, and decidable theories; these notions are well known from the literature.

CHAPTER 0

GENERAL THEORY OF ALGEBRAS

0. GENERAL THEORY OF ALGEBRAS

In this chapter we introduce and discuss a series of notions from the general theory of algebraic structures. Characteristic examples of such notions are those of subalgebra, isomorphism, homomorphism, and direct products. These notions can be rightly regarded as the most fundamental of modern algebra; they play a basic role in all branches of contemporary algebraic research, in particular in such highly developed theories as those of groups, rings, fields, and lattices. Some further notions, e.g., ultraproducts and free algebras, will be introduced here mainly because of their significance for model-theoretical discussions of algebraic structures. Finally, we shall discuss briefly a few notions, such as reducts, which are of lesser importance from a general viewpoint, but are relevant for the main purpose of this book, i.e., for the study of cylindric algebras.

Most of the notions discussed can be loosely characterized as procedures which apply to arbitrary algebraic structures (or systems of such structures) and yield new algebraic structures. A few concepts of the same general character — free products, direct and inverse limits, complex algebras — will be entirely omitted in the present discussion. These are concepts whose role in algebraic research is rather restricted and which have not yet been adequately studied in the general theory of algebras. They are also without broader significance for the theory developed in this book (although a special case of the notion of a complex algebra is introduced and used in Chapter 2, Section 7).

We shall state a number of theorems expressing simple properties of the notions introduced. Our purpose is to give a rather complete survey of the basic properties and interconnections of these notions. As a consequence, many theorems stated here will find no applications in the further chapters of this work.

Most of the theorems have a purely algebraic character. In addition we shall formulate some results with a metamathematical (model-theoretical) content. They will be found in the final portions of certain sections, and as a rule will not be applied in the proofs of purely algebraic results; however, we shall sometimes discuss informally the possibility of using them for such purposes.

The proofs of most of the theorems are straightforward. The material has been arranged in such an order that the reader will be able to reconstruct many of the less obvious proofs with a modest investment of effort. Frequently we indicate the principal theorems, stated earlier, from which a given result can be derived, and in more difficult cases we either supply sketches of proofs or refer to the literature.

The available literature in the general theory of algebraic structures is still rather scarce. To supplement some portions of this chapter the reader may consult Birkhoff [67*], P. M. Cohn [65*], Jónsson-Tarski [47*], and B. H. Neumann [62*], as well as Grätzer [68*] where, in particular, a comprehensive bibliography of the field can be found.

0.1. ALGEBRAS AND THEIR SUBALGEBRAS

DEFINITION 0.1.1. *By an* ***algebra*** *(or an* ***algebraic structure****) we understand a pair* $\mathfrak{A} = \langle A, Q \rangle$ *where A is a non-empty set and Q is a function which correlates with every element i of its domain a finitary operation* Q_i*, of positive rank, on and to elements of A, so that* $0 < \rho Q_i < \omega$ *and* $Q_i \in {}^{\rho Q_i}{}_A A$*. If* $\mathit{Do}\, Q = I$*, we also use the notation* $\mathfrak{A} = \langle A, Q_i \rangle_{i \in I}$*. A is called the* ***universe*** *of the algebra* $\mathfrak{A}$ *and is denoted by* $Uv\,\mathfrak{A}$*; I is the* ***index set*** *of* $\mathfrak{A}$*, symbolically* $In\,\mathfrak{A}$*; the operations* Q_i *are referred to as the* ***fundamental operations*** *of* $\mathfrak{A}$ *and are denoted by* $Op_i^{(\mathfrak{A})}$.

REMARKS 0.1.2. German capitals $\mathfrak{A}$, $\mathfrak{B}$, ... will be used to represent arbitrary algebras; if an algebra is denoted by a given German letter, say $\mathfrak{A}$, it will be assumed that the corresponding Roman letter, A, denotes the universe of $\mathfrak{A}$, $A = Uv\,\mathfrak{A}$. Moreover, the superscript $^{(\mathfrak{A})}$ in $Op_i^{(\mathfrak{A})}$ will be omitted whenever no confusion is likely to ensue. In addition to Op_i, some special symbols will be used to denote fundamental operations of algebras. In particular, the symbols $+, \cdot, \circ$, etc. will represent binary operations; thus, in general, these symbols will function as variables (unless they are used to denote the familiar arithmetical and set-theoretical operations in special algebraic structures). Of course, we may also represent the fundamental operations and the index set of an algebra by means of ordinary variables, e.g., Q_i and I (as in Definition 0.1.1).

Although (in opposition to a widespread custom) we shall distinguish between an algebra and its universe, we shall apply to an algebra various notions which are properly applicable to the universe of this algebra. For instance, we shall speak of elements of $\mathfrak{A}$ meaning the elements of A, we shall call $\mathfrak{A}$ finite if A is finite, etc.

REMARKS 0.1.3. One often deals with structures which formally are not algebras in the sense of Definition 0.1.1, but which can be subsumed under this definition in a most obvious and natural way. For example, one considers structures $\mathfrak{A} = \langle A, Q_i \rangle_{i \in I}$ where each Q_i is either an operation as described in 0.1.1, or is an element of A, a so-called *distinguished element*. To subsume such structures under algebras in the sense of 0.1.1, it suffices to identify each element $Q_i \in A$ with the constant operation Q'_i of rank 1 which assumes Q_i as its only value — or, speaking more pedantically, to replace each Q_i by the corresponding operation Q'_i. As another possibility we may concern ourselves with structures such as $\mathfrak{A} = \langle A, +, \cdot, - \rangle$ where $+$ and $\cdot$ are binary operations and $-$ is a unary operation (on and to elements of A); we identify this structure $\mathfrak{A}$ with the algebra $\langle A, Q_\nu \rangle_{\nu \in 3}$ where Q_0, Q_1, and Q_2 respectively coincide with $+$, $\cdot$, and $-$. Again, consider a structure $\mathfrak{A} = \langle A, Q_\xi, d_{\xi\eta} \rangle_{\xi, \eta < \alpha}$ formed by a non-empty set A, an α-termed sequence of finitary operations Q_ξ on and to elements of A, and a double sequence of distinguished elements $d_{\xi\eta} \in A$ indexed by ordered pairs $\langle \xi, \eta \rangle \in \alpha \times \alpha$, where α is an arbitrary ordinal. We identify $\mathfrak{A}$ with the algebra $\mathfrak{A}' = \langle A, Q'_i \rangle_{i \in I}$ where $I = \alpha \cup (\alpha \times \alpha)$, $Q'_\xi = Q_\xi$ for $\xi \in \alpha$, and $Q'_{\langle \xi, \eta \rangle}$ for $\langle \xi, \eta \rangle \in \alpha \times \alpha$ is the operation of rank 1 assuming $d_{\xi\eta}$ as the only value (recall that an ordered pair is never an ordinal). In all such cases and other analogous ones we shall treat the structures involved as algebras in the sense of 0.1.1 and, without any comment, we shall apply to them the general algebraic notions discussed in this chapter.

The fact that Definition 0.1.1 imposes no restrictions on index sets of algebras secures some flexibility in algebraic constructions. But it also has some undesirable consequences: it enlarges the variety of algebraic structures beyond necessity, and it sometimes forces us to make algebraically irrelevent distinctions. For example, in view of 0.1.1, two algebras $\mathfrak{A} = \langle A, Q_i \rangle_{i \in I}$ and $\mathfrak{A}' = \langle A, Q_{fj} \rangle_{j \in J}$ are always different unless $I = J$ and $Q_i = Q_{fi}$ for each $i \in I$. Thus an algebra $\mathfrak{A} = \langle A, Q_\nu \rangle_{\nu \in \{2,3\}} = \langle A, Q_2, Q_3 \rangle$ never coincides with $\mathfrak{A}' = \langle A, Q_{2+\mu} \rangle_{\mu \in \{0,1\}}$, although for an algebraist the passage from $\mathfrak{A}$ to $\mathfrak{A}'$ consists simply in a trivial "renaming" of indices of fundamental operations. (We do not claim that distinguishing $\mathfrak{A}$ from $\mathfrak{A}'$ in analogous situations is always algebraically irrelevant. For instance, by a similar "renaming" of indices we can obtain, from any given algebra $\mathfrak{A} = \langle A, +, \cdot \rangle$ with two binary operations, the so-called *dual algebra* $\mathfrak{A}' = \langle A, \cdot, + \rangle$. However, the distinction between an algebra and its dual and the discussion of mutual relations between these two algebras are quite important for the study of some special classes of algebras, e.g., lattices.)

To avoid the undesirable consequences of 0.1.1 we could consider a rather radical modification of the concept of an algebra which would eliminate entirely the notion of an index set: an algebra would be defined as an ordered pair $\langle A, S\rangle$ where A is an arbitrary non-empty set and S is any set (not system!) of finitary operations on A. It appears, however, that the new concept of algebras is not refined enough to provide an adequate framework for algebraic discussions. An algebraic structure $\langle A, S\rangle$ in the new sense would replace a whole class of algebras in the sense of 0.1.1, namely the class K of all those algebras $\langle A, Q_i\rangle_{i\in I}$ for which $S = \{Q_i : i \in I\}$. Thus, all algebras of K would be "identified", in spite of the fact that they may differ from each other in their basic algebraic properties. (In particular, under the new conception of an algebra we would lack a simple framework for discussing relations between an algebra $\mathfrak{A} = \langle A, +, \cdot\rangle$ and its dual $\mathfrak{A}' = \langle A, \cdot, +\rangle$ since these two algebras would always be replaced by one structure.)

A less radical method, which still removes many undesirable consequences of Definition 0.1.1, consists in providing this definition with an additional stipulation to the effect that the index set of an algebra is always a cardinal α (so that the fundamental operations are indexed by arbitrary ordinals $\xi < \alpha$). The notion of an algebra thus restricted is in principle fully adequate for algebraic discussions. In practice, however, the restriction leads to certain complications in discussing algebras with infinite index sets. The complications would arise primarily in those constructions in which we pass from one algebra to another by adjoining some new fundamental operations or removing some old ones, without necessarily changing the cardinality of the index sets. Thus, e.g., the discussion of reducts which will be outlined in the last section of this chapter would assume a more involved form.

As a kind of compromise, we could stipulate that the index sets of algebras are, not necessarily cardinals, but arbitrary ordinals (or, possibly, arbitrary sets of ordinals). However, even this less restrictive assumption proves not to be convenient for our purposes.

The term "algebra" is sometimes used in reference to relational structures which cannot easily be construed as algebras in the sense of Definition 0.1.1. This applies in particular to so-called *partial algebras*, that is, structures $\langle A, Q_i\rangle_{i\in I}$ where all Q_i's are finitary partial operations on A, i.e., functions with $Do\,Q_i \subseteq {}^{\kappa}A$ for some κ, $0 < \kappa < \omega$, and $Rg\,Q_i \subseteq A$, and also to *infinitary* (or *partial infinitary*) *algebras*, that is, structures $\langle A, Q_i\rangle_{i\in I}$ where all Q_i's are operations (or partial operations) on A, but not all of them are of finite rank. Many of the notions and results discussed in this chapter can be extended in a natural way to partial and infinitary algebras. For a discussion of partial and infinitary algebras see Bruck [66*] and Słomiński [59*], respectively.

REMARKS 0.1.4. Various special algebras and special classes of algebras have been thoroughly studied in modern mathematics. We shall mention here some of them, in particular those which are involved in the general theory of algebras or which will be used in this chapter to illustrate certain notions and results of the general theory and to construct examples.

From the point of view of the general theory the simplest algebraic structures are those in which all fundamental operations are unary (including the "degenerate" structures with no fundamental operations); these structures may be called *unary algebras.* Unary algebras have certain strong properties which do not extend to other algebras; these properties simplify the discussion of unary algebras but at the same time considerably reduce the heuristic value of the discussion for the general theory of algebraic structures. Nevertheless unary algebras can sometimes be used advantageously as a source of simple counterexamples. Probably the best known unary algebra is the algebra of natural numbers $\langle \omega, S \rangle$ where S is the successor operation, $S\xi = \xi+1$ for all $\xi < \omega$.

We turn to algebraic structures $\mathfrak{A} = \langle A, + \rangle$ with one binary operation; they are sometimes referred to in the literature as *groupoids.* Most of the structures of this kind which have been studied in modern algebra are *semigroups*, i.e., algebras in which the operation $+$, frequently referred to as (semigroup) composition, is *associative*: $x + (y + z) = (x + y) + z$ for any $x, y, z \in A$. Trivial examples of semigroups are algebras $\langle A, + \rangle$ with a constant operation as well as those in which $+$ coincides with ${}^2A \upharpoonright \mathsf{pj}_0$ (or ${}^2A \upharpoonright \mathsf{pj}_1$), i.e., $x+y = x$ (or $x+y = y$) for all $x, y \in A$.

If the operation $+$ in a semigroup is *commutative*, $x+y = y+x$ for all $x, y \in A$, $\mathfrak{A}$ is called a *commutative* or *Abelian semigroup*; if, in addition, $x+x = x$ for every $x \in A$, i.e., every element x is *idemmultiple*, $\mathfrak{A}$ is called a *semilattice*. A semigroup $\mathfrak{A}$ is a *cancellation semigroup* if $x+y = x+z$ always implies $y = z$, and so does $y+x = z+x$. A *group* can be defined as a semigroup $\mathfrak{A}$ in which, for any $x, y \in A$, there are $u, v \in A$ such that $x+u = y = v+x$. By a *zero* (*element*) of a binary operation $+$ on a set A we understand an element $z \in A$ such that $x+z = x = z+x$ for every $x \in A$; by a *unit* (*element*) or an *infinity element* we understand an element $u \in A$ such that $x+u = u = u+x$ for every $x \in A$. If such an element z, or u, exists, it is uniquely determined and is usually represented by 0, or 1, respectively; thus the symbols 0 and 1 sometimes function as variables. (The unit element is occasionally denoted by ∞ and referred to then as the infinity element.) If z is a zero element of $+$ and $x+y = y+x = z$, then y is called an *inverse* of x. If $+$ is a fundamental operation of the algebra $\mathfrak{A}$ and has a zero element 0, or a unit element 1, this element is often included in the definition of the algebras as a distinguished element. Thus groups are sometimes treated as structures $\langle A, +, 0\rangle$. We assume to be known what is meant by the νth *multiple* νx, $0 < \nu < \omega$, of an element x of a semigroup; if the zero element 0 exists, it is taken as the 0th multiple $0x$ of any element x. The *order* of an element x of a group is then the smallest number ν such that $0 < \nu < \omega$ and $\nu x = 0$, or else the number 0 if no such ν exists. A group $\mathfrak{A}$ is called *cyclic* if $\mathfrak{A} = \{\nu a : \nu < \omega\} \cup \{\nu b : \nu < \omega\}$ for some $a \in A$, where b is an inverse of a; it is called a *torsion-free group*, or a *torsion* (or *periodic*) *group*, if all its elements are of order 0, or none are of order 0, respectively. $\mathfrak{A}$ is *bounded* (or *of bounded period*) if there is a positive $\nu < \omega$ such that $\nu x = 0$ for every $x \in A$; if, in particular, $\nu = 2$, $\mathfrak{A}$ will be referred to as a *Boolean group*.

If the operation $+$ in a semigroup $\langle A, +\rangle$ is not assumed to be commutative, the multiplicative notation is usually applied rather than the additive one. Thus, the composition of two elements x and y is denoted by $x \cdot y$, the inverse of an element x in a group by x^{-1}, and we speak respectively of *idempotent elements* and of a *power* x^ν of an element rather than of idemmultiple elements and of a multiple of an element. The meanings of the terms "zero (element)" and "unit (element)" are reversed. Thus a zero 0 of a semigroup $\mathfrak{A} = \langle A, \cdot\rangle$ is an element z such that $x \cdot z = z = z \cdot x$ for every $x \in A$; it is also called an *annihilator* of the semigroup $\mathfrak{A}$ (or the operation $\cdot$).

We want to mention further some classes of algebras with two binary operations. $\mathfrak{A} = \langle A, +, \cdot\rangle$ is a *ring* if $\langle A, +\rangle$ is an Abelian group, $\langle A, \cdot\rangle$ is a semigroup, and two *distributive laws* hold: $x \cdot (y+z) = x \cdot y + x \cdot z$ and $(y+z) \cdot x = y \cdot x + z \cdot x$ for all $x, y, z \in A$. Consequently, the zero element of the group $\langle A, +\rangle$ is at the same time the zero (the annihilator) of the semigroup $\langle A, \cdot\rangle$, and is called the *zero* of the ring $\mathfrak{A}$; on the other hand, $\langle A, \cdot\rangle$ may have a unit, which is called the *unit* of the ring, while $\langle A, +\rangle$ as a group can never have a unit (unless $|A| = 1$). The ring $\mathfrak{A}$ is called *commutative* if $\cdot$ is commutative. The definitions of various important subclasses of the class of rings, such as the *integral domains* and the (*commutative*) *fields*, are assumed to be known.

$\mathfrak{A} = \langle A, +, \cdot\rangle$ is called a *lattice* if $\langle A, +\rangle$ and $\langle A, \cdot\rangle$ are commutative semigroups satisfying the *absorption law*: $x \cdot (x+y) = x = x + x \cdot y$. As a consequence, $\langle A, +\rangle$ and $\langle A, \cdot\rangle$ are both semilattices, the zero 0 of $\langle A, +\rangle$ (if it exists at all) coincides with the zero of $\langle A, \cdot\rangle$ and is called the *zero* of the lattice $\mathfrak{A}$, and analogously for the unit 1. The operations $+$ and $\cdot$ in a lattice are often referred to as *join* and *meet* and represented respectively by $\vee$ and $\wedge$. The class of lattices is closely related to the class of lattice ordering structures, i.e., structures $\langle A, \leqq \rangle$ in which $\leqq$ is a lattice ordering with field A (see the Preliminaries). With every lattice $\mathfrak{A} = \langle A, +, \cdot \rangle$ we can correlate a lattice ordering structure $\mathfrak{A}^* = \langle A, \leqq \rangle$ by defining $\leqq$ as the relation which holds between two elements $x, y \in A$ iff $x+y = y$ (or, equivalently, $x \cdot y = x$). Conversely, given a lattice ordering structure $\mathfrak{B} = \langle A, \leqq \rangle$ we construct the lattice $\mathfrak{A}$ for which $\mathfrak{A}^* = \mathfrak{B}$ by defining $x+y$ and $x \cdot y$ as the least upper bound and the greatest lower bound of elements $x, y \in A$ (under the partial ordering $\leqq$). Thus a natural one-one correspondence exists between the two classes of structures.

Given a lattice $\langle A, +, \cdot \rangle$ and any set $X \subseteq A$, we denote by ΣX the least upper bound and by ΠX the greatest lower bound of the set X under the relation $\leqq$ (assuming that these bounds exist). The lattice is called *α-complete* (α an infinite cardinal) if ΣX and ΠX exist for every set $X \subseteq A$ with $|X| < \alpha$. It is called *countably complete*, or *complete*, if it is ω_1-complete, or α-complete for every infinite cardinal α, respectively.

A lattice $\mathfrak{A} = \langle A, +, \cdot \rangle$ is called *modular* if for all elements $x, y, z \in A$ the *modular law* holds: $x \cdot (y + x \cdot z) = x \cdot y + x \cdot z$ (or, dually and equivalently, $x + y \cdot (x+z) = (x+y) \cdot (x+z)$); *distributive lattices* are defined analogously. The subclass of distributive lattices formed by the so-called Boolean algebras is of fundamental importance for the study of cylindric algebras. A *Boolean algebra* $\mathfrak{A} = \langle A, +, \cdot \rangle$ is a distributive lattice, with 0 and 1, in which for every $x \in A$ there is a (uniquely determined) element $-x$, called the *complement* of x, such that $x + (-x) = 1$ and $x \cdot (-x) = 0$. Boolean algebras will be discussed in some detail in Section 1.1 of Chapter 1; it will prove convenient for our purposes to define them, not simply as specialized lattices, but as lattices enriched by a unary operation of complementation and two distinguished elements, 0 and 1.

DEFINITION 0.1.5. *By the* ***similarity type*** *of an algebra $\mathfrak{A} = \langle A, Q_i^{(\mathfrak{A})} \rangle_{i \in I}$ we understand the system $\langle \rho(Q_i^{(\mathfrak{A})}) \rangle_{i \in I}$; two algebras are called* ***similar*** *if they have the same similarity type. The* ***similarity class*** *of $\mathfrak{A}$ is the class of all algebras similar to $\mathfrak{A}$. We speak of a similarity type of algebras, meaning the similarity type of some unspecified algebra $\mathfrak{A}$; analogously for the notion of a similarity class.*

By this definition, semigroups $\langle A, + \rangle$ discussed in 0.1.4 are of similarity type $\langle 2 \rangle$, rings $\langle A, +, \cdot \rangle$ of type $\langle 2, 2 \rangle$, and unary algebras $\langle A, Q_i \rangle_{i \in I}$ of type $\langle 1 \rangle_{i \in I}$. A similarity type is simply a function whose range is a subset of

$\omega \sim 1$. Note that a similarity class is always a proper class in Bernays' or Morse's set theory.

REMARK 0.1.6. In many discussions of special classes of algebraic structures, such as groups, rings, etc., the underlying class can be construed in several different ways, and both the fundamental operations and the similarity type of the structures discussed are not unambiguously determined. For example, we have defined groups in 0.1.4 as special algebras $\langle G, \cdot\rangle$ of type $\langle 2\rangle$; often, however, groups are treated as algebras $\langle G, \cdot, {}^{-1}\rangle$ of type $\langle 2, 1\rangle$, where $\cdot$ is the operation of composition and ${}^{-1}$ the operation of inversion. In many cases it is not even made clear which of the two notions of a group underlies the discussion.

The source of this ambiguity lies in the fact that the classes of algebras involved — the class K of groups $\langle A, \cdot\rangle$ in the first sense and the class K^* of groups $\langle A, \cdot, {}^{-1}\rangle$ in the second sense — are *definitionally equivalent*. By saying this we mean that it is possible to establish a one-one correspondence between structures in K and those in K^* so as to satisfy the following conditions: if $\mathfrak{G} = \langle G, \cdot\rangle \in \mathsf{K}$ and $\mathfrak{G}^* = \langle G^*, \cdot^*, {}^{-1^*}\rangle \in \mathsf{K}^*$ are two corresponding structures, then their universes G and G^* coincide and the fundamental operations of either of the structures are definable in terms of the fundamental operations of the other; moreover, the definitions of these operations can be chosen in such a way that they are formally the same for all the pairs of corresponding structures $\langle \mathfrak{G}, \mathfrak{G}^*\rangle$ involved. This notion of definitional equivalence has no precise meaning as long as we do not specify the formal language in which the mutual definitions of fundamental operations are supposed to be expressed. We arrive, for instance, at a precise (metamathematical) notion of first-order definitional equivalence by stipulating that the definitions should be formulated in the language of first-order predicate logic. When speaking in this work of definitional equivalence without further specification we shall always mean first-order definitional equivalence. In particular we call two structures $\mathfrak{A}$ and $\mathfrak{B}$ *definitionally equivalent* if the classes $\{\mathfrak{A}\}$ and $\{\mathfrak{B}\}$ are definitionally equivalent in the sense just described.

It is easily seen that the classes of groups in the two senses, $\mathfrak{G}$ and $\mathfrak{G}^*$, are indeed first-order definitionally equivalent. In fact, with every group $\mathfrak{A} = \langle A, \cdot\rangle \in \mathsf{K}$ we correlate a group $\mathfrak{A}^* = \langle A^*, \cdot^*, {}^{-1^*}\rangle \in \mathsf{K}^*$ by letting $A^* = A$, $x \cdot^* y = x \cdot y$ for all $x, y \in A$, and by stipulating that, for any given $x \in A$, x^{-1^*} is the inverse of x: the only element $z \in A$ such that, e.g., $(x \cdot z) \cdot x = x$. Conversely, given a group $\mathfrak{B} = \langle B, \cdot', {}^{-1'}\rangle \in \mathsf{K}^*$, we obtain a group $\mathfrak{A} = \langle A, \cdot\rangle \in \mathsf{K}$ for which $\mathfrak{A}^* = \mathfrak{B}$ simply by letting $\mathfrak{A} = \langle B, \cdot'\rangle$, i.e., by stipulating that $A = B$, and $x \cdot y = x \cdot' y$ for all $x, y \in B$. An elementary argument shows that in this way we actually establish a correspondence between K and K* which is one-one and has also all the other desired properties.

On the other hand, consider the algebras $\langle \omega, + \rangle$ and $\langle \omega, +, \cdot \rangle$, where $+$ and $\cdot$ are the ordinary arithmetical operations of addition and multiplication. Using the results in Presburger [30*] we can show that these two algebras are not first-order definitionally equivalent, while it is easily seen that they are second-order definitionally equivalent.

The notion of definitional equivalence naturally extends to relational structures of first and higher order, and to classes of such structures. For instance, from the discussion of the correspondence between lattices and lattice ordering structures in 0.1.4 we conclude at once that the class K of lattices $\langle A, +, \cdot \rangle$ and the class K* of lattice ordering structures $\langle A, \leqq \rangle$ are first-order definitionally equivalent. Furthermore, the two classes of second-order structures briefly discussed in the Preliminaries, the class of closure structures and that of closed-set structures, are easily seen to be second-order definitionally equivalent.

The practice of identifying two definitionally equivalent structures (or classes of structures) is frequently harmless. From the remarks in 0.1.13 below and especially from those which will be found in Section 0.4 under 0.4.14 it will clearly appear, however, that this practice should be abandoned since in certain situations it may lead to serious confusions.

REMARK 0.1.7. Throughout the work, when considering within a definition, theorem, or proof several individual algebras or classes of algebras, we shall tacitly assume that all of the algebras involved are of the same similarity type, unless the opposite is explicitly stated or obviously follows from the context.

Actually, to simplify the exposition, we shall concentrate from now on until the end of Section 0.4 on algebras of a special similarity type, namely of type $\langle 2 \rangle$. Nevertheless we shall sometimes discuss algebras of other similarity types in informal remarks, e.g., as examples. We believe that the extension of the definitions and theorems of this chapter to arbitrary algebras is a routine matter. In those exceptional cases in which the exact form of the extension may not be quite clear, we indicate the necessary modifications. We shall not concern ourselves with the few notions occasionally discussed in the literature (such as dual automorphisms and self-duality) which apply to algebras of type $\langle 2 \rangle$ and do not extend in a natural or useful way to arbitrary algebraic structures. In further chapters of the work we shall freely apply the definitions and theorems of this chapter to algebras of different similarity types which will be involved in our discussion.

In the remaining part of this section and in Section 0.2 we assume that $\mathfrak{A}$ is a fixed (but otherwise arbitrary) algebra with a binary operation $+$, $\mathfrak{A} = \langle A, +^{(\mathfrak{A})}\rangle = \langle A, +\rangle$. In addition to $\mathfrak{A}$ we shall frequently consider some other algebras, $\mathfrak{B} = \langle B, +^{(\mathfrak{B})}\rangle = \langle B, +\rangle$, $\mathfrak{C} = \langle C, +^{(\mathfrak{C})}\rangle = \langle C, +\rangle$, etc.

We now turn to the formation of subalgebras. This is the simplest process of "reducing the size" of an algebra, of forming smaller algebras from a larger one. Loosely speaking, the process consists in removing some elements from the universe of the algebra without changing the fundamental operations of the algebra.

DEFINITION 0.1.8. (i) *A set B is called a* ***subuniverse*** *of $\mathfrak{A}$, in symbols $B \in Su\,\mathfrak{A}$, if $B \subseteq A$ and if $x +^{(\mathfrak{A})} y \in B$ for any $x, y \in B$. An algebra $\mathfrak{B} = \langle B, +^{(\mathfrak{B})}\rangle$ is called a* ***subalgebra*** *of $\mathfrak{A}$, and $\mathfrak{A}$ is called a* ***superalgebra*** *or an* ***extension*** *of $\mathfrak{B}$, in symbols $\mathfrak{B} \subseteq \mathfrak{A}$ or $\mathfrak{A} \supseteq \mathfrak{B}$, if $B \in Su\,\mathfrak{A}$ and if $x +^{(\mathfrak{B})} y = x +^{(\mathfrak{A})} y$ for all $x, y \in B$. We put*

$$\mathsf{S}\mathfrak{A} = \{\mathfrak{B} : \mathfrak{B} \subseteq \mathfrak{A}\}$$

and, for every class K of algebras,

$$\mathsf{SK} = \bigcup\{\mathsf{S}\mathfrak{C} : \mathfrak{C} \in \mathsf{K}\}.$$

(ii) *A subuniverse B and (in case $B \neq 0$) the correlated subalgebra $\mathfrak{B}$ of $\mathfrak{A}$ are called* ***proper*** *if $B \neq A$. $\mathfrak{A}$ is called a* ***minimal*** *algebra if it has no proper subalgebra.*

REMARK 0.1.9. By 0.1.8(i) the subuniverses of an algebra are subsets of its universe. The only algebras $\langle A, +\rangle$ for which the converse holds are those in which $x + y \in \{x, y\}$ for any $x, y \in A$; obviously some of the "trivial" semigroups mentioned in 0.1.4 have this property.

The fundamental operations of an algebra $\mathfrak{A}$ and of any one of its proper subalgebras $\mathfrak{B} = \langle B, +^{(\mathfrak{B})}\rangle$ are not identical; $+^{(\mathfrak{B})}$ is the intersection of $+^{(\mathfrak{A})}$ and ${}^2B \times B$. Nevertheless, in agreement with 0.1.2, we shall generally use the same symbol for the fundamental operations of an algebra and all its subalgebras.

The reader will have noticed that we have also decided to use the symbol $\subseteq$ for both set-theoretical inclusion and the analogous relation between a subalgebra and an algebra. We feel confident that this symbolic convention, though formally incorrect, will not lead to any misunderstandings. The reader will observe several analogous instances of formally incorrect notations in our further discussion. We have, of course, been motivated by the desire to simplify the symbolism and in particular to avoid introducing an excessive number of new symbols.

THEOREM 0.1.10. (i) $\bigcup Su\mathfrak{A} = A \in Su\mathfrak{A}$.

(ii) *If* $K \subseteq Su\mathfrak{A}$, *then* $\bigcap K \in Su\mathfrak{A}$.

(iii) *If* K *is a non-empty subfamily of* $Su\mathfrak{A}$ *which is directed by the relation* $\subseteq$, *then* $\bigcup K \in Su\mathfrak{A}$.

(iv) $\bigcap Su\mathfrak{A} = 0 \in Su\mathfrak{A}$.

(v) *If* $B \in Su\mathfrak{A}$ *and* $B \subseteq D \subseteq A$, *then there is a maximal* $C \in Su\mathfrak{A}$ *such that* $B \subseteq C \subseteq D$.

0.1.10(ii) applies in particular to the case $K = 0$; by the intersection of the empty set of subuniverses of $\mathfrak{A}$ we understand the set A (see the Preliminaries).

REMARK 0.1.11. Parts (i)–(iii) of 0.1.10 express jointly the fact that $\langle A, Su\mathfrak{A}\rangle$ is an inductive closed-set structure (cf. the Preliminaries). This fact has many implications. In particular, 0.1.10(v) can easily be derived from it by means of Zorn's lemma. As another consequence of 0.1.10(ii), the partial ordering established in the family $Su\mathfrak{A}$ by the inclusion relation proves to be a complete lattice ordering. In fact, $Su\mathfrak{A}$ becomes the universe of a complete lattice if we take $X \cap Y$ for the meet and $\bigcap\{Z : X \cup Y \subseteq Z \in Su\mathfrak{A}\}$ (i.e., $Sg^{(\mathfrak{A})}(X \cup Y)$ according to Definition 0.1.15 below) for the join of any two sets $X, Y \in Su\mathfrak{A}$. In this lattice, the so-called *lattice of subuniverses* of $\mathfrak{A}$, the greatest lower bound of any family $L \subseteq Su\mathfrak{A}$ coincides with $\bigcap L$ and the least upper bound with $\bigcap\{Z : \bigcup L \subseteq Z \in Su\mathfrak{A}\}$; A is the unit and, in view of 0.1.10(iv), the empty set 0 is the zero of the lattice.

In a certain sense the converse of 0.1.10(i)–(iv) holds: for every inductive closed-set structure $\langle A, L\rangle$ with $0 \in L$ there is an algebra $\mathfrak{A}$ such that $Uv\mathfrak{A} = A$ and $Su\mathfrak{A} = L$; see Birkhoff-Frink [48*], p. 300, Theorem 10.

It can be shown by means of simple examples that the conclusion of 0.1.10(iii) in general fails if K is not directed by $\subseteq$, i.e., $\langle A, Su\mathfrak{A}\rangle$ is not in general a complete closed-set structure. Only in special algebras $\mathfrak{A}$ do we have $\langle A, Su\mathfrak{A}\rangle$ complete, e.g., in all unary algebras (cf. 0.1.4), and in those algebras $\mathfrak{A} = \langle A, +\rangle$ in which $x+y = x+z$ (or $y+x = z+x$) for all $x, y, z \in A$.

THEOREM 0.1.12. (i) $\mathfrak{A} \subseteq \mathfrak{A}$; *if* $\mathfrak{A} \subseteq \mathfrak{B} \subseteq \mathfrak{A}$, *then* $\mathfrak{A} = \mathfrak{B}$.

(ii) *If* $\mathfrak{B} \subseteq \mathfrak{A}$, *then* $Su\mathfrak{B} = Su\mathfrak{A} \cap Sb B$.

(iii) *If* $\mathfrak{B} \subseteq \mathfrak{A}$, *then* $\mathfrak{C} \subseteq \mathfrak{B}$ *iff* $\mathfrak{C} \subseteq \mathfrak{A}$ *and* $C \subseteq B$.

REMARK 0.1.13. By 0.1.12(i),(iii) the relation $\subseteq$ between algebras defined in 0.1.8(i) establishes a partial ordering in every class of algebras. As opposed, however, to observations in 0.1.11, this partial ordering is not in general a lattice ordering even if restricted to the class $\mathbf{S}\mathfrak{A}$ of subalgebras of a given algebra $\mathfrak{A}$. When so restricted, it is a lattice ordering, and actually a complete lattice ordering, in case $\mathfrak{A}$ has a least subalgebra or, what amounts to the same, the lattice of subuniverses of $\mathfrak{A}$ has a least non-zero element.

Notice that in an algebra $\mathfrak{A}$ with some distinguished elements every non-empty subuniverse of $\mathfrak{A}$ must contain all these distinguished elements, and hence, by 0.1.10(ii), $\mathfrak{A}$ has a least non-empty subuniverse and a least subalgebra. This may also be true, however, in case $\mathfrak{A}$ has no distinguished elements. A good illustration of these last remarks is provided by groups. We have discussed various ways in which the notion of a group is treated. If a group $\mathfrak{G}$ is regarded as an algebra with composition $\cdot$ as the only fundamental operation, $\mathfrak{G} = \langle G, \cdot \rangle$, then the subalgebras of $\mathfrak{G}$ are always cancellation semigroups but are not necessarily groups, and $\mathfrak{G}$ may have no least subalgebra; $\mathfrak{G}' = \langle \{1\}, \cdot \rangle$ is the unique minimal subalgebra but is not always the least subalgebra, i.e., is not necessarily included in every subalgebra of $\mathfrak{G}$. For instance, the additive group of integers has no least subalgebra. (It is easily seen that the following three conditions are equivalent: all the subalgebras of $\mathfrak{G}$ are groups; $\mathfrak{G}$ has a least subalgebra; $\mathfrak{G}$ is a torsion group (cf. 0.1.4).) If, however, 1 is included in the definition of $\mathfrak{G}$ as a distinguished element, $\mathfrak{G} = \langle G, \cdot, 1 \rangle$, then the subalgebras of $\mathfrak{G}$ are cancellation semigroups with 1 and $\mathfrak{G}' = \langle \{1\}, \cdot, 1 \rangle$ is the least subalgebra of $\mathfrak{G}$. On the other hand, if a group $\mathfrak{G}$ is regarded as an algebra with two fundamental operations, composition $\cdot$ and inversion $^{-1}$, i.e., $\mathfrak{G} = \langle G, \cdot, ^{-1} \rangle$, then all subalgebras of $\mathfrak{G}$ are groups (i.e., subalgebras coincide with subgroups), and $\mathfrak{G} = \langle \{1\}, \cdot, ^{-1} \rangle$ is again the least subalgebra of $\mathfrak{G}$; this remains true of course if, in addition, 1 is included in the definition of $\mathfrak{G}$ as a distinguished element. We see thus that in two definitionally equivalent algebras $\mathfrak{A}$ and $\mathfrak{B}$ the families $Su\,\mathfrak{A}$ and $Su\,\mathfrak{B}$ may differ very essentially.

THEOREM 0.1.14. (i) *For any class* K *of algebras,* $\mathsf{K} \subseteq \mathbf{S}\mathsf{K} = \mathbf{S}\mathbf{S}\mathsf{K}$.

(ii) *For any classes* K *and* L *of algebras, the formulas* $\mathsf{K} \subseteq \mathbf{S}\mathsf{L}$ *and* $\mathbf{S}\mathsf{K} \subseteq \mathbf{S}\mathsf{L}$ *are equivalent, and each of them is implied by* $\mathsf{K} \subseteq \mathsf{L}$.

(iii) *For any class* K *of algebras,* $\mathbf{S}\mathsf{K} = \bigcup_{\mathfrak{A} \in \mathsf{K}} \mathbf{S}\{\mathfrak{A}\}$.

Let M be any class of algebras such that $\mathsf{M} = \mathbf{S}\mathsf{M}$ (e.g., any similarity class of algebras). Let us assume the operation $\mathbf{S}$ is restricted to subclasses of M. Then, by 0.1.14, M and $\mathbf{S}$ satisfy all the conditions stated in the Preliminaries to characterize the notion of a complete closure structure. Disregarding certain difficulties related to the foundations of set theory, we could simply say that the ordered pair $\langle \mathsf{M}, \mathbf{S} \rangle$ is a complete closure structure.

DEFINITION 0.1.15. (i) *For every set* $X \subseteq A$ *the set* $\bigcap\{B : X \subseteq B \in Su\,\mathfrak{A}\}$ *is* *called the **subuniverse of** $\mathfrak{A}$ **generated by** X and is denoted by $Sg^{(\mathfrak{A})}X$, or SgX; in case $X \neq 0$, $\langle SgX, + \rangle$ is called the **subalgebra generated by** X and is denoted by $\mathfrak{Sg}^{(\mathfrak{A})}X$ or $\mathfrak{Sg}X$.*

(ii) *For any given cardinal α we put*

$$\mathbf{S}_\alpha\mathfrak{A} = \{\mathfrak{Sg}X : 0 \neq X \subseteq A \text{ and } |X| < \alpha\}$$

and, more generally, for every class K of algebras,

$$\mathbf{S}_\alpha\mathsf{K} = \bigcup\{\mathbf{S}_\alpha\mathfrak{C} : \mathfrak{C} \in \mathsf{K}\}.$$

*$\mathfrak{B}$ is called a **finitely generated subalgebra** of $\mathfrak{A}$ (and B a **finitely generated subuniverse** of $\mathfrak{A}$) if $\mathfrak{B} \in \mathbf{S}_\omega\mathfrak{A}$.*

REMARK 0.1.16. By saying that an algebra $\mathfrak{A}$ is finitely generated we mean of course that it is so generated as its own subalgebra, i.e., that $\mathfrak{A} \in \mathbf{S}_\omega\mathfrak{A}$.

The notation $\mathbf{S}_\alpha\mathsf{K}$ will be used mostly in Section 0.4 where it will help us to formulate concisely several theorems on free algebras. Many of the elementary properties of $\mathbf{S}_\alpha\mathsf{K}$ are analogous to those of $\mathbf{S}\mathsf{K}$.

THEOREM 0.1.17. *Let $X, Y \subseteq A$ and $K \subseteq SbA$. Then:*

(i) *$X \subseteq SgX = SgSgX \subseteq A$ and in particular $SgA = A$;*

(ii) *$SgX \subseteq SgY$ whenever $X \subseteq Y$;*

(iii) *if $\mathsf{K} \neq 0$ is directed by the relation $\subseteq$, then*
$Sg(\bigcup K) = \bigcup\{SgZ : Z \in K\}$;

(iv) *$Sg0 = 0$;*

(v) *$SgX \in Su\mathfrak{A}$; $X \in Su\mathfrak{A}$ iff $X = SgX$;*

(vi) *$Sg(X \cup Y) = Sg(X \cup SgY) = Sg(SgX \cup SgY)$;*

(vii) *$Sg(\bigcup \mathsf{K}) = Sg(\bigcup\{SgZ : Z \in K\})$;*

(viii) *$\bigcap\{SgZ : Z \in \mathsf{K}\} = Sg(\bigcap\{SgZ : Z \in \mathsf{K}\})$;*

(ix) *$SgX = \bigcup\{SgZ : Z \subseteq X, \ |Z| < \omega\}$;*

(x) *if X is a finitely generated subuniverse of $\mathfrak{A}$, then, for every $Z \subseteq A$ such that $X = SgZ$, there is a finite $Z' \subseteq Z$ such that $X = SgZ'$.*

Parts (i)–(iii) of 0.1.17 express jointly the fact that $\langle A, Sg^{(\mathfrak{A})}\rangle$ is an inductive closure structure; parts (vi)–(x) are simple consequences of this fact. In view of Definition 0.1.15(i), $\langle A, Sg^{(\mathfrak{A})}\rangle$ is the closure structure naturally correlated with the closed-set structure $\langle A, Su\mathfrak{A}\rangle$ which was discussed in 0.1.11; the correlation in the opposite direction is expressed in 0.1.17(v). (Compare here the Preliminaries, and see also 0.1.6.) 0.1.7(iv) is not a consequence of 0.1.17(i)–(iii), but depends on the specific property of $\langle A, Su\mathfrak{A}\rangle$ stated in 0.1.10(iv).

THEOREM 0.1.18. *If $\mathfrak{B} \subseteq \mathfrak{A}$ and $X \subseteq B$, then $Sg^{(\mathfrak{B})}X = Sg^{(\mathfrak{A})}X$, and $\mathfrak{Sg}^{(\mathfrak{B})}X = \mathfrak{Sg}^{(\mathfrak{A})}X$ in case $X \neq 0$.*

THEOREM 0.1.19. *For every* $X \subseteq A$ *we have* $|X| \leqq |Sg\,X| \leqq |X| \cup \omega$; *hence* $|Sg\,X| \leqq \omega$ *in case* $|X| \leqq \omega$, *and* $|Sg\,X| = |X|$ *in case* $|X| \geqq \omega$.

PROOF. Let $\langle Y_0, \ldots, Y_\kappa, \ldots \rangle_{\kappa<\omega}$ be the sequence such that $Y_0 = X$ and $Y_{\kappa+1} = Y_\kappa \cup \{x+y : x, y \in Y_\kappa\}$ for all $\kappa < \omega$. It is seen at once that $Sg\,X = \bigcup_{\kappa<\omega} Y_\kappa$, and hence the theorem can easily be derived by means of some familiar results from the general theory of sets.

REMARK 0.1.20. When extended to algebras $\mathfrak{A}$ of arbitrary similarity type, Theorem 0.1.19 undergoes a modification. We then have $|X| \leqq |Sg\,X| \leqq |X| \cup \beta \cup \omega$ where β is the cardinality of the set of fundamental operations of $\mathfrak{A}$, and hence $|Sg\,X| = |X|$ if both $|X| \geqq \beta$ and $|X| \geqq \omega$.

THEOREM 0.1.21. *Let* K *be a class of algebras and* α *a cardinal. Then*:

(i) $\mathbf{S}_0\mathsf{K} = \mathbf{S}_1\mathsf{K} = 0$;

(ii) $\mathbf{S}_\alpha\mathsf{K} = \mathbf{S}_\alpha\mathbf{S}_\alpha\mathsf{K} = \mathbf{S}_\alpha\mathbf{S}\mathsf{K} \subseteq \mathbf{S}\mathbf{S}_\alpha\mathsf{K} \subseteq \mathbf{S}\mathsf{K}$;

(iii) $\mathbf{S}_\alpha\mathsf{K} = \mathbf{S}\mathbf{S}_\alpha\mathsf{K} = \mathbf{S}\mathsf{K} \cap \{\mathfrak{A} : |A| < \alpha\}$ *in case* $\alpha > \omega$.

0.1.21(iii) is a simple consequence of 0.1.19; when extended to algebras of arbitrary classes, it requires the assumption $\alpha > \beta \cup \omega$ where β is the cardinality of the set of fundamental operations of $\mathfrak{A}$ (cf. 0.1.20).

By 0.1.21(i),(iii) the first inclusion symbol in 0.1.21(ii) can be replaced by an identity symbol in case $\alpha < 2$ or $\alpha > \omega$. It cannot be replaced in case $2 \leqq \alpha \leqq \omega$. In fact, let $\mathfrak{A} = \langle \omega, \cdot \rangle$ where $\cdot$ is the binary operation determined by the conditions: $0 \cdot \lambda = \lambda + 1$ and $\kappa \cdot \lambda = \kappa$ for $\kappa \neq 0$ $(\kappa, \lambda < \omega)$. As is easily seen, the set $\omega \sim 1$ (of positive integers) is a subuniverse of $\mathfrak{A}$ and hence $\mathfrak{B} = \langle \omega \sim 1, \cdot \rangle \subseteq \mathfrak{A}$. Moreover, $\mathfrak{A}$ is generated by one element, 0, while $\mathfrak{B}$ is not generated by any finite set of elements. Therefore, for every α with $2 \leqq \alpha \leqq \omega$, we have $\mathfrak{B} \in \mathbf{S}\mathbf{S}_\alpha\mathfrak{A}$ and $\mathfrak{B} \notin \mathbf{S}_\alpha\mathbf{S}\mathfrak{A}$. For no α can the second inclusion symbol in 0.1.21(ii) be replaced by an identity symbol; this can be shown by means of trivial counterexamples.

THEOREM 0.1.22. *If* $B \in Su\mathfrak{A}$, $B \subset D \subseteq A$, *and* $\mathfrak{Sg}D \in \mathbf{S}_\omega\mathfrak{A}$, *then there is a maximal* $C \in Su\mathfrak{A}$ *such that* $B \subseteq C \subset D$.

Theorem 0.1.22 is analogous to 0.1.10(v); again we are dealing with a particular case of a result which holds in all inductive closed-set structures and can easily be established by means of Zorn's lemma.

THEOREM 0.1.23. *The following conditions are equivalent*:

(i) $\mathfrak{A}$ *is a minimal algebra*;

(ii) $Su\mathfrak{A} = \{0, A\}$;

(iii) *for every non-empty* $X \subseteq A$, $\mathfrak{A} = \mathfrak{Sg}X$;
(iv) *for every* $a \in A$, $\mathfrak{A} = \mathfrak{Sg}\{a\}$.

DEFINITION 0.1.24. *By a* **union** *of a set* K *of algebras we understand an algebra which is the least common extension of all algebras of* K, *i.e., the least upper bound of* K *under the relation* $\subseteq$. *If such an algebra exists (and is uniquely determined), it is denoted by* $\mathsf{U}\mathsf{K}$.

It is obvious that a union of a set of algebras, if it exists, is always uniquely determined.

THEOREM 0.1.25. *For every set* K *of algebras and every algebra* $\mathfrak{B}$ *the following three conditions are equivalent*:
(i) $\mathfrak{B} = \mathsf{U}\mathsf{K}$;
(ii) $\mathsf{K} \subseteq \mathbf{S}\mathfrak{B}$ *and* ${}^2B = \mathsf{U}\{{}^2C : \mathfrak{C} \in \mathsf{K}\}$;
(iii) $+^{(\mathfrak{B})} = \mathsf{U}\{+^{(\mathfrak{C})} : \mathfrak{C} \in \mathsf{K}\}$, *in other words, for any* x, y, z *we have* $x, y \in B$ *and* $x +^{(\mathfrak{B})} y = z$ *iff* $x, y \in C$ *and* $x +^{(\mathfrak{C})} y = z$ *for some* $\mathfrak{C} \in \mathsf{K}$.

In this theorem only the assertion that (i) implies (ii) is perhaps not quite obvious. Assuming that (i) holds while (ii) fails, we can find elements $x_0, y_0 \in B$ such that $\langle x_0, y_0 \rangle \notin {}^2C$ for every $\mathfrak{C} \in \mathsf{K}$. Disregarding the obvious case $|B| = 1$, we pick in B an element $z \neq x_0 +^{(\mathfrak{B})} y_0$ and we define an algebra $\mathfrak{B}' = \langle B, +' \rangle$ by stipulating $x +' y = z$ if $\langle x, y \rangle = \langle x_0, y_0 \rangle$ and $x +' y = x +^{(\mathfrak{B})} y$ otherwise. Clearly $\mathsf{K} \subseteq \mathbf{S}\mathfrak{B}'$ while $\mathfrak{B} \subseteq \mathfrak{B}'$ does not hold, and this contradicts (i) in view of 0.1.24.

Conditions 0.1.25(ii),(iii) must be appropriately modified when applied to algebras of other similarity classes. If, e.g., in algebras of K all the fundamental operations are of rank $\leqq \nu$, $0 < \nu < \omega$, and one of them is actually of rank ν, we respectively replace in 0.1.25(ii) ${}^2B$ and ${}^2C$ by ${}^\nu B$ and ${}^\nu C$. If, however, such an ordinal ν does not exist, we replace the second formula of 0.1.25(ii) by the condition: ${}^\mu B = \mathsf{U}\{{}^\mu C : \mathfrak{C} \in \mathsf{K}\}$ for every $\mu \in \omega$.

REMARK 0.1.26. The empty set of algebras has no union. If $|\mathsf{K}| = 1$, then $\mathsf{U}\mathsf{K}$ obviously coincides with the only algebra in K. If $|\mathsf{K}| = 2$ and in fact $\mathsf{K} = \{\mathfrak{A}, \mathfrak{B}\}$, then $\mathsf{U}\mathsf{K}$ exists only in the trivial case when $\mathfrak{A} \subseteq \mathfrak{B}$ or $\mathfrak{B} \subseteq \mathfrak{A}$; $\mathsf{U}\mathsf{K}$ coincides then with the larger of the algebras $\mathfrak{A}$ and $\mathfrak{B}$, and hence belongs to K. On the other hand, for every cardinal $\alpha > 2$ we can construct a set K with $|\mathsf{K}| = \alpha$ such that $\mathsf{U}\mathsf{K}$ exists although K has no largest algebra, so that $\mathsf{U}\mathsf{K} \notin \mathsf{K}$. If, e.g., $\alpha = 3$ and K consists of algebras $\langle \{0, 1\}, \mathsf{u} \rangle$, $\langle \{0, 2\}, \mathsf{u} \rangle$, and $\langle \{1, 2\}, \mathsf{u} \rangle$, then $\mathsf{U}\mathsf{K} = \langle \{0, 1, 2\}, \mathsf{u} \rangle$. (More generally, for a given similarity class and cardinal α the following two conditions are equivalent: (I) there is a set K of algebras of the given similarity class such that $|\mathsf{K}| = \alpha$, $\mathsf{U}\mathsf{K}$ exists, and K has no largest algebra, (II) $\alpha > 1$ and all the fundamental operations of the algebras discussed are of rank $< \alpha$.)

The notion of a union plays a rather important role in the study of unary algebras. Every non-empty set of (similar) unary algebras with pairwise disjoint universes has a union. More generally, each of the following two conditions is both necessary and sufficient for a non-empty set K of unary algebras to have a union: (III) the algebras of K have a common extension; (IV) $O_i x = O'_i x$ for every couple of algebras $\mathfrak{A} = \langle A, O_i \rangle_{i \in I}$, $\mathfrak{A}' = \langle A', O'_i \rangle_{i \in I}$ in K, every $x \in A \cap A'$, and every $i \in I$. In the realm of non-unary algebras these two conditions (the second of them with appropriate changes) are still mutually equivalent and are necessary, though by no means sufficient, for K to have a union. In general, the existence of a union of arbitrary algebras is a rather exceptional phenomenon. When considering unions of sets K of algebras in the subsequent discussion, we shall restrict ourselves to sets directed by the relation $\subseteq$. For such sets we can prove

THEOREM 0.1.27. (i) *If* K *is a non-empty set of algebras directed by the relation* $\subseteq$, *then* $\mathsf{U}\mathsf{K}$ *exists.*

(ii) *In particular, for any algebra* $\mathfrak{A}$, *the set* $\mathbf{S}_\omega\mathfrak{A}$ *is directed by* $\subseteq$, *and* $\mathsf{U}\mathbf{S}_\omega\mathfrak{A} = \mathfrak{A}$.

In part (ii) of 0.1.27 we can replace ω by any cardinal $\alpha \geqq \omega$. If $2 < \nu < \omega$, then the set $\mathbf{S}_\nu\mathfrak{A}$ is not, in general, directed by $\subseteq$, but we still have $\mathsf{U}\mathbf{S}_\nu\mathfrak{A} = \mathfrak{A}$ in agreement with 0.1.24. (However, this last statement must be modified when applied to algebras of other similarity types.)

We do not introduce any special symbol (analogous to $\mathbf{S}$ or $\mathbf{S}_\alpha$) to denote the operation which, with any given class K of algebras, correlates the class of unions $\mathsf{U}\mathsf{L}$ with L ranging over all non-empty subsets of K directed by the relation $\subseteq$. This operation will be involved a few times in our further discussion. Some elementary facts concerning classes of algebras which are closed under the formation of unions are stated in the next theorem.

THEOREM 0.1.28. *Consider the following conditions concerning a class* K *of algebras.*

(i) $\mathsf{U}\mathsf{L} \in \mathsf{K}$ *whenever* L *is a non-empty subset of* K *directed by* $\subseteq$;

(ii) $\mathfrak{A} \in \mathsf{K}$ *whenever* $\mathbf{S}_\omega\mathfrak{A} \subseteq \mathsf{K}$ (*or, equivalently,* $\mathsf{L} \subseteq \mathsf{K}$ *whenever* $\mathbf{S}_\omega\mathsf{L} \subseteq \mathsf{K}$);

(iii) $\mathfrak{A} \in \mathsf{K}$ *whenever* $\mathbf{S}_\omega\mathfrak{A} \subseteq \mathbf{S}(\mathsf{K} \cap \mathbf{S}\mathfrak{A})$;

(iv) $\mathfrak{A} \in \mathsf{K}$ *whenever* $\mathbf{S}_\omega\mathfrak{A} \subseteq \mathbf{S}\mathsf{K}$ (*or, equivalently,* $\mathsf{L} \subseteq \mathsf{K}$ *whenever* $\mathbf{S}_\omega\mathsf{L} \subseteq \mathbf{S}\mathsf{K}$);

(v) $\mathfrak{A} \in \mathsf{K}$ *iff* $\mathbf{S}_\omega \mathfrak{A} \subseteq \mathsf{K}$ (*or, equivalently,* $\mathsf{L} \subseteq \mathsf{K}$ *iff* $\mathbf{S}_\omega \mathsf{L} \subseteq \mathsf{K}$);
(vi) $\mathbf{S}\mathsf{K} \subseteq \mathsf{K}$.

Condition (i) implies (ii) and is implied by (iii); under the assumption of (vi) the three conditions are equivalent. Conditions (iv) and (v) are equivalent to each other and are also equivalent to the conjunction of (vi) and any one of the conditions (i)–(iii).

The proof is easy. Notice that Theorem 0.1.28 remains valid if in (i) the word "directed" is replaced by "simply ordered" or "well ordered".

A class K of algebras satisfying condition 0.1.28(i) is sometimes called *local*, and so is the property of belonging to such a class. However, algebraists also use the term "local" in some other related senses corresponding to other conditions of 0.1.28.

Remark 0.1.29. To conclude this section we should like to mention some further notions which can be defined in terms of subalgebras, namely the important notions of *independence*. To define these notions conveniently, we consider again a fixed algebra $\mathfrak{A}$ and let

$$M = \bigcap\{X : 0 \neq X \in Su\mathfrak{A}\};$$

thus M is the least non-empty subuniverse of $\mathfrak{A}$ if such a subuniverse exists, and is empty otherwise. We can distinguish three notions of independence; a set $X \subseteq A$ is called *independent* in the sense (I), (II), or (III) if it satisfies respectively one of the following conditions:

(I) *for all* $Y, Z \subseteq X$, *if* $SgY = SgZ$, *then* $Y = Z$;
(II) $|X| = 1$, *or else* $X \cap M = 0$ *and for all non-empty* $Y, Z \subseteq X$, *if* $Y \cap Z = 0$, *then* $SgY \cap SgZ = M$;
(III) $|X| = 1$, *or else* $X \cap M = 0$ *and for all non-empty* $Y, Z \subseteq X$, $SgY \cap SgZ = M \cup Sg(Y \cap Z)$.

The following two conditions are easily seen each to be equivalent to (I):

(I′) $y \notin Sg(X \sim \{y\})$ *for all* $y \in X$;
(I″) *for all* $Y, Z \subseteq X$, *if* $SgY \subseteq SgZ$, *then* $Y \subseteq Z$.

Clearly (III) implies (II), and (II) implies (I); the implications in the opposite directions in general do not hold. With regard to any of the notions (I)–(III) the following can be established: 0 is independent; if X is independent and $Y \subseteq X$, then Y is independent; if every finite subset of X is independent, then X is independent; every independent set is included in a maximal independent set.

Among the three notions, the first one is probably the most important. If a set X of elements of an algebra $\mathfrak{A} = \langle A, +\rangle$ is independent in the sense (I) and $\mathfrak{A} = \mathfrak{S}\mathfrak{g}\, X$, we refer to X as an *independent* or *irredundant base* of $\mathfrak{A}$. Not every algebra has an irredundant base. On the other hand, it is easily seen that, if $\mathfrak{A}$ is finitely generated, then every set Y which generates $\mathfrak{A}$ includes a finite subset X which is an irredundant base of $\mathfrak{A}$. Hence every finitely generated algebra has a finite irredundant base, but no infinite irredundant base. The same algebra may have many finite irredundant bases with different cardinalities; however, two infinite irredundant bases of an algebra have the same cardinality. It has been shown that, if $\kappa < \lambda < \mu < \omega$ and $\mathfrak{A}$ has irredundant bases X and Z with $|X| = \kappa$ and $|Z| = \mu$, then it also has an irredundant base Y with $|Y| = \lambda$; cf. Tarski [68*], p. 282. This result applies to arbitrary algebras in which all fundamental operations are of rank $\leqq 2$. On the other hand, it is easy to construct a four-element algebra with a single ternary operation for which the result in its original formulation fails. (In a certain form the result can be extended to arbitrary algebras in which the ranks of all fundamental operations do not exceed a given natural number ν; the general result is, however, more involved and less interesting.)

It should be pointed out that there is still another notion of independence, and actually one which has been the most extensively discussed in the literature; this notion cannot be simply defined in terms of subalgebras and will be mentioned briefly in Section 0.4, Remark 0.4.56.

0.2. HOMOMORPHISMS, ISOMORPHISMS, CONGRUENCE RELATIONS, AND IDEALS

The formation of homomorphic images, which will be discussed in this section, is another process of reducing the size of an algebra. Roughly speaking, in this case, instead of removing elements from the universe of a given algebra, we identify some distinct elements, seeing to it, however, that the resulting structure preserves the character of an algebra. The important notion of an isomorphism is essentially a particular case of that of homomorphism. The significance of this notion is due to the fact that in algebraic discussions we consider almost exclusively those properties of algebraic structures which are invariant under the formation of isomorphic images, in the sense that when they apply to an algebra they apply to all isomorphic images as well. In particular, all the so-called *intrinsic* properties of algebras are invariant in this sense. By an intrinsic property of an algebra $\mathfrak{A}$ we understand, loosely speaking, a property which can be expressed entirely in terms of symbols denoting fundamental operations of $\mathfrak{A}$, the membership symbol $\in$, and variables ranging exclusively over elements of the universe A, subsets of A, relations between elements of A, sets of such subsets and relations, etc., and not, e.g., in case A is a family of sets, in terms of variables ranging over elements of $\bigcup A$. (Thus the notion of an intrinsic property is of metamathematical nature; to be made precise, it must be relativized to a well defined formal language.)

DEFINITION 0.2.1. *A function h is called a* ***homomorphism on*** $\mathfrak{A}$, *in symbols* $h \in Ho\,\mathfrak{A}$, *if* $Do\,h = A$ *and, for any* $x, x', y, y' \in A$, *the formulas* $hx = hx'$ *and* $hy = hy'$ *imply* $h(x+y) = h(x'+y')$; *h is called an* ***isomorphism on*** $\mathfrak{A}$, *in symbols* $h \in Is\,\mathfrak{A}$, *if in addition h is one-one.*

THEOREM 0.2.2. *If h is any one-one function with* $Do\,h = A$, *then* $h \in Is\,\mathfrak{A}$.

THEOREM 0.2.3. *If* $h \in Ho\,\mathfrak{A}$, *or* $h \in Is\,\mathfrak{A}$, *and* $\mathfrak{B} \subseteq \mathfrak{A}$, *then* $B \upharpoonright h \in Ho\,\mathfrak{B}$, *or* $B \upharpoonright h \in Is\,\mathfrak{B}$, *respectively.*

THEOREM 0.2.4. *If* $h \in Ho\,\mathfrak{A}$, *then there is a uniquely determined algebra* $\mathfrak{B}$ *satisfying the conditions*: (i) $B = Rg\,h$; (ii) *for any* $x, y \in A$, $h(x +^{(\mathfrak{A})} y) = hx +^{(\mathfrak{B})} hy$.

In view of 0.2.4 we define:

Definition 0.2.5. (i) *Let* $h \in Ho\,\mathfrak{A}$, *or* $h \in Is\,\mathfrak{A}$. *The algebra* $\mathfrak{B}$ *uniquely determined by conditions 0.2.4(i),(ii) is called the **h-image** of* $\mathfrak{A}$, *in symbols* $h^*\mathfrak{A}$, *and we say respectively that* h ***maps*** $\mathfrak{A}$ ***onto*** $\mathfrak{B}$ ***homomorphically,*** *or* ***isomorphically,*** *or that* h *is a* ***homomorphism,*** *or* ***isomorphism, from*** $\mathfrak{A}$ ***onto*** $\mathfrak{B}$, *in symbols* $h \in Ho(\mathfrak{A}, \mathfrak{B})$, *or* $h \in Is(\mathfrak{A}, \mathfrak{B})$.

(ii) *A homomorphism, or isomorphism, from* $\mathfrak{A}$ *onto a subalgebra of* $\mathfrak{B}$, *i.e., a function* $h \in Ho\,\mathfrak{A}$, *or* $h \in Is\,\mathfrak{A}$, *such that* $h^*\mathfrak{A} \subseteq \mathfrak{B}$, *is referred to respectively as a* ***homomorphism,*** *or* ***isomorphism, from*** $\mathfrak{A}$ ***into*** $\mathfrak{B}$, *in symbols* $h \in Hom(\mathfrak{A}, \mathfrak{B})$, *or* $h \in Ism(\mathfrak{A}, \mathfrak{B})$.

(iii) *A homomorphism from* $\mathfrak{A}$ *into* $\mathfrak{A}$, *i.e., a member of* $Hom(\mathfrak{A}, \mathfrak{A})$, *is called an* ***endomorphism*** *of* $\mathfrak{A}$. *An isomorphism from* $\mathfrak{A}$ *onto* $\mathfrak{A}$, *i.e., a member of* $Is(\mathfrak{A}, \mathfrak{A})$, *is called an* ***automorphism*** *of* $\mathfrak{A}$.

The symbolic expression $h:\mathfrak{A} \to \mathfrak{B}$ is sometimes used instead of $h \in Hom(\mathfrak{A}, \mathfrak{B})$, and there are analogous expressions for $h \in Ism(\mathfrak{A}, \mathfrak{B})$, $h \in Ho(\mathfrak{A}, \mathfrak{B})$, and $h \in Is(\mathfrak{A}, \mathfrak{B})$ (cf. MacLane [63*], p. 10).

We are frequently confronted with situations in which not the function h itself, but the restricted function $A1h$ is a homomorphism on $\mathfrak{A}$. In such situations we sometimes say (not quite correctly) that h is a homomorphism on $\mathfrak{A}$ and we use h instead of $A1h$ in the related notation. If, e.g., $h \in Ho\,\mathfrak{A}$ and $\mathfrak{B} \subseteq \mathfrak{A}$, we write $h^*\mathfrak{B}$ instead of $(B1h)^*\mathfrak{B}$ (cf. 0.2.3).

Definition 0.2.6. $\mathfrak{A}$ *is said to be* ***homomorphic to*** $\mathfrak{B}$, *in symbols* $\mathfrak{A} \succcurlyeq \mathfrak{B}$ *or* $\mathfrak{B} \preccurlyeq \mathfrak{A}$, *or* ***isomorphic to*** $\mathfrak{B}$, *in symbols* $\mathfrak{A} \cong \mathfrak{B}$, *if* $Ho(\mathfrak{A}, \mathfrak{B}) \neq 0$, *or* $Is(\mathfrak{A}, \mathfrak{B}) \neq 0$, *respectively. We put*

$$\mathbf{H}\mathfrak{A} = \{\mathfrak{C} : \mathfrak{A} \succcurlyeq \mathfrak{C}\},$$

$$\mathbf{I}\mathfrak{A} = \{\mathfrak{C} : \mathfrak{A} \cong \mathfrak{C}\},$$

and, for any class K *of algebras,*

$$\mathbf{H}\mathsf{K} = \bigcup\{\mathbf{H}\mathfrak{C} : \mathfrak{C} \in \mathsf{K}\},$$

$$\mathbf{I}\mathsf{K} = \bigcup\{\mathbf{I}\mathfrak{C} : \mathfrak{C} \in \mathsf{K}\}.$$

We do not introduce any special symbols to denote the relations between the algebras $\mathfrak{A}$ and $\mathfrak{B}$ which are expressed by the formulas $Hom(\mathfrak{A}, \mathfrak{B}) \neq 0$ and $Ism(\mathfrak{A}, \mathfrak{B}) \neq 0$. These relations, however, can be conveniently expressed using the symbols introduced in 0.2.6. In fact, we can treat $\succcurlyeq$, $\cong$, and $\subseteq$ as relations in the sense of set theory, i.e., as classes of ordered couples, and we can form the relative product of any two of these relations (see the Preliminaries). Then, obviously, $\mathfrak{A} \succcurlyeq | \subseteq \mathfrak{B}$ iff $Hom(\mathfrak{A}, \mathfrak{B}) \neq 0$, and $\mathfrak{A} \cong | \subseteq \mathfrak{B}$ iff $Ism(\mathfrak{A}, \mathfrak{B}) \neq 0$.

REMARKS 0.2.7. Many obvious theorems can be formulated which express the invariance under isomorphism of various properties of algebras; we have in mind theorems of the form "If a given property P applies to $\mathfrak{A}$, and $\mathfrak{A} \cong \mathfrak{B}$, then P applies to $\mathfrak{B}$" (or theorems of a related form, such as "If a given relation R holds between elements x and y of A and $h \in Is(\mathfrak{A}, \mathfrak{B})$, then R holds between the elements hx, hy of B"). No such theorems will be stated here explicitly.

It is important to observe that, according to Definition 0.2.6, no two algebras with distinct index sets are isomorphic, and this applies even to algebras such as $\mathfrak{A} = \langle A, Q_\nu \rangle_{\nu \in \{2,3\}}$ and $\mathfrak{A}' = \langle A, Q_{2+\mu} \rangle_{\mu \in \{0,1\}}$ which are obtainable from each other by trivial "renaming" of indices of fundamental operations. We could, of course, extend the notion of isomorphism to algebras with (possibly) distinct index sets by stipulating that two algebras $\mathfrak{A} = \langle A, Q_i \rangle_{i \in I}$ and $\mathfrak{B} = \langle B, P_j \rangle_{j \in J}$ are called isomorphic in the wider sense if there is a one-one function $f \in {}^I J$ such that the algebras $\mathfrak{A}$ and $\langle B, P_{fi} \rangle_{i \in I}$ are isomorphic in the sense of 0.2.6. This new notion, however, would not adequately serve as a substitute for the old notion since algebraic discussions are by no means restricted to those properties of algebras which are invariant under isomorphism in the wider sense. To illustrate the last remark, consider an arbitrary algebra $\mathfrak{A} = \langle A, +, \cdot \rangle$ of similarity type $\langle 2, 2 \rangle$ and its dual algebra $\mathfrak{A}' = \langle A, \cdot, + \rangle$. $\mathfrak{A}$ and $\mathfrak{A}'$ are trivially isomorphic in the wider sense. Nevertheless $\mathfrak{A}$ and $\mathfrak{A}'$ may differ from each other in their most elementary and basic properties, and therefore these properties are not invariant under isomorphism in the wider sense; e.g., if $\mathfrak{A}$ is a ring with more than one element, then $\mathfrak{A}'$ is certainly not a ring. As opposed to isomorphism in the wider sense, the notion of isomorphism in the sense of 0.2.6 can be fruitfully applied to the case of dual algebras. In general, of course, two dual algebras are not isomorphic in this sense. The property of being isomorphic (in the sense of 0.2.6) to its dual is an important property of algebraic structures; whenever this property is established for all members of a class of algebra, it simplifies considerably the study of this class. As is well known, the property discussed applies, e.g., to arbitrary Boolean algebras but not to arbitrary lattices and not even to arbitrary distributive lattices.

There is no doubt that our definitions of algebras and their isomorphisms, 0.1.1, 0.2.5, and 0.2.6, lead jointly to an inessential enlargement of the variety of algebraic isomorphism types (cf. 0.2.12). It appears, however, that a proper way of eliminating this undesirable consequence consists in imposing a restriction on index sets of algebras rather than in widening the notion of isomorphism (cf. 0.1.3).

THEOREM 0.2.8. (i) *$Is(\mathfrak{A}, \mathfrak{B}) \subseteq Ho(\mathfrak{A}, \mathfrak{B}) \subseteq Hom(\mathfrak{A}, \mathfrak{B})$ and $Is(\mathfrak{A}, \mathfrak{B}) \subseteq Ism(\mathfrak{A}, \mathfrak{B}) \subseteq Hom(\mathfrak{A}, \mathfrak{B})$.*

(ii) *$A 1 Id \in Is(\mathfrak{A}, \mathfrak{A})$; if $h = A \times \{b\}$ and $B = \{b\}$, then $h \in Ho(\mathfrak{A}, \mathfrak{B})$.*

THEOREM 0.2.9. *The following three conditions are equivalent:*

(i) *$h \in Is(\mathfrak{A}, \mathfrak{B})$,*

(ii) *$h^{-1} \in Is(\mathfrak{B}, \mathfrak{A})$,*

(iii) *both $h \in Ho(\mathfrak{A}, \mathfrak{B})$ and $h^{-1} \in Ho(\mathfrak{B}, \mathfrak{A})$.*

THEOREM 0.2.10. (i) *If $g \in Ho(\mathfrak{A}, \mathfrak{B})$ and $h \in Ho(\mathfrak{B}, \mathfrak{C})$, then $h \circ g \in Ho(\mathfrak{A}, \mathfrak{C})$; similarity with "Ho" replaced everywhere by "Hom", "Is", or "Ism".*

(ii) *If $g \in Ho(\mathfrak{A}, \mathfrak{B})$ and h is a function on B, then $h \circ g \in Ho(\mathfrak{A}, \mathfrak{C})$ implies $h \in Ho(\mathfrak{B}, \mathfrak{C})$; similarly with "Ho" replaced in its last two occurrences by "Hom", "Is", or "Ism".*

From 0.2.9 and 0.2.10(i) we easily conclude that the structure $\langle Is(\mathfrak{A}, \mathfrak{A}), \circ \rangle$ is a group. It is a subgroup of the group of all permutations of A and is called the *group of automorphisms* of $\mathfrak{A}$. A study of intrinsic properties of this group frequently leads to important conclusions concerning the algebra $\mathfrak{A}$ itself. It is known that every group is isomorphic to the group of automorphisms of some algebra $\mathfrak{A}$; cf. Birkhoff [46*].

One of the reasons why an identification of two definitionally equivalent algebras (see 0.1.6) in many cases does not lead to confusion is that the automorphism groups of these two algebras are identical. More generally, the following is true: if K and K^* are definitionally equivalent classes of algebras, and $\mathfrak{A}, \mathfrak{B}$ are any two members of K with $\mathfrak{A}^*, \mathfrak{B}^*$ the two corresponding members of K^*, then we always have $Is(\mathfrak{A}, \mathfrak{B}) = Is(\mathfrak{A}^*, \mathfrak{B}^*)$ and hence the formulas $\mathfrak{A} \cong \mathfrak{B}$ and $\mathfrak{A}^* \cong \mathfrak{B}^*$ are equivalent. This observation is rather frequently used in practice, namely, in those cases in which it is simpler to establish isomorphism for two members of one class rather than of another. For instance, in order to establish the isomorphism of two groups $\langle A, \cdot, ^{-1} \rangle$ and $\langle B, \cdot, ^{-1} \rangle$ it suffices to prove $\langle A, \cdot \rangle \cong \langle B, \cdot \rangle$. All that has just been said holds, not only for algebras, but also for any kind of relational structures, and applies, not only to first-order definitional equivalence, but to definitional equivalence with respect to any familiar system of logic.

THEOREM 0.2.11. (i) *$\mathfrak{A} \succcurlyeq \mathfrak{B}$ for every one-element algebra $\mathfrak{B}$.*

(ii) *$\mathfrak{A} \cong \mathfrak{A}$.*

(iii) $\mathfrak{A} \cong \mathfrak{B}$ *implies* $\mathfrak{B} \cong \mathfrak{A}$, $\mathfrak{A} \succcurlyeq \mathfrak{B}$, *and* $\mathfrak{B} \succcurlyeq \mathfrak{A}$.

(iv) $\mathfrak{A} \succcurlyeq \mathfrak{B} \succcurlyeq \mathfrak{C}$ *implies* $\mathfrak{A} \succcurlyeq \mathfrak{C}$; $\mathfrak{A} \cong \mathfrak{B} \cong \mathfrak{C}$ *implies* $\mathfrak{A} \cong \mathfrak{C}$.

REMARK 0.2.12. From 0.2.11 it follows in particular that $\cong$ is an equivalence relation between algebras. Hence, as was pointed out in the Preliminaries, there is a well defined function $\tau_{\cong}$, hereafter denoted simply by τ, on the class of all algebras which satisfies the following condition: for any algebras $\mathfrak{A}$ and $\mathfrak{B}$, $\tau\mathfrak{A} = \tau\mathfrak{B}$ iff $\mathfrak{A} \cong \mathfrak{B}$. The values of the function τ are referred to as *isomorphism types*; $\tau\mathfrak{A}$ is the *isomorphism type of the algebra* $\mathfrak{A}$.

$\mathfrak{A} \cong \mathfrak{B}$ obviously implies $|A| = |B|$. Only in some special classes of algebras does the converse hold (so that, within these classes, isomorphism types of algebras can be represented simply by cardinals). As examples we may mention the classes of "trivial" semigroups described in 0.1.4, furthermore the class of Boolean groups and, more generally, the class of all Abelian groups in which all non-zero elements have the given prime order π.

THEOREM 0.2.13. (i) *For every class* K *of algebras we have* $\mathsf{K} \subseteq \mathbf{I}\mathsf{K} = \mathbf{I}\mathbf{I}\mathsf{K} \subseteq \mathbf{H}\mathsf{K} = \mathbf{H}\mathbf{H}\mathsf{K} = \mathbf{H}\mathbf{I}\mathsf{K} = \mathbf{I}\mathbf{H}\mathsf{K}$.

(ii) *For any classes* K *and* L *of algebras, the formulas* $\mathbf{H}\mathsf{K} \subseteq \mathbf{H}\mathsf{L}$ *and* $\mathsf{K} \subseteq \mathbf{H}\mathsf{L}$ *are equivalent, and each of them is implied by* $\mathsf{K} \subseteq \mathsf{L}$; *similarly with* "$\mathbf{H}$" *replaced everywhere by* "$\mathbf{I}$".

(iii) *For any class* K *of algebras,* $\mathbf{H}\mathsf{K} = \bigcup_{\mathfrak{A}\in\mathsf{K}} \mathbf{H}\{\mathfrak{A}\}$ *and* $\mathbf{I}\mathsf{K} = \bigcup_{\mathfrak{A}\in\mathsf{K}} \mathbf{I}\{\mathfrak{A}\}$.

Since we are primarily interested in properties of algebras which are invariant under isomorphisms, most classes K of algebras involved in our discussion are closed under the formation of isomorphic images, i.e., satisfy the formula $\mathsf{K} = \mathbf{I}\mathsf{K}$. On the other hand, we shall often be concerned with classes K which are not closed under the formation of subalgebras or of homomorphic images, i.e., for which $\mathsf{K} \neq \mathbf{S}\mathsf{K}$ or $\mathsf{K} \neq \mathbf{H}\mathsf{K}$. For instance, referring to model-theoretic notions introduced in the Preliminaries, it is easily seen that a class K which is elementary (is an $\mathscr{E}\mathscr{C}$ or an $\mathscr{E}\mathscr{C}_\Delta$) always satisfies the formula $\mathsf{K} = \mathbf{I}\mathsf{K}$, but in general neither $\mathsf{K} = \mathbf{S}\mathsf{K}$ nor $\mathsf{K} = \mathbf{H}\mathsf{K}$; if K is universal (a $\mathscr{U}\mathscr{C}$ or a $\mathscr{U}\mathscr{C}_\Delta$), we have $\mathsf{K} = \mathbf{I}\mathsf{K} = \mathbf{S}\mathsf{K}$, but not in general $\mathsf{K} = \mathbf{H}\mathsf{K}$; finally, for a class K which is equational (an $\mathscr{E}\mathscr{Q}\mathscr{C}$ or an $\mathscr{E}\mathscr{Q}\mathscr{C}_\Delta$) we always have $\mathsf{K} = \mathbf{I}\mathsf{K} = \mathbf{S}\mathsf{K} = \mathbf{H}\mathsf{K}$.

THEOREM 0.2.14. *Assume that* $h \in Hom(\mathfrak{A}, \mathfrak{B})$. *Then*:

(i) $A' \in Su\,\mathfrak{A}$ *implies* $h^*A' \in Su\,\mathfrak{B}$;

(ii) $B' \in Su\,\mathfrak{B}$ *implies* $(h^{-1})^*B' \in Su\,\mathfrak{A}$;

(iii) *if, in addition,* $g \in Hom(\mathfrak{A}, \mathfrak{B})$, $X \subseteq A$, *and* $X1g = X1h$, *then* $(Sg\,X)1g = (Sg\,X)1h$.

THEOREM 0.2.15. *The following conditions are equivalent:*

(i) $h \in Hom(\mathfrak{A}, \mathfrak{B})$, *or* $h \in Ism(\mathfrak{A}, \mathfrak{B})$;

(ii) *there is an* $\mathfrak{A}' \supseteq \mathfrak{A}$ *and a* $k \in Ho(\mathfrak{A}', \mathfrak{B})$, *or* $k \in Is(\mathfrak{A}', \mathfrak{B})$, *such that* $A 1 k = h$.

PROOF. To obtain the assertion that (i) implies (ii), we assume $h \in Hom(\mathfrak{A}, \mathfrak{B})$, or $h \in Ism(\mathfrak{A}, \mathfrak{B})$, and we pick an element c such that for no x does $\langle x, c\rangle$ belong to A; we can take, e.g., A itself for c. We then let

$$C = (B \sim h^*A) \times \{c\}, \; A' = A \cup C, \text{ and } k = h \cup C 1 \mathsf{pj}_0$$

(where pj_0 is the 0th projection, so that $\mathsf{pj}_0\langle x, y\rangle = x$ for all x and y). Since $A \cap C = 0$, k is clearly a function and we have $Dok = A'$, $Rgk = B$, and $A 1 k = h$. By the axiom of choice there is a function $f \in {}^BA'$ such that $k \circ f = B 1 Id$. (If h is one-one, k is also one-one, and we can simply take k^{-1} for f without applying the axiom of choice.) We now define the algebra $\mathfrak{A}' = \langle A', +'\rangle$ by stipulating that

$$x +' y = x +^{(\mathfrak{A})} y \text{ if } \langle x, y\rangle \in {}^2A,$$

$$x +' y = f(kx +^{(\mathfrak{B})} ky) \text{ if } \langle x, y\rangle \in {}^2A' \sim {}^2A.$$

As is easily seen, $k \in Ho(\mathfrak{A}', \mathfrak{B})$ if $h \in Hom(\mathfrak{A}, \mathfrak{B})$, and $k \in Is(\mathfrak{A}', \mathfrak{B})$ if $h \in Ism(\mathfrak{A}, \mathfrak{B})$. In addition $\mathfrak{A}' \supseteq \mathfrak{A}$, and the proof that (i) implies (ii) is complete. The assertion that (ii) implies (i) follows directly from 0.2.14(i).

As direct consequences of 0.2.14 and 0.2.15 we obtain:

COROLLARY 0.2.16. *If* $\mathfrak{A} \subseteq | \preccurlyeq \mathfrak{B}$, *then* $\mathfrak{A} \preccurlyeq | \subseteq \mathfrak{B}$.

COROLLARY 0.2.17. (i) $\mathfrak{A} \subseteq | \succcurlyeq \mathfrak{B}$ *iff* $\mathfrak{A} \succcurlyeq | \subseteq \mathfrak{B}$.

(ii) $\mathfrak{A} \subseteq | \cong \mathfrak{B}$ *iff* $\mathfrak{A} \cong | \subseteq \mathfrak{B}$.

THEOREM 0.2.18. (i) *If* $h \in Hom(\mathfrak{A}, \mathfrak{B})$ *and* $X \subseteq A$, *then* $h^*(Sg^{(\mathfrak{A})}X) = Sg^{(\mathfrak{B})}h^*X$.

(ii) *If* $\mathfrak{A} \succcurlyeq \mathfrak{B}$, α *is any cardinal, and* $\mathfrak{A} \in \mathbf{S}_\alpha\mathfrak{A}$, *then* $\mathfrak{B} \in \mathbf{S}_\alpha\mathfrak{B}$.

(iii) *If* $g \in Ho(\mathfrak{A}, \mathfrak{B})$, $h \in Hom(\mathfrak{B}, \mathfrak{A})$, $A = Sg^{(\mathfrak{A})}X$, *and* $hgy = y$ *for every* $y \in X$, *then* $g \in Is(\mathfrak{A}, \mathfrak{B})$ *and* $h = g^{-1} \in Is(\mathfrak{B}, \mathfrak{A})$.

0.2.18(i) can be derived from 0.2.14(i),(ii); 0.2.18(ii) follows directly from 0.2.18(i). In proving 0.2.18(iii) we make use of 0.2.14(iii).

THEOREM 0.2.19. *For every class* K *of algebras and every cardinal* α, $\mathbf{SH}\mathsf{K} \subseteq \mathbf{HS}\mathsf{K}$, $\mathbf{S}_\alpha\mathbf{H}\mathsf{K} \subseteq \mathbf{HS}_\alpha\mathsf{K} = \mathbf{S}_\alpha\mathbf{HS}_\alpha\mathsf{K} = \mathbf{S}_\alpha\mathbf{HS}\mathsf{K}$, $\mathbf{SI}\mathsf{K} = \mathbf{IS}\mathsf{K}$, *and* $\mathbf{S}_\alpha\mathbf{I}\mathsf{K} = \mathbf{IS}_\alpha\mathsf{K}$.

In proving this theorem we make use of 0.2.16, 0.2.17(ii), and 0.2.18(i).

The converse of 0.2.16 does not hold and the inclusions in 0.2.19 cannot be replaced by equalities. In fact, let $\mathfrak{A}$ be a two-element field and $\mathfrak{B}$ be the field of rational numbers (with the ordinary operations $+$ and $\cdot$). We can easily show that $\mathfrak{A} \preccurlyeq | \subseteq \mathfrak{B}$ holds but $\mathfrak{A} \subseteq | \preccurlyeq \mathfrak{B}$ does not hold, and thus $\mathbf{HS}\mathfrak{B}$ is not included in $\mathbf{SH}\mathfrak{B}$; actually $\mathfrak{A} \in \mathbf{HS}_2\mathfrak{B}$, so that $\mathbf{HS}_2\mathfrak{B}$ is not included in either $\mathbf{SH}\mathfrak{B}$ or $\mathbf{S}_2\mathbf{H}\mathfrak{B}$.

Assuming that M is any class of algebras closed under the formation of homomorphic images we can say (disregarding certain set-theoretical difficulties) that the ordered pair $\langle \mathsf{M}, \mathbf{H} \rangle$ is, by 0.2.13, a complete closure structure. From 0.2.13 and 0.2.19 we further conclude that under appropriate assumptions the same applies to the pairs $\langle \mathsf{M}, \mathbf{I} \rangle$, $\langle \mathsf{M}, \mathbf{HS} \rangle$, and $\langle \mathsf{M}, \mathbf{IS} \rangle$, while the pair $\langle \mathsf{M}, \mathbf{SH} \rangle$ is not in general a closure structure. (Compare the Preliminaries, and the remarks following 0.1.14.)

We shall now discuss the important notion of a congruence relation and the related operation of forming quotient algebras. This operation may first appear to be a rather special case of the formation of homomorphic images. It turns out, however, that the general operation reduces up to isomorphism to the special one: every homomorphic image of an algebra can be isomorphically represented as a quotient algebra (cf. 0.2.23 below). The usefulness of the notion of a quotient algebra is largely due to its intrinsic character: while homomorphic images are built from arbitrary elements which may be completely extraneous to the original algebra, the elements of a quotient algebra are special subsets of the original algebra, namely equivalence classes under a congruence relation.

DEFINITION 0.2.20. *A binary relation R is called a* ***congruence relation*** *on $\mathfrak{A}$, in symbols $R \in Co\,\mathfrak{A}$, if R is an equivalence relation on A, and, for any $x, x', y, y' \in A$, the formulas xRx' and yRy' imply $(x+y)R(x'+y')$. The congruence relation R is called* ***proper*** *if $R \neq A \times A$.*

In the "trivial" semigroups mentioned in 0.1.4, as well as in every two-element algebra, every equivalence relation on the universe is a congruence relation. It is easily seen that no other algebras of similarity type $\langle 2 \rangle$ have this property. For algebras $\mathfrak{A} = \langle A, Q_i \rangle_{i \in I}$ of arbitrary similarity type we have $Co\,\mathfrak{A} = \bigcap_{i \in I} Co\langle A, Q_i \rangle$.

We recall that, for every binary relation R and every $x \in Do\,R$, $R^{\star}x = \{y : xRy\}$; h being any function, $h|h^{-1}$ is a relation which holds between x and y iff $x, y \in Do\,h$ and $hx = hy$.

THEOREM 0.2.21. (i) *$R \in Co\,\mathfrak{A}$ iff R is an equivalence relation on A and $R^{\bigstar} \in Ho\,\mathfrak{A}$.*

(ii) *$h \in Ho\,\mathfrak{A}$ iff h is a function on A and $h|h^{-1} \in Co\,\mathfrak{A}$; if $R \in Co\,\mathfrak{A}$ and $h = A 1 R^{\bigstar}$, then $h|h^{-1} = R$.*

(iii) *The function $\langle h|h^{-1} : h \in Ho\,\mathfrak{A}\rangle$ maps $Ho\,\mathfrak{A}$ onto $Co\,\mathfrak{A}$ (in a many-one way).*

In view of 0.2.21(i) we define:

DEFINITION 0.2.22. *If $R \in Co\,\mathfrak{A}$, the algebra $(R^{\bigstar})^*\mathfrak{A}$ is called the* ***quotient algebra*** *of $\mathfrak{A}$ over R, symbolically $\mathfrak{A}/R$.*

THEOREM 0.2.23. (i) *If $h \in Ho\,\mathfrak{A}$ and $R = h|h^{-1}$, then $h^*\mathfrak{A} \cong \mathfrak{A}/R$, and in fact $h^{-1}|R^{\bigstar} \in Is(h^*\mathfrak{A}, \mathfrak{A}/R)$.*

(ii) *$\mathfrak{A} \succcurlyeq \mathfrak{B}$ iff $\mathfrak{A}/R \cong \mathfrak{B}$ for some $R \in Co\,\mathfrak{A}$.*

(iii) *If $h \in Ho\,\mathfrak{A}$, $R \in Co\,\mathfrak{A}$, and $R \subseteq h|h^{-1}$, then $\{\langle a/R, ha\rangle : a \in A\} \in Ho(\mathfrak{A}/R)$.*

THEOREM 0.2.24. *Theorem 0.1.10 remains valid if 0, A, and $Su\,\mathfrak{A}$ are respectively replaced by $A 1 Id$, $A \times A$, and $Co\,\mathfrak{A}$.*

The proof of 0.2.24 is straight-forward. If we want, we can also derive this theorem as an immediate corollary from 0.1.10, using the following observation which shows that the notion of congruence relation can be subsumed under that of subuniverse. For any given algebra $\mathfrak{A} = \langle A, +\rangle$ we form a new algebra $\mathfrak{A}^*$ with the universe $A \times A$, two binary operations, $+'$ and $|$, one unary operation, $^{-1}$, and the system of distinguished elements $\langle x, x\rangle$ where x is any element of A. The operations $+'$, $|$, and $^{-1}$ are defined by the formulas:

$$\langle x, y\rangle +' \langle x', y'\rangle = \langle x+x', y+y'\rangle,$$

$$\langle x, y\rangle \,|\, \langle x', y'\rangle = \langle x, y'\rangle \text{ in case } y = x',$$

$$\langle x, y\rangle \,|\, \langle x', y'\rangle = \langle x, y\rangle \text{ otherwise},$$

$$\langle x, y\rangle^{-1} = \langle y, x\rangle.$$

($\langle A \times A, +'\rangle$ is what will be called in Section 0.3 the direct square of $\mathfrak{A}$ and denoted by $\mathfrak{A} \times \mathfrak{A}$.) It is easily seen that congruence relations of the original algebra $\mathfrak{A}$ are just non-empty subuniverses of $\mathfrak{A}^*$.

The above observation can be used to check some remarks and establish some facts stated in the subsequent discussion. It permits us for instance to derive 0.2.29 below as a direct corollary of 0.1.17.

Note that the part of 0.2.24 corresponding to 0.1.10(ii), namely, that $\bigcap K \in Co\,\mathfrak{A}$ whenever $K \subseteq Co\,\mathfrak{A}$, applies in particular to $K = 0$; in accordance with a general convention described in the Preliminaries, by the intersection of the empty set of congruence relations on $\mathfrak{A}$ we understand the set $A \times A$ (cf. the remark immediately following 0.1.10).

In connection with Theorem 0.2.24 we can repeat, with obvious changes, all the remarks in 0.1.11 concerning Theorem 0.1.10 and its implications. Thus, by 0.2.24, $\langle A \times A, Co\,\mathfrak{A}\rangle$ is an inductive closed-set structure. As a consequence, the inclusion relation establishes a complete lattice ordering in $Co\,\mathfrak{A}$. Hence $Co\,\mathfrak{A}$ becomes the universe of a complete lattice, the so-called *lattice of congruence relations on* $\mathfrak{A}$, under appropriate definitions of join and meet.

An intrinsic characterization of all those lattices which are isomorphic to lattices of congruence relations on algebras can be found in the literature; it turns out that the same characterization applies to those lattices which are isomorphic to lattices of subuniverses. For the result concerning the lattices of subuniverses cf. Birkhoff [67*], p. 187 ff. where further bibliographic references can be found; cf. in addition Hanf [56*]. For the result concerning the lattices of congruence relations see Grätzer-Schmidt [63*].

THEOREM 0.2.25. *$\mathfrak{A}/(A \times A)$ is a one-element algebra while $\mathfrak{A}/(A 1 Id) \cong \mathfrak{A}$.*

THEOREM 0.2.26. *If $\mathfrak{B} \subseteq \mathfrak{A}$ and $R \in Co\,\mathfrak{A}$, then*

(i) $(B \times B) \cap R \in Co\,\mathfrak{B}$;

(ii) *$R^*B \in Su\,\mathfrak{A}$ and $B/R = R^*B/R \in Su(\mathfrak{A}/R)$;*

(iii) $\mathfrak{B}/((B \times B) \cap R) \cong \langle B/R, +^{(\mathfrak{A}/R)}\rangle \in \mathbf{S}(\mathfrak{A}/R)$.

THEOREM 0.2.27. *Let $R \in Co\,\mathfrak{A}$, $K = \{S : R \subseteq S \in Co\,\mathfrak{A}\}$, and $F = \langle S/R : S \in K\rangle$. Then:*

(i) *F maps K onto $Co(\mathfrak{A}/R)$ in a one-one way;*

(ii) *$F^{-1}T = \{\langle x, y\rangle : x, y \in A, (x/R)T(y/R)\}$ for every $T \in Co(\mathfrak{A}/R)$;*

(iii) *if $L \subseteq K$, then $F(\bigcap L) = \bigcap F^*L$;*

(iv) *$\mathfrak{A}/S \cong (\mathfrak{A}/R)/FS = (\mathfrak{A}/R)/(S/R)$ for every $S \in K$.*

0.2.26(iii) and 0.2.27(iv) are generalizations of the so-called first and second isomorphism theorems of group theory.

From 0.2.27 we can conclude that, for every algebra $\mathfrak{A}$ and every $R \in Co\,\mathfrak{A}$, the lattice of congruence relations on $\mathfrak{A}/R$ is isomorphic to a sublattice of the lattice of congruence relations on $\mathfrak{A}$ (cf. the remarks following 0.2.24); the universe of this sublattice is just the set K defined in 0.2.27.

DEFINITION 0.2.28. *For every relation $X \subseteq A \times A$, the relation $R = \bigcap\{Y : X \subseteq Y \in Co\,\mathfrak{A}\}$ is called the* **congruence relation generated by** *X and is* *denoted by* $Cg^{(\mathfrak{A})}X$ *or* $Cg\,X$. *We say that* R *is a* ***finitely generated congruence relation*** *on* $\mathfrak{A}$ *if* $R = Cg\,X$ *for some finite* $X \subseteq A \times A$.

THEOREM 0.2.29. *Theorem 0.1.17 remains true if "A", "Su", and "Sg" are respectively replaced by "$A \times A$", "Co", and "Cg". However, part (iv) of* 0.1.17 *goes over into*

$$Cg0 = A1Id.$$

With obvious changes, we can repeat here the remarks immediately following 0.1.17.

THEOREM 0.2.30. *If* $0 \neq K \subseteq Co\mathfrak{A}$, *then* $Cg(\bigcup K) = \bigcup\{R_0|\ldots|R_{\nu-1}: 0 < \nu < \omega$ *and* $R \in {}^{\nu}K\}$.

THEOREM 0.2.31. *For any* $R, S \in Co\mathfrak{A}$ *the following five conditions are equivalent*: (i) $R|S$ *is symmetric*, (ii) $R|S = S|R$, (iii) $R|S$ *is transitive*, (iv) $R|S \in Co\mathfrak{A}$, *and* (v) $R|S = Cg(R \cup S)$.

PROOF: by 0.2.20 and 0.2.30.

If all the congruence relations on an algebra $\mathfrak{A}$ satisfy the formulas of 0.2.31, then $\mathfrak{A}$ is called an *algebra with commuting congruence relations*. Many familiar algebras are of this kind, e.g., all groups and rings. If $\mathfrak{A}$ is an algebra with commuting congruence relations and $\mathfrak{L}$ the lattice of congruence relations on $\mathfrak{A}$, then, as a consequence of 0.2.31, the join of any two relations R, S in $\mathfrak{L}$ is simply $R|S$ (while their meet is, as always, $R \cap S$). Hence we can derive some further properties of $\mathfrak{L}$; in particular, we can conclude that the lattice $\mathfrak{L}$ is modular. Cf. Birkhoff [67*], p. 162, Theorem 4.

THEOREM 0.2.32. (i) *If* $\mathfrak{A}$ *is a finitely generated algebra, then* $A \times A$ *is a finitely generated congruence relation on* $\mathfrak{A}$.

(ii) *If* $R \in Co\mathfrak{A}$, $R \subset T \subseteq A \times A$, *and* CgT *is a finitely generated congruence relation on* $\mathfrak{A}$, *then there is a maximal* $S \in Co\mathfrak{A}$ *such that* $R \subseteq S \subset T$.

PROOF: To obtain (i) it suffices to notice that $\mathfrak{A} = \mathfrak{S}\mathfrak{g}\,X$ implies

$$A \times A = Cg\{\langle x, y+z\rangle : x, y, z \in X\};$$

if X is finite, then obviously so is the set $\{\langle x, y+z\rangle : x, y, z \in X\}$. 0.2.32(ii) is an analogue of 0.1.22 and can easily be derived from 0.2.29(iii) by means of Zorn's lemma.

REMARK 0.2.33. It should be mentioned that 0.2.32(i) can be extended, not to arbitrary algebras $\mathfrak{A} = \langle A, Q_i\rangle_{i\in I}$, but only to those with finitely many fundamental operations. For example, let $\mathfrak{A} = \langle \omega, \omega \times \{\nu\}\rangle_{\nu<\omega}$; obviously, for every $\nu < \omega$, $\omega \times \{\nu\}$ is a constant operation with ν as its only value. As is easily seen, $\mathfrak{A}$ is minimal and hence is finitely generated (cf. 0.1.23) while the congruence relation $A \times A$ is not.

DEFINITION 0.2.34. *An algebra $\mathfrak{A}$ is called **simple** if $|A| \geqq 2$ and $Co\,\mathfrak{A} = \{A \upharpoonright Id, A \times A\}$ (or, equivalently, if $|Co\,\mathfrak{A}| = 2$).*

THEOREM 0.2.35. (i) *If $\mathfrak{A}$ is not simple, then there is an algebra $\mathfrak{B} \in \mathbf{S}_4\mathfrak{A}$ which is not simple, and which actually is not a subalgebra of any simple subalgebra of $\mathfrak{A}$.*[1]

(ii) *If K is a set of simple algebras directed by the relation $\subseteq$, then UK is simple.*

PROOF. By the hypothesis of (i) there is an $R \in Co\,\mathfrak{A}$ such that $A \upharpoonright Id \subset R \subset A \times A$ and hence there must be three elements a, b, c such that $\langle a, c\rangle \in R \sim Id$ and $\langle b, c\rangle \in (A \times A) \sim R$. We have $\mathfrak{B} = \mathfrak{Sg}\{a, b, c\} \in \mathbf{S}_4\mathfrak{A}$ and, by 0.2.26(i), $\mathfrak{B}$ is not simple, as required in the conclusion of (i). More generally, no algebra $\mathfrak{C}$ such that $\mathfrak{B} \subseteq \mathfrak{C} \subseteq \mathfrak{A}$ is simple. (ii) follows by 0.1.28.

The observations made in 0.2.35 seem to be new. In connection with 0.2.35(i) it may be interesting to mention as a curiosity that, if a unary algebra $\mathfrak{A}$ (cf. 0.1.4) is simple, then either $|A| = 2$ or $\mathfrak{A} \in \mathbf{S}_2\mathfrak{A}$.

THEOREM 0.2.36. *The following two conditions are equivalent:*

(i) *$\mathfrak{A}$ is simple;*

(ii) *$|A| \geqq 2$ and, for every $h \in Ho\,\mathfrak{A}$, either $h \in Is\,\mathfrak{A}$ or h is a constant function.*

Each of these conditions implies:

(iii) *$|A| \geqq 2$ and, whenever $\mathfrak{A} \succcurlyeq \mathfrak{B}$, we have either $\mathfrak{A} \cong \mathfrak{B}$ or else $|B| = 1$.*

REMARK 0.2.37. Algebras satisfying condition 0.2.36(iii) may be called *pseudo-simple*. Every finite pseudo-simple algebra is simple. In general, however, this does not apply to infinite algebras. For instance, given a prime number π, let $\mathfrak{A}_\pi = \langle A_\pi, +'\rangle$ be the Abelian group such that A_π is the set of all rational numbers of the form ν/π^κ where $0 \leqq \nu < \pi^\kappa$ and $0 < \kappa < \omega$, and $+'$ is the arithmetical addition modulo 1 (i.e., $x+'y = x+y$ if $x+y < 1$, and $x+'y = (x+y)-1$ otherwise). Then $\mathfrak{A}_\pi$ proves to be pseudo-simple but not simple. It has also been shown that every pseudo-simple Abelian group which is not simple is isomorphic to one of the groups $\mathfrak{A}_\pi$; see Szélpál [49*]. A necessary, though not sufficient, condition for an algebra $\mathfrak{A}$ to be pseudo-simple is that the set $Co\,\mathfrak{A}$ be well-ordered by the relation $\subseteq$; cf. Monk [62*].

[1]) This theorem was originally proved by the authors in a somewhat weaker form, with "$\mathbf{S}_4\mathfrak{A}$" replaced by "$\mathbf{S}_5\mathfrak{A}$". The present form is due to Bjarni Jónsson.

THEOREM 0.2.38. (i) *If* $R \in Co\mathfrak{A}$, *then* $\mathfrak{A}/R$ *is simple iff* R *is a maximal proper congruence relation.*

(ii) *If* $|A| \geqq 2$ *and* $A \times A$ *is a finitely generated congruence relation on* $\mathfrak{A}$, *then for each proper* $S \in Co\mathfrak{A}$ *there is an* R *such that* $S \subseteq R \in Co\mathfrak{A}$ *and* $\mathfrak{A}/R$ *is simple.*

(iii) *The conclusion of (ii) holds in particular if* $|A| \geqq 2$ *and* $\mathfrak{A}$ *is a finitely generated algebra.*

PROOF: (i) by 0.2.27(i),(iii); (ii) by 0.2.32(ii); (iii) by (ii) and 0.2.32(i).

From 0.2.38(iii) it follows that every pseudo-simple finitely generated algebra is simple (see 0.2.37). This last observation and Theorem 0.2.38(iii) itself extend only to algebras with finitely many operations (compare 0.2.33).[1]

DEFINITION 0.2.39. *For any* $z \in A$ *a set* $I \subseteq A$ *is called a* ***z-ideal*** *of* $\mathfrak{A}$, *in symbols* $I \in Il_z\mathfrak{A}$, *if* $I = z/R$ *for some* $R \in Co\mathfrak{A}$. *The z-ideal* I *is called* ***proper*** *if* $I \neq A$.

REMARK 0.2.40. The z-ideals play a rather restricted role in the general theory of algebras; their use is essentially confined to those cases when the z-ideals function properly in a given algebra $\mathfrak{A}$ — in the sense of Definition 0.2.47 below; cf. Remark 0.2.48.

THEOREM 0.2.41. *For every* $z \in A$, *Theorem 0.1.10 remains valid if* 0 *and* $Su\mathfrak{A}$ *are replaced everywhere by* $\{z\}$ *and* $Il_z\mathfrak{A}$, *respectively.*

To show, for example, that $\bigcup K \in Il_z\mathfrak{A}$ whenever $0 \neq K \subseteq Il_z\mathfrak{A}$ and K is directed by $\subseteq$, we let, for each $I \in K$,

$$S_I = \bigcap\{R : R \in Co\mathfrak{A}, \quad I \subseteq z/R\}.$$

Hence, by 0.2.24, $S_I \in Co\mathfrak{A}$ and, as is easily seen, $z/S_I = I$. Moreover, $\{S_J : J \in K\}$ is directed by $\subseteq$. Therefore, by applying 0.2.24 again, $\bigcup\{S_J : J \in K\} \in Co\mathfrak{A}$ and

$$\bigcup K = \bigcup\{z/S_J : J \in K\} = z/\bigcup\{S_J : J \in K\} \in Il_z\mathfrak{A}.$$

As in the case of 0.2.24 we can repeat here again, with obvious changes, all that was said in 0.1.11. In particular, Theorem 0.2.41 leads, for any given $z \in A$, to the construction of the complete *lattice of z-ideals of* $\mathfrak{A}$.

THEOREM 0.2.42. (i) *If* $\{z\} \in Su\mathfrak{A}$, *then* $Il_z\mathfrak{A} \subseteq Su\mathfrak{A}$.
(ii) *If* $\mathfrak{B} \subseteq \mathfrak{A}$, $z \in B$, *and* $I \in Il_z\mathfrak{A}$, *then* $B \cap I \in Il_z\mathfrak{B}$.
PROOF: by 0.2.26(i).

[1]) 0.2.38(ii),(iii) and the subsequent remarks are due to Tarski.

DEFINITION 0.2.43. *For any* $z \in A$ *and any* $X \subseteq A$, *the set* $\bigcap\{Y : X \subseteq Y \in Il_z\mathfrak{A}\}$ *is called the* **z-ideal generated by** *X and is denoted by* $Ig_z^{(\mathfrak{A})}X$ *or* Ig_zX. *We say that I is a* **finitely generated** **z-ideal**, *or a* **principal** **z-ideal**, *of* $\mathfrak{A}$ *if* $I = Ig_zX$ *for some finite* $X \subseteq A$, *or for some* $X \subseteq A$ *with* $|X| = 1$, *respectively.*

THEOREM 0.2.44. *For every* $z \in A$, *Theorem 0.1.17 remains valid if "Su" and "Sg" are respectively replaced by "*Il_z*" and "*Ig_z*". However, 0.1.17(iv) goes over into*

$$Ig_z0 = \{z\} = Ig_z\{z\}.$$

Compare here the remarks immediately following 0.1.17.

THEOREM 0.2.45. *For every* $z \in A$ *we have*:

(i) *If* $X \subseteq A$, *then* $\bigcap\{R : R \in Co\mathfrak{A},\ z/R \supseteq X\} = Cg(\{z\} \times X)$ *and* $Ig_zX = z/Cg(\{z\} \times X)$;

(ii) $I \in Il_z\mathfrak{A}$ *iff* $I \subseteq A$ *and* $I = z/Cg(\{z\} \times I)$.

PROOF. For any $R \in Co\mathfrak{A}$, $z/R \supseteq X$ iff $R \supseteq \{z\} \times X$. This gives the first equality of (i), and using this equality we have

$$\begin{aligned} Ig_zX &= \bigcap\{z/R : R \in Co\mathfrak{A},\ z/R \supseteq X\} \\ &= z/\bigcap\{R : R \in Co\mathfrak{A},\ z/R \supseteq X\} \\ &= z/Cg(\{z\} \times X). \end{aligned}$$

To prove (ii) let us assume $I \in Il_z\mathfrak{A}$. Then $Ig_zI = I$ and, hence, taking $X = I$ in the second equation of (i) we get $I = z/Cg(\{z\} \times I)$. Thus $I \in Il_z\mathfrak{A}$ implies $I \subseteq A$ and $I = z/Cg(\{z\} \times I)$. The implication in the opposite direction is obvious.

If I is a z-ideal, then, in general, there may be many congruence relations R such that $z/R = I$. From 0.2.45 we conclude that $Cg(\{z\} \times I)$ is the least congruence relation with this property. It can also be easily shown that there is a largest congruence relation R with $z/R = I$.

THEOREM 0.2.46. *For every* $z \in A$ *we have*:

(i) *I is a finitely generated z-ideal iff* $I \in Il_z\mathfrak{A}$ *and* $Cg(\{z\} \times I)$ *is a finitely generated congruence relation*;

(ii) *A is a finitely generated z-ideal iff* $A \times A$ *is a finitely generated congruence relation*;

(iii) *if* $I \in Il_z\mathfrak{A}$, $I \subset L \subseteq A$, *and* Ig_zL *is a finitely generated z-ideal, then there is a maximal* $J \in Il_z\mathfrak{A}$ *such that* $I \subseteq J \subset L$.

PROOF. If $I = Ig_z X$, then, for any $R \in Co\mathfrak{A}$, $z/R \supseteq X$ iff $z/R \supseteq I$; thus, by 0.2.45(i), $Cg(\{z\} \times I) = Cg(\{z\} \times X)$. Hence $Cg(\{z\} \times I)$ is a finitely generated congruence relation whenever I is a finitely generated z-ideal. Now assume $I \in Il_z\mathfrak{A}$, and $Cg(\{z\} \times I)$ is finitely generated. Then by 0.2.29(x) there is an $X \subseteq I$ with $|X| < \omega$ such that $Cg(\{z\} \times X) = Cg(\{z\} \times I)$, and thus, by 0.2.45,

$$I = z/Cg(\{z\} \times I) = z/Cg(\{z\} \times X) = Ig_z X;$$

therefore, I is finitely generated. This proves (i). Since $A \times A = Cg(\{z\} \times A)$, (ii) follows directly from (i). (iii) can easily be derived from 0.2.44 using Zorn's lemma.

We now turn to the discussion of those situations when, for every z-ideal I, there is just one congruence relation R such that $I = z/R$.

DEFINITION 0.2.47. *We say that the z-**ideals function properly** in $\mathfrak{A}$ if, for any $R, S \in Co\mathfrak{A}$, $z/R = z/S$ implies $R = S$. If the z-ideals function properly in $\mathfrak{A}$, then by the **quotient algebra of** $\mathfrak{A}$ **over a z-ideal** I, symbolically $\mathfrak{A}/(I, z)$, we understand the algebra $\mathfrak{A}/R$ where R is the unique congruence relation on $\mathfrak{A}$ such that $z/R = I$. We use analogously the symbols $x/(I, z)$ for $x \in A$, $Y/(I, z)$ for $Y \subseteq A$, $S/(I, z)$ for $S \subseteq A \times A$, and $(I, z)^\bigstar$ (instead of $R^\bigstar$).*

REMARK 0.2.48. When applying the notion of an ideal to a particular algebra, we usually fix an element $z \in A$ such that the z-ideals function properly in $\mathfrak{A}$, and we refer to these z-ideals simply as ideals; we write, e.g., $Il\mathfrak{A}$ and $\mathfrak{A}/I$ instead of $Il_z\mathfrak{A}$ and $\mathfrak{A}/(I, z)$. An element z with this property cannot always be found and, if it can be found, it is by no means uniquely determined; we usually pick for z an element which is easily definable in terms of fundamental operations of $\mathfrak{A}$. Once z has been fixed and the z-ideals have been shown to function properly, we have a one-one correspondence between ideals and congruence relations (cf. Theorem 0.2.49 below), and we can replace the latter by the former in the subsequent development. This frequently simplifies the discussion. In fact, ideals, as subsets of A, can in principle be handled more easily than congruence relations which are subsets of $A \times A$; moreover, in all practically important cases the z-ideals, under a proper choice of z, admit a simple characterization in terms of fundamental operations on their elements without referring to the notion of a congruence relation. Actually a metamathematical result is known by which the z-ideals, in case they function properly, can always be characterized by means of first-order conditions involving only elements and fundamental operations of the algebra; cf. here Vaught [66*].

A few examples may illustrate the above remarks. In the algebra $\mathfrak{A} = \langle A, +\rangle$ where $A = \omega$ and $+$ is the ordinary addition of natural numbers, the z-ideals do not function properly for any z. If A is the set of integers (positive, negative, and 0) with the additional infinity element ∞, and $+$ is the addition of integers with the supplementary stipulation $x+\infty = \infty+x = \infty$ for every $x \in A$, then the z-ideals ($z \in A$) function properly iff $z \neq \infty$. If $\mathfrak{G} = \langle G, \cdot\rangle$ is a group, the z-ideals function properly for every $z \in G$; if we take z to be the unit element of $\mathfrak{G}$, then the z-ideals coincide with normal subgroups. Similarly, if $\mathfrak{R}$ is a ring, the z-ideals function properly for every $z \in R$; taking z to be the zero element 0, the z-ideals then coincide with ordinary (two-sided) ring ideals. Thus, in particular, we recognize in 0.2.46 a generalization of the well-known theorem by which every proper ideal in a ring with unit can be extended to a maximal proper ideal.

It would be interesting to find a simple characterization of those algebras in which there is an element z such that the z-ideals function properly, as well as those in which this happens for every element z. In this connection compare Jakubík [54*], Mal'cev [54*], and Valucè [63*].

THEOREM 0.2.49. *Let $z \in A$ and assume that the z-ideals function properly in $\mathfrak{A}$; let*

$$F = \langle z/R : R \in Co\,\mathfrak{A}\rangle.$$

Then:

(i) *F maps $Co\,\mathfrak{A}$ onto $Il_z\mathfrak{A}$ in a one-one way*;

(ii) *if $L \subseteq Co\,\mathfrak{A}$, then $F(\bigcap L) = \bigcap F^*L$.*

Hence we see that, under the hypothesis of 0.2.49, the lattice of z-ideals of $\mathfrak{A}$ is isomorphic to the lattice of congruence relations on $\mathfrak{A}$. Notice that 0.2.49(ii) does not require the assumption that the z-ideals function properly.

THEOREM 0.2.50. *Assume that $z \in A$.*

(i) *The following three conditions are equivalent*:

(i′) *the z-ideals function properly in $\mathfrak{A}$*;

(i″) *for any $R \in Co\,\mathfrak{A}$ and $X \subseteq A$, $z/R = Ig_z X$ iff $R = Cg(\{z\} \times X)$*;

(i‴) *for any $R \in Co\,\mathfrak{A}$, $R = Cg(\{z\} \times z/R)$.*

(ii) *If the z-ideals function properly in $\mathfrak{A}$, then R is a finitely generated congruence relation iff $R \in Co\,\mathfrak{A}$ and z/R is a finitely generated z-ideal.*

PROOF. That (i′) implies (i″) follows immediately from 0.2.45(i). Taking $X = z/R$ in (i″) we get (i‴), and (i‴) obviously implies (i′). This proves (i).

To obtain (ii) take $I = z/R$ in 0.2.46(i) and use the fact that (i′) implies (i‴).[1]

Without the assumption that the z-ideals function properly the congruence relation R may be finitely generated even though the ideal z/R is not; an example of such an R was pointed out by Joel Karnofsky.

THEOREM 0.2.51. *If $z \in A$, $R \in Co\mathfrak{A}$ (or $I \in Il_z\mathfrak{A}$), and the z-ideals function properly in $\mathfrak{A}$, then the z/R-ideals (or $z/(I, z)$-ideals) function properly in $\mathfrak{A}/R$ (or $\mathfrak{A}/(I, z)$). If $R \in Co\mathfrak{A}$ and $z/R \subseteq I \in Il_z\mathfrak{A}$, then $I/R \in Il_{z/R}(\mathfrak{A}/R)$.*

PROOF: by 0.2.27.

REMARK 0.2.52. Using 0.2.49 (and in some cases 0.2.51), we can automatically express various theorems on congruence relations in the terminology of z-ideals; we restrict ourselves, of course to those algebras in which the z-ideals function properly. Thus we can establish a direct correspondence between homomorphisms and z-ideals (cf. 0.2.21 and 0.2.23) as well as between the z-ideals of $\mathfrak{A}$ which include a given z-ideal I and the $z/(I, z)$-ideals of $\mathfrak{A}/(I, z)$ (cf. 0.2.27), and we can characterize simple algebras in terms of z-ideals (cf. 0.2.34 and 0.2.38(i)). We shall not formulate explicitly the theorems thus obtained.

[1]) The fact that the ideal z/R is finitely generated whenever R is finitely generated (under the assumption that the z-ideals function properly) was pointed out by Bjarni Jónsson. We do not give here Jónsson's proof of this implication since we derive it as a direct consequence from a more general result, 0.2.46(i), which was subsequently found.

In this section we shall concern ourselves with an operation, called direct multiplication, which when performed on a system $\langle \mathfrak{A}_i : i \in I \rangle$ of similar algebras yields an algebra $\mathfrak{B}$ similar to the $\mathfrak{A}_i$'s. As opposed to the formation of subalgebras and homomorphic images this operation leads from smaller, simpler algebras to larger, more complicated ones. In this respect direct multiplication resembles the operation of forming the union of algebras (which we introduced in passing at the end of Section 0.1), but its range of application in the theory of algebras is much wider.

DEFINITION 0.3.1. (i) *By the **direct** (or **cardinal**) **product** of a system* $\mathfrak{A}$ *of algebras* $\mathfrak{A}_i = \langle A_i, +_i \rangle$ *indexed by a set* I, *symbolically* $\mathsf{P}\mathfrak{A}$ *or* $\mathsf{P}\langle \mathfrak{A}_i : i \in I \rangle$ *or* $\mathsf{P}_{i \in I} \mathfrak{A}_i$, *we understand the algebra* $\mathfrak{B} = \langle B, + \rangle$ *such that* $B = \mathsf{P}_{i \in I} A_i$ *and* $+$ *is the operation defined by the formula*

$$f + g = \langle f_i +_i g_i : i \in I \rangle$$

for any $f, g \in B$. *The product* $\mathsf{P}_{i \in I} \mathfrak{A}_i$ *in which* $I = 2$, $\mathfrak{A}_0 = \mathfrak{C}$, *and* $\mathfrak{A}_1 = \mathfrak{D}$ *is denoted by* $\mathfrak{C} \times \mathfrak{D}$; *analogous notations are used for* $I = 3, 4, \ldots$. *If* $\mathfrak{A}_i = \mathfrak{C}$ *for every* $i \in I$, *the product* $\mathsf{P}_{i \in I} \mathfrak{A}_i$ *is called the* I^{th} ***direct power*** *of the algebra* $\mathfrak{C}$ *and is denoted by* ${}^I\mathfrak{C}$.

(ii) *An algebra* $\mathfrak{C}$ *is said to be a **direct factor** of the algebra* $\mathfrak{B}$ (*or to **divide directly** the algebra* $\mathfrak{B}$), *in symbols* $\mathfrak{C} \mid \mathfrak{B}$, *if there is a system* $\langle \mathfrak{A}_i : i \in I \rangle$ *of algebras and an element* $j \in I$ *such that* $\mathfrak{B} \cong \mathsf{P}_{i \in I} \mathfrak{A}_i$ *and* $\mathfrak{C} = \mathfrak{A}_j$.[1)]

(iii) *By* $\mathbf{P}\mathfrak{C}$ *we denote the class of all algebras isomorphic to direct powers of* $\mathfrak{C}$. *More generally, for any class* K *of algebras we put*

$$\mathbf{P}\mathsf{K} = \mathbf{I}\{\mathsf{P}\mathfrak{A} : \mathfrak{A} \in {}^I\mathsf{K} \textit{ for some } I\}.$$

Notice a difference between the definition of $\mathbf{P}\mathsf{K}$ and the definitions of the analogous notions $\mathbf{S}\mathsf{K}$ and $\mathbf{H}\mathsf{K}$ (0.1.8(i) and 0.2.6): we include in $\mathbf{P}\mathsf{K}$, not only the direct products of algebras in K, but also all isomorphic images of

1) Distinguish the symbol "∣" just introduced and the symbol "|" denoting the product of relations. (Cf. 0.3.10(v) below.)

such products. The reason is that we are mostly interested in classes K of algebras which are invariant under isomorphism, i.e., for which $\mathbf{I}\mathsf{K} = \mathsf{K}$. If K is such a class, the same applies to $\mathbf{S}\mathsf{K}$ and $\mathbf{H}\mathsf{K}$ (actually $\mathbf{H}\mathsf{K}$ is invariant under isomorphism for every class K); however, in view of the special structure of elements of direct products, this would not apply to $\mathbf{P}\mathsf{K}$ if this class consisted exclusively of the direct products of algebras in K.

THEOREM 0.3.2. (i) $\mathsf{P}_{i\in 0}\mathfrak{A}_i = \langle 1, +\rangle$ *where* $+$ *is the unique binary operation on* 1.

(ii) $\mathsf{P}_{i\in\{j\}}\mathfrak{A}_i \cong \mathfrak{A}_j$.

(iii) *If* $j \neq k$, *then* $\mathsf{P}_{i\in\{j,k\}}\mathfrak{A}_i \cong \mathfrak{A}_j \times \mathfrak{A}_k$.

In connection with 0.3.2(ii),(iii), $\mathsf{P}_{i\in\{j\}}\mathfrak{A}_i$ and $\mathsf{P}_{i\in\{j,k\}}\mathfrak{A}_i$ for $j \neq k$ are usually identified with $\mathfrak{A}_j$ and $\mathfrak{A}_j \times \mathfrak{A}_k$.

THEOREM 0.3.3. (i) *If* f *is any permutation of the set* I, *then*

$$\mathsf{P}_{i\in I}\mathfrak{A}_i \cong \mathsf{P}_{i\in I}\mathfrak{A}_{fi}.$$

(ii) *Let* $J = \langle J_i : i \in I\rangle$ *be a system of sets*, $K = \{\langle i,j\rangle : i \in I, j \in J_i\}$, *and* $\langle \mathfrak{A}_{ij} : \langle i,j\rangle \in K\rangle$ *be a system of algebras. Then*

$$\mathsf{P}_{i\in I}\mathsf{P}_{j\in J_i}\mathfrak{A}_{ij} \cong \mathsf{P}_{\langle i,j\rangle\in K}\mathfrak{A}_{ij}.$$

Theorem 0.3.3(i) is the *general commutative law* and Theorem 0.3.3(ii) the *general associative law* for direct products.

THEOREM 0.3.4. (i) $\mathsf{P}_{i\in I}\mathfrak{A}_i$ *is a one-element algebra iff all* $\mathfrak{A}_i$*'s with* $i \in I$ *are one-element algebras.*

(ii) *If* $J \subseteq I$ *and* $\mathsf{P}_{i\in I\sim J}\mathfrak{A}_i$ *is a one-element algebra, then* $\mathsf{P}_{i\in I}\mathfrak{A}_i \cong \mathsf{P}_{i\in J}\mathfrak{A}_i$.

THEOREM 0.3.5. $\mathsf{P}_{i\in I}\mathfrak{B}_i \subseteq \mathsf{P}_{i\in I}\mathfrak{A}_i$ *iff* $\mathfrak{B}_i \subseteq \mathfrak{A}_i$ *for every* $i \in I$.

We recall that pj_i (the ith projection) is the function whose domain consists of all functions f with $i \in Dof$ and which is defined by the formula

$$\mathsf{pj}_i f = f_i.$$

We also recall the convention made in the remark following 0.2.5.

THEOREM 0.3.6. (i) $\mathsf{pj}_j \in Ho(\mathsf{P}_{i\in I}\mathfrak{A}_i, \mathfrak{A}_j)$ *for every* $j \in I$.

(ii) *If* $g_i \in Hom(\mathfrak{C}, \mathfrak{A}_i)$ *for each* $i \in I$ *and* h *is the function on* $\mathfrak{C}$ *such that*

$$hx = \langle g_i x : i \in I\rangle$$

for every $x \in C$, *then* $h \in Hom(\mathfrak{C}, \mathsf{P}_{i\in I}\mathfrak{A}_i)$, *and* $\mathsf{pj}_i \circ h = g_i$ *for every* $i \in I$. *If, moreover,* g_j *is one-one for some* $j \in I$, *then* $h \in Ism(\mathfrak{C}, \mathsf{P}_{i\in I}\mathfrak{A}_i)$.

(iii) *If* $g_i \in Ho(\mathfrak{A}_i, \mathfrak{B}_i)$ *for every* $i \in I$ *and* h *is the function on* $\mathsf{P}_{i\in I}\mathfrak{A}_i$ *such that*

$$hf = \langle g_i(f_i) : i \in I \rangle$$

for every $f \in \mathsf{P}_{i\in I}\mathfrak{A}_i$, *then* $h \in Ho(\mathsf{P}_{i\in I}\mathfrak{A}_i, \mathsf{P}_{i\in I}\mathfrak{B}_i)$, *and* $\mathsf{pj}_i \circ h = g_i \circ \mathsf{pj}_i$ *for every* $i \in I$; *similarly with "Ho" replaced by "Hom", "Is", or "Ism".*

(iv) $h \in Ho(\mathfrak{C}, \mathsf{P}_{i\in I}\mathfrak{A}_i)$ *iff* h *is a function from* $\mathfrak{C}$ *onto* $\mathsf{P}_{i\in I}\mathfrak{A}_i$, *and* $\mathsf{pj}_i \circ h \in Ho(\mathfrak{C}, \mathfrak{A}_i)$ *for every* $i \in I$.

Corollary 0.3.7. (i) $\mathsf{P}_{i\in I}\mathfrak{A}_i \succcurlyeq \mathfrak{A}_j$ *for every* $j \in I$; *more generally,* $\mathsf{P}_{i\in I}\mathfrak{A}_i \succcurlyeq \mathsf{P}_{i\in J}\mathfrak{A}_i$ *whenever* $J \subseteq I$.

(ii) *If* $\mathfrak{C} \succcurlyeq|\subseteq \mathfrak{A}_i$ *for every* $i \in I$, *then* $\mathfrak{C} \succcurlyeq|\subseteq \mathsf{P}_{i\in I}\mathfrak{A}_i$. *If, moreover,* $\mathfrak{C} \cong|\subseteq \mathfrak{A}_j$ *for some* $j \in J$, *then* $\mathfrak{C} \cong|\subseteq \mathsf{P}_{i\in I}\mathfrak{A}_i$.

(iii) *If* $\mathfrak{A}_i \succcurlyeq \mathfrak{B}_i$ *for every* $i \in I$, *then* $\mathsf{P}_{i\in I}\mathfrak{A}_i \succcurlyeq \mathsf{P}_{i\in I}\mathfrak{B}_i$; *similarly with* "$\succcurlyeq$" *replaced by* "$\succcurlyeq|\subseteq$", "$\cong$", *or* "$\cong|\subseteq$".

Theorem 0.3.8. *If* $J \subseteq I$, *the following two conditions are equivalent:*

(i) $\mathsf{P}_{i\in J}\mathfrak{A}_i \cong|\subseteq \mathsf{P}_{i\in I}\mathfrak{A}_i$;

(ii) $\mathsf{P}_{i\in J}\mathfrak{A}_i \succcurlyeq|\subseteq \mathfrak{A}_j$ *for every* $j \in I \sim J$.

Both conditions are satisfied if each algebra $\mathfrak{A}_i$ *with* $i \in I \sim J$ *has a one-element subalgebra.*

The proof of this theorem is based primarily on 0.2.17 and 0.3.7(i),(ii); we also use 0.2.11(i)–(iv) and a special form of 0.3.3(ii).

Theorem 0.3.9. (i) *If* $|I| = |J|$, *then* ${}^I\mathfrak{B} \cong {}^J\mathfrak{B}$.

(ii) *If* $K = \bigcup\{J_i : i \in I\}$ *and* $J_i \cap J_{i'} = 0$ *for any two distinct* $i, i' \in I$, *then* ${}^K\mathfrak{B} \cong \mathsf{P}_{i\in J}{}^{J_i}\mathfrak{B}$.

(iii) ${}^J(\mathsf{P}_{i\in I}\mathfrak{A}_i) \cong \mathsf{P}_{i\in I}{}^J\mathfrak{A}_i$.

(iv) ${}^J({}^I\mathfrak{B}) \cong {}^I({}^J\mathfrak{B}) \cong {}^{I\times J}\mathfrak{B}$.

(v) *If* $I \neq 0$, *then* ${}^I\mathfrak{A} \succcurlyeq \mathfrak{A}$ *and* ${}^I\mathfrak{A} \supseteq|\cong \mathfrak{A}$.

(vi) *If* $J \subseteq I$, *then* ${}^I\mathfrak{A} \succcurlyeq {}^J\mathfrak{A}$; *if, in addition,* $J \neq 0$, *then* ${}^I\mathfrak{A} \supseteq|\cong {}^J\mathfrak{A}$.

Theorem 0.3.10. (i) $\mathfrak{A}|\mathfrak{B}$ *iff* $\mathfrak{A} \times \mathfrak{C} \cong \mathfrak{B}$ *for some* $\mathfrak{C}$.

(ii) *If* $|A| = 1$ *or* $\mathfrak{A} \cong \mathfrak{B}$, *then* $\mathfrak{A}|\mathfrak{B}$.

(iii) $\mathfrak{A}|\mathfrak{A}$; *if* $\mathfrak{A}|\mathfrak{B}|\mathfrak{C}$, *then* $\mathfrak{A}|\mathfrak{C}$.

(iv) *If* $\mathfrak{A}|\mathfrak{B}$, *then* $\mathfrak{A} \preccurlyeq \mathfrak{B}$; *if, in addition,* $\mathfrak{A} \succcurlyeq|\subseteq \mathfrak{B}$ *(or, in particular, if* $\mathfrak{B}$ *has a one-element subalgebra), then* $\mathfrak{A} \cong|\subseteq \mathfrak{B}$.

(v) *If* $\mathfrak{A} \subseteq||\mathfrak{B}$, *then* $\mathfrak{A} ||\subseteq \mathfrak{B}$; *similarly with* "$\subseteq$" *replaced by* "$\supseteq$".

(vi) *If* $J \subseteq I$, *then* $\mathsf{P}_{i\in J}\mathfrak{A}_i | \mathsf{P}_{i\in I}\mathfrak{A}_i$.

(vii) *If* $\mathfrak{A}_i|\mathfrak{B}_i$ *for every* $i \in I$, *then* $\mathsf{P}_{i\in I}\mathfrak{A}_i | \mathsf{P}_{i\in I}\mathfrak{B}_i$.

REMARKS 0.3.11. Some properties of direct products of algebras follow directly from purely set-theoretical properties of Cartesian products of sets. E.g., it is easily seen that a denumerable algebra is not isomorphic to any direct product of finite algebras; more generally, an algebra whose cardinality is cofinal with ω is not isomorphic to a direct product of algebras with smaller cardinality.

Various notions which we have introduced so far for arbitrary algebras can be extended in a most natural way to isomorphism types of algebras (cf. 0.2.12). For instance, given a system $\alpha = \langle \alpha_i : i \in I \rangle$ of isomorphism types, by the direct product $\mathsf{P}\alpha = \mathsf{P}_{i \in I} \alpha_i$ of this system we understand the uniquely determined isomorphism type β satisfying the following condition: there is a system $\langle \mathfrak{A}_i : i \in I \rangle$ of algebras such that $\tau \mathfrak{A}_i = \alpha_i$ for each $i \in I$ and $\tau(\mathsf{P}_{i \in I} \mathfrak{A}_i) = \beta$. As particular cases of this notion we obtain the direct product $\alpha \times \beta$ of two isomorphism types and the direct power ${}^{\kappa}\alpha$ of an isomorphism type (in view of 0.3.9(i) we can restrict ourselves to the case when the exponent κ in ${}^{\kappa}\alpha$ is a cardinal). Analogously the relations $\subseteq$, $\preccurlyeq$, and $|$ between algebras induce the corresponding relations between isomorphism types, which we may denote respectively by the same symbols. From 0.3.2–0.3.10 we can immediately derive various corollaries concerning the corresponding operations on and relations between isomorphism types; actually these corollaries have as a rule a somewhat simpler form than the original theorems. For illustration, consider any class $\mathsf{K} \neq 0$ of algebras such that for any algebras $\mathfrak{A}$, $\mathfrak{B}$ and $\mathfrak{C}$ with $\mathfrak{A} \cong \mathfrak{B} \times \mathfrak{C}$ we have $\mathfrak{A} \in \mathsf{K}$ iff $\mathfrak{B}, \mathfrak{C} \in \mathsf{K}$; let $T = \{\tau \mathfrak{A} : \mathfrak{A} \in \mathsf{K}\}$. Clearly $\mathfrak{K} = \langle \mathsf{K}, \times \rangle$ and $\mathfrak{T} = \langle T, \times \rangle$ are again algebras (we disregard some inessential set-theoretical difficulties which arise in case the classes K and T are not sets); the isomorphism relation $\cong$ restricted to K is a congruence relation on $\mathfrak{K}$, and $\mathfrak{T} \cong \mathfrak{K}/{\cong}$. However, $\mathfrak{K}$ does not belong to any of the familiar classes of algebras discussed in 0.1.4 since the operation $\times$ on algebras is neither commutative nor associative. On the other hand, from 0.3.3(i) and 0.3.3(ii) (i.e., from what we call, not quite properly, the general commutative and associative laws for direct product of algebras) we see at once that the corresponding operation $\times$ on isomorphism types is commutative and associative and that consequently $\mathfrak{T}$ is a commutative semigroup. By 0.3.4(ii) this algebra has a unit. By 0.3.10(i) $|$ is a relation between elements $\alpha, \beta \in T$ such that $\alpha | \beta$ iff $\alpha \times \gamma = \beta$ for some $\gamma \in T$. By 0.3.10(i),(iii) this relation is reflexive and transitive; from the remarks in 0.3.33 below it will be seen that, in general, this relation is not antisymmetric and hence does not establish a partial ordering in the class of all isomorphism types.

THEOREM 0.3.12. *For any classes* K *and* L *of algebras and any cardinal* α,

(i) $\mathsf{K} \subseteq \mathbf{I}\mathsf{K} \subseteq \mathbf{P}\mathsf{K} = \mathbf{PP}\mathsf{K} = \mathbf{PI}\mathsf{K} = \mathbf{IP}\mathsf{K}$,

(ii) $\mathbf{PS}\mathsf{K} \subseteq \mathbf{SP}\mathsf{K}$ *and* $\mathbf{PH}\mathsf{K} \subseteq \mathbf{HP}\mathsf{K}$,

(iii) $\mathbf{S}_\alpha\mathbf{P}\mathsf{K} = \mathbf{S}_\alpha\mathbf{PS}_\alpha\mathsf{K} \subseteq \mathbf{SPS}_\alpha\mathsf{K} \subseteq \mathbf{SP}\mathsf{K}$,

(iv) *the formulas* $\mathsf{K} \subseteq \mathbf{P}\mathsf{L}$ *and* $\mathbf{P}\mathsf{K} \subseteq \mathbf{P}\mathsf{L}$ *are equivalent, and each of them is implied by* $\mathsf{K} \subseteq \mathsf{L}$.

In proving 0.3.12(ii), 0.3.5 and 0.3.7(iii) may be used.

The inequalities of 0.3.12(ii),(iii) cannot in general be replaced by equalities. For example, if K consists exclusively of finite algebras, and $|A| > 1$ for some $\mathfrak{A} \in \mathsf{K}$, then all the algebras in $\mathbf{PS}\mathsf{K}$ and $\mathbf{PH}\mathsf{K}$ are finite or non-denumerable while $\mathbf{SP}\mathsf{K}$ certainly contains denumerable algebras; under some suitable further restrictions on K the same applies to $\mathbf{HP}\mathsf{K}$. With certain trivial exceptions $\mathbf{S}_\alpha\mathbf{PS}_\alpha\mathsf{K}$ is obviously different from $\mathbf{SPS}_\alpha\mathsf{K}$. If $\mathfrak{A} = \langle A, +, \cdot \rangle$ is a field with $|A| \geqq \alpha \cup \omega_1$, then $\mathbf{SPS}_\alpha\mathfrak{A} \neq \mathbf{SP}\mathfrak{A}$.

THEOREM 0.3.13. *Let* K *be a class of algebras.*

(i) *The following three conditions are equivalent:*

$$\mathsf{K} = \mathbf{S}\mathsf{K} = \mathbf{H}\mathsf{K} = \mathbf{P}\mathsf{K},$$

$$\mathsf{K} = \mathbf{HSP}\mathsf{K},$$

$$\mathsf{K} = \mathbf{HSP}\mathsf{L} \textit{ for some class } \mathsf{L} \textit{ of algebras.}$$

(ii) $\mathbf{HSP}\mathsf{K}$ *is the intersection of all classes* L *such that* $\mathsf{K} \subseteq \mathsf{L} = \mathbf{HSP}\mathsf{L}$.

PROOF: by 0.2.19 and 0.3.12.

REMARK 0.3.14. A class L of algebras may be called *algebraically closed* if it is closed under all the basic algebraic operations discussed in this and the preceding two sections, i.e., under the formation of subalgebras, homomorphic (and isomorphic) images, and direct products. As examples we may mention the class of all semilattices $\mathfrak{A} = \langle A, + \rangle$ and all groups $\mathfrak{G} = \langle G, \cdot, ^{-1} \rangle$ (cf. here 0.1.4 and 0.1.6); from 0.1.13 it is seen that, under the alternative conception of a group discussed in 0.1.4, the class of all groups is not algebraically closed.

From 0.3.13 we see that $\mathbf{HSP}\mathsf{K}$ is the least algebraically closed class which includes K; hence we can call $\mathbf{HSP}\mathsf{K}$ the class *algebraically generated by* K. For instance, the class of all one-element algebras (of a given similarity type) is algebraically closed and is algebraically generated by the empty class. We shall see in 0.3.17 that $\mathbf{HSP}\mathsf{K}$ is also closed under the formation of unions if this operation is restricted to sets of algebras directed by the relation $\subseteq$ (cf. 0.1.24 and 0.1.27).

THEOREM 0.3.15. *Let* K *and* L *be any classes of algebras, and let* Q, Q', *and* Q'' *be any finite compositions of all or some of the operations* $\mathbf{S}$, $\mathbf{H}$, *and* $\mathbf{P}$ *(including compositions with repeating terms as well as the* 0*-termed composition, i.e., the identity operation). Then the two formulas* $\mathbf{HSP}\mathsf{K} \subseteq \mathbf{HSP}\mathsf{L}$ *and* $\mathsf{Q}\mathsf{K} \subseteq \mathbf{HSP}\mathsf{L}$ *are equivalent, and each of them is implied by* $\mathsf{Q}'\mathsf{K} \subseteq \mathsf{Q}''\mathsf{L}$. *Furthermore, the four formulas* $\mathbf{HSPS}_\alpha\mathsf{K} \subseteq \mathbf{HSP}\mathsf{L}$, $\mathsf{Q}\mathbf{S}_\alpha\mathsf{K} \subseteq \mathbf{HSP}\mathsf{L}$, $\mathbf{HSPS}_\alpha\mathsf{K} \subseteq \mathbf{HSPS}_\alpha\mathsf{L}$, *and* $\mathsf{Q}\mathbf{S}_\alpha\mathsf{K} \subseteq \mathbf{HSPS}_\alpha\mathsf{L}$ *are equivalent for every cardinal* α, *and each of them is implied by* $\mathsf{Q}'\mathbf{S}_\alpha\mathsf{K} \subseteq \mathsf{Q}''\mathsf{L}$ *as well as by* $\mathsf{Q}'\mathbf{S}_\alpha\mathsf{K} \subseteq \mathsf{Q}''\mathbf{S}_\alpha\mathsf{L}$.

The formulation of 0.3.15 may raise some doubts from the viewpoint of axiomatic set theory: we speak of finite sequences of the operations $\mathbf{S}$, $\mathbf{H}$, and $\mathbf{P}$, while, for instance, in Bernays' set theory no such entities as operations on arbitrary classes and finite sequences of such operations exist. Without too much trouble we could adjust the formulation of 0.3.15 to Bernays' set theory by eliminating any reference to "operations" and using the definition of a finite sequence of classes given in the Preliminaries, but we abstain from doing so since the new formulation would be more complicated and the whole question is rather irrelevant for our purposes.

REMARK 0.3.16. It should be observed that both parts of Theorem 0.3.13 as well as Theorem 0.3.15 cease to be valid under any non-identical permutation of $\mathbf{H}$, $\mathbf{S}$, and $\mathbf{P}$ in $\mathbf{HSP}$. More generally, we cannot replace $\mathbf{HSP}$ in 0.3.13 and 0.3.15 by any finite composition Q of $\mathbf{S}$, $\mathbf{H}$, $\mathbf{P}$, unless some occurrence of $\mathbf{H}$ in Q precedes some occurrence of $\mathbf{S}$ which in turn precedes some occurrence of $\mathbf{P}$. Compositions which satisfy the condition just mentioned are the only ones for which $\mathsf{QK} = \mathbf{HSP}\mathsf{K}$ for all K.

The variety of compositions Q of $\mathbf{S}$, $\mathbf{H}$, and $\mathbf{P}$ have been studied in detail by Don Pigozzi. He has arrived at the following conclusions, partly announced in Pigozzi [66*]. (i) There are eighteen operations Q_ν, $\nu < 18$, such that every composition Q coincides with one of them, i.e., $\mathsf{QK} = \mathsf{Q}_\nu\mathsf{K}$ for some $\nu < 18$ and for every class K of algebras; these are the sixteen compositions without repetition, Q_0–Q_{15} (the identity operation, $\mathbf{S}$, $\mathbf{H}$, $\mathbf{P}$, $\mathbf{SH}$, $\mathbf{HS}$, etc.), and the two operations $\mathsf{Q}_{16} = \mathbf{SPHS}$ and $\mathsf{Q}_{17} = \mathbf{SHPS}$. (ii) No two operations Q_μ and Q_ν with $\mu, \nu < 18$ and $\mu \neq \nu$ coincide, i.e., no equality $\mathsf{Q}_\mu\mathsf{K} = \mathsf{Q}_\nu\mathsf{K}$ is satisfied by all classes K. (iii) Also no inclusion $\mathsf{Q}_\mu\mathsf{K} \subseteq \mathsf{Q}_\nu\mathsf{K}$, with μ and ν as before, is satisfied by all classes K unless it is one of the inclusions which trivially follow from 0.1.14, 0.2.13, 0.2.19, and 0.3.12. (iv) All the examples needed for establishing (ii) and (iii) are provided by two classes of algebras, K_0 and K_1, each of which consists of a single Abelian group, $\mathsf{K}_0 = \{\mathfrak{A}_0\}$

and $\mathsf{K}_1 = \{\mathfrak{A}_1\}$. We can take for $\mathfrak{A}_0$ a torsion-free cyclic group and for $\mathfrak{A}_1$ any torsion Abelian group which is not bounded (cf. 0.1.4); the situation does not change if we include in K_0 and K_1 all isomorphic images of $\mathfrak{A}_0$ and $\mathfrak{A}_1$, respectively. Notice that $\mathbf{HSP}\mathsf{K}_0 = \mathbf{HSP}\mathsf{K}_1$ is the class of all commutative semigroups. Actually, a class K consisting of a single algebra $\mathfrak{A}$ is known which alone suffices for establishing (ii) and (iii). The definition of this algebra is, however, rather artificial; no single Abelian group can be used for this purpose.

Continuing the remarks which follow 0.1.14 and 0.2.19 we can ask the question: for which operations Q of the kind discussed is the pair $\langle \mathsf{M}, \mathsf{Q}\rangle$ (where M is any class of algebras with $\mathsf{M} = \mathsf{QM}$) a closure structure? From 0.1.14, 0.2.19, and 0.3.12 it easily follows that $\mathsf{K} \subseteq \mathsf{QK}$ for every class K, and $\mathsf{QK} \subseteq \mathsf{QL}$ whenever $\mathsf{K} \subseteq \mathsf{L}$; hence our problem reduces to the question: for which operations Q do we always have $\mathsf{QK} = \mathsf{QQK}$? Since the operations $\mathbf{S}$, $\mathbf{H}$, and their compositions were previously discussed, we restrict ourselves now to those compositions Q which have $\mathbf{P}$ as a factor. The answer now easily follows from 0.3.12, 0.3.15, and our recent remarks. It turns out that the only operations Q of this kind for which $\langle \mathsf{M}, \mathsf{Q}\rangle$ is always a closure structure are $\mathbf{P}$, $\mathbf{SP}$, $\mathbf{HP}$, and $\mathbf{HSP}$; moreover, it can be shown that for no such Q is $\langle \mathsf{M}, \mathsf{Q}\rangle$ always a complete or even always an inductive closure structure.

THEOREM 0.3.17. (i) *If* L *is a non-empty set of algebras directed by the relation* $\subseteq$, *then* $\bigcup\mathsf{L} \in \mathbf{HSP}\mathsf{L}$; *if, moreover,* K *is any class of algebras, then the formulas* $\mathsf{L} \subseteq \mathbf{HSP}\mathsf{K}$ *and* $\bigcup\mathsf{L} \in \mathbf{HSP}\mathsf{K}$ *are equivalent.*

(ii) *In particular, for every algebra* $\mathfrak{A}$ *and every class* K *of algebras, we have* $\mathfrak{A} \in \mathbf{HSPS}_\omega\mathfrak{A}$, *and the formulas* $\mathbf{S}_\omega\mathfrak{A} \subseteq \mathbf{HSP}\mathsf{K}$ *and* $\mathfrak{A} \in \mathbf{HSP}\mathsf{K}$ *are equivalent.*

PROOF. To prove the first half of (i), let A be the set of all $f \in \mathsf{P}\langle B : \mathfrak{B} \in \mathsf{L}\rangle$ for which there is a $\mathfrak{B} \in \mathsf{L}$ such that $f_\mathfrak{B} = f_\mathfrak{C}$ for every $\mathfrak{C}$ with $\mathfrak{B} \subseteq \mathfrak{C} \in \mathsf{L}$. Clearly, $0 \neq A \in Su\mathsf{P}\langle \mathfrak{B} : \mathfrak{B} \in \mathsf{L}\rangle$, whence $\mathfrak{A} = \langle A, +\rangle \in \mathbf{SP}\mathsf{L}$. From the fact that L is directed by $\subseteq$ we easily conclude that for every $f \in A$ there is just one $c \in \bigcup\{B : \mathfrak{B} \in \mathsf{L}\}$ such that, for some $\mathfrak{B} \in \mathsf{L}$ and for every $\mathfrak{C}$ with $\mathfrak{B} \subseteq \mathfrak{C} \in \mathsf{L}$, $f_\mathfrak{C} = c$. Therefore we can define a function h with $Doh = A$ by stipulating that, for every $f \in A$, hf is the only element c just described. We show without difficulty that $h \in Ho(\mathfrak{A}, \bigcup\mathsf{L})$, and hence $\bigcup\mathsf{L} \in \mathbf{HSP}\mathsf{L}$. The second half of (i) follows from the first by 0.1.24 and 0.3.15. (ii) is simply a particular case of (i) in view of 0.1.27(ii).[1)]

[1)] Theorem 0.3.17 can also be derived as an immediate corollary from a model-theoretical result, Theorem 0.4.63, which will be given in the next section; this was first pointed out several years ago by Tarski. Subsequently Saunders MacLane suggested a purely mathematical proof of 0.3.17, which was essentially the same as the one just outlined in the text.

THEOREM 0.3.18. *For every class* K *of algebras and every cardinal* $\alpha \geqq \omega$, $\mathbf{HSP}\mathsf{K} = \mathbf{HSPS}_\alpha\mathsf{K}$.

PROOF. We have $\mathsf{K} \subseteq \mathbf{HSPS}_\omega\mathsf{K}$ by 0.3.17, whence $\mathbf{HSP}\mathsf{K} \subseteq \mathbf{HSPS}_\omega\mathsf{K}$ by 0.3.15; on the other hand, $\mathbf{HSPS}_\omega\mathsf{K} \subseteq \mathbf{HSPS}_\alpha\mathsf{K} \subseteq \mathbf{HSP}\mathsf{K}$ by 0.1.15(ii) and 0.3.15.

This theorem does not extend to any cardinal $\alpha < \omega$.

THEOREM 0.3.19. (i) *If* K *is a set of algebras, then*

$$\mathbf{HSP}\mathsf{K} = \mathbf{HSP}(\mathsf{P}\langle\mathfrak{B} : \mathfrak{B} \in \mathsf{K}\rangle).$$

(ii) *Let* K *be any class of algebras,* $\mathsf{L} = \{\mathfrak{B} : \mathfrak{B} \in \mathbf{IS}\mathsf{K}, B \subseteq \omega\}$, *and* $\mathsf{M} = \{\mathfrak{B} : \mathfrak{B} \in \mathbf{IS}_\omega\mathsf{K}, B \subseteq \omega\}$. *Then*

$$\mathbf{HSP}\mathsf{K} = \mathbf{HSP}(\mathsf{P}\langle\mathfrak{B} : \mathfrak{B} \in \mathsf{L}\rangle) = \mathbf{HSP}(\mathsf{P}\langle\mathfrak{B} : \mathfrak{B} \in \mathsf{M}\rangle).$$

PROOF. Since clearly $\mathsf{K} \subseteq \mathbf{H}(\mathsf{P}\langle\mathfrak{B} : \mathfrak{B} \in \mathsf{K}\rangle)$ and $\mathsf{P}\langle\mathfrak{B} : \mathfrak{B} \in \mathsf{K}\rangle \in \mathbf{P}\mathsf{K}$, (i) follows by 0.3.15. Since, by 0.1.19, $\mathbf{S}_\omega\mathsf{K} \subseteq \mathbf{I}\mathsf{M} \subseteq \mathbf{I}\mathsf{L} \subseteq \mathbf{IS}\mathsf{K}$, we obtain $\mathbf{HSP}\mathsf{K} = \mathbf{HSP}\mathsf{L} = \mathbf{HSP}\mathsf{M}$ by 0.3.15 and 0.3.18, and hence (ii) follows by (i).

When extending 0.3.19(ii) to algebras of arbitrary similarity types, we replace ω by $\beta \cup \omega$ where β is the cardinality of the index set in each of the algebras involved.

REMARK 0.3.20. From 0.3.19 we see that every algebraically closed class of algebras is algebraically generated by a single algebra (cf. 0.3.14). Notice that 0.3.19(i) cannot be applied to very comprehensive classes K of algebras, such as the class of all groups, which are not sets in the sense of familiar axiomatic systems of set theory and for which, therefore, the direct products $\mathsf{P}\langle\mathfrak{B} : \mathfrak{B} \in \mathsf{K}\rangle$ do not exist. On the other hand, the more involved construction given in 0.3.19(ii) can always be carried through, since the class of all algebras whose universes consist exclusively of natural numbers is certainly a set, and so is every subclass of this class.

We turn to the discussion of connections between direct products and congruence relations.

THEOREM 0.3.21. *Let* $R_i \in Co\,\mathfrak{A}_i$ *for every* $i \in I$ *and let* $S = \{\langle f, g\rangle : f, g \in \mathsf{P}_{i\in I}\mathfrak{A}_i$ *and* $f_i R_i g_i$ *for each* $i \in I\}$. *Then* $S \in Co(\mathsf{P}_{i\in I}\mathfrak{A}_i)$ *and* $\mathsf{P}_{i\in I}(\mathfrak{A}_i/R_i) \cong (\mathsf{P}_{i\in I}\mathfrak{A}_i)/S$.

0.3.21 is merely a translation of a part of 0.3.7(iii) in terms of congruence relations.

THEOREM 0.3.22. *For* $\mathfrak{A} \cong \mathsf{P}_{i\in I}\mathfrak{B}_i$ *it is necessary and sufficient that there exists a system* $R = \langle R_i : i \in I\rangle$ *satisfying the following conditions*:

(i) $R \in {}^I Co\,\mathfrak{A}$;

(ii) $\bigcap_{i\in I} R_i = A 1 Id$;

(iii) $\bigcap_{i\in I}(x_i/R_i) \neq 0$ *for every* $x \in {}^I A$;

(iv) $\mathfrak{A}/R_i \cong \mathfrak{B}_i$ *for every* $i \in I$.

Applying 0.2.27 we immediately obtain a generalization of Theorem 0.3.22 which gives a necessary and sufficient condition for $\mathfrak{A}/S \cong \mathsf{P}_{i\in I}\mathfrak{B}_i$, where S is any given congruence relation on $\mathfrak{A}$; for this purpose it suffices to replace $A1Id$ by S in 0.3.22(ii).

THEOREM 0.3.23. *Every system R satisfying conditions 0.3.22(i),(iii) also satisfies the following conditions*:

(v) *if* $J \subseteq I$, $K \subseteq I$, *and* $J \cap K = 0$, *then* $(\bigcap_{i\in J} R_i) \mid (\bigcap_{i\in K} R_i) = A \times A$;

(vi) *if* $j \in I$, *then* $R_j \mid (\bigcap_{i\in I\sim\{j\}} R_i) = A \times A$;

(vii) *if* $j, k \in I$ *and* $j \neq k$, *then* $R_j \mid R_k = A \times A$.

In case $|I| < \omega$ *every system R satisfying conditions 0.3.22(i), and 0.3.23(v) or 0.3.23(vi) satisfies 0.3.22(iii) as well.*

The final statement of Theorem 0.3.23 can be easily proven by induction on $|I|$. We may note that, even if $0 < |I| < \omega$, condition 0.3.23(vii) together with 0.3.22(i),(ii) is not sufficient to assure that 0.3.22(iii) holds.

Theorem 0.3.22 shows that the study of direct products of algebras largely reduces to that of systems of congruence relations satisfying certain special conditions. Such a reduction proves especially efficient in discussions concerning direct decompositions of algebras. These are discussions in which we consider the problem whether a given algebra is isomorphically representable as a direct product of algebras of a certain kind, i.e., whether it is isomorphic to, and not necessarily identical with, a direct product of those algebras.[1)]

REMARK 0.3.24. A system $\langle R_i : i \in I\rangle$ satisfying conditions 0.3.22(i)–(iii) may be referred to as a *system of complementary congruence relations* on $\mathfrak{A}$. A relation S is called a *factor congruence relation* on $\mathfrak{A}$ if it occurs in some system of complementary congruence relations on $\mathfrak{A}$. It can easily be shown that, for S to be a factor congruence relation on $\mathfrak{A}$, it is necessary (and obviously sufficient) that S occur in some system of just two complementary congruence

[1)] The possibility of representing direct decompositions of algebras in terms of congruence relations was first noticed by Birkhoff for the case of a finite system of factors; see Birkhoff [67*], p. 164, Theorem 5. An extension to arbitrary systems of factors can be found in Hashimoto [57*].

relations; in other words, S is a factor congruence relation on $\mathfrak{A}$ iff $S \in Co\,\mathfrak{A}$ and there is a $T \in Co\,\mathfrak{A}$ such that $S \cap T = A1Id$ and $S|T = A \times A$. Obviously, $A1Id$ and $A \times A$ are factor congruence relations on $\mathfrak{A}$.

Keeping these observations in mind we define:

DEFINITION 0.3.25. *An algebra* $\mathfrak{A}$ *is called* ***directly indecomposable*** *if* $|A| \geqq 2$ *and if the formulas* $R, S \in Co\,\mathfrak{A}$, $R \cap S = A1Id$, *and* $R|S = A \times A$ *always imply that either* $R = A \times A$ (*i.e.*, $S = A1Id$) *or else* $S = A \times A$ (*i.e.*, $R = A1Id$).

In other words, following 0.3.24, $\mathfrak{A}$ is directly indecomposable iff $|A| \geqq 2$ and $\mathfrak{A}$ has no factor congruence relation different from $A \times A$ and $A1Id$. This formulation, when compared with Definition 0.2.34, exhibits at once a close connection between the notions of simplicity and direct indecomposability; cf. also Theorem 0.3.58 below. We may note that, in contrast to 0.2.35(ii), a directed union of directly indecomposable algebras is not necessarily directly indecomposable.

In view of 0.3.24 we can re-formulate 0.3.25 using arbitrary systems of complementary relations:

COROLLARY 0.3.26. $\mathfrak{A}$ *is directly indecomposable iff* $|A| \geqq 2$ *and if, for every system* R *satisfying conditions 0.3.22(i)–(iii) with* $|I| \geqq 2$, *there is a* $j \in I$ *such that* $R_j = A \times A$ (*or, equivalently, there is a* $k \in I$ *such that* $R_k = A1Id$ *while* $R_i = A \times A$ *for every* $i \in I \sim \{k\}$).

Using 0.3.22, we can, of course, express the content of 0.3.25 and 0.3.26 equivalently in terms of isomorphisms and direct products:

COROLLARY 0.3.27. *The following three conditions are equivalent*:

(i) $\mathfrak{A}$ *is directly indecomposable*;

(ii) $|A| \geqq 2$, *and* $h \in Is(\mathfrak{A}, \mathfrak{B} \times \mathfrak{C})$ *always implies that either* $\mathsf{pj}_0 \circ h$ *is constant and* $\mathsf{pj}_1 \circ h \in Is(\mathfrak{A}, \mathfrak{C})$ *or else* $\mathsf{pj}_0 \circ h$ *is constant and* $\mathsf{pj}_1 \circ h \in Is(\mathfrak{A}, \mathfrak{B})$;

(iii) $|A| \geqq 2$, *and* $h \in Is(\mathfrak{A}, \mathsf{P}_{i \in I}\,\mathfrak{B}_i)$ *with* $|I| \geqq 2$ *always implies the existence of a* $k \in I$ *such that* $\mathsf{pj}_k \circ h \in Is(\mathfrak{A}, \mathfrak{B}_k)$ *while* $\mathsf{pj}_i \circ h$ *is constant for every* $i \in I \sim \{k\}$.

Somewhat more interesting is the following corollary in which the relations of isomorphism between algebras, but not the actual isomorphic mappings, are involved.

COROLLARY 0.3.28. *The following three conditions are equivalent*:

(i) $\mathfrak{A}$ *is directly indecomposable*;

(ii) $|A| \geqq 2$, *and* $\mathfrak{A} \cong \mathfrak{B} \times \mathfrak{C}$ *always implies that* $|B| = 1$ *or* $|C| = 1$;

(iii) $|A| \geqq 2$, *and* $\mathfrak{A} \cong \mathsf{P}_{i \in I} \mathfrak{B}_i$ *with* $|I| \geqq 2$ *always implies that* $|B_j| = 1$ *for some* $j \in I$.

An obvious consequence of 0.3.28 is that every finite algebra whose cardinality is a prime number is directly indecomposable. In some classes of algebras the converse also holds. This applies, e.g., to all the "trivial" semigroups mentioned in 0.1.4. It applies as well as to Boolean groups since the only directly indecomposable Boolean groups are two-element groups; an analogous remark applies to Boolean algebras. On the other hand, the converse fails in the class of all groups; it is well known, e.g., that directly indecomposable finite Abelian groups coincide with cyclic groups whose cardinality is a positive power of a prime.

REMARK 0.3.29. Consider still the following conditions closely related to 0.3.28(ii),(iii):

(ii′) $|A| \geqq 2$, *and* $\mathfrak{A} \cong \mathfrak{B} \times \mathfrak{C}$ *always implies that* $|B| = 1$ *or* $\mathfrak{A} \cong \mathfrak{B}$;

(ii″) $|A| \geqq 2$, *and* $\mathfrak{A} \cong \mathfrak{B} \times \mathfrak{C}$ *always implies that* $\mathfrak{A} \cong \mathfrak{B}$ *or* $\mathfrak{A} \cong \mathfrak{C}$;

(iii′) $|A| \geqq 2$, *and* $\mathfrak{A} \cong \mathsf{P}_{i \in I} \mathfrak{B}_i$ *always implies that, for every* $i \in I$, *either* $|B_i| = 1$ *or* $\mathfrak{A} \cong \mathfrak{B}_i$;

(iii″) $|A| \geqq 2$, *and* $\mathfrak{A} \cong \mathsf{P}_{i \in I} \mathfrak{B}_i$ *always implies that* $\mathfrak{A} \cong \mathfrak{B}_j$ *for some* $j \in I$.

Each of these conditions is necessary but, in general, not sufficient for $\mathfrak{A}$ to be directly indecomposable; each is sufficient in case $\mathfrak{A}$ is finite. We are confronted here with various notions of *direct pseudo-indecomposability* analogous to the notion of pseudo-simplicity (cf. 0.2.37). It is easily seen that conditions (ii′) and (iii′) are equivalent and that each of them implies (iii″), which in turn implies (ii″); the implications in the opposite direction do not hold. Thus, e.g., every Boolean group of power 2^{ω} satisfies (ii″) but not (iii″). A denumerable Boolean group satisfies (iii″) but neither (ii′) nor (iii′). Finally, a denumerable atomless Boolean algebra satisfies (ii′) but is not directly indecomposable.

By comparing 0.2.27 with 0.3.22, 0.3.23, and 0.3.25 we easily obtain various conditions formulated entirely in terms of congruence relations on $\mathfrak{A}$ which are necessary and sufficient for $\mathfrak{A}/R$ to be isomorphic to $\mathsf{P}_{i \in I} \mathfrak{B}_i$ or to be directly indecomposable. We may note that the ring of integers furnishes an example of a directly indecomposable algebra $\mathfrak{A}$ such that, for some $R \in Co\,\mathfrak{A}$, $\mathfrak{A}/R$ is not directly indecomposable.

DEFINITION 0.3.30. (i) *An algebra* $\mathfrak{A}$ *is called* ***totally decomposable*** *if there* *is a system* $\langle \mathfrak{B}_i : i \in I \rangle$ *of directly indecomposable algebras such that* $\mathfrak{A} \cong \mathsf{P}_{i \in I} \mathfrak{B}_i$.

(ii) $\mathfrak{A}$ *is said to be* ***uniquely totally decomposable,*** *or to have the* ***unique decomposition property,*** *if* $\mathfrak{A}$ *has a decomposition* $\mathfrak{A} \cong \mathsf{P}_{i \in I} \mathfrak{B}_i$ *into directly indecomposable factors* $\mathfrak{B}_i$ *and if, for any other decomposition* $\mathfrak{A} \cong \mathsf{P}_{j \in J} \mathfrak{C}_j$ *of this kind, there is a one-one function* f *from* I *onto* J *such that* $\mathfrak{B}_i \cong \mathfrak{C}_{fi}$ *for every* $i \in I$.

In case an algebra $\mathfrak{A}$ is uniquely totally decomposable we sometimes say that the decomposition of $\mathfrak{A}$ into indecomposable factors is *unique up to isomorphism.* There is a stronger variant of this notion, unicity up to order, which is formulated in terms of systems of complementary congruence relations.

THEOREM 0.3.31. (i) *Every finite algebra is totally decomposable.*

(ii) *If a finite algebra* $\mathfrak{A}$ *has the unique decomposition property, the same applies to all direct factors of* $\mathfrak{A}$.

In connection with 0.3.31(i) it should be mentioned that not every finite algebra is uniquely totally decomposable. A simple example of a finite algebra without the unique decomposition property is provided by the four-element algebra $\langle 4, \circ \rangle$ in which $\kappa \circ \lambda = 3 - \kappa$ for $\kappa, \lambda = 0, 1, 2, 3$ (cf. Jónsson-Tarski [47*], pp. 61 f.). Nonetheless, unicity of decompositions does hold for many important finite algebras. A fairly general result in this direction is the following: If K is the class of all finite algebras $\mathfrak{A} = \langle A, +, O_i \rangle_{i \in I}$ which have an element 0 such that 0 is a zero for $+$ and $\{0\} \in Su\,\mathfrak{A}$, then every algebra in K has the unique decomposition property (op. cit., p. 49, Theorem 3.10). Particular cases of this general result are the well-known unique decomposition theorems for finite groups and for finite lattices (and semilattices) with a zero. Infinite algebras are not, in general, totally decomposable as is seen by considering arbitrary denumerable Boolean groups and denumerable Boolean algebras. Moreover, those infinite algebras which are totally decomposable are not necessarily uniquely totally decomposable; in fact, examples to this effect can be found among Abelian groups. (Cf. Jónsson [57*].)

REMARK 0.3.32. With regard to 0.3.31(ii) let us say that an algebra $\mathfrak{A}$ *hereditarily* has a given property P if $\mathfrak{A}$ as well as every direct factor of $\mathfrak{A}$ has the property P. Thus, by 0.3.31(ii), every finite algebra which has the unique decomposition property has this property hereditarily. A Boolean group $\mathfrak{A}$ with $|A| = 2^{\omega}$ can serve as an example of an algebra which is totally decomposable, but not hereditarily totally decomposable, since it has denumerable direct factors which are not totally decomposable. Under the assumption of the continuum hypothesis, $\mathfrak{A}$ is also an example of an algebra which has the unique decomposition property, but not hereditarily; no such example independent of some analogous set-theoretical hypothesis is known.

Some related questions are: is there an algebra $\mathfrak{A}$ with a direct factor $\mathfrak{B}$ such that $\mathfrak{A}$ is uniquely totally decomposable and $\mathfrak{B}$ is totally but not uniquely totally decomposable? Is there an algebra $\mathfrak{A}$ which is hereditarily totally decomposable and has the unique decomposition property, but does not have this property hereditarily? The answer to these questions is negative in case $\mathfrak{A}$ is decomposable into finitely many indecomposable factors. On the other hand, under certain set-theoretical hypotheses the answer to the first question turns out to be affirmative; assuming, e.g., that $2^{\omega} = 2^{\omega_1} = \omega_2$ and $2^{\omega_2} = \omega_3$, we can take for $\mathfrak{A}$ and $\mathfrak{B}$ two Boolean groups with cardinalities ω_3 and ω_2, respectively. (These assumptions, just as the continuum hypothesis, are now known to be compatible with ordinary axioms of set theory; cf. Solovay [65*].) The second question remains open.

REMARKS 0.3.33. Whenever the unique decomposition property is established for the algebras of a given class K, the result has far reaching implications and considerably simplifies the study of this class of algebras. Numerous problems concerning direct products are known whose formulations are very simple and which nevertheless present considerable difficulties when applied to arbitrary algebras, but which become almost trivial when the algebras involved are assumed to have the unique decomposition property hereditarily. For illustration consider the following two properties of an algebra $\mathfrak{A}$.

(α) *For any algebras* $\mathfrak{B}$ *and* $\mathfrak{C}$, *if* $\mathfrak{A} \cong \mathfrak{A}\times\mathfrak{B}\times\mathfrak{C}$, *then* $\mathfrak{A} \cong \mathfrak{A}\times\mathfrak{B} \cong \mathfrak{A}\times\mathfrak{C}$ (*or, equivalently, for every algebra* $\mathfrak{B}$, *if* $\mathfrak{A}\,|\,\mathfrak{B}\,|\,\mathfrak{A}$, *then* $\mathfrak{A} \cong \mathfrak{B}$).

(β) *For any algebras* $\mathfrak{B}$ *and* $\mathfrak{C}$, *if* $\mathfrak{A} \cong \mathfrak{B}\times\mathfrak{B} \cong \mathfrak{C}\times\mathfrak{C}$, *then* $\mathfrak{B} \cong \mathfrak{C}$.

We can show very easily that (α) and (β) hold for every algebra $\mathfrak{A}$ which is hereditarily uniquely totally decomposable. The discussion of (α) and (β) for arbitrary algebras is difficult and is far from being completed. Examples of groups, even of Abelian groups, as well as of Boolean algebras in which (α) and (β) fail can be found in the literature but are by no means simple; cf. Kinoshita [53*], Hanf [57*], Tarski [57*], Jónsson [57*], [57a*], Sąsiada [61*], and Corner [64*]. A simple example of an algebra $\mathfrak{A}$ in which (β) fails, due to J. C. C. McKinsey, is obtained by letting $\mathfrak{B} = \langle\omega, S\rangle$ (the algebra of natural numbers with the successor operation) and $\mathfrak{A} = \mathfrak{C} = \mathfrak{B}\times\mathfrak{B}$; we have then $\mathfrak{A} \cong \mathfrak{B}\times\mathfrak{B} \cong \mathfrak{C}\times\mathfrak{C}$ but not $\mathfrak{B} \cong \mathfrak{C}$. No comparably simple counterexample for (α) is known. (α) trivially holds if the algebra $\mathfrak{A}$ is finite; the corresponding result for (β) has been established in Lovász [67*].

The notions of total and unique total decomposability extend automatically from algebras to their isomorphism types. Problems referring to these notions can be treated as problems concerning the algebra $\mathfrak{T} = \langle T, \times \rangle$, where T is the class of all isomorphism types (of a given similarity type), and various subalgebras of $\mathfrak{T}$; compare here 0.3.11. In particular, the properties of isomorphism types corresponding to (α) and (β) (but not, e.g., the unique decomposition property) are elementary in the metamathematical sense, i.e., they are expressible in the discourse language of $\mathfrak{T}$; see the Preliminaries. (If $\mathfrak{U} = \langle U, \times \rangle$ is a subalgebra of $\mathfrak{T}$ such that $\gamma \times \delta \in U$ implies $\gamma, \delta \in U$, then to say that (α) applies to all types of U amounts to saying that the relation $|$ restricted to U is antisymmetric and hence establishes a partial order in U.) Some remarks previously made can now be restated in the following way: there are numerous problems in the elementary theories of the algebra $\mathfrak{T}$ and its subalgebras whose study, although presenting considerable difficulty in general, becomes almost trivial when restricted to subalgebras $\mathfrak{U} = \langle U, \times \rangle$ in which U consists exclusively of types with the unique decomposition property and is closed under the passage to direct factors. Under some additional assumptions on $\mathfrak{U}$ we can easily show, using the results in Presburger [30*], Mostowski-Tarski [49*], and Mostowski [52*], that the elementary theory of $\mathfrak{U}$ is decidable. For instance, the following assumption is sufficient for this purpose: there is a cardinal $\kappa \geqq \omega$ such that, for every system $\langle \alpha_j : j \in J \rangle$ of indecomposable types, $\mathsf{P}_{j \in J} \alpha_j \in U$ iff $\{\alpha_j : j \in J\} \subseteq U$ and $|J| < \kappa$. The structure of $\mathfrak{U}$ is especially simple in case $\kappa = \omega$ and U is denumerable (thus, e.g., in case U consists exclusively of types of finite algebras); $\mathfrak{U}$ is then isomorphic either to the algebra $\langle \omega \sim \{0\}, \cdot \rangle$, the multiplicative semigroup of positive integers, or to some subalgebra of $\langle \omega \sim \{0\}, \cdot \rangle$ generated by a finite set of prime numbers. On the other hand, by an unpublished result of Ralph McKenzie, the elementary theory of the whole algebra $\mathfrak{T} = \langle T, \times \rangle$ is undecidable, and the same applies to the subalgebra $\mathfrak{T}' = \langle T', \times \rangle$ of $\mathfrak{T}$ in which T' is the set of all types of finite algebras.

A property of algebras which is frequently involved in discussing direct decompositions and is closely related to the unique decomposition property is the so-called refinement property.

DEFINITION 0.3.34. (i) *An algebra $\mathfrak{A}$ is said to have the* ***refinement property****, or to be* ***refinable****, if for any two direct decompositions of $\mathfrak{A}$, $\mathfrak{A} \cong \mathsf{P}_{i \in I} \mathfrak{B}_i \cong \mathsf{P}_{j \in J} \mathfrak{C}_j$, there are algebras $\mathfrak{D}_{ij}$ with $\langle i, j \rangle \in I \times J$ such that $\mathfrak{B}_i \cong \mathsf{P}_{j \in J} \mathfrak{D}_{ij}$ for each $i \in I$ and $\mathfrak{C}_j \cong \mathsf{P}_{i \in I} \mathfrak{D}_{ij}$ for each $j \in J$.*

(ii) *An algebra* $\mathfrak{A}$ *is said to have the* ***finite refinement property,*** *or to be* ***finitely refinable,*** *if for any two direct decompositions of* $\mathfrak{A}$, $\mathfrak{A} \cong \mathfrak{B} \times \mathfrak{C} \cong \mathfrak{D} \times \mathfrak{E}$, *there are algebras* $\mathfrak{F}_0, \mathfrak{F}_1, \mathfrak{F}_2, \mathfrak{F}_3$ *such that* $\mathfrak{B} \cong \mathfrak{F}_0 \times \mathfrak{F}_1$, $\mathfrak{C} \cong \mathfrak{F}_2 \times \mathfrak{F}_3$, $\mathfrak{D} \cong \mathfrak{F}_0 \times \mathfrak{F}_2$, *and* $\mathfrak{E} \cong \mathfrak{F}_1 \times \mathfrak{F}_3$.

Note that the $\mathfrak{B}_i$'s and $\mathfrak{C}_j$'s in 0.3.34(i) and $\mathfrak{B}$, $\mathfrak{C}$, $\mathfrak{D}$, and $\mathfrak{E}$ in 0.3.34(ii) are not assumed to be indecomposable. As in the case of unique total decomposability, there is a stronger variant of the refinement property formulated in terms of congruence relations.

THEOREM 0.3.35. *For every algebra* $\mathfrak{A}$, *the following two conditions are equivalent:*

(i) $\mathfrak{A}$ *has the refinement property and is totally decomposable;*

(ii) $\mathfrak{A}$ *is uniquely totally decomposable and all direct factors of* $\mathfrak{A}$ *are totally decomposable.*

If, in particular, $\mathfrak{A}$ *is finite, then the refinement property, the finite refinement property, and the unique decomposition property are equivalent.*

By this theorem an algebra $\mathfrak{A}$ which is totally decomposable cannot have the refinement property if it is not hereditarily totally decomposable or does not have the unique decomposition property. Hence, e.g., a Boolean group $\mathfrak{A}$ with $|A| = 2^\omega$ does not have the refinement property although, under the continuum hypothesis, it has the unique decomposition property; cf. 0.3.32. On the other hand, we know many examples of infinite algebras which have the refinement property but are not totally decomposable; for instance, all "trivial" semigroups mentioned in 0.1.4 which are denumerable are of this kind. Also all rings with unit, all lattices, and in particular all Boolean algebras have the refinement property, while only some of them (e.g., only those Boolean algebras which are complete and atomic) are totally decomposable. A Boolean group $\mathfrak{A}$ with $|A| = \omega + 2^\omega + 2^{2^\omega} + \dots$ provides an example of an algebra which has the refinement property, but not hereditarily. As a consequence of the fact that $|A|$ is cofinal with ω, $\mathfrak{A}$ proves to be directly pseudo-indecomposable in the sense of condition (iii″) of 0.3.29 (cf. 0.3.11), and it is easily seen that every algebra satisfying this condition has the refinement property; on the other hand, Boolean groups with cardinalities $2^\omega, 2^{2^\omega}, \dots$, which are direct factors of $\mathfrak{A}$, do not have this property, although they satisfy 0.3.29(ii″) and, as a consequence, have the finite refinement property. From 0.3.35 it follows that an algebra is totally decomposable and hereditarily refinable iff it hereditarily has the unique decomposition property. It is not known whether there exists an algebra which is totally decomposable and refinable, but not hereditarily refinable; this is an equivalent formulation of a problem mentioned in 0.3.32.

The significance of the notion of the refinement property results from the following observations. Some general methods are known which enable one to establish this property for many important classes of algebras; when so established and applied to algebras which are totally decomposable, it directly implies the unique decomposition property. On the other hand, this property can be meaningfully and fruitfully discussed also for those algebras which are not totally decomposable. In many discussions only the finite refinement property is needed.

In the following theorem we find an example of useful conclusions which can be drawn from the refinement property for algebras not assumed to be totally decomposable.

THEOREM 0.3.36. (i) *If* $\mathfrak{A}$ *has the refinement property,* $\mathfrak{B}$ *is a direct product of finitely many directly indecomposable algebras, and* $\mathfrak{A} \cong \mathfrak{B} \times \mathfrak{C}$, *then* $\mathfrak{C}$ *also has the refinement property.*

(ii) *Under the assumptions of* (*i*), *if in addition* $\mathfrak{A} \cong \mathfrak{B} \times \mathfrak{C}'$, *then* $\mathfrak{C} \cong \mathfrak{C}'$.

(iii) *Parts* (*i*) *and* (*ii*) *remain valid if "refinement property" is replaced everywhere by "finite refinement property".*

The proof can be most conveniently carried through by induction on the number of factors in $\mathfrak{B}$.

For more information on the refinement property see Jónsson-Tarski [47*], Chang-Jónsson-Tarski [64*], Crawley-Jónsson [64*], and Jónsson [66*].

When applied to algebras which are not assumed to be totally decomposable, the refinement property leads to especially interesting conclusions if it is combined with the so-called remainder property.

DEFINITION 0.3.37. *An algebra* $\mathfrak{A}$ *is said to have the* ***remainder***, *or* ***infinite chain, property*** *if, for any two infinite sequences of algebras* $\langle \mathfrak{B}_\nu : \nu < \omega \rangle$ *and* $\langle \mathfrak{C}_\nu : \nu < \omega \rangle$ *such that* $\mathfrak{B}_0 = \mathfrak{A}$ *and* $\mathfrak{B}_\nu \cong \mathfrak{B}_{\nu+1} \times \mathfrak{C}_\nu$ *for every* $\nu < \omega$, *there is an algebra* $\mathfrak{D}$ *satisfying the formula*

$$\mathfrak{B}_\nu \cong \mathfrak{D} \times \mathsf{P}\langle \mathfrak{C}_{\nu+\pi} : \pi < \omega \rangle$$

for every $\nu < \omega$.

THEOREM 0.3.38. (i) *Every finite algebra has the remainder property.*

(ii) *If an algebra is totally decomposable and if it together with all its factors has the refinement property, then it also has the remainder property.*

(iii) *If an algebra has the remainder property, then all its direct factors have this property.*

REMARK 0.3.39. The class R of those algebras $\mathfrak{A}$ which both have the remainder property and are hereditarily refinable is especially interesting from our point of view. (It is sufficient to assume for our purposes that the algebras in R have the refinement property restricted to finite and denumerable decompositions.) R contains all the algebras which hereditarily have the unique decomposition property, but it also contains many other algebras, for example, all countably complete Boolean algebras (which in general are not totally decomposable). If $\mathfrak{A}$ is any algebra in R and $T\mathfrak{A} = \{\tau\mathfrak{B} : \mathfrak{B} \mid \mathfrak{A}\}$, then the structure $\langle T\mathfrak{A}, \times, \mathsf{P}\rangle$ (with the operation P restricted to ω-termed sequences of isomorphism types) is what is called in the literature a *generalized cardinal algebra*; cf. Tarski [49*], Part II, Section 5. This has far reaching implications for the study of direct decompositions of algebras in R. In fact, generalized cardinal algebras are known to possess a great variety of interesting arithmetical properties; all these properties automatically apply to the isomorphism types in $T\mathfrak{A}$ and carry with them interesting consequences for algebras $\mathfrak{A}$ in R and their direct decompositions. In particular, it turns out that every algebra $\mathfrak{A} \in \mathsf{R}$ has the properties (α) and (β) stated above in 0.3.33. We do not assume in this monograph that the reader is familiar with the theory of cardinal algebras. On the other hand, the class R is too special from the viewpoint of the general theory of algebraic structures to be discussed with details in the present chapter. In Section 2.4, however, we shall establish (α), (β), and several related properties for a comprehensive class of cylindric algebras using exclusively the fact that the cylindric algebras involved belong to the class R. It will be seen that the arguments applied there are much more intricate than those which lead to analogous conclusions for the (more restricted) class of algebras with the unique decomposition property.

We shall now discuss a special kind of subalgebra of a direct product called a subdirect product; later on we consider a special kind of homomorphic image called a reduced product, and a particular case of this notion, that of an ultraproduct.

DEFINITION 0.3.40. *An algebra $\mathfrak{B}$ is called a* **subdirect product** *of a system of algebras $\mathfrak{A} = \langle \mathfrak{A}_i : i \in I\rangle$ if $\mathfrak{B} \subseteq \mathsf{P}_{i\in I}\mathfrak{A}_i$ and $\mathsf{pj}_i \in Ho(\mathfrak{B}, \mathfrak{A}_i)$ for each $i \in I$. Given two algebras $\mathfrak{B}$ and $\mathfrak{C}$ we shall write $\mathfrak{B} \subseteq_{\mathsf{d}} \mathfrak{C}$ or $\mathfrak{C} \supseteq_{\mathsf{d}} \mathfrak{B}$ just in case there is a system $\mathfrak{A}$ of algebras such that $\mathfrak{B}$ is a subdirect product of $\mathfrak{A}$ and $\mathfrak{C}$ is the direct product of $\mathfrak{A}$.*

When using the notation just introduced, we should bear in mind that if, for a given algebra $\mathfrak{C}$, there is a system $\mathfrak{A}$ of algebras for which $\mathfrak{C} = \mathsf{P}\mathfrak{A}$, then $\mathfrak{A}$ is uniquely determined.

In most cases we are interested in knowing whether an algebra $\mathfrak{B}$ can be isomorphically represented as (and not whether it actually coincides with) a subdirect product of algebras $\mathfrak{A}_i$. In this connection we note the following immediate consequence of Definition 0.3.40.

COROLLARY 0.3.41. *If* $\mathfrak{B} \cong|\subseteq_{\mathsf{d}} \mathsf{P}_{i\in I}\mathfrak{A}_i$, *then* $\mathfrak{B} \cong|\subseteq \mathsf{P}_{i\in I}\mathfrak{A}_i$ *and* $\mathfrak{B} \succcurlyeq \mathfrak{A}_i$ *for each* $i \in I$.

The symbol $\subseteq_{\mathsf{d}}$ will occur most frequently in such a context as the one in the hypothesis of 0.3.41. Notice in this connection that the formulas $\mathfrak{B} \cong|\subseteq_{\mathsf{d}} \mathsf{P}_{i\in I}\mathfrak{A}_i$ and $\mathsf{P}_{i\in I}\mathfrak{A}_i \cong \mathfrak{C}$ do not imply in general $\mathfrak{B} \cong|\subseteq_{\mathsf{d}} \mathfrak{C}$. The implication may fail even if $I = \{0\}$ and $\mathfrak{C} = \mathfrak{A}_0$; in fact, $\mathfrak{B} \cong|\subseteq_{\mathsf{d}} \mathsf{P}_{i\in\{0\}}\mathfrak{A}_i$ holds iff $\mathfrak{B} \cong \mathsf{P}_{i\in\{0\}}\mathfrak{A}_i$, while $\mathfrak{B} \cong|\subseteq_{\mathsf{d}} \mathfrak{C}$ never holds in case $\mathfrak{C}$ is not of the form $\mathfrak{C} = \mathsf{P}\mathfrak{D}$ for some system $\mathfrak{D}$ of algebras.

The converse of Corollary 0.3.41 does not hold. To show this we take, e.g., for $\mathfrak{B}$ the finite algebra $\langle 4, \circ\rangle$ described following Theorem 0.3.31, and we define two congruence relations R and S on $\mathfrak{B}$ for which $\mathfrak{B} \cong|\subseteq (\mathfrak{B}/R) \times (\mathfrak{B}/S)$ holds while $\mathfrak{B} \cong|\subseteq_{\mathsf{d}} (\mathfrak{B}/R) \times (\mathfrak{B}/S)$ fails. The details are left to the reader.

THEOREM 0.3.42. (i) $\mathsf{P}\mathfrak{A} \subseteq_{\mathsf{d}} \mathsf{P}\mathfrak{A}$ *for every system* $\mathfrak{A}$ *of algebras.*

(ii) *If f is a permutation of the set I, then the formulas* $\mathfrak{B} \cong|\subseteq_{\mathsf{d}} \mathsf{P}_{i\in I}\mathfrak{A}_i$ *and* $\mathfrak{B} \cong|\subseteq_{\mathsf{d}} \mathsf{P}_{i\in I}\mathfrak{A}_{fi}$ *are equivalent.*

(iii) *Let* $\langle J_i : i \in I\rangle$ *be a system of sets,* $K = \{\langle i,j\rangle : i \in I, j \in J_i\}$, *and* $\langle \mathfrak{A}_{ij} : \langle i,j\rangle \in K\rangle$ *be a system of algebras. If* $\mathfrak{B}_i \cong|\subseteq_{\mathsf{d}} \mathsf{P}_{j\in J_i}\mathfrak{A}_{ij}$ *for each* $i \in I$ *and* $\mathfrak{C} \cong|\subseteq_{\mathsf{d}} \mathsf{P}_{i\in I}\mathfrak{B}_i$, *then* $\mathfrak{C} \cong|\subseteq_{\mathsf{d}} \mathsf{P}_{\langle i,j\rangle\in K}\mathfrak{A}_{ij}$.

THEOREM 0.3.43. $\mathfrak{C} \subseteq \mathsf{P}_{i\in I}\mathfrak{A}_i$ *iff there is a system of algebras* $\langle \mathfrak{B}_i : i \in I\rangle$ *such that* $\mathfrak{B}_i \subseteq \mathfrak{A}_i$ *for each* $i \in I$ *and* $\mathfrak{C} \subseteq_{\mathsf{d}} \mathsf{P}_{i\in I}\mathfrak{B}_i$.

THEOREM 0.3.44. *Under the hypothesis of 0.3.6(ii) we have* $h^*\mathfrak{C} \subseteq_{\mathsf{d}} \mathsf{P}_{i\in I}g_i^*\mathfrak{C}$; *hence,* if $h \in Is\mathfrak{C}$, *then* $\mathfrak{C} \cong|\subseteq_{\mathsf{d}} \mathsf{P}_{i\in I}g_i^*\mathfrak{C}$.

THEOREM 0.3.45. *If* $J \subseteq I$, *the following two conditions are equivalent:*

(i) $\mathsf{P}_{i\in J}\mathfrak{A}_i \cong|\subseteq_{\mathsf{d}} \mathsf{P}_{i\in I}\mathfrak{A}_i$;

(ii) $\mathsf{P}_{i\in J}\mathfrak{A}_i \succcurlyeq \mathfrak{A}_i$ *for every* $i \in I \sim J$.

In particular, if $0 \neq J \subseteq I$, *we have* ${}^J\mathfrak{B} \cong|\subseteq_{\mathsf{d}} {}^I\mathfrak{B}$ *for every algebra* $\mathfrak{B}$.

THEOREM 0.3.46. *For* $\mathfrak{A} \cong|\subseteq_{\mathsf{d}} \mathsf{P}_{i\in I}\mathfrak{B}_i$ *it is necessary and sufficient that there exist a system R of relations satisfying conditions 0.3.22(i),(ii),(iv).*

We obtain a generalization of 0.3.46 by changing $\mathfrak{A}$ to $\mathfrak{A}/S$ at the beginning of the theorem (where S is any congruence relation on $\mathfrak{A}$) and replacing $A1Id$ by S in 0.3.22(ii). Compare the remark following 0.3.22.

The following elementary observations may be of some interest. Every subset of the universe of $\mathfrak{A} \times \mathfrak{B}$ is of course a binary relation R with $Do\,R \subseteq A$ and $Rg\,R \subseteq B$, and conversely. Some of these relations R are subuniverses of $\mathfrak{A} \times \mathfrak{B}$, and the corresponding algebras $\langle R, +\rangle$ are subalgebras of $\mathfrak{A} \times \mathfrak{B}$ and, in some cases, subdirect products of the system $\langle \mathfrak{A}, \mathfrak{B}\rangle$. Among such relations we find various structures discussed in the preceding section. This is seen from the next theorem.

THEOREM 0.3.47. (i) *The following two conditions are equivalent*:

(i′) $h \in Hom(\mathfrak{A}, \mathfrak{B})$;

(i″) $h \in Su(\mathfrak{A} \times \mathfrak{B})$ *and h is a function with* $Do\,h = A$.

Each of these conditions implies that $\mathfrak{A} \cong \langle h, +\rangle \succcurlyeq\!\mid\, \subseteq \mathfrak{B}$.

(ii) *The following two conditions are equivalent*:

(ii′) $h \in Ho(\mathfrak{A}, \mathfrak{B})$;

(ii″) $h \in Su(\mathfrak{A} \times \mathfrak{B})$ *and h is a function with* $Do\,h = A$ *and* $Rg\,h = B$.

Each of these conditions implies that $\mathfrak{A} \cong \langle h, +\rangle \succcurlyeq \mathfrak{B}$ *and* $\langle h, +\rangle \subseteq_d \mathfrak{A} \times \mathfrak{B}$.

(iii) *The above statements (i) and (ii) remain valid if we replace "Hom" by "Ism" in (i′), "Ho" by "Is" in (ii′), "function" by "one-one function" in (i″) and (ii″), and "$\succcurlyeq$" by "$\cong$" in the last parts of (i) and (ii).*

(iv) *The following two conditions are equivalent*:

(iv′) $R \in Co\,\mathfrak{A}$;

(iv″) $R \in Su(\mathfrak{A} \times \mathfrak{A})$ *and R is an equivalence relation on A.*

Each of these conditions implies that $\langle R, +\rangle \succcurlyeq \mathfrak{A}$ *and* $\langle R, +\rangle \subseteq_d \mathfrak{A} \times \mathfrak{A}$.

DEFINITION 0.3.48. *An algebra $\mathfrak{A}$ is called* ***semisimple*** *if it is isomorphic to a subdirect product of simple algebras.*

THEOREM 0.3.49. *$\mathfrak{A}$ is semisimple iff $A \upharpoonright Id$ is the intersection of all maximal proper congruence relations on $\mathfrak{A}$.*

PROOF: by 0.2.38(i) and 0.3.46.

From 0.3.49 we see that every semisimple algebra with more than one element has at least one maximal proper congruence relation.

DEFINITION 0.3.50. (i) *An algebra $\mathfrak{A}$ is called* ***weakly subdirectly indecomposable*** *if $|A| \geqq 2$ and if the formulas $R, S \in Co\,\mathfrak{A}$ and $R \cap S = A \upharpoonright Id$ always imply that $R = A \upharpoonright Id$ or $S = A \upharpoonright Id$.*

(ii) *An algebra $\mathfrak{A}$ is called* ***subdirectly indecomposable*** *if $|A| \geqq 2$ and if, for every system R of relations satisfying conditions 0.3.22(i),(ii), there is an $i \in I$ such that $R_i = A \upharpoonright Id$.*

By comparing 0.3.50 with 0.3.25 and 0.3.26 we notice immediately a close connection between the notions of direct and subdirect indecomposability.

On the other hand, we see that the conditions stated in 0.3.25 and 0.3.26 are equivalent and characterize the same notion, while for the analogous conditions in 0.3.50(i) and 0.3.50(ii) two different terms have been introduced. We shall see below, in the remark following 0.3.58, that the two notions defined in 0.3.50 are actually different when applied to infinite algebras; it is easily seen, however, that they coincide for finite algebras. The subdirectly indecomposable algebras in the sense of 0.3.50(ii) play a much larger role in algebraic discussions than the weakly subdirectly indecomposable algebras in the sense of 0.3.50(i).

THEOREM 0.3.51. (i) *If* $\mathfrak{A}$ *is not weakly subdirectly indecomposable, then some subalgebra of* $\mathfrak{A}$ *with four generators is not weakly subdirectly indecomposable, and actually is not a subalgebra of any weakly subdirectly indecomposable subalgebra of* $\mathfrak{A}$.

(ii) *The union of a set of weakly subdirectly indecomposable algebras which is directed by the relation* $\subseteq$ *is weakly subdirectly indecomposable.*

PROOF. (i) Suppose $R, S \in Co\mathfrak{A}$, $R \cap S = A 1 Id$, and $R, S \neq A 1 Id$. Choose $\langle a, b\rangle \in R \sim (A 1 Id)$, $\langle c, d\rangle \in S \sim (A 1 Id)$. By 0.2.26(i), the algebra $\mathfrak{B} = \mathfrak{Sg}\{a, b, c, d\}$ is not weakly subdirectly indecomposable, as required in the conclusion of (i). More generally, no algebra $\mathfrak{C}$ such that $\mathfrak{B} \subseteq \mathfrak{C} \subseteq \mathfrak{A}$ is weakly subdirectly indecomposable. Hence (i) is proved, and (ii) follows from (i) by 0.1.28.

Thus the property of being weakly subdirectly indecomposable, just as that of being simple, is local (cf. the remark following 0.1.28). In contrast, the properties of being directly and subdirectly indecomposable are not local. In fact, let $\mathfrak{A} = \langle A, +\rangle$ be any infinite "trivial" semigroup (cf. 0.1.4) with $x+y = x$ for any $x, y \in A$. Since every subset of A is a subuniverse of $\mathfrak{A}$, the set L of all finite subalgebras of $\mathfrak{A}$ whose cardinality is a prime number is obviously directed by $\subseteq$; moreover, all algebras in L are directly indecomposable while $\mathfrak{A} = \bigcup \mathsf{L}$ is not. On the other hand, let $\mathfrak{A} = \langle \omega, \circ\rangle$ where $\kappa \circ \kappa = \kappa$ and $\kappa \circ \lambda = \lambda \circ \kappa = \kappa + 1$ for $\kappa < \lambda < \omega$. Then, for each $\kappa < \omega$, $\kappa \neq 0, 1$, the subalgebra $\mathfrak{B}_\kappa = \langle \kappa, \circ\rangle$ is subdirectly indecomposable, but $\mathfrak{A} = \bigcup_{\kappa<\omega} \mathfrak{B}_\kappa$ is not. The last example is due to Ralph McKenzie, who called our attention to the fact that the property of being subdirectly indecomposable is not local.

A more concise form of 0.3.50(ii) is

COROLLARY 0.3.52. $\mathfrak{A}$ *is subdirectly indecomposable iff* $A 1 Id \subset \bigcap\{R: A 1 Id \subset R \in Co\mathfrak{A}\}$, *or, equivalently, iff there is a (uniquely determined) least relation* R *such that* $A 1 Id \subset R \in Co\mathfrak{A}$.

COROLLARY 0.3.53. *Let* $R \in Co\mathfrak{A}$. $\mathfrak{A}/R$ *is subdirectly indecomposable iff* $R \subset \bigcap\{S : R \subset S \in Co\mathfrak{A}\}$, *or, in other words, iff there are two elements* $x, y \in A$ *such that* R *is a maximal congruence relation on* $\mathfrak{A}$ *for which* xRy *does not hold.*

THEOREM 0.3.54. *For every algebra* $\mathfrak{A}$ *there is a system* $\langle \mathfrak{B}_i : i \in I \rangle$ *of subdirectly indecomposable algebras (and, in fact, of algebras of the form* $\mathfrak{B}_i = \mathfrak{A}/S_i$ *where* $S_i \in Co\mathfrak{A}$ *for each* $i \in I$*) such that* $\mathfrak{A} \cong|\subseteq_{\mathsf{d}} \mathsf{P}_{i\in I}\, \mathfrak{B}_i$.

PROOF. We let $I = (A \times A) \sim (A1Id)$ and, by applying Zorn's lemma, we pick for each $\langle x, y \rangle \in I$ a congruence relation S_{xy} on $\mathfrak{A}$ which is maximal among congruence relations not containing $\langle x, y \rangle$. To complete the proof we apply 0.3.46 and 0.3.53.

Theorem 0.3.54, which is due to Birkhoff [44*], accounts for much of the usefulness of the notion of subdirect indecomposability. In opposition to the case of direct decomposability we see that every algebra has a subdirect decomposition into subdirectly indecomposable factors.

THEOREM 0.3.55. *The following three conditions are equivalent*:

(i) $\mathfrak{A}$ *is subdirectly indecomposable*;

(ii) $|A| \geqq 2$, *and, if* h *is an isomorphism from* $\mathfrak{A}$ *onto a subdirect product of* $\langle \mathfrak{B}_i : i \in I \rangle$, *then* $\mathsf{pj}_i \circ h \in Is(\mathfrak{A}, \mathfrak{B}_i)$ *for some* $i \in I$;

(iii) $|A| \geqq 2$, *and the formula* $\mathfrak{A} \cong|\subseteq_{\mathsf{d}} \mathsf{P}_{i\in I}\, \mathfrak{B}_i$ *always implies that* $\mathfrak{A} \cong \mathfrak{B}_i$ *for some* $j \in I$.

Theorem 0.3.54 may be used to show that 0.3.55(iii) implies 0.3.55(i).

THEOREM 0.3.56. *Every weakly subdirectly indecomposable algebra* $\mathfrak{A}$ *satisfies the following condition*:

(i) $|A| \geqq 2$, *and the formula* $\mathfrak{A} \cong|\subseteq_{\mathsf{d}} \mathfrak{B} \times \mathfrak{C}$ *always implies that* $\mathfrak{A} \cong \mathfrak{B}$ *or* $\mathfrak{A} \cong \mathfrak{C}$.

REMARK 0.3.57. The converse of 0.3.56 does not hold. For example, every infinite "trivial" semigroup satisfies 0.3.56(i) but is not weakly subdirectly indecomposable; this is a consequence of those properties of "trivial" semigroups which were pointed out in 0.2.12 and in the remark following 0.2.20. We see thus that the notion of weak subdirect indecomposability leads to a new "pseudo-notion" (cf. 0.2.37 and 0.3.29). On the other hand, in view of Theorem 0.3.55, we do not know of any pseudo-notion induced in a natural way by the proper notion of subdirect indecomposability.

THEOREM 0.3.58. (i) *Every simple algebra is subdirectly indecomposable.*

(ii) *Every subdirectly indecomposable algebra is weakly subdirectly indecomposable.*

(iii) *Every weakly subdirectly indecomposable algebra is directly indecomposable.*

The converses of all three parts of 0.3.58 fail. In fact, the algebra $\mathfrak{C} = \langle 3, P \rangle$, where P is the unary operation on 3 such that $P0 = P1 = 0$ and $P2 = 1$, is subdirectly indecomposable but not simple; the ring of integers is weakly subdirectly indecomposable but not subdirectly indecomposable; and no "trivial" semigroup $\mathfrak{A}$ with $2 < |A| < \omega$ is weakly subdirectly indecomposable although every such algebra for which $|A|$ is a prime is directly indecomposable.

Remark 0.3.59. In addition to the four notions involved in 0.3.58 we have briefly considered five "pseudo-notions": pseudo-simplicity (0.2.37), three variants of direct pseudo-indecomposability (0.3.29), and weak subdirect pseudo-indecomposability (0.3.57). There are still two valid implications between these notions which do not follow from our previous remarks: every pseudo-simple algebra is subdirectly indecomposable, and every algebra satisfying 0.3.56(i) satisfies also 0.3.29(ii″), i.e., the weakest condition of direct pseudo-indecomposability. Using as counterexamples the special algebras mentioned in 0.2.37, 0.3.29, 0.3.57, and in the remark following 0.3.58, we can easily convince ourselves that no further implications hold between any two of the nine notions discussed (disregarding, of course, all the implications which follow by transitivity from those explicitly stated).

Remark 0.3.60. By 0.3.42(i) direct products can be treated as particular instances of subdirect products. Another important class of particular instances of subdirect products is formed by the so-called *weak direct products* (known also in the literature as *direct sums*). When introducing this notion in the general theory of algebras it seems natural to restrict oneself to algebras of the form $\mathfrak{B} = \langle B, b, Q_i \rangle_{i \in I}$ where b is a distinguished element and $\{b\}$ is a subuniverse of $\mathfrak{B}$. (It will be obvious that the discussion can be extended to any class K of algebras provided there is a uniform method which permits us to single out a one-element subalgebra in each algebra of K; thus, e.g., the notion of weak direct product can be applied to the class of all algebras $\mathfrak{B} = \langle B, + \rangle$ where $+$ is a binary operation with zero.) Given a system $\mathfrak{B} = \langle \mathfrak{B}^{(j)} : j \in J \rangle$ of algebras $\mathfrak{B}^{(j)} = \langle B^{(j)}, b^{(j)}, Q_i^{(j)} \rangle_{i \in I}$, by the weak direct product of $\mathfrak{B}$ we understand the subalgebra $\mathfrak{C}$ of $\mathsf{P}\mathfrak{B}$ whose universe C consists of all those functions $f \in \mathsf{P}_{j \in J} B^{(j)}$ for which the set $\{j : j \in J, fj \neq b^{(j)}\}$ is finite; thus C is what we have called in the Preliminaries the weak Cartesian product of B relative to $b = \langle b^{(j)} : j \in J \rangle$. The formation of weak direct products, just as that of ordinary direct products, is an operation which (whenever applicable) correlates a well determined algebra with a given system of algebras. The two operations have many properties in common; e.g., the general commutative and associative laws, 0.3.3(i) and 0.3.3(ii), hold for both kinds of direct products. Also various notions defined in terms of direct products and results concerning these notions extend in an obvious way to weak direct products; we have in mind here such notions as direct power, unique decomposition property, and refinement property. For any finite system of algebras the direct product and the weak direct product obviously coincide. Hence, for instance, the notion of a directly indecomposable algebra is the same for both kinds of direct products. Properties (α) and (β) discussed in 0.3.33 and 0.3.39 apply to all algebras which hereditarily have the unique weak decomposition property and, more generally, to those which both are hereditarily weakly refinable and have the weak remainder property.

It should be pointed out that in discussions of various special classes of algebras, in particular in the theories of groups and rings, weak direct products actually play a more essential role than ordinary direct products. On the other hand, weak products are of no importance for the discussion of some other classes of algebras, for instance, of lattices with zero and unit. If we treat such lattices as algebras with two distinguished elements, 0 and 1, which in general do not coincide, our construction of weak products does not apply since the algebras involved do not have one-element subalgebras. If we treat them as algebras with one distinguished element, say 0, and only postulate the existence of 1, then the weak direct product of an infinite system of such algebras (each with at least two different elements) is never itself an algebra of this kind. In particular a weak direct product of an infinite system of Boolean algebras is not a Boolean algebra. For analogous reasons weak products will not be involved in the discussion of cylindric algebras.

The notions of reduced product and ultraproduct, to which we now turn, were introduced in the theory of algebras only a few years ago. During these few years they have come to serve as a powerful instrument of research in the theory of models — the study of relationships between mathematical properties of algebras (and of other mathematical structures) and the form of sentences expressing these properties in some formalism. With the help of reduced products and ultraproducts several basic model-theoretical notions have been characterized in mathematical terms and further consequences have been derived from these characterizations by means of mathematical methods; this has led in some cases to the solution of purely mathematical problems whose metamathematical connections are quite remote.

The use of reduced products and ultraproducts in this work will be rather restricted. Here we define both notions and state some elementary consequences of these definitions. For a detailed discussion see Frayne-Morel-Scott [62*]; further bibliographic references, in particular to Łoś [55a*], can also be found there.

THEOREM 0.3.61. *If* $\langle \mathfrak{A}_i : i \in I \rangle$ *is a system of algebras and F is a filter on I, then* $\{\langle f, g \rangle : f, g \in \mathsf{P}_{i\in I} A_i,\ \{i : i \in I, f_i = g_i\} \in F\}$ *is a congruence relation on* $\mathsf{P}_{i\in I} \mathfrak{A}_i$.

In view of this theorem we define:

DEFINITION 0.3.62. (i) *For any system of algebras* $\mathfrak{A} = \langle \mathfrak{A}_i : i \in I \rangle$ *and any filter F on I we let*

$$\bar{F}^{(\mathfrak{A})} = \bar{F} = \{\langle f, g \rangle : f, g \in \mathsf{P}_{i\in I} A_i \text{ and } \{i : i \in I, f_i = g_i\} \in F\}.$$

The algebra $\mathfrak{B} = \mathsf{P}\mathfrak{A}/\bar{F} = \mathsf{P}_{i\in I} \mathfrak{A}_i/\bar{F}$ *is called the* ***direct product of the system*** $\mathfrak{A}$ ***reduced by the filter*** *F or, briefly, the F-**reduced product** of* $\mathfrak{A}$; *in case F is an ultrafilter,* $\mathfrak{B}$ *is called the F-**ultraproduct** of* $\mathfrak{A}$. *The terminology extends in an obvious way to direct powers.*

(ii) *For any algebra* $\mathfrak{A}$ *we let* $\mathsf{Up}\,\mathfrak{A}$ *be the class of all algebras isomorphic to ultrapowers of* $\mathfrak{A}$; *more generally, for any class* K *of algebras we put*

$$\mathsf{Up}\,\mathsf{K} = \mathsf{I}\{\mathsf{P}_{i\in I} \mathfrak{B}_i/\bar{F} : \mathfrak{B} \in {}^I\mathsf{K},\ F \text{ is an ultrafilter on } I\}.$$

Notice that, just as in the case of direct products, we include in Up K not only all the ultraproducts of algebras in the class K but also all the isomorphic images of these ultraproducts; cf. the comment following 0.3.1. We do not introduce any special symbol to denote the class of all reduced products of algebras in K (and their isomorphic images) since such a symbol is not needed for the later discussion.

THEOREM 0.3.63. *Let* $\mathfrak{A} = \langle \mathfrak{A}_i : i \in I \rangle$ *be a system of algebras and* $A = \langle A_i : i \in I \rangle$ *be the system of universes of these algebras.*

(i) *If* $F = \{I\}$, *then* $\bar{F} = \mathsf{P}A \upharpoonright Id$ *and* $\mathsf{P}\mathfrak{A}/\bar{F} \cong \mathsf{P}\mathfrak{A}$.

(ii) *If* $F = SbI$, *then* $\bar{F} = \mathsf{P}A \times \mathsf{P}A$ *and* $\mathsf{P}\mathfrak{A}/\bar{F}$ *is a one-element algebra.*

(iii) *If F is a principal filter on I and, in fact,* $F = \{X : J \subseteq X \subseteq I\}$ *where* $J \subseteq I$, *then* $\mathsf{P}\mathfrak{A}/\bar{F} \cong \mathsf{P}_{i\in J} \mathfrak{A}_i$.

(iv) *If F is a principal ultrafilter on I and, in fact,* $F = \{X : j \in X \subseteq I\}$ *where* $j \in I$, *then* $\mathsf{P}\mathfrak{A}/\bar{F} \cong \mathfrak{A}_j$.

Since every filter on a finite set is principal, it follows from 0.3.63(iii),(iv) that the notions of reduced product and ultraproduct present no interest when applied to finite systems of algebras.

THEOREM 0.3.64. *Let* $J = \langle J_i : i \in I \rangle$ *be a system of sets,* $K = \{\langle i, j \rangle : i \in I, j \in J_i\}$, *and* $\mathfrak{A} = \langle \mathfrak{A}_{ij} : \langle i, j \rangle \in K \rangle$ *be a system of algebras. For each* $i \in I$ *let* F_i *be a filter, or an ultrafilter, on* J_i, G *be a filter, or an ultrafilter, on* I, *and let* $H = \{X : X \subseteq K$ *and* $\{i : i \in I, X^{\star} i \in F_i\} \in G\}$. *Then* H *is respectively a filter, or an ultrafilter, on* K, *and*

$$\mathsf{P}_{i \in I}(\mathsf{P}_{j \in J_i} \mathfrak{A}_{ij} / \bar{F}_i) / \bar{G} \cong \mathsf{P}_{\langle i, j \rangle \in K} \mathfrak{A}_{ij} / \bar{H}.$$

PROOF. The desired isomorphism is the relation

$$\{\langle \langle \langle f_{ij} : j \in J_i \rangle / \bar{F}_i : i \in I \rangle / \bar{G}, f / \bar{H} \rangle : f \in \mathsf{P}_{\langle i, j \rangle \in K} \mathfrak{A}_{ij}\}.$$

COROLLARY 0.3.65. *Let* $\langle \mathfrak{A}_i : i \in I \rangle$ *be a system of algebras,* F *a filter on* I, J *a non-empty subset of* I, *and let* $G = \{X \cap J : X \in F\}$. *Then* G *is a filter on* J *satisfying the following conditions*:

(i) *if* $\mathfrak{A}_i$ *has a one-element subalgebra for every* $i \in I \sim J$, *then*

$$\mathsf{P}_{i \in J} \mathfrak{A}_i / \bar{G} \cong | \subseteq \mathsf{P}_{i \in I} \mathfrak{A}_i / \bar{F};$$

(ii) *if* $J \in F$, *then* $\mathsf{P}_{i \in J} \mathfrak{A}_i / \bar{G} \cong \mathsf{P}_{i \in I} \mathfrak{A}_i / \bar{F}$;

(iii) *if* $\{i \in I : |A_i| \geq 2\} \subseteq J$, *then* $\mathsf{P}_{i \in J} \mathfrak{A}_i / \bar{G} \cong \mathsf{P}_{i \in I} \mathfrak{A}_i / \bar{F}$.

Moreover, in case F is an ultrafilter on I, G is an ultrafilter on J or else is identical with $Sb\,J$ *(i.e., is the improper filter on J).*

THEOREM 0.3.66. *If* $\langle \mathfrak{A}_i : i \in I \rangle$, $\langle \mathfrak{B}_i : i \in I \rangle$ *are systems of algebras,* F *is a filter on* I, *and* $\mathfrak{A}_i \subseteq \mathfrak{B}_i$ *for each* $i \in I$, *then*

$$\mathsf{P}_{i \in I} \mathfrak{A}_i / \bar{F} \cong | \subseteq \mathsf{P}_{i \in I} \mathfrak{B}_i / \bar{F}.$$

PROOF. The relation

$$\{\langle f / \bar{F}^{(\mathfrak{A})}, f / \bar{F}^{(\mathfrak{B})} \rangle : f \in \mathsf{P}_{i \in I} A_i\}$$

is easily seen to be a function mapping $\mathsf{P}_{i \in I} \mathfrak{A}_i / \bar{F}$ isomorphically into $\mathsf{P}_{i \in I} \mathfrak{B}_i / \bar{F}$.

THEOREM 0.3.67. *For every algebra* $\mathfrak{A}$, *if* F *is a proper filter on* I, *then* $\mathfrak{A} \cong | \subseteq {}^I \mathfrak{A} / \bar{F}$.

PROOF. The function $\langle (I \times \{a\}) / \bar{F} : a \in A \rangle$ maps $\mathfrak{A}$ isomorphically into ${}^I \mathfrak{A} / \bar{F}$.

THEOREM 0.3.68. *If* $\langle \mathfrak{A}_i : i \in I \rangle$, $\langle \mathfrak{B}_i : i \in I \rangle$ *are systems of algebras,* F *is a filter on* I, *and* $\mathfrak{A}_i \succcurlyeq \mathfrak{B}_i$ *for each* $i \in I$, *then*

$$\mathsf{P}_{i \in I} \mathfrak{A}_i / \bar{F} \succcurlyeq \mathsf{P}_{i \in I} \mathfrak{B}_i / \bar{F}.$$

PROOF. Suppose $f_i \in Ho(\mathfrak{A}_i \mathfrak{B}_i)$ for each $i \in I$. For any given $g \in \mathsf{P}_{i \in I} A_i$ we let

$$g^+ = \langle f_i(g_i) : i \in I \rangle.$$

Then, as is easily seen, the relation

$$\{\langle g/\bar{F}, g^+/\bar{F}\rangle : g \in \mathsf{P}_{i\in I} A_i\}$$

is a function which maps $\mathsf{P}_{i\in I}\mathfrak{A}_i/\bar{F}$ homomorphically onto $\mathsf{P}_{i\in I}\mathfrak{B}_i/\bar{F}$.

THEOREM 0.3.69. *For any classes* K *and* L *of algebras we have*:

(i) $\mathsf{K} \subseteq \mathbf{I}\mathsf{K} \subseteq \mathsf{Up}\,\mathsf{K} = \mathsf{Up}\,\mathsf{Up}\,\mathsf{K} = \mathsf{Up}\mathbf{I}\mathsf{K} = \mathbf{I}\mathsf{Up}\,\mathsf{K} \subseteq \mathbf{HP}\mathsf{K}$,

(ii) $\mathsf{Up}\mathbf{S}\mathsf{K} \subseteq \mathbf{S}\mathsf{Up}\,\mathsf{K}$ *and* $\mathsf{Up}\mathbf{H}\mathsf{K} \subseteq \mathbf{H}\mathsf{Up}\,\mathsf{K}$,

(iii) *the formulas* $\mathsf{K} \subseteq \mathsf{Up}\,\mathsf{L}$ *and* $\mathsf{Up}\,\mathsf{K} \subseteq \mathsf{Up}\,\mathsf{L}$ *are equivalent, and each of them is implied by* $\mathsf{K} \subseteq \mathsf{L}$.

Part (ii) of this theorem follows from 0.3.66 and 0.3.68.

Regarding the possibility of replacing the inclusions (ii) of 0.3.69 by equalities see Remark 0.3.76 below. It has been observed by Ralph McKenzie that neither of the inclusions $\mathsf{Up}\mathbf{P}\mathsf{K} \subseteq \mathbf{P}\mathsf{Up}\,\mathsf{K}$ and $\mathbf{P}\mathsf{Up}\,\mathsf{K} \subseteq \mathsf{Up}\mathbf{P}\mathsf{K}$ holds for all classes K; both inclusions fail, e.g., for $\mathsf{K} = \{\mathfrak{A}\}$ where $\mathfrak{A}$ is the field of rational numbers.

Theorem 0.3.69 permits us to supplement 0.3.15: the conclusion of this theorem proves to hold if Q, Q', Q'' are assumed to be any compositions of four operations, $\mathbf{S}$, $\mathbf{H}$, $\mathbf{P}$, and Up. Another consequence of 0.3.69, to which we shall refer later in this section, is

COROLLARY 0.3.70. *Let* K *be any class of algebras.*

(i) *The following three conditions are equivalent*:

$\mathsf{K} = \mathbf{S}\mathsf{K} = \mathsf{Up}\,\mathsf{K}$,

$\mathsf{K} = \mathbf{S}\mathsf{Up}\,\mathsf{K}$,

$\mathsf{K} = \mathbf{S}\mathsf{Up}\,\mathsf{L}$ *for some class* L *of algebras.*

(ii) $\mathbf{S}\mathsf{Up}\,\mathsf{K}$ *is the intersection of all classes* L *such that* $\mathsf{K} \subseteq \mathsf{L} = \mathbf{S}\mathsf{Up}\,\mathsf{L}$.

(iii) *Parts (i) and (ii) remain true if* "$\mathbf{S}$" *is replaced everywhere by* "$\mathbf{H}$".

THEOREM 0.3.71. *Theorem 0.3.17 remains true if* "$\mathbf{HSP}$" *is replaced everywhere by* "$\mathbf{S}\mathsf{Up}$".

PROOF. Consider any non-empty set L of algebras directed by the relation $\subseteq$. By the axiom of choice $\mathsf{P}\langle A : \mathfrak{A} \in \mathsf{L}\rangle \neq 0$; let $f \in \mathsf{P}\langle A : \mathfrak{A} \in \mathsf{L}\rangle$. For each $a \in \mathsf{U}\mathsf{L}$ we put

$$\mathsf{M}_a = \{\mathfrak{A} : a \in A,\ \mathfrak{A} \in \mathsf{L}\},\ g_a = \mathsf{M}_a \times \{a\} \cup (\mathsf{L} \sim \mathsf{M}_a)1f;$$

clearly,

$$g_a \in \mathsf{P}\langle A : \mathfrak{A} \in \mathsf{L}\rangle.$$

Since L is directed by $\subseteq$, we easily see that the set

$$G = \{X : \textit{for some } \mathfrak{B} \in \mathsf{L},\ \{\mathfrak{A} : \mathfrak{B} \subseteq \mathfrak{A} \in \mathsf{L}\} \subseteq X \subseteq \mathsf{L}\}$$

is a proper filter on L; hence, by a well-known set-theoretical result, G can be extended to an ultrafilter F on L. For every such ultrafilter F (or, more generally, for every proper filter $F \supseteq G$) we show without difficulty that the function

$$\langle g_a/\bar{F} : a \in \mathsf{UL} \rangle$$

is an isomorphism from UL into $\mathsf{P}\langle \mathfrak{A} : \mathfrak{A} \in \mathsf{L} \rangle / \bar{F}$. Consequently,

$$\mathsf{UL} \in \mathbf{S}\mathsf{Up}\,\mathsf{L}.$$

We have thus obtained the first part of 0.3.17(i) with "**HSP**" replaced by "**S**Up"; hence, with the help of 0.3.70(i), 0.1.24, and 0.1.27(ii), we easily derive both the second part of 0.3.17(i) and 0.3.17(ii), again with "**HSP**" replaced by "**S**Up".

Theorem 0.3.71 is clearly an improvement of 0.3.17 (since every class of the form **HSP**K is also of the form **S**Up L); an intermediate statement between 0.3.17 and 0.3.71 is one in which "**HSP**" in 0.3.17 is replaced by "**SHP**". By comparing the proofs of 0.3.17 and 0.3.71 we notice that they are based on closely related ideas and that the proof of 0.3.17 implicitly involves the notion of a reduced product.

THEOREM 0.3.72. (i) *If* $\mathfrak{A} = \langle \mathfrak{A}_i : i \in I \rangle$ *is a system of algebras and*

$$\mathsf{L} = \{\mathsf{P}_{i \in J}\mathfrak{A}_i : J \subseteq I,\ |J| < \omega\},$$

then $\mathsf{P}\mathfrak{A} \in \mathbf{S}\mathsf{Up}\,\mathsf{L}$; *hence, for any class* K *of algebras,* $\mathsf{L} \subseteq \mathbf{S}\mathsf{Up}\,\mathsf{K}$ *implies* $\mathsf{P}\mathfrak{A} \in \mathbf{S}\mathsf{Up}\,\mathsf{K}$.

(ii) *For any class* K *of algebras, if* $\mathfrak{B} \times \mathfrak{C} \in \mathbf{S}\mathsf{Up}\,\mathsf{K}$ *whenever* $\mathfrak{B}, \mathfrak{C} \in \mathbf{S}\mathsf{Up}\,\mathsf{K}$, *then* $\mathbf{PS}\mathsf{Up}\,\mathsf{K} = \mathbf{S}\mathsf{Up}\,\mathsf{K}$.

PROOF. Under the assumptions of (i) let

$$M = \{J : J \subseteq I,\ |J| < \omega\},$$

and

$$g_f = \langle J 1 f : J \in M \rangle \textit{ for every } f \in \mathsf{P}_{i \in I} A_i;$$

clearly g_f belongs to the universe of $\mathsf{P}_{J \in M}(\mathsf{P}_{j \in J}\mathfrak{A}_i)$. Also let

$$G = \{X : \textit{for some } J \in M,\ \{N : J \subseteq N \in M\} \subseteq X \subseteq M\};$$

G is a proper filter on M and hence can be extended to an ultrafilter F. For every such ultrafilter F (and, more generally, for every proper filter F on M which includes G) the function $\langle g_f/\bar{F} : f \in \mathsf{P}_{i \in I} A_i \rangle$ turns out to be an isomorphism from $\mathsf{P}\mathfrak{A}$ into $\mathsf{P}_{J \in M}(\mathsf{P}_{i \in J}\mathfrak{A}_i)/\bar{F}$. Consequently, $\mathsf{P}\mathfrak{A} \in \mathbf{S}\mathsf{Up}\,\mathsf{L}$. Hence the remaining part of (i) follows immediately, and (ii) is an easy consequence of (i).

The notions of reduced product and ultraproduct can be applied, not only to algebras, but also to sets treated as universes of some (unspecified) algebras. In fact, from Definition 0.3.62 it is seen that the universe of the F-reduced product $\mathsf{P}_{i\in I}\mathfrak{A}_i/\bar{F}$ is fully determined by the system $\langle A_i : i\in I\rangle$ of universes and by the filter F, and does not depend at all on the fundamental operations of algebras $\mathfrak{A}_i$. We can thus speak of the *F-reduced product of any system $\langle A_i : i\in I\rangle$ of (non-empty) sets*; in agreement with our general symbolic conventions in the Preliminaries referring to equivalence relations, we shall of course denote the reduced product by $\mathsf{P}_{i\in I}A_i/\bar{F}$. In this context we come across various purely set-theoretical problems concerning in particular the cardinality of reduced products and ultraproducts. Many problems of this kind are still open; in the next three theorems we state some of the results which are known at present.

THEOREM 0.3.73. *If $|B| \geqq \omega$, then for every non-empty set I there is an ultrafilter F on I such that $|{}^IB/\bar{F}| = |{}^IB|$.*

PROOF. By 0.3.63(iv) the conclusion is obvious in case $|I| < \omega$. If $|I| \geqq \omega$, we argue as in the proof of 0.3.72, choosing for the $\mathfrak{A}_i$ any algebras such that $A_i = B$ for every $i\in I$. We obtain

$$\mathsf{P}_{i\in I}\mathfrak{A}_i \cong | \subseteq \mathsf{P}_{J\in M}(\mathsf{P}_{i\in J}\mathfrak{A}_i)/\bar{F}$$

whence, passing to universes,

$$|{}^IB| \leqq |\mathsf{P}_{J\in M}{}^JB/\bar{F}|;$$

here F is an ultrafilter on M. Since $|M| = |I|$, $|{}^JB| = |B|$ for every non-empty $J\in M$, and $|{}^0B| = 1 \leqq |B|$, we conclude that there is an ultrafilter F' on I for which $|{}^IB/\bar{F}'| = |{}^IB|$.

THEOREM 0.3.74. *If F is an ultrafilter on I, then $|\mathsf{P}_{i\in I}A_i/\bar{F}| \geqq \omega$ or else $\{i : |A_i| \leqq \kappa\}\in F$ for some $\kappa < \omega$.*

PROOF. Assume $\{i : |A_i| \leqq \kappa\}\notin F$ for each $\kappa < \omega$. Let X be any finite subset of $\mathsf{P}_{i\in I}A_i/\bar{F}$; thus $X = \{f^{(\lambda)}/\bar{F} : \lambda < \kappa\}$ for some $\kappa < \omega$ and $f = \langle f^{(\lambda)} : \lambda < \kappa\rangle \in {}^{\kappa}(\mathsf{P}_{i\in I}A_i)$. By the axiom of choice we can find a $g\in \mathsf{P}_{i\in I}A_i$ such that $g_i\in A_i \sim \{f_i^{(\lambda)} : \lambda < \kappa\}$ whenever $i\in I$ and $|A_i| > \kappa$. Our assumption implies then that $g/\bar{F} \neq f^{(\lambda)}$ for each $\lambda < \kappa$ and hence $g/\bar{F}\in(\mathsf{P}_{i\in I}A_i/\bar{F}) \sim X$. Consequently the set $\mathsf{P}_{i\in I}A_i/\bar{F}$ is infinite.

THEOREM 0.3.75. *If F is an ultrafilter on I, then $|\mathsf{P}_{i\in I}A_i/\bar{F}| \geqq 2^{\omega}$ or else $|\mathsf{P}_{i\in I}A_i/\bar{F}| = |A_j|$ for some $j\in I$.*

The proof of this theorem is more difficult; compare Frayne-Morel-Scott [62*], pp. 208 f. (Theorem 1.29).

REMARK 0.3.76. By 0.3.75, neither of the inclusions of 0.3.69(ii) can be replaced by an equality. Indeed, 0.3.75 implies that for any class K of finite algebras, every algebra in $\mathsf{Up}\,\mathsf{K}$ is either finite or of cardinality 2^{ω}. Thus if all algebras in K are finite, the same holds for $\mathbf{S}\mathsf{K}$ and $\mathbf{H}\mathsf{K}$, and therefore no algebra belonging to $\mathsf{Up}\mathbf{S}\mathsf{K}$ or $\mathsf{Up}\mathbf{H}\mathsf{K}$ is denumerable. On the other hand, if we take for K, e.g., the class of all finite algebras $\mathfrak{A} = \langle A, {}^2A \upharpoonright \mathsf{pj}_0 \rangle$ (or $\mathfrak{A} = \langle A, A \upharpoonright Id \rangle$), we easily see that both classes $\mathbf{S}\mathsf{Up}\,\mathsf{K}$ and $\mathbf{H}\mathsf{Up}\,\mathsf{K}$ contain denumerable algebras among their members.

To conclude this section we shall discuss briefly model-theoretical properties of ultraproducts (using various metalogical notions and symbols which were introduced in the Preliminaries). The next theorem is a basic result in this direction; practically all of the more profound applications of ultraproducts in mathematics and metamathematics are based upon it.

THEOREM 0.3.77. *Let $\langle \mathfrak{A}_i : i \in I \rangle$ be a system of algebras, F be an ultrafilter on I, and $\mathfrak{B} = \mathsf{P}_{i \in I} \mathfrak{A}_i / \bar{F}$. Furthermore, let $x \in {}^{\omega}(\mathsf{P}_{i \in I} A_i)$, and let φ be any formula in the discourse language of $\mathfrak{B}$ (and hence also in the language of $\mathfrak{A}_i$ for each $i \in I$). Then*

(i) $\mathfrak{B} \models \varphi[\bar{F}^{\star} \circ x]$ *iff* $\{i \in I : \mathfrak{A}_i \models \varphi[\mathsf{pj}_i \circ x]\} \in F$.

Hence, in case φ is a sentence,

(ii) $\mathfrak{B} \models \varphi$ *iff* $\{i \in I : \mathfrak{A}_i \models \varphi\} \in F$; *in particular, we have* $\mathfrak{B} \models \varphi$ *whenever* $\mathfrak{A}_i \models \varphi$ *for every* $i \in I$.

Theorem 0.3.77 is essentially due to Łoś [55a*], p. 105, 2.6; for a proof of it see Frayne-Morel-Scott [62*], pp. 213 f. An immediate corollary of 0.3.77(ii) is

COROLLARY 0.3.78. *If $\mathfrak{A}$ is any algebra, I any set, and F any ultrafilter on I, then $\mathfrak{A}$ and ${}^I\mathfrak{A}/\bar{F}$ are elementarily equivalent.*

As a consequence of 0.3.77 (using the well-known theorem on the extension of filters to ultrafilters) we obtain the model-theoretical compactness theorem:

If Σ is any set of sentences (in the language of the elementary theory of some class of algebras) and every finite subset of Σ has a model, then Σ also has a model.

The compactness theorem can be fruitfully applied in algebraic investigations; its importance consists in the fact that in many cases it enables one to assert the existence of an algebra satisfying some prescribed condition, even though no effective method for constructing such an algebra is available. Instead of proving the compactness theorem by means of 0.3.77 we can derive it directly from the completeness theorem for predicate logic; while neither of these proofs is effective, the one based upon 0.3.77 provides us with much more insight into the structure of algebras which can be claimed to be models of the set Σ.

To illustrate the applicability of 0.3.77 (and also of the compactness theorem) to specialized situations, consider the case when all the algebras $\mathfrak{A}_i$ are integral domains. Then, by 0.3.77, their ultraproduct $\mathfrak{B}$ is also an integral domain. On the other hand, it is well known that the direct product of $\mathfrak{A}_i$'s is never an integral domain (disregarding the trivial case when all $\mathfrak{A}_i$'s with at most one exception are one-element algebras). Since, by 0.3.63(i), the direct product is isomorphic to a reduced product, we conclude that 0.3.77 does not extend to direct products nor *a fortiori* to arbitrary reduced products.

Assume now that I consists of all natural numbers which are prime powers and that, for each $i \in I$, $\mathfrak{A}_i$ is a finite (Galois) field with exactly i elements. Let F be any non-principal ultrafilter on I; thus all complements of finite subsets of I are members of F. By 0.3.77 the ultraproduct $\mathfrak{B} = \mathsf{P}_{i\in I}\mathfrak{A}_i/\bar{F}$ is again a field which, in addition, exhibits the following particularities: every property expressible in the language of the elementary theory of fields and which holds in all finite fields holds in $\mathfrak{B}$ as well; the same applies to every property which holds in "almost all" finite fields, i.e., in all finite fields with at least κ elements, where κ is some natural number given in advance. Thus, for instance, every element of $\mathfrak{B}$ is a sum of two squares, and any homogeneous polynomial in κ variables ($\kappa > 1$) of degree $< \kappa$, with coefficients in $\mathfrak{B}$, has a non-trivial zero in $\mathfrak{B}$. Furthermore, for every $\kappa < \omega$, $\mathfrak{B}$ has at least κ elements (since "almost all" finite fields have at least κ elements); hence $\mathfrak{B}$ is infinite and, in view of 0.3.75, actually has cardinality 2^ω. On the other hand, nothing definite can be said about the characteristic of $\mathfrak{B}$. By choosing F appropriately, $\mathfrak{B}$ can be forced to have any characteristic given in advance, in particular characteristic zero, and this in spite of the fact that all the fields $\mathfrak{A}_i$ have positive characteristic. If, in constructing $\mathfrak{B}$, we replace the set of all powers of primes by the set of all primes (and thus restrict ourselves to finite prime fields), then $\mathfrak{B}$ always has characteristic zero. The observations just made concerning the cardinality and characteristic of $\mathfrak{B}$ by no means contradict 0.3.77; they simply show that, as a consequence of 0.3.77, the properties of being finite, or infinite, and of having positive characteristic, or characteristic zero, turn out not to be expressible in the language of the elementary theory of fields. (All the observations in this paragraph can be established as easily by means of the compactness theorem.)

With a slight modification in the definition of ultraproducts Theorem 0.3.77 extends to arbitrary (first order) relational structures; cf. Frayne-Morel-Scott [62*]. By applying this extended theorem (or, instead, the compactness theorem) to ordered fields $\mathfrak{F} = \langle F, +, \cdot, \leqq \rangle$, thus to structures which are not algebras in our sense, we obtain, e.g., the following result: for every ordered field $\mathfrak{F}$ there is a non-Archimedean ordered field $\mathfrak{F}'$ which is elementarily equivalent with $\mathfrak{F}$, i.e., which has all the properties of $\mathfrak{F}$ expressible in the discourse language of $\mathfrak{F}$; cf. Tarski [52]. The familiar methods of constructing non-Archimedean ordered fields do not yield a field $\mathfrak{F}'$ elementarily equivalent with $\mathfrak{F}$.

For the discussion to follow recall from the Preliminaries that "$\mathscr{GCH}$" is an abbreviation for "the generalized continuum hypothesis".

THEOREM 0.3.79. *Let* $\mathfrak{A}$ *and* $\mathfrak{B}$ *be any two algebras.*

(i) *The condition* $\mathsf{Up}\,\mathfrak{A} \cap \mathsf{Up}\,\mathfrak{B} \neq 0$ *is sufficient for* $\mathfrak{A}$ *and* $\mathfrak{B}$ *to be elementarily equivalent.*

(ii) *Under the assumption of* $\mathscr{GCH}$, *this condition is also necessary.*

0.3.79(i) is a direct consequence of 0.3.77(ii). The proof of 0.3.79(ii) is much more difficult; it is due to Keisler [61a*]. The question to what extent the role of $\mathscr{GCH}$ in 0.3.79(ii) is essential has not yet been cleared up.[1)]

REMARK 0.3.80. Let us agree to write $\mathfrak{A} \equiv \mathfrak{B}$ in case the algebras $\mathfrak{A}$ and $\mathfrak{B}$ are elementarily equivalent, and $\mathfrak{A} \triangleq \mathfrak{B}$ in case $\mathsf{Up}\,\mathfrak{A} \cap \mathsf{Up}\,\mathfrak{B} \neq 0$. Thus the relation $\triangleq$ has a purely mathematical character, while the relation $\equiv$ has a model-theoretical one. 0.3.79 shows that, assuming $\mathscr{GCH}$, the two relations coincide. (If the algebras involved are finite, each of these two relations coincide with $\cong$.) Thus this theorem enables us, in principle, to establish the relation $\triangleq$ between two given algebras $\mathfrak{A}$ and $\mathfrak{B}$ by means of a metamathematical argument, or to establish the elementary equivalence of these algebras using purely mathematical methods. So far, however, in its application to particular algebras Theorem 0.3.79 has proved useful mostly in the first direction; in many cases the only proof of the formula $\mathfrak{A} \triangleq \mathfrak{B}$ which is now available is the one based upon the formula $\mathfrak{A} \equiv \mathfrak{B}$ which was previously established by a metamathematical argument. Some other mathematical cha-

[1)] [*Added in proof.*] Recently Saharon Shelah has shown that the conclusion of 0.3.79 (ii) holds without the assumption of $\mathscr{GCH}$. In view of this result various statements and remarks in our further discussion require adjustment.

racterizations of elementary equivalence, in addition to the one given in 0.3.79, are also known (cf. Fraïssé [55*], Ehrenfeucht [61*], Taĭmanov [61*], Kochen [61*], and Keisler [63*]). They are mathematically less elegant and more involved, but have two virtues: they do not depend on $\mathscr{GCH}$ and turn out to be more readily applicable in practice. In various particular situations they have actually made it possible to establish the elementary equivalence of given algebras by means of mathematical methods.

With the continuous use of 0.3.79 we can develop a whole theory of the relation $\triangleq$. However (and rather unfortunately), this theory appears at present to depend essentially on $\mathscr{GCH}$. This applies even to the results expressing the simplest and most basic properties of the relation $\triangleq$, namely to the statements that $\triangleq$ is an equivalence relation and that the partition induced by $\triangleq$ in any similarity class of algebras (with non-empty and at most denumerable systems of operations) has cardinality 2^ω. The corresponding properties of $\equiv$ are obvious, but we do not know how to establish these properties for $\triangleq$ without the help of 0.3.79. Another important property of $\triangleq$ is expressed in the following statement: For every infinite algebra $\mathfrak{A}$ (with at most denumerably many fundamental operations) and every infinite cardinal β, there is an algebra $\mathfrak{B}$ such that $\mathfrak{A} \triangleq \mathfrak{B}$ and $|B| = \beta$. The corresponding model-theoretical result concerning the relation $\equiv$ is the well-known Löwenheim-Skolem theorem.

As corollaries of various results concerning the elementary equivalence of algebras which are known from the literature, we can establish a number of theorems connecting $\triangleq$ with direct products and reduced products. Thus, using the results in Mostowski [52*] and Feferman-Vaught [59*], we can show that, for any two systems $\mathfrak{A} = \langle \mathfrak{A}_i : i \in I \rangle$ and $\mathfrak{B} = \langle \mathfrak{B}_i : i \in I \rangle$ of algebras and every filter F on I, the condition $\mathfrak{A}_i \triangleq \mathfrak{B}_i$ for every $i \in I$ implies $\mathsf{P}\mathfrak{A}/\bar{F} \triangleq \mathsf{P}\mathfrak{B}/\bar{F}$ and in particular $\mathsf{P}\mathfrak{A} \triangleq \mathsf{P}\mathfrak{B}$; also that, for every algebra $\mathfrak{C}$ and any infinite sets I and J, ${}^I\mathfrak{C} \triangleq {}^J\mathfrak{C}$. By a recent result of Fred Galvin we have ${}^J({}^I\mathfrak{C}/\bar{F})/\bar{G} \triangleq {}^I({}^J\mathfrak{C}/\bar{G})/\bar{F}$ for very algebra $\mathfrak{C}$. By another of his results, for any algebras $\mathfrak{A}$, $\mathfrak{B}$, and $\mathfrak{C}$, the formula $\mathfrak{A} \triangleq \mathfrak{A} \times \mathfrak{B} \times \mathfrak{C}$ implies $\mathfrak{A} \triangleq \mathfrak{A} \times \mathfrak{B} \triangleq \mathfrak{A} \times \mathfrak{C}$; in other words, the property (α) discussed in 0.3.33, but with "$\cong$" replaced by "$\triangleq$", holds for every algebra $\mathfrak{A}$. Cf. here Galvin [67*], p. 63. On the other hand, using the same example as in 0.3.33, we see that the property (β), again with "$\cong$" replaced by "$\triangleq$", fails for some algebras. Other, more complicated, properties involving $\triangleq$ and analogous to those discussed in 0.3.30–0.3.39, such as the unique factorization and the refinement properties, have not yet been studied.

In formulating the next few theorems we shall use the abbreviations $\mathscr{EC}$, $\mathscr{EC}_\Delta$, etc. introduced in the Preliminaries.

THEOREM 0.3.81. *Let* K *be any class of algebras included in a similarity class* L.

(i) *The condition*

$$\mathsf{K} = \mathsf{Up}\,\mathsf{K} \text{ and } \mathsf{L} \sim \mathsf{K} = \mathsf{Up}(\mathsf{L} \sim \mathsf{K})$$

is necessary for K *to be an* $\mathscr{EC}$.

(ii) *The condition*

$$\mathsf{K} = \mathsf{Up}\,\mathsf{K} = \{\mathfrak{A} : \mathsf{Up}\,\mathfrak{A} \cap \mathsf{K} \neq 0\}$$

is necessary for K *to be an* $\mathscr{EC}_\Delta$.

(iii) *Under the assumption of* $\mathscr{GCH}$, *the conditions in* (*i*) *and* (*ii*) *are also sufficient.*

Parts (i) and (ii) of 0.3.81 are easy consequences of 0.3.77(ii) and 0.3.78. Part (iii) can be established with the essential help of 0.3.79(ii); see Keisler [61a*].

REMARK 0.3.82. If we agree to denote by $\mathsf{Up}'\mathsf{K}$ the class of all algebras isomorphic to an ultrapower of some algebra in K, then the condition in 0.3.81(ii) can be expressed analogously to that in 0.3.81(i):

$$\mathsf{K} = \mathsf{Up}\,\mathsf{K} \text{ and } \mathsf{L} \sim \mathsf{K} = \mathsf{Up}'(\mathsf{L} \sim \mathsf{K}).$$

On the other hand, let $\mathsf{Uf}\,\mathsf{K} = \{\mathfrak{A} : \mathsf{Up}\,\mathfrak{A} \cap \mathsf{K} \neq 0\}$. The condition in 0.3.81(ii) then assumes the form $\mathsf{K} = \mathsf{Up}\,\mathsf{K} = \mathsf{Uf}\,\mathsf{K}$ or, equivalently, $\mathsf{K} = \mathsf{Uf}\,\mathsf{Up}\,\mathsf{K}$; another equivalent condition is: $\mathsf{K} = \mathsf{Uf}\,\mathsf{Up}\,\mathsf{L}$ *for some class* L. Theorem 0.3.81(ii),(iii) implies that (assuming $\mathscr{GCH}$) $\mathsf{Uf}\,\mathsf{Up}\,\mathsf{K}$ is the least class which includes K and is an $\mathscr{EC}_\Delta$, in other words, $\mathsf{Uf}\,\mathsf{Up}\,\mathsf{K} = \mathsf{Md}\,\theta\rho\,\mathsf{K}$ (cf. Keisler [61a*]).

The formula $\mathsf{K} = \mathsf{Uf}\,\mathsf{Up}\,\mathsf{K}$ expresses a purely mathematical property of a class K of algebras, while the property of being an $\mathscr{EC}_\Delta$ is of a model-theoretical nature. Regarding the relation between these two properties various observations can be made which are analogous to those in 0.3.80 concerning the relation between $\triangleq$ and $\equiv$. The difference is that, in the present case, the model-theoretical property is the stronger one: it easily implies the mathematical property, while the implication in the opposite direction depends on $\mathscr{GCH}$. At any rate, Theorem 0.3.81 gives an interesting and relatively simple mathematical characterization of elementary classes of algebras (both in the narrower and in the wider sense); this is important since most familiar classes of algebras, such as groups, rings, integral domains, and fields, are actually $\mathscr{EC}$'s or $\mathscr{EC}_\Delta$'s.

We know various properties of classes of algebras which are stronger than the one expressed by the formula $\mathsf{K} = \mathsf{Uf}\,\mathsf{Up}\,\mathsf{K}$, and which therefore imply that K is an $\mathscr{EC}_\Delta$. From 0.3.67 and 0.3.69(i) it easily follows that such is, e.g., the property of being an algebraically closed class, $\mathsf{K} = \mathbf{HSP}\mathsf{K}$ (cf. 0.3.14). In Section 0.4 we shall discuss a model-theoretical characterization of this property; as we shall see there, the proof that every algebraically closed class is an $\mathscr{EC}_\Delta$ does not depend on $\mathscr{GCH}$. The same applies to the property $\mathsf{K} = \mathbf{S}\mathsf{Up}\mathsf{K}$, which is weaker than $\mathsf{K} = \mathbf{HSP}\mathsf{K}$. This property was discussed in 0.3.70 and 0.3.71; its model-theoretical characterization is given in the following

THEOREM 0.3.83. *Given any class* K *of algebras, the condition* $\mathsf{K} = \mathbf{S}\mathsf{Up}\mathsf{K}$ *is necessary and sufficient for* K *to be a* $\mathscr{UC}_\Delta$.

This is a result of Łoś [55a*], p. 105, 2.5 (obtained without the help of $\mathscr{GCH}$).

REMARK 0.3.84. Using 0.3.83 we can of course "translate" every mathematical statement concerning classes K of algebras with $\mathsf{K} = \mathbf{S}\mathsf{Up}\mathsf{K}$ into a model-theoretical statement on $\mathscr{UC}_\Delta$'s, and conversely. For instance, it is obvious that every $\mathscr{UC}_\Delta$ is an $\mathscr{EC}_\Delta$ and hence, by 0.3.81(ii) and 0.3.83, we obtain the mathematical statement: $\mathsf{K} = \mathbf{S}\mathsf{Up}\mathsf{K}$ always implies $\mathsf{K} = \mathsf{Uf}\,\mathsf{Up}\,\mathsf{K}$, where Uf is the operation defined in 0.3.82. (This implication can also be easily derived from 0.3.67 and 0.3.69(i).) On the other hand, using 0.3.83 we can derive from 0.3.71 and 0.3.72 the following properties of $\mathscr{UC}_\Delta$'s: (1) if K is a $\mathscr{UC}_\Delta$ and L is a subset of K ordered by the relation $\subseteq$, then $\mathbf{U}\mathsf{L} \in \mathsf{K}$; (2) if K is a $\mathscr{UC}_\Delta$, $\mathfrak{A} = \langle \mathfrak{A}_i : i \in I\rangle$, and $\{\mathsf{P}_{i\in J}\mathfrak{A}_i : J \subseteq I, |J| < \omega\} \subseteq \mathsf{K}$, then $\mathsf{P}\mathfrak{A} \in \mathsf{K}$. Both these statements, however, can be established as easily by means of direct model-theoretical arguments. Actually (1) is almost obvious; a model-theoretical proof of (2) (in a somewhat weaker form) is outlined in Tarski [54*], p. 587. Moreover, by a result of Vaught in Feferman-Vaught [59*], p. 84, Corollary 6.7.2, the property (2) applies not only to $\mathscr{UC}_\Delta$'s but to arbitrary $\mathscr{EC}_\Delta$'s; hence, by making use of 0.3.81(iii), we conclude that, under the assumption of $\mathscr{GCH}$, 0.3.72 remains valid if "$\mathbf{S}\mathsf{Up}$" is replaced everywhere by "Uf Up".

By rearranging the succession of theorems we could use (1) and (2) to derive 0.3.71 and 0.3.72 (as well as 0.3.17) as simple corollaries from 0.3.83.

REMARK 0.3.85. We may mention here still another mathematical characterization of $\mathscr{UC}_\Delta$'s (announced in Tarski [53*] and established in Tarski [54*], p. 578, Theorem 1.2). It is conceptually simpler than the one given in 0.3.83 since it does not involve the notion of an ultraproduct; it involves instead the notions of a partial subalgebra and of isomorphism of such subalgebras. By a partial subalgebra of an algebra $\mathfrak{A} = \langle A, + \rangle$ we understand any structure $\langle B, +' \rangle$ where B is any non-empty subset of A (not necessarily a subuniverse of $\mathfrak{A}$) and $+'$ is the operation $+$ with the domain restricted to the set $\{\langle x, y \rangle : x, y, x+y \in B\}$. Thus $\langle B, +' \rangle$ is in general not an algebra in the sense of 0.1.1, but is one of the partial algebras which were mentioned briefly in 0.1.3. Let $\mathsf{Ps}\,\mathfrak{A}$ and $\mathsf{Ps}_\omega\mathfrak{A}$ be respectively the classes of all partial subalgebras and all finite partial subalgebras of an algebra $\mathfrak{A}$; the meaning of $\mathsf{Ps}\,\mathsf{K}$ and $\mathsf{Ps}_\omega\mathsf{K}$ for a class K of algebras is obvious. It now turns out that the following condition is necessary and sufficient for K to be a $\mathscr{UC}_\Delta$:

(I) $\mathsf{IK} \subseteq \mathsf{K}$, *and, for every* $\mathfrak{A}$, $\mathfrak{A} \in \mathsf{K}$ *whenever* $\mathsf{Ps}_\omega\mathfrak{A} \subseteq \mathsf{Ps}\,\mathsf{K}$.

It may be noticed that the closely related condition

(II) $\mathsf{IK} \subseteq \mathsf{K}$, *and, for every* $\mathfrak{A}$, $\mathfrak{A} \in \mathsf{K}$ *whenever* $\mathbf{S}_\omega\mathfrak{A} \subseteq \mathbf{S}\mathsf{K}$ (*or, equivalently by* 0.1.28, $\mathfrak{A} \in \mathsf{K}$ *iff* $\mathbf{S}_\omega\mathfrak{A} \subseteq \mathsf{K}$)

is only necessary, but not sufficient, for K to be a $\mathscr{UC}_\Delta$; this condition will be satisfied, e.g., by the class K of all torsion groups $\mathfrak{G} = \langle G, \cdot, {}^{-1} \rangle$ although K is not a $\mathscr{UC}_\Delta$ and not even an $\mathscr{EC}_\Delta$.[1)] Using either 0.3.81(ii) and 0.3.83 or else the second characterization of $\mathscr{UC}_\Delta$'s we can easily establish the following rather important and by no means obvious model-theoretical result (cf. Tarski [54*], p. 583, Theorem 1.6, and Łoś [55*], p. 49, Theorem 7).

THEOREM 0.3.86. *If* K *is an* $\mathscr{EC}_\Delta$, *then* $\mathbf{S}\mathsf{K}$ *is a* $\mathscr{UC}_\Delta$.

REMARK 0.3.87. To obtain a mathematical characterization of $\mathscr{UC}$'s, we first notice that a class K is a $\mathscr{UC}$ iff it is both an $\mathscr{EC}$ and a $\mathscr{UC}_\Delta$, and then we apply 0.3.81(i),(iii) and 0.3.83; the resulting characterization is rather complicated ($\mathsf{K} = \mathbf{S}\mathsf{Up}\,\mathsf{K}$ and $\mathsf{L} \sim \mathsf{K} = \mathsf{Up}(\mathsf{L} \sim \mathsf{K})$) but turns out to be independent of $\mathscr{GCH}$. However, as indicated in Vaught [54*], a simple variant of the second characterization of $\mathscr{UC}_\Delta$'s proves to be adequate for this purpose as well.

REMARK 0.3.88. Theorems 0.3.81 and 0.3.83 are samples of a rather long series of related results which have been obtained in the theory of models (cf. Lyndon [59*], [59a*], Łoś-Suszko [57*], Chang [59*], and Keisler [65*]). As was already mentioned, another result of the same kind will be discussed in the next section.

[1)] The close relation between (I) and (II) is made even more evident by the following observations: A class of algebras satisfies (I) iff it can be characterized by a set of universal sentences in ordinary predicate logic. On the other hand, a class satisfies (II) iff it can be characterized by a set of universal sentences in predicate logic with infinitely long formulas, and in fact of sentences each containing only finitely many distinct variables. Cf. Tarski [58*].

0.4. POLYNOMIALS AND FREE ALGEBRAS

In the first half of this section we discuss polynomials over a given algebra $\mathfrak{A}$. Polynomials are, loosely speaking, all the operations, of finite or infinite rank, which can be obtained from purely set-theoretical operations, namely projections pj_ξ, and the fundamental operations of $\mathfrak{A}$ by composition. The most convenient way of defining the set of polynomials of a given rank α over an algebra $\mathfrak{A}$ consists in introducing this set as a special subuniverse of the direct power ${}^{^\alpha A}\mathfrak{A}$. Hence this set induces a subalgebra of ${}^{^\alpha A}\mathfrak{A}$ called the *polynomial algebra of rank* α *over* $\mathfrak{A}$. Thus the polynomial algebras are constructed by essentially the same method as subdirect products (discussed in the preceding section), i.e., by forming subalgebras of direct products; in the present case, however, this method assumes a highly specialized character.

Throughout this section α and β will represent arbitrary cardinals different from 0. On the other hand, by ξ we shall represent, as usual, an arbitrary ordinal.

DEFINITION 0.4.1. *For every algebra* $\mathfrak{A}$ *and every cardinal* $\alpha \neq 0$, *the set* $Pl_\alpha\mathfrak{A}$ *of* ***polynomials in*** α ***variables***, *or* ***polynomial operations with rank*** α, ***over*** $\mathfrak{A}$ *is the subuniverse of* ${}^{^\alpha A}\mathfrak{A}$ *generated by the set* $\{{}^\alpha A \restriction \mathsf{pj}_\xi : \xi < \alpha\}$; *the correlated algebra* $\mathfrak{Pl}_\alpha\mathfrak{A} = \langle Pl_\alpha\mathfrak{A}, +\rangle$ (*where* $+$ *is, of course, the fundamental operation of* ${}^{^\alpha A}\mathfrak{A}$) *is called the* ***polynomial algebra of rank*** α ***over*** $\mathfrak{A}$. *The class* $Pl\mathfrak{A}$ *of all polynomials over* $\mathfrak{A}$ *is defined by*

$$Pl\mathfrak{A} = \bigcup\{Pl_\beta\mathfrak{A} : \beta \neq 0\}.$$

Thus $Pl_1\mathfrak{A}$ contains as elements $A \restriction Id$, $\langle x+x : x \in A\rangle$, $\langle x+(x+x) : x \in A\rangle$, etc. Among elements of $Pl_2\mathfrak{A}$ we find ${}^2A \restriction \mathsf{pj}_0 = \langle x : \langle x, y\rangle \in {}^2A\rangle$, ${}^2A \restriction \mathsf{pj}_1 = \langle y : \langle x, y\rangle \in {}^2A\rangle$, $+ = \langle x+y : \langle x, y\rangle \in {}^2A\rangle$, $\langle y+x : \langle x, y\rangle \in {}^2A\rangle$, $\langle x+(y+x) : \langle x, y\rangle \in {}^2A\rangle$, etc.

We could define the set $Pl_\alpha\mathfrak{A}$, not only for a cardinal, but for every ordinal α; more generally, we could define $Pl_I\mathfrak{A}$ for every index set I. These extensions of the notion of polynomials do not appear to be either essential or useful.

REMARK 0.4.2. For some purposes an extension of the notion of polynomials in a different direction proves to be useful. The set of *polynomials* (in α variables over $\mathfrak{A}$) *in the wider sense* is the subuniverse of ${}^{\alpha A}\mathfrak{A}$ generated by the set consisting of all projections ${}^{\alpha}A 1 \mathsf{pj}_{\xi}$, $\xi < \alpha$, and of all constant functions ${}^{\alpha}A \times \{c\}$ where c is any element of A.

The notion of a polynomial originated with the discussion of various number fields and has been subsequently extended first to abstract rings and fields, and then to arbitrary algebras. The notion had originally a metamathematical character and was related to what is now technically called a "term" in the elementary theory of algebras. It has been reconstrued in mathematical terms and in this process has undergone some modifications. It is easily seen that what are normally called polynomials with coefficients from a field $\mathfrak{F} = \langle F, +, \cdot \rangle$ can be identified with polynomials over $\mathfrak{F}$ in the wider sense just defined, at least in the case of an infinite field; if $\mathfrak{F}$ is a field of characteristic 0, then polynomials with integral coefficients can be identified with polynomials over $\mathfrak{F}$ in the sense of 0.4.1 (including for this purpose -1 as a distinguished element in the definition of $\mathfrak{F}$).

As almost immediate consequences of 0.4.1 we obtain 0.4.3 and 0.4.4:

THEOREM 0.4.3. $\mathfrak{Pl}_{\alpha}\mathfrak{A} \in \mathbf{S}_{\alpha^+}\mathbf{P}\mathfrak{A}$.

THEOREM 0.4.4. (i) *If* $|A| = 1$, *then* $|\{{}^{\alpha}A 1 \mathsf{pj}_{\xi} : \xi < \alpha\}| = 1 = |Pl_{\alpha}\mathfrak{A}|$ (*for every* α).

(ii) *If* $|A| > 1$, *then* $|\{{}^{\alpha}A 1 \mathsf{pj}_{\xi} : \xi < \alpha\}| = \alpha \leqq |Pl_{\alpha}\mathfrak{A}| \leqq \alpha \cup \omega$; *if, in addition,* $\alpha \geqq \omega$, *then* $|Pl_{\alpha}\mathfrak{A}| = \alpha$.

(iii) *The fundamental operation* $+$ *of* $\mathfrak{A}$ *is a member of* $Pl_2\mathfrak{A}$.

PROOF. (i) is obvious, and (ii) follows from 0.1.19. Since $+ = ({}^{2}A 1 \mathsf{pj}_0) +' ({}^{2}A 1 \mathsf{pj}_1)$, $+'$ being the fundamental operation of ${}^{2A}\mathfrak{A}$, (iii) follows.

0.4.4(i) holds for algebras of any similarity classes. When extending 0.4.4(ii) to other similarity classes, we apply 0.1.20 with $|X| = \alpha$. Note that, if f is a κ-ary fundamental operation of an algebra $\mathfrak{A}$, then $f \in Pl_{\kappa}\mathfrak{A}$.

THEOREM 0.4.5. *If* $O \in Pl_{\alpha}\mathfrak{A}$, $Q = \langle Q_{\xi} : \xi < \alpha \rangle \in {}^{\alpha}Pl_{\beta}\mathfrak{A}$, *and* $C(O, Q) = \langle O\langle Q_{\xi}x : \xi < \alpha \rangle : x \in {}^{\beta}A \rangle$, *then* $C(O, Q) \in Pl_{\beta}\mathfrak{A}$.

PROOF. For a given $Q \in {}^{\alpha}Pl_{\beta}\mathfrak{A}$, let

$$B = \{O : O \in {}^{\alpha A}A,\ C(O, Q) \in Pl_{\beta}\mathfrak{A}\}.$$

It is easily seen that $B \in Su({}^{\alpha A}\mathfrak{A})$ and that $\{{}^{\alpha}A 1 \mathsf{pj}_{\xi} : \xi < \alpha\} \subseteq B$. Hence $Pl_{\alpha}\mathfrak{A} \subseteq B$, and the theorem follows.

THEOREM 0.4.6. *$Pl\mathfrak{A}$ is the least class K satisfying the following conditions:*

(i) *${}^{\alpha}A \upharpoonright \mathsf{pj}_{\xi} \in K$ for every α and every ordinal $\xi < \alpha$;*

(ii) *the fundamental operation $+$ of $\mathfrak{A}$ belongs to K;*

(iii) *for any two cardinals $\alpha, \beta \neq 0$, if $O \in K \cap {}^{{}^{\alpha}A}A$, $Q \in {}^{\alpha}(K \cap {}^{{}^{\beta}A}A)$, and $C(O, Q)$ is defined as in 0.4.5, then $C(O, Q) \in K$.*

PROOF. By 0.4.1, 0.4.4(iii), and 0.4.5, $K = Pl\mathfrak{A}$ satisfies (i)–(iii), and hence, if K is the smallest class satisfying (i)–(iii), we have $K \subseteq Pl\mathfrak{A}$. Also, for any α, ${}^{{}^{\alpha}A}A \cap K$ is a subuniverse of ${}^{{}^{\alpha}A}\mathfrak{A}$ which includes $\{{}^{\alpha}A \upharpoonright \mathsf{pj}_{\xi} : \xi < \alpha\}$, and so $Pl_{\alpha}\mathfrak{A} \subseteq {}^{{}^{\alpha}A}\mathfrak{A} \cap K$; thus $Pl\mathfrak{A} \subseteq K$, and the proof is complete.

REMARK 0.4.7. The operation on functions described in 0.4.5 can be called *generalized composition.* When performed on a function O in α variables and an α-termed sequence Q of functions in β variables, it yields the function $C(O, Q)$ in β variables; in case $\alpha = \beta = 1$ it reduces to ordinary composition. Thus, 0.4.6 justifies the remark made at the beginning of this section regarding the possibility of defining polynomials over $\mathfrak{A}$ as operations obtained from projections and fundamental operations of $\mathfrak{A}$ by composition. Besides generalized composition, there is another related operation, called *superposition,* under which $Pl\mathfrak{A}$ is closed; cf. Jónsson-Tarski [51], p. 895.

With an obvious modification of condition (ii), Theorem 0.4.6 extends to algebras $\mathfrak{A}$ of arbitrary similarity classes.

THEOREM 0.4.8. (i) *If $0 \neq \Gamma \subseteq \alpha$ and $O \in Pl_{\alpha}\mathfrak{A}$, then the following conditions are equivalent:*

(i′) *$O \in Sg^{(\mathfrak{Pl}_{\alpha}\mathfrak{A})}\{{}^{\alpha}A \upharpoonright \mathsf{pj}_{\xi} : \xi \in \Gamma\}$;*

(i″) *$Ox = Oy$ whenever $x, y \in {}^{\alpha}A$ and $\Gamma \upharpoonright x = \Gamma \upharpoonright y$.*

(ii) *If $0 \neq \Gamma \subseteq \alpha$ and $|\Gamma| = \beta$, then*

$$\mathfrak{Sg}^{(\mathfrak{Pl}_{\alpha}\mathfrak{A})}\{{}^{\alpha}A \upharpoonright \mathsf{pj}_{\xi} : \xi \in \Gamma\} \cong \mathfrak{Pl}_{\beta}\mathfrak{A}.$$

(iii) *If $\alpha \geqq \beta$, $\mathfrak{Pl}_{\beta}\mathfrak{A} \cong | \subseteq \mathfrak{Pl}_{\alpha}\mathfrak{A}$.*

PROOF. (i) The set of all $O \in Pl_{\alpha}\mathfrak{A}$ satisfying (i″) is clearly a subuniverse of $\mathfrak{Pl}_{\alpha}\mathfrak{A}$ which includes $\{{}^{\alpha}A \upharpoonright \mathsf{pj}_{\xi} : \xi \in \Gamma\}$; thus (i′) implies (i″). Turning to the implication in the opposite direction we let $B = Sg^{(\mathfrak{Pl}_{\alpha}\mathfrak{A})}\{{}^{\alpha}A \upharpoonright \mathsf{pj}_{\xi} : \xi \in \Gamma\}$. We will need the following lemma whose proof is almost identical to that of 0.4.5:

(1) If $O \in Pl_{\alpha}\mathfrak{A}$, $Q = \langle Q_{\xi} : \xi < \alpha \rangle \in {}^{\alpha}B$, and $C(O, Q) = \langle O\langle Q_{\xi}x : \xi < \alpha \rangle : x \in {}^{\alpha}A \rangle$, then $C(O, Q) \in B$.

Now suppose $O \in Pl_{\alpha}\mathfrak{A}$ satisfies condition (i″). Since $\Gamma \neq 0$, there is a $\sigma \in {}^{\alpha}\Gamma$ such that

$$\Gamma \upharpoonright \sigma = \Gamma \upharpoonright Id. \tag{2}$$

Let $Q = \langle {}^{\alpha}A 1 \mathsf{pj}_{\sigma\xi} : \xi < \alpha \rangle$. Then

$$(3) \qquad Q \in {}^{\alpha}B$$

and by (2) we have

$$\Gamma 1 x = \Gamma 1 \langle ({}^{\alpha}A 1 \mathsf{pj}_{\sigma\xi}) x : \xi < \alpha \rangle$$

for every $x \in {}^{\alpha}A$. Consequently, if $C(O, Q)$ is defined as in (1), from (i″) we get

$$Ox = O\langle ({}^{\alpha}A 1 \mathsf{pj}_{\sigma\xi}) x : \xi < \alpha \rangle = C(O, Q)x$$

for every $x \in {}^{\alpha}A$, and thus $O = C(O, Q)$. Therefore $O \in B$ by (1) and (3), i.e., O satisfies condition (i′).

(ii) Let h be a one-one function from β onto Γ. Let $F = \langle \langle Q(x \circ h) : x \in {}^{\alpha}A \rangle : Q \in {}^{\beta A}A \rangle$. Clearly $F \in Ism({}^{\beta A}\mathfrak{A}, {}^{\alpha A}\mathfrak{A})$. Since $F^*\{{}^{\beta}A 1 \mathsf{pj}_{\xi} : \xi < \beta\} = \{{}^{\alpha}A 1 \mathsf{pj}_{\xi} : \xi \in \Gamma\}$, the desired conclusion follows by 0.2.18(i).

(iii): by (ii).

The implication from (i″) to (i′) in Theorem 0.4.8 was pointed out by Don Pigozzi.

A consequence of 0.4.8(i) and 0.1.17(ix) is the following theorem which shows that every polynomial in infinitely many variables essentially depends only on finitely many of its arguments:

THEOREM 0.4.9. *For every $Q \in Pl_{\alpha}\mathfrak{A}$ there is a finite $\Gamma \subseteq \alpha$ such that $Qx = Qy$ whenever $x, y \in {}^{\alpha}A$ and $\Gamma 1 x = \Gamma 1 y$.*

In terms of polynomials we obtain a simple description of the subuniverse generated by a given set X:

THEOREM 0.4.10. (i) *If $x \in {}^{\alpha}A$, then*

$$Sg^{(\mathfrak{A})}\{x_{\xi} : \xi \in \alpha\} = \{Qx : Q \in Pl_{\alpha}\mathfrak{A}\}$$

or, more concisely,

$$Sg^{(\mathfrak{A})}Rgx = \mathsf{pj}_x^*(Pl_{\alpha}\mathfrak{A}).$$

(ii) *If $X \subseteq A$, then*

$$Sg^{(\mathfrak{A})}X = \bigcup\{Q^*({}^{\kappa}X) : \kappa < \omega, Q \in Pl_{\kappa}\mathfrak{A}\}$$

and, for every $\alpha \geqq \omega$,

$$Sg^{(\mathfrak{A})}X = \bigcup\{Q^*({}^{\alpha}X) : Q \in Pl_{\alpha}\mathfrak{A}\}.$$

PROOF. (i) By 0.2.14(i) and 0.3.6(i), $\mathsf{pj}_x^*(Pl_{\alpha}\mathfrak{A})$ is a subuniverse of $\mathfrak{A}$; it clearly includes Rgx, so $SgRgx \subseteq \mathsf{pj}_x^*(Pl_{\alpha}\mathfrak{A})$. Also, by 0.2.14(ii), $(({}^{\alpha A}A 1$

$\mathsf{pj}_x)^{-1})^*Sg\,Rgx$ is a subuniverse of ${}^{\alpha A}\mathfrak{A}$ including $\{{}^{\alpha}A1\,\mathsf{pj}_\xi : \xi < \alpha\}$, and hence $\mathsf{pj}_x^*(Pl_\alpha\mathfrak{A}) \subseteq Sg\,Rgx$ easily follows.

(ii): by (i) and 0.1.17(ix).

THEOREM 0.4.11. (i) *If* $\mathfrak{A} \supseteq \mathfrak{B}$, *then* $\mathfrak{Pl}_\alpha\mathfrak{A} \succcurlyeq \mathfrak{Pl}_\alpha\mathfrak{B}$.

(ii) *If* $\mathfrak{A} \succcurlyeq \mathfrak{B}$, *then* $\mathfrak{Pl}_\alpha\mathfrak{A} \succcurlyeq \mathfrak{Pl}_\alpha\mathfrak{B}$.

(iii) $\mathfrak{Pl}_\alpha\mathfrak{A} \cong \mathfrak{Pl}_\alpha({}^I\mathfrak{A})$ *for any non-empty set I.*

PROOF. (i) Let

$$F = \langle {}^{\alpha}B1\,Q : Q \in Pl_\alpha\mathfrak{A}\rangle.$$

Then clearly $F \in Hom(\mathfrak{Pl}_\alpha\mathfrak{A}, {}^{\alpha B}\mathfrak{B})$. It is easily checked that $\{{}^{\alpha}A1\,\mathsf{pj}_\xi : \xi < \alpha\} \subseteq (F^{-1})^*(Pl_\alpha\mathfrak{B}) \in Su({}^{\alpha A}\mathfrak{A})$. Hence $Pl_\alpha\mathfrak{A} \subseteq (F^{-1})^*(Pl_\alpha\mathfrak{B})$. Similarly $Pl_\alpha\mathfrak{B} \subseteq F^*(Pl_\alpha\mathfrak{A})$. Thus $F \in Ho(Pl_\alpha\mathfrak{A}, Pl_\alpha\mathfrak{B})$, and (i) holds.

(ii) By hypothesis there is an $h \in Ho(\mathfrak{A}, \mathfrak{B})$. We let

$$G = \langle\{\langle h \circ x, hQx\rangle : x \in {}^{\alpha}A\} : Q \in Pl_\alpha\mathfrak{A}\rangle,$$

and argue as in (i) to show that $G \in Ho(\mathfrak{Pl}_\alpha\mathfrak{A}, \mathfrak{Pl}_\alpha\mathfrak{B})$.

(iii) Let

$$H = \langle\langle\langle Q\langle x_\xi i : \xi < \alpha\rangle : i \in I\rangle : x \in {}^{\alpha}({}^{I}A)\rangle : Q \in Pl_\alpha\mathfrak{A}\rangle.$$

By arguing again as in (i) we show that $H \in Ho(\mathfrak{Pl}_\alpha\mathfrak{A}, \mathfrak{Pl}_\alpha({}^I\mathfrak{A}))$. Suppose now $Q, Q' \in Pl_\alpha\mathfrak{A}$ and $Q \neq Q'$. Say, $Qy \neq Q'y$ for a given $y \in {}^{\alpha}A$. Let $x = \langle\langle y_\xi : i \in I\rangle : \xi < \alpha\rangle$. Then, for any $i \in I$, $(HQx)i = Qy \neq Q'y = (HQ'x)i$, so that $HQ \neq HQ'$. Thus $H \in Is(\mathfrak{Pl}_\alpha\mathfrak{A}, \mathfrak{Pl}_\alpha({}^I\mathfrak{A}))$, and the proof is complete.

From 0.4.10(i) we see that every subuniverse of an algebra $\mathfrak{A}$ is closed under all polynomial operations over $\mathfrak{A}$; by analyzing the proof of 0.4.11(ii) we easily conclude that every homomorphism on $\mathfrak{A}$ preserves all these operations.

The two algebras, $\mathfrak{A}$ and $\mathfrak{B}$, involved in the next two theorems are not assumed to be similar.

THEOREM 0.4.12. *For any two algebras $\mathfrak{A}$ and $\mathfrak{B}$ with non-empty index sets, the following five conditions are equivalent:*

(i) *every fundamental operation of $\mathfrak{B}$ is a polynomial over $\mathfrak{A}$;*

(ii) $Pl_\kappa\mathfrak{B} \subseteq Pl_\kappa\mathfrak{A}$ *for every κ with* $0 < \kappa < \omega$;

(iii) $Pl_\alpha\mathfrak{B} \subseteq Pl_\alpha\mathfrak{A}$ *for some* $\alpha \geqq \omega$;

(iv) $Pl_\alpha\mathfrak{B} \subseteq Pl_\alpha\mathfrak{A}$ *for every* α;

(v) $Pl\mathfrak{B} \subseteq Pl\mathfrak{A}$.

PROOF. Note that each condition (i)–(v) implies that $A = B$. Clearly (iv) implies (v), (v) implies (iii), and (ii) implies (i); also, (i) implies (iv) since, by 0.4.5,

$$\{{}^{\alpha}B1\,\mathsf{pj}_\xi : \xi < \alpha\} \subseteq Pl_\alpha\mathfrak{A} \in Su({}^{\alpha B}\mathfrak{B}),$$

so that $Pl_\alpha\mathfrak{B} \subseteq Pl_\alpha\mathfrak{A}$. It remains to show that (iii) implies (ii). Since $\{{}^\kappa B 1 \mathrm{pj}_\xi : \xi < \kappa\} \subseteq Pl_\kappa\mathfrak{A}$, it suffices to show that $Pl_\kappa\mathfrak{A} \in Su({}^{\kappa B}\mathfrak{B})$. Let Q_i be any fundamental operation of $\mathfrak{B}$, say Q_i is λ-ary, and let $T = \langle T_\xi : \xi < \lambda\rangle \in {}^\lambda Pl_\kappa\mathfrak{A}$. Let $R = \langle Q_i(\lambda 1 x) : x \in {}^\alpha B\rangle$. Now $Q_i \in Pl_\lambda\mathfrak{B}$ by 0.4.4(iii), and hence $R \in Pl_\alpha\mathfrak{B} \subseteq Pl_\alpha\mathfrak{A}$ (cf. proof of 0.4.8(ii)). Let $h \in {}^\alpha Pl_\kappa\mathfrak{A}$ be such that $T \subseteq h$. Then, by 0.4.5, $C(R, h) \in Pl_\kappa\mathfrak{A}$ (we use here the notation of 0.4.5). It is easily checked that $C(R, h) = \hat{Q}_i\langle T_\xi : \xi < \lambda\rangle$, $\hat{Q}_i$ being the fundamental operation of ${}^{\kappa B}\mathfrak{B}$ corresponding to Q_i; this completes the proof of (ii).

THEOREM 0.4.13. *For any two algebras $\mathfrak{A}$ and $\mathfrak{B}$ with non-empty index sets, each of the conditions 0.4.12(i)–0.4.12(v) implies:*

(vi) $A = B$;

(vii) $Su\mathfrak{A} \subseteq Su\mathfrak{B}$;

(viii) *if* $\mathfrak{A}' \subseteq \mathfrak{A}$, $\mathfrak{B}' \subseteq \mathfrak{B}$, *and* $A' = B'$, *then* $Pl\mathfrak{B}' \subseteq Pl\mathfrak{A}'$;

(ix) $Sg^{(\mathfrak{B})}X \subseteq Sg^{(\mathfrak{A})}X$ *for every* $X \subseteq A$;

(x) $Ho\mathfrak{A} \subseteq Ho\mathfrak{B}$, *and* $Plh^*\mathfrak{B} \subseteq Plh^*\mathfrak{A}$ *for every* $h \in Ho\mathfrak{A}$;

(xi) $Co\mathfrak{A} \subseteq Co\mathfrak{B}$, *and* $Pl(\mathfrak{B}/R) \subseteq Pl(\mathfrak{A}/R)$ *for every* $R \in Co\mathfrak{A}$;

(xii) $Pl({}^I\mathfrak{B}) \subseteq Pl({}^I\mathfrak{A})$ *for every set* I;

(xiii) *if* $\mathfrak{A} = \mathsf{P}_{i\in I}\mathfrak{A}'_i$ *and* $\mathfrak{B} = \mathsf{P}_{i\in I}\mathfrak{B}'_i$, *then* $Pl\mathfrak{B}'_i \subseteq Pl\mathfrak{A}'_i$ *for every* $i \in I$.

PROOF. Clearly (i) implies (vi), and (iii) implies (vii) by 0.4.10(ii). To see that (iv) implies (viii) observe that

$$Pl\mathfrak{A}' = \bigcup_{\alpha\neq 0}\{{}^\alpha A' 1 Q : Q \in Pl_\alpha\mathfrak{A}\}$$

and

$$Pl\mathfrak{B}' = \bigcup_{\alpha\neq 0}\{{}^\alpha B' 1 Q : Q \in Pl_\alpha\mathfrak{B}\}$$

(cf. the proof of 0.4.11(i)). (i) implies the first inclusion of (x), and, since

$$Plh^*\mathfrak{A} = \bigcup_{\alpha\neq 0}\{\{\langle h\circ x, hQx\rangle : x \in {}^\alpha A\} : Q \in Pl_\alpha\mathfrak{A}\}$$

and

$$Plh^*\mathfrak{B} = \bigcup_{\alpha\neq 0}\{\{\langle h\circ x, hQx\rangle : x \in {}^\alpha B\} : Q \in Pl_\alpha\mathfrak{B}\}$$

(cf. the proof of 0.4.11(ii)), we see that (iv) implies the second inclusion of (x). Furthermore, (vii) obviously implies (ix), and (x) implies (xi) by 0.2.21.

For any α and any set I, $Q \in Pl_\alpha({}^I\mathfrak{A})$ iff there is an $O \in Pl_\alpha\mathfrak{A}$ such that $(Qx)i = O\langle x_\xi i : \xi < \alpha\rangle$ for every $x \in {}^\alpha({}^I\mathfrak{A})$ and $i \in I$; similarly with $\mathfrak{B}$ in place of $\mathfrak{A}$ (cf. the proof of 0.4.11(iii)). Thus (iv) implies (xii). Finally, it is obvious that (x) implies (xiii).

REMARKS 0.4.14. Theorems 0.4.12 and 0.4.13 clearly remain valid if all the inclusions are replaced in them by equations. (Of course, condition 0.4.12(i) should be formulated then in both directions; in (viii) the formulas $\mathfrak{A}' \subseteq \mathfrak{A}$

and $\mathfrak{B}' \subseteq \mathfrak{B}$ should be left unchanged.) We thus obtain a number of conditions which are necessary and sufficient for two algebras, possibly of different similarity classes, to have the same sets of polynomials.

We shall refer to two algebras $\mathfrak{A}$ and $\mathfrak{B}$ for which $Pl\mathfrak{A} = Pl\mathfrak{B}$ as *polynomially equivalent*. It can easily be realized that the notion of polynomial equivalence is a special case of the notion of (first-order) definitional equivalence, which was discussed in 0.1.6. In fact, the algebras $\mathfrak{A}$ and $\mathfrak{B}$ are polynomially equivalent if they are definitionally equivalent and, moreover, the fundamental operations of either of these algebras can be defined in terms of the fundamental operations of the other by means of special (first-order) sentences having the form of identities. If, for instance, $+$ is a constant in the discourse language for $\mathfrak{A}$ denoting a binary fundamental operation of $\mathfrak{A}$, this operation must be definable by means of a formula

$$\forall_{xy} x+y = \tau$$

where τ is a term in the discourse language of $\mathfrak{B}$ containing no variables different from $\boldsymbol{x}$ and $\boldsymbol{y}$. From 0.4.12 and 0.4.1 it is seen that for every algebra $\mathfrak{A}$ we can construct many different algebras polynomially equivalent with $\mathfrak{A}$, simply by adjoining any polynomials of finite rank over $\mathfrak{A}$ to the system of fundamental operations of $\mathfrak{A}$ (or else by removing some suitable operations, e.g., any projections ${}^{\kappa}A1\,\mathsf{pj}_{\xi}$, $\xi < \kappa < \omega$, which may occur in this system).

Following the lines of Remark 0.1.6, we can extend the notion of polynomial equivalence to classes of algebras. Two classes K and K^*, each consisting of similar algebras, are called polynomially equivalent if it is possible to establish a one-one correspondence between algebras in K and those in K^* in such a way that any two corresponding algebras $\mathfrak{A} \in \mathsf{K}$ and $\mathfrak{A}^* \in \mathsf{K}^*$ have the same universe, and the fundamental operations of either of these algebras are definable in terms of fundamental operations of the other by means of identities (of the form prescribed above); moreover, a fixed system of such identities can be chosen which can serve as a system of mutual definitions for fundamental operations of all pairs $\langle \mathfrak{A}, \mathfrak{A}^* \rangle$ of corresponding algebras.[1] For illustration consider the class K of all groups $\mathfrak{G} = \langle G, \cdot, {}^{-1} \rangle$. With each group $\mathfrak{G} \in \mathsf{K}$ we correlate the algebra $\mathfrak{G}^* = \langle G, : \rangle$ where : is the righthand division, i.e., $x:y$ is the (uniquely determined) element $z \in G$ such that $x = y \cdot z$. Letting

[1]) The notion of polynomial equivalence was introduced in Mal'cev [58*] under the name *rational equivalence*. For a simple purely mathematical characterization of polynomially equivalent classes of algebras, applying to the case when the classes involved are equational, see op. cit. p. 32, Theorem 6. Cf. also Felscher [68*].

$\mathsf{K}^* = \{\mathfrak{G}^* : \mathfrak{G} \in \mathsf{K}\}$, we easily see that the classes K and K^* are polynomially equivalent. As mutual definitions of fundamental operations for any pair $\langle \mathfrak{G}, \mathfrak{G}^* \rangle$ of corresponding algebras we can use in one direction the formula

$$\forall_{xy} x : y = y^{-1} \cdot x$$

and in the other direction the formulas

$$\forall_{xy} x \cdot y = y : ((x : x) : x), \quad \forall_x x^{-1} = (x : x) : x.$$

Another well known example of polynomially equivalent classes is provided by the class of Boolean algebras $\mathfrak{B} = \langle B, \vee, \wedge, - \rangle$ treated as distributive lattices with join $\vee$, meet $\wedge$, and the additional unary operation $-$ of forming complements, and the class of Boolean rings $\mathfrak{B}^* = \langle B, +, \cdot, 1 \rangle$, i.e., rings with unit in which every element is idempotent.

Two definitionally equivalent algebras or classes of algebras may differ considerably in many important algebraic aspects. For instance (as is seen from 0.1.6 and the remarks following 0.1.13) two groups $\mathfrak{G} = \langle G, \cdot, ^{-1} \rangle$ and $\mathfrak{G}' = \langle G, \cdot \rangle$, with the same universe and the same group composition, are definitionally equivalent but, in general, $Su\, \mathfrak{G} \neq Su\, \mathfrak{G}'$ and, in fact, $Su\, \mathfrak{G} \subset Su\, \mathfrak{G}'$; if K is the class of all groups $\langle G, \cdot, ^{-1} \rangle$ and K' the class of all groups $\langle G, \cdot \rangle$, then K is closed under the formation of subalgebras while K' is not. In this example, if H is a common subuniverse of two corresponding groups $\mathfrak{G}$ and $\mathfrak{G}'$, then the induced subalgebras $\mathfrak{H} = \langle H, \cdot, ^{-1} \rangle$ and $\mathfrak{H}' = \langle H, \cdot \rangle$ are still definitionally equivalent; we also have $Ho\, \mathfrak{G} = Ho\, \mathfrak{G}'$, $Co\, \mathfrak{G} = Co\, \mathfrak{G}'$, and the groups $\mathfrak{G}/R$ and $\mathfrak{G}'/R$ are definitionally equivalent for every $R \in Co\, \mathfrak{G}$; finally, both classes K and K' are closed under the formation of homomorphic images and direct products. However, these similarities between K and K' and between their corresponding members do not extend to all definitionally equivalent classes of algebras, as is easily shown by examples.

On the other hand, the connections between two polynomially equivalent algebras (or classes of algebras) are much closer. By 0.4.13 any two such algebras, $\mathfrak{A}$ and $\mathfrak{B}$, have the same subuniverses, the same homomorphisms, and the same congruence relations; two corresponding subalgebras, or homomorphic images, of $\mathfrak{A}$ and $\mathfrak{B}$ are again polynomially equivalent; in case $\mathfrak{A}$ and $\mathfrak{B}$ are direct products of two systems of algebras $\langle \mathfrak{A}'_i : i \in I \rangle$ and $\langle \mathfrak{B}'_i : i \in I \rangle$, any two corresponding factors, $\mathfrak{A}'_i$ and $\mathfrak{B}'_i$, are polynomially equivalent as well. If K and K^* are two polynomially equivalent classes of algebras, and $\mathfrak{A} = \langle \mathfrak{A}_i : i \in I \rangle$ and $\mathfrak{A}^* = \langle \mathfrak{A}^*_i : i \in I \rangle$ are two systems of correlated algebras in these classes, then one can show that $\mathsf{P}\mathfrak{A}$ and $\mathsf{P}\mathfrak{A}^*$ are again polynomially equivalent. The fact that $\langle \mathsf{K}, \mathsf{K}^* \rangle$ is a pair of polynomially equivalent classes implies that the same is true of the pairs $\langle \mathbf{S}\mathsf{K}, \mathbf{S}\mathsf{K}^* \rangle$, $\langle \mathbf{H}\mathsf{K}, \mathbf{H}\mathsf{K}^* \rangle$, and $\langle \mathbf{P}\mathsf{K}, \mathbf{P}\mathsf{K}^* \rangle$. Moreover, the formulas $\mathsf{K} = \mathbf{S}\mathsf{K}$ and $\mathsf{K}^* = \mathbf{S}\mathsf{K}^*$ are equivalent, and so are the formulas $\mathsf{K} = \mathbf{H}\mathsf{K}$ and $\mathsf{K}^* = \mathbf{H}\mathsf{K}^*$, as well as $\mathsf{K} = \mathbf{P}\mathsf{K}$ and $\mathsf{K}^* = \mathbf{P}\mathsf{K}^*$; thus, in particular, if one of the classes K and K^* is algebraically closed, then so is the other. We can sum up most of the facts just mentioned by saying that the function $F = \langle \mathfrak{A}^* : \mathfrak{A} \in \mathsf{K} \rangle$ isomorphically maps the structure $\langle \mathsf{K}, \subseteq, \preccurlyeq, \mathsf{P} \rangle$ onto the structure $\langle \mathsf{K}^*, \subseteq, \preccurlyeq, \mathsf{P} \rangle$, and that this isomorphism extends from classes K and K^* to their algebraic closures $\mathbf{HSP}\mathsf{K}$ and $\mathbf{HSP}\mathsf{K}^*$. K and K^* also have many model-theoretical properties in common; e.g., if one of these classes is an $\mathscr{E}\mathscr{C}$, a $\mathscr{U}\mathscr{C}$, or an $\mathscr{E}\mathscr{Q}\mathscr{C}$, then so is the other.

From these remarks it is clearly seen that two polynomially equivalent algebras (or classes of algebras) can be used interchangeably in various algebraic arguments. In ordinary algebraic discussions two such algebras are often identified, and such an identification — as opposed to that of two arbitrary algebras which are definitionally, but not polynomially, equivalent — does not lead in general to any confusion.

Polynomials find an application in the discussion of ideals. As we shall see below, in terms of polynomials we can formulate a useful, though rather special, condition under which, for every element z of an algebra $\mathfrak{A}$, the z-ideals function properly in $\mathfrak{A}$ (cf. 0.2.40 and 0.2.48).

DEFINITION 0.4.15. *A binary operation Q is called a* ***group-forming polynomial over*** $\mathfrak{A}$ *if $Q \in Pl_2\mathfrak{A}$ and $\langle A, Q \rangle$ is a group.*

For instance, in every Boolean algebra $\mathfrak{A} = \langle A, +, \cdot, - \rangle$ the operation $Q = \langle x \cdot -y + y \cdot -x : x, y \in A \rangle$, known as *symmetric difference*, is a group-forming polynomial.

THEOREM 0.4.16. *If $\mathfrak{A}$ has a group-forming polynomial and $R, S \in Co\,\mathfrak{A}$, then $Cg(R \cup S) = R|S = S|R$.*

PROOF. Let $\cdot$ be a group-forming polynomial over $\mathfrak{A}$, so that $\mathfrak{B} = \langle A, \cdot \rangle$ is a group. By 0.4.13 we have $Co\,\mathfrak{A} \subseteq Co\,\mathfrak{B}$ and hence $R, S \in Co\,\mathfrak{B}$. Therefore, by a familiar result from group theory, $R|S = S|R$; an application of 0.2.31 completes the proof.

In connection with 0.4.16 cf. 0.4.48 below.

THEOREM 0.4.17. *Assume that $\mathfrak{A}$ has a group-forming polynomial $\cdot$; let $^{-1}$ be, as usual, the operation of forming inverses in the group $\langle A, \cdot \rangle$, and let z be any element of A. Then the following conditions hold:*

(i) $I \in Il_z\mathfrak{A}$ *and* $R = \{\langle x, y\rangle : x, y \in A$ *and* $x \cdot y^{-1} \cdot z \in I\}$ *iff* $R \in Co\,\mathfrak{A}$ *and* $z/R = I$;

(ii) $I \in Il_z\mathfrak{A}$ *iff*

(ii′) $z \in I \subseteq A$,

(ii″) $z \cdot x^{-1} \cdot z \in I$ *whenever* $x \in I$,

and

(ii‴) $(x+x') \cdot (y+y')^{-1} \cdot z \in I$ *whenever* $x, x', y, y' \in A$ *and* $x \cdot y^{-1} \cdot z,\ x' \cdot (y')^{-1} \cdot z \in I$;

(iii) *the z-ideals function properly in* $\mathfrak{A}$;

(iv) *if* $I, J \in Il_z\mathfrak{A}$, *then* $Ig_z(I \cup J) = \{x : x \in A$, *and* $x \cdot y^{-1} \cdot z \in I$ *for some* $y \in J\}$.

PROOF. Let $\mathfrak{B} = \langle A, \cdot\rangle$ and 1 be the unit of the group $\mathfrak{B}$.

(i) If $R \in Co\,\mathfrak{A}$ and $z/R = I$, then, by 0.4.13, $R \in Co\,\mathfrak{B}$. Hence, by using elements of group theory, we show that, for any $x, y \in A$, the formulas xRy, $(x \cdot y^{-1})R1$, and $(x \cdot y^{-1} \cdot z)Rz$, i.e., $x \cdot y^{-1} \cdot z \in I$, are mutually equivalent. Therefore

(1) $$R = \{\langle x, y\rangle : x, y \in A \text{ and } x \cdot y^{-1} \cdot z \in I\}.$$

Also, $I \in Il_z\mathfrak{A}$ by 0.2.39. If, conversely, $I \in Il_z\mathfrak{A}$ and (1) holds, we have $z/S = I$ for some $S \in Co\,\mathfrak{A}$. Hence, by what has just been proved, we obtain (1) with R replaced by S. Therefore $R = S$, so that $R \in Co\,\mathfrak{A}$ and $z/R = I$.

(ii) Assume first $I \in Il_z\mathfrak{A}$. This implies (ii′) at once. Furthermore, $z/R = I$ for some $R \in Co\,\mathfrak{A}$. Hence we have (1) by (i). Since zRx whenever $x \in I$, and $(x+x')R(y+y')$ whenever xRy and $x'Ry'$, we obtain (ii″) and (ii‴) by applying (1).

Assume now, conversely, that (ii′)–(ii‴) hold, and let R be the relation defined by (1). By (ii′) and (1) we have xRx for every $x \in A$, so that the field of R is A. By applying (1), (ii″) with x replaced by $x \cdot y^{-1} \cdot z$, and then (1) again, we conclude that R is symmetric. From (ii‴) and (1) we see that R preserves $+$, i.e., xRy and $x'Ry'$ imply $(x+x'')R(y+y')$. By an easy induction this extends from $+$ to every polynomial operation over $\mathfrak{A}$ and hence, in particular, to $\cdot$. Hence, if xRy and yRz, we obtain $(x \cdot y^{-1} \cdot y)R(y \cdot y^{-1} \cdot z)$, i.e., xRz. Thus R is transitive and therefore $R \in Co\,\mathfrak{A}$. Finally, (1) implies: xRz iff $x \in I$, so that $I = z/R$ and $I \in Il_z\mathfrak{A}$.

(iii) follows immediately from (i) and 0.2.47; (iv) is a consequence of (iii), 0.2.49 and 0.4.16.

Conclusions (i), (ii), and (iv) of this theorem become simpler, of course, if we take as z the unit element of the group $\langle A, \cdot\rangle$. When referring 0.4.17 to algebras of other similarity types, condition 0.4.17(ii‴) must be modified or supplemented; for example, for algebras $\langle A, +, \circ, -\rangle$ of the similarity type $\langle 2, 2, 1\rangle$, we must supplement (ii‴) in the following way:

$$(x+x')\cdot(y+y')^{-1}\cdot z,\ (x\circ x')\cdot(y\circ y')^{-1}\cdot z,\ (-x)\cdot(-y)^{-1}\cdot z \in I$$
$$\textit{whenever } x\cdot y^{-1}\cdot z,\ x'\cdot(y')^{-1}\cdot z \in I.$$

In connection with 0.4.17(ii) recall the result of Vaught [66*] mentioned previously in 0.2.48.

REMARK 0.4.18. From 0.4.16 and 0.4.17 we see that every algebra with a group-forming polynomial has the following two important properties: all its congruence relations commute ($R|S = S|R$) and its z-ideals function properly (for every element z of the algebra). It is a common phenomenon that these two properties go together, that is, either both hold or both fail in given algebras. In general, however, it is not true that the two properties are equivalent, and not even that one of them implies the other. In fact, if $\mathfrak{A} = \langle A, +\rangle$ is the algebra of integers with an additional infinity element ∞ discussed in 0.2.48, then the congruence relations on $\mathfrak{A}$ commute, but the ∞-ideals do not function properly. In the other direction, let R and S be the equivalence relations on the set 6 associated respectively with the partitions $\{\{0, 1\}, \{2, 3\}, \{4, 5\}\}$ and $\{\{1, 2\}, \{3, 4\}, \{5, 0\}\}$; let K be the set of all unary operations Q such that $\{R, S\} \subseteq Co\langle 6, Q\rangle$; finally let $\mathfrak{B} = \langle 6, Q\rangle_{Q\in K}$. It can be shown that $Co\,\mathfrak{B} = \{6 1 Id, R, S, 6\times 6\}$. Hence the z-ideals of $\mathfrak{B}$ function properly (for $z = 0, \ldots, 5$), but $R|S \neq S|R$. (Cf. Mal'cev [54*] and Valucè [63*].)

Some further properties of polynomials and polynomial algebras can or actually will be obtained as simple consequences from the discussion of free algebras, to which we turn now. Compare here, e.g., 0.4.51.

The notion of a free algebra is of model-theoretical origin. We can characterize *a free algebra with α generators over a class* K of similar algebras as an algebra (of the same similarity class) which is generated by a set G with cardinality α such that an equation (in the discourse language of K) with κ variables, $\kappa \leqq \alpha$, is satisfied by any given κ distinct elements of G iff it is identically satisfied in every algebra of K. This characterization implies that any two free algebras over the same class K and with the same number α of generators are isomorphic. Actually, we shall choose, for each K and each α, a well defined algebra of this kind with a fixed set of generators and refer to it as *the free algebra over* K *with α generators*, in symbols $\mathfrak{Fr}_\alpha\mathsf{K}$; to express the fact that an algebra $\mathfrak{A}$ is a free algebra over K with α generators we shall, of course, use the formula $\mathfrak{A} \cong \mathfrak{Fr}_\alpha\mathsf{K}$.

The model-theoretical characterization of free algebras leads to a simple metamathematical construction of $\mathfrak{Fr}_\alpha \mathsf{K}$, which will be described at the end of this section. It seems preferable, however, to base the discussion of free algebras on a purely mathematical construction, for, in consequence, we acquire in free algebras an instrument for studying some model-theoretical notions by means of mathematical methods. The role of free algebras becomes thus analogous to that of reduced products and ultraproducts (cf. the remarks preceding 0.3.61 as well as Remarks 0.3.80 and 0.3.82), although their range of applications is much narrower: reduced products and ultraproducts are used in the study of arbitrary elementary properties of algebras (i.e., properties which can be formulated in the language of first-order logic), while free algebras are applied to the discussion of those properties which can be expressed by means of identities.

Several different mathematical methods of constructing free algebras are known. We could use, e.g., polynomials and polynomial algebras for this purpose. In fact, as we shall see below, $\mathfrak{Fr}_\alpha \mathsf{K}$ could simply be defined as the α-polynomial algebra over the direct product of algebras of K (although this construction has to be modified in case K is a class of algebras which is not a set; cf. 0.4.50 below). However, the definition of $\mathfrak{Fr}_\alpha \mathsf{K}$ which we shall actually accept here is based upon a different, conceptually simpler idea; it can be viewed as an exact mathematical paraphrase of the metamathematical construction mentioned above.[1]

DEFINITION 0.4.19. (i) *For every (non-zero cardinal) α we let*

$$Fr_\alpha = \bigcap\{X : \alpha \cup {}^2X \subseteq X\} \text{ and } \mathfrak{Fr}_\alpha = \langle Fr_\alpha, {}^2Fr_\alpha \upharpoonright Id \rangle;$$

$\mathfrak{Fr}_\alpha$ is called the ***absolutely free algebra*** *(of similarity type $\langle 2 \rangle$)* ***with*** *α* ***generators.*** *The ordinals $\xi < \alpha$ are referred to as* ***letters*** *and the elements $x \in Fr_\alpha$ as* ***words*** *of the algebra $\mathfrak{Fr}_\alpha$.*

(ii) *For every class K of algebras (of similarity type $\langle 2 \rangle$) and for every α we let*

$$Cr_\alpha \mathsf{K} = \bigcap\{R : R \in Co\mathfrak{Fr}_\alpha \text{ and } \mathfrak{Fr}_\alpha/R \in \mathbf{ISK}\};$$

[1]) The notion of a free algebra was first studied for the special case of groups; indeed, the definition of an abstract group first given by Cayley in the 1850's was metamathematical — using generators and defining relations — and clearly implied the notion of a free group as a special case, although free groups were evidently first studied and named by Jakob Nielson in 1921. The general definition, patterned after that for groups and hence metamathematical in nature, was given in Birkhoff [35*]. The equivalent characterization in terms of polynomial algebras can be found in McKinsey-Tarski [44*], while the characterization in terms of extending to homomorphisms is due to the Bourbaki school (cf. Samuel [48*]). The present mathematical formulation, while closely related to the metamathematical definition, appears to be new; it is related to constructions given in Felscher [65*] and Kerkhoff [65*].

we also let

$$Fr_\alpha \mathsf{K} = Fr_\alpha / Cr_\alpha \mathsf{K} \text{ and } \mathfrak{Fr}_\alpha \mathsf{K} = \mathfrak{Fr}_\alpha / Cr_\alpha \mathsf{K}.$$

$\mathfrak{Fr}_\alpha \mathsf{K}$ *is called the* ***free algebra over*** K ***with*** α ***generators.***

In case $\mathsf{K} = \{\mathfrak{A}\}$ *we shall write* $Cr_\alpha \mathfrak{A}$, $Fr_\alpha \mathfrak{A}$, $\mathfrak{Fr}_\alpha \mathfrak{A}$ *instead of* $Cr_\alpha\{\mathfrak{A}\}$, *etc.*

REMARKS 0.4.20. According to 0.4.19(i), Fr_α is the smallest set which contains all ordinals $\xi < \alpha$ and is closed under the operation of forming ordered pairs; $\mathfrak{Fr}_\alpha$ is the algebra $\langle A, +\rangle$ where $A = Fr_\alpha$ and $x+y = \langle x, y\rangle$ for all $x, y \in A$.

The definitions of Fr_α and $\mathfrak{Fr}_\alpha$ must of course be modified when applied to algebras of other similarity types. In general, for algebras of the similarity type $\langle \rho_i : i \in I\rangle$ (cf. 0.1.5) we define Fr_α as the intersection of all sets X such that $\alpha \subseteq X$, and $\langle i, y\rangle \in X$ whenever $i \in I$ and $y \in {}^{\rho_i}X$; we let $\mathfrak{Fr}_\alpha = \langle Fr_\alpha, Q_i\rangle_{i \in I}$ where Q_i is a ρ_i-termed operation on Fr_α such that $Q_i y = \langle i, y\rangle$ for all $i \in I$, $y \in {}^{\rho_i}Fr_\alpha$. For the special case of algebras of similarity type $\langle 2\rangle$ we have simplified this definition, but the reader should have no trouble in carrying through the development in the general case.

From the formula defining $Cr_\alpha \mathsf{K}$ in 0.4.19(ii) it follows immediately by 0.2.24(ii) that $Cr_\alpha \mathsf{K} \in Co\,\mathfrak{Fr}_\alpha$. This justifies the definition of $Fr_\alpha \mathsf{K}$ and $\mathfrak{Fr}_\alpha \mathsf{K}$ given in the later part of 0.4.19(ii). Just as Fr_α and $\mathfrak{Fr}_\alpha$, the notation $Fr_\alpha \mathsf{K}$ and $\mathfrak{Fr}_\alpha \mathsf{K}$ will be used for classes K of algebras of arbitrary similarity types.

THEOREM 0.4.21. *If* K *is the similarity class of* $\mathfrak{Fr}_\alpha$, *then* $Cr_\alpha \mathsf{K} = Fr_\alpha 1 Id$ *and* $\mathfrak{Fr}_\alpha \mathsf{K} \cong \mathfrak{Fr}_\alpha$.

THEOREM 0.4.22. *For every class* K *of algebras and every* α *we have*

$$\mathfrak{Fr}_\alpha \mathsf{K} \in \mathbf{SPS}_{\alpha^+} \mathsf{K} \subseteq \mathbf{SP}\mathsf{K}.$$

PROOF: by 0.3.12 and 0.3.46.

DEFINITION 0.4.23. *Given an algebra* $\mathfrak{A}$ *and a class* K *of algebras, we say that a set* X K-***freely generates*** $\mathfrak{A}$ *if* $\mathfrak{A} = Sg\,X$ *and, for every* $\mathfrak{B} \in \mathsf{K}$ *and* $f \in {}^X B$, *there is an* $h \in Hom(\mathfrak{A}, \mathfrak{B})$ *such that* $h \supseteq f$.

THEOREM 0.4.24. *If* X K-*freely generates* $\mathfrak{A}$, *then, for every* $\mathfrak{B} \in \mathsf{K}$ *and* $f \in {}^X B$, *there is just one* $h \in Hom(\mathfrak{A}, \mathfrak{B})$ *such that* $h \supseteq f$.

PROOF: by 0.2.14(iii).

THEOREM 0.4.25. (i) *If* $\{X_i : i \in I\}$ *and* $\{\mathfrak{A}_i : i \in I\}$ *are each directed by the relation* $\subseteq$ *and* X_i K-*freely generates* $\mathfrak{A}_i$ *for each* $i \in I$, *then* $\bigcup\{X_i : i \in I\}$ K-*freely generates* $\bigcup\{\mathfrak{A}_i : i \in I\}$.

(ii) *In particular, if* $\mathfrak{Sg}\, X = \mathfrak{A}$ *and, for each finite subset* Y *of* X, Y K*-freely generates* $\mathfrak{Sg}\, Y$, *then* X K*-freely generates* $\mathfrak{A}$.

PROOF: by 0.1.27, 0.4.24.

THEOREM 0.4.26. (i) *If* X K*-freely generates* $\mathfrak{A}$ *and* $\mathsf{K} \supseteq \mathsf{L}$, *then* X L*-freely generates* $\mathfrak{A}$.

(ii) *Let* Q *be any composition of some or all of the operations* **S**, **H**, *and* **P**. *Then* X K*-freely generates* $\mathfrak{A}$ *iff* X QK*-freely generates* $\mathfrak{A}$.

PROOF. (i) is obvious. To prove (ii), in view of (i), 0.2.18(i), and 0.4.23 it suffices to show that, if X K-freely generates $\mathfrak{A}$, then X **H**K- and **P**K-freely generates $\mathfrak{A}$. Suppose first that $g \in Ho(\mathfrak{B}, \mathfrak{C})$, $\mathfrak{B} \in \mathsf{K}$, and $f \in {}^X C$. By the axiom of choice there is an $f' \in {}^X B$ such that $g \circ f' = f$. Since X K-freely generates $\mathfrak{A}$, f' can be extended to a (uniquely determined) $h \in Hom(\mathfrak{A}, \mathfrak{B})$. Then $g \circ h \in Hom(\mathfrak{A}, \mathfrak{C})$ and $g \circ h \supseteq f$. Suppose now that $\mathfrak{C} \in {}^I\mathsf{K}$, $\mathfrak{B} = \mathsf{P}\mathfrak{C}$, and $f \in {}^X B$. Then, for each $i \in I$, we have $\mathsf{pj}_i \circ f \in {}^X C_i$, and hence there exists a $g_i \in Hom(\mathfrak{A}, \mathfrak{C}_i)$ such that $g_i \supseteq \mathsf{pj}_i \circ f$. Therefore, by 0.3.6(ii),

$$h = \langle \langle g_i x : i \in I \rangle : x \in A \rangle \in Hom(\mathfrak{A}, \mathfrak{B}),$$

and clearly $h \supseteq f$.

THEOREM 0.4.27. (i) *If* α *is any non-zero cardinal and* K *is the similarity class of algebras of type* $\langle 2 \rangle$, *then* α K*-freely generates* $\mathfrak{Fr}_\alpha$.

(ii) *More generally, if* K *is any class of algebras, then* $\alpha/Cr_\alpha\mathsf{K}$ (*i.e.*, $\{\xi/Cr_\alpha\mathsf{K} : \xi < \alpha\}$) K*-freely generates* $\mathfrak{Fr}_\alpha\mathsf{K}$.

PROOF. (i) Assume $\mathfrak{B} = \langle B, + \rangle \in \mathsf{K}$ and $f \in {}^\alpha B$. Since $\alpha \subseteq Fr_\alpha$, f is a subset of $Fr_\alpha \times B$. Therefore, let $h = Sg^{(\mathfrak{Fr}_\alpha \times \mathfrak{B})} f$, so that $h \supseteq f$. Clearly $z \in h$ iff either $z \in f$ or else for some $x', x'' \in Fr_\alpha$ and $y', y'' \in B$ we have $z = \langle \langle x', x'' \rangle, y' + y'' \rangle$ and $\langle x', y' \rangle, \langle x'', y'' \rangle \in h$. Using this and letting

$$A = \{x : x \in Fr_\alpha \text{ and there is just one } y \in B \text{ such that } \langle x, y \rangle \in h\},$$

we easily show that $\alpha \subseteq A$, and $\langle x', x'' \rangle \in A$ whenever $x', x'' \in A$. Hence $A = Fr_\alpha$ and therefore h is a function with $Do\,h = Fr_\alpha$. By applying 0.3.47(i) we obtain at once $h \in Hom(\mathfrak{Fr}_\alpha, \mathfrak{B})$.[1)]

(ii) Assume now $\mathfrak{B} \in \mathsf{K}$ and $f \in {}^{\alpha/Cr_\alpha\mathsf{K}} B$. Then $\alpha 1 (f \circ (Cr_\alpha\mathsf{K})^\star) \in {}^\alpha B$ whence, by (i), there is an $h \in Hom(\mathfrak{Fr}_\alpha, \mathfrak{B})$ such that $h \supseteq \alpha 1 (f \circ (Cr_\alpha\mathsf{K})^\star)$. Then, by 0.2.23(i), $\mathfrak{Fr}_\alpha/(h|h^{-1}) \in \mathbf{ISK}$ and hence $Cr_\alpha\mathsf{K} \subseteq h|h^{-1}$; by 0.2.23(iii), the relation $k = \{\langle a/Cr_\alpha\mathsf{K}, ha \rangle : a \in Fr_\alpha\}$ is a member of $Ho(\mathfrak{Fr}_\alpha\mathsf{K})$. Clearly $k \supseteq f$ and $k \in Hom(\mathfrak{Fr}_\alpha\mathsf{K}, \mathfrak{B})$.

1) The original proof of 0.4.27(i) was somewhat more involved. The present version, based upon 0.3.47(i), was suggested to the authors by Karl-Heinz Diener.

THEOREM 0.4.28. *If X K-freely generates $\mathfrak{A}$, Y K-freely generates $\mathfrak{B}$, $f \in {}^XY$, $f^{-1} \in {}^YX$, and $\mathfrak{A}, \mathfrak{B} \in$ **HSP**K, then there is just one $h \in Is(\mathfrak{A}, \mathfrak{B})$ such that $h \supseteq f$.*

PROOF. Choose $h \in Hom(\mathfrak{A}, \mathfrak{B})$ and $k \in Hom(\mathfrak{B}, \mathfrak{A})$ such that $h \supseteq f$, $k \supseteq f^{-1}$. By 0.2.14(iii) we have $k \circ h = A \upharpoonright Id$ and $h \circ k = B \upharpoonright Id$. Hence $h \in Is(\mathfrak{A}, \mathfrak{B})$. The uniqueness of h follows from 0.4.24.

THEOREM 0.4.29. *The following two conditions are equivalent:*

(i) $\mathfrak{A} \cong \mathfrak{Fr}_\alpha \mathsf{K}$;

(ii) $\mathfrak{A} \in$ **HSP**K *and there exists a set X such that X K-freely generates $\mathfrak{A}$ and $|X| = |\alpha/Cr_\alpha \mathsf{K}|$.*

In (ii) **HSP**K *can be replaced by* **SP**K.

PROOF. (i) implies (ii) by 0.4.22 and 0.4.27(ii). (ii) implies (i) because of the implication just established, together with 0.4.28. The same argument can be used after the indicated replacement.

Theorem 0.4.29 may be called the *fundamental theorem on free algebras.* It provides a very useful and easily applicable characteristic property of free algebras and their isomorphic images, which yields interesting and important consequences. In particular, this theorem leads to various methods of constructing free algebras which are different from the method used in the original definition 0.4.19.

THEOREM 0.4.30. *Let* K *be the similarity class containing the algebra $\mathfrak{A}$ and let $X \subseteq A$. In order for X to* K*-freely generate $\mathfrak{A}$ it is necessary and sufficient that it satisfy the conditions*

(i) $\mathfrak{A} = \mathfrak{Sg} X$,

(ii) *if $x, y \in A$, then $x+y \notin X$,*

and, moreover, that $\mathfrak{A}$ satisfy the condition

(iii) *if $x, y, x', y' \in A$ and $x+y = x'+y'$, then $x = x'$ and $y = y'$.*

PROOF. Assume X K-freely generates $\mathfrak{A}$. Let $\alpha = |X|$ and f be an α-termed sequence without repetitions and with $Rgf = X$. Then by 0.4.27(i) and 0.4.28 there is an $h \in Is(\mathfrak{Fr}_\alpha, \mathfrak{A})$ such that $f \subseteq h$. Thus $h^*\alpha = X$ and hence conditions (i)–(iii) hold.

Now assume (i)–(iii) hold and take $\mathfrak{B} = \langle B, + \rangle \in \mathsf{K}$ and $f \in {}^XB$. Since $X \subseteq A$, f is a subset of $A \times B$. Therefore, let $h = Sg^{(\mathfrak{A} \times \mathfrak{B})}f$, so that $h \supseteq f$. Clearly, $z \in h$ iff either $z \in f$ or else for some $a, a' \in A$ and $b, b' \in B$ we have $z = \langle a +^{(\mathfrak{A})} a', b +^{(\mathfrak{B})} b' \rangle$ and $\langle a, b \rangle, \langle a', b' \rangle \in h$. Using this fact and letting

$$C = \{a : a \in A \text{ and there is just one } b \in B \text{ such that } \langle a, b \rangle \in h\},$$

we conclude respectively from (ii) and (iii) that $X \subseteq C$ and $C \in Su\mathfrak{A}$. Hence

$C = A$ by (i), and thus h is a function with $Doh = A$. By applying 0.3.47(i) we obtain at once $h \in Hom(\mathfrak{A}, \mathfrak{B})$. Therefore, X K-freely generates $\mathfrak{A}$.

Compare the second half of this proof with the proof of 0.4.27(i).

It is easily seen that, if K is the whole similarity class of an algebra $\mathfrak{A}$, then there is at most one subset X of A which K-freely generates $\mathfrak{A}$; in fact, X if it exists must be the set of all those elements of A which do not belong to the range of $+$.

THEOREM 0.4.31. *$\mathfrak{A} \cong \mathfrak{Fr}_\alpha$ iff there is a set X with $|X| = \alpha$ satisfying conditions 0.4.30(i)–(iii).*

PROOF: by 0.4.27(i), 0.4.29, and 0.4.30.

Theorems 0.4.30 and 0.4.31 must be appropriately modified when applied to other similarity classes. For example, when applying it to algebras $\langle A, +, -\rangle$ of similarity type $\langle 2, 1\rangle$, we have to supplement 0.4.30(i)–(iii) by the following conditions:

(ii′) *if $x \in A$, then $-x \notin X$;*
(iii′) *if $x, x' \in A$ and $-x = -x'$, then $x = x'$;*
(iv) *if $x, y, z \in A$, then $x+y \neq -z$.*

As an interesting application of 0.4.30 we give

THEOREM 0.4.32. *If $1 \leqq \beta \leqq \omega$ and $2 \leqq \gamma < \omega$, then ${}^\gamma\mathfrak{Fr}_\beta \cong \mathfrak{Fr}_\omega$. More generally, if $2 \leqq \gamma$ and $(\beta \cup \omega)^\gamma = \alpha$, then ${}^\gamma\mathfrak{Fr}_\beta \cong \mathfrak{Fr}_\alpha$.*

PROOF. We apply 0.4.30 with $\mathfrak{A} = {}^\gamma\mathfrak{Fr}_\beta$. Denoting by $+$ the fundamental operation of $\mathfrak{A}$, we see that 0.4.30(iii) holds. The set X satisfying 0.4.30(i),(ii) is defined by the formula

$$X = \{x : x \in {}^\gamma Fr_\beta, \text{ and } x_\xi \in \beta \text{ for some } \xi \in \gamma\}.$$

To verify 0.4.30(i) let B be the set of all $b \in Fr_\beta$ such that, for every $a \in A$, $b \in Rga$ implies $a \in Sg X$; then notice that $\beta \subseteq B \in Su\mathfrak{Fr}_\beta$. Finally, it is easy to show that $|X| = \alpha$.

When applying this theorem to algebras of an arbitrary similarity type $\langle \rho_i : i \in I\rangle$ with $I \neq 0$, we have to replace ω (in all its occurrences except the one in the formula $2 \leqq \gamma < \omega$) by $|I| \cup \omega$ or, what amounts to the same, by $|Fr_1|$.

Theorem 0.4.32 provides examples of algebras $\mathfrak{A}$ which do not possess property (β) discussed in 0.3.33, i.e., for which there exist non-isomorphic algebras $\mathfrak{B}$, $\mathfrak{C}$ such that $\mathfrak{A} \cong \mathfrak{B} \times \mathfrak{B} \cong \mathfrak{C} \times \mathfrak{C}$. In fact, to obtain such examples we let $\mathfrak{A} = \mathfrak{B} = \mathfrak{Fr}_\beta \times \mathfrak{Fr}_\beta$ and $\mathfrak{C} = \mathfrak{Fr}_\beta$, where $1 \leqq \beta < \omega$. For algebras of type $\langle 1\rangle$ and $\beta = 1$ the first part of 0.4.32 was found by J. C. C. McKinsey who used it just for the purpose of constructing an algebra for which (β) fails to hold. (He actually considered, not $\mathfrak{Fr}_1$, but the algebra of natural numbers with the successor operation, $\langle\omega, S\rangle$, which is isomorphic with $\mathfrak{Fr}_1$ by 0.4.30; cf. 0.3.33.) The generalization of McKinsey's result to algebras of arbitrary types and to arbitrary β with $1 \leqq \beta \leqq \omega$ was recently obtained and communicated to the authors by Karl-Heinz Diener.

THEOREM 0.4.33. (i) *If* $|A| = 1$ *for every* $\mathfrak{A} \in \mathsf{K}$, *or if* $\alpha = 1$ *and* $\{a\} \in Su\mathfrak{A}$ *for every* $\mathfrak{A} \in \mathsf{K}$ *and every* $a \in A$, *then* $Cr_\alpha\mathsf{K} = Fr_\alpha \times Fr_\alpha$ *and* $|\alpha/Cr_\alpha\mathsf{K}| = |Fr_\alpha\mathsf{K}| = 1$.

(ii) *If* $|A| > 1$ *for some* $\mathfrak{A} \in \mathsf{K}$ *and* $\alpha > 1$, *or else if* $\{a\} \notin Su\mathfrak{A}$ *for some* $\mathfrak{A} \in \mathsf{K}$ *and some* $a \in A$, *then* $Cr_\alpha\mathsf{K} \neq Fr_\alpha \times Fr_\alpha$, *the function* $\langle\xi/Cr_\alpha\mathsf{K} : \xi \in \alpha\rangle$ *is one-one*, $|\alpha/Cr_\alpha\mathsf{K}| = \alpha$, *and* $\alpha \leqq |Fr_\alpha\mathsf{K}| \leqq \alpha \cup \omega$ (*whence* $|Fr_\alpha\mathsf{K}| = \alpha$ *in case* $\alpha \geqq \omega$).

PROOF. (i) is clear. To prove (ii), first consider the case when $\alpha > 1$ and $|A| > 1$ for a given $\mathfrak{A} \in \mathsf{K}$. Then, obviously, for any two distinct $\xi, \eta < \alpha$ there is an $f \in {}^\alpha A$ such that $f\xi \neq f\eta$. In view of 0.4.27(i), f can be extended to a (uniquely determined) $h \in Hom(\mathfrak{Fr}_\alpha, \mathfrak{A})$. Clearly $Cr_\alpha\mathsf{K} \subseteq h|h^{-1}$ and $\langle\xi, \eta\rangle \notin h|h^{-1}$. Since ξ and η are arbitrary, we conclude that $\langle\xi/Cr_\alpha\mathsf{K} : \xi \in \alpha\rangle$ is one-one. An application of 0.1.19 completes the proof.

Now assume that $\{a\} \notin Su\mathfrak{A}$ for a given $\mathfrak{A} \in \mathsf{K}$ and a given $a \in A$. Letting $f = (\alpha/Cr_\alpha\mathsf{K}) \times \{a\}$ we obtain $f \in {}^{\alpha/Cr_\alpha\mathsf{K}}A$ and $Rgf = \{a\}$. By applying 0.4.27(ii) we extend f to an $h \in Hom(\mathfrak{Fr}_\alpha\mathsf{K}, \mathfrak{A})$. Since $h^*Fr_\alpha\mathsf{K} \in Su\mathfrak{A}$, we have $h^*Fr_\alpha\mathsf{K} \supset \{a\}$; therefore $\mathfrak{Fr}_\alpha\mathsf{K}$ is not a one-element algebra, and hence $Cr_\alpha\mathsf{K} \neq Fr_\alpha \times Fr_\alpha$. Furthermore, the function $\langle\xi/Cr_\alpha\mathsf{K} : \xi \in \alpha\rangle$ is one-one (obviously in case $\alpha = 1$, and by what was shown above in case $\alpha > 1$); together with 0.1.19 this implies the remaining conclusions of (ii).

REMARK 0.4.34. Notice that the formula $\{a\} \in Su\mathfrak{A}$ when applied to an algebra $\mathfrak{A} = \langle A, +\rangle$ means simply that $a+a = a$. To extend 0.4.33(ii) to algebras of arbitrary similarity types apply 0.1.20 with $|X| = \alpha$.

THEOREM 0.4.35. *If* $f \in Ho(\mathfrak{Fr}_\alpha\mathsf{K}, \mathfrak{B})$, $g \in Ho\,(\mathfrak{A}, \mathfrak{B})$, *and* $\mathfrak{A} \in \mathbf{HSP}\mathsf{K}$, *then there is an* $h \in Hom(\mathfrak{Fr}_\alpha\mathsf{K}, \mathfrak{A})$ *such that* $g \circ h = f$.

PROOF. By the axiom of choice there is a $k \in {}^{\alpha/Cr_\alpha\mathsf{K}}A$ such that $g \circ k = (\alpha/Cr_\alpha\mathsf{K})1f$. The desired homomorphism h is obtained by applying 0.4.26(ii) and 0.4.27(ii) (cf. 0.2.14(iii)).

THEOREM 0.4.36. *For every class* K *of algebras and every* α *we have*

$$Cr_\alpha\mathsf{K} = Cr_\alpha\mathsf{QK} \text{ and } \mathfrak{Fr}_\alpha\mathsf{K} = \mathfrak{Fr}_\alpha\mathsf{QK}$$

where Q *is any composition of some or all of the operations* $\mathbf{S}$, $\mathbf{H}$, $\mathbf{P}$, *and* $\mathbf{S}_\beta$ *with* $\beta > \alpha$.

PROOF. It clearly suffices to establish two special formulas:

(1) $Cr_\alpha\mathsf{K} \subseteq Cr_\alpha\mathbf{H}\mathsf{K}$ and (2) $Cr_\alpha\mathsf{K} \subseteq Cr_\alpha\mathbf{P}\mathsf{K}$.

To obtain (1) notice first that the formula amounts to

$$\bigcap\{R: R \in Co\mathfrak{Fr}_\alpha, \mathfrak{Fr}_\alpha/R \in \mathbf{ISH}\mathsf{K}\} \supseteq \bigcap\{S: S \in Co\mathfrak{Fr}_\alpha, \mathfrak{Fr}_\alpha/S \in \mathbf{IS}\mathsf{K}\}.$$

Thus consider an $R \in Co\mathfrak{Fr}_\alpha$ with $\mathfrak{Fr}_\alpha/R \in \mathbf{ISH}\mathsf{K} \subseteq \mathbf{HS}\mathsf{K}$. Hence, for some g, $\mathfrak{A}$, and $\mathfrak{B}$ we have $\mathfrak{A} \subseteq \mathfrak{B} \in \mathsf{K}$ and $g \in Ho(\mathfrak{A}, \mathfrak{Fr}_\alpha/R)$. By 0.4.35 there is an $h \in Hom(\mathfrak{Fr}_\alpha, \mathfrak{A})$ such that $g \circ h = R^\bigstar$. By letting $S = h|h^{-1}$ we obtain $S \subseteq R$, $S \in Co\mathfrak{Fr}_\alpha$, and $\mathfrak{Fr}_\alpha/S \cong h^*\mathfrak{Fr}_\alpha \subseteq \mathfrak{B} \in \mathsf{K}$, whence $\mathfrak{Fr}_\alpha/S \in \mathbf{IS}\mathsf{K}$. Since the argument holds for an arbitrary R, the proof of (1) is complete.

To prove (2) consider any $R \in Co\mathfrak{Fr}_\alpha$ such that $\mathfrak{Fr}_\alpha/R \in \mathbf{ISP}\mathsf{K}$. Hence, for some f we have $f \in Ism(\mathfrak{Fr}_\alpha/R, \mathsf{P}\mathfrak{A})$ where $\mathfrak{A} \in {}^I\mathsf{K}$. By letting

$$S_i = (\mathsf{pj}_i \circ f \circ R^\bigstar) \mid (\mathsf{pj}_i \circ f \circ R^\bigstar)^{-1},$$

we obtain $R = \bigcap\{S_i : i \in I\}$, $\{S_i : i \in I\} \subseteq Co\mathfrak{Fr}_\alpha$, and $\mathfrak{Fr}_\alpha/S_i \cong (\mathsf{pj}_i \circ f)^*(\mathfrak{Fr}_\alpha/R) \subseteq \mathfrak{A}_i$ whence $\mathfrak{Fr}_\alpha/S_i \in \mathbf{IS}\mathsf{K}$, for each $i \in I$. The observation that, again, this argument holds for an arbitrary R completes the proof of (2).

THEOREM 0.4.37. *Let* K *and* L *be any classes of algebras,* α *any non-zero cardinal, and* Q_1, Q_2, Q_3, *and* Q_4 *any compositions of some or all of the operations* $\mathbf{S}$, $\mathbf{H}$, $\mathbf{P}$, *and* $\mathbf{S}_\beta$ *with* $\beta > \alpha$.

(i) *If* $\mathsf{Q}_1\mathsf{K} \supseteq \mathsf{Q}_2\mathsf{L}$, *then* $Cr_\alpha\mathsf{K} \subseteq Cr_\alpha\mathsf{L}$ *and* $\mathfrak{Fr}_\alpha\mathsf{K} \succcurlyeq \mathfrak{Fr}_\alpha\mathsf{L}$.

(ii) *If* $\mathsf{Q}_1\mathsf{K} = \mathsf{Q}_2\mathsf{L}$ *or, more generally, if* $\mathsf{Q}_1\mathsf{K} \supseteq \mathsf{Q}_2\mathsf{L}$ *and* $\mathsf{Q}_3\mathsf{L} \supseteq \mathsf{Q}_4\mathsf{K}$, *then* $Cr_\alpha\mathsf{K} = Cr_\alpha\mathsf{L}$ *and* $\mathfrak{Fr}_\alpha\mathsf{K} = \mathfrak{Fr}_\alpha\mathsf{L}$.

THEOREM 0.4.38. (i) *If* K *is any set of algebras, then*

$$\mathfrak{Fr}_\alpha\mathsf{K} = \mathfrak{Fr}_\alpha(\mathsf{P}\langle\mathfrak{B} : \mathfrak{B} \in \mathsf{K}\rangle).$$

(ii) *If* K *is any class of algebras, and* L *and* M *are the sets defined in 0.3.19(ii), then*

$$\mathfrak{Fr}_\alpha\mathsf{K} = \mathfrak{Fr}_\alpha(\mathsf{P}\langle\mathfrak{B} : \mathfrak{B} \in \mathsf{L}\rangle) = \mathfrak{Fr}_\alpha(\mathsf{P}\langle\mathfrak{B} : \mathfrak{B} \in \mathsf{M}\rangle).$$

PROOF: (i) directly by 0.4.37(ii), since $\mathsf{P}\langle\mathfrak{B} : \mathfrak{B} \in \mathsf{K}\rangle \in \mathbf{P}\mathsf{K}$ and $\mathsf{K} \subseteq \mathbf{H}(\mathsf{P}\langle\mathfrak{B} : \mathfrak{B} \in \mathsf{K}\rangle)$; (ii) by 0.3.19(ii) and 0.4.37(ii).

THEOREM 0.4.39. $\mathbf{S}_{\alpha^+}\mathsf{K} \subseteq \mathbf{H}\mathfrak{Fr}_\alpha\mathsf{K}$ *for every class* K *of algebras and every* α.

PROOF: by 0.4.27 and 0.4.33.

THEOREM 0.4.40. *For every class* K *of algebras and every* α *we have*

(i) $\mathbf{HSP}\mathfrak{Fr}_\alpha\mathsf{K} = \mathbf{HSPS}_{\alpha^+}\mathsf{K} \subseteq \mathbf{HSP}\mathsf{K}$;

(ii) $\mathbf{HSP}\mathfrak{Fr}_\alpha\mathsf{K} = \mathbf{HSP}\mathsf{K}$ *in case* $\alpha \geqq \omega$.

PROOF: (i) by 0.3.15, 0.4.22, and 0.4.39; (ii) by (i) and 0.3.18.

THEOREM 0.4.41. *For every class* K *of algebras,*

$$\mathsf{K} \subseteq \bigcup_{\beta \neq 0} \mathbf{H}\mathfrak{Fr}_\beta\mathsf{K} = \mathbf{HSP}\mathsf{K}.$$

PROOF: by 0.4.26, 0.4.27, 0.4.39 and 0.4.40(i).

THEOREM 0.4.42. *For any classes* K *and* L *of algebras and every* α, *the following conditions are equivalent:*

(i) $Cr_\alpha\mathsf{K} \subseteq Cr_\alpha\mathsf{L}$,

(ii) $\mathfrak{Fr}_\alpha\mathsf{K} \succcurlyeq \mathfrak{Fr}_\alpha\mathsf{L}$,

(iii) $\mathbf{HSP}\mathsf{K} \supseteq \mathbf{S}_{\alpha^+}\mathsf{L}$.

In case $\alpha \geqq \omega$, *these conditions are also equivalent to*

(iv) $\mathbf{HSP}\mathsf{K} \supseteq \mathsf{L}$.

PROOF. (i)–(iii) are equivalent by 0.4.22, 0.4.37(i), and 0.4.39. Clearly (iv) implies (iii). If (iii) holds and $\alpha \geqq \omega$, then, by 0.3.18, $\mathsf{L} \subseteq \mathbf{HSP}\mathsf{L} = \mathbf{HSPS}_{\alpha^+}\mathsf{L} \subseteq \mathbf{HSP}\mathsf{K}$.

THEOREM 0.4.43. *For any classes* K *and* L *of algebras and every* α, *the following conditions are equivalent:*

(i) $Cr_\alpha\mathsf{K} = Cr_\alpha\mathsf{L}$,

(ii) $\mathfrak{Fr}_\alpha\mathsf{K} = \mathfrak{Fr}_\alpha\mathsf{L}$,

(iii) $\mathfrak{Fr}_\alpha\mathsf{K} \cong \mathfrak{Fr}_\alpha\mathsf{L}$,

(iv) $\mathfrak{Fr}_\alpha\mathsf{K} \succcurlyeq \mathfrak{Fr}_\alpha\mathsf{L} \succcurlyeq \mathfrak{Fr}_\alpha\mathsf{K}$,

(v) $\mathbf{HSP}\mathsf{K} \supseteq \mathbf{S}_{\alpha^+}\mathsf{L}$ *and* $\mathbf{HSP}\mathsf{L} \supseteq \mathbf{S}_{\alpha^+}\mathsf{K}$.

In case $\alpha \geqq \omega$, *these conditions are also equivalent to*

(vi) $\mathbf{HSP}\mathsf{K} = \mathbf{HSP}\mathsf{L}$.

PROOF. (i), (iv), and (v) are equivalent by 0.4.42. Obviously (i) implies (ii), (iii) implies (iv), and (ii) implies (iii). If $\alpha \geqq \omega$, (vi) is equivalent to (i)–(v) by 0.4.42 and 0.3.15.

Concerning the possibility of replacing 0.4.40(i),(ii), 0.4.42(iii),(iv), and 0.4.43(v),(vi) by various related formulas compare 0.3.15–0.3.16.

THEOREM 0.4.44. *If any one of the conditions 0.4.42(i)–(iii) holds for a given* α, *it also holds for every* $\beta \leqq \alpha$ *and, in case* $\alpha \geqq \omega$, *for any* β *whatsoever. The same applies to conditions 0.4.43(i)–(v).*

THEOREM 0.4.45. *For every algebra* $\mathfrak{A}$ *and every* α,

$$\mathfrak{Fr}_\alpha \succcurlyeq \mathfrak{Pl}_\alpha\mathfrak{A}.$$

More specifically, there is just one $h \in Ho(\mathfrak{Fr}_\alpha, \mathfrak{Pl}_\alpha\mathfrak{A})$ *such that* $h\xi = {}^\alpha A 1\, \mathsf{pj}_\xi$ *for every* $\xi < \alpha$.

PROOF: by 0.4.1, 0.4.27, and 0.4.24.

In view of 0.4.45 we introduce the following notation:

DEFINITION 0.4.46. *Given an algebra* $\mathfrak{A}$ *and a cardinal* $\alpha \neq 0$, $Pd_\alpha^{(\mathfrak{A})}$ *denotes the unique* $h \in Ho(\mathfrak{Fr}_\alpha, \mathfrak{Pl}_\alpha\mathfrak{A})$ *such that* $h\xi = {}^\alpha A 1\, \mathsf{pj}_\xi$ *for every* $\xi < \alpha$. *For any* $u \in Fr_\alpha$, *we refer to* $Pd_\alpha^{(\mathfrak{A})}u$ *as the* ***polynomial in*** α ***variables determined by the word*** u.

REMARKS 0.4.47. The notation $Pd_\alpha^{(\mathfrak{A})}$ enables us to relate polynomials over different algebras. Two polynomials over two (similar) algebras, Q over $\mathfrak{A}$ and Q' over $\mathfrak{A}'$, can be called *conjugated polynomials* if there is an α and a word $u \in Fr_\alpha$ for which $Q = Pd_\alpha^{(\mathfrak{A})}u$ and $Q' = Pd_\alpha^{(\mathfrak{A}')}u$ (so that, consequently, both Q and Q' are polynomials in α variables). For example let $\mathfrak{A} = \langle A, +\rangle$, $\mathfrak{A}' = \langle A', +'\rangle$, and let Q and Q' be polynomials determined by the formulas

$$Q(x, y, z) = (x+z)+(y+z) \text{ for } x, y, z \in A,$$

$$Q'(x, y, z) = (x+'z)+'(y+'z) \text{ for } x, y, z \in A'.$$

Q and Q' are indeed conjugated polynomials since by letting $\alpha = 3$ and

$$u = \langle\langle 0, 2\rangle, \langle 1, 2\rangle\rangle$$

we obtain $u \in Fr_\alpha$, $Q = Pd_\alpha^{(\mathfrak{A})}u$, and $Q' = Pd_\alpha^{(\mathfrak{A}')}u$. Two such polynomials are frequently referred to as *polynomials of the same form*, or even as *identical polynomials* (over different algebras); we prefer the term "conjugated polynomials" since it is more neutral and not misleading.

The notion of conjugated polynomials is implicitly involved in the notion of polynomially equivalent classes of algebras, which was discussed in 0.4.14. Actually, we can use 0.4.46 to formulate a precise definition of the latter notion in mathematical terms; the formulation of such a definition is left to the reader.

We could generalize the notation introduced in 0.4.46 by relativizing it to an arbitrary class K containing $\mathfrak{A}$: $Pd_\alpha^{(\mathfrak{A},\mathsf{K})}$ would be defined as the unique $h \in Ho(\mathfrak{Fr}_\alpha\mathsf{K}, \mathfrak{Pl}_\alpha\mathfrak{A})$ such that $h(\xi/Cr_\alpha\mathsf{K}) = {}^\alpha A 1\, \mathsf{pj}_\xi$ for every $\xi < \alpha$.

It should be pointed out that in many cases a condition which is imposed on all algebras of a class K and is formulated in terms of the notion $Pd_\alpha^{(\mathfrak{A})}$ or $Pd_\alpha^{(\mathfrak{A},\mathsf{K})}$ can be simplified by eliminating this notion and using polynomials over free algebras instead. Consider for instance the following condition in which K is assumed to be an algebraically closed class of similar algebras:

(I) *there is a* $u \in Fr_3$ *such that* $(Pd_3^{(\mathfrak{A})}u)(x, y, y) = x$ *and* $(Pd_3^{(\mathfrak{A})}u)(x, x, y) = y$ *for all* $\mathfrak{A} \in \mathsf{K}$ *and* $x, y \in A$.

Clearly, this condition expresses the following fact: with every algebra $\mathfrak{A} \in \mathsf{K}$ we can correlate a polynomial $Q_{\mathfrak{A}}$ in 3 variables over $\mathfrak{A}$ so that all polynomials so correlated are of the same form and that, for all $\mathfrak{A} \in \mathsf{K}$ and $x, y \in A$, we have $Q_{\mathfrak{A}}(x, y, y) = x$ and $Q_{\mathfrak{A}}(x, x, y) = y$. It is easily seen that the same fact can be expressed somewhat more simply, without using $Pd_3^{(\mathfrak{A})}$:

(II) *there is a* $Q \in Pl_3\mathfrak{Fr}_\alpha\mathsf{K}$ *such that* $Q(x, y, y) = x$ *and* $Q(x, x, y) = y$ *for all* $x, y \in Fr_\alpha\mathsf{K}$.

α in (II) represents an arbitrary cardinal $\geqq 3$.

We can use either (I) or (II) to formulate an interesting result stated and proved in Mal'cev [54*]. We shall actually use (II):

THEOREM 0.4.48. *If* $\mathsf{K} = \mathbf{HSP}\mathsf{K}$ *and* $\alpha \geqq 3$, *then the following two conditions are equivalent*:

(i) $R|S = S|R$ *for all* $R, S \in Co\,\mathfrak{A}$ *and all* $\mathfrak{A} \in \mathsf{K}$;

(ii) *there is a* $Q \in Pl_3\mathfrak{Fr}_\alpha\mathsf{K}$ *such that* $Q(x, y, y) = x$ *and* $Q(x, x, y) = y$ *for all* $x, y \in Fr_\alpha\mathsf{K}$.

This theorem can be applied to the class K of all groups treated as algebras $\mathfrak{G} = \langle G, \cdot, ^{-1}\rangle$ of type $\langle 2, 1\rangle$. It is known that $\mathsf{K} = \mathbf{HSP}\mathsf{K}$. Furthermore, in every group $\mathfrak{G} \in \mathsf{K}$, and in particular in $\mathfrak{Fr}_\alpha\mathsf{K}$, we can construct a polynomial Q in 3 variables such that $Q(x, y, y) = x$ and $Q(x, x, y) = y$ for any $x, y \in G$; in fact, it suffices to set $Q(x, y, z) = (x \cdot y^{-1}) \cdot z$. Hence, as an immediate consequence of 0.4.48, we obtain the well-known fact that the congruence relations in every group commute. 0.4.48 is closely connected with 0.4.16, but we will not analyze here the connection in detail.

THEOREM 0.4.49. *For every algebra* $\mathfrak{A}$ *and every* α, *the set* $\{{}^{\alpha}A1\,\mathsf{pj}_\xi : \xi < \alpha\}$ $\{\mathfrak{A}\}$-*freely generates* $\mathfrak{Pl}_\alpha\mathfrak{A}$ *and* $|\{{}^{\alpha}A1\,\mathsf{pj}_\xi : \xi < \alpha\}| = |\alpha/Cr_\alpha\mathfrak{A}|$.

PROOF. If $|A| = 1$, the desired conclusion follows from 0.4.4(i) and 0.4.33(i). Assume $|A| > 1$. Then by 0.4.4(ii) and 0.4.33 we obtain $|\{{}^{\alpha}A1\,\mathsf{pj}_\xi : \xi < \alpha\}| = \alpha = |\alpha/Cr_\alpha\mathfrak{A}|$. Set $h = \langle {}^{\alpha}A1\,\mathsf{pj}_\xi : \xi < \alpha\rangle$ and let f be any function from $\{{}^{\alpha}A1\,\mathsf{pj}_\xi : \xi < \alpha\}$ into A. By 0.3.6(i) we have

$$g = {}^{\alpha A}A1\,\mathsf{pj}_{f \circ h} \in Ho({}^{\alpha A}\mathfrak{A}, \mathfrak{A}).$$

Hence $Pl_\alpha\mathfrak{A}1g \in Hom(\mathfrak{Pl}_\alpha\mathfrak{A}, \mathfrak{A})$, and $f \subseteq Pl_\alpha\mathfrak{A}1g$. Thus $\{{}^\alpha A1\,\mathsf{pj}_\xi : \xi < \alpha\}$ $\{\mathfrak{A}\}$-freely generates $\mathfrak{Pl}_\alpha\mathfrak{A}$, as was to be shown.

THEOREM 0.4.50. *Let* α *be any non-zero cardinal.*

(i) *For every algebra* $\mathfrak{A}$, $\mathfrak{Fr}_\alpha\mathfrak{A} \cong \mathfrak{Pl}_\alpha\mathfrak{A}$.

(ii) *For every set* K *of algebras,*

$$\mathfrak{Fr}_\alpha\mathsf{K} \cong \mathfrak{Pl}_\alpha(\mathsf{P}\langle\mathfrak{B} : \mathfrak{B} \in \mathsf{K}\rangle).$$

(iii) *For every class* K *of algebras, if* L *and* M *are defined as in 0.3.19(ii), then*

$$\mathfrak{Fr}_\alpha\mathsf{K} \cong \mathfrak{Pl}_\alpha(\mathsf{P}\langle\mathfrak{B} : \mathfrak{B} \in \mathsf{L}\rangle) \cong \mathfrak{Pl}_\alpha(\mathsf{P}\langle\mathfrak{B} : \mathfrak{B} \in \mathsf{M}\rangle).$$

PROOF: by 0.4.29, 0.4.38, 0.4.49.

This theorem, and in particular its part (iii), shows that free algebras can be constructed as special polynomial algebras, in agreement with the remarks preceding Definition 0.4.19. The construction of a free algebra over a class K described in this definition could be loosely characterized as a construction "from above" or "from the outside", as opposed to the construction described in 0.4.50 — a construction "from below" or "from the inside".

THEOREM 0.4.51. *For every algebra* $\mathfrak{A}$ *and any* α *and* β *we have*

(i) $\mathfrak{Fr}_\alpha\mathfrak{A} \cong \mathfrak{Pl}_\alpha\mathfrak{A} \succcurlyeq \mathfrak{Fr}_\alpha\mathfrak{Fr}_\beta\mathfrak{A} = \mathfrak{Fr}_\alpha\mathfrak{Pl}_\beta\mathfrak{A} \cong \mathfrak{Pl}_\alpha\mathfrak{Fr}_\beta\mathfrak{A} \cong \mathfrak{Pl}_\alpha\mathfrak{Pl}_\beta\mathfrak{A}$.

If, moreover, $\alpha \leqq \beta$ *or else if* $\omega \leqq \beta$ *(and* α *is arbitrary), then*

(ii) $\mathfrak{Fr}_\alpha\mathfrak{A} = \mathfrak{Fr}_\alpha\mathfrak{Fr}_\beta\mathfrak{A} = \mathfrak{Fr}_\alpha\mathfrak{Pl}_\beta\mathfrak{A} \cong \mathfrak{Pl}_\alpha\mathfrak{A} \cong \mathfrak{Pl}_\alpha\mathfrak{Fr}_\beta\mathfrak{A} \cong \mathfrak{Pl}_\alpha\mathfrak{Pl}_\beta\mathfrak{A}$.

PROOF: by 0.4.40(ii), 0.4.42, 0.4.43, 0.4.50(i).

The parts of this theorem which concern only polynomial algebras could be proved with no special difficulties in the first half of this section, before introducing the notion of a free algebra.

0.4.51(ii) may fail in case $\beta < \alpha$ and $\beta < \omega$. Consider, e.g., any non-commutative semigroup $\mathfrak{A}$. Then, as is easily seen, $\mathfrak{Pl}_1\mathfrak{A}$ and hence also $\mathfrak{Pl}_2\mathfrak{Pl}_1\mathfrak{A}$ is commutative while $\mathfrak{Pl}_2\mathfrak{A}$ is not; therefore $\mathfrak{Pl}_2\mathfrak{A}$ and $\mathfrak{Pl}_2\mathfrak{Pl}_1\mathfrak{A}$ are not isomorphic.

THEOREM 0.4.52. (i) *If* X K*-freely generates* $\mathfrak{A}$ *and* $X \supseteq Y \neq 0$, *then* Y K*-freely generates* $\mathfrak{Sg}^{(\mathfrak{A})}Y$.

(ii) *If* $|B| > 1$ *for some* $\mathfrak{B} \in \mathsf{K}$, X K*-freely generates* $\mathfrak{A}$, $\mathfrak{A} = \mathfrak{Sg}Y$, *and* $X \supseteq Y$, *then* $X = Y$.

PROOF. (i): by 0.2.14(i) and 0.4.23.

(ii) Suppose on the contrary that $x \in X \sim Y$. Choose $\mathfrak{B} \in \mathsf{K}$ such that $|B| > 1$. From 0.4.23 we conclude that there are $f, g \in Hom(\mathfrak{A}, \mathfrak{B})$ such that

$fx \neq gx$ and $Y1f = Y1g$. Hence by 0.2.14(iii) we obtain $f = g$, which is impossible.

From 0.4.52(ii) we see that a set X which K-freely generates an algebra $\mathfrak{A}$ is an irredundant base of $\mathfrak{A}$ in the sense of 0.1.29.

THEOREM 0.4.53. (i) *If* $0 \neq \Gamma \subseteq \alpha$ *and* $|\Gamma| = \beta$, *then* $\mathfrak{Sg}^{(\mathfrak{Fr}_\alpha\mathsf{K})}(\Gamma/Cr_\alpha\mathsf{K})$ *is* K*-freely generated by* $\Gamma/Cr_\alpha\mathsf{K}$ *and is isomorphic to* $\mathfrak{Fr}_\beta\mathsf{K}$.

(ii) *If* $\alpha \geqq \beta$, *then* $\mathfrak{Fr}_\beta\mathsf{K} \cong | \subseteq \mathfrak{Fr}_\alpha\mathsf{K}$ *and* $\mathfrak{Fr}_\beta\mathsf{K} \preccurlyeq \mathfrak{Fr}_\alpha\mathsf{K}$.

PROOF: (i) by 0.4.27(ii), 0.4.33, 0.4.29, 0.4.52(i); (ii) by (i), 0.4.22, 0.4.39.

By comparing 0.4.51 with 0.4.53(ii), we obtain some further formulas analogous to 0.4.51(i),(ii). Thus, if $\alpha \leqq \beta$, we have

$$\mathfrak{Pl}_\beta\mathfrak{A} \succcurlyeq \mathfrak{Pl}_\beta\mathfrak{Pl}_\alpha\mathfrak{A} \succcurlyeq \mathfrak{Pl}_\alpha\mathfrak{A} \cong \mathfrak{Pl}_\alpha\mathfrak{Pl}_\beta\mathfrak{A}.$$

THEOREM 0.4.54. *Assuming that* $|B| > 1$ *for some* $\mathfrak{B} \in \mathsf{K}$, *and* $\alpha \geqq \omega$, *or else that* $\omega > |B| > 1$ *for some* $\mathfrak{B} \in \mathsf{K}$ (*and* α *is arbitrary*), *we have*:

(i) *if* $\mathfrak{Fr}_\alpha\mathsf{K} = \mathfrak{Sg}X$, *then* $|X| \geqq \alpha$;
(ii) *if* X K*-freely generates* $\mathfrak{Fr}_\alpha\mathsf{K}$, *then* $|X| = \alpha$;
(iii) *if* $\mathfrak{Fr}_\alpha\mathsf{K} \succcurlyeq \mathfrak{Fr}_\beta\mathsf{K}$, *then* $\alpha \geqq \beta$;
(iv) *if* $\mathfrak{Fr}_\beta\mathsf{K} \succcurlyeq \mathfrak{Fr}_\alpha\mathsf{K}$, *then* $\beta \geqq \alpha$;
(v) *if* $\mathfrak{Fr}_\alpha\mathsf{K} \cong \mathfrak{Fr}_\beta\mathsf{K}$, *then* $\alpha = \beta$.

For $\alpha \geqq \omega$ this theorem follows almost immediately, using 0.1.17(ix),(x), 0.4.27, and 0.4.52(ii) (cf. also Fujiwara [55*]). For $\alpha < \omega$ the proof is essentially given in Jónsson-Tarski [61*], where some related results can also be found.

A well-known example of a class of algebras which satisfies conditions (i)–(v) of 0.4.54 are the groups. On the other hand, a class K of infinite algebras is constructed in op. cit. for which these conditions fail; actually in this class any two algebras $\mathfrak{Fr}_\alpha\mathsf{K}$ and $\mathfrak{Fr}_\beta\mathsf{K}$ with $\alpha, \beta < \omega$ are isomorphic. (Consequently, in opposition to 0.4.43(ii),(iii), the formulas $\mathfrak{Fr}_\alpha\mathsf{K} \cong \mathfrak{Fr}_\beta\mathsf{K}$ and $\mathfrak{Fr}_\alpha\mathsf{K} = \mathfrak{Fr}_\beta\mathsf{K}$ are, in general, not equivalent; as is easily seen, $\mathfrak{Fr}_\alpha\mathsf{K} = \mathfrak{Fr}_\beta\mathsf{K}$ never holds unless $\alpha = \beta$.)

THEOREM 0.4.55 *The assumption*

$$\mathbf{HSP}\mathsf{K} = \mathbf{HSP}\{\mathfrak{A} : \mathfrak{A} \in \mathsf{K}, |A| < \omega\}$$

implies the following conclusion:

$\mathfrak{Fr}_\alpha\mathsf{K}$ *is* K*-freely generated by every set* X *such that* $|X| = \alpha < \omega$, $X \subseteq Fr_\alpha\mathsf{K}$, *and* $\mathfrak{Sg}X = \mathfrak{Fr}_\alpha\mathsf{K}$.

The proof can be found in Jónsson-Tarski [61*]. For a more suggestive form of the assumption of this theorem see 0.4.64 below.

Theorems 0.4.54 and 0.4.55 extend to algebras of arbitrary similarity types. In the really interesting case, however, that of finite α, the two theorems are of relatively little value when directly applied to algebras with infinite index sets, since, in this case, the hypotheses of both of them frequently fail (so that the theorems hold only vacuously). Concerning this situation compare Remark 0.5.18 in the next section.

In 0.4.42, 0.4.43, 0.4.53, and 0.4.54 we have concerned ourselves with relations between $\mathfrak{Fr}_\alpha\mathsf{K}$ and $\mathfrak{Fr}_\alpha\mathsf{L}$ as well as between $\mathfrak{Fr}_\alpha\mathsf{K}$ and $\mathfrak{Fr}_\beta\mathsf{K}$. More generally, one can study relations between $\mathfrak{Fr}_\alpha\mathsf{K}$ and $\mathfrak{Fr}_\beta\mathsf{L}$ where K, L are any two classes of algebras and α, β are any two cardinals. In particular, one can look for conditions under which these two free algebras are isomorphic. This problem, however, easily reduces to those previously discussed. In fact, as is easily seen from 0.4.42, 0.4.43 and 0.4.53, the formula

$$\mathfrak{Fr}_\alpha\mathsf{K} \cong \mathfrak{Fr}_\beta\mathsf{L}$$

is equivalent to

$$\mathfrak{Fr}_\alpha\mathsf{K} = \mathfrak{Fr}_\alpha\mathsf{L} \cong \mathfrak{Fr}_\beta\mathsf{L},$$

or else to

$$\mathfrak{Fr}_\alpha\mathsf{K} \cong \mathfrak{Fr}_\beta\mathsf{K} = \mathfrak{Fr}_\beta\mathsf{L},$$

dependent on whether $\alpha \leqq \beta$ or $\alpha \geqq \beta$.

REMARK 0.4.56. An important notion which admits a simple definition in terms of freely generating sets is that of a new notion of an independent set of elements of an algebra, which is closely related to the three notions of independence briefly discussed in Remark 0.1.29. It is, however, stronger than all of them. Given an algebra $\mathfrak{A}$, a set $X \subseteq A$ is *independent* in the new sense iff either X is empty or X $\{\mathfrak{A}\}$-freely generates $\mathfrak{Sg}^{(\mathfrak{A})}X$. All the properties of independent sets listed in 0.1.29 apply to the new notion of independence as well. The Abelian groups $\langle G, +\rangle$ in which all elements are of the same prime order $\pi > 0$ provide an interesting example of a class of algebras for which all the four notions of independence coincide. For further information concerning independent sets see Marczewski [66*] where in particular a comprehensive bibliography of the subject can be found.

In the last portion of this section we state a few model-theoretical results which involve free algebras. The first of them is of the same general character as Theorems 0.4.53(ii) and 0.4.54 (iii)–(iv), i.e., it concerns the relationship between two free algebras over the same class but with different numbers of generators.

THEOREM 0.4.57. *For every class* K *of algebras and any* $\alpha, \beta \geqq \omega$, *the algebras* $\mathfrak{Fr}_\alpha \mathsf{K}$ *and* $\mathfrak{Fr}_\beta \mathsf{K}$ *are elementarily equivalent.*

For the proof see Tarski-Vaught [57*], p. 98, Theorem 3.5.

Assuming $\mathscr{GCH}$ and applying 0.3.79 we conclude from 0.4.57 that, for every class K of algebras and any infinite α and β,

$$\mathsf{Up}\mathfrak{Fr}_\alpha \mathsf{K} \cap \mathsf{Up}\mathfrak{Fr}_\beta \mathsf{K} \neq 0.$$

Thus we have here a new result concerning the relation $\triangleq$ which was discussed in 0.3.80.

Theorem 0.4.57 does not extend to finite cardinals. For instance, in case K is the class of all groups, the free groups $\mathfrak{Fr}_1 \mathsf{K}$ and $\mathfrak{Fr}_2 \mathsf{K}$ are not elementarily equivalent since the former is commutative while the latter is not (cf. the remarks following 0.4.51). It seems very likely that any two free groups $\mathfrak{Fr}_\alpha \mathsf{K}$ and $\mathfrak{Fr}_\beta \mathsf{K}$ with $\alpha, \beta \geqq 2$ are elementarily equivalent; however, no proof of this statement has yet been found.

We want now to describe the metamathematical construction of free algebras which was mentioned in the remarks preceding 0.4.19. To this end consider the discourse language of the similarity class of algebras of type $\langle 2 \rangle$ (cf. 0.1.5). We assume that this language has been provided with α different variables arranged as usual in the sequence $\langle \mathbf{v}_\xi : \xi < \alpha \rangle$; here α is an arbitrary non-zero cardinal which is regarded as fixed in each construction of a free algebra, but varies from one construction to another. As we know, the language has a binary operation symbol, say $+$, as the only non-logical constant; this operation symbol induces a binary operation on expressions also denoted by $+$. We recall that the set $T\mu^{(\alpha)}$ of all α-terms (or simply terms) is the least set containing all variables and closed under the operation $+$, and that two α-terms σ and τ are called equivalent with respect to a class K of algebras, in symbols $\sigma \equiv_{\mathsf{K}}^{(\alpha)} \tau$, if $\mathsf{K} \models \sigma = \tau$.

THEOREM 0.4.58. (i) $\langle T\mu^{(\alpha)}, + \rangle$ *is an algebra and* $\langle T\mu^{(\alpha)}, + \rangle = \mathfrak{Sg}\{\mathbf{v}_\xi : \xi < \alpha\}$.

(ii) *For every class* K *of algebras,* $\equiv_{\mathsf{K}}^{(\alpha)} \in Co\langle T\mu^{(\alpha)}, + \rangle$ *and hence* $\langle T\mu^{(\alpha)}, + \rangle / \equiv_{\mathsf{K}}^{(\alpha)}$ *is also an algebra.*

DEFINITION 0.4.59. *The algebra* $\mathfrak{Tm}_\alpha = \langle T\mu^{(\alpha)}, + \rangle$ *is called the* ***free term algebra.*** *For every class* K *of algebras, the algebra* $\mathfrak{Tm}_\alpha \mathsf{K} = \mathfrak{Tm}_\alpha / \equiv_{\mathsf{K}}^{(\alpha)}$ *is called the* ***term algebra of*** K.

THEOREM 0.4.60. (i) *There is a unique f such that $f \in Is(\mathfrak{Tm}_\alpha, \mathfrak{Fr}_\alpha)$ and $fv_\xi = \xi$ for every $\xi < \alpha$.*

The unique f of (i) satisfies also the following conditions:

(ii) $\langle\sigma, \tau\rangle \in \equiv_{\mathsf{K}}^{(\alpha)}$ *iff* $\langle f\sigma, f\tau\rangle \in Cr_\alpha\mathsf{K}$, *for any α-terms σ, τ and every class* K *of algebras;*

(iii) $\tilde{\sigma}^{(\mathfrak{A})} = Pd_\alpha^{(\mathfrak{A})}(f\sigma)$ *for every term σ and every algebra* $\mathfrak{A}$.

PROOF. It follows immediately from 0.4.30 and 0.4.59 that, if K is a similarity class, then $\{v_\xi : \xi < \alpha\}$ K-freely generates $\mathfrak{Tm}_\alpha$. This fact combined with 0.4.27(i) and 0.4.28 gives (i). To show (ii) we first prove the following result:

(1) if K is any class of algebras, then $\{v_\xi/\equiv_{\mathsf{K}}^{(\alpha)} : \xi < \alpha\}$ K-freely generates $\mathfrak{Tm}_\alpha\mathsf{K}$.

Assume $\mathfrak{B} \in \mathsf{K}$ and g is a function from $\{v_\xi/\equiv_{\mathsf{K}}^{(\alpha)} : \xi < \alpha\}$ into B, and set

(2) $$h = \langle \tilde{\sigma}^{(\mathfrak{B})}x : \sigma \in T\mu^{(\alpha)}\rangle$$

where x is the α-termed sequence of elements of $\mathfrak{B}$ such that $x_\xi = g(v_\xi/\equiv_{\mathsf{K}}^{(\alpha)})$ for every $\xi < \alpha$. It is easily seen that

(3) $$g\circ(\equiv_{\mathsf{K}}^{(\alpha)})^\star \subseteq h \in Hom(\mathfrak{Tm}_\alpha, \mathfrak{B}).$$

From (2) we get $\equiv_{\mathsf{K}}^{(\alpha)} \subseteq h|h^{-1}$; hence, by (3) and 0.2.23(iii),

$$k = \{\langle \sigma/\equiv_{\mathsf{K}}^{(\alpha)}, h\sigma\rangle : \sigma \in T\mu^{(\alpha)}\} \in Hom(\mathfrak{Tm}_\alpha\mathsf{K}, \mathfrak{B}).$$

Since it is clear that $g \subseteq k$, (1) is proved.

Let K be any class of algebras. Then for any $\mathfrak{B} \in \mathsf{K}$ and $g \in {}^{\{v_\xi : \xi < \alpha\}}B$ we have

$$g \subseteq \langle \tilde{\sigma}^{(\alpha)}x : \sigma \in T\mu^{(\alpha)}\rangle \in Hom(\mathfrak{Tm}_\alpha, \mathfrak{B})$$

where $x \in {}^\alpha B$ and $x_\xi = gv_\xi$ for every $\xi < \alpha$. Thus we easily conclude that

$$\equiv_{\mathsf{K}}^{(\alpha)} = \bigcap\{h|h^{-1} : \mathfrak{B} \in \mathsf{K}, h \in Hom(\mathfrak{Tm}_\alpha, \mathfrak{B})\}.$$

Hence, by 0.2.23(i), 0.3.12(ii), and 0.3.46 along with the remark just following it,

$$\mathfrak{Tm}_\alpha\mathsf{K} \in \mathbf{SP}\mathsf{K};$$

furthermore, in view of 0.4.33 it is seen that $\{\langle v_\xi/\equiv_{\mathsf{K}}^{(\alpha)}, \xi/Cr_\alpha\mathsf{K}\rangle : \xi < \alpha\}$ is a one-one function. Combining these two facts with (1), 0.4.27(ii), and 0.4.28 we obtain a function $k \in Is(\mathfrak{Tm}_\alpha\mathsf{K}, \mathfrak{Fr}_\alpha\mathsf{K})$ such that $k(v_\xi/\equiv_{\mathsf{K}}^{(\alpha)}) = \xi/Cr_\alpha\mathsf{K}$ for every $\xi < \alpha$. Therefore, by 0.2.14(iii),

$$(Cr_\alpha\mathsf{K})^\star \circ f = k\circ(\equiv_{\mathsf{K}}^{(\alpha)})^\star.$$

Hence, since f and k are both one-one, we immediately get (ii).

To complete the proof of the theorem observe that, if Γ is the set of all α-

terms σ such that the equality of (iii) holds, then $\{\mathfrak{v}_\xi : \xi < \alpha\} \subseteq \Gamma \in Su\mathfrak{Tm}_\alpha$; thus we have (iii).

THEOREM 0.4.61. $\mathfrak{Fr}_\alpha \cong \mathfrak{Tm}_\alpha$, *and* $\mathfrak{Fr}_\alpha\mathsf{K} \cong \mathfrak{Tm}_\alpha\mathsf{K}$ *for every class* K *of algebras.*

Proof: by 0.4.60.

From 0.4.60(ii) we can easily derive the model-theoretical characterization of free algebras. We give here a precise formulation of this result:

Let K *be a class of algebras containing at least one algebra with two or more elements. For* $\mathfrak{A} \cong \mathfrak{Fr}_\alpha\mathsf{K}$ *it is necessary and sufficient that there exists a set* X *with cardinality* α *which generates* $\mathfrak{A}$ *and satisfies the following two conditions:*

(I) *every equation* φ *which is satisfied in* $\mathfrak{A}$ *under some assignment of distinct elements of* X *to the variables in* φ *is satisfied identically in all algebras of* K;

(II) *every equation* φ *which is satisfied identically in all algebras of* K *is satisfied in* $\mathfrak{A}$ *under every such assignment.*

It may also be noticed that the above condition (I) is necessary and sufficient for a given set X generating $\mathfrak{A}$ to K-freely generate $\mathfrak{A}$.

The metamathematical construction of free algebras which has just been outlined (in 0.4.58–0.4.61) leads to several interesting model-theoretical results; we will state one of them below as Theorem 0.4.63. The validity of these results depends on the assumption that $\alpha \geqq \omega$, i.e., that the underlying elementary theory of algebras is provided with infinitely many variables; we simply let $\alpha = \omega$.

We recall that a class K of algebras is an equational class, K is an $\mathscr{E}\mathscr{Q}\mathscr{C}_\Delta$, if there is a set Γ of identities such that $\mathsf{K} = \mathsf{Md}\,\Gamma$.

THEOREM 0.4.62. K *is an* $\mathscr{E}\mathscr{Q}\mathscr{C}_\Delta$ *iff, for every algebra* $\mathfrak{A}$, $\equiv^{(\omega)}_{\mathsf{K}} \subseteq \equiv^{(\omega)}_{\mathfrak{A}}$ *implies* $\mathfrak{A} \in \mathsf{K}$.

The most important consequence of 0.4.60 and, more generally, the most important application of the theory of free algebras to model theory is the following

THEOREM 0.4.63. *For every class* K *of algebras the following conditions are equivalent:*

(i) K *is an* $\mathscr{E}\mathscr{Q}\mathscr{C}_\Delta$,

(ii) $\mathsf{K} = \mathbf{S}\mathsf{K} = \mathbf{H}\mathsf{K} = \mathbf{P}\mathsf{K}$,

(iii) $\mathsf{K} = \mathbf{HSP}\mathsf{K}$.

PROOF. (ii) and (iii) are equivalent by 0.3.13. Consider the additional condition:

(iv) $\mathfrak{A} \in \mathsf{K}$ *for all* $\mathfrak{A}$ *such that* $Cr_\omega \mathsf{K} \subseteq Cr_\omega \mathfrak{A}$.

Then (i) and (iv) are equivalent by 0.4.60(ii) and 0.4.62; moreover, (iii) and (iv) are equivalent by 0.4.42(i),(iv).

This result, without condition (iii), is the well-known theorem of Birkhoff [35*], p. 441, Theorem 10. It shows that the mathematical notion of an algebraically closed class (cf. 0.3.14) and the model-theoretical notion of an equational class (in the wider sense) extensionally coincide. Several variants of 0.4.63 are known. Cf. Tarski [55*], pp. 58 ff., and Šaîn [65*]; in both these papers further bibliographic references are given. (A correction to Tarski [55*] can be found in Tarski [58*], p. 175.)

REMARKS 0.4.64. In view of 0.3.13, Theorem 0.4.63 immediately implies that, for every class K of algebras, $\mathbf{HSP}\mathsf{K}$ is the least equational class which includes K. This permits us, for instance, to give the assumption of 0.4.55 the following equivalent form: *every identity which holds in all finite algebras of* K *holds also in all algebras of* K.

REMARKS 0.4.65. To conclude this section we wish to discuss briefly a certain generalization of the notion of a free algebra. We have here in mind the notion of a conditionally free algebra, more specifically, a free algebra over a given class K of similar algebras, with generators subject to certain gonditions. These conditions are expressed by equations in the discourse lancuage Λ of K or, what amounts to the same, of the similarity class L which includes K; they are called *defining relations*, and the algebra itself is referred to as a *free algebra with defining relations*.

To formulate a model-theoretical characterization of this notion, we denote by Σ the set of equations in terms of which the defining relations are expressed, and we agree to call a sequence $x = \langle x_\xi \rangle_{\xi<\alpha}$ of elements in an algebra $\mathfrak{B}$ of L *Σ-appropriate* if it satisfies all the equations of Σ. (It is assumed here that the sequence of variables $\mathbf{v}_\xi$ in the discourse language Λ of L is of length α where α is a cardinal; whether the sequence $x = \langle x_\xi \rangle_{\xi<\alpha}$ satisfies an equation of Σ with the variables $\mathbf{v}_{\xi_0}, \ldots, \mathbf{v}_{\xi_{\kappa-1}}$ depends, of course, exclusively on the terms $x_{\xi_0}, \ldots, x_{\xi_{\kappa-1}}$ of this sequence.) We can now say that $\mathfrak{A}$ is a free algebra over L, with an α-termed sequence $g = \langle g_\xi \rangle_{\xi<\alpha}$ of generators and with defining relations expressed by equations of a set Σ, if (I) $\mathfrak{A}$ is an algebra in L and g is a Σ-appropriate sequence of elements of $\mathfrak{A}$; (II) $\mathfrak{A}$ is generated by the range of g; and (III) whenever an equation in Λ is satisfied by g, it is also satisfied by every Σ-appropriate sequence of elements of every algebra in L, and conversely. Notice that equations of the set Σ may imply that certain terms in every Σ-appropriate sequence are equal; as a consequence, the cardinality of the range of the sequence g may be in some cases smaller than α.

Just as in the case of ordinary free algebras, we can replace this characterization by a purely mathematical construction, in which, in addition, the sequence of generators is fixed, so that the algebra $\mathfrak{A}$ becomes uniquely determined by the class K of algebras, the cardinal α, and the set of defining relations. To this end we use ordered pairs of words $\langle w, u\rangle$, i.e., of elements $w, u \in Fr_\alpha$, as mathematical substitutes for equations in Σ; we can simply identify these pairs with defining relations. If, for instance, K is a class of algebras $\langle B, +\rangle$ and a defining relation is expressed by the equation

$$\mathbf{v}_\xi + (\mathbf{v}_\eta + \mathbf{v}_\zeta) = (\mathbf{v}_\xi + \mathbf{v}_\eta) + \mathbf{v}_\zeta,$$

we identify the defining relation with the pair $\langle w, u\rangle$ where $w = \langle \xi, \langle \eta, \zeta\rangle\rangle$ and $u = \langle\langle \xi, \eta\rangle, \zeta\rangle$. Let $\mathfrak{Fr}_\alpha^{(S)}\mathsf{K}$ be the free algebra over K with α generators and with the set S of defining relations. The desired definition of $\mathfrak{Fr}_\alpha^{(S)}\mathsf{K}$ is obtained by a simple modification of Definition 0.4.19(ii) of $\mathfrak{Fr}_\alpha\mathsf{K}$; in fact, we let

$$Cr_\alpha^{(S)}\mathsf{K} = \bigcap\{R : R \in Co\mathfrak{Fr}_\alpha, \mathfrak{Fr}_\alpha/R \in \mathbf{IS}\mathsf{K}, S \subseteq R\}$$

and

$$\mathfrak{Fr}_\alpha^{(S)}\mathsf{K} = \mathfrak{Fr}_\alpha/Cr_\alpha^{(S)}\mathsf{K}.$$

On the basis of this definition we could develop the whole theory of $\mathfrak{Fr}_\alpha^{(S)}\mathsf{K}$ and carry over to it most of the results obtained for $\mathfrak{Fr}_\alpha\mathsf{K}$; the formulations of results are, of course, more complicated. In particular, we can extend the notion of a set K-freely generating a given algebra, which was defined in 0.4.23. In fact, given an algebra $\mathfrak{A} \in \mathsf{L}$ we say that a sequence $x \in {}^\alpha A$ *K-freely generates $\mathfrak{A}$ under a set $S \subseteq Fr_\alpha \times Fr_\alpha$* (of defining relations) if (I) x is S-appropriate; (II) $\mathfrak{A}$ is generated by Rgx; and (III) for any $\mathfrak{B} \in \mathsf{K}$ and any S-appropriate $y \in {}^\alpha B$ there exists a (unique) homomorphism h from $\mathfrak{A}$ into $\mathfrak{B}$ such that $hx_\xi = y_\xi$ for every $\xi < \alpha$. (We can formally define the term "S-appropriate" by saying that an α-termed sequence x of elements of $\mathfrak{A}$ is S-appropriate if $S \subseteq h|h^{-1}$ where h is the unique homomorphism from $\mathfrak{Fr}_\alpha$ into A such that $h\xi = x_\xi$ for every $\xi < \alpha$.) Theorem 0.4.29, the fundamental theorem on free algebras, assumes now the following form: $\mathfrak{A} \cong \mathfrak{Fr}_\alpha^{(S)}\mathsf{K}$ iff $\mathfrak{A} \in \mathbf{SP}\mathsf{K}$ and there is a sequence $x \in {}^\alpha A$ which K-freely generates $\mathfrak{A}$ under S.

It should be emphasized that, while the notion of a free algebra with defining relations is closely related in content to the original notion of free algebra, it is far more comprehensive in scope. In fact, it can easily be shown that for any congruence relation R on the free algebra $\mathfrak{Fr}_\alpha\mathsf{K}$ we have

$$\mathfrak{Fr}_\alpha\mathsf{K}/R \cong \mathfrak{Fr}_\alpha^{(R)}\mathsf{K};$$

consequently, any algebra is isomorphic to some free algebra with defining relations.

It is seen from the above remarks that the notion discussed is conceptually rather involved. The notion is now intensely used in the discussion of groups and semigroups, but has not found so far many applications elsewhere; see, for instance, Magnus-Karrass-Solitar [66*] and Adjan [66*]. In the theory of cylindric algebras only a very special case of this notion will be involved (see Chapter 2, Definition 2.5.31). These are the reasons why free algebras with defining relations are not discussed here in a systematic way. Nevertheless it seems likely that this notion will become a useful tool in the general theory of algebras, especially in the study of some metamathematical aspects of the theory; cf. here Lyndon [62*].

We may mention still another notion of a general algebraic character which is closely related to that of a free algebra (and essentially can be subsumed under the notion of a free algebra with defining relations). It is the notion of a free product of algebras. In this case again a study of the notion within the framework of the general theory of algebras seems desirable and promises to be fruitful. For work in this direction see Łoś [65*].

Like the formation of subalgebras and homomorphic images, the formation of reducts is an operation "reducing the size of an algebra". Loosely speaking, when taking a reduct of an algebra we leave the universe unchanged but discard some of the fundamental operations. In consequence, the formation of reducts (as opposed to all the operations on algebras previously discussed) leads, as a rule, from algebras of a given similarity class to algebras of a different similarity class.

Some particular cases of the notion of a reduct are known from the discussion of special classes of algebras. If, for instance, we form a reduct of a ring by discarding the multiplication, we arrive at what is called the additive group of a ring. In its general form the notion of a reduct was introduced in Tarski [54*], p. 580. The notion is much less known and less frequently discussed than the notions introduced in the previous sections. It proves to be really useful when applied to algebras with infinitely many operations, for it enables one to reduce some problems concerning such algebras to analogous problems for algebras with finitely many operations.

DEFINITION 0.5.1. *Let $\mathfrak{A} = \langle A, Q_i \rangle_{i \in I}$ be an algebra, J be any set, and r be a one-one function from J into I. The algebra $\mathfrak{B} = \langle A, Q_{rj} \rangle_{j \in J}$ is called the r-**reduct** of $\mathfrak{A}$ and denoted by $\mathfrak{Rd}_J^{(r)}\mathfrak{A}$ or, simply, by $\mathfrak{Rd}^{(r)}\mathfrak{A}$. In case $r = J 1 Id$ (and thus $J \subseteq I$), we call $\mathfrak{B}$ the J-**reduct** of $\mathfrak{A}$ and denote it by $\mathfrak{Rd}_J\mathfrak{A}$. If J is finite, $\mathfrak{Rd}_J\mathfrak{A}$ is called a **finite reduct** of $\mathfrak{A}$.*

For any class K of similar algebras we set

$$\mathsf{Rd}_J^{(r)}\mathsf{K} = \{\mathfrak{Rd}_J^{(r)}\mathfrak{A} : \mathfrak{A} \in \mathsf{K}\}$$

and, if $J \subseteq I$,

$$\mathsf{Rd}_J\mathsf{K} = \{\mathfrak{Rd}_J\mathfrak{A} : \mathfrak{A} \in \mathsf{K}\}.$$

REMARK 0.5.2. In the discussion of special classes of algebras it sometimes appears desirable to use the term "reduct" in a more restricted sense. Assume, e.g., that we are interested in groups with operators, i.e., in algebras $\mathfrak{A} =$

$\langle A, +, O_i \rangle_{i \in I}$ where $\langle A, + \rangle$ is a group and all O_i's are unary operations distributive with respect to $+$. If one undertakes a systematic study of reducts of such algebras, it seems reasonable to restrict oneself to reducts of the form $\langle A, +, O_{rj} \rangle_{j \in J}$ (which continue to be groups with operators), disregarding entirely those reducts in which the operation $+$ has been removed. The way in which the notion of a reduct should be restricted varies from one particular case to another and is determined by the actual interests of algebraists.

The reducts occasionally prove useful in establishing the isomorphism of structures. Thus to prove the isomorphism of two groups $\langle A, \cdot, ^{-1} \rangle$ and $\langle B, \cdot, ^{-1} \rangle$ it suffices to do so for their reducts $\langle A, \cdot \rangle$ and $\langle B, \cdot \rangle$, since the class of groups $\langle A, \cdot, ^{-1} \rangle$ and that of corresponding reducts $\langle A, \cdot \rangle$ are definitionally equivalent (cf. 0.1.6 and the remarks following 0.2.10). For the same reason, two Boolean algebras $\langle A, +, \cdot, -, 0, 1 \rangle$ and $\langle B, +, \cdot, -, 0, 1 \rangle$ (cf. 0.1.4) are isomorphic whenever their reducts $\langle A, + \rangle$ and $\langle B, + \rangle$ (or $\langle A, \cdot \rangle$ and $\langle B, \cdot \rangle$) are isomorphic.

All the theorems on reducts stated in this section are simple and similar in character: they establish basic connections between properties of algebras and properties of their reducts.

THEOREM 0.5.3. (i) *If r is a one-one function from J into $In\,\mathfrak{A}$, then $Uv\,\mathfrak{Rd}^{(r)}\mathfrak{A} = Uv\,\mathfrak{A}$, $In\,\mathfrak{Rd}^{(r)}\mathfrak{A} = J$, and $Op_j\mathfrak{Rd}^{(r)}\mathfrak{A} = Op_{rj}\mathfrak{A}$ for every $j \in J$.*

(ii) *If $J \subseteq In\,\mathfrak{A}$, then $Uv\,\mathfrak{Rd}_J\mathfrak{A} = Uv\,\mathfrak{A}$, $In\,\mathfrak{Rd}_J\mathfrak{A} = J$, and $Op_j\mathfrak{Rd}_J\mathfrak{A} = Op_j\mathfrak{A}$ for every $j \in J$.*

From now on until the end of this section the variable r is assumed to represent an arbitrary one-one function whose range is included in the index set of a given algebra (or in the common index set of all algebras of a given class or a given system of algebras). An analogous stipulation applies to the variable J.

THEOREM 0.5.4. (i) *$Su\,\mathfrak{A} \subseteq Su\,\mathfrak{Rd}^{(r)}\mathfrak{A}$.*

(ii) *If $\mathfrak{A} \subseteq \mathfrak{B}$, then $\mathfrak{Rd}^{(r)}\mathfrak{A} \subseteq \mathfrak{Rd}^{(r)}\mathfrak{B}$.*

(iii) $\mathsf{Rd}^{(r)}\mathbf{S}\mathsf{K} \subseteq \mathbf{S}\mathsf{Rd}^{(r)}\mathsf{K}$.

(iv) *$Sg^{(\mathfrak{Rd}^{(r)}\mathfrak{A})}X \subseteq Sg^{(\mathfrak{A})}X$ and $\mathfrak{Sg}^{(\mathfrak{Rd}^{(r)}\mathfrak{A})}X \subseteq \mathfrak{Rd}^{(r)}\mathfrak{Sg}^{(\mathfrak{A})}X$, for every $X \subseteq A$.*

(v) *$Sg^{(\mathfrak{A})}X = \bigcup\{Sg^{(\mathfrak{Rd}_J\mathfrak{A})}X : J \subseteq In\,\mathfrak{A}, |J| < \omega\}$ for every $X \subseteq A$.*

(vi) *If $\mathfrak{Rd}^{(r)}\mathfrak{A}$ is minimal, or finitely generated, then so is $\mathfrak{A}$.*

(vii) *If K is a non-empty set of algebras directed by the relation $\subseteq$, then so is $\mathsf{Rd}^{(r)}\mathsf{K}$, and we have $\mathsf{URd}^{(r)}\mathsf{K} = \mathsf{Rd}^{(r)}\mathsf{UK}$; in particular, $\mathsf{URd}^{(r)}\mathbf{S}_\omega\mathfrak{A} = \mathfrak{Rd}^{(r)}\mathfrak{A}$ for every algebra $\mathfrak{A}$.*

One can show by simple examples that the implications and inclusions stated in 0.5.4 cannot in general be replaced by equivalences and equalities. The same applies to 0.5.12 and 0.5.13 below.

THEOREM 0.5.5. (i) *If* r' *is a one-one function with* $Rgr' \subseteq Rgr$, *then* $\mathfrak{Rd}^{(r')}\mathfrak{Rd}^{(r)}\mathfrak{A} = \mathfrak{Rd}^{(r \circ r')}\mathfrak{A}$.

(ii) *If* $J' \subseteq J$, *then* $\mathfrak{Rd}_{J'}\mathfrak{Rd}_J\mathfrak{A} = \mathfrak{Rd}_{J'}\mathfrak{A}$.

DEFINITION 0.5.6. *An algebra* $\mathfrak{B}$ *is called a* ***subreduct*** *of an algebra* $\mathfrak{A}$, *in symbols* $\mathfrak{B} \subseteq^{\mathsf{r}} \mathfrak{A}$, *or* $\mathfrak{A} \supseteq^{\mathsf{r}} \mathfrak{B}$, *if* $In\mathfrak{B} \subseteq In\mathfrak{A}$ *and* $\mathfrak{B} \subseteq \mathfrak{Rd}_{In\mathfrak{B}}\mathfrak{A}$.

If, e.g., $\mathfrak{F} = \langle F, +, \cdot \rangle$ is a field with zero element 0, then the multiplicative group of $\mathfrak{F}$, $\langle F \sim \{0\}, \cdot \rangle$, is a subreduct of $\mathfrak{F}$.

THEOREM 0.5.7. (i) *If* $\mathfrak{A} \subseteq \mathfrak{B}$, *then* $\mathfrak{A} \subseteq^{\mathsf{r}} \mathfrak{B}$; *in case* $\mathfrak{A}$ *and* $\mathfrak{B}$ *are similar, the converse also holds.*

(ii) *If* $\mathfrak{A} \subseteq^{\mathsf{r}} \mathfrak{B} \subseteq^{\mathsf{r}} \mathfrak{A}$, *then* $\mathfrak{A} = \mathfrak{B}$.

(iii) *If* $\mathfrak{A} \subseteq^{\mathsf{r}} \mathfrak{B} \subseteq^{\mathsf{r}} \mathfrak{C}$, *then* $\mathfrak{A} \subseteq^{\mathsf{r}} \mathfrak{C}$.

By 0.5.7 the relation $\subseteq^{\mathsf{r}}$ establishes a partial ordering in every class of algebras. Hence we can speak of the least upper bound of algebras under this relation and formulate the following

DEFINITION 0.5.8. *By a* ***reduct union*** *of a set* K *of (not necessarily similar) algebras we understand an algebra which is the least upper bound of* K *under the relation* $\subseteq^{\mathsf{r}}$. *If such an algebra exists, it is denoted by* $\mathsf{U}^{\mathsf{r}}\mathsf{K}$.

REMARK 0.5.9. U^{r} is the only operation considered in this chapter which can be applied to a set of non-similar algebras. It is closely related in its properties to the operation U defined in 0.1.24 and, as will be stated in 0.5.10(iii), it reduces to the latter when applied to a set of similar algebras.

THEOREM 0.5.10. *Assuming that* $\mathsf{U}^{\mathsf{r}}\mathsf{K}$ *exists, the following conditions hold*:

(i) $\mathsf{U}\{Uv\mathfrak{A} : \mathfrak{A} \in \mathsf{K}\} = Uv\,\mathsf{U}^{\mathsf{r}}\mathsf{K}$;

(ii) $\mathsf{U}\{In\mathfrak{A} : \mathfrak{A} \in \mathsf{K}\} = In\,\mathsf{U}^{\mathsf{r}}\mathsf{K}$;

(iii) *if* K *is a set of similar algebras, then* $\mathsf{U}^{\mathsf{r}}\mathsf{K} = \mathsf{U}\mathsf{K}$.

THEOREM 0.5.11. (i) *If* K *is a non-empty set of algebras directed by the relation* $\subseteq^{\mathsf{r}}$, *then* $\mathsf{U}^{\mathsf{r}}\mathsf{K}$ *exists.*

(ii) *In particular, for any algebra* $\mathfrak{A}$ *we have* $\mathsf{U}^{\mathsf{r}}\mathsf{K} = \mathfrak{A}$ *if* K *is any one of the following three sets*:

(ii′) $\{\mathfrak{Rd}_J\mathfrak{A} : J \subseteq In\mathfrak{A}, |J| < \omega\}$,

(ii″) $\mathsf{U}\{\mathsf{Rd}_J\mathbf{S}_\omega\mathfrak{A} : J \subseteq In\mathfrak{A}, |J| < \omega\}$,

(ii‴) $\mathsf{U}\{\mathbf{S}_\omega\mathfrak{Rd}_J\mathfrak{A} : J \subseteq In\mathfrak{A}, |J| < \omega\}$.

THEOREM 0.5.12. (i) *$Ho\mathfrak{A} \subseteq Ho\,\mathfrak{Rd}^{(r)}\mathfrak{A}$ and $Is\mathfrak{A} = Is\,\mathfrak{Rd}^{(r)}\mathfrak{A}$.*

(ii) *$Ho(\mathfrak{A}, \mathfrak{B}) \subseteq Ho(\mathfrak{Rd}^{(r)}\mathfrak{A}, \mathfrak{Rd}^{(r)}\mathfrak{B})$ and $Is(\mathfrak{A}, \mathfrak{B}) \subseteq Is(\mathfrak{Rd}^{(r)}\mathfrak{A}, \mathfrak{Rd}^{(r)}\mathfrak{B})$.*

(iii) *For every class K of similar algebras, $\mathsf{Rd}^{(r)}\mathbf{H}\mathsf{K} \subseteq \mathbf{H}\mathsf{Rd}^{(r)}\mathsf{K}$ and $\mathsf{Rd}^{(r)}\mathbf{I}\mathsf{K} = \mathbf{I}\mathsf{Rd}^{(r)}\mathsf{K}$.*

(iv) *$Co\mathfrak{A} \subseteq Co\,\mathfrak{Rd}^{(r)}\mathfrak{A}$.*

(v) *$Cg^{(\mathfrak{Rd}^{(r)}\mathfrak{A})}X \subseteq Cg^{(\mathfrak{A})}X$ for every $X \subseteq A$.*

(vi) *If $\mathfrak{Rd}^{(r)}\mathfrak{A}$ is simple, then $\mathfrak{A}$ is simple.*

(vii) *$Il_z\mathfrak{A} \subseteq Il_z\mathfrak{Rd}^{(r)}\mathfrak{A}$ for every $z \in A$.*

(viii) *If the z-ideals function properly in $\mathfrak{Rd}^{(r)}\mathfrak{A}$, then the z-ideals function properly in $\mathfrak{A}$ as well.*

THEOREM 0.5.13. (i) $\mathfrak{Rd}^{(r)}(\mathsf{P}_{i\in I}\mathfrak{B}_i) = \mathsf{P}_{i\in I}\mathfrak{Rd}^{(r)}\mathfrak{B}_i$.

(ii) $\mathfrak{Rd}^{(r)}({}^I\mathfrak{A}) = {}^I\mathfrak{Rd}^{(r)}\mathfrak{A}$.

(iii) $\mathsf{Rd}^{(r)}\mathbf{P}\mathsf{K} = \mathbf{P}\mathsf{Rd}^{(r)}\mathsf{K}$.

(iv) *If $\mathfrak{A} \subseteq_{\mathsf{d}} \mathsf{P}_{i\in I}\mathfrak{B}_i$, then $\mathfrak{Rd}^{(r)}\mathfrak{A} \subseteq_{\mathsf{d}} \mathsf{P}_{i\in I}\mathfrak{Rd}^{(r)}\mathfrak{B}_i$.*

(v) *If $\mathfrak{Rd}^{(r)}\mathfrak{A}$ is directly indecomposable, weakly subdirectly indecomposable, or subdirectly indecomposable, then so is $\mathfrak{A}$.*

(vi) *$\mathfrak{Rd}^{(r)}(\mathsf{P}_{i\in I}\mathfrak{B}_i/\bar{F}) = \mathsf{P}_{i\in I}\mathfrak{Rd}^{(r)}\mathfrak{B}_i/\bar{F}$ for every filter F on I.*

(vii) *$\mathfrak{Rd}^{(r)}({}^I\mathfrak{A}/\bar{F}) = {}^I\mathfrak{Rd}^{(r)}\mathfrak{A}/\bar{F}$ for every filter F on I.*

(viii) $\mathsf{Rd}^{(r)}\mathsf{Up}\,\mathsf{K} = \mathsf{Up}\,\mathsf{Rd}^{(r)}\mathsf{K}$.

REMARK 0.5.14. It is easy to construct an algebra $\mathfrak{A}$ such that $\mathfrak{A}$ is semisimple but $\mathfrak{Rd}^{(r)}\mathfrak{A}$ is not, or such that $\mathfrak{Rd}^{(r)}\mathfrak{A}$ is semisimple but $\mathfrak{A}$ is not. The same applies to the notion of pseudo-simplicity and the various notions of pseudo-indecomposability (cf. 0.3.59).

THEOREM 0.5.15. *Let K be any class of similar algebras; put $\mathsf{K}' = \mathbf{S}\mathsf{Up}\,\mathsf{K}$.*

(i) *If L is a non-empty set of algebras directed by the relation $\subseteq^{\mathsf{r}}$ such that $\mathsf{U}^{\mathsf{r}}\mathsf{L}$ is similar to algebras in K, then the following two conditions are equivalent:*

(i′) *$\mathfrak{A} \in \mathbf{S}\mathsf{Rd}_{In\,\mathfrak{A}}\mathsf{K}'$ for each $\mathfrak{A} \in \mathsf{L}$;*

(i″) *$\mathsf{U}^{\mathsf{r}}\mathsf{L} \in \mathsf{K}'$.*

(ii) *In particular, for any algebra $\mathfrak{A}$ similar to algebras in K, the following conditions are equivalent:*

(ii′) *$\mathfrak{A} \in \mathsf{K}'$;*

(ii″) *$\mathfrak{Rd}_J\mathfrak{A} \in \mathsf{Rd}_J\mathsf{K}'$ for every finite $J \subseteq In\mathfrak{A}$;*

(ii‴) *$\mathfrak{Sg}^{(\mathfrak{Rd}_J\mathfrak{A})}X \in \mathbf{S}\mathsf{Rd}_J\mathsf{K}'$ for every finite $J \subseteq In\mathfrak{A}$ and finite $X \subseteq A$.*

This theorem clearly extends and improves 0.3.71. The proof of 0.5.15(i) is entirely analogous to that of 0.3.71(i); 0.5.15(ii) follows directly from 0.5.15(i) and 0.5.11(ii).

THEOREM 0.5.16. *For every cardinal* $\alpha \neq 0$,

(i) $Pl_\alpha \mathfrak{Rd}^{(r)}\mathfrak{A} \subseteq Pl_\alpha \mathfrak{A}$, $\mathfrak{Pl}_\alpha \mathfrak{Rd}^{(r)}\mathfrak{A} \subseteq \mathfrak{Rd}^{(r)}\mathfrak{Pl}_\alpha \mathfrak{A}$, *and* $Pl\,\mathfrak{Rd}^{(r)}\mathfrak{A} \subseteq Pl\,\mathfrak{A}$;

(ii) $\mathfrak{Fr}_\alpha \mathsf{Rd}^{(r)}\mathsf{K} \cong | \subseteq \mathfrak{Rd}^{(r)}\mathfrak{Fr}_\alpha \mathsf{K}$.

THEOREM 0.5.17. *Let* K *be a class of similar algebras and* I *their common index set. If, for each finite* $J \subseteq I$, *the class* $\mathsf{Rd}_J\mathsf{K}$ *satisfies 0.4.54(i) (with* K *replaced by* $\mathsf{Rd}_J\mathsf{K}$), *or the conclusion of 0.4.55 (with the same replacement), then so does* K; *similarly for 0.4.54(ii)–(v).*

For a proof of 0.5.17 see Keisler [61*].

REMARK 0.5.18. Theorem 0.5.17 essentially extends the range of applicability of 0.4.54, and 0.4.55. As was pointed out in the remark immediately following 0.4.55, Theorems 0.4.54 and 0.4.55 in the case of finite α have relatively little value when applied to classes K of algebras with infinite index sets, since many such classes contain no finite algebras. However, it can well happen that, for every finite subset J of the index set, the class $\mathbf{HSP}\mathsf{Rd}_J\mathsf{K}$ contains a finite algebra, with more than one element, among its members. Then, by applying 0.4.36 and 0.4.54, we infer that conditions 0.4.54(i)–(v) hold for each of the classes $\mathsf{Rd}_J\mathsf{K}$ and that consequently, by 0.5.17, they hold for the class K itself. A similar remark applies to 0.4.55.

REMARKS 0.5.19. We want still to discuss briefly some model-theoretical properties of reducts; a more detailed discussion of these properties can be found in Tarski [54*], [55*]. The most important questions here are of the following type: is a given property preserved when we pass from any class K of similar algebras to the class $\mathsf{Rd}^{(r)}\mathsf{K}$ or $\mathsf{Rd}_J\mathsf{K}$ of its reducts (for any given function r or set J)? With respect to all model-theoretical properties of K defined in the Preliminaries — thus, the properties of being elementary, universal, or equational (in the narrower or wider sense) — the answers to these questions prove to be negative. As instructive examples consider the classes K_1 of all commutative rings and K_2 of all integral domains, both treated as classes of algebraic structures $\mathfrak{A} = \langle A, +, -, \cdot \rangle$ of the type $\langle 2, 1, 2 \rangle$ with the common index set $3 = \{0, 1, 2\}$; let $J = 2 = \{0, 1\}$. As is well known, K_1 and K_2 are $\mathscr{UC}$'s, hence also $\mathscr{EC}$'s, and K_1 is actually an $\mathscr{EQC}$. $\mathsf{Rd}_J\mathsf{K}_1$, the class of all additive groups of rings, turns out to coincide with that of all Abelian groups, and hence it preserves the properties of K_1 just mentioned. On the other hand, $\mathsf{Rd}_J\mathsf{K}_2$ does not preserve the properties of K_2. In fact, let $\mathfrak{A}'$ be the additive group of integers and $\mathfrak{A}'' = \mathfrak{A}' \times \mathfrak{B}$ where $\mathfrak{B}$ is the additive group of rationals. $\mathfrak{A}'$ and $\mathfrak{A}''$ are known to be elementarily equivalent; cf. Szmielew [55*], p. 268, Theorem 6.6. Clearly $\mathfrak{A}' \in \mathsf{Rd}_J\mathsf{K}_2$; it can easily be shown, however, that $\mathfrak{A}'' \notin \mathsf{Rd}_J\mathsf{K}_2$. Hence we conclude at once that $\mathsf{Rd}_J\mathsf{K}_2$ is not a $\mathscr{UC}$ and not even an $\mathscr{EC}$ or an $\mathscr{EC}_\Delta$.

From these negative observations we can draw by means of 0.3.83 some algebraic conclusions. Consider the algebraic properties of K expressed by the formula

$$\mathsf{K} = \mathbf{S}\mathsf{K},\ \mathsf{K} = \mathbf{H}\mathsf{K},\ \mathsf{K} = \mathbf{P}\mathsf{K},\ \mathsf{K} = \mathsf{Up}\,\mathsf{K}.$$

By 0.5.13(iii),(viii) the last two properties are preserved by the classes of reducts; but the first two are not. In fact, the class K_2 of integral domains provides an example of a class K such that $\mathsf{K} = \mathbf{S}\mathsf{K} = \mathsf{Up}\,\mathsf{K}$ but, for some r, $\mathsf{Rd}^{(r)}\mathsf{K} \neq \mathbf{S}\mathsf{Rd}^{(r)}\mathsf{K}$. Also it is easy to construct a class K which is equational and hence satisfies the formulas $\mathsf{K} = \mathbf{S}\mathsf{K} = \mathbf{H}\mathsf{K} = \mathbf{P}\mathsf{K}$ while, for some r, $\mathbf{S}\mathsf{Rd}^{(r)}\mathsf{K} \neq \mathsf{Rd}^{(r)}\mathsf{K} \neq \mathbf{H}\mathsf{Rd}^{(r)}\mathsf{K}$.

In view of the above remarks the following result, which improves 0.3.86, may appear somewhat unexpected.

THEOREM 0.5.20. *If* K *is an* $\mathscr{EC}_\Delta$, *then* $\mathbf{S}\mathsf{Rd}^{(r)}\mathsf{K}$ *is a* $\mathscr{UC}_\Delta$.

This is a consequence of 0.3.70(i), 0.3.83, and 0.5.13(viii). The result was first proved by a different method in Tarski [54*], p. 585, Theorem 1.9.

Applying 0.3.71 and 0.3.83 we can obtain a model-theoretical version of 0.5.15 which states various properties of $\mathscr{UC}_\Delta$'s involving reducts and reduct unions. This version is, however, obvious and almost trivial; in a different arrangement of material it could be used to give an alternative proof of the original Theorem 0.5.15.

REMARK 0.5.21. Since there are classes of algebras which are elementary while their corresponding classes of reducts are not, it appears natural to single out those classes K of (similar) algebras which can be represented in the form $\mathsf{K} = \mathsf{Rd}^{(r)}\mathsf{L}$ where L is an $\mathscr{EC}$, or an $\mathscr{EC}_\Delta$. Such a class K is respectively called *pseudo-elementary in the narrower sense, or pseudo-elementary* (*in the wider sense*), for brevity a $\mathscr{PC}$, or a $\mathscr{PC}_\Delta$. Obviously, every $\mathscr{EC}$ is a $\mathscr{PC}$ and every $\mathscr{EC}_\Delta$ a $\mathscr{PC}_\Delta$; by 0.5.19 the converses do not hold. 0.5.13(viii) implies that $\mathscr{PC}_\Delta$'s share an important property with $\mathscr{EC}_\Delta$'s — the closure under the formation of ultraproducts. By 0.5.20, Theorem 0.3.86 also extends from $\mathscr{EC}_\Delta$'s to $\mathscr{PC}_\Delta$'s.

REMARK 0.5.22. The notion of reduct, like many other notions discussed in this chapter, extends in a natural way to arbitrary relational structures, and many results stated here apply with obvious changes to the extended notion. This is important for the purposes of the general theory of algebras since it frequently happens that a class K of algebras can be naturally represented as the class of reducts of a class L of relational structures; from known properties of the class L we may then be able to draw some conclusions concerning the class K. In principle, we could always construe L as a class of algebras. This is a consequence of the simple observation that a relation can always be replaced by its characteristic function; cf. the Preliminaries. Such a procedure, however, would be rather unnatural and would present some disadvantages from a heuristic point of view.

If $\mathfrak{A}$ is a relational structure $\langle A, R_i, O_j\rangle_{i\in I, j\in J}$ with relations R_i and operations O_j, each of its reducts is determined by two functions, r and s with ranges included in I and J, respectively, or by two sets, a subset I' of I and a subset J' of J. Thus we have to use the symbolic expressions $\mathfrak{Rd}^{(r,s)}\mathfrak{A}$ and $\mathfrak{Rd}_{I',J'}\mathfrak{A}$ to denote reducts of $\mathfrak{A}$, and similarly $\mathsf{Rd}^{(r,s)}\mathsf{L}$ and $\mathsf{Rd}_{I',J'}\mathsf{L}$ for classes of reducts. In case r or I' is empty, the resulting reducts are algebras. If a class K of algebras (or, more generally, of relational structures) can be represented in the form $\mathsf{K} = \mathsf{Rd}^{(r,s)}\mathsf{L}$ where L is an elementary class of relational structures, K can still be called pseudo-elementary. Since, e.g., Theorems 0.5.13(viii) and 0.5.20 apply to the extended reducts (see Frayne-Morel-Scott [62*] for the necessary changes in definitions of reduced product and ultraproduct), a class K of algebras which is pseudo-elementary in the extended sense is still closed under the formation of ultraproducts, and the class **S**K is universal.

To give a rather interesting example, consider a class K of algebras $\mathfrak{A} = \langle A, O_j\rangle_{j\in J}$. Let L be the class of all relational structures $\mathfrak{R} = \langle A, R, O_j\rangle_{j\in J}$ such that $\mathfrak{A} = \langle A, O_j\rangle_{j\in J}$ is an algebra in K and R is a congruence relation of $\mathfrak{A}$ different from $A1Id$ and $A\times A$. As is easily seen, the class $\mathsf{K}' = \mathsf{Rd}_{0,J}\mathsf{L}$ is the class of all those algebras $\mathfrak{A}\in\mathsf{K}$ which are not simple. Thus, if K is elementary, the class K′ of all non-simple algebras in K is pseudo-elementary. It is known that, in general, K′ is not elementary. (For the case of the class K of groups this is stated in Tarski [52a], p. 717, footnote 17.) We mention here some further classes K′ which by means of an analogous argument can be shown to be pseudo-elementary in case K itself is elementary: the class of all algebras in K which have a proper subalgebra, or are directly decomposable, or are weakly subdirectly decomposable; the class of all algebras which have a subalgebra, or have a quotient algebra, belonging to K; the class of all algebras which are isomorphic to direct products of two, or of any other fixed finite number, of algebras in K. Many other examples of this kind are known.

PROBLEMS

PROBLEM 0.1. *Let $\mathfrak{A}$ be a finite algebra such that every subset of A consisting of a single element is a subuniverse of $\mathfrak{A}$. Does $\mathfrak{A}$ necessarily have the unique decomposition property?*

For related results see McKenzie [70*]. In particular it is stated there that Problem 0.1 admits an affirmative solution when restricted to semigroups.

PROBLEM 0.2. *Does there exist an algebra which is uniquely totally decomposable and all factors of which are totally decomposable but not all are uniquely totally decomposable?*

Compare the solution of a weaker problem, obtained moreover on the basis of a special set-theoretical hypothesis, in 0.3.32.

PROBLEM 0.3. *Can one prove without the help of any set-theoretical hypothesis like the continuum hypothesis that there is an algebra which is uniquely totally decomposable but does not have the refinement property?*

PROBLEM 0.4. *Is there an algebra which has the remainder property and the finite refinement property but fails to have the general refinement property (possibly restricted to the case of denumerably many factors)?*

This problem was raised by Robert Bradford.

PROBLEM 0.5. *Let $F\mathfrak{A}$ denote the set of all factor congruence relations on a given algebra $\mathfrak{A}$. By a complement of a congruence relation $R \in F\mathfrak{A}$ we understand any congruence relation $S \in F\mathfrak{A}$ such that $R \cap S = A \upharpoonright Id$ and $R|S = A \times A$. Consider the following two properties of $\mathfrak{A}$:* (i) *every $R \in F\mathfrak{A}$ has just one complement*; (ii) *the system $\mathfrak{F}\mathfrak{A} = \langle F\mathfrak{A}, |, \cap \rangle$ is a distributive lattice or — what amounts to the same in this case — a Boolean algebra (in the sense of Remarks 0.1.4). Does (i) always imply (ii)? If not, does (i) imply (ii) in case $\mathfrak{F}\mathfrak{A}$ is a lattice or $A \times A$ is a finitely generated congruence relation or both?*

It is known that condition (ii) is equivalent to: (iii) $\mathfrak{A}$ *has the strict refinement property*. (This is the stronger variant of the refinement property mentioned in connection with 0.3.34. Cf. Chang-Jónsson-Tarski [64*].) Notice that in general $\mathfrak{F}\mathfrak{A}$ is not an algebra at all, since the set $F\mathfrak{A}$ is not necessarily closed under the operations | and $\mathbf{\cap}$; if, however, $\mathfrak{F}\mathfrak{A}$ is an algebra, then it is a lattice and actually a modular lattice.

PROBLEM 0.6. *For* $|A| = \omega$, *does there exist a cardinal* $\alpha \geqq \omega$ *and an ultrafilter* F *on* α *such that* $\omega < |{}^{\alpha}A/F| < 2^{\alpha}$?

This problem is formulated in Frayne-Morel-Scott [62*], p. 208.

PROBLEM 0.7. *Consider the infinite sequence of operations (from and to classes of algebras)* **P**, **PUp**, **PUpP**, ... *Are all the operations in this sequence distinct?*

PROBLEM 0.8. *Is it possible to prove without* $\mathscr{GCH}$ *that any two elementarily equivalent algebras have isomorphic ultrapowers?*

PROBLEM 0.9. *Let the formula* $\mathfrak{A} \triangleq \mathfrak{B}$ *express the fact that the algebras* $\mathfrak{A}$ *and* $\mathfrak{B}$ *have some isomorphic ultrapowers. Can one prove without the help of* $\mathscr{GCH}$ *that* $\triangleq$ *is an equivalence relation and that the partition induced by this relation in any similarity class of algebras, with non-empty and at most denumerable systems of fundamental operations, has the cardinality of the set of real numbers?*

In connection with the last two problems compare 0.3.79 and 0.3.80. Both problems are formulated in Keisler [61a*], p. 487.[1)]

[1)] [*Added in proof.*] The last two problems have recently been solved affirmatively by Saharon Shelah; cf. footnote[1)] on p. 113.

CHAPTER 1

ELEMENTARY PROPERTIES OF CYLINDRIC ALGEBRAS

1. ELEMENTARY PROPERTIES OF CYLINDRIC ALGEBRAS

In this chapter we shall define the class of cylindric algebras and develop the most basic portion of its elementary (arithmetical) theory. In the course of the development we shall have occasion to introduce various concepts which play a fundamental role in the later, more specialized discussion.

1.1. CYLINDRIC ALGEBRAS

A cylindric algebra is a Boolean algebra enriched by certain additional operations. Before beginning the discussion of cylindric algebras we establish the terminology and notation to be used in dealing with the Boolean part of the algebra. As a general reference for the theory of Boolean algebras see Halmos [63*] or Sikorski [64*].

A Boolean algebra is an algebraic structure $\mathfrak{A} = \langle A, +, \cdot, -, 0, 1\rangle$ in which $+$ and $\cdot$ are binary operations and $-$ is a unary operation on A, while 0 and 1 are distinguished elements of A, and which satisfies certain well-known postulates. We may mention, e.g., the following postulates which are known to be adequate for this purpose when satisfied for arbitrary $x, y, z \in A$:

(B_0) $x+y = y+x$, $\quad x\cdot y = y\cdot x$,

(B_1) $x+(y\cdot z) = (x+y)\cdot(x+z)$, $\quad x\cdot(y+z) = x\cdot y+x\cdot z$,

(B_2) $x+0 = x$, $\quad x\cdot 1 = x$,

(B_3) $x+-x = 1$, $\quad x\cdot -x = 0$.[1)]

Given a Boolean algebra $\mathfrak{A}$, for any $x, y \in A$ we let $x-y = x\cdot -y$ and $x\oplus y = (x-y)+(y-x)$; we also let $x \leqq y$, or $y \geqq x$, iff $x+y = y$, and $x < y$, or $y > x$, iff both $x \leqq y$ and $x \neq y$. $x-y$ is called the *difference* of x and y, $x\oplus y$ the *symmetric difference* of x and y, and $\leqq$ is called the *inclusion* relation. The class of all Boolean algebras is denoted by BA. We also use

1) This postulate system is essentially due to Huntington [04*], p. 292 f. but contains a modification suggested by Bernstein [50*].

"BA" just as an abbreviation for "Boolean algebra", saying, e.g., that a given algebra $\mathfrak{A}$ is a BA. We use the same convention with respect to various classes introduced below, such as the class CA of all cylindric algebras. If $\mathfrak{A} = \langle A, +, \cdot, -, 0, 1\rangle$ is a BA, then $\langle A, +, \cdot\rangle$ is a lattice, and we apply to $\mathfrak{A}$ the terminology for lattices introduced in Remarks 0.1.4; in particular, $\sum^{(\mathfrak{A})}$ and $\prod^{(\mathfrak{A})}$ (or simply $\sum$ and $\prod$) are the infinitary generalizations of $+$ and $\cdot$, respectively. We assume as known the notion of an *atom.* $\mathfrak{A}$ is *atomic* if for every non-zero $x \in A$ there is an atom y such that $y \leqq x$. $At\mathfrak{A}$ is the set of all atoms of $\mathfrak{A}$.

A *field of sets* is a non-empty family F of sets such that F is closed under $\cup$, $\cap$, and complementation with respect to $\bigcup F$, and such that $0, \bigcup F \in F$. Note that $\{0\}$ is a field of sets. A *Boolean set algebra* is a structure $\langle A, \cup, \cap, \sim, 0, \bigcup A\rangle$ where A is a field of sets; $\bigcup A$ is the *unit set* of both the field of sets and the Boolean set algebra just mentioned. Clearly such a structure is a Boolean algebra.

We also assume as known the notions of *ideal* and *filter* (or *dual ideal*) in a Boolean algebra. A brief discussion of these notions will be found in Chapter 2; see the remarks preceding 2.3.4. Other more special notions will be introduced when needed in the further discussion.

DEFINITION 1.1.1. *By a* ***cylindric algebra of dimension*** α, *where* α *is any ordinal number, we mean an algebraic structure*

$$\mathfrak{A} = \langle A, +, \cdot, -, 0, 1, \mathsf{c}_\kappa, \mathsf{d}_{\kappa\lambda}\rangle_{\kappa,\lambda<\alpha}$$

such that 0, 1, *and* $\mathsf{d}_{\kappa\lambda}$ *are distinguished elements of* A (*for all* $\kappa, \lambda < \alpha$), $-$ *and* c_κ *are unary operations on* A (*for all* $\kappa < \alpha$), $+$ *and* $\cdot$ *are binary operations on* A, *and such that the following postulates are satisfied for any* $x, y \in A$ *and any* $\kappa, \lambda, \mu < \alpha$:

(C_0) *the structure* $\langle A, +, \cdot, -, 0, 1\rangle$ *is a* BA;
(C_1) $\mathsf{c}_\kappa 0 = 0$;
(C_2) $x \leqq \mathsf{c}_\kappa x$ (*i.e.*, $x + \mathsf{c}_\kappa x = \mathsf{c}_\kappa x$);
(C_3) $\mathsf{c}_\kappa(x \cdot \mathsf{c}_\kappa y) = \mathsf{c}_\kappa x \cdot \mathsf{c}_\kappa y$;
(C_4) $\mathsf{c}_\kappa \mathsf{c}_\lambda x = \mathsf{c}_\lambda \mathsf{c}_\kappa x$;
(C_5) $\mathsf{d}_{\kappa\kappa} = 1$;
(C_6) *if* $\kappa \neq \lambda, \mu$, *then* $\mathsf{d}_{\lambda\mu} = \mathsf{c}_\kappa(\mathsf{d}_{\lambda\kappa} \cdot \mathsf{d}_{\kappa\mu})$;
(C_7) *if* $\kappa \neq \lambda$, *then* $\mathsf{c}_\kappa(\mathsf{d}_{\kappa\lambda} \cdot x) \cdot \mathsf{c}_\kappa(\mathsf{d}_{\kappa\lambda} \cdot -x) = 0$.[1)]

[1)] A definition with essentially the same postulate system was published in Tarski-Thompson [52]; see also Thompson [52], p. 10. It may be mentioned that Postulate (C_7) was used

The class of all cylindric algebras is denoted by CA, *and the class of all cylindric algebras of dimension* α *by* CA_α. *The elements* $\mathsf{d}_{\kappa\lambda}$ *are called* ***diagonal elements,*** *and the operations* c_κ *are called* ***cylindrifications.*** *The structure* $\mathfrak{B}\mathfrak{l}\,\mathfrak{A} = \langle A, +, \cdot, -, 0, 1\rangle$ *is the* ***Boolean part*** *of* $\mathfrak{A}$.

The meanings of the symbols $+$, ..., c_κ, and $\mathsf{d}_{\kappa\lambda}$ of course depend on (and are uniquely determined by) the cylindric algebra $\mathfrak{A}$ which is formed from the operations and elements denoted by these symbols. Theoretically we should make this dependence explicit, e.g., by providing each of these symbols with the superscript $^{(\mathfrak{A})}$. In practice, however, the superscript will be omitted whenever there is no possibility of confusion. Analogous remarks apply to all symbols and technical terms which will be introduced in our further discussion and which designate notions relative to a given algebra $\mathfrak{A}$; cf. here 0.1.2.

According to 1.1.1 cylindric algebras are Boolean algebras with additional operations indexed by ordinals. The notion of a cylindric algebra as a structure with operations indexed by an arbitrary set could also be introduced; the generality thereby gained is slight, however, and the Definition 1.1.1 allows more ease of expression. Elementary notions specific for the Boolean part of an $\mathfrak{A} \in \mathsf{CA}_\alpha$ will be applied also to $\mathfrak{A}$ itself. Thus, e.g., $\mathfrak{A}$ is *atomic*, or *countably complete*, iff $\mathfrak{B}\mathfrak{l}\,\mathfrak{A}$ is atomic, or countably complete.

Sometimes we shall have to deal, not with CA's, but with algebras similar to them in the sense of 0.1.5. As was pointed out in 0.1.3, we have to change Definition 1.1.1 slightly in order to apply to CA's the terminology of the general theory of algebraic structures. The only modification of any substance consists in replacing the distinguished elements 0, 1, $\mathsf{d}_{\kappa\lambda}$ by the constant unary operations $0'$, $1'$, $\mathsf{d}'_{\kappa\lambda}$ respectively, assuming these elements as their unique values. With this modification an algebra $\mathfrak{B}'$ is similar to CA_α's in the sense of 0.1.5 iff

$$\mathfrak{B}' = \langle B, +, \cdot, -, 0', 1', \mathsf{c}_\kappa, \mathsf{d}'_{\kappa\lambda}\rangle_{\kappa,\lambda<\alpha}$$

where $+$ and $\cdot$ are arbitrary binary operations while $-$, $0'$, $1'$, c_κ, $\mathsf{d}'_{\kappa\lambda}$ are arbitrary unary operations on B; we do not require the operations $0'$, $1'$, $\mathsf{d}'_{\kappa\lambda}$ to be constant since the similarity type of an algebra depends exclusively on the ranks of its fundamental operations. It proves, however, more convenient for our purposes to use the term "similar" in this context in a more restrictive

originally in the following form:

$$\text{if } \kappa \neq \lambda \text{ then } \mathsf{d}_{\kappa\lambda} \cdot \mathsf{c}_\kappa(\mathsf{d}_{\kappa\lambda} \cdot x) \leq x.$$

The modified formulation actually adopted both here and in op. cit. is due to a suggestion of Roger Lyndon.

sense, and to call $\mathfrak{B}'$ similar to CA_α's just in case the operations $0'$, $1'$, $\mathsf{d}'_{\kappa\lambda}$ are actually constant. We can now return to our original definition of CA_α's and refer to a structure $\mathfrak{B}$ as *similar* to a CA_α if

$$\mathfrak{B} = \langle B, +, \cdot, -, 0, 1, \mathsf{c}_\kappa, \mathsf{d}_{\kappa\lambda}\rangle_{\kappa,\lambda<\alpha}$$

where $+, \cdot, \ldots, \mathsf{d}_{\kappa\lambda}$ represent objects of the same kinds as they do in 1.1.1.

DEFINITION 1.1.2. *$\mathfrak{B}$ is a* ***diagonal-free cylindric algebra of dimension*** *α, in symbols $\mathfrak{B} \in \mathsf{Df}_\alpha$, if $\mathfrak{B}$ is an algebraic structure $\langle B, +, \cdot, -, 0, 1, \mathsf{c}_\kappa\rangle_{\kappa<\alpha}$ where $+, \cdot, -, 0, 1$, and c_κ are as in 1.1.1, and Postulates (C_0)–(C_4) of 1.1.1 hold.*

If $\mathfrak{A} = \langle A, +, \cdot, -, 0, 1, \mathsf{c}_\kappa, \mathsf{d}_{\kappa\lambda}\rangle_{\kappa,\lambda<\alpha} \in \mathsf{CA}_\alpha$, the ***diagonal-free part of*** *$\mathfrak{A}$ is the structure $\mathfrak{Df}\,\mathfrak{A} = \langle A, +, \cdot, -, 0, 1, \mathsf{c}_\kappa\rangle_{\kappa<\alpha}$.*

Various definitions and conventions introduced for CA's will be automatically carried over to Df's without explicit statement. Thus if $\mathfrak{A} \in \mathsf{Df}_\alpha$, then $\mathfrak{Bl}\,\mathfrak{A}$ is its Boolean part.

Our main concern here is with cylindric algebras (with diagonal elements), and the class Df_α will play only a minor role in our discussion.

Obviously we have

THEOREM 1.1.3. *If $\mathfrak{A} \in \mathsf{CA}_\alpha$, then $\mathfrak{Df}\,\mathfrak{A} \in \mathsf{Df}_\alpha$.*

Clearly, if $\mathfrak{A}$ is a CA, then $\mathfrak{Df}\,\mathfrak{A}$ is a reduct of $\mathfrak{A}$ (cf. 0.5.1); however, it is easily seen from 1.5.24 below that not every diagonal-free cylindric algebra is a reduct of a CA.

Singly the postulates (C_1)–(C_4) are well known. (C_1) expresses the normality of each of the operations c_κ (in the sense of Jónsson-Tarski [51], p. 877). Postulate (C_3) may be called the modular law in view of its close analogy to the modular laws in lattices (see the definition of modular lattices in 0.1.4). (C_4) expresses the commutativity of the operations c_κ.

To obtain the simplest (and rather trivial) examples of CA's, we notice that every Boolean algebra $\mathfrak{A} = \langle A, +, \cdot, -, 0, 1\rangle$ can be transformed into a CA_α (of any prescribed dimension α) just by letting $\mathsf{c}_\kappa x = x$ and $\mathsf{d}_{\kappa\lambda} = 1$ for every $x \in A$ and all $\kappa, \lambda < \alpha$. The algebras thus obtained present little interest as cylindric algebras; the discussion of these algebras reduces entirely to that of their Boolean parts. In Section 1.3 we shall say something more about these algebras, and in particular we shall agree to refer to them as *discrete* CA's; cf. Definition 1.3.10 and the immediately following theorems.

REMARKS 1.1.4. We now discuss briefly two construction methods which lead to two important classes of CA's — the *cylindric set algebras* and the *cylindric algebras of formulas*. In the second part of this work these two classes of CA's will be intensively studied in their own right and in their relation to arbitrary CA's. It seems desirable, however, to introduce the two notions at this early stage since their discussion throws much light on our motives in designing the general notion of cylindric algebras and in selecting the postulate system; moreover, the two classes of CA's will serve as a rich source of examples and counterexamples, and will help to illustrate the content of particular results. We shall sometimes refer informally to algebras of the two classes as *special cylindric algebras*, occasionally including also the discrete algebras under this term. (We do not use the term "special cylindric algebra" here in the same sense as in the Foreword.)

The discussion of the first class of special CA's, that of cylindric set algebras, leads naturally to most of our terminology adopted in the general theory. The construction of these is set-theoretical or — in some very general and abstract sense — geometrical in character. The basic ideas are found in the analytical geometry of real three-dimensional space. If R is the set of real numbers, we apply customary geometric terminology to the space ${}^3R$, except that we speak of the 0-, 1-, and 2-coordinates of a point rather than its x-, y-, and z-coordinates. On point sets of ${}^3R$ we can of course perform the ordinary set-theoretical operations of union, intersection, and complementation (with respect to ${}^3R$). We are also interested in the less familiar operation of cylindrification parallel to an axis: if $\kappa < 3$ and $X \subseteq {}^3R$, let

$$\mathsf{C}_\kappa X = \{y : y \in {}^3R \text{ and } y_\lambda = x_\lambda \text{ for some } x \in X \text{ and all } \lambda \neq \kappa\}.$$

Thus $\mathsf{C}_\kappa X$ is the cylinder generated by moving the point set X parallel to the κ-axis. Moreover, certain special point sets of ${}^3R$ are of interest to us. These are, first of all, the empty point set 0 and the whole space ${}^3R$. Then, if $\kappa, \lambda < 3$, let

$$\mathsf{D}_{\kappa\lambda} = \{x : x \in {}^3R,\ x_\kappa = x_\lambda\}.$$

Thus $\mathsf{D}_{\kappa\lambda}$ is a diagonal plane in ${}^3R$ if $\kappa \neq \lambda$, and $\mathsf{D}_{\kappa\kappa} = {}^3R$. Now suppose A is a family of subsets of ${}^3R$ which is closed under the Boolean operations and under $\mathsf{C}_\kappa(\kappa < 3)$, and contains all the sets $\mathsf{D}_{\kappa\lambda}$ $(\kappa, \lambda < 3)$. Let $\mathfrak{A} = \langle A, \cup, \cap, {}_{{}^3R}\sim, 0, {}^3R, \mathsf{C}_\kappa, \mathsf{D}_{\kappa\lambda}\rangle_{\kappa,\lambda<3}$. Then it is easily verified that $\mathfrak{A}$ is a three-dimensional cylindric algebra. For example, (C_6) holds since, if $\kappa \neq \lambda$, μ and $\lambda \neq \mu$, the diagonal plane $\mathsf{D}_{\lambda\mu}$ can be obtained by cylindrifying the diagonal line $\mathsf{D}_{\lambda\kappa} \cap \mathsf{D}_{\kappa\mu}$, while in case $\lambda = \mu$ the space ${}^3R$ can be obtained by cylindrifying the plane $\mathsf{D}_{\lambda\kappa}$; (C_7) holds since cylindrification by C_κ or C_λ of disjoint subsets of the plane $\mathsf{D}_{\kappa\lambda}$ for $\kappa \neq \lambda$ leads to disjoint subsets of the space. Further elementary properties and notions are suggested upon considering this CA_3.

For example we clearly have $\mathsf{C}_\kappa(\sim\mathsf{C}_\kappa X) = \sim\mathsf{C}_\kappa X$ and $\mathsf{C}_\kappa(X \cup Y) = \mathsf{C}_\kappa X \cup \mathsf{C}_\kappa Y$ whenever $\kappa < 3$ and $X, Y \subseteq {}^3R$. Much of Chapter 1 will be devoted to the proof of such elementary laws directly from the postulates.

This construction can obviously be generalized by taking an arbitrary set U instead of R and an arbitrary ordinal α instead of 3:[1)]

DEFINITION 1.1.5. (i) *For any set U, any ordinal α, and any $\kappa < \alpha$ we let $\mathsf{C}_\kappa^{(U,\alpha)}$, or $\mathsf{C}_\kappa^{(U)}$, or even simply C_κ, be the function from $Sb^\alpha U$ into $Sb^\alpha U$ such that*

$$\mathsf{C}_\kappa X = \{y : y \in {}^\alpha U \text{ and for some } x \in X \text{ we have } y_\lambda = x_\lambda \text{ for all } \lambda \neq \kappa\}$$

for every $X \subseteq {}^\alpha U$.

(ii) *For any set U, any ordinal α, and any $\kappa, \lambda < \alpha$ we denote the set*

$$\{y : y \in {}^\alpha U, y_\kappa = y_\lambda\}$$

by $\mathsf{D}_{\kappa\lambda}^{(U,\alpha)}$, or $\mathsf{D}_{\kappa\lambda}^{(U)}$, or even simply $\mathsf{D}_{\kappa\lambda}$.

(iii) *A is an **α-dimensional cylindric field of sets** iff there is a set U, called the **base** of A, such that A is a non-empty subset of $Sb^\alpha U$ closed under all the operations $\cup$, $\cap$, $\sim$ (i.e., ${}_{^\alpha U}\!\sim$), and C_κ (for $\kappa < \alpha$) and containing all the sets $\mathsf{D}_{\kappa\lambda}$.*

(iv) *$\mathfrak{A}$ **is a cylindric set algebra of dimension α with base** U iff $\mathfrak{A} = \langle A, \cup, \cap, \sim, 0, {}^\alpha U, \mathsf{C}_\kappa, \mathsf{D}_{\kappa\lambda}\rangle_{\kappa,\lambda<\alpha}$ where A is an α-dimensional cylindric field of sets with base U. In case $A = Sb^\alpha U$, A and $\mathfrak{A}$ are respectively called a **full cylindric field of sets** and a **full cylindric set algebra.***

When referring to cylindric set algebras in informal remarks, we shall use geometrical terminology (extending the usual terminology of analytic geometry). Thus the set ${}^\alpha U$, the αth Cartesian power of U in the sense of the Preliminaries, is also called the *α-dimensional* (*Cartesian*) *space with base U* (or *over U*). Elements of this set are referred to as *points* and subsets as *point sets*. If x is a point of ${}^\alpha U$ and hence a sequence $\langle x_\xi : \xi < \alpha\rangle$, we refer to the ξth projection of x, $\mathsf{pj}_\xi x = x_\xi$, as the ξth *coordinate* of x. A point set X such that $X = \mathsf{C}_\kappa X$ is called a *cylinder parallel to the κth axis* or, for brevity, a *κ-cylinder*. An α-dimensional cylindric field of sets A, or cylindric set algebra $\mathfrak{A}$, with the base U will sometimes be referred to as a field, or an algebra, *in the space ${}^\alpha U$*; this space is clearly the unit set of both the space A and the algebra $\mathfrak{A}$.

In application to set algebras we shall use the symbol "C" and "D" to replace "c" and "d", not only in the basic situation referred to in 1.1.5, but also in some related contexts introduced below in 1.7.1, 1.8.1, and 1.9.1.

THEOREM 1.1.6. *Every cylindric set algebra of dimension α is a CA_α.*

[1)] The construction in its essential parts was first carried out in Tarski [31]; see in particular the last paragraph of the paper.

The proof presents no difficulty.

For the purpose of graphic illustration, we shall frequently use 2-dimensional cylindric set algebras, since in this case the base U can be pictured as a segment and ${}^2U$ as a square. For instance, in Fig. 1.1.7 we exhibit a cylindric field of subsets of ${}^2U$, a member X of this field, and the cylinders C_0X and C_1X. Fig. 1.1.8 helps to check the validity of Postulate (C_3).

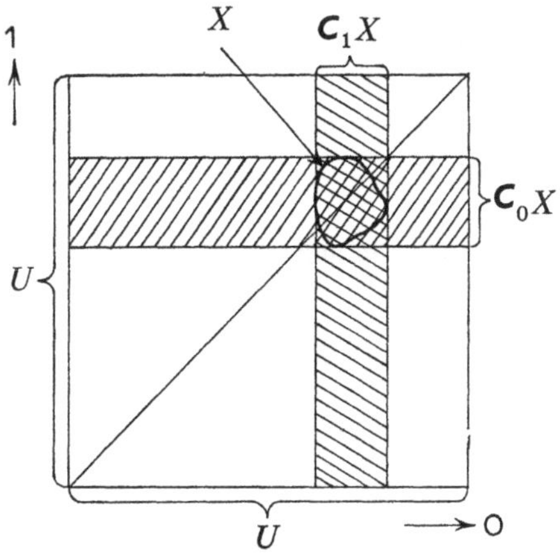

Fig. 1.1.7

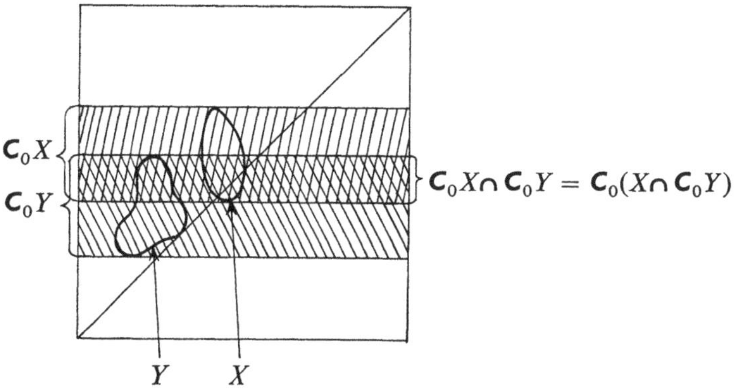

Fig. 1.1.8

It is easily seen how Definition 1.1.5 should be modified to obtain the notions of an *α-dimensional diagonal-free cylindric field of sets* and of a *diagonal-free cylindric set algebra of dimension* α. One has only to disregard everything in 1.1.5 having to do with diagonal elements, and then to replace the Cartesian power ${}^{\alpha}U$ by the more general Cartesian product $\mathsf{P}_{\xi<\alpha}U_\xi$ where $\langle U_\xi : \xi < \alpha\rangle$ is any α-termed sequence of sets. Clearly, every diagonal-free cylindric set algebra of dimension α is a Df_α.

We now turn to the second class of special algebras, the cylindric algebras of formulas. The construction of these algebras has a metamathematical or rather metalogical character and resembles the metamathematical construction of free algebras (cf. Section 0.4).

DEFINITION 1.1.9. *For each language* Λ *of predicate logic we let* $\mathfrak{Fm}^{(\Lambda)}$ *be the algebra*

$$\langle \Phi\mu^{(\Lambda)}, \vee, \wedge, \neg, F, T, \exists_{v_\kappa}, v_\kappa = v_\lambda\rangle_{\kappa,\lambda<\alpha}$$

where α *is the length of the sequence of variables of* Λ. $\mathfrak{Fm}^{(\Lambda)}$ *is called the* ***free algebra of formulas*** ***(in*** Λ).

Recall here that $\vee$, $\wedge$ and $\neg$ are treated as operations on formulas, as is $\exists_{v_\kappa}$ for each $\kappa < \alpha$; cf. the Preliminaries. To justify the terminology introduced in 1.1.9 notice that, K being the class of all algebras similar to CA's, the algebra $\mathfrak{Fm}^{(\Lambda)}$ is seen to be K-freely generated, in the sense of 0.4.23, by the set A of all those atomic formulas in Λ which contain some non-logical constants (assuming $A \neq 0$).

THEOREM 1.1.10. *Let* Λ *be a language of predicate logic in which the sequence of variables has length* α. *We then have*:

(i) $\mathfrak{Fm}^{(\Lambda)}$ *is similar to* CA_α's;

(ii) *if* Σ *is a set of sentences of* Λ, *then* $\equiv_\Sigma \in Co\mathfrak{Fm}^{(\Lambda)}$ (*i.e.*, $\equiv_\Sigma$ *is a congruence relation of* $\mathfrak{Fm}^{(\Lambda)}$) *and* $\mathfrak{Fm}^{(\Lambda)}/\equiv_\Sigma \in \mathsf{CA}_\alpha$.

The proof of 1.1.10 is straightforward. To prove the last part of 1.1.10(ii) we have to check that postulates (C_0)–(C_7) hold in $\mathfrak{Fm}^{(\Lambda)}/\equiv_\Sigma$. Indeed, in the process of checking these postulates the logical origin of cylindric algebras comes to light. We shall illustrate this by considering two postulates, (C_3) and (C_7).

It is well known that, for any formulas φ and ψ and any variable v_κ with $\kappa < \alpha$, the formulas

$$\exists_{v_\kappa}(\varphi \wedge \exists_{v_\kappa}\psi) \text{ and } \exists_{v_\kappa}\varphi \wedge \exists_{v_\kappa}\psi$$

are logically equivalent. In fact, this equivalence is one of the basic laws by means of which any formula can be equivalently expressed in the prenex normal form. The fact that (C_3) holds in $\mathfrak{Fm}^{(\Lambda)}/\equiv_\Sigma$ is an immediate consequence of this equivalence.

On the other hand, (C_7) expresses a simple property of the operation of substituting one variable for another in a given formula. This metalogical operation is well known and need not be described here in detail. Roughly speaking, to substitute $\mathbf{v}_\lambda$ for $\mathbf{v}_\kappa$ in φ we simply replace $\mathbf{v}_\kappa$ in all its free occurrences by $\mathbf{v}_\lambda$; if, however, $\mathbf{v}_\lambda$ occurs bound in φ, then to avoid "collisions" we may have to begin by picking a "non-colliding" variable $\mathbf{v}_\mu$ (e.g., a variable not occurring in φ at all) and replacing $\mathbf{v}_\lambda$ by $\mathbf{v}_\mu$ wherever the former is bound in φ. If ψ and ψ' are respectively the formulas obtained from φ and $\neg\varphi$ by substituting $\mathbf{v}_\lambda$ for $\mathbf{v}_\kappa$, then clearly $\psi' = \neg\psi$ (assuming that the same variable $\mathbf{v}_\mu$ was used in eliminating the bound occurrences of $\mathbf{v}_\lambda$ from φ and from $\neg\varphi$). On the other hand, it is known (and can easily be checked) that, in case $\kappa \neq \lambda$, the formulas $\exists_{\mathbf{v}_\kappa}(\mathbf{v}_\kappa = \mathbf{v}_\lambda \wedge \varphi)$ and $\exists_{\mathbf{v}_\kappa}(\mathbf{v}_\kappa = \mathbf{v}_\lambda \wedge \neg\varphi)$ are respectively equivalent to ψ and ψ'.[1] Hence the formula

$$\neg[\exists_{\mathbf{v}_\kappa}(\mathbf{v}_\kappa = \mathbf{v}_\lambda \wedge \varphi) \wedge \exists_{\mathbf{v}_\kappa}(\mathbf{v}_\kappa = \mathbf{v}_\lambda \wedge \neg\varphi)]$$

is logically valid, and consequently (C_7) holds in $\mathfrak{Fm}^{(\Lambda)}/\equiv_\Sigma$. The role of (C_7) in discussion of properties of the substitution operation will be seen further in Sections 1.5 and 1.11.

DEFINITION 1.1.11. *Given a language* Λ *of predicate logic and a set* Σ *of sentences in* Λ, *the algebra* $\mathfrak{Fm}^{(\Lambda)}/\equiv_\Sigma$ *is referred to as a* ***cylindric algebra of formulas*** *or, more specifically, as the* ***(cylindric) algebra of formulas (in*** Λ***) associated with*** Σ.[2]

REMARKS 1.1.12. At many places in the later discussion it will prove convenient to use cylindric algebras of formulas associated with a theory Θ rather

[1]) This observation was made in Tarski [51*]; its significance for the formulation of predicate logic is discussed in Tarski [65*].

[2]) The method of constructing algebras of formulas (correlated with a set of sentences Σ) was originally applied to languages containing sentential connectives but not necessarily quantifiers, so that the construction led to Boolean algebras rather than cylindric algebras. In this form, but with the use of somewhat different terminology, the method was outlined in Tarski [35*]; see, in particular, Theorem 4 on p. 510, as well as the footnote on pp. 511 f. where references to some earlier papers of B. A. Bernstein, E. V. Huntington, and Tarski are also given. In its application to languages with quantifiers the method was discussed at some length in Henkin-Tarski [61]; cf. pp. 85 ff., as well as historical remarks on pp. 111 f. For misinformation concerning the history of this method see the footnote on p. 245 f. in Rasiowa-Sikorski [63*] and compare it with the article of J. C. C. McKinsey quoted there.

than with an arbitrary set of sentences Σ. Actually no generality would be lost even if we agreed to use the notion of an algebra of formulas exclusively in this restricted context. This follows from the simple fact that, for any set of sentences Σ in a language Λ, we have $\mathfrak{Fm}^{(\Lambda)}/\equiv_\Sigma = \mathfrak{Fm}^{(\Lambda)}/\equiv_\Theta$ where Θ is the set of all formulas of Λ which are consequences of Σ, i.e., $\Theta = \Theta\rho\,\mathsf{Md}\,\Sigma$ (cf. the Preliminaries).

The metalogical method of constructing CA's suggests an alternative terminology for abstract cylindric algebras. Thus recalling that the operation on formulas correlated with the (existential) quantifier is termed (existential) quantification (with respect to a variable), we could extend this terminology to arbitrary CA's and refer to the operations c_κ as *quantifications* or, more specifically, *existential quantifications*. Similarly, the elements $\mathsf{d}_{\kappa\lambda}$ could be called *identity elements*, and the CA's themselves could be referred to as *quantifier algebras*. We shall occasionally indicate how this alternative terminology could be extended to other notions which will be introduced in our discussion.[1)]

REMARKS 1.1.13. We refer to cylindric set algebras and cylindric algebras of formulas as "special CA's" not only because of the specialized structure of their elements. The class of CA's obtained from these algebras by adjoining all their isomorphic images turns out to be still rather restricted and in this sense "special". However, the special algebras include those CA's for whose study the general theory of CA's was originally designed and which greatly influenced its subsequent development; they continue to play a predominant part in applications of the general theory. Hence, naturally, the problems of characterizing those algebras which can be isomorphically represented as special CA's (of either of the two classes) have become outstanding problems of the theory of CA's. These problems, which are known as the *representation problems*, will be studied exhaustively in Part II of our work. Some preliminary remarks seem, however, to be appropriate at this point.

We concentrate first on cylindric set algebras of finite dimension. It is easily seen that every cylindric set algebra of given dimension $\alpha < \omega$ is simple and therefore subdirectly (and directly) indecomposable in the sense of the general theory of algebras; cf. 0.3.58. Hence, when discussing the problem which CA_α's are isomorphic to cylindric set algebras, we restrict ourselves to subdirectly indecomposable algebras. (As we shall see in Chapter 2, for all CA's of finite dimension the notions of simplicity, subdirect indecomposability, and

1) The algebraic structures referred to as quantifier algebras in Halmos [56b] roughly correspond to what we call here diagonal-free CA's.

direct indecomposability coincide.) On the other hand, as a consequence of a theorem of Birkhoff, 0.3.54, every CA_α is isomorphic to a subdirect product of subdirectly indecomposable CA_α's. Thus we are naturally led to the problem of characterizing those CA_α's which are isomorphic to subdirect products of cylindric set algebras, and we agree to call such algebras *representable*. An immediate consequence of this definition is that, for $\alpha < \omega$, CA_α's isomorphic to set algebras are just those representable CA_α's which are subdirectly indecomposable.

We extend our definition of representability, without any change in its formulation, to algebras of infinite dimension. In this case, however, an intuitive justification of the definition is less clear, since cylindric set algebras of infinite dimension are not, in general, subdirectly indecomposable. For $\alpha \geqq \omega$, no intrinsic property is known which singles out the algebras isomorphic to set algebras from among all representable CA_α's.

The definition of representability can be formulated equivalently in a different, more geometrical way, by generalizing the notion of a set algebra given in 1.1.5. The generalization consists in taking for the unit set V of the algebra, not a Cartesian space ${}^\alpha U$, but a union of pairwise disjoint Cartesian spaces, i.e., a set of the form $\bigcup_{i\in I}{}^\alpha U_i$ where $U_i \cap U_j = 0$ for any two distinct $i, j \in I$. The defining formulas for the operations $\mathsf{C}_\kappa = \mathsf{C}_\kappa^{[V]}$ from SbV into SbV and the sets $\mathsf{D}_{\kappa\lambda}$ are obtained from those in 1.1.5 by changing ${}^\alpha U$ to V. Then a family A of subsets of V is called an (*α-dimensional*) *generalized cylindric field of sets* if A is closed under all the operations $\cup, \cap, {}_V\!\sim, \mathsf{C}_\kappa^{[V]}$ (for $\kappa < \alpha$) and contains all the sets $\mathsf{D}_{\kappa\lambda}^{[V]}$; in this case the algebra $\langle A, \cup, \cap, {}_V\!\sim, 0, V, \mathsf{C}_\kappa^{[V]}, \mathsf{D}_{\kappa\lambda}^{[V]}\rangle_{\kappa,\lambda<\alpha}$ is called a *generalized cylindric set algebra* (*of dimension* α). A generalized set algebra is easily seen to be a CA, and it turns out that a CA_α with $\alpha \neq 0$ is representable iff it is isomorphic to some generalized set algebra.

This, by the way, is not the case for $\alpha = 0$. In fact, while CA_0's coincide with arbitrary Boolean algebras, there is just one generalized set algebra of dimension 0 and it is a 2-element Boolean set algebra.

A simple consequence of the definition of representability is that all CA_α's with $\alpha = 0, 1$ and all discrete CA's are representable. On the other hand, we shall see still in this part of our work (cf. 2.6.44) that there exist non-representable CA_α's for every $\alpha \geqq 2$.

An important goal of the representation theory of CA's is to provide a workable criterion, i.e., a necessary and sufficient condition, for the representability of a CA. The criterion should be formulated, if possible, in terms of intrinsic properties of the CA's involved. It should be simple and powerful enough to be applicable to the problems of representability of various individually defined CA's as well as to various general problems concerning the class of representable CA's. No such simple criterion has yet been found. The theory of representation developed in Part II will provide, however, in addition to some complicated necessary and sufficient conditions, a number of interesting partial criteria, i.e., sufficient conditions, for representability of a CA; we shall learn that several rather comprehensive classes of CA's, with simple intrinsic characterizations, are included in the class of representable CA's. Thus, e.g., we shall see that all CA's of infinite dimension which are semisimple in the sense of 0.3.48 are representable.

One of the main results of the representation theory is that, for any given α, the class of all representable CA_α's, just as the class of all CA_α's, is equational, i.e., can be characterized by a set of identities. From 1.1.1 we see that the class of all CA_α's is characterized by a set Γ of identities obtained from one of its finite subsets Δ by replacing all the indices of cylindrifications c_κ and diagonal elements $d_{\kappa\lambda}$ occurring in the identities of Δ by arbitrary ordinals $<\alpha$; thus, in case $\alpha < \omega$, the set Γ itself is finite. In contrast to this, it will be shown that a set of identities characterizing the class of representable CA_α's for any $\alpha > 2$ is always infinite and cannot be obtained from any of its finite subsets in the way just indicated. Except for the cases $\alpha = 0, 1, 2$, we do not know any simple description of a set of identities characterizing the class of representable CA_α's; in Part II however a rather complicated set of such identities for an arbitrary $\alpha > 2$ will be explicitly described.

The notions discussed so far in 1.1.13 extend in a natural way to Df's. Again a Df is said to be *representable* if it is isomorphic to a subdirect product of diagonal-free set algebras. By generalizing the original notion of a cylindric set algebra we have obtained the notion of a generalized cylindric set algebra. In essentially the same way we pass from the original diagonal-free cylindric set algebras to *generalized diagonal-free cylindric set algebras*; a unit set of such an algebra is a set V of the form $\bigcup_{i\in I}(\mathsf{P}_{\xi<\alpha} U_\xi^{(i)})$ where, for each $\xi < \alpha$, $U_\xi^{(i)} \cap U_\xi^{(j)} = 0$ for any two distinct $i, j \in I$. As before, a Df_α with $\alpha \neq 0$ is representable iff it is isomorphic to a generalized diagonal-free cylindric set algebra.

In opposition to CA_α's, every Df_α, not only with $\alpha = 0, 1$, but also with $\alpha = 2$ proves to be representable. On the other hand, for every $\alpha \geqq 3$ there are non-representable Df_α's. Just as the class of representable CA_α's, the class of representable Df_α's is equational for every given α. All the resulst just mentioned will be established in Part II.

We now consider the second class of special CA's, i.e., algebras of formulas (associated with sets of sentences in first-order languages). The representation problem involving these special algebras largely reduces to the problem previously discussed. In fact, all the CA's which are isomorphically representable as algebras of formulas can easily be shown to be representable in the original sense, i.e., as subdirect products of set algebras. (For this purpose we could make use of 0.3.46, or rather the remark immediately following it.) The converse, however, is by no means true except for CA's of finite dimension. As regards CA's of infinite dimension, it is not difficult to provide an intrinsic characterization of all those algebras isomorphic to algebras of formulas. Indeed, they can be singled out from among all CA's and all representable CA's (of the same dimension) by means of a simple, very restrictive property of a finitary character. More specifically, the class of these CA's proves to coincide with the class of *locally finite-dimensional* cylindric algebras, which will be defined and discussed in Section 1.11.

With every set of sentences in a first-order language it proves convenient to associate, in addition to a cylindric algebra of formulas, a closely related Boolean algebra of sentences. This can be carried through in the following way.

THEOREM 1.1.14. *Let* Λ *be a language of predicate logic and* $\mathfrak{A}$ *be the algebra*

$$\langle \Sigma v^{(\Lambda)}, \vee, \wedge, \neg, F, T \rangle.$$

Then $\mathfrak{A}$ *is similar to* BA's, *and for any* $\Sigma \subseteq \Sigma v^{(\Lambda)}$ *we have* $({}^2\Sigma v^{(\Lambda)} \cap \equiv_\Sigma) \in Co\,\mathfrak{A}$ *and* $\mathfrak{A}/({}^2\Sigma v^{(\Lambda)} \cap \equiv_\Sigma) \in$ BA.

In analogy to $\mathfrak{Fm}^{(\Lambda)}$ the algebra $\mathfrak{A}$ from 1.1.14 may be called the *free algebra of sentences* (*in* Λ).

DEFINITION 1.1.15. *Let* Λ *and* $\mathfrak{A}$ *be as in 1.1.14. If* $\Sigma \subseteq \Sigma v^{(\Lambda)}$, *we refer to* $\mathfrak{A}/({}^2\Sigma v^{(\Lambda)} \cap \equiv_\Sigma)$ *as a* ***Boolean algebra of sentences*** *or, more specifically, as the* ***(Boolean) algebra of sentences (in*** Λ***) associated with*** Σ.

REMARK 1.1.16. There is of course a close connection between the Boolean algebra $\mathfrak{B} = \mathfrak{A}/({}^2\Sigma v^{(\Lambda)} \cap \equiv_\Sigma)$ just defined and the cylindric algebra $\mathfrak{C} = \mathfrak{Fm}^{(\Lambda)}/\equiv_\Sigma$. In fact, as a direct consequence of 0.2.26(iii) (the first isomorphism theorem), we conclude that $\mathfrak{B}$ is isomorphic to the algebra $\mathfrak{B}' = \langle \Sigma v^{(\Lambda)}/\equiv_\Sigma, +, \cdot, -, 0, 1 \rangle$ where $+, \cdot, -, 0, 1$ are the operations and distinguished elements of $\mathfrak{C}$ induced by the corresponding notions $\vee, \wedge, \neg, F, T$ of $\mathfrak{Fm}^{(\Lambda)}$ ($\mathfrak{B}'$ is thus a subalgebra of $\mathfrak{Bl}\,\mathfrak{C}$). For this reason either of the two algebras $\mathfrak{B}$ and $\mathfrak{B}'$ will be referred to in informal remarks as the algebra of sentences associated with Σ. The remark at the beginning of 1.1.12 applies to Boolean algebras of sentences as well. Notice that in case $\alpha = 0$ the algebras $\mathfrak{A}$ and $\mathfrak{Fm}^{(\Lambda)}$ coincide, and so do $\mathfrak{B}$ and $\mathfrak{C}$.

1.2. CYLINDRIFICATIONS

In this section we present the most basic laws which hold for cylindrifications in arbitrary CA's. Since the diagonal elements play no role in the discussion, we can assume in this section that $\mathfrak{A}$ is just an arbitrary Df_α. κ, λ will range over ordinals less than α, and x, y, z will range over elements of A or possibly systems of elements of A.

From (C_1) and (C_2) we obtain

THEOREM 1.2.1. $\mathsf{c}_\kappa x = 0$ *iff* $x = 0$.

Taking $x = 1$ in (C_2) we arrive at

THEOREM 1.2.2. $\mathsf{c}_\kappa 1 = 1$.

THEOREM 1.2.3. $\mathsf{c}_\kappa \mathsf{c}_\kappa y = \mathsf{c}_\kappa y$.

PROOF. Put $x = 1$ in (C_3) and use 1.2.2.

COROLLARY 1.2.4. $x = \mathsf{c}_\kappa x$ *iff* $x = \mathsf{c}_\kappa y$ *for some* $y \in A$.

Thus the set of so-called fixed points of the function c_κ coincides with the range of c_κ.

THEOREM 1.2.5. $x \cdot \mathsf{c}_\kappa y = 0$ *iff* $y \cdot \mathsf{c}_\kappa x = 0$.

PROOF. By (C_3) we have $\mathsf{c}_\kappa(x \cdot \mathsf{c}_\kappa y) = \mathsf{c}_\kappa(y \cdot \mathsf{c}_\kappa x)$. Hence the theorem follows by 1.2.1.

In the terminology of the theory of Boolean algebras, Theorem 1.2.5 expresses the fact that c_κ is a *self-conjugate* operation on $\mathfrak{B}\mathfrak{l}\mathfrak{A}$; cf. Jónsson-Tarski [51], p. 903. Every such operation is known to be *completely additive* in the sense of op. cit., and indeed this will be evident from the fact that in the following proof only 1.2.5 is employed.

THEOREM 1.2.6. (i) *If* $\sum_{i \in I} z_i$ *exists, then so does* $\sum_{i \in I} \mathsf{c}_\kappa z_i$ *and we have* $\mathsf{c}_\kappa(\sum_{i \in I} z_i) = \sum_{i \in I} \mathsf{c}_\kappa z_i$.

(ii) *In particular,* $\mathsf{c}_\kappa(x+y) = \mathsf{c}_\kappa x + \mathsf{c}_\kappa y$.

PROOF. It suffices, of course, to prove (i). Let $w = \sum_{i \in I} z_i$. Since $\mathsf{c}_\kappa w \cdot -\mathsf{c}_\kappa w = 0$, we see by 1.2.5 that $w \cdot \mathsf{c}_\kappa - \mathsf{c}_\kappa w = 0$; hence $z_i \cdot \mathsf{c}_\kappa - \mathsf{c}_\kappa w = 0$ for each $i \in I$. Another application of 1.2.5 then gives $\mathsf{c}_\kappa z_i \cdot -\mathsf{c}_\kappa w = 0$, or $\mathsf{c}_\kappa z_i \leqq \mathsf{c}_\kappa w$, for each $i \in I$. That is, $\mathsf{c}_\kappa w$ is an upper bound for the set of all elements $\mathsf{c}_\kappa z_i$ with $i \in I$.

Now suppose that t is any other upper bound of these elements: $\mathsf{c}_\kappa z_i \leqq t$ for all $i \in I$. Then $\mathsf{c}_\kappa z_i \cdot -t = 0$, and so $z_i \cdot \mathsf{c}_\kappa - t = 0$, by 1.2.5, for each $i \in I$. It follows that $w \cdot \mathsf{c}_\kappa - t = 0$ and hence, by a final application of 1.2.5, $\mathsf{c}_\kappa w \cdot -t = 0$, or $\mathsf{c}_\kappa w \leqq t$.

The following two corollaries follow just from the fact that c_κ is *additive*, i.e., from 1.2.6 (ii).

COROLLARY 1.2.7. *If* $x \leqq y$, *then* $\mathsf{c}_\kappa x \leqq \mathsf{c}_\kappa y$.

COROLLARY 1.2.8. (i) $\mathsf{c}_\kappa x - \mathsf{c}_\kappa y \leqq \mathsf{c}_\kappa(x-y)$;
(ii) $\mathsf{c}_\kappa x \oplus \mathsf{c}_\kappa y \leqq \mathsf{c}_\kappa(x \oplus y)$.

PROOF. (i) By additivity we have $\mathsf{c}_\kappa x = \mathsf{c}_\kappa(x \cdot y) + \mathsf{c}_\kappa(x \cdot -y)$, and hence by 1.2.7 it follows that $\mathsf{c}_\kappa x \leqq \mathsf{c}_\kappa y + \mathsf{c}_\kappa(x-y)$. The desired result is now obvious.

(ii) By (i) and additivity we get

$$\mathsf{c}_\kappa x \oplus \mathsf{c}_\kappa y = (\mathsf{c}_\kappa x - \mathsf{c}_\kappa y) + (\mathsf{c}_\kappa y - \mathsf{c}_\kappa x) \leqq \mathsf{c}_\kappa(x-y) + \mathsf{c}_\kappa(y-x) = \mathsf{c}_\kappa(x \oplus y).$$

Notice that, in view of (C_1), (C_2), 1.2.3, and 1.2.6(ii), for each $\mathfrak{A} \in \mathsf{Df}_\alpha$ and each particular $\kappa < \alpha$, $\langle A, +, \cdot, -, 0, 1, \mathsf{c}_\kappa \rangle$ is a closure algebra in the sense of McKinsey-Tarski [44*], pp. 145 f., Definition 1.1. For this reason we shall sometimes say that x is *closed under* c_κ, or simply *κ-closed*, when $\mathsf{c}_\kappa x = x$, and x is *open under* c_κ, or simply *κ-open*, when $-x$ is κ-closed, i.e., $\mathsf{c}_\kappa - x = -x$. Occasionally the κ-closed elements will also be called *κ-cylinders*; this seems to be especially appropriate when they appear as members of a cylindric set algebra.

The following two theorems are consequences of the fact that $\langle A, +, \cdot, -, 0, 1, \mathsf{c}_\kappa \rangle$ is a closure algebra.

THEOREM 1.2.9. $x \leqq \mathsf{c}_\kappa y$ *iff* $\mathsf{c}_\kappa x \leqq \mathsf{c}_\kappa y$.

PROOF. If $\mathsf{c}_\kappa x \leqq \mathsf{c}_\kappa y$, then $x \leqq \mathsf{c}_\kappa y$ by (C_2). Conversely, if $x \leqq \mathsf{c}_\kappa y$, then $\mathsf{c}_\kappa x \leqq \mathsf{c}_\kappa y$ by 1.2.3 and 1.2.7.

From 1.2.4 and 1.2.9 we see that $\mathsf{c}_\kappa x$ can be characterized as the least κ-cylinder which includes x. Thus the operation c_κ could be defined in terms of the set of κ-cylinders, and could be replaced by this set in the list of primitive notions of CA's. (Of course, after this replacement the CA's would no longer be algebras in the sense of 0.1.1.)

THEOREM 1.2.10. *If* $\prod_{i\in I} c_\kappa z_i$ *exists, then* $c_\kappa(\prod_{i\in I} c_\kappa z_i) = \prod_{i\in I} c_\kappa z_i$; *in particular,* $c_\kappa(c_\kappa x \cdot c_\kappa y) = c_\kappa x \cdot c_\kappa y$.

PROOF. Let $w = \prod_{i\in I} c_\kappa z_i$. Then $w \leqq c_\kappa z_i$, and so $c_\kappa w \leqq c_\kappa z_i$, by 1.2.9, for each $i \in I$. But then $c_\kappa w \leqq w$, and the desired conclusion follows by (C_2).

THEOREM 1.2.11. $c_\kappa - c_\kappa x = -c_\kappa x$.

PROOF. Since $-c_\kappa x \cdot c_\kappa c_\kappa x = 0$ by 1.2.3, we get $c_\kappa x \cdot c_\kappa - c_\kappa x = 0$ by 1.2.5, so that $c_\kappa - c_\kappa x \leqq -c_\kappa x$. The opposite inclusion is a particular case of (C_2).

COROLLARY 1.2.12. (i) $c_\kappa x = x$ *iff* $c_\kappa - x = -x$.

(ii) $c_\kappa(x - c_\kappa y) = c_\kappa x - c_\kappa y$.

PROOF: (i) directly by 1.2.11; (ii) by (C_3) and 1.2.11.

1.2.12(ii) has a form closely related to that of (C_3); both are modular laws for c_κ, the latter under multiplication and the former under subtraction.

In the terminology of closure algebras (cf. the remarks following 1.2.8) either of the statements 1.2.11 and 1.2.12(i) expresses the fact that, in the closure algebra $\langle A, +, \cdot, -, 0, 1, c_\kappa\rangle$ which is the reduct of an $\mathfrak{A} \in \mathsf{CA}$, every closed element is open (and conversely).

The statements 1.2.1–1.2.12 have been derived exclusively from Postulates (C_1), (C_2), and (C_3) (in addition, of course, to (C_0)). Some of the statements can be used to replace equivalently (C_1)–(C_3) in the original postulate system. For example, it can be shown that we can replace (C_1)–(C_3) by 1.2.2, 1.2.3, and 1.2.5; alternatively, retaining (C_2), we can replace (C_1) and (C_3) by 1.2.12(ii) or else by 1.2.6(ii) and 1.2.11.

When applied to the case $\alpha = 1$ the observations just made provide new characterizations of CA_1's and Df_1's. It turns out, for instance, that Df_1's coincide with those closure algebras in which each closed element is open. (In connection with such closure algebras cf. Halmos [56], pp. 222 ff. and Rubin [56].)

The following two theorems, which are easy to prove, furnish simple and rather trivial methods for constructing Df_1's and CA_1's.

THEOREM 1.2.13. *If* $\langle A, +, \cdot, -, 0, 1\rangle \in \mathsf{BA}$, *then* $\langle A, +, \cdot, -, 0, 1, A 1 Id, 1\rangle \in \mathsf{CA}_1$ (*and* $\langle A, +, \cdot, -, 0, 1, A 1 Id\rangle \in \mathsf{Df}_1$).

THEOREM 1.2.14. *If* $\langle A, +, \cdot, -, 0, 1\rangle \in \mathsf{BA}$ *and if* f *is a function on* A *such that, for every* $x \in A$, $fx = 0$ *or* $fx = 1$ *according as* $x = 0$ *or* $x \neq 0$, *then* $\langle A, +, \cdot, -, 0, 1, f, 1\rangle \in \mathsf{CA}_1$ (*and* $\langle A, +, \cdot, -, 0, 1, f\rangle \in \mathsf{Df}_1$).

Concerning the Df_1's and CA_1's just described compare Remark 2.3.15 in the next chapter.

We now turn to laws in which more than one cylindrification is involved, and which depend on (C_4).

THEOREM 1.2.15. $c_\kappa x \cdot c_\lambda y = 0$ *iff* $c_\lambda x \cdot c_\kappa y = 0$.
PROOF: by 1.2.5 and (C_4).

THEOREM 1.2.16. $c_\kappa - c_\lambda x \cdot c_\lambda - c_\kappa - x = 0$.
PROOF. From (C_2) we see that $c_\lambda x + c_\kappa - x = 1$, and hence $-c_\lambda x \cdot - c_\kappa - x = 0$. Then by 1.2.11 we have $c_\lambda - c_\lambda x \cdot c_\kappa - c_\kappa - x = 0$, and an application of 1.2.15 gives the desired conclusion.

A more intuitive form of the preceding theorem is given as 1.4.4(iv) below·

THEOREM 1.2.17. *If* $\kappa, \lambda < \omega$, $\sigma \in {}^\kappa\alpha$, $\tau \in {}^\lambda\alpha$, *and* $Rg\sigma = Rg\tau$, *then* $c_{\sigma 0} \ldots c_{\sigma_{\kappa-1}} x = c_{\tau 0} \ldots c_{\tau_{\lambda-1}} x$.
PROOF. An inductive proof can be based on 1.2.3 and the following lemma:

(1) if $\mu < \omega$, $\rho \in {}^\mu\alpha$, and π is a permutation of μ, then

$$c_{\rho 0} \ldots c_{\rho_{\mu-1}} x = c_{\rho_{\pi 0}} \ldots c_{\rho_{\pi\mu-1}} x.$$

The lemma follows from (C_4), since π can be expressed as a product of transpositions $(\nu, \nu+1)$ where $\nu+1 < \mu$.

Theorem 1.2.17 will be used in Section 1.7 to define a generalized cylindrification, $c_{(\Gamma)}$, for any finite subset Γ of α.

1.3. DIAGONAL ELEMENTS

We present now the most basic laws involving diagonal elements. Throughout this section and the remainder of this chapter, unless otherwise stated, it is tacitly assumed that $\mathfrak{A}$ is a fixed but arbitrary CA_α with any given dimension α; $\kappa, \lambda, \mu, \nu, \xi, \rho$ are any ordinals less than α; and x, y, z, w are arbitrary elements of A. In further chapters of our work these and similar assumptions will not always be explicitly stated. Of course they will be omitted only when the possibility of confusion seems negligible.

THEOREM 1.3.1. $\mathsf{d}_{\kappa\lambda} = \mathsf{d}_{\lambda\kappa}$.

PROOF. If $\kappa = \lambda$, there is nothing to prove. Otherwise, we apply (C_7), taking $x = \mathsf{d}_{\lambda\kappa}$, to get $\mathsf{c}_\kappa(\mathsf{d}_{\kappa\lambda}\cdot\mathsf{d}_{\lambda\kappa})\cdot\mathsf{c}_\kappa(\mathsf{d}_{\kappa\lambda}\cdot-\mathsf{d}_{\lambda\kappa}) = 0$. But by (C_5) and (C_6) we have $\mathsf{c}_\kappa(\mathsf{d}_{\kappa\lambda}\cdot\mathsf{d}_{\lambda\kappa}) = \mathsf{d}_{\lambda\lambda} = 1$. Hence using 1.2.1 we obtain $\mathsf{d}_{\kappa\lambda}\cdot-\mathsf{d}_{\lambda\kappa} = 0$, i.e. $\mathsf{d}_{\kappa\lambda} \leqq \mathsf{d}_{\lambda\kappa}$. The inclusion in the opposite direction is obtained by symmetry.

THEOREM 1.3.2. $\mathsf{c}_\kappa\mathsf{d}_{\kappa\lambda} = 1$.

PROOF. If $\kappa = \lambda$, the result is immediate from (C_5) and 1.2.2. If $\kappa \neq \lambda$, then by (C_5), (C_6), and 1.3.1 we have $\mathsf{c}_\kappa\mathsf{d}_{\kappa\lambda} = \mathsf{c}_\kappa(\mathsf{d}_{\kappa\lambda}\cdot\mathsf{d}_{\lambda\kappa}) = \mathsf{d}_{\lambda\lambda} = 1$.

The proofs of 1.3.1 and 1.3.2 make no use of Axiom (C_1). We can, therefore, apply these theorems to derive (C_1) by putting $x = 1$ in (C_7) — provided $\alpha > 1$. But (C_1) is independent of the remaining axioms if $\alpha = 1$.

THEOREM 1.3.3. *If $\kappa \neq \lambda, \mu$, then* $\mathsf{c}_\kappa\mathsf{d}_{\lambda\mu} = \mathsf{d}_{\lambda\mu}$.

PROOF. Apply c_κ to each side of the equation (C_6). The result then follows from 1.2.3.

THEOREM 1.3.4. *If $\kappa \neq \lambda$, then* $\mathsf{c}_\kappa(\mathsf{d}_{\kappa\lambda}\cdot-x) = -\mathsf{c}_\kappa(\mathsf{d}_{\kappa\lambda}\cdot x)$.

PROOF. The product of $\mathsf{c}_\kappa(\mathsf{d}_{\kappa\lambda}\cdot-x)$ and $\mathsf{c}_\kappa(\mathsf{d}_{\kappa\lambda}\cdot x)$ is 0 by (C_7), and their sum is 1 by 1.2.6 and 1.3.2. The theorem follows by (C_0).

COROLLARY 1.3.5. *If $\kappa \neq \lambda, \mu$, then* $\mathsf{c}_\kappa(\mathsf{d}_{\lambda\kappa}\cdot-\mathsf{d}_{\kappa\mu}) = -\mathsf{d}_{\lambda\mu}$.

THEOREM 1.3.6. *If $\kappa \neq \lambda$, then* $\mathsf{c}_\kappa(\mathsf{d}_{\kappa\lambda}\cdot x\cdot y) = \mathsf{c}_\kappa(\mathsf{d}_{\kappa\lambda}\cdot x)\cdot\mathsf{c}_\kappa(\mathsf{d}_{\kappa\lambda}\cdot y)$.

PROOF. We have $c_\kappa(d_{\kappa\lambda}\cdot(x+y)) = c_\kappa(d_{\kappa\lambda}\cdot x)+c_\kappa(d_{\kappa\lambda}\cdot y)$ by 1.2.6. Thus by 1.3.4 the operation on A which assigns $c_\kappa(d_{\kappa\lambda}\cdot x)$ to x is an endomorphism of $\mathfrak{Bl}\mathfrak{A}$. Hence the desired result follows.

The endomorphism $\langle c_\kappa(d_{\kappa\lambda}\cdot x):x\in A\rangle$ for $\kappa \neq \lambda$ involved in the proof of 1.3.6 will be studied in greater detail in Section 1.5. Using (C_1) we see that (C_7) is just a special case of 1.3.6. Hence 1.3.6 could replace (C_7) in Definition 1.1.1.

THEOREM 1.3.7. $d_{\kappa\lambda}\cdot d_{\lambda\mu} = d_{\kappa\lambda}\cdot d_{\kappa\mu}$.

PROOF. If $\lambda = \kappa$ or $\lambda = \mu$, the result is immediate. If $\lambda \neq \kappa$ and $\lambda \neq \mu$, then by (C_2) and (C_6) we have $d_{\kappa\lambda}\cdot d_{\lambda\mu} \leqq d_{\kappa\mu}$, and the theorem follows by 1.3.1 and symmetry.

THEOREM 1.3.8. $d_{\kappa\lambda}\cdot c_\kappa x = 0$ *iff* $x = 0$.

PROOF: by 1.2.5 and 1.3.2.

THEOREM 1.3.9. *If* $\kappa \neq \lambda$, *then* $d_{\kappa\lambda}\cdot c_\kappa(d_{\kappa\lambda}\cdot x) = d_{\kappa\lambda}\cdot x$.

PROOF. We have $d_{\kappa\lambda}\cdot -x \leqq c_\kappa(d_{\kappa\lambda}\cdot -x) \leqq -c_\kappa(d_{\kappa\lambda}\cdot x)$ by (C_2) and (C_7); hence $d_{\kappa\lambda}\cdot c_\kappa(d_{\kappa\lambda}\cdot x) \leqq d_{\kappa\lambda}\cdot x$. The opposite inequality follows from (C_2).

It is easy to see that 1.3.9 could also replace (C_7) in Definition 1.1.1.

1.3.1–1.3.9 give the most useful properties of diagonal elements. We now discuss cylindric algebras in which all the diagonal elements and cylindrifications are trivial.

DEFINITION 1.3.10. $\mathfrak{A}$ *is* ***discrete*** *if* $d_{\kappa\lambda} = 1$ *and* $c_\kappa x = x$ *for all* $\kappa, \lambda < \alpha$ *and all* $x \in A$.

Discrete CA's are in a sense degenerate and in some cases will explicitly be ruled out of consideration.

We now state formally a fact noted in the remarks preceding 1.1.4.

THEOREM 1.3.11. *Suppose* $\mathfrak{A} = \langle A, +, \cdot, -, 0, 1\rangle$ *is a* BA. *For each* $\kappa < \alpha$ *let* $c_\kappa = A \upharpoonright Id$, *and for all* $\kappa, \lambda < \alpha$ *let* $d_{\kappa\lambda} = 1$. *Then* $\mathfrak{A}' = \langle A, +, \cdot, -, 0, 1, c_\kappa, d_{\kappa\lambda}\rangle_{\kappa,\lambda<\alpha}$ *is a discrete* CA_α.

The term "discrete" derives of course from the fact that each closure operation c_κ is discrete. Although by 1.3.11 discrete CA's exist in profusion, it is easily seen that a cylindric set algebra on a base with at least two elements is never discrete if $\alpha > 1$. A cylindric algebra of formulas associated with a theory Θ is discrete iff the sentence $\forall_x \forall_y x = y$ is valid in Θ. Such theories Θ are rather trivial.

THEOREM 1.3.12. *For any given* κ *the following two conditions are equivalent:*

(i) $\mathfrak{A}$ *is a discrete* CA_α;

(ii) *for all* x, $\mathsf{c}_\kappa x = x$.

Moreover, for any given $\lambda \neq \kappa$, *(i) is also equivalent to the condition*

(iii) $\mathsf{d}_{\kappa\lambda} = 1$.

PROOF. Obviously (i) implies (ii); if $\alpha \leqq 1$, it is also clear that (ii) implies (i). Assuming $\alpha > 1$ and $\lambda \neq \kappa$ it is easily seen using 1.3.2 that (ii) implies (iii).

Finally, we show that (iii) implies (i). Suppose $\kappa \neq \lambda$ and $\mathsf{d}_{\kappa\lambda} = 1$. First we establish:

$$\text{(1)} \qquad \text{if } \mu \neq \kappa, \text{ then } \mathsf{d}_{\kappa\mu} = 1.$$

To prove (1) we may assume that $\mu \neq \lambda$. Then $\mathsf{d}_{\kappa\mu} = \mathsf{c}_\lambda(\mathsf{d}_{\kappa\lambda} \cdot \mathsf{d}_{\lambda\mu}) = \mathsf{c}_\lambda \mathsf{d}_{\lambda\mu} = 1$ by (C_6) and 1.3.2. Now suppose $\mu, \nu < \alpha$. By 1.3.1 and 1.3.7 we have $\mathsf{d}_{\kappa\mu} \cdot \mathsf{d}_{\kappa\nu} \leqq \mathsf{d}_{\mu\nu}$. Hence, by (1) and (C_5), $\mathsf{d}_{\mu\nu} = 1$. Thus

$$\text{(2)} \qquad \mathsf{d}_{\mu\nu} = 1 \text{ for all } \mu, \nu < \alpha.$$

Suppose $\mu < \alpha$. Choose $\nu \in \alpha \sim \{\mu\}$. Then by (2) and (C_7) we have $\mathsf{c}_\mu x \cdot \mathsf{c}_\mu - x = \mathsf{c}_\mu(\mathsf{d}_{\mu\nu} \cdot x) \cdot \mathsf{c}_\mu(\mathsf{d}_{\mu\nu} \cdot -x) = 0$. Hence from 1.2.3 and 1.2.5 we infer that $\mathsf{c}_\mu x \cdot -x = 0$, and then by (C_2) that $\mathsf{c}_\mu x = x$. In view of (2) this completes the proof.

THEOREM 1.3.13. *If* $\alpha > 2$, *then* $\mathfrak{A}$ *is a discrete* CA_α *iff there exist distinct* κ, λ *such that* $\mathsf{c}_\kappa = \mathsf{c}_\lambda$.

PROOF. The necessity is immediate from 1.3.10. Now assume $\mathsf{c}_\kappa = \mathsf{c}_\lambda$ with κ, λ distinct. Choose $\mu \neq \kappa, \lambda$. Then, by 1.3.2 and 1.3.3, $\mathsf{d}_{\kappa\mu} = \mathsf{c}_\lambda \mathsf{d}_{\kappa\mu} = 1$, so $\mathfrak{A}$ is discrete by 1.3.12.

The conclusion of Theorem 1.3.13 is easily seen to fail for $\alpha = 2$.

THEOREM 1.3.14. *If* $\alpha > 2$, *then* $\mathfrak{A}$ *is a discrete* CA_α *iff there exist* $\kappa, \lambda, \mu, \nu$ *with* κ, λ, μ *distinct such that* $\mathsf{d}_{\kappa\lambda} = \mathsf{d}_{\mu\nu}$.

PROOF. The existence of such $\kappa, \lambda, \mu, \nu$ is obviously necessary for $\mathfrak{A}$ to be discrete when $\alpha > 2$. Now assume that κ, λ, μ are distinct and $\mathsf{d}_{\kappa\lambda} = \mathsf{d}_{\mu\nu}$. Applying c_μ to both sides of the equality and using 1.3.2 and 1.3.3, we get $\mathsf{d}_{\kappa\lambda} = 1$. Hence, by 1.3.12, $\mathfrak{A}$ is discrete.

COROLLARY 1.3.15. *If* $\mathfrak{A}$ *is non-discrete, then* $|\alpha| \leqq |A|$. *In particular, every non-discrete* CA_α *with* $\alpha \geqq \omega$ *is infinite.*

In case $\alpha < \omega$, we can obtain a better estimate for the lower bound of the cardinalities of non-discrete CA_α's; cf. 2.4.72 and 2.4.73 below.

THEOREM 1.3.16. (i) *Every* CA_0 *is discrete.*
(ii) *Every* $\mathfrak{A} \in \mathsf{CA}_\alpha$ *with* $|A| \leqq 2$ *is discrete.*
PROOF: (i) by 1.3.10; (ii) by (C_1), 1.2.2, and 1.3.12.

We conclude this section with three theorems about complements of diagonal elements. The first theorem will be generalized and the discussion continued in Section 1.9.

DEFINITION 1.3.17. *A complement of a diagonal element is called a* ***co-diagonal element.***

THEOREM 1.3.18. *Assume that* $\kappa \neq \lambda$. *Then*
(i) $\mathsf{c}_\kappa \mathsf{c}_\lambda[-\mathsf{d}_{\kappa\lambda} \cdot \mathsf{c}_\kappa x \cdot \mathsf{c}_\lambda(\mathsf{d}_{\kappa\lambda} \cdot \mathsf{c}_\kappa x)] \leqq \mathsf{c}_\lambda(-\mathsf{d}_{\kappa\lambda} \cdot \mathsf{c}_\kappa x)$;
(ii) $\mathsf{c}_\kappa \mathsf{c}_\lambda - \mathsf{d}_{\kappa\lambda} = \mathsf{c}_\kappa - \mathsf{d}_{\kappa\lambda} = \mathsf{c}_\lambda - \mathsf{d}_{\kappa\lambda}$;
(iii) *if in addition* $\mu \neq \nu$, *then* $\mathsf{c}_\kappa \mathsf{c}_\lambda - \mathsf{d}_{\kappa\lambda} = \mathsf{c}_\mu \mathsf{c}_\nu - \mathsf{d}_{\mu\nu}$.
PROOF. Notice first of all that $-\mathsf{c}_\lambda(-\mathsf{d}_{\kappa\lambda} \cdot \mathsf{c}_\kappa x) \leqq \mathsf{d}_{\kappa\lambda} + -\mathsf{c}_\kappa x$ by (C_2). From this is follows that

$$\mathsf{d}_{\kappa\lambda} \cdot \mathsf{c}_\kappa x \cdot -\mathsf{c}_\lambda(-\mathsf{d}_{\kappa\lambda} \cdot \mathsf{c}_\kappa x) = \mathsf{c}_\kappa x \cdot -\mathsf{c}_\lambda(-\mathsf{d}_{\kappa\lambda} \cdot \mathsf{c}_\kappa x). \tag{1}$$

Using the hypothesis $\kappa \neq \lambda$ we have

$$\begin{aligned}
\mathsf{c}_\kappa[\mathsf{d}_{\kappa\lambda} \cdot \mathsf{c}_\lambda(-\mathsf{d}_{\kappa\lambda} \cdot \mathsf{c}_\kappa x)] &= -\mathsf{c}_\kappa[\mathsf{d}_{\kappa\lambda} \cdot -\mathsf{c}_\lambda(-\mathsf{d}_{\kappa\lambda} \cdot \mathsf{c}_\kappa x)] && \text{by 1.3.4}\\
&\leqq -\mathsf{c}_\kappa[\mathsf{d}_{\kappa\lambda} \cdot \mathsf{c}_\kappa x \cdot -\mathsf{c}_\lambda(-\mathsf{d}_{\kappa\lambda} \cdot \mathsf{c}_\kappa x)] && \text{by 1.2.7}\\
&= -\mathsf{c}_\kappa[\mathsf{c}_\kappa x \cdot -\mathsf{c}_\lambda(-\mathsf{d}_{\kappa\lambda} \cdot \mathsf{c}_\kappa x)] && \text{by (1)}\\
&= -[\mathsf{c}_\kappa x \cdot \mathsf{c}_\kappa - \mathsf{c}_\lambda(-\mathsf{d}_{\kappa\lambda} \cdot \mathsf{c}_\kappa x)] && \text{by } (C_3)\\
&= -\mathsf{c}_\kappa x + -\mathsf{c}_\kappa - \mathsf{c}_\lambda(-\mathsf{d}_{\kappa\lambda} \cdot \mathsf{c}_\kappa x)\\
&\leqq -\mathsf{c}_\kappa x + \mathsf{c}_\lambda(-\mathsf{d}_{\kappa\lambda} \cdot \mathsf{c}_\kappa x) && \text{by } (C_2).
\end{aligned}$$

It follows that $\mathsf{c}_\lambda(-\mathsf{d}_{\kappa\lambda} \cdot \mathsf{c}_\kappa x) \geqq \mathsf{c}_\kappa x \cdot \mathsf{c}_\kappa[\mathsf{d}_{\kappa\lambda} \cdot \mathsf{c}_\lambda(-\mathsf{d}_{\kappa\lambda} \cdot \mathsf{c}_\kappa x)]$, and hence, by 1.2.9,

$$\begin{aligned}
\mathsf{c}_\lambda(-\mathsf{d}_{\kappa\lambda} \cdot \mathsf{c}_\kappa x) &\geqq \mathsf{c}_\lambda\{\mathsf{c}_\kappa x \cdot \mathsf{c}_\kappa[\mathsf{d}_{\kappa\lambda} \cdot \mathsf{c}_\lambda(-\mathsf{d}_{\kappa\lambda} \cdot \mathsf{c}_\kappa x)]\}\\
&= \mathsf{c}_\kappa \mathsf{c}_\lambda[\mathsf{c}_\kappa x \cdot \mathsf{d}_{\kappa\lambda} \cdot \mathsf{c}_\lambda(-\mathsf{d}_{\kappa\lambda} \cdot \mathsf{c}_\kappa x)] && \text{by } (C_3), (C_4)\\
&= \mathsf{c}_\kappa \mathsf{c}_\lambda[\mathsf{d}_{\kappa\lambda} \cdot \mathsf{c}_\lambda(\mathsf{d}_{\kappa\lambda} \cdot \mathsf{c}_\kappa x) \cdot \mathsf{c}_\lambda(-\mathsf{d}_{\kappa\lambda} \cdot \mathsf{c}_\kappa x)] && \text{by 1.3.9}\\
&= \mathsf{c}_\kappa[\mathsf{c}_\lambda \mathsf{d}_{\kappa\lambda} \cdot \mathsf{c}_\lambda(\mathsf{d}_{\kappa\lambda} \cdot \mathsf{c}_\kappa x) \cdot \mathsf{c}_\lambda(-\mathsf{d}_{\kappa\lambda} \cdot \mathsf{c}_\kappa x)] && \text{by } (C_3)\\
&= \mathsf{c}_\kappa \mathsf{c}_\lambda[-\mathsf{d}_{\kappa\lambda} \cdot \mathsf{c}_\kappa x \cdot \mathsf{c}_\lambda(\mathsf{d}_{\kappa\lambda} \cdot \mathsf{c}_\kappa x)] && \text{by } (C_3), \text{1.3.2}.
\end{aligned}$$

Thus part (i) of the theorem is proved. Setting $x = 1$ in (i) and using (C_2) and 1.3.2 we easily obtain $\mathsf{c}_\kappa \mathsf{c}_\lambda - \mathsf{d}_{\kappa\lambda} = \mathsf{c}_\lambda - \mathsf{d}_{\kappa\lambda}$, and with the help of (C_4) and 1.3.1 we infer in a similar manner that $\mathsf{c}_\kappa \mathsf{c}_\lambda - \mathsf{d}_{\kappa\lambda} = \mathsf{c}_\kappa - \mathsf{d}_{\kappa\lambda}$. Thus (ii) holds.

Suppose now that $\kappa \neq \mu$ (as well as $\kappa \neq \lambda$). Then by (C_4), 1.2.12(i), and 1.3.3 we get

$$(2) \qquad c_\mu c_\lambda - d_{\kappa\lambda} = c_\lambda - d_{\kappa\lambda},$$

and hence

$$\begin{aligned} c_\kappa c_\lambda - d_{\kappa\lambda} &= c_\kappa(c_\mu d_{\mu\kappa} \cdot c_\lambda - d_{\kappa\lambda}) && \text{by 1.3.2} \\ &= c_\mu c_\kappa(d_{\mu\kappa} \cdot c_\lambda - d_{\kappa\lambda}) && \text{by (2), } (C_3), (C_4) \\ &= c_\mu - c_\kappa(d_{\mu\kappa} \cdot - c_\lambda - d_{\kappa\lambda}) && \text{by 1.3.4} \\ &\geqq c_\mu - c_\kappa(d_{\mu\kappa} \cdot d_{\kappa\lambda}) && \text{by } (C_2), \text{ 1.2.7} \\ &= c_\mu - d_{\mu\lambda} && \text{by } (C_6). \end{aligned}$$

Therefore, by (C_4) and 1.2.9 we conclude that $c_\kappa c_\lambda - d_{\kappa\lambda} \geqq c_\lambda c_\mu - d_{\lambda\mu}$, and it is clear that this result holds whenever $\kappa \neq \lambda$, regardless of whether $\kappa = \mu$ or $\kappa \neq \mu$. By iterated use of this result we get

$$c_\kappa c_\lambda - d_{\kappa\lambda} \geqq c_\mu c_\nu - d_{\mu\nu} \geqq c_\kappa c_\lambda - d_{\kappa\lambda}$$

provided $\kappa \neq \lambda$ and $\mu \neq \nu$, and hence (iii) is proved.

1.3.18(ii),(iii) provide us with a variety of symbolic expressions which can be used to represent one and the same element $c_0 - d_{01}$ (in a given CA_α with $\alpha \geqq 2$). In fact, we can use for this purpose any one of the expressions $c_1 - d_{01}$, $c_0 c_1 - d_{01}$, as well as $c_\kappa - d_{\kappa\lambda}$, $c_\lambda - d_{\kappa\lambda}$, and $c_\kappa c_\lambda - d_{\kappa\lambda}$ where κ and λ are any two distinct ordinals $< \alpha$.

The two equations in 1.3.18(ii),(iii) correspond to the following two logically valid formulas:

$$\exists_x x \neq y \leftrightarrow \exists_x \exists_y x \neq y;$$
$$\exists_x \exists_y x \neq y \leftrightarrow \exists_z \exists_w z \neq w.$$

The inclusion 1.3.18(i) is also related to certain logically valid formulas. Let, for instance, Γ be any predicate of rank 1 in a language of predicate logic. Then the implication

$$\exists_x \exists_y (x \neq y \wedge \Gamma x \wedge \Gamma y) \rightarrow \forall_x \exists_y (x \neq y \wedge \Gamma y)$$

and hence also the implication

$$\exists_x \exists_y (x \neq y \wedge \Gamma x \wedge \Gamma y) \rightarrow \exists_y (x \neq y \wedge \Gamma y)$$

is easily seen to be logically valid. The second implication is just one of the formulas corresponding to 1.3.18(i).

THEOREM 1.3.19. *If* $\kappa \neq \lambda$, *the following conditions are equivalent*:

(i) $x \cdot c_\kappa - d_{\kappa\lambda} = 0$;

(ii) $x \leqq d_{\kappa\lambda}$ *and* $c_\kappa x = x$;

(iii) $x \leqq d_{\mu\nu}$ *and* $c_\mu x = x$ *for all* μ, ν.

PROOF. To show that (i) implies (iii) notice that the hypothesis implies $\alpha \geqq 2$; hence by (C_5) we may assume that $\mu \neq \nu$. Since $x \cdot c_\mu - d_{\mu\nu} = 0$ by 1.3.18 (cf. the remark immediately following 1.3.18), we infer by 1.2.5 that $c_\mu x \cdot - d_{\mu\nu} = 0$; so by (C_2) $x \leqq c_\mu x \leqq d_{\mu\nu}$. Hence

$$\begin{aligned} x &= d_{\mu\nu} \cdot x \\ &= d_{\mu\nu} \cdot c_\mu (d_{\mu\nu} \cdot x) \qquad \text{by 1.3.9} \\ &= d_{\mu\nu} \cdot c_\mu x \\ &= c_\mu x. \end{aligned}$$

(iii) obviously implies (ii). Assume finally that (ii) holds. Then $c_\kappa x \cdot - d_{\kappa\lambda} = x \cdot - d_{\kappa\lambda} = 0$; thus, by 1.2.5, $x \cdot c_\kappa - d_{\kappa\lambda} = 0$, and (i) holds. The proof is now complete.

THEOREM 1.3.20. *If* $\kappa \neq \lambda$, *then* $c_\kappa x \cdot - c_\kappa - d_{\kappa\lambda} = x \cdot - c_\kappa - d_{\kappa\lambda}$.

PROOF. By taking $x \cdot - c_\kappa - d_{\kappa\lambda}$ for x in 1.3.19 we obtain

$$c_\kappa (x \cdot - c_\kappa - d_{\kappa\lambda}) = x \cdot - c_\kappa - d_{\kappa\lambda}.$$

However, by (C_3) and 1.2.11 we have

$$c_\kappa (x \cdot - c_\kappa - d_{\kappa\lambda}) = c_\kappa x \cdot - c_\kappa - d_{\kappa\lambda}.$$

The conclusion follows immediately.

As is well known, whenever the algebra $\mathfrak{B} = \langle B, +, \cdot, -, 0, 1\rangle$ is a BA, the algebra $\mathfrak{B}^{\partial} = \langle B, \cdot, +, -, 1, 0\rangle$ is also a BA called the *dual* of $\mathfrak{B}$, and the operation $-$ is an isomorphism from $\mathfrak{B}$ onto $\mathfrak{B}^{\partial}$ as well as from $\mathfrak{B}^{\partial}$ onto $\mathfrak{B}$ (in the sense of 0.2.5). This notion of duality can be extended to CA_α's:

DEFINITION 1.4.1. *We define the operation* $\mathsf{c}_\kappa^\partial$ *on the* CA_α $\mathfrak{A}$ *by letting* $\mathsf{c}_\kappa^\partial x = -\mathsf{c}_\kappa - x$ *for all* $x \in A$. *The operations* $\mathsf{c}_\kappa^\partial$ *thus defined are called* ***inner cylindrifications.*** *We also let* $\mathfrak{A}^\partial = \langle A, \cdot, +, -, 1, 0, \mathsf{c}_\kappa^\partial, -\mathsf{d}_{\kappa\lambda}\rangle_{\kappa,\lambda<\alpha}$. *Similar notation is applied to diagonal-free cylindric algebras.*

To illustrate this definition consider the special classes of non-trivial CA's discussed in Section 1.1. First suppose $\mathfrak{A}$ is a cylindric set algebra of dimension α with base U; naturally the κth inner cylindrification of $\mathfrak{A}$ will be denoted by $\mathbf{C}_\kappa^\partial$. For any $X \in A$, $\mathbf{C}_\kappa^\partial X$ is the set of all $y \in {}^\alpha U$ such that for every $x \in {}^\alpha U$, if $x_\lambda = y_\lambda$ whenever $\lambda \neq \kappa$, then $x \in X$. In the geometric terminology of 1.1.4 we see that $\mathbf{C}_\kappa^\partial X$ is the largest cylinder with axis parallel to the κ-axis which is included in X. Hence the name "inner cylindrification"; in this context the old operations c_κ could be referred to as *outer cylindrifications.* The application of $\mathsf{c}_\kappa^\partial$ in a simple situation is illustrated in Fig. 1.4.3.

On the other hand, let Σ be a set of sentences and φ a formula of a language Λ. Then, in the CA of formulas associated with Σ, $\mathsf{c}_\kappa^\partial(\varphi/\equiv_\Sigma)$ is easily seen to be the equivalence class of the formula $\forall_{v_\kappa}\varphi$. Thus, e.g., $\mathsf{c}_0^\partial\mathsf{c}_1^\partial\mathsf{d}_{01}$ is the equivalence class of the formula $\forall_x\forall_y x = y$. For this reason the operations $\mathsf{c}_\kappa^\partial$ could also be called *universal quantifications.* The logically valid formula $\forall_{v_\kappa}\varphi \rightarrow \exists_{v_\kappa}\varphi$ gives rise to the inequality $\mathsf{c}_\kappa^\partial x \leq \mathsf{c}_\kappa x$, valid in any CA_α (see 1.4.4(i) below).

THEOREM 1.4.2. $\mathfrak{A}^\partial$ *is a* CA_α, *and the operation* $-$ *is an isomorphism from* $\mathfrak{A}$ *onto* $\mathfrak{A}^\partial$. *If* $\mathfrak{B} \in \mathsf{Df}_\alpha$, *then* $\mathfrak{B}^\partial \in \mathsf{Df}_\alpha$ *and* $-$ *is an isomorphism from* $\mathfrak{B}$ *onto* $\mathfrak{B}^\partial$.

PROOF. The fact that $-$ is a one-one function from A onto A, and indeed an isomorphism from the Boolean part $\mathfrak{Bl}\mathfrak{A}$ onto $\mathfrak{Bl}\mathfrak{A}^\partial$, is well known. The

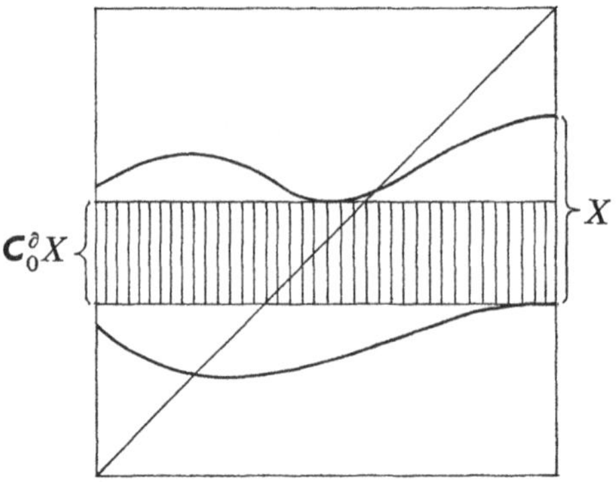

Fig. 1.4.3

fact that $-$ carries c_κ into $\mathsf{c}_\kappa^\partial$ (i.e., that $-\mathsf{c}_\kappa x = \mathsf{c}_\kappa^\partial - x$ for all $x \in A$) and $\mathsf{d}_{\kappa\lambda}$ into $-\mathsf{d}_{\kappa\lambda}$ is completely obvious. Thus $-$ is an isomorphism from $\mathfrak{A}$ onto $\mathfrak{A}^\partial$. Hence, as observed in the remarks beginning Section 0.2, $\mathfrak{A}^\partial$ must satisfy all of the algebraic laws (in particular, all identities) which are satisfied by $\mathfrak{A}$. Therefore, $\mathfrak{A}^\partial$ must be a CA_α by 1.1.1. This can also be shown by checking directly that $\mathfrak{A}^\partial$ satisfies (C_0)–(C_7).

If a notion introduced in this work is explicitly or implicitly relativized to a given cylindric algebra $\mathfrak{A}$, then the same notion relativized to $\mathfrak{A}^\partial$ is referred to as the *dual* of the original notion. If the symbol for the original notion is provided with the superscript $^{(\mathfrak{A})}$, then the dual symbol is obtained by replacing $^{(\mathfrak{A})}$ by $^{(\mathfrak{A}^\partial)}$; if, however, an explicit reference to $\mathfrak{A}$ is not made in the original symbol (cf. the remarks immediately following 1.1.1), then in forming the dual symbol we use the simpler superscript $^\partial$ in place of $^{(\mathfrak{A}^\partial)}$. From 1.4.1 and 1.4.2 we see that the symbol $\mathsf{c}_\kappa^\partial$ has been introduced consistently with this convention, and that $+^\partial$, $\cdot^\partial$, $-^\partial$, 0^∂, 1^∂, and $\mathsf{d}_{\kappa\lambda}^\partial$ are respectively synonymous with $\cdot$, $+$, $-$, 1, 0 and $-\mathsf{d}_{\kappa\lambda}$. Analogous conclusions can easily be obtained for other notions using 1.4.2 and the relevant definitions; e.g., the duals of the notions $\leqq$, $\sum$, and $\prod$ are respectively $\geqq$, $\prod$, and $\sum$. (Notice an exception to our convention: the symbol $\mathsf{C}_\kappa^\partial$ casually introduced in remarks following 1.4.1 to represent the κth inner cylindrification of a cylindric set algebra $\mathfrak{A}$ cannot be interpreted as an abbreviation for $\mathsf{C}_\kappa^{(\mathfrak{A}^\partial)}$. In fact, $\mathfrak{A}^\partial$ is not a cylindric set algebra according to our Definition 1.1.5 and $\mathsf{C}_\kappa^{(\mathfrak{A}^\partial)}$ has no meaning whatever. We can only say that $\mathsf{C}_\kappa^\partial$ is an alternative symbol for denoting $\mathsf{c}_\kappa^{(\mathfrak{A}^\partial)}$ in case $\mathfrak{A}$ is a cylindric set algebra.) From 1.1.1 and 1.4.2 we easily see that the duals of dual notions are the original notions, e.g., $(\mathsf{c}_\kappa^\partial)^\partial = \mathsf{c}_\kappa$. A notion is called *self-dual* if it coincides with its dual; thus complementation is self-dual.

If, in any theorem, we replace all the notions involved by their duals, we obtain what we call the *dual* of the theorem. The dual of a theorem will be denoted by affixing the superscript $^\partial$ to the number of the theorem. Thus, e.g., the dual of 1.2.1, denoted by $1.2.1^\partial$, is the following statement:

$$\mathsf{c}_\kappa^\partial x = 1 \text{ iff } x = 1$$

(for all $\kappa < \alpha$ and $x \in A$). We shall rarely, if ever, formulate the dual theorems explicitly, but we shall use them and refer to them in our arguments.

THEOREM 1.4.4. (i) $\mathsf{c}_\kappa^\partial x \leqq \mathsf{c}_\kappa x$;

(ii) $\mathsf{c}_\kappa \mathsf{c}_\kappa^\partial x = \mathsf{c}_\kappa^\partial x$;

(iii) $\mathsf{c}_\kappa^\partial \mathsf{c}_\kappa x = \mathsf{c}_\kappa x$;

(iv) $\mathsf{c}_\kappa \mathsf{c}_\lambda^\partial x \leqq \mathsf{c}_\lambda^\partial \mathsf{c}_\kappa x$;

(v) $\mathsf{c}_\kappa x \leqq \mathsf{c}_\lambda^\partial y$ *iff* $\mathsf{c}_\lambda x \leqq \mathsf{c}_\kappa^\partial y$;

(vi) *if* $\kappa \neq \lambda$, *then* $\mathsf{c}_\kappa(\mathsf{d}_{\kappa\lambda} \cdot x) = \mathsf{c}_\kappa^\partial(-\mathsf{d}_{\kappa\lambda} + x)$;

(vii) *if* $\kappa \neq \lambda$, *then* $\mathsf{c}_\kappa \mathsf{c}_\lambda^\partial \mathsf{d}_{\kappa\lambda} \leqq \mathsf{d}_{\mu\nu}$ *for all* μ, ν;

(viii) $\mathsf{c}_\kappa^\partial \mathsf{c}_\lambda - \mathsf{d}_{\kappa\lambda} = \mathsf{c}_\kappa \mathsf{c}_\lambda - \mathsf{d}_{\kappa\lambda}$.

PROOF. Statements (i)–(iii) follow from the most elementary laws of Section 1.2. Formula (iv) is a direct consequence of 1.2.16, and condition (v) is a restatement of 1.2.15. As to (vi), we have $\mathsf{c}_\kappa^\partial(-\mathsf{d}_{\kappa\lambda} + x) = -\mathsf{c}_\kappa - (-\mathsf{d}_{\kappa\lambda} + x) = -\mathsf{c}_\kappa(\mathsf{d}_{\kappa\lambda} \cdot -x) = \mathsf{c}_\kappa(\mathsf{d}_{\kappa\lambda} \cdot x)$. To prove (vii), we may assume that $\mu \neq \nu$. Then

$$\begin{aligned}
\mathsf{c}_\kappa \mathsf{c}_\lambda^\partial \mathsf{d}_{\kappa\lambda} &= \mathsf{c}_\kappa - \mathsf{c}_\lambda - \mathsf{d}_{\kappa\lambda} \\
&= \mathsf{c}_\kappa - \mathsf{c}_\kappa \mathsf{c}_\lambda - \mathsf{d}_{\kappa\lambda} && \text{by 1.3.18} \\
&= -\mathsf{c}_\kappa \mathsf{c}_\lambda - \mathsf{d}_{\kappa\lambda} \\
&= -\mathsf{c}_\mu \mathsf{c}_\nu - \mathsf{d}_{\mu\nu} && \text{by 1.3.18} \\
&= \mathsf{c}_\mu^\partial \mathsf{c}_\nu^\partial \mathsf{d}_{\mu\nu} \\
&\leqq \mathsf{d}_{\mu\nu} && \text{by } (C_2)^\partial.
\end{aligned}$$

Finally, formula (viii) follows directly from 1.3.18.

REMARK 1.4.5. By comparing 1.4.4(viii) with 1.3.18(ii),(iii) we obtain new expressions representing the element $\mathsf{c}_0 - \mathsf{d}_{01}$ (cf. the remark following 1.3.18). Instances of such expressions are $\mathsf{c}_0^\partial \mathsf{c}_1 - \mathsf{d}_{01}$ and $\mathsf{c}_1^\partial \mathsf{c}_0 - \mathsf{d}_{01}$; to represent the complement $-\mathsf{c}_0 - \mathsf{d}_{01}$ we can use $\mathsf{c}_0^\partial \mathsf{d}_{01}$, $\mathsf{c}_1^\partial \mathsf{d}_{01}$, $\mathsf{c}_0 \mathsf{c}_1^\partial \mathsf{d}_{01}$, and $\mathsf{c}_0^\partial \mathsf{c}_1^\partial \mathsf{d}_{01}$. In all these expressions 0 and 1 may be replaced respectively by any two distinct ordinals κ and λ.

1.5. SUBSTITUTIONS

In this section we consider certain operations, defined in terms of cylindrifications and diagonal elements, which are suggested by the discussion of the special CA's introduced in Section 1.1, and which play a useful role in developing the general theory of CA's.

DEFINITION 1.5.1. *For any $\kappa, \lambda < \alpha$ and $x \in A$ we set*

$$\mathsf{s}_\lambda^\kappa x = \begin{cases} x & \text{if } \kappa = \lambda, \\ \mathsf{c}_\kappa(\mathsf{d}_{\kappa\lambda} \cdot x) & \text{if } \kappa \neq \lambda. \end{cases}$$

In case $\mathfrak{A}$ is a discrete CA_α we obviously have $\mathsf{s}_\lambda^\kappa x = x$ for any $\kappa, \lambda < \alpha$ and any $x \in A$.

The metalogical interpretation of the operation $\mathsf{s}_\lambda^\kappa$ is simple. Let Σ be a set of sentences in a language Λ and $\mathfrak{A}$ be the CA of formulas associated with Σ. Any given element x of $\mathfrak{A}$ is an equivalence class $\varphi/\equiv_\Sigma$ where φ is a formula in Λ. By comparing 1.5.1 with the remarks following 1.1.10 we see at once that $\mathsf{s}_\lambda^\kappa x$ is the equivalence class $\psi/\equiv_\Sigma$ where ψ is the formula obtained from φ by substituting the variable $\mathfrak{v}_\lambda$ for $\mathfrak{v}_\kappa$.

Turning to the geometrical (or set-theoretical) interpretation, we consider an arbitrary set algebra $\mathfrak{A}$ of dimension α with base U, a subset X of ${}^\alpha U$, and two distinct ordinals $\kappa, \lambda < \alpha$. We let X' be the set of all points $x \in X$ such that $x_\kappa = x_\lambda$. Then $\mathsf{s}_\lambda^\kappa X$ is simply the cylinder obtained by moving X' parallel to the κ-axis. (In Fig. 1.5.2 this process is represented for $\alpha = 2$.) There seems to be no standard geometric term for this operation; if, however, $\mathbf{C}_\lambda X = X$ (and $\kappa \neq \lambda$), the operation is just that of reflection in the hyperplane $\mathbf{D}_{\kappa\lambda}$. In real three-dimensional space the set $\mathsf{s}_1^0 \mathbf{C}_1 X$ can also be obtained from $\mathbf{C}_1 X$ by a rotation of 90° about the 2-axis.

Since the intuitive meaning of the operation $\mathsf{s}_\lambda^\kappa$ is simpler and clearer in its metalogical rather than geometrical interpretation, we shall use a term of metalogical origin to refer to this operation in informal remarks; in fact, we shall call it *substitution* or, more specifically the *λ-for-κ substitution*.

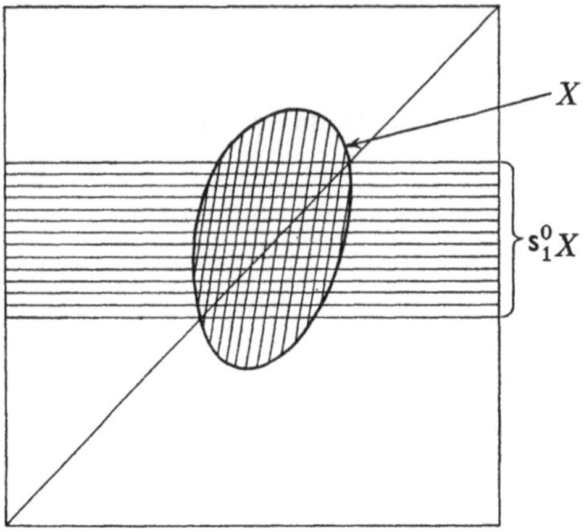

Fig. 1.5.2

THEOREM 1.5.3. (i) *If* $\sum_{i\in I} z_i$ *exists, then* $\sum_{i\in I} s^\kappa_\lambda z_i$ *exists, and* $s^\kappa_\lambda \sum_{i\in I} z_i = \sum_{i\in I} s^\kappa_\lambda z_i$; *in particular* $s^\kappa_\lambda(x+y) = s^\kappa_\lambda x + s^\kappa_\lambda y$;

(ii) $s^\kappa_\lambda - x = -s^\kappa_\lambda x$.

PROOF: by 1.2.6 and 1.3.4.

In view of part (ii) of this theorem, part (i) remains valid if Σ and $+$ are respectively replaced by Π and $\cdot$; (i) and (ii) jointly imply that $s^\kappa_\lambda 0 = 0$ and $s^\kappa_\lambda 1 = 1$. With these supplements Theorem 1.5.3 shows the function s^κ_λ to be an endomorphism, and in fact what is called a *complete endomorphism*, of the Boolean algebra $\mathfrak{Bl}\,\mathfrak{A}$.

THEOREM 1.5.4. (i) *If* $\kappa \neq \mu$, *then* $s^\kappa_\lambda d_{\kappa\mu} = d_{\lambda\mu}$;

(ii) *If* $\kappa \neq \mu, \nu$ *then* $s^\kappa_\lambda d_{\mu\nu} = d_{\mu\nu}$.

PROOF. Both statements are obvious if $\kappa = \lambda$. If $\kappa \neq \lambda$, then (i) follows from (C_6) and 1.3.1, and (ii) follows from (C_3), 1.3.2, and 1.3.3.

By 1.3.9 we have

THEOREM 1.5.5. $d_{\kappa\lambda} \cdot s^\kappa_\lambda x = d_{\kappa\lambda} \cdot x$.

THEOREM 1.5.6. $d_{\kappa\lambda} \cdot s^\mu_\kappa x = d_{\kappa\lambda} \cdot s^\mu_\lambda x$.

PROOF. If $\mu = \kappa$ or $\mu = \lambda$ the theorem reduces to 1.5.5. If $\mu \neq \kappa$ and $\mu \neq \lambda$, then $d_{\kappa\lambda} \cdot s^\mu_\kappa x = c_\mu(d_{\kappa\lambda} \cdot d_{\mu\kappa} \cdot x) = c_\mu(d_{\kappa\lambda} \cdot d_{\mu\lambda} \cdot x) = d_{\kappa\lambda} \cdot s^\mu_\lambda x$ by (C_3), 1.3.1, 1.3.3, and 1.3.7.

THEOREM 1.5.7. *$\mathsf{d}_{\kappa\lambda}$ is the least element x such that $\mathsf{s}^\kappa_\lambda x = 1$.*

PROOF. By 1.5.4(i), $\mathsf{d}_{\kappa\lambda}$ does satisfy the equation $\mathsf{s}^\kappa_\lambda \mathsf{d}_{\kappa\lambda} = 1$. Now suppose $\mathsf{s}^\kappa_\lambda x = 1$. Then $\mathsf{d}_{\kappa\lambda} \cdot x = \mathsf{d}_{\kappa\lambda} \cdot \mathsf{s}^\kappa_\lambda x = \mathsf{d}_{\kappa\lambda}$ by 1.5.5, and hence $\mathsf{d}_{\kappa\lambda} \leqq x$.

From 1.5.7 we see that, in principle, the operations $\mathsf{s}^\kappa_\lambda$ could be taken as fundamental operations in 1.1.1 in place of the elements $\mathsf{d}_{\kappa\lambda}$.

THEOREM 1.5.8. (i) $\mathsf{s}^\kappa_\lambda \mathsf{c}_\kappa x = \mathsf{c}_\kappa x$;

(ii) *if $\mu \neq \kappa, \lambda$, then* $\mathsf{s}^\kappa_\lambda \mathsf{c}_\mu x = \mathsf{c}_\mu \mathsf{s}^\kappa_\lambda x$.

PROOF. For both (i) and (ii) we may assume that $\kappa \neq \lambda$. Then (i) follows by (C_3) and 1.3.2, and (ii) follows from (C_3), (C_4), and 1.3.3.

THEOREM 1.5.9. (i) $\mathsf{c}_\lambda \mathsf{s}^\kappa_\lambda x = \mathsf{c}_\kappa \mathsf{s}^\lambda_\kappa x$;

(ii) *if $\kappa \neq \lambda$, then* $\mathsf{c}_\kappa \mathsf{s}^\kappa_\lambda x = \mathsf{s}^\kappa_\lambda x$.

PROOF. The first equation follows immediately from (C_4); the second is implied by 1.2.3.

THEOREM 1.5.10. (i) *If $\kappa \neq \mu$, then* $\mathsf{s}^\kappa_\lambda \mathsf{s}^\kappa_\mu x = \mathsf{s}^\kappa_\mu x$;

(ii) $\mathsf{s}^\kappa_\lambda \mathsf{s}^\mu_\kappa x = \mathsf{s}^\kappa_\lambda \mathsf{s}^\mu_\lambda x$;

(iii) *if $\lambda \neq \mu \neq \kappa \neq \nu$, then* $\mathsf{s}^\mu_\nu \mathsf{s}^\kappa_\lambda x = \mathsf{s}^\kappa_\lambda \mathsf{s}^\mu_\nu x$;

(iv) $\mathsf{s}^\kappa_\lambda \mathsf{s}^\mu_\lambda x = \mathsf{s}^\mu_\lambda \mathsf{s}^\kappa_\lambda x$;

(v) $\mathsf{s}^\kappa_\lambda \mathsf{s}^\lambda_\kappa x = \mathsf{s}^\kappa_\lambda x$;

(vi) $\mathsf{s}^\kappa_\lambda \mathsf{s}^\mu_\kappa x = \mathsf{s}^\mu_\lambda \mathsf{s}^\kappa_\mu x$.

PROOF. Equation (i) is immediate by 1.5.1 and 1.5.8(i). Applying c_κ to both sides of 1.5.6 gives (ii) if $\kappa \neq \lambda$; the case $\kappa = \lambda$ is trivial. (iii) follows from (C_3), (C_4), 1.3.3, and 1.5.1. If $\kappa = \lambda$, $\kappa = \mu$, or $\lambda = \mu$, then (iv) is trivial by 1.5.1. If $\kappa \neq \lambda$, $\kappa \neq \mu$, and $\lambda \neq \mu$, then (iv) is a special case of (iii). From (ii) we easily deduce (v), and (vi) follows from (ii) and (iv).

THEOREM 1.5.11. (i) $\mathsf{s}^\kappa_\lambda \mathsf{s}^\mu_\kappa \mathsf{c}_\kappa x = \mathsf{s}^\mu_\lambda \mathsf{c}_\kappa x$;

(ii) $\mathsf{s}^\kappa_\lambda \mathsf{s}^\mu_\kappa \mathsf{c}_\kappa \mathsf{c}_\nu x = \mathsf{s}^\nu_\lambda \mathsf{s}^\mu_\nu \mathsf{c}_\kappa \mathsf{c}_\nu x$.

PROOF. Equation (i) follows from 1.5.8(i) and 1.5.10(vi), and (ii) is obtained from (i) using (C_4).

DEFINITION 1.5.12. *For any $\kappa, \lambda, \mu < \alpha$ and $x \in A$ we set*

$$_\mu\mathsf{s}(\kappa, \lambda)x = \mathsf{s}^\mu_\kappa \mathsf{s}^\kappa_\lambda \mathsf{s}^\lambda_\mu x.$$

The operation $_\mu\mathsf{s}(\kappa, \lambda)$ acquires a simple meaning when applied to the special CA's introduced in Section 0.1, and specifically to those elements of the algebras which are μ-closed.

First, consider the metalogical interpretation. Let $\mathfrak{A}$ be a CA_α of formulas associated with a set of sentences Σ in a language Λ; thus the elements of $\mathfrak{A}$ are the equivalence classes $\varphi/\equiv_\Sigma$ of formulas φ in Λ. The operation ${}_\mu\mathsf{s}(\kappa, \lambda)$ is closely related to the familiar metalogical operation of interchanging the variables $\mathbf{v}_\kappa$ and $\mathbf{v}_\lambda$ (or rather the free occurrences of these variables) in a formula φ. The latter operation does not have to be described here in detail. It should only be pointed out that, in case the language is provided with sufficiently many distinct variables, the operation reduces to a succession of simple substitutions of one variable for another. In fact, to interchange $\mathbf{v}_\kappa$ and $\mathbf{v}_\lambda$ in φ we can first pick any variable $\mathbf{v}_\mu$ which does not occur in φ, and then substitute successively $\mathbf{v}_\mu$ for $\mathbf{v}_\lambda$ in φ, $\mathbf{v}_\lambda$ for $\mathbf{v}_\kappa$ in the resulting formula φ', and finally $\mathbf{v}_\kappa$ for $\mathbf{v}_\mu$ in the formula obtained from φ' by this second substitution. Hence if $\mathbf{v}_\mu$ does not occur in φ and therefore the element $x = \varphi/\equiv_\Sigma$ is μ-closed, then by 1.5.12, ${}_\mu\mathsf{s}(\kappa, \lambda)x$ is the equivalence class of the formula obtained from φ by interchanging $\mathbf{v}_\kappa$ and $\mathbf{v}_\lambda$.

It should be mentioned that both the operation of substituting $\mathbf{v}_\lambda$ for $\mathbf{v}_\kappa$ and that of interchanging $\mathbf{v}_\kappa$ and $\mathbf{v}_\lambda$ are particular cases of a general metalogical operation: that of simultaneously substituting several variables $\mathbf{v}_{\lambda_0}, \ldots, \mathbf{v}_{\lambda_{\nu-1}}$ for given variables $\mathbf{v}_{\kappa_0}, \ldots, \mathbf{v}_{\kappa_{\nu-1}}$. Interchanging $\mathbf{v}_\kappa$ and $\mathbf{v}_\lambda$ means the same as simultaneously substituting $\mathbf{v}_\kappa$ for $\mathbf{v}_\lambda$ and $\mathbf{v}_\lambda$ for $\mathbf{v}_\kappa$. The operation of substituting many variables simultaneously, just as that of interchanging two variables, can be reduced to a succession of substitutions of single variables in case the language is provided with sufficiently many distinct variables. In Section 1.11 we shall introduce and study an operation in CA's corresponding to that of simultaneous substitution; just as in the case of ${}_\mu\mathsf{s}(\kappa, \lambda)$, this operation will be defined as a composition of operations $\mathsf{s}^\kappa_\lambda$ (cf. 1.11.9). However, to obtain satisfactory results about this operation we shall have to restrict ourselves in that discussion to special classes of CA's which will be defined in an earlier part of Section 1.11.

Turning now to the set-theoretical interpretation, consider a CA $\mathfrak{A}$ of sets, say of dimension 3, and let X be any 2-cylinder of $\mathfrak{A}$. From 1.5.12 we easily conclude that

$$ {}_2\mathsf{s}(0, 1)X = \{x : x = \langle x_0, x_1, x_2\rangle \in {}^3U,\ \langle x_1, x_0, x_2\rangle \in X\}. $$

Thus in this case ${}_2\mathsf{s}(0, 1)X$ is obtained from X by reflection in the plane $\mathbf{D}_{01}$ (and hence coincides with $\mathsf{s}^0_1 X$ if $\mathsf{C}_1 X = X$). If binary relations on U are identified with 2-cylinders included in ${}^3U$ (and this seems to be a natural procedure in discussing three-dimensional set algebras), then the operation ${}_2\mathsf{s}(0, 1)$ when applied to binary relations coincides with the familiar operation of forming converses.

THEOREM 1.5.13. (i) ${}_{\kappa}\mathsf{s}(\kappa, \lambda)x = \mathsf{s}^{\kappa}_{\lambda}x$;
(ii) ${}_{\lambda}\mathsf{s}(\kappa, \lambda)x = \mathsf{s}^{\lambda}_{\kappa}x$;
(iii) ${}_{\mu}\mathsf{s}(\kappa, \kappa)x = \mathsf{s}^{\mu}_{\kappa}x$.
PROOF: by 1.5.10(v).

THEOREM 1.5.14. *If* $\kappa, \lambda, \mu \neq \nu$, *then*

$$ {}_{\mu}\mathsf{s}(\kappa, \lambda)\mathsf{c}_{\mu}\mathsf{c}_{\nu}x = {}_{\mu}\mathsf{s}(\lambda, \kappa)\mathsf{c}_{\mu}\mathsf{c}_{\nu}x. $$

PROOF. By 1.5.13 we may assume that $\kappa, \lambda, \mu, \nu$ are distinct. Then

$$
\begin{aligned}
\mathsf{s}^{\lambda}_{\mu}\mathsf{s}^{\mu}_{\kappa}\mathsf{s}^{\kappa}_{\lambda}\mathsf{s}^{\lambda}_{\mu}\mathsf{c}_{\mu}\mathsf{c}_{\nu}x &= \mathsf{s}^{\lambda}_{\mu}\mathsf{s}^{\mu}_{\kappa}\mathsf{s}^{\kappa}_{\lambda}\mathsf{s}^{\nu}_{\mu}\mathsf{s}^{\lambda}_{\nu}\mathsf{c}_{\mu}\mathsf{c}_{\nu}x && \text{by } (\mathrm{C}_4), 1.5.11(\mathrm{i}) \\
&= \mathsf{s}^{\lambda}_{\mu}\mathsf{s}^{\mu}_{\kappa}\mathsf{s}^{\nu}_{\mu}\mathsf{c}_{\mu}\mathsf{s}^{\kappa}_{\lambda}\mathsf{s}^{\lambda}_{\nu}\mathsf{c}_{\nu}x && \text{by } 1.5.8(\mathrm{ii}), 1.5.10(\mathrm{iii}) \\
&= \mathsf{s}^{\lambda}_{\mu}\mathsf{s}^{\nu}_{\kappa}\mathsf{s}^{\kappa}_{\lambda}\mathsf{s}^{\lambda}_{\nu}\mathsf{c}_{\mu}\mathsf{c}_{\nu}x && \text{by } 1.5.8(\mathrm{ii}), 1.5.11(\mathrm{i}) \\
&= \mathsf{s}^{\nu}_{\kappa}\mathsf{s}^{\lambda}_{\mu}\mathsf{s}^{\kappa}_{\lambda}\mathsf{c}_{\lambda}\mathsf{s}^{\lambda}_{\nu}\mathsf{c}_{\mu}\mathsf{c}_{\nu}x && \text{by } 1.5.9(\mathrm{ii}), 1.5.10(\mathrm{iii}) \\
&= \mathsf{s}^{\nu}_{\kappa}\mathsf{s}^{\kappa}_{\mu}\mathsf{s}^{\lambda}_{\nu}\mathsf{c}_{\mu}\mathsf{c}_{\nu}x && \text{by } 1.5.9(\mathrm{ii}), 1.5.11(\mathrm{i}) \\
&= \mathsf{s}^{\nu}_{\kappa}\mathsf{s}^{\lambda}_{\nu}\mathsf{c}_{\nu}\mathsf{s}^{\kappa}_{\mu}\mathsf{c}_{\mu}x && \text{by } (\mathrm{C}_4), 1.5.8(\mathrm{ii}), 1.5.10(\mathrm{iii}) \\
&= \mathsf{s}^{\lambda}_{\kappa}\mathsf{s}^{\kappa}_{\mu}\mathsf{c}_{\mu}\mathsf{c}_{\nu}x && \text{by } (\mathrm{C}_4), 1.5.8(\mathrm{ii}), 1.5.11(\mathrm{i}).
\end{aligned}
$$

Hence, applying $\mathsf{s}^{\mu}_{\lambda}$ to both sides and using 1.5.9(ii) and 1.5.11(i), the desired result follows.

The result just obtained gives rise to a natural question: can Theorem 1.5.14 be improved by omitting all references to ν (and hence by replacing $\mathsf{c}_{\nu}x$ by x in the conclusion)? It is easily seen that the improved theorem holds for the special algebras of Section 1.1 (and hence, in view of 0.4.63, for all representable algebras in the sense of 1.1.13). In particular, it becomes obviously true when applied to CA's of formulas. As was previously pointed out, the operation of interchanging two variables, $\boldsymbol{v}_{\kappa}$ and $\boldsymbol{v}_{\lambda}$, in a formula φ reduces to a succession of simple substitutions of one variable for another. This process is not uniquely determined; for instance, after having picked a variable $\boldsymbol{v}_{\mu}$ not occurring in φ, we can substitute successively either $\boldsymbol{v}_{\mu}$ for $\boldsymbol{v}_{\lambda}$, $\boldsymbol{v}_{\lambda}$ for $\boldsymbol{v}_{\kappa}$, and $\boldsymbol{v}_{\kappa}$ for $\boldsymbol{v}_{\mu}$, or else $\boldsymbol{v}_{\mu}$ for $\boldsymbol{v}_{\kappa}$, $\boldsymbol{v}_{\kappa}$ for $\boldsymbol{v}_{\lambda}$, and $\boldsymbol{v}_{\lambda}$ for $\boldsymbol{v}_{\mu}$. The outcome is the same in both cases, and this is just the content of the improved Theorem 1.5.14. In Part II of this work we shall see, however, that Theorem 1.5.14 in the improved form does not hold in the general theory of CA's: for each $\alpha \geqq 3$ there is a CA_{α} in which it fails.

THEOREM 1.5.15. *If* $\{\kappa, \lambda\} \cap \{\mu, \nu\} = 0$, *then*

$$ {}_{\mu}\mathsf{s}(\kappa, \lambda)\mathsf{c}_{\mu}\mathsf{c}_{\nu}x = {}_{\nu}\mathsf{s}(\kappa, \lambda)\mathsf{c}_{\mu}\mathsf{c}_{\nu}x. $$

PROOF. By 1.5.13(iii) we may assume that κ, λ, μ, ν are distinct. Then

$$\begin{aligned}
{}_\mu s(\kappa, \lambda) c_\mu c_\nu x &= s^\mu_\kappa s^\kappa_\lambda s^\lambda_\mu c_\nu c_\mu x && \text{by } (C_4) \text{ and } 1.5.12 \\
&= s^\mu_\kappa s^\kappa_\lambda s^\nu_\mu s^\lambda_\nu c_\nu c_\mu x && \text{by } 1.5.11(\text{i}) \\
&= s^\mu_\kappa s^\nu_\mu c_\mu s^\kappa_\lambda s^\lambda_\nu c_\nu x && \text{by } (C_4),\ 1.5.8(\text{ii}),\ 1.5.10(\text{iii}) \\
&= {}_\nu s(\kappa, \lambda) c_\mu c_\nu x && \text{by } 1.5.8(\text{ii}),\ 1.5.11(\text{i}).
\end{aligned}$$

THEOREM 1.5.16. (i) *If* $\sum_{i \in I} z_i$ *exists, then* $\sum_{i \in I}\, {}_\mu s(\kappa, \lambda) z_i$ *exists, and* ${}_\mu s(\kappa, \lambda) \sum_{i \in I} z_i = \sum_{i \in I}\, {}_\mu s(\kappa, \lambda) z_i$; *in particular*, ${}_\mu s(\kappa, \lambda)(x+y) = {}_\mu s(\kappa, \lambda)x + {}_\mu s(\kappa, \lambda)y$;
(ii) ${}_\mu s(\kappa, \lambda) -x = -{}_\mu s(\kappa, \lambda)x$.

PROOF: by 1.5.3.

Thus ${}_\mu s(\kappa, \lambda)$, like s^κ_λ, is a complete endomorphism of $\mathfrak{Bl}\,\mathfrak{A}$; cf. the remark immediately following 1.5.3.

THEOREM 1.5.17. *If* $\kappa, \lambda, \mu \neq \nu$, *then*

$${}_\mu s(\kappa, \lambda)\, {}_\mu s(\kappa, \lambda) c_\mu c_\nu x = c_\mu c_\nu x.$$

PROOF. We have

$$\begin{aligned}
{}_\mu s(\kappa, \lambda)\, {}_\mu s(\kappa, \lambda) c_\mu c_\nu x &= {}_\mu s(\lambda, \kappa)\, {}_\mu s(\kappa, \lambda) c_\mu c_\nu x && \text{by } 1.5.14 \\
&= s^\mu_\lambda s^\lambda_\kappa s^\kappa_\mu s^\mu_\kappa s^\kappa_\lambda s^\lambda_\mu c_\mu c_\nu x \\
&= c_\mu c_\nu x && \text{by } 1.5.11(\text{i}).
\end{aligned}$$

Theorem 1.5.17 implies that, in case $\kappa, \lambda, \mu \neq \nu$, the function ${}_\mu s(\kappa, \lambda)$ is a permutation of the set of all elements $c_\mu c_\nu x$ (and in fact a permutation of a very special kind usually called an *inversion*). Thus, in particular, when restricted to this set the function ${}_\mu s(\kappa, \lambda)$ is one-one.

THEOREM 1.5.18. *If* $\kappa, \lambda, \mu, \nu, \xi$ *are distinct, then*

$${}_\nu s(\kappa, \lambda)\, {}_\nu s(\kappa, \mu) c_\nu c_\xi x = {}_\nu s(\lambda, \mu)\, {}_\nu s(\kappa, \lambda) c_\nu c_\xi x.$$

PROOF. We have

$$\begin{aligned}
{}_\nu s(\kappa, \lambda)\, {}_\nu s(\kappa, \mu) c_\nu c_\xi x &= s^\nu_\kappa s^\kappa_\lambda s^\lambda_\nu s^\xi_\kappa s^\kappa_\mu s^\mu_\xi c_\nu c_\xi x && \text{by } 1.5.15 \\
&= s^\nu_\kappa s^\kappa_\lambda s^\xi_\kappa s^\lambda_\nu s^\kappa_\mu s^\mu_\xi c_\nu c_\xi x && \text{by } 1.5.10(\text{iii}) \\
&= s^\nu_\kappa s^\xi_\lambda s^\lambda_\nu s^\kappa_\mu s^\mu_\xi c_\nu c_\xi x && \text{by } 1.5.11(\text{i}) \\
&= s^\xi_\lambda s^\nu_\kappa s^\kappa_\mu s^\mu_\xi s^\lambda_\nu c_\nu c_\xi x && \text{by } 1.5.10(\text{iii}) \\
&= s^\xi_\lambda s^\nu_\kappa s^\lambda_\mu s^\kappa_\lambda s^\mu_\xi s^\lambda_\nu c_\nu c_\xi x && \text{by } 1.5.11(\text{i}) \\
&= s^\xi_\lambda s^\lambda_\mu s^\mu_\xi s^\nu_\kappa s^\kappa_\lambda s^\lambda_\nu c_\nu c_\xi x && \text{by } 1.5.10(\text{iii}) \\
&= {}_\nu s(\lambda, \mu)\, {}_\nu s(\kappa, \lambda) c_\nu c_\xi x && \text{by } 1.5.15.
\end{aligned}$$

In connection with 1.5.17 and 1.5.18 the question arises whether these theorems can be improved by omitting all references to ν in the former and to ξ in the latter. The situation turns out to be exactly the same as in the case of Theorem 1.5.14, and we can repeat here with minor changes the remarks concerning that theorem.

THEOREM 1.5.19. *If* $\kappa, \lambda, \mu, \nu$ *are distinct, then*
(i) ${}_\nu\mathsf{s}(\kappa, \lambda)\mathsf{s}^\mu_\kappa x = \mathsf{s}^\mu_\lambda {}_\nu\mathsf{s}(\kappa, \lambda)x$;
(ii) ${}_\nu\mathsf{s}(\kappa, \lambda)\mathsf{s}^\mu_\lambda x = \mathsf{s}^\mu_\kappa {}_\nu\mathsf{s}(\kappa, \lambda)x$.
PROOF. To prove (i) we compute:

$$\begin{aligned}
{}_\nu\mathsf{s}(\kappa, \lambda)\mathsf{s}^\mu_\kappa x &= \mathsf{s}^\nu_\kappa \mathsf{s}^\kappa_\lambda \mathsf{s}^\lambda_\nu \mathsf{s}^\mu_\kappa x \\
&= \mathsf{s}^\nu_\kappa \mathsf{s}^\kappa_\lambda \mathsf{s}^\mu_\kappa \mathsf{s}^\lambda_\nu x && \text{by 1.5.10(iii)} \\
&= \mathsf{s}^\nu_\kappa \mathsf{s}^\mu_\lambda \mathsf{s}^\kappa_\lambda \mathsf{s}^\lambda_\nu x && \text{by 1.5.10(ii), (iv)} \\
&= \mathsf{s}^\mu_\lambda {}_\nu\mathsf{s}(\kappa, \lambda)x && \text{by 1.5.10(iii).}
\end{aligned}$$

A similar computation gives (ii):

$$\begin{aligned}
{}_\nu\mathsf{s}(\kappa, \lambda)\mathsf{s}^\mu_\lambda x &= \mathsf{s}^\nu_\kappa \mathsf{s}^\kappa_\lambda \mathsf{s}^\lambda_\nu \mathsf{s}^\mu_\lambda x \\
&= \mathsf{s}^\nu_\kappa \mathsf{s}^\kappa_\lambda \mathsf{s}^\mu_\nu \mathsf{s}^\lambda_\nu x && \text{by 1.5.10(ii), (iv)} \\
&= \mathsf{s}^\nu_\kappa \mathsf{s}^\mu_\nu \mathsf{s}^\kappa_\lambda \mathsf{s}^\lambda_\nu x && \text{by 1.5.10(iii)} \\
&= \mathsf{s}^\mu_\kappa \mathsf{s}^\nu_\kappa \mathsf{s}^\kappa_\lambda \mathsf{s}^\lambda_\nu x && \text{by 1.5.10(ii), (iv)} \\
&= \mathsf{s}^\mu_\kappa {}_\nu\mathsf{s}(\kappa, \lambda)x.
\end{aligned}$$

THEOREM 1.5.20. *If* κ, λ, μ *are distinct, then*

$${}_\mu\mathsf{s}(\kappa, \lambda)\mathsf{s}^\lambda_\kappa \mathsf{c}_\mu x = \mathsf{s}^\kappa_\lambda \mathsf{c}_\mu x.$$

PROOF. We have

$$\begin{aligned}
{}_\mu\mathsf{s}(\kappa, \lambda)\mathsf{s}^\lambda_\kappa \mathsf{c}_\mu x &= \mathsf{s}^\mu_\kappa \mathsf{s}^\kappa_\lambda \mathsf{s}^\lambda_\kappa \mathsf{c}_\mu x && \text{by 1.5.10(i)} \\
&= \mathsf{s}^\mu_\kappa \mathsf{s}^\kappa_\lambda \mathsf{c}_\mu x && \text{by 1.5.10(v)} \\
&= \mathsf{s}^\kappa_\lambda \mathsf{c}_\mu x && \text{by 1.5.8(i), (ii).}
\end{aligned}$$

THEOREM 1.5.21. *If* κ, λ, μ *are distinct, then*

$$\mathsf{c}_\kappa {}_\mu\mathsf{s}(\kappa, \lambda)\mathsf{c}_\mu x = {}_\mu\mathsf{s}(\kappa, \lambda)\mathsf{c}_\lambda \mathsf{c}_\mu x.$$

PROOF. We have

$$\begin{aligned}
c_{\kappa\mu}s(\kappa,\lambda)c_\mu x &= c_\kappa s^\mu_\kappa s^\kappa_\lambda s^\lambda_\mu c_\mu x && \text{by 1.5.12}\\
&= c_\mu s^\kappa_\mu s^\kappa_\lambda s^\lambda_\mu c_\mu x && \text{by 1.5.9(i)}\\
&= c_\mu s^\kappa_\lambda s^\lambda_\mu c_\mu x && \text{by 1.5.10(i)}\\
&= s^\kappa_\lambda c_\mu s^\lambda_\mu c_\mu x && \text{by 1.5.8(ii)}\\
&= s^\kappa_\lambda c_\lambda s^\mu_\lambda c_\mu x && \text{by 1.5.9(i)}\\
&= s^\kappa_\lambda c_\lambda c_\mu x && \text{by 1.5.8(i)}\\
&= c_\mu s^\kappa_\lambda c_\lambda c_\mu x && \text{by 1.5.8(ii)}\\
&= s^\mu_\kappa s^\kappa_\lambda c_\lambda c_\mu x && \text{by 1.5.8(i), (ii)}\\
&= s^\mu_\kappa s^\kappa_\lambda s^\lambda_\mu c_\lambda c_\mu x && \text{by 1.5.8(i)}\\
&= {}_\mu s(\kappa,\lambda)c_\lambda c_\mu x.
\end{aligned}$$

This completes our exposition of the basic properties of the operations s^κ_λ and ${}_\mu s(\kappa, \lambda)$. As was previously mentioned, in Section 1.11 we shall introduce and study (for a special class of CA's) the general notion of substitution operator, corresponding to the metalogical operation of the simultaneous substitution of several variables. In discussing this notion we shall make essential use of various properties of s^λ_κ. On the other hand, the results concerning ${}_\mu s(\kappa, \lambda)$ will find applications in Part II of this work. We note in passing that, by 1.4.4(vi), both s^κ_λ and ${}_\mu s(\kappa, \lambda)$ are self-dual (see the remark following 1.4.2).

REMARK 1.5.22. The operation s^κ_λ can sometimes be applied successfully in proving general laws of the theory of CA's, the formulation of which do not involve, implicitly or explicitly, either the operation itself or, more generally, the notion of diagonal elements. The proof of our next theorem provides an interesting example.

Consider the following inequality (first discussed in Thompson [52]):

$$c_\kappa x\cdot c_\lambda y\cdot c_\mu z \leqq c_\kappa c_\lambda c_\mu[c_\mu(c_\lambda x\cdot c_\kappa y)\cdot c_\lambda(c_\mu x\cdot c_\kappa z)\cdot c_\kappa(c_\mu y\cdot c_\lambda z)].$$

This inequality (for distinct κ, λ, μ) is again an instance of formulas which are identically satisfied in all special CA's but which can be shown (and will be shown in Part II) to fail in the general case. From the following theorem[1)] it is seen, however, that in a somewhat weaker formulation the inequality does hold for arbitrary CA's.

[1)] Due to Henkin [67], p. 33, where the theorem is given a metalogical rather than algebraic formulation.

THEOREM 1.5.23. $c_\kappa c_\nu x \cdot c_\lambda c_\nu y \cdot c_\mu c_\nu z \leqq c_\kappa c_\lambda c_\mu [c_\mu(c_\lambda c_\nu x \cdot c_\kappa c_\nu y) \cdot c_\lambda(c_\mu c_\nu x \cdot c_\kappa c_\nu z) \cdot c_\kappa(c_\mu c_\nu y \cdot c_\lambda c_\nu z)]$.

PROOF. The inequality can easily be checked in case any two of the ordinals κ, λ, μ, and ν are identical; therefore we assume that they are all distinct. For brevity we let

$$x' = c_\nu x, \quad y' = c_\nu y, \text{ and } z' = c_\nu z.$$

Let also

$$w = c_\mu(c_\lambda x' \cdot c_\kappa y') \cdot c_\lambda(c_\mu x' \cdot c_\kappa z') \cdot c_\kappa(c_\mu y' \cdot c_\lambda z').$$

Then

$$\begin{aligned} s_\nu^\lambda x' \cdot c_\kappa(y' \cdot s_\nu^\lambda c_\mu z') &\leqq c_\lambda x' \cdot c_\kappa y' \cdot s_\nu^\lambda x' \cdot c_\kappa(c_\mu y' \cdot s_\nu^\lambda c_\mu z') \\ &\leqq c_\mu(c_\lambda x' \cdot c_\kappa y') \cdot c_\mu s_\nu^\lambda x' \cdot c_\mu c_\kappa(c_\mu y' \cdot s_\nu^\lambda z') \\ &\leqq c_\mu[c_\mu(c_\lambda x' \cdot c_\kappa y') \cdot s_\nu^\lambda c_\mu x' \cdot s_\nu^\lambda c_\kappa z' \cdot c_\kappa(c_\mu y' \cdot c_\lambda z')] \\ &\leqq c_\mu[c_\mu(c_\lambda x' \cdot c_\kappa y') \cdot s_\nu^\lambda(c_\mu x' \cdot c_\kappa z') \cdot c_\kappa(c_\mu y' \cdot c_\lambda z')] \\ &\leqq c_\mu w. \end{aligned}$$

Hence we successively obtain

$$\begin{aligned} c_\kappa c_\lambda[s_\nu^\lambda x' \cdot c_\kappa(y' \cdot s_\nu^\lambda c_\mu z')] &\leqq c_\kappa c_\lambda c_\mu w; \\ c_\kappa[s_\nu^\lambda x' \cdot c_\kappa(c_\lambda y' \cdot s_\nu^\lambda c_\mu z')] &\leqq c_\kappa c_\lambda c_\mu w; \\ s_\nu^\lambda c_\kappa x' \cdot c_\lambda y' \cdot s_\nu^\lambda c_\mu z' &\leqq c_\kappa c_\lambda c_\mu w; \\ s_\nu^\lambda(c_\kappa x' \cdot c_\lambda y' \cdot c_\mu z') &\leqq c_\kappa c_\lambda c_\mu w. \end{aligned}$$

Applying s_λ^ν to both sides of this inequality and using 1.5.8(i) and 1.5.10(v), we easily obtain $c_\kappa x' \cdot c_\lambda y' \cdot c_\mu z' \leqq c_\kappa c_\lambda c_\mu w$ as desired.

REMARKS 1.5.24. The inequality in 1.5.23 is formulated entirely in terms of fundamental operations of Df's. Since, however, in the proof of 1.5.23 we have made essential use of the operations s_λ^κ, and hence implicitly of diagonal elements, we have not established the fact that this inequality is identically satisfied in arbitrary Df's. Actually it will be shown in Part II that the inequality (with say $\kappa = 0$, $\lambda = 1$, $\mu = 2$, and $\nu = 3$) fails in some Df_α for each $\alpha \geqq 4$ and that, consequently, in every proof of the inequality diagonal elements, and some of the postulates (C_5)–(C_7), must be essentially involved. It will also be shown that no such inequality exists in case $\alpha \leqq 2$: in this case every equation (and hence, obviously, every inequality) not involving diagonal elements which holds in all CA_α's holds in all Df_α's as well. For $\alpha = 3$ the analogous problem is still open.

On the other hand, for every $\alpha \geqq 2$ there are Df_α's which are not diagonal-free parts of CA_α's in the sense of 1.1.2. In fact, for any given $\alpha \geqq 2$ it is easy to construct a Df_α $\mathfrak{A}$ such that, in opposition to 1.3.12, the equation $\mathsf{c}_0 x = x$ is identically satisfied in $\mathfrak{A}$ while $\mathsf{c}_1 x = x$ is not.

1.6. DIMENSION SETS

DEFINITION 1.6.1. *By the **dimension set** of x, in symbols $\Delta^{(\mathfrak{A})}x$, or simply Δx, is meant the set of all κ for which $c_\kappa x \neq x$.*

This notion (introduced in Tarski-Thompson [52], where the term "dimension index" was used) will be employed later in this chapter to define some important classes of cylindric algebras. In the case of cylindric set algebras, ΔX is the set of all $\kappa < \alpha$ such that X does not form a cylinder parallel to the κth axis.

In the case of a cylindric algebra of formulas associated with a theory Θ, $\Delta(\varphi/\equiv_\Theta)$ consists of all $\kappa < \alpha$ such that $\exists_{v_\kappa} \varphi \leftrightarrow \varphi$ — or, equivalently, $\forall_{v_\kappa} \varphi \leftrightarrow \varphi$ — is not valid in Θ. Hence, $\Delta(\varphi/\equiv_\Theta)$ contains only κ for which v_κ is free in φ. But it does not necessarily contain all such κ. For instance, if ψ is any formula in which v_κ does not occur free and if we let $\varphi = \psi \wedge v_\kappa = v_\kappa$, then $\forall_{v_\kappa} \varphi \leftrightarrow \varphi$ is logically valid and hence valid in Θ, so that $\kappa \notin \Delta(\varphi/\equiv_\Theta)$. It is easily seen however that the following holds: $\Delta(\varphi/\equiv_\Theta)$ contains all those and only those κ for which v_κ is free in every formula ψ belonging to the equivalence class $\varphi/\equiv_\Theta$. If we agree to call v_κ *essentially free in* φ (*with respect to* Θ) just in case v_κ is free in every formula belonging to $\varphi/\equiv_\Theta$ — or, what amounts to the same, in case $\forall_{v_\kappa} \varphi \leftrightarrow \varphi$ is not valid in Θ — then we can also say that $\Delta(\varphi/\equiv_\Theta)$ is the set of all those $\kappa < \alpha$ for which v_κ is essentially free in φ. If, in particular, $\Delta(\varphi/\equiv_\Theta) = 0$, then no variable v_κ occurs essentially free in φ with respect to Θ, whence $\varphi \leftrightarrow [\varphi]$ is in Θ and therefore $(\varphi/\equiv_\Theta) = ([\varphi]/\equiv_\Theta)$; we recall here that $[\varphi]$ is the sentence which is the closure of φ (cf. the Preliminaries). If, conversely, there is a sentence σ such that $\varphi \equiv \sigma$ is valid in Θ or, what amounts to the same, $(\varphi/\equiv_\Theta) = ([\varphi]/\equiv_\Theta)$, then, as is easily seen, $\Delta(\varphi/\equiv_\Theta) = 0$.

THEOREM 1.6.2. *$\Delta 0$ and $\Delta 1$ are empty.*

THEOREM 1.6.3. *$\mathfrak{A}$ is discrete iff $\Delta x = 0$ for all $x \in A$.*
PROOF: by 1.3.12.

THEOREM 1.6.4. *If $d_{\kappa\lambda} \neq 1$, then $\Delta d_{\kappa\lambda} = \{\kappa, \lambda\}$.*
PROOF: by 1.3.2 and 1.3.3.

THEOREM 1.6.5. *If* $\sum_{i\in I} z_i$ *exists, then* $\Delta(\sum_{i\in I} z_i) \subseteq \bigcup_{i\in I} \Delta z_i$; *in particular,* $\Delta(x+y) \subseteq \Delta x \cup \Delta y$.

PROOF. If $\kappa \notin \bigcup_{i\in I} \Delta z_i$, then $c_\kappa z_i = z_i$ for every $i \in I$, and it follows that $\kappa \notin \Delta(\sum_{i\in I} z_i)$ by 1.2.6.

THEOREM 1.6.6. *If* $\prod_{i\in I} z_i$ *exists, then* $\Delta(\prod_{i\in I} z_i) \subseteq \bigcup_{i\in I} \Delta z_i$; *in particular,* $\Delta(x\cdot y) \subseteq \Delta x \cup \Delta y$.

PROOF. The proof is analogous to that of 1.6.5, 1.2.10 being used instead of 1.2.6.

It is easy to obtain examples to show that equality cannot replace inclusion in Theorems 1.6.5 and 1.6.6; for example, we may take any non-discrete CA_α $\mathfrak{A}$, an element $x \in A$, and an index $\kappa < \alpha$ for which $c_\kappa x \neq x$, and let $y = -x$.

THEOREM 1.6.7. (i) $\Delta(-x) = \Delta x$.

(ii) *If* $\Delta y = 0$, *then*

$$\Delta x = \Delta(x\cdot y) \cup \Delta(x\cdot -y) = \Delta(x+y) \cup \Delta(x+-y).$$

PROOF. (i) follows directly from 1.2.12. Since $x = x\cdot y + x\cdot -y$, we have, by 1.6.5, $\Delta x \subseteq \Delta(x\cdot y) \cup \Delta(x\cdot -y)$. The inclusion in the opposite direction follows from 1.6.6 and part (i) of our theorem, using the fact that $\Delta y = 0$. Thus, $\Delta x = \Delta(x\cdot y) \cup \Delta(x\cdot -y)$. The remaining part of the conclusion is then obtained by means of (i).

THEOREM 1.6.8. $\Delta(c_\kappa x) \subseteq \Delta x \sim \{\kappa\}$.

PROOF. If $\lambda \notin \Delta x \sim \{\kappa\}$, then $\lambda \notin \Delta x$ or $\lambda = \kappa$. In either case $c_\lambda c_\kappa x = c_\kappa x$ by (C_4) or 1.2.3.

It is easily seen that the inclusion symbol in 1.6.8 can be replaced by the equality symbol in case $\alpha = 1$. In general, however, this cannot be done. If, e.g., $\mathfrak{A}$ is a non-discrete CA_α with $\alpha \geqq 2$ and if $x = d_{01}$, then $\Delta x = \{0, 1\}$ by 1.3.12 and 1.6.4; hence $\Delta x \sim \{0\} = \{1\}$, while $\Delta c_0 x$ is empty by 1.3.2 and 1.6.2.

The remaining theorems in this section are less obvious and some of them are more difficult to prove.

THEOREM 1.6.9. (i) *For any* κ *and* λ,

$$\Delta(c_\kappa - d_{\kappa\lambda}) = 0.$$

(ii) *More generally, for any element* x *we have*

$$\Delta(x\cdot c_\kappa - d_{\kappa\lambda}) = \Delta x$$

provided $\kappa \neq \lambda$.

PROOF. (i) is obtained directly from 1.3.18. From 1.6.7(ii) we get $\Delta x = \Delta(x \cdot c_\kappa - d_{\kappa\lambda}) \cup \Delta(x \cdot c_\kappa^\partial d_{\kappa\lambda})$. However, $\Delta(x \cdot c_\kappa^\partial d_{\kappa\lambda}) = 0$ by 1.3.19. This proves (ii).

Theorem 1.6.9 (ii) was pointed out by Don Pigozzi.

THEOREM 1.6.10. *If* $\kappa \neq \lambda$, $\{\kappa, \lambda\} \nsubseteq \Delta x$, *and* $\Delta x \neq 0$ *or, more generally,* $x \cdot c_\kappa - d_{\kappa\lambda} \neq 0$, *then*

$$\Delta(x \cdot d_{\kappa\lambda}) = \Delta(x \cdot - d_{\kappa\lambda}) = \Delta x \cup \{\kappa, \lambda\}.$$

PROOF. In view of 1.6.9(ii) we can disregard the premiss $\Delta x \neq 0$ and assume $x \cdot c_\kappa - d_{\kappa\lambda} \neq 0$. We begin by proving the following lemma:

(1) For any element y with $\{\kappa, \lambda\} \nsubseteq \Delta y$, the following conditions are equivalent:
(a) $y \cdot c_\kappa - d_{\kappa\lambda} = 0$, (b) $y \cdot - d_{\kappa\lambda} = 0$, (c) $y \cdot d_{\kappa\lambda} \cdot c_\kappa - d_{\kappa\lambda} = 0$.

As is easily seen from 1.3.18(ii), we can assume, without loss of generality, that $\kappa \notin \Delta y$. Then the equivalence of (a) and (b) follows immediately by 1.2.5. (a) clearly implies (c). The converse is obtained by applying c_κ to both sides of (c) and using (C_3), 1.3.2, and the fact that $c_\kappa y = y$. This proves (1).

In view of the premiss $x \cdot c_\kappa - d_{\kappa\lambda} \neq 0$ we have, taking x for y in (1),

(2) $$x \cdot - d_{\kappa\lambda} \neq 0 \text{ and } x \cdot d_{\kappa\lambda} \cdot c_\kappa - d_{\kappa\lambda} \neq 0.$$

In establishing the conclusion of the theorem we first want to show

(3) $$\{\kappa, \lambda\} \subseteq \Delta(x \cdot d_{\kappa\lambda}) \text{ and } \{\kappa, \lambda\} \subseteq \Delta(x \cdot - d_{\kappa\lambda}).$$

Suppose $\{\kappa, \lambda\} \nsubseteq \Delta(x \cdot d_{\kappa\lambda})$ and apply (1), taking y to be $x \cdot d_{\kappa\lambda}$. Then $x \cdot d_{\kappa\lambda} \cdot c_\kappa - d_{\kappa\lambda} = 0$, which contradicts (2). Thus the first inclusion of (3) holds. Now suppose $\{\kappa, \lambda\} \nsubseteq \Delta(x \cdot - d_{\kappa\lambda})$ and again apply (1), this time taking $x \cdot - d_{\kappa\lambda}$ for y. We get $x \cdot - d_{\kappa\lambda} = 0$, which contradicts (2), and thus the second inclusion of (3) holds.

We next show

(4) $$\Delta x \subseteq \Delta(x \cdot d_{\kappa\lambda}) \text{ and } \Delta x \subseteq \Delta(x \cdot - d_{\kappa\lambda}).$$

To prove (4) we will use the premiss $\{\kappa, \lambda\} \nsubseteq \Delta x$; say $\kappa \notin \Delta x$. Assume, first of all, that $\mu \notin \Delta(x \cdot d_{\kappa\lambda})$. We then have $c_\mu(x \cdot d_{\kappa\lambda}) = x \cdot d_{\kappa\lambda}$; applying c_κ to both sides of this equation and using (C_3), (C_4), and 1.3.2 we get $c_\mu x = x$. Thus $\mu \notin \Delta x$ and the first inclusion of (4) is proved. Now assume $\mu \notin \Delta(x \cdot - d_{\kappa\lambda})$. We then have $c_\mu(x \cdot - d_{\kappa\lambda}) = x \cdot - d_{\kappa\lambda}$; again applying c_κ to both sides of this equation we obtain $c_\mu(x \cdot c_\kappa - d_{\kappa\lambda}) = x \cdot c_\kappa - d_{\kappa\lambda}$. Thus $\mu \notin \Delta x$ by 1.6.9(ii), and the second inclusion of (4) is proved.

Finally, by 1.6.4, 1.6.6, and 1.6.7 we have immediately

$$\Delta(x\cdot \mathsf{d}_{\kappa\lambda}) \subseteq \Delta x \cup \{\kappa, \lambda\} \text{ and } \Delta(x\cdot -\mathsf{d}_{\kappa\lambda}) \subseteq \Delta x \cup \{\kappa, \lambda\};$$

these inclusions along with those of (3) and (4) establish the desired conclusions.

COROLLARY 1.6.11. *Suppose* $\mathfrak{A}$ *is not discrete. If* $\alpha < \omega$, *or, more generally, if* $|\alpha \sim \Delta x| < \omega$ *for some* $x \in A$, *then there is a* $y \in A$ *such that* $\Delta y = \alpha$.

THEOREM 1.6.12. *If* $\kappa \notin \Delta x$, $\lambda \notin \Delta y$, *and* $\kappa \neq \lambda$, *then* $\Delta x \cup \Delta y \subseteq \Delta(x\cdot \mathsf{d}_{\kappa\lambda} + y\cdot -\mathsf{d}_{\kappa\lambda}) \cup \{\kappa, \lambda\}$.

PROOF. Suppose $\mu \notin \Delta(x\cdot \mathsf{d}_{\kappa\lambda} + y\cdot -\mathsf{d}_{\kappa\lambda}) \cup \{\kappa, \lambda\}$. Then $\mu \neq \kappa, \lambda$, and

$$\mathsf{c}_\mu x\cdot \mathsf{d}_{\kappa\lambda} + \mathsf{c}_\mu y\cdot -\mathsf{d}_{\kappa\lambda} = \mathsf{c}_\mu(x\cdot \mathsf{d}_{\kappa\lambda} + y\cdot -\mathsf{d}_{\kappa\lambda}) = x\cdot \mathsf{d}_{\kappa\lambda} + y\cdot -\mathsf{d}_{\kappa\lambda}.$$

Hence $\mathsf{c}_\mu x\cdot \mathsf{d}_{\kappa\lambda} = x\cdot \mathsf{d}_{\kappa\lambda}$ and $\mathsf{c}_\mu y\cdot -\mathsf{d}_{\kappa\lambda} = y\cdot -\mathsf{d}_{\kappa\lambda}$. Applying c_κ to both sides of the first of these two equations, we obtain $\mathsf{c}_\mu x = x$, i.e., $\mu \notin \Delta x$. Applying c_λ to both sides of the second equation we get, using 1.3.18 and 1.3.20, $\mathsf{c}_\mu y = y$, or $\mu \notin \Delta y$. This completes the proof.[1]

THEOREM 1.6.13. $\Delta(\mathsf{s}^\kappa_\lambda x) \subseteq (\Delta x \sim \{\kappa\}) \cup \{\lambda\}$.

PROOF: by 1.6.4, 1.6.6, and 1.6.8.

THEOREM 1.6.14. $\mathsf{s}^\kappa_\lambda x = x$ *iff* $\kappa = \lambda$ *or* $\kappa \notin \Delta x$.

PROOF: by 1.5.8(i) and 1.5.9(ii).

THEOREM 1.6.15. *If* $\kappa \in \Delta x$ *and* $\lambda \notin \Delta x$, *then* $\Delta(\mathsf{s}^\kappa_\lambda x) = (\Delta x \sim \{\kappa\}) \cup \{\lambda\}$.

PROOF. By 1.6.13 we have $\Delta(\mathsf{s}^\kappa_\lambda x) \subseteq (\Delta x \sim \{\kappa\}) \cup \{\lambda\}$. We claim that

$$(1) \qquad \lambda \in \Delta(\mathsf{s}^\kappa_\lambda x).$$

For otherwise we have $\mathsf{s}^\kappa_\lambda x = \mathsf{c}_\lambda \mathsf{s}^\kappa_\lambda x = \mathsf{c}_\kappa \mathsf{s}^\lambda_\kappa x = \mathsf{c}_\kappa x$ by 1.5.9(i) and the fact that $\lambda \notin \Delta x$. Multiplying by $\mathsf{d}_{\kappa\lambda}$ on both sides of the resulting equation $\mathsf{s}^\kappa_\lambda x = \mathsf{c}_\kappa x$ we get $\mathsf{d}_{\kappa\lambda}\cdot x = \mathsf{d}_{\kappa\lambda}\cdot \mathsf{c}_\kappa x$ by 1.5.5. Applying c_λ we then infer that $x = \mathsf{c}_\kappa x$, which contradicts the assumption $\kappa \in \Delta x$. Thus (1) holds.

Now using 1.6.13 and 1.5.11(i) we see that $\Delta x = \Delta(\mathsf{s}^\lambda_\kappa \mathsf{s}^\kappa_\lambda x) \subseteq (\Delta(\mathsf{s}^\kappa_\lambda x) \sim \{\lambda\}) \cup \{\kappa\}$. Hence $\Delta x \sim \{\kappa\} \subseteq \Delta(\mathsf{s}^\kappa_\lambda x)$, and the desired conclusion follows by (1).

THEOREM 1.6.16. *Suppose* $\mu \neq \kappa, \lambda$. *Then*:

(i) *if* $\kappa, \lambda \notin \Delta \mathsf{c}_\mu x$ *or* $\kappa, \lambda \in \Delta \mathsf{c}_\mu x$, *then* $\Delta({}_\mu\mathsf{s}(\kappa, \lambda)\mathsf{c}_\mu x) = \Delta \mathsf{c}_\mu x$;

(ii) *if* $\kappa \in \Delta \mathsf{c}_\mu x$ *and* $\lambda \notin \Delta \mathsf{c}_\mu x$, *then* $\Delta({}_\mu\mathsf{s}(\kappa, \lambda)\mathsf{c}_\mu x) = (\Delta \mathsf{c}_\mu x \sim \{\kappa\}) \cup \{\lambda\}$;

(iii) *if* $\kappa \notin \Delta \mathsf{c}_\mu x$ *and* $\lambda \in \Delta \mathsf{c}_\mu x$, *then* $\Delta({}_\mu\mathsf{s}(\kappa, \lambda)\mathsf{c}_\mu x) = (\Delta \mathsf{c}_\mu x \sim \{\lambda\}) \cup \{\kappa\}$.

[1]) 1.6.10 and 1.6.11 were proved by Tarski, while 1.6.12 is due to Jack Silver; 1.6.10–1.6.12 are lemmas leading to Theorem 1.11.4 below.

PROOF. If $\kappa, \lambda \notin \Delta c_\mu x$, then ${}_\mu s(\kappa, \lambda) c_\mu x = c_\mu x$; if $\kappa, \lambda \in \Delta c_\mu x$, then $\Delta({}_\mu s(\kappa, \lambda) c_\mu x) = \Delta c_\mu x$ by 1.6.15. Thus (i) holds. That (ii) and (iii) hold follows from 1.6.15 and the following observations: under the hypothesis of (ii) we have ${}_\mu s(\kappa, \lambda) c_\mu x = s^\kappa_\lambda c_\mu x$, while under the hypothesis of (iii) we get ${}_\mu s(\kappa, \lambda) c_\mu x = s^\lambda_\kappa c_\mu x$.

THEOREM 1.6.17. (i) $\Delta c^\partial_\kappa x = \Delta c_\kappa - x$;
(ii) $\Delta^\partial x = \Delta x$.
PROOF: by 1.4.1, 1.4.2, and 1.6.7 (cf. the remarks following 1.4.2).

To complete this section we introduce some notions which are directly defined in terms of the dimension set.

DEFINITION 1.6.18. (i) *Let Γ be any (not necessarily finite) subset of α. An element x is called a (Γ)-**closed** element or a (Γ)-**cylinder** if $\Delta x \cap \Gamma = 0$. We denote the set of (Γ)-closed elements by $Cl_\Gamma \mathfrak{A}$ and let $\mathfrak{Cl}_\Gamma \mathfrak{A} = \langle Cl_\Gamma \mathfrak{A}, +, \cdot, -, 0, 1 \rangle$.*
(ii) *An element x is called **zero-dimensional** if $\Delta x = 0$. We denote the set of zero-dimensional elements by $Zd\mathfrak{A}$ and let $\mathfrak{Zd}\mathfrak{A} = \langle Zd\mathfrak{A}, +, \cdot, -, 0, 1 \rangle$.*

A metalogical interpretation of $Zd\mathfrak{A}$ is clear: if $\mathfrak{A}$ is a cylindric algebra of formulas, then $Zd\mathfrak{A}$ is the set of all those elements of $\mathfrak{A}$ which are equivalence classes of sentences, i.e., of formulas without free variables; compare the remarks following 1.6.1. Thus, if $\mathfrak{A}$ is the cylindric algebra of formulas associated with a theory Θ, $\mathfrak{Zd}\mathfrak{A}$ is the Boolean algebra of sentences associated with Θ; cf. 1.1.15 and 1.1.16.

THEOREM 1.6.19. *$\mathfrak{Cl}_\Gamma \mathfrak{A}$ for every $\Gamma \subseteq \alpha$ (Γ not necessarily finite) and $\mathfrak{Zd}\mathfrak{A}$ are* BA's. *In fact, $\mathfrak{Cl}_\Gamma \mathfrak{A}$ is a subalgebra of $\mathfrak{Bl}\mathfrak{A}$, and $\mathfrak{Zd}\mathfrak{A}$ is a subalgebra of $\mathfrak{Cl}_\Gamma \mathfrak{A}$.*
PROOF. The proof reduces to showing that the sets $Cl_\Gamma \mathfrak{A}$ and $Zd\mathfrak{A}$ are closed under $+, \cdot, -$ and contain 0 and 1 as elements. This follows immediately from 1.6.2 and 1.6.5–1.6.7.

It will be seen in Section 2.6 that $\mathfrak{Cl}_\Gamma \mathfrak{A}$ can be represented in a natural way as the Boolean part of a subreduct of $\mathfrak{A}$ (cf. 0.5.6).

By 1.6.3 a CA $\mathfrak{A}$ is discrete iff all its elements are zero-dimensional, i.e., if $Zd\mathfrak{A} = Bl\mathfrak{A}$. It may be pointed out that a non-discrete CA may contain many zero-dimensional elements in addition to 0 and 1. For instance, in the cylindric algebra of formulas associated with a theory Θ, the zero-dimensional elements coincide with equivalence classes of sentences; if the theory Θ is not complete, there are sentences in the language of Θ which are not valid in Θ and whose negations are also not valid, and the equivalence classes of such sentences are different from 0 and 1. As is easily seen, a finite-dimensional cylindric set algebra has no zero-dimensional members different from 0 (the empty set) and 1 (the whole space). It is not difficult, however, to exhibit infinite-dimensional cylindric set algebras which have such members.

As was mentioned in the remarks following 1.3.18, every CA_α $\mathfrak{A}$ with $\alpha \geqq 2$ has just one element of the form $\mathsf{c}_\kappa - \mathsf{d}_{\kappa\lambda} = \mathsf{c}_\kappa \mathsf{c}_\lambda - \mathsf{d}_{\kappa\lambda}$ for some $\kappa, \lambda < \alpha$, $\kappa \neq \lambda$; from what we have already seen in 1.6.9 and 1.6.10, this element appears to play a distinguished role in the algebra $\mathfrak{A}$. The following theorem, which was brought to our attention by Don Pigozzi, throws further light on this point.

THEOREM 1.6.20. *For any $\kappa, \lambda < \alpha$ with $\kappa \neq \lambda$, $\mathsf{c}_\kappa - \mathsf{d}_{\kappa\lambda}$ is the least element $a \in A$ such that $x \in Zd\mathfrak{A}$ for every $x \geqq a$.*

PROOF. By 1.6.9, for every $x \geqq \mathsf{c}_\kappa - \mathsf{d}_{\kappa\lambda}$ we have $\Delta x = \Delta(\mathsf{c}_\kappa - \mathsf{d}_{\kappa\lambda}) = 0$. Conversely, suppose $x \in Zd\mathfrak{A}$ for all $x \geqq a$. Then, in particular, $\{\kappa, \lambda\} \nsubseteq \Delta a$ and $\Delta(a + \mathsf{d}_{\kappa\lambda}) = 0$; hence $a \geqq \mathsf{c}_\kappa - \mathsf{d}_{\kappa\lambda}$ by $1.6.10^\partial$ and 1.6.17(ii).

We shall see in Section 2.4 that zero-dimensional elements play an essential role in the study of direct decompositions of CA's. In particular, we shall state a result concerning direct decompositions which is closely related to 1.6.20; cf. Theorem 2.4.37.

1.7. GENERALIZED CYLINDRIFICATIONS

We now turn to generalizations of the elementary notions introduced so far. In this section we consider direct generalizations of the notions of cylindrification and inner cylindrification.

DEFINITION 1.7.1. *We let* $c_{(0)}x = x$, *and, if* Γ *is any non-empty finite subset of* α, *we let* $c_{(\Gamma)}x = c_{\kappa_0} \ldots c_{\kappa_{\lambda-1}}x$ *for all* $x \in A$ *where* $|\Gamma| = \lambda$ *and* $\langle \kappa_0, \ldots, \kappa_{\lambda-1} \rangle$ *is the strictly increasing sequence whose range is* Γ. *The operations* $c_{(\Gamma)}$ *are called* ***generalized cylindrifications.***

Notice the difference between c_κ and $c_{(\kappa)}$ where κ is any finite ordinal. c_κ is of course one of the fundamental operations of the algebra $\mathfrak{A} \in \mathsf{CA}_\alpha$ (provided $\kappa < \alpha$). On the other hand, κ as a finite ordinal is also a finite set of ordinals, $\kappa = \{\xi : \xi < \kappa\}$ (cf. the Preliminaries), and hence the meaning of $c_{(\kappa)}$ is determined by 1.7.1: $c_{(0)}x = x$ and $c_{(\kappa)}x = c_0 \ldots c_{\kappa-1}x$ when $\kappa > 0$.

There seems to be no natural way of extending 1.7.1 to infinite sets Γ for arbitrary CA's.

COROLLARY 1.7.2. *If* Γ *is a finite subset of* α, $\lambda < \omega$, *and* κ *is a function from* λ *onto* Γ, *then* $c_{(\Gamma)}x = c_{\kappa_0} \ldots c_{\kappa_{\lambda-1}}x$.

PROOF: by 1.2.17.

In a cylindric set algebra $\mathfrak{A}$ with base U, $\mathsf{C}_{(\Gamma)}X$ is the set of all $y \in {}^\alpha U$ such that there is an $x \in X$ with $x_\kappa = y_\kappa$ for all $\kappa \in \alpha \sim \Gamma$; thus $\mathsf{C}_{(\Gamma)}X$ is obtained from X by cylindrifying along all κ-axes for $\kappa \in \Gamma$. In the cylindric algebra of formulas associated with a set Σ of sentences, $c_{(\Gamma)}(\varphi/{\equiv_\Sigma})$ is the equivalence class of the formula $\exists_{u_0} \ldots \exists_{u_{\lambda-1}} \varphi$ where $\Gamma = \{\kappa_0, \ldots, \kappa_{\lambda-1}\}$ and $u_0 = \mathbf{v}_{\kappa_0}, \ldots, u_{\lambda-1} = \mathbf{v}_{\kappa_{\lambda-1}}$. Thus $c_{(\Gamma)}$ could have been termed alternatively a *generalized (existential) quantification.*

Through the rest of this chapter Γ, Δ, and Θ will represent finite subsets of α (or sometimes sequences of such subsets) unless otherwise indicated.

THEOREM 1.7.3. $c_{(\Gamma)}c_{(\Delta)}x = c_{(\Gamma \cup \Delta)}x$.

PROOF: by 1.2.17.

THEOREM 1.7.4. $\mathsf{c}_{(\{\kappa\})} = \mathsf{c}_\kappa$.

THEOREM 1.7.5. *For any sequence* $\langle \Gamma_\kappa : \kappa < \beta \rangle$ *of finite subsets of* α, *the structure* $\langle A, +, \cdot, -, 0, 1, \mathsf{c}_{(\Gamma_\kappa)} \rangle_{\kappa<\beta}$ *is a diagonal-free cylindric algebra.*

PROOF. By 1.1.2 the proof of the theorem amounts to showing that the following four equations hold for any Δ and Θ (finite subsets of α):

(1) $$\mathsf{c}_{(\Delta)} 0 = 0;$$

(2) $$x \leqq \mathsf{c}_{(\Delta)} x;$$

(3) $$\mathsf{c}_{(\Delta)}(x \cdot \mathsf{c}_{(\Delta)} y) = \mathsf{c}_{(\Delta)} x \cdot \mathsf{c}_{(\Delta)} y;$$

(4) $$\mathsf{c}_{(\Delta)} \mathsf{c}_{(\Theta)} x = \mathsf{c}_{(\Theta)} \mathsf{c}_{(\Delta)} x.$$

The first three equations are easily proved by induction on $|\Delta|$, and (4) follows from 1.7.3.

An important consequence of 1.7.5 is that all the theorems of Section 1.2 extend automatically to generalized cylindrifications, and the same applies to parts of 1.4.4; cf. the initial remarks of Section 1.2 and the proof of 1.4.4(i)–(v).

THEOREM 1.7.6. *If* $\Gamma \neq 0$, *then the following two conditions are equivalent:*

(i) $\mathfrak{A}$ *is discrete;*

(ii) *for all* x, $\mathsf{c}_{(\Gamma)} x = x$.

PROOF: by (C_2) and 1.3.12.

In contrast with 1.3.13, we may have a non-discrete CA_α with $\alpha > 2$ and distinct Γ and Δ such that $\mathsf{c}_{(\Gamma)} x = \mathsf{c}_{(\Delta)} x$ for all x. This is the case, for example, with certain CA_3's considered in Part II, where $\mathsf{c}_{(2)} x = \mathsf{c}_{(3)} x = 1$ for all $x \neq 0$, while, of course, $\mathsf{c}_{(2)} 0 = \mathsf{c}_{(3)} 0 = 0$.

THEOREM 1.7.7. (i) $\Delta(\mathsf{c}_{(\Gamma)} x) \subseteq \Delta x \sim \Gamma$.

(ii) $\mathsf{c}_{(\Gamma)} x = \mathsf{c}_{(\Gamma \cap \Delta x)} x$.

PROOF. (i) is proved by induction on $|\Gamma|$, using 1.6.8. To prove (ii) notice that, by 1.7.3, $\mathsf{c}_{(\Gamma)} x = \mathsf{c}_{(\Gamma \sim \Delta x)} \mathsf{c}_{(\Gamma \cap \Delta x)} x$, and then argue by induction on $|\Gamma \sim \Delta x|$, using (i).

THEOREM 1.7.8. $\mathsf{c}^{\partial}_{(\Gamma)} x = -\mathsf{c}_{(\Gamma)} - x$.

PROOF: by 1.7.1 and the remarks following 1.4.3.

$\mathsf{c}^{\partial}_{(\Gamma)}$ can of course be called a *generalized inner cylindrification.* In a cylindric set algebra $\mathbf{C}^{\partial}_{(\Gamma)} X$ is the largest $Y \subseteq X$ which for each $\kappa \in \Gamma$ forms a cylinder parallel to the κ-axis. In the cylindric algebra associated with a set of sentences Σ, $\mathsf{c}^{\partial}_{(\Gamma)}(\varphi / \equiv_\Sigma)$ is the equivalence class of $\forall_{u_0} \dots \forall_{u_{\lambda-1}} \varphi$ where $\Gamma = \{\kappa_0, \dots,$

$\kappa_{\lambda-1}\}$ and $u_0 = v_{\kappa_0}, \ldots, u_{\lambda-1} = v_{\kappa_{\lambda-1}}$. For this reason $c_{(\Gamma)}$ may alternatively be called a *generalized universal quantification.*

REMARK 1.7.9. We would like to note here in connection with 1.7.5 and the remark immediately following it that, given any CA_α $\mathfrak{A}$ and any sequence $\langle \Gamma_\kappa : \kappa < \beta \rangle$ of finite subsets of α, and letting $\mathfrak{B} = \langle A, +, \cdot, -, 0, 1, c_{(\Gamma_\kappa)} \rangle_{\kappa<\beta}$, we have

$$c_{(\Gamma_\kappa)}^{(\mathfrak{A})} = c_\kappa^{(\mathfrak{B})} \text{ and } c_{(\Gamma_\kappa)}^{(\mathfrak{A}\partial)} = c_\kappa^{(\mathfrak{B}\partial)}.$$

Thus, for example, from 1.4.4(iv) (which is easily seen to hold for all Df's) and 1.7.5 we obtain

$$c_{(\Gamma)} c^{\partial}_{(\Delta)} x \leqq c^{\partial}_{(\Delta)} c_{(\Gamma)} x.$$

In the last two theorems of this section generalized cylindrification is applied to the discussion of the notions of (Γ)-closed and zero-dimensional elements introduced at the end of Section 1.6 (see Definition 1.6.18).

THEOREM 1.7.10. (i) *Let Γ be any (not necessarily finite) subset of α. Then $Cl_\Gamma \mathfrak{A} = \{x : x \in A,\ c_{(\Delta)} x = x$ for every finite $\Delta \subseteq \Gamma\}$. If $|\Gamma| < \omega$, then $Cl_\Gamma \mathfrak{A} = \{x : x \in A,\ c_{(\Gamma)} x = x\} = Rg\, c_{(\Gamma)}$.*

(ii) *$Zd\mathfrak{A} = Cl_\alpha \mathfrak{A} = \{x : x \in A,\ c_{(\Delta)} x = x$ for every finite $\Delta \subseteq \alpha\}$. If $\alpha < \omega$, then $Zd\mathfrak{A} = \{x : x \in A,\ c_{(\alpha)} x = x\} = Rg\, c_{(\alpha)}$.*

PROOF. 1.2.4, 1.7.5, and 1.7.7 together give the second part of (i); the first part is obvious. (ii) follows easily from (i).

THEOREM 1.7.11. (i) *If $|\Delta x| < \omega$, then $c_{(\Delta x)} x \in Zd\mathfrak{A}$; if, moreover, $x \leqq y \in Zd\mathfrak{A}$, then $c_{(\Delta x)} x \leqq y$.*

(ii) *If $\alpha < \omega$ and $x \in A$, then $c_{(\Delta x)} x = c_{(\alpha)} x$.*

PROOF. The first part of (i) is an immediate consequence of 1.7.7(ii), and (ii) is an immediate consequence of 1.7.7(ii); the second part of (i) follows from 1.2.7 and 1.7.10(ii).

Theorem 1.7.11(i) can clearly be expressed in the following way: $c_{(\Delta x)} x$ when Δx is finite is the least zero-dimensional element $\geqq x$; similarly, $c^{\partial}_{(\Delta x)} x$ is the largest zero-dimensional element of $\mathfrak{A}$ $\leqq x$.

Let $\mathfrak{A}$ be a CA of formulas and x an element of $\mathfrak{A}$ which is the equivalence class of a formula φ. It is easily seen that Δx is finite; therefore, $c_{(\Delta x)} x$ and $c^{\partial}_{(\Delta x)} x$ exist, and as zero-dimensional elements they must be the equivalence classes of sentences (cf. the remark following 1.6.18). If in fact $u_0, \ldots, u_{\kappa-1}$ are all the variables occurring free in φ, or, at least, all the variables essentially free in φ, then $c_{(\Delta x)} x$ corresponds to $\exists_{u_0 \cdots u_{\kappa-1}} \varphi$ and $c^{\partial}_{(\Delta x)} x$ corresponds to $\forall_{u_0 \cdots u_{\kappa-1}} \varphi$, i.e., to $[\varphi]$.

1.8. GENERALIZED DIAGONAL ELEMENTS

In this section we generalize the notion of a diagonal element. The relations between these generalized diagonal elements and generalized cylindrifications are frequently analogous to those between the diagonal elements $\mathsf{d}_{\kappa\lambda}$ and the cylindrifications c_{κ}.

DEFINITION 1.8.1. *We let* $\mathsf{d}_{\Gamma} = \prod_{\kappa,\lambda\in\Gamma}\mathsf{d}_{\kappa\lambda}$. *The elements* d_{Γ} *are called* ***generalized diagonal elements.***

In agreement with remarks following 1.1.5, we shall use $\mathbf{D}_{\Gamma}$ to denote generalized diagonal elements in cylindric set algebras; analogously for generalized co-diagonal elements which will be introduced in the next section.

COROLLARY 1.8.2. $\mathsf{d}_0 = 1$.

In the cylindric algebra of formulas associated with a set of sentences Σ, d_{Γ} is the element $\varphi/\equiv_{\Sigma}$ where φ is a conjunction of all formulas $\mathfrak{v}_{\kappa} = \mathfrak{v}_{\lambda}$ for $\kappa, \lambda \in \Gamma$; if $\Gamma = 0$, we may take $\boldsymbol{T}$ for φ. On the other hand, in a cylindric set algebra $\mathbf{D}_{\Gamma}$ is the linear subspace characterized by the system of equations $x_{\kappa} = x_{\lambda}$ for all $\kappa, \lambda \in \Gamma$.

THEOREM 1.8.3. (i) $\mathsf{d}_{\{\kappa\}} = 1$;
(ii) $\mathsf{d}_{\{\kappa,\lambda\}} = \mathsf{d}_{\kappa\lambda}$;
(iii) *if* $\Gamma \subseteq \Delta$, *then* $\mathsf{d}_{\Gamma} \geqq \mathsf{d}_{\Delta}$.

THEOREM 1.8.4. *If* $\Gamma \cap \Delta \neq 0$, *then* $\mathsf{d}_{\Gamma}\cdot\mathsf{d}_{\Delta} = \mathsf{d}_{\Gamma\cup\Delta}$.

PROOF. Clearly we have $\mathsf{d}_{\Gamma\cup\Delta} \leqq \mathsf{d}_{\Gamma}\cdot\mathsf{d}_{\Delta}$ by 1.8.3(iii). To establish the inequality in the other direction, and hence the theorem, it is sufficient to show that, if $\kappa \in \Gamma$ and $\lambda \in \Delta$, then $\mathsf{d}_{\Gamma}\cdot\mathsf{d}_{\Delta} \leqq \mathsf{d}_{\kappa\lambda}$. To this end, choose $\mu \in \Gamma \cap \Delta$ (which is possible by the hypothesis). Thus $\mathsf{d}_{\Gamma} \leqq \mathsf{d}_{\kappa\mu}$ and $\mathsf{d}_{\Delta} \leqq \mathsf{d}_{\mu\lambda}$, so that $\mathsf{d}_{\Gamma}\cdot\mathsf{d}_{\Delta} \leqq \mathsf{d}_{\kappa\mu}\cdot\mathsf{d}_{\mu\lambda} \leqq \mathsf{d}_{\kappa\lambda}$ by 1.3.7, and this completes the proof.

THEOREM 1.8.5. *Suppose B is a finite set of diagonal elements of $\mathfrak{A}$, each different from* 1. *Let R be the least equivalence relation including the set*

$\{\langle\kappa, \lambda\rangle : \mathsf{d}_{\kappa\lambda} \in B\}$ *and let E be the set of all R-equivalence classes (i.e., E = FdR/R). Then E and all members of E are finite, and we have*

$$\prod B = \prod_{\Gamma \in E} \mathsf{d}_\Gamma.$$

PROOF. Let

$$S = \{\langle\kappa, \lambda\rangle : \mathsf{d}_{\kappa\lambda} \in B\} \cup \{\langle\kappa, \kappa\rangle : \text{for some } \lambda,\ \mathsf{d}_{\kappa\lambda} \in B\};$$

then $R = \bigcup_{\mu \in \omega \sim 1} S^{[\mu]}$, and $S^{[\mu]} \subseteq S^{[\nu]}$ if $1 \leqq \mu \leqq \nu < \omega$. (Recall that $S^{[\mu]}$ is the μth relative power of S; see the Preliminaries.) By 1.3.14, the assumption that B is finite implies that S and $S^{[\mu]}$ are finite for each $\mu \in \omega \sim 1$. By an easy induction on μ we prove that $\prod B = \prod\{\mathsf{d}_{\kappa\lambda} : \langle\kappa, \lambda\rangle \in S^{[\mu]}\}$ for all $\mu \in \omega \sim 1$. Since S is finite, we must have $R = S^{[\mu]}$ for some $\mu \in \omega \sim 1$. Therefore, $\prod B = \prod\{\mathsf{d}_{\kappa\lambda} : \langle\kappa, \lambda\rangle \in R\}$, whence the conclusion easily follows.

THEOREM 1.8.6. *If* $0 < \kappa < \omega$ *and* $\langle\Theta_\lambda : \lambda < \kappa\rangle$ *is a sequence of mutually disjoint finite subsets of* α, *then* $\mathsf{c}_{(\Gamma)}(\mathsf{d}_{\Theta_0} \cdot \ldots \cdot \mathsf{d}_{\Theta_{\kappa-1}}) = \mathsf{d}_{\Theta_0 \sim \Gamma} \cdot \ldots \cdot \mathsf{d}_{\Theta_{\kappa-1} \sim \Gamma}$; *in particular,* $\mathsf{c}_{(\Gamma)}\mathsf{d}_\Delta = \mathsf{d}_{\Delta \sim \Gamma}$.

PROOF. It clearly suffices to treat the case in which Γ has exactly one element, say ν. If $\nu \notin \Theta_\lambda$, then $\mathsf{c}_\nu \mathsf{d}_{\Theta_\lambda} = \mathsf{d}_{\Theta_\lambda}$ by 1.2.10 and 1.3.3. Thus the desired equation follows in case $\nu \notin \Theta_\lambda$ for all $\lambda < \kappa$. Now assume that $\nu \in \Theta_\lambda$ for a certain $\lambda < \kappa$. Then

(1) $$\mathsf{c}_\nu(\mathsf{d}_{\Theta_0} \cdot \ldots \cdot \mathsf{d}_{\Theta_{\kappa-1}}) = \mathsf{c}_\nu \mathsf{d}_{\Theta_\lambda} \cdot \prod_{\mu \in \kappa \sim \{\lambda\}} \mathsf{d}_{\Theta_\mu}.$$

If ν is the only member of Θ_λ, the desired conclusion follows from 1.8.2 and 1.8.3(i). If, on the other hand, $\nu, \xi \in \Theta_\lambda$ with $\nu \neq \xi$, we can set $\Lambda = \Theta_\lambda \sim \{\nu\}$ and apply 1.8.4 to get $\mathsf{d}_{\Theta_\lambda} = \mathsf{d}_\Lambda \cdot \mathsf{d}_{\nu\xi}$. Since $\nu \notin \Lambda$, we have (as noted above) $\mathsf{c}_\nu \mathsf{d}_\Lambda = \mathsf{d}_\Lambda$, whence $\mathsf{c}_\nu \mathsf{d}_{\Theta_\lambda} = \mathsf{c}_\nu(\mathsf{d}_\Lambda \cdot \mathsf{d}_{\nu\xi}) = \mathsf{d}_\Lambda \cdot \mathsf{c}_\nu \mathsf{d}_{\nu\xi} = \mathsf{d}_\Lambda$ by 1.3.2, and the desired conclusion follows by (1).

COROLLARY 1.8.7. $\mathsf{c}_{(\Gamma)}\mathsf{d}_\Gamma = 1$ *and* $\mathsf{c}_{(\Gamma)}\mathsf{d}_{\Gamma \cup \{\kappa\}} = 1$.

THEOREM 1.8.8. *Let* $\langle\Gamma_\kappa : \kappa < \lambda\rangle$ *and* $\langle\Delta_\kappa : \kappa < \mu\rangle$ *be sequences of finite subsets of* α *satisfying the following conditions*:

(i) $0 < \lambda, \mu < \omega$;

(ii) $\mathsf{d}_{\Gamma_\kappa}, \mathsf{d}_{\Delta_\nu} \neq 1$ *for all* $\kappa < \lambda$ *and* $\nu < \mu$;

(iii) $\Gamma_\kappa \cap \Gamma_\nu = 0$ *if* $\kappa < \nu < \lambda$, *and* $\Delta_\kappa \cap \Delta_\nu = 0$ *if* $\kappa < \nu < \mu$;

(iv) $\prod_{\kappa<\lambda} \mathsf{d}_{\Gamma_\kappa} = \prod_{\kappa<\mu} \mathsf{d}_{\Delta_\kappa}$.

Then $\lambda = \mu$, *and there is a permutation* σ *of* λ *such that* $\Gamma_{\sigma\kappa} = \Delta_\kappa$ *for all* $\kappa < \lambda$.

PROOF. By symmetry it clearly suffices to prove the following statement:

(1) if $\kappa < \lambda$, then there is a $\nu < \mu$ such that $\Gamma_\kappa \subseteq \Delta_\nu$.

To prove (1), first assume that $\Gamma_\kappa \nsubseteq \bigcup_{\nu<\mu}\Delta_\nu$; say $\xi \in \Gamma_\kappa \sim \bigcup_{\nu<\mu}\Delta_\nu$. Since $\mathsf{d}_{\Gamma_\kappa} \neq 1$, by 1.8.3(i) we may choose $\pi \in \Gamma_\kappa \sim \{\xi\}$. Let $\Theta = \bigcup_{\nu<\mu}\Delta_\nu \sim \{\pi\}$. Then

$$\begin{aligned} 1 &= \mathsf{c}_{(\Theta)}\textstyle\prod_{\nu<\mu}\mathsf{d}_{\Delta_\nu} && \text{by 1.8.2, 1.8.3, 1.8.6} \\ &= \mathsf{c}_{(\Theta)}\textstyle\prod_{\nu<\lambda}\mathsf{d}_{\Gamma_\nu} && \text{by (iv)} \\ &\leqq \mathsf{d}_{\xi\pi} && \text{by 1.8.6.} \end{aligned}$$

But this implies that $\mathfrak{A}$ is discrete, which clearly contradicts (ii). Thus the assumption $\Gamma_\kappa \nsubseteq \bigcup_{\nu\sim\mu}\Delta_\nu$ is false; we have

(2) $$\Gamma_\kappa \subseteq \bigcup_{\nu<\mu}\Delta_\nu.$$

Now if (1) is false, then by (2) there exist distinct $\nu', \nu'' < \mu$ such that $\Gamma_\kappa \cap \Delta_{\nu'}, \Gamma_\kappa \cap \Delta_{\nu''} \neq 0$. Say $\xi' \in \Gamma_\kappa \cap \Delta_{\nu'}$ and $\xi'' \in \Gamma_\kappa \cap \Delta_{\nu''}$. Let $\Lambda = \bigcup_{\nu<\mu}\Delta_\nu \sim \{\xi', \xi''\}$. Then, applying $\mathsf{c}_{(\Lambda)}$ to both sides of equation (iv), we easily infer that $1 = \mathsf{d}_{\xi'\xi''}$ by 1.8.6, which leads to a contradiction as before. Thus (1) is true, and the proof is complete.

THEOREM 1.8.9. *If Γ has at least two elements and $\Gamma \neq \Delta$, then the following conditions are equivalent*:

(i) $\mathfrak{A}$ *is discrete*;
(ii) $\mathsf{d}_\Gamma = 1$;
(iii) $\mathsf{d}_\Gamma = \mathsf{d}_\Delta$.

PROOF. (i) is equivalent to (ii) by 1.3.12, and (ii) is equivalent to (iii) by 1.8.8.

THEOREM 1.8.10. *If $\Gamma \subset \Delta$, then $\mathsf{c}_{(\Gamma)}(\mathsf{d}_\Delta \cdot x)\cdot\mathsf{c}_{(\Gamma)}(\mathsf{d}_\Delta \cdot -x) = 0$.*

PROOF. First we prove:

(1) if $\kappa \notin \Gamma$, then $\mathsf{c}_{(\Gamma)}(\mathsf{d}_{\Gamma\cup\{\kappa\}}\cdot x)\cdot\mathsf{c}_{(\Gamma)}(\mathsf{d}_{\Gamma\cup\{\kappa\}}\cdot -x) = 0$.

To prove (1) we proceed by induction on $|\Gamma|$. If $\Gamma = 0$, (1) is trivial. Suppose then that $\lambda \notin \Gamma \cup \{\kappa\}$ and the conclusion of (1) holds for every x. We want to obtain the same conclusion with Γ replaced by $\Gamma \cup \{\lambda\}$. Bearing in mind that $\mathsf{d}_{\Gamma\cup\{\kappa,\lambda\}} = \mathsf{d}_{\Gamma\cup\{\kappa\}}\cdot\mathsf{d}_{\kappa\lambda}$ by 1.8.4, we see that, for every $x \in A$,

$$\begin{aligned} \mathsf{c}_{(\Gamma\cup\{\lambda\})}(\mathsf{d}_{\Gamma\cup\{\kappa,\lambda\}}\cdot x) &= \mathsf{c}_{(\Gamma)}\mathsf{c}_\lambda(\mathsf{d}_{\Gamma\cup\{\kappa\}}\cdot\mathsf{d}_{\kappa\lambda}\cdot x) \\ &= \mathsf{c}_{(\Gamma)}(\mathsf{d}_{\Gamma\cup\{\kappa\}}\cdot\mathsf{c}_\lambda(\mathsf{d}_{\kappa\lambda}\cdot x)) \\ &\leqq -\mathsf{c}_{(\Gamma)}(\mathsf{d}_{\Gamma\cup\{\kappa\}}\cdot -\mathsf{c}_\lambda(\mathsf{d}_{\kappa\lambda}\cdot x)) && \text{by induction hypothesis} \\ &= -\mathsf{c}_{(\Gamma)}(\mathsf{d}_{\Gamma\cup\{\kappa\}}\cdot\mathsf{c}_\lambda(\mathsf{d}_{\kappa\lambda}\cdot -x)) && \text{by 1.3.4} \\ &= -\mathsf{c}_{(\Gamma\cup\{\lambda\})}(\mathsf{d}_{\Gamma\cup\{\kappa,\lambda\}}\cdot -x). \end{aligned}$$

This completes the inductive proof of (1). Turning to the proof of the theorem itself, choose $\kappa \in \Delta \sim \Gamma$. Since, for any $x \in A$, $c_{(\Gamma)}(d_\Delta \cdot x) \leqq c_{(\Gamma)}(d_{\Gamma \cup \{\kappa\}} \cdot x)$ by 1.8.3(iii), the desired result follows at once from (1).

Corollary 1.8.11. *Suppose* $\kappa \notin \Gamma$. *Let* $c'_0 = c'_1 = c_{(\Gamma)}$, $d'_{00} = d'_{11} = 1$, *and* $d'_{01} = d'_{10} = d_{\Gamma \cup \{\kappa\}}$. *Then* $\langle A, +, \cdot, -, 0, 1, c'_\kappa, d'_{\kappa\lambda} \rangle_{\kappa,\lambda<2}$ *is a* CA_2.

Proof: by 1.7.5, 1.8.7, and 1.8.10.

If $\Gamma \subset \Delta$ and in fact $\Delta \sim \Gamma = \{\kappa\}$, we see from 1.8.11 that the function on A which correlates $c_{(\Gamma)}(d_\Delta \cdot x)$ with every $x \in A$ is a complete endomorphism of $\mathfrak{Bl}\mathfrak{A}$; cf. the remark following 1.5.3. From 1.8.6 we see that this result cannot be extended to the case when $|\Delta \sim \Gamma| \geqq 2$.

Theorem 1.8.12. *If* $\Gamma \subset \Delta$, *we have*:

(i) $c_{(\Gamma)}(d_\Delta \cdot (x+y)) = c_{(\Gamma)}(d_\Delta \cdot x) + c_{(\Gamma)}(d_\Delta \cdot y)$;

(ii) $c_{(\Gamma)}(d_\Delta \cdot x \cdot y) = c_{(\Gamma)}(d_\Delta \cdot x) \cdot c_{(\Gamma)}(d_\Delta \cdot y)$;

(iii) $c_{(\Gamma)}(d_\Delta \cdot -x) = c_{(\Gamma)} d_\Delta - c_{(\Gamma)}(d_\Delta \cdot x)$;

(iv) $d_\Delta \cdot c_{(\Gamma)}(d_\Delta \cdot x) = d_\Delta \cdot x$.

Proof. (i) is obvious. For the proof of (ii) it is convenient to denote $c_{(\Gamma)}(d_\Delta \cdot z)$ by $f(z)$ for each $z \in A$. Then, by (i) and 1.8.10,

$$f(x) \cdot f(y) = f(x) \cdot [f(x \cdot y) + f(-x \cdot y)] = f(x) \cdot f(x \cdot y).$$

Since $f(x \cdot y) \leqq f(x)$, (ii) is proved. (iv) is an easy consequence of (i) and 1.8.10; this is the case also for (iii), as is readily seen by observing that only (C_2) and (C_7) were used in the proof of 1.3.9.

Theorem 1.8.13. (i) *If* $\Gamma \cap \Delta = 0$ *or* $|\Delta| < 2$, *then* $c^\partial_{(\Gamma)} d_\Delta = d_\Delta$.

(ii) *If* $\Gamma \cap \Delta \neq 0$ *and* $|\Delta| \geqq 2$, *then* $c^\partial_{(\Gamma)} d_\Delta = c^\partial_0 d_{01}$.

Proof. (i) follows from 1.4.4(iii) and 1.7.5 (cf. Remark 1.7.9) along with 1.8.2, 1.8.3, and 1.8.6. We now turn to (ii). Let $\mu \in \Gamma \cap \Delta$ and $\nu \in \Delta \sim \{\mu\}$. Then by (i), $1.2.6^\partial$, $1.7.3^\partial$, and 1.8.4 we have

$$c^\partial_{(\Gamma)} d_\Delta = c^\partial_{(\Gamma)} c^\partial_\mu (d_{\Delta \sim \{\mu\}} \cdot d_{\mu\nu}) = c^\partial_{(\Gamma)} (d_{\Delta \sim \{\mu\}} \cdot c^\partial_\mu d_{\mu\nu}).$$

Hence $c^\partial_{(\Gamma)} d_\Delta = c^\partial_0 d_{01}$ by 1.3.19 and $1.6.9(\mathrm{i})^\partial$; cf. also Remark 1.4.5. This completes the proof.

Theorem 1.8.14. (i) *If* $\kappa \notin \Gamma$, *then* $s^\kappa_\lambda d_\Gamma = d_\Gamma$;

(ii) *if* $\kappa \in \Gamma$, *then* $s^\kappa_\lambda d_\Gamma = d_{(\Gamma \sim \{\kappa\}) \cup \{\lambda\}}$.

Proof. (i) follows from 1.5.8(i) and 1.8.6, and (ii) results from 1.8.4 and 1.8.6.

THEOREM 1.8.15. *If* $\lambda < \omega$ *and* $\mathsf{d}_{\Delta_\kappa} \neq 1$ *for each* $\kappa < \lambda$, *then* $\Delta(\prod_{\kappa<\lambda} \mathsf{d}_{\Delta_\kappa}) = \bigcup_{\kappa<\lambda} \Delta_\kappa$; *in particular, if* $\mathsf{d}_\Gamma \neq 1$, *then* $\Delta(\mathsf{d}_\Gamma) = \Gamma$.

PROOF. By 1.6.4 and 1.6.6 it is clear that $\Delta(\prod_{\kappa<\lambda} \mathsf{d}_{\Delta_\kappa}) \subseteq \bigcup_{\kappa<\lambda} \Delta_\kappa$. To prove the converse inclusion we may, by 1.8.4, assume that $\Delta_\kappa \cap \Delta_\mu = 0$ for $\kappa < \mu < \lambda$. The converse inclusion then follows by 1.8.6 and 1.8.8.

1.9. GENERALIZED CO-DIAGONAL ELEMENTS

The notion of a co-diagonal element can be generalized in much the same way as that of a diagonal element. The usefulness of the generalization is somewhat limited, however: it will be applied primarily in the detailed investigation of some rather restricted classes of cylindric algebras.[1)]

The reader may find it difficult to grasp the intuitive source and content of many of the theorems in this section; since, in addition, some proofs are complicated, he may not see any justification for including these results in the work. In most cases a metalogical interpretation of the results proves helpful. It turns out that in this interpretation the results discussed correspond to certain logical statements which have a clear intuitive content and are obviously true, although their formal derivations from logical axioms are not quite trivial. It seems to be of some interest that the simple postulates underlying the general theory of CA's are adequate for an algebraic reconstruction of all these derivations, and it becomes clear that an essential simplification of many proofs in the present section can hardly be expected.

Independently of the above observations it should be borne in mind that the theorems stated here function primarily as lemmas which are useful in certain later portions of our discussion. The reader may therefore find it convenient to postpone the study of the material given in this section until it is actually needed in Chapter 2.

Throughout this section R, S, and T will represent finite relations $\subseteq \alpha \times \alpha$, and Γ, Δ, and Θ will continue to represent finite subsets of α.

DEFINITION 1.9.1. *We let* $\bar{\mathsf{d}}R = \prod_{\langle \kappa, \lambda \rangle \in R \sim Id} - \mathsf{d}_{\kappa\lambda}$. *The elements* $\bar{\mathsf{d}}R$ *are called* ***generalized co-diagonal elements.***

In a cylindric algebra of formulas, $\bar{\mathsf{d}}R$ is the equivalence class of the conjunction of all the formulas $\mathbf{v}_\kappa \neq \mathbf{v}_\lambda$ with $\langle \kappa, \lambda \rangle \in R$ and $\kappa \neq \lambda$. If $R \subseteq \alpha \upharpoonright Id$, we may take $\boldsymbol{T}$ as the formula determining $\bar{\mathsf{d}}R$.

[1)] The material given in this section was first presented in Monk [64a], where it formed an essential part of the proof that all the monadic-generated CA's defined in 2.2.16 below are representable.

THEOREM 1.9.2. (i) $\bar{\mathsf{d}}(R^{-1}) = \bar{\mathsf{d}}R$.

(ii) $\bar{\mathsf{d}}(R \sim Id) = \bar{\mathsf{d}}R$.

(iii) $\bar{\mathsf{d}}R = 1$ *iff* $R \subseteq Id$ *or* $0 = 1$ (*i.e., A is a singleton*).

(iv) $\bar{\mathsf{d}}(R \cup S) = \bar{\mathsf{d}}R \cdot \bar{\mathsf{d}}S$.

(v) *If* $R \subseteq S$, *then* $\bar{\mathsf{d}}R \geqq \bar{\mathsf{d}}S$.

PROOF. (ii), (iv), and (v) follow directly from 1.9.1, while (i) is an easy consequence of 1.3.1. In order to see that (iii) holds we need only observe that, by 1.3.2, $-\mathsf{d}_{\kappa\lambda}$ for $\kappa \neq \lambda$ is never equal to 1 unless $1 = 0$.

If a CA_α $\mathfrak{A}$ is discrete, then we have either $\bar{\mathsf{d}}R = 1$ or $\bar{\mathsf{d}}R = 0$ for every R, according as $R \subseteq Id$ or $R \nsubseteq Id$. If, conversely, we have $\bar{\mathsf{d}}R = 0$ or $\bar{\mathsf{d}}R = 1$ for every R (or, equivalently, if we have $\bar{\mathsf{d}}R = 0$ for every $R \nsubseteq Id$), then $\mathfrak{A}$ is discrete provided $\alpha > 1$. However, we may have $\bar{\mathsf{d}}R = 0$ even if $\mathfrak{A}$ is not discrete. Thus, if $\alpha \geqq \kappa$ where $1 < \kappa < \omega$, and if $\mathfrak{A}$ is a cylindric set algebra whose base has less than κ elements, then $\bar{\mathsf{d}}(\kappa \times \kappa) = 0$; cf. 2.4.61 and the remark following 2.4.62 below.

THEOREM 1.9.3. *If* $S = R \cap [(\Gamma \times \alpha) \cup (\alpha \times \Gamma)]$, *then* $\mathsf{c}_{(\Gamma)}\bar{\mathsf{d}}R = \mathsf{c}_{(\Gamma)}\bar{\mathsf{d}}S \cdot \bar{\mathsf{d}}(R \sim S)$.

PROOF. From 1.2.11, 1.3.3, and 1.7.1 we conclude that $\mathsf{c}_{(\Gamma)} - \mathsf{d}_{\kappa\lambda} = -\mathsf{d}_{\kappa\lambda}$ if $\kappa, \lambda \notin \Gamma$. The proof is completed by an easy induction on $|R|$.

The next two theorems are trivially true when applied to cylindric algebras of formulas; we have only to keep in mind that the operation $\mathsf{c}^{\partial}_{(\Gamma)}$ corresponds to universal quantification and $\mathsf{s}^{\kappa}_{\lambda}$ to substitution (see the remarks following 1.5.1 and 1.7.8).

THEOREM 1.9.4. *Let* $S = R \cap [(\Gamma \times \alpha) \cup (\alpha \times \Gamma)]$.

(i) *If* $S \subseteq Id$, *then* $\mathsf{c}^{\partial}_{(\Gamma)}\bar{\mathsf{d}}R = \bar{\mathsf{d}}R = \mathsf{c}_{(\Gamma)}\bar{\mathsf{d}}R$.

(ii) *If* $S \nsubseteq Id$, *then* $\mathsf{c}^{\partial}_{(\Gamma)}\bar{\mathsf{d}}R = 0$.

PROOF. (i) By applying 1.9.2(ii),(iii) and 1.9.3 we obtain immediately $\mathsf{c}_{(\Gamma)}\bar{\mathsf{d}}R = \bar{\mathsf{d}}R$. Hence the dual formula $\mathsf{c}^{\partial}_{(\Gamma)}\bar{\mathsf{d}}R = \bar{\mathsf{d}}R$ can be derived by 1.4.4(iii); cf. Remark 1.7.9.

(ii) By 1.2.6(i) and 1.7.8 we have

$$\mathsf{c}^{\partial}_{(\Gamma)}\bar{\mathsf{d}}R = -\sum\{\mathsf{c}_{(\Gamma)}\mathsf{d}_{\kappa\lambda} : \langle\kappa, \lambda\rangle \in R \sim Id\}.$$

Since $S \nsubseteq Id$ implies the existence of $\langle\kappa, \lambda\rangle \in R \sim Id$ such that $\kappa \in \Gamma$ or $\lambda \in \Gamma$, the conclusion of (ii) follows by 1.3.2.

THEOREM 1.9.5. (i) *If* $\langle\kappa, \lambda\rangle \in (R \cup R^{-1}) \sim Id$, *then* $\mathsf{s}^{\kappa}_{\lambda}\bar{\mathsf{d}}R = 0$.

(ii) *Suppose* $\langle\kappa, \lambda\rangle \notin (R \cup R^{-1}) \sim Id$ *and let f be the function on* $R \sim Id$ *determined by the conditions*

$$f\langle\mu, \nu\rangle = \langle\mu, \nu\rangle \text{ if } \mu, \nu \neq \kappa,$$

$$f\langle\mu, \kappa\rangle = \langle\mu, \lambda\rangle \text{ and } f\langle\kappa, \nu\rangle = \langle\lambda, \nu\rangle.$$

Then

$$\mathsf{s}^{\kappa}_{\lambda}\bar{\mathsf{d}}R = \bar{\mathsf{d}}(Rgf).$$

PROOF. (i) follows easily from the definitions of the notions involved; (ii) is obtained with the help of 1.5.3 and 1.5.4.

In the next few theorems we deal with products of diagonal and co-diagonal elements, i.e., with elements of the form $\prod_{\langle\kappa,\lambda\rangle\in T}\mathsf{d}_{\kappa\lambda}\cdot\prod_{\langle\mu,\nu\rangle\in R}-\mathsf{d}_{\mu\nu}$; they can sometimes be represented more simply as $\mathsf{d}_{\Gamma}\cdot\bar{\mathsf{d}}R$. Such products are of interest to us primarily in the case when $T\cap R$ or $(\Gamma\times\Gamma)\cap(R\sim Id)$ is empty (for otherwise they are equal to 0). In this case the products can be subsumed under the general Boolean-algebraic notion of constituents of a given set, and in fact they can be treated as constituents of the set of all diagonal elements. We want to explain here the meaning of this notion, especially since it will play a role in some portions of our further discussion.

Given a Boolean algebra $\mathfrak{A}$, and a set B of its elements, by a *constituent associated with B*, or, more briefly, a *constituent of B*, we understand an arbitrary element of the form

$$\prod_{x\in X}x\cdot\prod_{y\in Y}-y = \prod X - \sum Y$$

where X and Y are any disjoint finite subsets of B. It is well known that, if the set B generates $\mathfrak{A}$ (in the sense of 0.1.15(i)), so does the set of constituents of B; in fact, every element of A is then a sum of finitely many constituents of B. If, in particular, B is finite, then the constituents $\prod X - \sum(B\sim X)$ can be called *minimal constituents* of B. It is known that the sum of all minimal constituents is always 1, and that any two distinct minimal constituents are disjoint. In case the finite set B generates $\mathfrak{A}$, the set of non-zero minimal constituents also generates $\mathfrak{A}$ and coincides with the set of all atoms of $\mathfrak{A}$. As in other analogous cases we shall apply the notion of a constituent not only to BA's, but also to CA's.

THEOREM 1.9.6. *Let* $S = R\cup\{\langle\kappa, \lambda\rangle : \kappa\in\Gamma$, *and* $\langle\mu, \lambda\rangle\in R\cup R^{-1}$ *for some* $\mu\in\Gamma\sim\{\lambda\}\}$. Then

$$\mathsf{d}_{\Gamma}\cdot\bar{\mathsf{d}}R = \mathsf{d}_{\Gamma}\cdot\bar{\mathsf{d}}S.$$

PROOF. By 1.9.2(v) we have $\mathsf{d}_{\Gamma}\cdot\bar{\mathsf{d}}S \leqq \mathsf{d}_{\Gamma}\cdot\bar{\mathsf{d}}R$. For the inclusion in the opposite direction it suffices to note that if $\kappa\in\Gamma$, $\mu\in\Gamma\sim\{\lambda\}$, and $\langle\mu, \lambda\rangle\in$

$R \cup R^{-1}$, we obtain with the help of 1.3.7

$$\mathsf{d}_\Gamma \cdot \bar{\mathsf{d}}R \leqq \mathsf{d}_{\kappa\mu} \cdot -\mathsf{d}_{\mu\lambda} \leqq -\mathsf{d}_{\kappa\lambda}.$$

THEOREM 1.9.7. *Let P be a partition of a set Γ. For every $\Delta \subseteq \Gamma$ such that $|\Delta \cap \Theta| = 1$ for each $\Theta \in P$ we have*

$$\prod_{\Theta \in P} \mathsf{d}_\Theta \cdot \bar{\mathsf{d}}(\Delta \times \Delta) = \prod_{\Theta \in P} [\mathsf{d}_\Theta \cdot \bar{\mathsf{d}}(\Theta \times (\Gamma \sim \Theta))].$$

PROOF. Consider the equation to be proved as the conjunction of two inclusions. By 1.9.2(v) it is clear that $\prod_{\Theta \in P} \bar{\mathsf{d}}(\Theta \times (\Gamma \sim \Theta)) \leqq \bar{\mathsf{d}}(\Delta \times \Delta)$, and this directly implies one inclusion. Now suppose that $\langle \kappa, \lambda \rangle \in \Theta \times (\Gamma \sim \Theta)$ for some $\Theta \in P$. There exists an $\Omega \in P$ such that $\lambda \in \Omega$ and $\Theta \cap \Omega = 0$. There exist also a $\mu \in \Delta \cap \Theta$ and a $\nu \in \Delta \cap \Omega$. With the help of 1.3.7 we obtain

$$\prod_{\Theta \in P} \mathsf{d}_\Theta \cdot \bar{\mathsf{d}}(\Delta \times \Delta) \leqq \mathsf{d}_{\kappa\mu} \cdot \mathsf{d}_{\lambda\nu} \cdot -\mathsf{d}_{\mu\nu} \leqq -\mathsf{d}_{\kappa\lambda},$$

and hence the other inclusion easily follows.

THEOREM 1.9.8. *Suppose $\Gamma \neq 0$.*

(i) *If $B = \{\mathsf{d}_{\kappa\lambda} : \kappa, \lambda \in \Gamma\}$, $X \subseteq B$, and $\prod X - \sum(B \sim X) \neq 0$, then there is a unique partition P of Γ such that*

$$\prod X - \sum(B \sim X) = \prod_{\Theta \in P} [\mathsf{d}_\Theta \cdot \bar{\mathsf{d}}(\Theta \times (\Gamma \sim \Theta))].$$

(ii) *Let K be the set of all ordered pairs $\langle P, \Delta \rangle$ such that P is a partition of Γ, $\Delta \subseteq \Gamma$, and $|\Delta \cap \Theta| = 1$ for each $\Theta \in P$. We then have*

$$\sum_{\langle P, \Delta \rangle \in K} [\prod_{\Theta \in P} \mathsf{d}_\Theta \cdot \bar{\mathsf{d}}(\Delta \times \Delta)] = 1.$$

PROOF. Let $R = \{\langle \kappa, \lambda \rangle : \mathsf{d}_{\kappa\lambda} \in X\}$. If $\langle \kappa, \kappa \rangle \notin R$ for some $\kappa \in \Gamma$, then $1 = \mathsf{d}_{\kappa\kappa} \in B \sim X$; hence $\prod X \sim \sum(B \sim X) = 0$, which contradicts the hypothesis of (i). Therefore R is reflexive over Γ, and for a similar reason R is symmetric and transitive and thus is an equivalence relation over Γ. Let $P = \Gamma / R$ be the associated partition of Γ. We then have that $\prod X - \sum(B \sim X)$ equals

(1) $$\prod_{\Theta \in P} \mathsf{d}_\Theta \cdot \bar{\mathsf{d}}(\Theta \times (\Gamma \sim \Theta)).$$

In order to complete the proof of (i) observe that, if P and P' are distinct partitions of Γ, then the product (1) and the corresponding product associated with P' are disjoint, and hence distinct if different from 0.

(ii) follows easily from (i), 1.9.7, and the fact that the sum of all minimal constituents of a finite set B is the unit element.

THEOREM 1.9.9. *If* $\Gamma \subseteq \Delta \subseteq \Theta$, *then* $\mathsf{c}_{(\Delta)}\bar{\mathsf{d}}(\Delta\times\Theta) \leqq \mathsf{c}_{(\Gamma)}\bar{\mathsf{d}}(\Gamma\times\Theta)$.

PROOF. By an easy inductive argument it is seen that we may confine ourselves to the case where $\Delta = \Gamma \cup \{\kappa\}$ for a given $\kappa \notin \Gamma$. We now distinguish three possible cases.

(I) $|\Gamma| = 0$. This case is obvious in view of 1.9.2(iii).

(II) $|\Gamma| = 1$, say $\Gamma = \{\lambda\}$. Then, with $\Theta' = \Theta \sim \{\kappa, \lambda\}$,

$$\begin{aligned}\mathsf{c}_{(\Delta)}\bar{\mathsf{d}}(\Delta\times\Theta) &= \mathsf{c}_\kappa\mathsf{c}_\lambda\bar{\mathsf{d}}(\{\kappa,\lambda\}\times\Theta)\\ &= \mathsf{c}_\kappa\mathsf{c}_\lambda[-\mathsf{d}_{\kappa\lambda}\cdot\bar{\mathsf{d}}(\{\lambda\}\times\Theta')\cdot\bar{\mathsf{d}}(\{\kappa\}\times\Theta')]\\ &= \mathsf{c}_\kappa\mathsf{c}_\lambda[-\mathsf{d}_{\kappa\lambda}\cdot\bar{\mathsf{d}}(\{\lambda\}\times\Theta')\cdot\mathsf{s}^\lambda_\kappa\bar{\mathsf{d}}(\{\lambda\}\times\Theta')] && \text{by 1.9.5(ii)}\\ &\leqq \mathsf{c}_\lambda[-\mathsf{d}_{\kappa\lambda}\cdot\bar{\mathsf{d}}(\{\lambda\}\times\Theta')] && \text{by 1.3.18(i)}\\ &= \mathsf{c}_{(\Gamma)}\bar{\mathsf{d}}(\Gamma\times\Theta).\end{aligned}$$

(III) $|\Gamma| > 1$. For each $\lambda\in\Gamma$ we have

$$\begin{aligned}\mathsf{c}_{(\Gamma)}\bar{\mathsf{d}}(\Gamma\times\Theta) &= \mathsf{c}_{(\Gamma)}[\bar{\mathsf{d}}(\{\lambda\}\times\Theta)\cdot\bar{\mathsf{d}}((\Gamma\sim\{\lambda\})\times(\Theta\sim\{\lambda\}))]\\ &= \mathsf{c}_{(\Gamma\sim\{\lambda\})}[\mathsf{c}_\lambda\bar{\mathsf{d}}(\{\lambda\}\times\Theta)\cdot\bar{\mathsf{d}}((\Gamma\sim\{\lambda\})\times(\Theta\sim\{\lambda\}))] && \text{by 1.9.3}\\ &\geqq \mathsf{c}_{(\Gamma\sim\{\lambda\})}[\mathsf{c}_\kappa\mathsf{c}_\lambda\bar{\mathsf{d}}(\{\kappa,\lambda\}\times\Theta)\cdot\bar{\mathsf{d}}((\Gamma\sim\{\lambda\})\times(\Theta\sim\{\lambda\}))] && \text{by (II)}\\ &= \mathsf{c}_{(\Gamma)}[\mathsf{c}_\kappa\bar{\mathsf{d}}(\Delta\times\Theta)\cdot\bar{\mathsf{d}}((\Gamma\sim\{\lambda\})\times\{\kappa\})] && \text{by 1.9.3.}\end{aligned}$$

Hence

$$\mathsf{c}_{(\Gamma)}\bar{\mathsf{d}}(\Gamma\times\Theta) \geqq \textstyle\sum_{\lambda\in\Gamma}\mathsf{c}_{(\Gamma)}[\mathsf{c}_\kappa\bar{\mathsf{d}}(\Delta\times\Theta)\cdot\bar{\mathsf{d}}((\Gamma\sim\{\lambda\})\times\{\kappa\})]$$

and

$$\mathsf{c}_{(\Gamma)}\bar{\mathsf{d}}(\Gamma\times\Theta) \geqq \mathsf{c}_{(\Gamma)}[\mathsf{c}_\kappa\bar{\mathsf{d}}(\Delta\times\Theta)\cdot\textstyle\sum_{\lambda\in\Gamma}\bar{\mathsf{d}}((\Gamma\sim\{\lambda\})\times\{\kappa\})]. \tag{2}$$

If $\lambda\in\Gamma$, then $\mathsf{d}_{\kappa\lambda}\cdot\bar{\mathsf{d}}(\Gamma\times\Gamma) \leqq \bar{\mathsf{d}}((\Gamma\sim\{\lambda\})\times\{\kappa\})$ by 1.9.2(v) and 1.9.6. Therefore

$$\begin{aligned}\bar{\mathsf{d}}(\Gamma\times\Gamma) &\leqq \bar{\mathsf{d}}(\Gamma\times\{\kappa\})+\textstyle\sum_{\lambda\in\Gamma}\bar{\mathsf{d}}((\Gamma\sim\{\lambda\})\times\{\kappa\})\\ &= \textstyle\sum_{\lambda\in\Gamma}\bar{\mathsf{d}}((\Gamma\sim\{\lambda\})\times\{\kappa\}) && \text{by 1.9.2(v).}\end{aligned}$$

Hence by (2), 1.9.2(v), and 1.9.3 we have

$$\mathsf{c}_{(\Gamma)}\bar{\mathsf{d}}(\Gamma\times\Theta) \geqq \mathsf{c}_{(\Gamma)}[\mathsf{c}_\kappa\bar{\mathsf{d}}(\Delta\times\Theta)\cdot\bar{\mathsf{d}}(\Gamma\times\Gamma)] = \mathsf{c}_{(\Delta)}\bar{\mathsf{d}}(\Delta\times\Theta).$$

This completes the proof.

COROLLARY 1.9.10. *If* $\Gamma\subseteq\Delta$, *then* $\mathsf{c}_{(\Delta)}\bar{\mathsf{d}}(\Delta\times\Delta) \leqq \mathsf{c}_{(\Gamma)}\bar{\mathsf{d}}(\Gamma\times\Gamma)$.

PROOF: by 1.9.9. (There is a simple direct proof of this corollary; cf. the proof of Lemma 2.2.20 in the next chapter.)

The following theorem is a generalization of 1.3.18.

THEOREM 1.9.11. (i) $\mathsf{c}_{(\Gamma\sim\{\kappa\})}\mathsf{d}(\Gamma\times\Gamma) = \mathsf{c}_{(\Gamma)}\mathsf{d}(\Gamma\times\Gamma)$.

(ii) *If* $|\Gamma| = |\Delta|$, *then* $\mathsf{c}_{(\Gamma)}\mathsf{d}(\Gamma\times\Gamma) = \mathsf{c}_{(\Delta)}\mathsf{d}(\Delta\times\Delta)$.

PROOF. (i) We have

$$\begin{aligned}\mathsf{c}_{(\Gamma\sim\{\kappa\})}\mathsf{d}(\Gamma\times\Gamma) &\leqq \mathsf{c}_{(\Gamma)}\mathsf{d}(\Gamma\times\Gamma)\\ &\leqq \mathsf{c}_{(\Gamma\sim\{\kappa\})}\mathsf{d}((\Gamma\sim\{\kappa\})\times\Gamma) && \text{by 1.9.9}\\ &= \mathsf{c}_{(\Gamma\sim\{\kappa\})}\mathsf{d}(\Gamma\times\Gamma).\end{aligned}$$

Thus (i) holds.

(ii) It suffices to note that, if $\kappa\in\Gamma$ and $\lambda\notin\Gamma$, then

$$\begin{aligned}\mathsf{c}_{(\Gamma)}\mathsf{d}(\Gamma\times\Gamma) &= \mathsf{s}^{\kappa}_{\lambda}\mathsf{c}_{(\Gamma)}\mathsf{d}(\Gamma\times\Gamma) && \text{by 1.5.8(i)}\\ &= \mathsf{s}^{\kappa}_{\lambda}\mathsf{c}_{(\Gamma\sim\{\kappa\})}\mathsf{d}(\Gamma\times\Gamma) && \text{by (i)}\\ &= \mathsf{c}_{(\Gamma\sim\{\kappa\})}\mathsf{s}^{\kappa}_{\lambda}\mathsf{d}(\Gamma\times\Gamma) && \text{by 1.5.8(ii)}\\ &= \mathsf{c}_{(\Gamma\sim\{\kappa\})}\mathsf{d}[((\Gamma\sim\{\kappa\})\cup\{\lambda\})\times((\Gamma\sim\{\kappa\})\cup\{\lambda\})] && \text{by 1.9.5(ii)}\\ &= \mathsf{c}_{((\Gamma\sim\{\kappa\})\cup\{\lambda\})}\mathsf{d}[((\Gamma\sim\{\kappa\})\cup\{\lambda\})\times((\Gamma\sim\{\kappa\})\cup\{\lambda\})] && \text{by (i).}\end{aligned}$$

The desired result then follows by induction.

Corollary 1.9.10 and Theorem 1.9.11(ii) become self-evident when applied to cylindric algebras of formulas. In this application the elements of the form $\mathsf{c}_{(\Gamma)}\mathsf{d}(\Gamma\times\Gamma)$ acquire an especially simple meaning; indeed, it is easily seen that each such element is the equivalence class of a sentence which has no non-logical constants and expresses in a natural way the fact that there are at least $|\Gamma|$ elements. E.g., in case $|\Gamma| = 3$, $\mathsf{c}_{(\Gamma)}\mathsf{d}(\Gamma\times\Gamma)$ corresponds to the sentence $\exists_{xyz}(x\neq y\wedge x\neq z\wedge y\neq z)$. Hence, in the application discussed, the meaning of $\mathsf{c}_{(\Gamma)}\mathsf{d}(\Gamma\times\Gamma)$ depends only on the cardinality of Γ, and this makes 1.9.11(ii) obvious. Also, assuming that σ_Γ and σ_Δ are sentences corresponding respectively to $\mathsf{c}_{(\Gamma)}\mathsf{d}(\Gamma\times\Gamma)$ and $\mathsf{c}_{(\Delta)}\mathsf{d}(\Delta\times\Delta)$, it is clearly seen that σ_Δ implies σ_Γ whenever $\Gamma\subseteq\Delta$ (or, more generally, $|\Gamma|\leqq|\Delta|$); this is just the content of Corollary 1.9.10.

THEOREM 1.9.12. (i) *If* $0\neq\Delta\subseteq\Gamma$ *and* $\lambda\in\Gamma\sim\Delta$, *then*

$$\mathsf{c}^{\partial}_{\lambda}\mathsf{c}_{(\Delta)}\mathsf{d}(\Delta\times\Gamma) = \mathsf{c}_{(\Delta\cup\{\lambda\})}\mathsf{d}((\Delta\cup\{\lambda\})\times\Gamma).$$

(ii) *If* $\kappa\in\Gamma$, *then*

$$\mathsf{c}_{(\Gamma\sim\{\kappa\})}^{\partial}\mathsf{c}_{\kappa}\mathsf{d}(\{\kappa\}\times\Gamma) = \mathsf{c}_{(\Gamma)}\mathsf{d}(\Gamma\times\Gamma).$$

PROOF. (i) From 1.9.9 we get $\mathsf{c}_{(\Delta\cup\{\lambda\})}\mathsf{d}((\Delta\cup\{\lambda\})\times\Gamma)\leqq\mathsf{c}_{(\Delta)}\mathsf{d}(\Delta\times\Gamma)$, so, by 1.2.7$^{\partial}$ and 1.4.4(iii),

$$\mathsf{c}_{(\Delta\cup\{\lambda\})}\mathsf{d}((\Delta\cup\{\lambda\})\times\Gamma)\leqq\mathsf{c}^{\partial}_{\lambda}\mathsf{c}_{(\Delta)}\mathsf{d}(\Delta\times\Gamma).$$

We now proceed to prove the inclusion in the opposite direction. Let $\kappa \in \Delta$ and $\Gamma' = \Gamma \sim (\Delta \cup \{\lambda\})$; then

$$\begin{aligned} \mathsf{c}_\lambda^\partial \mathsf{c}_{(\Delta)} \bar{\mathsf{d}}(\Delta \times \Gamma) &\leqq \mathsf{c}_\lambda^\partial \mathsf{c}_\kappa \bar{\mathsf{d}}(\{\kappa\} \times \Gamma') \\ &= \mathsf{c}_\lambda^\partial \mathsf{c}_\kappa \mathsf{s}_\kappa^\lambda \bar{\mathsf{d}}(\{\kappa\} \times \Gamma') && \text{by 1.5.8(i)} \\ &= \mathsf{c}_\lambda \mathsf{s}_\lambda^\kappa \bar{\mathsf{d}}(\{\kappa\} \times \Gamma') && \text{by 1.4.4(iii), 1.5.9(i)} \\ &= \mathsf{c}_\lambda \bar{\mathsf{d}}(\{\lambda\} \times \Gamma') && \text{by 1.9.5(ii).} \end{aligned}$$

Therefore,

$$\begin{aligned} \mathsf{c}_\lambda^\partial \mathsf{c}_{(\Delta)} \bar{\mathsf{d}}(\Delta \times \Gamma) &\leqq \mathsf{c}_\lambda^\partial \mathsf{c}_{(\Delta)} \bar{\mathsf{d}}(\Delta \times \Gamma) \cdot \mathsf{c}_\lambda \bar{\mathsf{d}}(\{\lambda\} \times \Gamma') \\ &= \mathsf{c}_\lambda [\mathsf{c}_\lambda^\partial \mathsf{c}_{(\Delta)} \bar{\mathsf{d}}(\Delta \times \Gamma) \cdot \bar{\mathsf{d}}(\{\lambda\} \times \Gamma')] && \text{by } (C_3) \text{ and 1.4.4(ii)} \\ &\leqq \mathsf{c}_\lambda [\mathsf{c}_{(\Delta)} \bar{\mathsf{d}}(\Delta \times \Gamma) \cdot \bar{\mathsf{d}}(\{\lambda\} \times \Gamma')] && \text{by } (C_2^\partial) \\ &= \mathsf{c}_\lambda \mathsf{c}_{(\Delta)} [\bar{\mathsf{d}}(\Delta \times \Gamma) \cdot \bar{\mathsf{d}}(\{\lambda\} \times \Gamma')] && \text{by 1.9.3} \\ &= \mathsf{c}_{(\Delta \cup \{\lambda\})} \bar{\mathsf{d}}((\Delta \cup \{\lambda\}) \times \Gamma) && \text{by 1.9.2(iv).} \end{aligned}$$

This completes the argument.

(ii) It is enough to prove the following lemma by induction on $|\Theta|$:

(1) If $\kappa \in \Gamma$ and $\Theta \subseteq \Gamma \sim \{\kappa\}$, then

$$\mathsf{c}_{(\Theta)}^\partial \mathsf{c}_\kappa \bar{\mathsf{d}}(\{\kappa\} \times \Gamma) = \mathsf{c}_{(\Theta \cup \{\kappa\})} \bar{\mathsf{d}}((\Theta \cup \{\kappa\}) \times \Gamma).$$

The case $\Theta = 0$ is trivial. Assume (1), and suppose that $\lambda \in \Gamma \sim \{\kappa\}$ and $\lambda \notin \Theta$. Then

$$\begin{aligned} \mathsf{c}_{(\Theta \cup \{\lambda\})}^\partial \mathsf{c}_\kappa \bar{\mathsf{d}}(\{\kappa\} \times \Gamma) &= \mathsf{c}_\lambda^\partial \mathsf{c}_{(\Theta)}^\partial \mathsf{c}_\kappa \bar{\mathsf{d}}(\{\kappa\} \times \Gamma) && \text{by } 1.7.3^\partial \\ &= \mathsf{c}_\lambda^\partial \mathsf{c}_{(\Theta \cup \{\kappa\})} \bar{\mathsf{d}}((\Theta \cup \{\kappa\}) \times \Gamma) && \text{by (1)} \\ &= \mathsf{c}_{(\Theta \cup \{\lambda\} \cup \{\kappa\})} \bar{\mathsf{d}}((\Theta \cup \{\lambda\} \cup \{\kappa\}) \times \Gamma) && \text{by (i),} \end{aligned}$$

which completes the inductive proof.

The content of Theorem 1.9.12, just as with 1.9.11, becomes self-evident when applied to cylindric algebras of formulas. For instance, 1.9.12(ii) when so applied with $\Gamma = \{0, 1, 2\}$ and $\kappa = 2$ expresses the well-known fact that the sentences $\forall_{xy} \exists_z (x \neq z \wedge y \neq z)$ and $\exists_{xyz} (x \neq y \wedge x \neq z \wedge y \neq z)$ are logically equivalent. It may be noted that, although this equivalence is intuitively clear, its formal derivation from the usual axioms of predicate logic is not completely trivial, and the proof becomes even more complicated when analogous sentences with a larger number of variables are involved; this reflects itself to some extent in the proof of 1.9.12.

Our next theorem, 1.9.13, when combined with some earlier and more elementary results (1.8.5, 1.8.14, 1.9.3, and 1.9.5) reduces the cylindrification of any product of diagonal and co-diagonal elements to a succession of Boolean operations performed on diagonal elements and elements of the form $\mathsf{c}_{(\Gamma)}\bar{\mathsf{d}}(\Gamma\times\Gamma)$. The significance of the whole procedure will become clear only in the next chapter. The final conclusions will be drawn and presented in a simple form in Theorem 2.1.17, for which 1.9.13 will serve as the principal lemma. Some further comments, concerning, in particular, a metalogical interpretation of the results, will be found in the Remarks 2.1.18.

THEOREM 1.9.13. *If* $\kappa \notin \Gamma$, *then*

$$\mathsf{c}_\kappa\bar{\mathsf{d}}(\{\kappa\}\times\Gamma) = \prod_{\Delta\subseteq\Gamma}[-\bar{\mathsf{d}}(\Delta\times\Delta)+\mathsf{c}_{(\Delta\cup\{\kappa\})}\bar{\mathsf{d}}((\Delta\cup\{\kappa\})\times(\Delta\cup\{\kappa\}))].$$

PROOF. If $\Delta\subseteq\Gamma$, then

$$\begin{aligned}\mathsf{c}_\kappa\bar{\mathsf{d}}(\{\kappa\}\times\Gamma)\cdot\bar{\mathsf{d}}(\Delta\times\Delta) &\leqq \mathsf{c}_\kappa[\bar{\mathsf{d}}(\{\kappa\}\times\Delta)\cdot\bar{\mathsf{d}}(\Delta\times\Delta)] && \text{by 1.9.2(v), 1.9.3}\\ &= \mathsf{c}_\kappa\bar{\mathsf{d}}((\Delta\cup\{\kappa\})\times(\Delta\cup\{\kappa\})) && \text{by 1.9.2(iv)}\\ &\leqq \mathsf{c}_{(\Delta\cup\{\kappa\})}\bar{\mathsf{d}}((\Delta\cup\{\kappa\})\times(\Delta\cup\{\kappa\})).\end{aligned}$$

Hence

$$\mathsf{c}_\kappa\bar{\mathsf{d}}(\{\kappa\}\times\Gamma)\leqq -\bar{\mathsf{d}}(\Delta\times\Delta)+\mathsf{c}_{(\Delta\cup\{\kappa\})}\bar{\mathsf{d}}((\Delta\cup\{\kappa\})\times(\Delta\cup\{\kappa\})).$$

Since this inequality holds for every $\Delta\subseteq\Gamma$, we have

$$\mathsf{c}_\kappa\mathsf{d}(\{\kappa\}\times\Gamma)\leqq\prod_{\Delta\subseteq\Gamma}[-\bar{\mathsf{d}}(\Delta\times\Delta)+\mathsf{c}_{(\Delta\cup\{\kappa\})}\bar{\mathsf{d}}((\Delta\cup\{\kappa\})\times(\Delta\cup\{\kappa\}))].$$

To prove the opposite inequality we will make use of 1.9.8. Assume, then, that P is an arbitrary partition of Γ, and that Ω is any subset of Γ such that $|\Omega\cap\Theta| = 1$ for each $\Theta\in P$. Then

$$\begin{aligned}&\bar{\mathsf{d}}(\Omega\times\Omega)\cdot\prod_{\Theta\in P}\mathsf{d}_\Theta\cdot\prod_{\Delta\subseteq\Gamma}[-\bar{\mathsf{d}}(\Delta\times\Delta)+\mathsf{c}_{(\Delta\cup\{\kappa\})}\bar{\mathsf{d}}((\Delta\cup\{\kappa\})\times(\Delta\cup\{\kappa\}))]\\ &\qquad\leqq\prod_{\Theta\in P}\mathsf{d}_\Theta\cdot\mathsf{c}_{(\Omega\cup\{\kappa\})}\bar{\mathsf{d}}((\Omega\cup\{\kappa\})\times(\Omega\cup\{\kappa\}))\\ &\qquad\leqq\prod_{\Theta\in P}\mathsf{d}_\Theta\cdot\mathsf{c}_\kappa\bar{\mathsf{d}}(\{\kappa\}\times\Omega) && \text{by 1.9.9}\\ &\qquad=\mathsf{c}_\kappa(\prod_{\Theta\in P}\mathsf{d}_\Theta\cdot\bar{\mathsf{d}}(\{\kappa\}\times\Omega)) && \text{by 1.8.6}\\ &\qquad\leqq\mathsf{c}_\kappa\bar{\mathsf{d}}(\{\kappa\}\times\Gamma) && \text{by 1.9.6.}\end{aligned}$$

An application of 1.9.8 completes the proof.

The following theorem is a generalization of 1.8.15.

THEOREM 1.9.14. *Let* $\mathfrak{A}$ *be a non-discrete* CA_α *and* B *a finite subset of diagonal and co-diagonal elements of* $\mathfrak{A}$ *such that* $\prod B\neq 0$. *Then*

$$\Delta(\prod B)=\bigcup_{x\in B}\Delta x=\{\kappa:\mathsf{d}_{\kappa\lambda}\in B \text{ or } -\mathsf{d}_{\kappa\lambda}\in B \text{ for some } \lambda\neq\kappa\}.$$

PROOF. We may, of course, assume $\alpha \geqq 2$. Let $\Gamma = \bigcup_{x \in B} \Delta x$, and notice that, by 1.6.4 and 1.6.7, $\Gamma = \{\kappa : \mathsf{d}_{\kappa\lambda} \in B$ or $-\mathsf{d}_{\kappa\lambda} \in B$ for some $\lambda \neq \kappa\}$. Also observe that $\prod B \cdot \mathsf{c}_0 - \mathsf{d}_{01} \neq 0$ (cf. the remark after 1.3.18); for, if $\prod B \cdot \mathsf{c}_0 - \mathsf{d}_{01} = 0$, then, by 1.3.19, $\prod B \leqq \mathsf{d}_{\kappa\lambda}$ for all $\kappa, \lambda < \alpha$. This implies that B consists exclusively of diagonal elements; hence, using 1.6.9(i) and 1.8.7 we have

$$0 = \mathsf{c}_{(\Gamma)}(\prod B \cdot \mathsf{c}_0 - \mathsf{d}_{01}) = \mathsf{c}_{(\Gamma)} \prod B \cdot \mathsf{c}_0 - \mathsf{d}_{01} = \mathsf{c}_0 - \mathsf{d}_{01}.$$

This is impossible since $\mathfrak{A}$ is assumed to be non-discrete. Therefore, $\prod B \cdot \mathsf{c}_0 - \mathsf{d}_{01} \neq 0$.

Consider any $\kappa \in \Gamma$ and assume, by way of contradiction, that $\kappa \notin \Delta(\prod B)$. Then we must have either $\mathsf{d}_{\kappa\lambda} \in B$ or $-\mathsf{d}_{\kappa\lambda} \in B$ for some $\lambda \neq \kappa$; since also $\{\kappa, \lambda\} \nsubseteq \Delta(\prod B)$ by assumption, 1.6.10 gives us

$$\Delta(\prod B) = \Delta(\prod B) \cup \{\kappa, \lambda\}.$$

This contradicts the assumption $\kappa \notin \Delta(\prod B)$, and thus, in fact, κ is contained in $\Delta(\prod B)$. This shows $\Gamma \subseteq \Delta(\prod B)$. The inclusion in the opposite direction is an immediate consequence of 1.6.4, 1.6.5, and 1.6.7, which completes the proof.

COROLLARY 1.9.15. *If B is a finite set of diagonal and co-diagonal elements, C is another such set, and $\prod B \cdot \prod C \neq 0$, then*

$$\Delta(\prod B \cdot \prod C) = \Delta(\prod B) \cup \Delta(\prod C).$$

The proof is trivial in case $\mathfrak{A}$ is discrete. In the other case apply 1.9.14 directly.

COROLLARY 1.9.16. (i) *If $\bar{\mathsf{d}}R \neq 0$, then $\Delta(\bar{\mathsf{d}}R) = Fd(R \sim Id)$.*
(ii) *If $0 \neq \bar{\mathsf{d}}(\Gamma \times \Delta) \neq 1$, then $\Delta(\bar{\mathsf{d}}(\Gamma \times \Delta)) = \Gamma \cup \Delta$.*

In connection with 1.9.16(i) and its proof see the remark following 1.9.2.

We recall that by atoms of a cylindric algebra $\mathfrak{A}$ we understand atoms of its Boolean part $\mathfrak{Bl}\mathfrak{A}$, and that the set of these atoms is denoted by $At\mathfrak{A}$. Remembering that $\mathfrak{Cl}_\Gamma\mathfrak{A}$ is a BA (see 1.6.19) we now define:

DEFINITION 1.10.1. *By a **Γ-atom** of $\mathfrak{A}$ we mean an atom of $\mathfrak{Cl}_\Gamma\mathfrak{A}$, i.e., an element of $At\mathfrak{Cl}_\Gamma\mathfrak{A}$.*

According to our convention the set Γ in 1.10.1 is assumed to be finite. This restriction, however, is not essential; in view of 1.6.18 and 1.6.19 we could consider Ω-atoms for an arbitrary set $\Omega \subseteq \alpha$.

In an atomic cylindric set algebra whose atoms are singletons, the $\{\kappa\}$-atoms are just lines parallel to the κ-axis; Γ-atoms are linear subspaces parallel to the subspace spanned by the set of all λ-axes for $\lambda \in \Gamma$.

COROLLARY 1.10.2. *The 0-atoms of $\mathfrak{A}$ coincide with the atoms of $\mathfrak{A}$, i.e., $At\mathfrak{Cl}_0\mathfrak{A} = At\mathfrak{A}$.*

THEOREM 1.10.3. (i) *If $\Gamma \subseteq \Delta$ and x is a Γ-atom, then $c_{(\Delta)}x$ is a Δ-atom.*

(ii) *If $\Gamma \subseteq \Delta$, $\kappa \notin \Delta$, and x is a Δ-atom, then $x \cdot d_{\Gamma\cup\{\kappa\}}$ is a $(\Delta \sim \Gamma)$-atom.*

PROOF. (i) Under the hypothesis of (i) we have $0 < x \leqq c_{(\Delta)}x$ by 1.7.5, so $c_{(\Delta)}x \neq 0$. Now suppose $c_{(\Delta)}x \cdot y \neq 0$ where $y \in Cl_\Delta\mathfrak{A}$. Then, by 1.7.5, $x \cdot y \neq 0$. Since $y \in Cl_\Gamma\mathfrak{A}$ and x is a Γ-atom, it follows that $x \leqq y$, and hence, by 1.7.5 again, that $c_{(\Delta)}x \leqq y$. This shows that $c_{(\Delta)}x$ is a Δ-atom.

(ii) Since $c_{(\Gamma)}(x \cdot d_{\Gamma\cup\{\kappa\}}) = x \cdot d_{\{\kappa\}} = x$ by 1.8.3(i) and 1.8.6, we have $x \cdot d_{\Gamma\cup\{\kappa\}} \neq 0$. Also, $x \cdot d_{\Gamma\cup\{\kappa\}} \in Cl_{\Delta\sim\Gamma}\mathfrak{A}$ by 1.8.6.

Now suppose $x \cdot d_{\Gamma\cup\{\kappa\}} \cdot y \neq 0$ where $y \in Cl_{\Delta\sim\Gamma}\mathfrak{A}$. Then $c_{(\Delta)}(x \cdot d_{\Gamma\cup\{\kappa\}} \cdot y) \neq 0$, and, since $c_{(\Delta)}x = x$, we obtain $x \cdot c_{(\Delta)}(d_{\Gamma\cup\{\kappa\}} \cdot y) \neq 0$. Since x is a Δ-atom, it follows that

(1) $$x \leqq c_{(\Delta)}(d_{\Gamma\cup\{\kappa\}} \cdot y).$$

Now

$$c_{(\Delta)}(d_{\Gamma\cup\{\kappa\}} \cdot y) = c_{(\Gamma)}c_{(\Delta\sim\Gamma)}(d_{\Gamma\cup\{\kappa\}} \cdot y) = c_{(\Gamma)}(d_{\Gamma\cup\{\kappa\}} \cdot y).$$

Hence, by (1) and 1.8.12(iv),

$$x \cdot \mathsf{d}_{\Gamma \cup \{\kappa\}} \leqq \mathsf{d}_{\Gamma \cup \{\kappa\}} \cdot \mathsf{c}_{(\Gamma)}(\mathsf{d}_{\Gamma \cup \{\kappa\}} \cdot y) = \mathsf{d}_{\Gamma \cup \{\kappa\}} \cdot y \leqq y.$$

This proves that $x \cdot \mathsf{d}_{\Gamma \cup \{\kappa\}}$ is a $(\Delta \sim \Gamma)$-atom.

COROLLARY 1.10.4. (i) *If* $\Gamma, \Delta \subseteq \Theta$, $\kappa \notin \Theta$, *and* x *is a* Γ*-atom, then* $\mathsf{c}_{(\Theta)} x \cdot \mathsf{d}_{\Delta \cup \{\kappa\}}$ *is a* $(\Theta \sim \Delta)$*-atom.*

(ii) *If, in particular,* x *is an atom and* $\kappa \notin \Theta$, *then* $\mathsf{c}_{(\Theta)} x \cdot \mathsf{d}_{\Theta \cup \{\kappa\}}$ *is an atom.*

THEOREM 1.10.5. (i) *If* x *is a* Γ*-atom, then either* $\Delta x = 0$ *or else* $\Delta x = \alpha \sim \Gamma$; *in case* $|\alpha \sim \Gamma| \geqq 2$ *we have* $\Delta x = 0$ *iff* $x \leqq \mathsf{c}_0^\partial \mathsf{d}_{01}$, *and* $\Delta x = \alpha \sim \Gamma$ *iff* $x \leqq \mathsf{c}_0 - \mathsf{d}_{01}$.

(ii) *If, in particular,* x *is an atom, then either* $\Delta x = 0$ *or else* $\Delta x = \alpha$; *in case* $\alpha \geqq 2$ *we have* $\Delta x = 0$ *iff* $x \leqq \mathsf{c}_0^\partial \mathsf{d}_{01}$ *and* $\Delta x = \alpha$ *iff* $x \leqq \mathsf{c}_0 - \mathsf{d}_{01}$.

(iii) *If* x *is an atom, then* $\mathsf{c}_\kappa^\partial x = 0$ *for every* $\kappa \in \Delta x$, *and* $\Delta(\mathsf{c}_{(\Gamma)} x) = \Delta x \sim \Gamma$ *for every* $\Gamma \subseteq \alpha$ *such that* $|\alpha \sim \Gamma| \geqq 2$.

PROOF. For the proof of (i) we may suppose

(1) $$|\alpha \sim \Gamma| \geqq 2$$

since, otherwise, (i) obviously holds. Assume $\Delta x \neq \alpha \sim \Gamma$, i.e., $\Delta x \subset \alpha \sim \Gamma$ ($\Delta x \cap \Gamma = 0$ by the hypothesis of (i)). Then because of (1) we may choose $\kappa, \lambda < \alpha$ such that

(2) $$\kappa, \lambda \in \alpha \sim \Gamma, \quad \kappa \neq \lambda,$$

and

(3) $$\mathsf{c}_\kappa x = x.$$

Using (3) and 1.3.2 we obtain $\mathsf{c}_\kappa(x \cdot \mathsf{d}_{\kappa\lambda}) = x$; hence $x \cdot \mathsf{d}_{\kappa\lambda} \neq 0$. Therefore, since x is a Γ-atom, and $\mathsf{d}_{\kappa\lambda} \in Cl_\Gamma \mathfrak{A}$ by (2) and 1.6.4, we have $x \leqq \mathsf{d}_{\kappa\lambda}$. This inclusion when combined with (2) and (3) gives $\Delta x = 0$ and $x \leqq \mathsf{c}_0^\partial \mathsf{d}_{01}$, by 1.3.19 (cf. Remark 1.4.5). The proof of (i) is completed by observing that, if $\Delta x = \alpha \sim \Gamma$, then $x \cdot \mathsf{c}_0 - \mathsf{d}_{01} \neq 0$ by 1.6.9(ii), and hence $x \leqq \mathsf{c}_0 - \mathsf{d}_{01}$ since $\mathsf{c}_0 - \mathsf{d}_{01} \in Cl_\Gamma \mathfrak{A}$. Taking Γ to be empty in (i) we obtain (ii).

We now turn to the proof of (iii). Assume x is an atom. Then the first part of the conclusion of (iii) is an easy consequence of 1.6.17(ii). In order to see that the second part of the conclusion holds, assume, on the contrary, that $|\alpha \sim \Gamma| \geqq 2$ and $\Delta(\mathsf{c}_{(\Gamma)} x) \subset \Delta x \sim \Gamma$. Then, *a fortiori,*

(4) $$\Delta x \neq 0,$$

and hence we know by part (ii) of the present theorem that $\Delta x = \alpha$. Therefore, $\Delta(\mathsf{c}_{(\Gamma)} x) \subset \alpha \sim \Gamma$, and since $\mathsf{c}_{(\Gamma)} x$ is a Γ-atom by 1.10.3(i), we have

$\Delta(\mathsf{c}_{(\Gamma)}x) = 0$ upon applying part (i). Furthermore, in light of the assumption $|\alpha \sim \Gamma| \geqq 2$ we can conclude from (i) that $\mathsf{c}_{(\Gamma)}x \leqq \mathsf{c}_0^\partial \mathsf{d}_{01}$. Hence $x \leqq \mathsf{c}_0^\partial \mathsf{d}_{01}$, and $\Delta x = 0$ by 1.3.19. This contradicts (4), and the proof is complete.

It will be seen in Chapter 2 that the assumption $|\alpha \sim \Gamma| \geqq 2$ is necessary both in (i) and (iii); cf. the remark after 2.1.22.

From 1.10.5(ii) we see that two very different kinds of atoms occur in cylindric algebras.

The notion of an atom is a purely Boolean concept. In the theory of CA's a special role is played by certain atoms, called rectangular atoms, which are distinguished by means of a condition expressed in terms of cylindrifications. This condition may be applied to non-atomic elements as well and leads to the following

DEFINITION 1.10.6. *An element x is said to be* **rectangular** *iff* $\mathsf{c}_{(\Gamma)}x \cdot \mathsf{c}_{(\Delta)}x = \mathsf{c}_{(\Gamma \cap \Delta)}x$ *for all Γ and Δ.*

Clearly all zero-dimensional elements are rectangular. In a non-discrete CA the element $\mathsf{d}_{\kappa\lambda}$ with $\kappa \neq \lambda$ is not rectangular, as is easily seen from 1.10.6 by letting $\Gamma = \{\kappa\}$ and $\Delta = \{\lambda\}$. In a cylindric set algebra of dimension 2 an element x is rectangular iff it is a "generalized rectangle" in the sense that whenever $\langle u_0, u_1\rangle, \langle v_0, v_1\rangle \in x$ we also have $\langle u_0, v_1\rangle, \langle v_0, u_1\rangle \in x$ (see Fig. 1.10.7). A generalization to higher finite dimensions is given in 1.10.12 below; from this it results that a rectangular element of a finite-dimensional cylindric set algebra can be regarded as the Cartesian product of its projections on each coordinate axis.

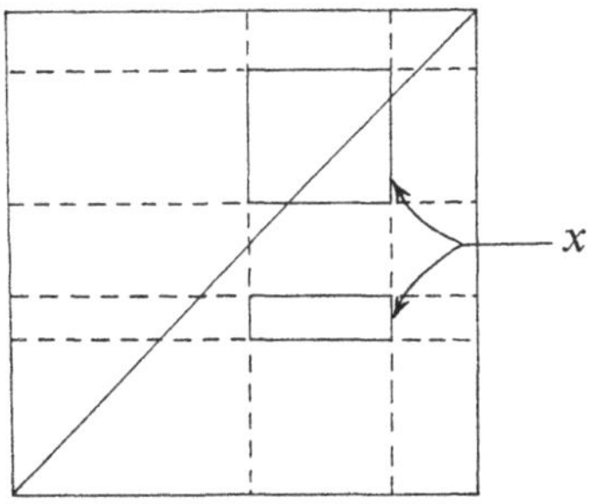

Fig. 1.10.7

COROLLARY 1.10.8. *If x is a rectangular element, then so is $\mathsf{c}_{(\Gamma)}x$.*

THEOREM 1.10.9. *The following two conditions are equivalent:*

(i) *x is rectangular;*

(ii) *for every Γ and for any κ, λ such that $\kappa \neq \lambda$, we have $\mathsf{c}_{(\Gamma\cup\{\kappa\})}x\cdot\mathsf{c}_{(\Gamma\cup\{\lambda\})}x = \mathsf{c}_{(\Gamma)}x$.*

PROOF. The necessity of (ii) is clear from 1.10.6. Now assume that (ii) holds. We shall show that $\mathsf{c}_{(\Gamma)}x\cdot\mathsf{c}_{(\Delta)}x = \mathsf{c}_{(\Gamma\cap\Delta)}x$ by induction on the number of elements of $\Theta = (\Gamma \sim \Delta)\cup(\Delta \sim \Gamma)$.

Indeed, if Θ is empty, then $\Gamma = \Delta$, and there is nothing to prove. Therefore, assume inductively that Θ is non-empty. If either of the sets Γ or Δ is included in the other, the desired result is immediate. Hence we can assume that there are elements $\kappa \in \Gamma \sim \Delta$ and $\lambda \in \Delta \sim \Gamma$. Now $(\Gamma \sim (\Delta \cup \{\kappa\}))\cup((\Delta\cup\{\kappa\}) \sim \Gamma) = \Theta \sim \{\kappa\}$, and thus, by the induction hypothesis, $\mathsf{c}_{(\Gamma)}x\cdot\mathsf{c}_{(\Delta\cup\{\kappa\})}x = \mathsf{c}_{((\Gamma\cap\Delta)\cup\{\kappa\})}x$. Similarly, $\mathsf{c}_{(\Gamma\cup\{\lambda\})}x\cdot\mathsf{c}_{(\Delta)}x = \mathsf{c}_{((\Gamma\cap\Delta)\cup\{\lambda\})}x$. Hence

$$\begin{aligned}\mathsf{c}_{(\Gamma)}x\cdot\mathsf{c}_{(\Delta)}x &\leqq \mathsf{c}_{((\Gamma\cap\Delta)\cup\{\kappa\})}x\cdot\mathsf{c}_{((\Gamma\cap\Delta)\cup\{\lambda\})}x\\ &= \mathsf{c}_{(\Gamma\cap\Delta)}x \qquad\qquad \text{by (ii)}\\ &\leqq \mathsf{c}_{(\Gamma)}x\cdot\mathsf{c}_{(\Delta)}x,\end{aligned}$$

and the inductive proof is complete.

THEOREM 1.10.10. *If x is a rectangular element and $0 < |I| < \omega$, then*

$$\prod_{i\in I}\mathsf{c}_{(\Gamma_i)}x = \mathsf{c}_{(\Delta)}x$$

where $\Delta = \bigcap_{i\in I}\Gamma_i$.

PROOF: by induction on the number of elements of I, using 1.10.6.

THEOREM 1.10.11. *The following two conditions are equivalent:*

(i) *x is rectangular,*

(ii) *$x = \prod_{\lambda\in\Gamma}\mathsf{c}_{(\Gamma\sim\{\lambda\})}x$ for every non-empty $\Gamma \subseteq \alpha$.*

If $0 < |\Delta x| < \omega$, (i) is also equivalent to

(iii) *$x = \prod_{\lambda\in\Delta x}\mathsf{c}_{(\Delta x\sim\{\lambda\})}x$;*

hence, if $0 < \alpha < \omega$, (i) is equivalent to

(iv) *$x = \prod_{\lambda<\alpha}\mathsf{c}_{(\alpha\sim\{\lambda\})}x$.*

PROOF. The implication from (i) to (ii) is a special case of 1.10.10. To show the implication in the opposite direction consider any Δ and Θ and set $\Gamma = \Delta\cup\Theta$. Then using (i) and ($C_2$) we get

$$\begin{aligned}\mathsf{c}_{(\Delta)}x &= \mathsf{c}_{(\Delta)}\prod_{\lambda\in\Gamma}\mathsf{c}_{(\Gamma\sim\{\lambda\})}x\\ &= \prod_{\lambda\in\Gamma\sim\Delta}\mathsf{c}_{(\Gamma\sim\{\lambda\})}x\cdot\mathsf{c}_{(\Delta)}\prod_{\lambda\in\Delta}\mathsf{c}_{(\Gamma\sim\{\lambda\})}x\\ &= \prod_{\lambda\in\Gamma\sim\Delta}\mathsf{c}_{(\Gamma\sim\{\lambda\})}x\cdot\prod_{\lambda\in\Delta}\mathsf{c}_{(\Gamma)}x\end{aligned}$$

and hence

(1) $$\mathsf{c}_{(\Delta)}x = \prod_{\lambda\in\Gamma\sim\Delta}\mathsf{c}_{(\Gamma\sim\{\lambda\})}x.$$

Similarly we obtain

(2) $$\mathsf{c}_{(\Theta)}x = \prod_{\lambda\in\Gamma\sim\Theta}\mathsf{c}_{(\Gamma\sim\{\lambda\})}x \text{ and } \mathsf{c}_{(\Delta\cap\Theta)}x = \prod_{\lambda\in\Gamma\sim(\Delta\cap\Theta)}\mathsf{c}_{(\Gamma\sim\{\lambda\})}x.$$

(1) and (2) immediately imply

$$\mathsf{c}_{(\Delta)}x\cdot\mathsf{c}_{(\Theta)}x = \mathsf{c}_{(\Delta\cap\Theta)}x,$$

and thus x is rectangular by 1.10.6.

Now we assume Δx is finite and non-empty. Trivially (ii) implies (iii). To show, conversely, that (iii) implies (ii) we take any non-empty $\Gamma\subseteq\alpha$ and compute:

$$\begin{aligned}\prod_{\lambda\in\Gamma}\mathsf{c}_{(\Gamma\sim\{\lambda\})}x &= \prod_{\lambda\in\Gamma}\mathsf{c}_{((\Gamma\cap\Delta x)\sim\{\lambda\})}x && \text{by 1.7.7(ii)}\\ &= \prod_{\lambda\in\Delta x}\mathsf{c}_{((\Gamma\cap\Delta x)\sim\{\lambda\})}x && \text{by } (C_2)\text{, 1.7.3}\\ &\leqq \prod_{\lambda\in\Delta x}\mathsf{c}_{(\Delta x\sim\{\lambda\})}x && \text{by } (C_2)\text{, 1.7.3, 1.7.5}\\ &= x && \text{by (iii).}\end{aligned}$$

On the other hand, $x\leqq\prod_{\lambda\in\Gamma}\mathsf{c}_{(\Gamma\sim\{\lambda\})}x$ by (C_2) and 1.7.5. Since Γ was an arbitrary (finite) subset of α, we obtain (iii).

Under the hypothesis $0<\alpha<\omega$, (iv) is a trivial consequence of (ii) and is easily seen to imply (ii). This completes the proof.

REMARK 1.10.12. By applying 1.10.11 to cylindric set algebras of finite dimension α we see that the rectangular sets X in such algebras can be characterized as "generalized parallelepipeds" (or "generalized boxes"), i.e., as sets which are Cartesian products of their projections on all coordinate axes:

$$X = \mathsf{P}_{\kappa<\alpha}\mathsf{pj}_\kappa{}^*X.$$

The next theorem is rather special and has a restricted scope of application.

THEOREM 1.10.13. *Let $\alpha<\omega$ and let $\mathfrak{A}$ be a CA_α such that for every $x\neq 0$ we have $\mathsf{c}_{(\alpha)}x=1$.*

(i) *If $|\Gamma|\geqq 2$, z is rectangular, and $0\neq z\leqq\mathsf{d}_\Gamma$, then $\mathsf{c}_{(\alpha\sim\Gamma)}z$ is a $(\alpha\sim\Gamma)$-atom.*

(ii) *If in particular $\alpha\geqq 2$, z is rectangular, and $0\neq z\leqq\mathsf{d}_{\kappa\lambda}$ for all $\kappa,\lambda<\alpha$, then z is an atom.*

PROOF. Let $\Delta=\alpha\sim\Gamma$. Assume that $\mathsf{c}_{(\Delta)}z\cdot y\neq 0$ for some $y\in A$ such that $\mathsf{c}_{(\Delta)}y=y$. We wish to show that $\mathsf{c}_{(\Delta)}z\leqq y$. Since $|\Gamma|\geqq 2$, there are $\Theta,\Phi\neq\Gamma$

such that $\Theta \cup \Phi = \Gamma$ and $\Theta \cap \Phi = 0$. Then

$$\begin{aligned}
\mathsf{c}_{(\Theta)}(\mathsf{c}_{(\Delta)}z \cdot -y) &= \mathsf{c}_{(\alpha)}(\mathsf{c}_{(\Delta)}z \cdot y) \cdot \mathsf{c}_{(\Theta)}(\mathsf{c}_{(\Delta)}z \cdot -y) && \text{by hypothesis of theorem} \\
&= \mathsf{c}_{(\Theta)}[\mathsf{c}_{(\Phi)}(\mathsf{c}_{(\Delta)}z \cdot y) \cdot \mathsf{c}_{(\Theta)}(\mathsf{c}_{(\Delta)} \cdot -y)] && \text{by } (C_3), 1.7.5 \\
&= \mathsf{c}_{(\Theta)}[\mathsf{c}_{(\Phi)}(\mathsf{c}_{(\Delta)}z \cdot \mathsf{d}_\Gamma \cdot y) \cdot \mathsf{c}_{(\Theta)}(\mathsf{c}_{(\Delta)}z \cdot \mathsf{d}_\Gamma \cdot -y)] \\
& && \text{by hypothesis of (i)} \\
&= \mathsf{c}_{(\Theta)}[\mathsf{c}_{(\Phi)}(\mathsf{c}_{(\Delta)}z \cdot \mathsf{d}_\Gamma) \cdot \mathsf{c}_{(\Phi)}(\mathsf{d}_\Gamma \cdot y) \cdot \mathsf{c}_{(\Theta)}(\mathsf{c}_{(\Delta)}z \cdot \mathsf{d}_\Gamma) \cdot \mathsf{c}_{(\Theta)}(\mathsf{d}_\Gamma \cdot -y)] \\
& && \text{by 1.8.12(ii)} \\
&= \mathsf{c}_{(\Theta)}[\mathsf{c}_{(\Delta \cup \Phi)}z \cdot \mathsf{c}_{(\Delta \cup \Theta)}z \cdot \mathsf{c}_{(\Phi)}(\mathsf{d}_\Gamma \cdot y) \cdot \mathsf{c}_{(\Theta)}(\mathsf{d}_\Gamma \cdot -y)] \\
& && \text{by hypothesis of (i)} \\
&= \mathsf{c}_{(\Theta)}[\mathsf{c}_{(\Delta)}z \cdot \mathsf{d}_\Gamma \cdot \mathsf{c}_{(\Phi)}(\mathsf{d}_\Gamma \cdot y) \cdot \mathsf{c}_{(\Theta)}(\mathsf{d}_\Gamma \cdot -y)] \\
& && \text{by hypothesis of (i), 1.10.6} \\
&= \mathsf{c}_{(\Theta)}(\mathsf{c}_{(\Delta)}z \cdot \mathsf{d}_\Gamma \cdot y \cdot \mathsf{d}_\Gamma \cdot -y) && \text{by 1.8.12(iv)} \\
&= 0.
\end{aligned}$$

Thus $\mathsf{c}_{(\Delta)}z \cdot -y = 0$, i.e., $\mathsf{c}_{(\Delta)}z \leqq y$ as desired. Hence we obtain the conclusion of (i). Since (ii) is a particular case of (i), the proof is complete.

The premisses of 1.10.13 are rather restrictive, but it can be shown by means of simple examples that none of them can be dropped. This applies in the first place to the premiss: $\mathsf{c}_{(\alpha)}x = 1$ *for all* $x \neq 0$. It can easily be shown that an equivalent form of this condition (for $\alpha < \omega$) is: the only two zero-dimensional elements are 0 and 1; cf. 1.7.10(ii). It will be seen from the discussion in Section 2.4 that the condition in the second form characterizes those CA_α's which are directly indecomposable in the sense of 0.3.25. As regards the premiss $\alpha < \omega$, a generalization of 1.10.13 to infinite-dimensional algebras is known, but this requires imposing on $\mathfrak{A}$ additional conditions which deprive the result of much of its interest. The premisses $|\Gamma| \geqq 2$ in 1.10.13(i) and $\alpha \geqq 2$ in 1.10.13(ii) prove also to be essential.

1.11. LOCALLY FINITE-DIMENSIONAL AND DIMENSION-COMPLEMENTED CYLINDRIC ALGEBRAS

In this section we discuss elementary properties of two important classes of cylindric algebras — the *locally finite-dimensional* CA_α's and the *dimension-complemented* CA_α's. Both these notions are of interest only when applied to CA_α's with $\alpha \geqq \omega$ since, as is seen from 1.11.3 below, they prove to be trivial in case $\alpha < \omega$.

The metalogical aspect of the theory of CA's leads in a most natural way to the first of these notions; for some elaboration of this remark see 1.11.2. The second notion is more general than the first (disregarding the trivial case $\alpha < \omega$). It turns out, however, that most of the interesting results concerning the first notion easily extend to the second. This is the reason why in the present section we shall deal primarily with dimension-complemented algebras. However, the intuitive background of the second notion is less clear and its connection with metalogical representations of CA's is much looser than in the case of the first notion. Again we shall have something to add to this point in 1.11.2.[1]

DEFINITION 1.11.1. (i) $\mathfrak{A}$ *is called a* ***locally finite-dimensional,*** *or simply a* ***locally finite,*** *cylindric algebra of dimension* α, *in symbols* $\mathfrak{A} \in \mathsf{Lf}_\alpha$, *if* $|\Delta x| < \omega$ *for every* $x \in A$.

(ii) $\mathfrak{A}$ *is a* ***dimension-complemented*** *cylindric algebra of dimension* α, *in symbols* $\mathfrak{A} \in \mathsf{Dc}_\alpha$, *if* $\Delta x \neq \alpha$ *for every* $x \in A$.

Examples of dimension-complemented cylindric algebras which are not locally finite are easily constructed. They can be found, e.g., among subalgebras of any infinite-dimensional full cylindric set algebra whose base contains at least two different elements.

REMARKS 1.11.2. In any language Λ of predicate logic in which the sequence of variables is of length α all the cylindric algebras of formulas are easily

[1]) The class of locally finite-dimensional algebras was first defined in Tarski-Thompson [52]. The class of dimension-complemented algebras was first introduced by a different though equivalent definition in Henkin [55a], pp. 90 f.; cf. here footnote 1 on p. 233.

seen to be Lf_α's. This is a direct consequence of the trivial fact that every formula in Λ is a finite string of symbols and hence contains only finitely many variables. In case $\alpha \geqq \omega$ it turns out that, conversely, every Lf_α is isomorphic with an algebra of formulas in some such language Λ. Assume for the moment that the construction of cylindric algebras of formulas described in 1.1.11 (or, more specifically, the definition of the relation $\equiv_\Sigma$ underlying this construction) is based upon the syntactical notion of derivability rather that the semantical notion of consequence; cf. the Preliminaries. In this case the converse theorem is pretty trivial and its proof is straightforward; loosely speaking, we simply check that the postulates characterizing CA's form an adequate algebraic transcription of axioms of predicate logic. Moreover, if we extend literally an appropriate version of the syntactical notion of derivability (e.g., the notion defined in Monk [65a]) to languages Λ with $\alpha < \omega$, then the converse theorem also extends to CA_α's with $\alpha < \omega$ and implies that every CA of finite dimension is isomorphic with an algebra of formulas.

Actually, however, our construction of cylindric algebras of formulas is implicitly based upon the semantical notion of consequence. As a result, the converse theorem acquires a somewhat deeper character and its proof depends, explicitly or implicitly (according as to the method of proof chosen), on the completeness theorem for predicate logic; such a proof will be given in detailed form in Part II of our work. Since, moreover, the completeness theorem in its usual form fails for languages with finitely many variables, the converse theorem applies only to Lf's of infinite dimension.

In a certain portion of Part II we shall discuss a comprehensive class of formal languages which, in addition to languages Λ described in the Preliminaries, contains also analogously constructed languages with infinitely long expressions. In such infinitary languages we can form disjunctions and conjunctions of any sequence of formulas whose length is smaller than some fixed infinite (regular) cardinal $\kappa > \omega$; relation symbols and operation symbols may be of finite or infinite rank, but at any rate of rank $< \kappa$; the length α of the sequence of variables is usually assumed to be at least equal to κ. Our construction of cylindric algebras of formulas easily extends to these infinitary languages. However, in opposition to the case of finitary languages, the resulting algebras are no longer locally finite-dimensional. They still prove to be dimension-complemented. In fact, they all belong to a certain class of CA_α's which is intermediate between Lf_α and Dc_α and is characterized by the following condition: the cardinality of the dimension set of every element x in such an algebra is smaller than the cardinality of the dimension α of the whole algebra.

Since all cylindric algebras of formulas of infinite dimension α are representable in the sense of 1.1.13, the same applies to all Lf_α's. In Part II it will be shown that all Dc_α's are also representable. Actually, it will be proved there that the class of representable CA_α's (with $\alpha \geqq \omega$) is algebraically generated in the sense of 0.3.14 by either of the classes Lf_α or Dc_α.

THEOREM 1.11.3. (i) *If* $\alpha \geqq \omega$, *then* $\mathsf{Lf}_\alpha \subseteq \mathsf{Dc}_\alpha$.
(ii) *If* $\alpha < \omega$, *then* $\mathsf{Lf}_\alpha = \mathsf{CA}_\alpha$.
(iii) *If* $0 < \alpha < \omega$, *then* $\mathfrak{A}$ *is dimension-complemented iff it is discrete.*
(iv) *Every discrete* CA_α *is locally finite.*
PROOF: (i), (ii), and (iv) by the definitions involved; (iii) by 1.6.11.

THEOREM 1.11.4. *For* $\alpha \geqq \omega$ *the following conditions are equivalent*:
(i) $\mathfrak{A} \in \mathsf{Dc}_\alpha$;
(ii) $\alpha \sim \bigcup_{x \in X} \Delta x$ *is non-empty for every finite* $X \subseteq A$;
(iii) $\alpha \sim \bigcup_{x \in X} \Delta x$ *is infinite for every finite* $X \subseteq A$.[1]
PROOF. Obviously (iii) implies (ii) and (ii) implies (i). It remains to show that (i) implies (iii); to this end assume $\mathfrak{A} \in \mathsf{Dc}_\alpha$. Then, by 1.6.11,

(1) $$\alpha \sim \Delta x \text{ is infinite for each } x \in A.$$

We shall prove (iii) by induction on $|X|$. The case $|X| = 0$ is trivial, and the case $|X| = 1$ is covered by (1). Now assume inductively that $|X| > 1$, and, to derive a contradiction, that $\alpha \sim \bigcup_{x \in X} \Delta x$ is finite. Choose $y, z \in X$, $y \neq z$. By (1) there exist κ, λ such that $\kappa \notin \Delta x$, $\lambda \notin \Delta z$, and $\kappa \neq \lambda$. Then

$$\alpha \sim [(\bigcup_{x \in X \sim \{y,z\}} \Delta x) \cup \Delta(y \cdot \mathsf{d}_{\kappa\lambda} + z \cdot - \mathsf{d}_{\kappa\lambda})]$$

is finite by 1.6.12. This contradicts the induction hypothesis. Hence (iii) holds.

REMARK 1.11.5. Theorem 1.11.4 suggests a generalization of the notion of dimension-complementedness. A CA_α $\mathfrak{A}$ may be called κ-*dimension-complemented* (κ any cardinal) if $\alpha \neq \bigcup_{x \in X} \Delta x$ whenever $X \subseteq A$ and $|X| < \kappa$. By 1.11.4, the class of κ-dimension-complemented CA's for $2 \leqq \kappa \leqq \omega$ coincides with that of dimension-complemented CA's in the sense of 1.11.1. On the other hand, one can easily show by means of examples that the class of κ- and λ-dimension-complemented CA_α's are distinct whenever $\kappa < \lambda \leqq |\alpha|$ and $\lambda > \omega$. The class of κ-dimension-complemented CA's contains in particular the CA's of formulas in infinitary languages mentioned in 1.11.2.

[1]) Henkin [55a] originally defined the notion of a Dc_α by means of the condition 1.11.4(iii). Later Tarski showed that this condition was equivalent to 1.11.4(ii). The final simplification, leading to Definition 1.11.1(ii) and Theorem 1.11.4, is due to Jack Silver.

THEOREM 1.11.6. *For every* $\mathfrak{A} \in \mathsf{Dc}_\alpha$ *and* $\kappa < \alpha$ *we have:*

(i) $$\mathsf{c}_\kappa x = \sum_{\lambda < \alpha} \mathsf{s}^\kappa_\lambda x$$

and, more generally,

$$\mathsf{c}_\kappa x = \sum_{\lambda \in \Gamma} \mathsf{s}^\kappa_\lambda x$$

for any set Γ *such that* $\alpha \sim \Delta x \subseteq \Gamma \subseteq \alpha$;

(ii) *if* $\alpha \geqq 2$ *and* $\kappa < \alpha$, *then*

$$\sum_{\kappa \neq \lambda < \alpha} \mathsf{d}_{\kappa\lambda} = \sum_{\mu < \lambda < \alpha} \mathsf{d}_{\mu\lambda} = 1$$

and

$$\prod_{\lambda < \alpha} \mathsf{d}_{\kappa\lambda} = \prod_{\mu, \lambda < \alpha} \mathsf{d}_{\mu\lambda} = \mathsf{c}^\partial_0 \mathsf{d}_{01}.$$

PROOF. (i) In view of 1.11.3(iii) and the remark immediately following 1.5.1 we can assume that $\alpha \geqq \omega$. Clearly $\mathsf{s}^\kappa_\lambda x \leqq \mathsf{c}_\kappa x$ for every $\lambda \in \Gamma$. Now assume that $\mathsf{s}^\kappa_\lambda x \leqq y$ for all $\lambda \in \Gamma$. By 1.11.4 there is a $\lambda \in \alpha \sim (\Delta x \cup \Delta y)$; hence $\lambda \in \Gamma$, $\mathsf{c}_\lambda x = x$, and $\mathsf{c}_\lambda y = y$. Then $\mathsf{c}_\lambda \mathsf{s}^\kappa_\lambda x \leqq y$ by 1.2.9, and therefore $\mathsf{c}_\kappa x \leqq y$ by 1.5.8(i) and 1.5.9(i). This completes the proof of (i).

(ii) We first choose $\kappa, \mu < \alpha$ with $\kappa \neq \mu$. To obtain the formula for $\sum$, we respectively replace κ, x, and Γ in (i) by μ, $\mathsf{d}_{\kappa\mu}$, and $\alpha \sim \{\kappa\}$; to obtain the formula for $\prod$, we replace κ, x, and Γ by μ, $-\mathsf{d}_{\kappa\mu}$, and α, and then use Remark 1.4.5.

Thus in a Dc_α a cylindrification can be expressed in terms of the operations $\mathsf{s}^\kappa_\lambda$. We have already seen in 1.5.7 that the diagonal elements can also be so expressed. This paves the way for a possible new treatment of Dc_α's. The Dc_α's in this new sense would be algebraic structures

$$\mathfrak{A}^* = \langle A, +, \cdot, -, 0, 1, \mathsf{s}^\kappa_\lambda \rangle_{\kappa, \lambda < \alpha}$$

each correlated with a Dc_α in the old sense, with $A, +, \ldots, 1$ taken from $\mathfrak{A}$ and $\mathsf{s}^\kappa_\lambda$ defined as in 1.5.1. The relation between the classes of Dc_α's in the old and the new sense can be described as a variant of definitional equivalence discussed in 0.1.6. This would not be, however, the ordinary first-order definitional equivalence. The problem naturally arises of providing a simple and elegant characterization of Dc_α's in the new sense. A solution of this problem can be found in Preller [70].[1)]

In Section 1.5 we came across various simple equations which do not hold in every CA_α but which do hold in all the special CA_α's discussed in Section 1.1.

[1)] Earlier work in this direction was done in Galler [57] and LeBlanc [61]. See also Preller [68] where algebraic structures related to $\mathfrak{A}^*$ but of a more general nature are studied.

Consequently, these equations hold in all representable CA's and hence, in particular, in all Dc_α's; cf. the remarks following 1.5.14 and the concluding remark of 1.11.2. Actually, we shall see later (cf. 2.6.53) that all the conditions of the form "a given equation holds in all algebras of the class K" are equivalent if we take for K any one of the following classes: Lf_α, Dc_α, the class of all cylindric set algebras of dimension α, and the class of all representable CA_α's, where α is any given infinite ordinal.

The next theorem provides a new example of an equation (given in the form of an inclusion) which will be shown in 2.6.42 not to hold in arbitrary CA_α's; we shall show here by a simple direct argument that it holds in every Dc_α.

THEOREM 1.11.7. *If* $\mathfrak{A} \in \mathsf{Dc}_\alpha$, $\kappa, \lambda < \alpha$, $\kappa \neq \lambda$, *and* $x, y \in A$, *then* $\mathsf{c}_\kappa(x \cdot y \cdot \mathsf{c}_\lambda(x \cdot -y)) \leqq \mathsf{c}_\lambda(\mathsf{c}_\kappa x \cdot -\mathsf{d}_{\kappa\lambda})$.

PROOF. By 1.11.4 choose $\mu \in \alpha \sim (\Delta x \cup \Delta y \cup \{\kappa, \lambda\})$. Now by 1.3.9 we have $x \cdot y \cdot \mathsf{c}_\lambda(x \cdot -y \cdot \mathsf{d}_{\mu\lambda}) \cdot \mathsf{d}_{\mu\lambda} = 0$. Hence

$$x \cdot y \cdot \mathsf{c}_\lambda(x \cdot -y \cdot \mathsf{d}_{\mu\lambda}) \leqq x \cdot -\mathsf{d}_{\mu\lambda} \leqq \mathsf{c}_\lambda(\mathsf{c}_\kappa x \cdot -\mathsf{d}_{\mu\lambda}).$$

Also, $x \cdot y \cdot \mathsf{c}_\lambda(x \cdot -y \cdot -\mathsf{d}_{\mu\lambda}) \leqq \mathsf{c}_\lambda(\mathsf{c}_\kappa x \cdot -\mathsf{d}_{\mu\lambda})$, so $x \cdot y \cdot \mathsf{c}_\lambda(x \cdot -y) \leqq \mathsf{c}_\lambda(\mathsf{c}_\kappa x \cdot -\mathsf{d}_{\mu\lambda})$. By 1.2.9 it follows that $\mathsf{c}_\kappa(x \cdot y \cdot \mathsf{c}_\lambda(x \cdot -y)) \leqq \mathsf{c}_\lambda(\mathsf{c}_\kappa x \cdot -\mathsf{d}_{\mu\lambda})$. Applying s^μ_κ to both sides of this inequality, the desired result follows.

THEOREM 1.11.8. *Suppose* $\mathfrak{A} \in \mathsf{Dc}_\alpha$.

(i) *If* $x \in At\mathfrak{A}$, *then* $\Delta x = 0$.

(ii) *If* $\alpha \geqq 2$ *and* $\mathsf{c}^\partial_0 \mathsf{d}_{01} = 0$, *then* $\mathfrak{A}$ *is atomless.*

PROOF: by 1.10.5(ii) and 1.11.1(ii).

In the remainder of this section we shall be concerned with a generalization of the operation $\mathsf{s}^\kappa_\lambda$ (and also ${}_\mu\mathsf{s}(\kappa, \lambda)$) in certain CA_α's $\mathfrak{A}$. The new operations will be correlated not only with a pair of ordinals $\kappa, \lambda < \alpha$, but with each pair of finite sequences $\langle \kappa_0, \ldots, \kappa_{\nu-1} \rangle$, $\langle \lambda_0, \ldots, \lambda_{\nu-1} \rangle$ of ordinals $< \alpha$. Just as the operation $\mathsf{s}^\kappa_\lambda$ in its metalogical interpretation corresponds to the substitution of a variable $\mathbf{v}_\lambda$ for $\mathbf{v}_\kappa$ in a formula φ, so the generalized operation corresponds to the simultaneous substitution of variables $\mathbf{v}_{\lambda_0}, \ldots, \mathbf{v}_{\lambda_{\nu-1}}$ for $\mathbf{v}_{\kappa_0}, \ldots, \mathbf{v}_{\kappa_{\nu-1}}$; cf. the remarks following 1.5.12. We shall refer to the generalized operation as the substitution operator, and we shall denote it by s. (There will be no conflict with the use of s in the expressions $\mathsf{s}^\kappa_\lambda$ and ${}_\mu\mathsf{s}(\kappa, \lambda)$ since the letter s will now appear in a different symbolic context.) It proves more convenient to relativize the substitution operator to an arbitrary finite transformation of α rather than to two finite sequences of ordinals $< \alpha$. Thus s will correlate with every finite transformation τ of α a function s_τ from A to A

(which, in fact, will be a Boolean endomorphism of $\mathfrak{A}$); s_τ generalizes the operation $\mathsf{s}_\lambda^\kappa$ in the sense that if τ is the replacement $[\kappa/\lambda]$, then s_τ coincides with $\mathsf{s}_\lambda^\kappa$. The difference between these two approaches may seem to be rather insignificant and purely formal since, as we recall from the Preliminaries, every finite transformation τ of α has a unique canonical representation $\tau = [\xi_0/\eta_0, \ldots, \xi_{\nu-1}/\eta_{\nu-1}]$ in terms of two finite sequences of ordinals. However, the relativization of the substitution operator to finite transformations suggests the idea of further generalizing this notion, and indeed of extending it to arbitrary transformations of α. We shall see that to some extent this idea can be materialized.[1]

In the whole discussion the cylindric algebras involved will be assumed to be Dc_α's, and the part of the discussion relating to substitutions for arbitrary (not necessarily finite) transformations will be restricted to Lf_α's. It is known that these restrictions are essential. In particular, it will be seen from results in Part II of this work that it is impossible to define a substitution operator for arbitrary CA's which would agree in its basic properties with the notions we will introduce for Dc's and Lf's.

DEFINITION 1.11.9. *Let $\mathfrak{A}$ be a Dc_α with $\alpha \geqq \omega$. By the* ***substitution operator*** s *(or $\mathsf{s}^{(\mathfrak{A})}$) we understand the function which correlates with every finite transformation τ of α a unary operation s_τ on A determined by the following stipulation:*

If $\tau = [\mu_0/\nu_0, \ldots, \mu_{\kappa-1}/\nu_{\kappa-1}]$ is the canonical representation of τ ($\mu, \nu \in {}^\kappa\alpha$, $\mu_0 < \ldots < \mu_{\kappa-1}$), if x is any element of A, and if $\pi_0, \ldots, \pi_{\kappa-1}$ are in this order the first κ ordinals in $\alpha \sim (\Delta x \cup Rg\mu \cup Rg\nu)$, then

$$\mathsf{s}_\tau x = \mathsf{s}_{\nu_0}^{\pi_0} \ldots \mathsf{s}_{\nu_{\kappa-1}}^{\pi_{\kappa-1}} \mathsf{s}_{\pi_0}^{\mu_0} \ldots \mathsf{s}_{\pi_{\kappa-1}}^{\mu_{\kappa-1}} x.$$ [2]

REMARK 1.11.10. Definition 1.11.9 can be applied in particular to any CA of formulas in a language Λ with variables arranged in a transfinite sequence of length α. In fact, we know that every such algebra $\mathfrak{A}$ is an Lf and hence also a Dc. The elements of $\mathfrak{A}$ are equivalence classes correlated with formulas

[1] The idea of relativizing the substitution operator to arbitrary transformations of α originates with Halmos, and underlies his development of the theory of polyadic algebras; cf. Halmos [57]. In this theory the substitution operator is not defined in terms of cylindrifications and diagonal elements, but is treated as a primitive notion, and its fundamental properties are simply postulated. Details will be found in Part II.

[2] The theory of the substitution operator based upon Definition 1.11.9 was developed early in the study of cylindric algebras; it provided an apparatus used in the proof of the representation theorem for locally finite algebras. The representation theorem was first announced in Tarski [52], but the theory of the substitution operator was not published at that time. A few years later this theory was independently developed by Galler [57].

in Λ. Let $\tau = [\mu_0/\nu_0, \dots, \mu_{\kappa-1}/\nu_{\kappa-1}]$ be a finite transformation of α, φ be any formula in Λ, and ψ be the formula obtained by simultaneously substituting the variables $\mathbf{v}_{\nu_0}, \dots, \mathbf{v}_{\nu_{\kappa-1}}$ for $\mathbf{v}_{\mu_0}, \dots, \mathbf{v}_{\mu_{\kappa-1}}$, respectively; cf. the remarks after 1.5.12. From Definition 1.11.9 it follows then that, if x is the equivalence class of φ, then $\mathsf{s}_\tau x$ is the equivalence class of ψ.

The intuitive meaning of the operator s as applied to cylindric set algebras is less clear. If $\mathfrak{A}$ is a dimension-complemented cylindric set algebra, say with base U and dimension $\alpha \geqq \omega$, if τ is a finite transformation of α, and if X is any member of A, then 1.11.9 yields

$$\mathsf{s}_\tau X = \{x : x \in {}^\alpha U \text{ and } x \circ \tau \in X\}.$$

The following theorem expresses some basic properties of the operator s. It will be seen that the proof of this theorem, which is based directly on Definition 1.11.9, is somewhat complicated in details. The theorem could be easily derived as a corollary from a general result in the theory of semigroups established in Jónsson [62]; however, we do not wish to assume here knowledge of Jónsson's paper. Our further discussion will be based exclusively on Theorem 1.11.11 and will not revert to Definition 1.11.9.

THEOREM 1.11.11. *Let $\mathfrak{A}$ be a Dc_α with $\alpha \geqq \omega$. We then have:*

(i) *For all $\kappa, \lambda < \alpha$, $\mathsf{s}_{[\kappa/\lambda]} = \mathsf{s}^\kappa_\lambda$.*

(ii) *For all $\kappa, \lambda, \mu < \alpha$ with $\kappa \neq \lambda$, if $\mu \notin \Delta x$, then $\mathsf{s}_{[\kappa/\lambda, \lambda/\kappa]} x = {}_\mu\mathsf{s}(\kappa, \lambda)x$.*

(iii) *For every finite transformation τ of α and every $x, y \in A$, $\mathsf{s}_\tau(x+y) = \mathsf{s}_\tau x + \mathsf{s}_\tau y$, $\mathsf{s}_\tau(x \cdot y) = \mathsf{s}_\tau x \cdot \mathsf{s}_\tau y$, $\mathsf{s}_\tau - x = -\mathsf{s}_\tau x$, and also $\mathsf{s}_\tau 0 = 0$ and $\mathsf{s}_\tau 1 = 1$; in other words, s_τ is an endomorphism of $\mathfrak{B}\mathfrak{l}\mathfrak{A}$.*

(iv) *If σ and τ are any two finite transformations of α, then $\mathsf{s}_{\sigma\circ\tau} = \mathsf{s}_\sigma \circ \mathsf{s}_\tau$.*

PROOF. (i) follows from 1.5.11(i), and (ii) follows from 1.5.10(iii), 1.5.11(i), and 1.5.15.

For the proof of (iii) and (iv) we will need some lemmas. In these lemmas, and throughout the remaining part of the proof, it will be convenient to represent the function $\mathsf{s}^{\mu_{\pi 0}}_{\nu_{\pi 0}} \circ \mathsf{s}^{\mu_{\pi 1}}_{\nu_{\pi 1}} \circ \dots \circ \mathsf{s}^{\mu_{\pi(\kappa-1)}}_{\nu_{\pi(\kappa-1)}}$ simply as s^μ_ν, whenever μ and ν are functions from the same finite subset of ω into α and π is the unique strictly increasing sequence such that $Rg\pi = Do\mu = Do\nu$.

(1) If $\kappa < \omega$, if $\mu, \nu, \pi, \rho \in {}^\kappa\alpha$, if μ, π, ρ are one-one, and if $(Rg\pi \cup Rg\rho) \cap (\Delta x \cup Rg\mu \cup Rg\nu) = 0$, then

$$s^\pi_\nu s^\mu_\pi x = s^\rho_\nu s^\mu_\rho x.$$

Indeed, (1) is easily established if $Rg\pi \cap Rg\rho = 0$ by making use of 1.5.10(iii) and 1.5.11(ii). To prove (1) in general, note that there is a $\xi \in {}^\kappa\alpha$ with ξ one-

one, $Rg\xi \cap (\Delta x \cup Rg\mu \cup Rg\nu) = 0$, and $Rg\xi \cap (Rg\pi \cup Rg\rho) = 0$, and then use the special case just established.

(2) Let τ be a finite transformation of α. If $\kappa < \omega$, if $\mu, \pi \in {}^{\kappa}\alpha$, if μ and π are one-one, if $\{\lambda : \tau\lambda \neq \lambda\} \subseteq Rg\mu$, and if $Rg\pi \cap (\Delta x \cup Rg\mu \cup \tau^* Rg\mu) = 0$, then

$$\mathsf{s}_{\tau} x = s^{\pi}_{\tau\circ\mu} s^{\mu}_{\pi} x.$$

(2) is an easy consequence of (1), 1.5.8(i), and 1.5.10(iii),(v).

Now we turn to the proof of (iii). Let x and y be given; let τ be a finite transformation and let $\kappa < \omega$. Let μ be a one-one function in ${}^{\kappa}\alpha$ such that $\{\lambda : \tau\lambda \neq \lambda\} \subseteq Rg\mu$. Choose $\pi \in {}^{\kappa}\alpha$ such that π is one-one and $Rg\pi \cap (\Delta x \cup \Delta y \cup Rg\mu \cup \tau^* Rg\mu) = 0$. Then

$$\begin{aligned}
\mathsf{s}_{\tau}(x+y) &= s^{\pi}_{\tau\circ\mu} s^{\mu}_{\pi}(x+y) && \text{by (2)}\\
&= s^{\pi}_{\tau\circ\mu} s^{\mu}_{\pi} x + y s^{\pi}_{\tau\circ\mu} s^{\mu}_{\pi} y && \text{by 1.5.3(i)}\\
&= \mathsf{s}_{\tau} x + \mathsf{s}_{\tau} y && \text{by (2).}
\end{aligned}$$

Furthermore, it is clear that $\mathsf{s}_{\tau} - x = -\mathsf{s}_{\tau} x$; hence the remaining formulas of (iii) are established.

For (iv), let σ and τ be finite transformations of α. Then there is a $\kappa < \omega$ and a $\mu \in {}^{\kappa}\alpha$ such that μ is one-one, $\{\lambda : \tau\lambda \neq \lambda\} \cup \{\lambda : \sigma\lambda \neq \lambda\} \subseteq Rg\mu$, and $\sigma^* Rg\mu \cup \tau^* Rg\mu \subseteq Rg\mu$. There exist $\pi, \rho \in {}^{\kappa}\alpha$ such that π and ρ are one-one, $Rg\pi \cap (\Delta x \cup Rg\mu) = 0$, and $Rg\rho \cap (\Delta \mathsf{s}_{\tau} x \cup Rg\mu \cup Rg\pi) = 0$. Thus, by (2),

(3) $$\mathsf{s}_{\sigma}\mathsf{s}_{\tau} x = s^{\rho}_{\sigma\circ\mu} s^{\mu}_{\rho} s^{\pi}_{\tau\circ\mu} s^{\mu}_{\pi} x.$$

We now claim that, for any $\lambda \leqq \kappa$,

(4) $$\mathsf{s}_{\sigma}\mathsf{s}_{\tau} x = s^{\lambda 1\pi}_{\lambda 1\sigma\circ\tau\circ\mu} s^{\rho}_{\sigma\circ\mu} s^{\mu}_{\rho} s^{(\kappa\sim\lambda)1\pi}_{(\kappa\sim\lambda)1\tau\circ\mu} s^{\mu}_{\pi} x.$$

For $\lambda = 0$, (4) reduces to (3). Assuming that (4) holds for $\lambda < \kappa$ we prove it for $\lambda+1$. Let

(5) $$\tau\mu_{\lambda} = \mu_{\eta};$$

we then have

$$\begin{aligned}
\mathsf{s}_{\sigma}\mathsf{s}_{\tau} x &= s^{\lambda 1\pi}_{\lambda 1\sigma\circ\tau\circ\mu} s^{\rho}_{\sigma\circ\mu} s^{\mu}_{\rho} \mathsf{s}^{\mu_{\eta}}_{\rho_{\eta}} s^{(\kappa\sim\lambda)1\pi}_{(\kappa\sim\lambda)1\tau\circ\mu} s^{\mu}_{\pi} x && \text{by (4), 1.5.10(iii)}\\
&= s^{\lambda 1\pi}_{\lambda 1\sigma\circ\tau\circ\mu} s^{\rho}_{\sigma\circ\mu} s^{\mu}_{\rho} \mathsf{s}^{\mu_{\eta}}_{\rho_{\eta}} \mathsf{s}^{\pi_{\lambda}}_{\rho_{\eta}} s^{(\kappa\sim(\lambda+1))1\pi}_{(\kappa\sim(\lambda+1))1\tau\circ\mu} s^{\mu}_{\pi} x && \text{by 1.5.10(ii), (5)}\\
&= s^{\lambda 1\pi}_{\lambda 1\sigma\circ\tau\circ\mu} s^{\rho}_{\sigma\circ\mu} \mathsf{s}^{\rho_{\eta}}_{\sigma\mu_{\eta}} \mathsf{s}^{\pi_{\lambda}}_{\rho_{\eta}} s^{\mu}_{\rho} s^{(\kappa\sim(\lambda+1))1\pi}_{(\kappa\sim(\lambda+1))1\tau\circ\mu} s^{\mu}_{\pi} x && \text{by 1.5.10(iii)}\\
&= s^{\lambda 1\pi}_{\lambda 1\sigma\circ\tau\circ\mu} s^{\rho}_{\sigma\circ\mu} \mathsf{s}^{\rho_{\eta}}_{\sigma\mu_{\eta}} \mathsf{s}^{\pi_{\lambda}}_{\sigma\mu_{\eta}} s^{\mu}_{\rho} s^{(\kappa\sim(\lambda+1))1\pi}_{(\kappa\sim(\lambda+1))1\tau\circ\mu} s^{\mu}_{\pi} x && \text{by 1.5.10(ii)}\\
&= s^{(\lambda+1)1\pi}_{(\lambda+1)1\sigma\circ\tau\circ\mu} s^{\rho}_{\sigma\circ\mu} s^{\mu}_{\rho} s^{(\kappa\sim(\lambda+1))1\pi}_{(\kappa\sim(\lambda+1))1\tau\circ\mu} s^{\mu}_{\pi} x && \text{by (5), 1.5.10(iii).}
\end{aligned}$$

This completes the inductive proof of (4). By taking $\lambda = \kappa$ in (4) and using (2), 1.5.8(i), and 1.5.10(i) we get

$$\mathsf{s}_\sigma \mathsf{s}_\tau x = s^{\pi}_{\sigma\circ\tau\circ\mu} s^{\mu}_{\pi} = \mathsf{s}_{\sigma\circ\tau} x.$$

The theorem is thus proved.

THEOREM 1.11.12. *Let $\mathfrak{A} \in \mathsf{Dc}_\alpha$ with $\alpha \geqq \omega$ and let σ, τ be finite transformations of α; suppose $\kappa, \lambda < \alpha$ and, as usual, let Γ be a finite subset of α, R a finite subset of $\alpha \times \alpha$, and x any element of A. We then have:*

(i) $\mathsf{s}_{\alpha 1 Id} = A 1 Id$.

(ii) *If τ is one-one, then s_τ is an automorphism of $\mathfrak{Bl}\mathfrak{A}$.*

(iii) *If $y \in {}^I A$ and $\sum_{i\in I} y_i$ exists, then $\sum_{i\in I} \mathsf{s}_\tau y_i$ also exists and we have*

$$\mathsf{s}_\tau \textstyle\sum_{i\in I} y_i = \sum_{i\in I} \mathsf{s}_\tau y_i;$$

similarly, with "$\sum$" replaced by "$\prod$". In other words, s_τ is a complete endomorphism of $\mathfrak{Bl}\mathfrak{A}$.

(iv) *If $(\alpha \sim \Gamma) 1 \sigma = (\alpha \sim \Gamma) 1 \tau$, then $\mathsf{s}_\sigma \mathsf{c}_{(\Gamma)} x = \mathsf{s}_\tau \mathsf{c}_{(\Gamma)} x$.*

(v) *If $\Delta x 1 \sigma = \Delta x 1 \tau$, then $\mathsf{s}_\sigma x = \mathsf{s}_\tau x$.*

(vi) *If $(\tau^{-1})^* \Gamma = \Delta$ and $\Delta 1 \tau$ is one-one, then $\mathsf{c}_{(\Gamma)} \mathsf{s}_\tau x = \mathsf{s}_\tau \mathsf{c}_{(\Delta)} x$.*

(vii) $\mathsf{s}_\tau \mathsf{d}_\Gamma = \mathsf{d}_{\tau^*\Gamma}$.

(viii) *If $\langle \kappa, \lambda \rangle \in R \sim Id$ and $\tau\kappa = \tau\lambda$, then $\mathsf{s}_\tau \bar{\mathsf{d}} R = 0$.*

(ix) *If $\tau\kappa \neq \tau\lambda$ whenever $\langle \kappa, \lambda \rangle \in R \sim Id$, then $\mathsf{s}_\tau \bar{\mathsf{d}} R = \bar{\mathsf{d}}\{\langle \tau\kappa, \tau\lambda \rangle : \langle \kappa, \lambda \rangle \in R\}$.*

(x) $\Delta(\mathsf{s}_\tau x) \subseteq \tau^* \Delta x$.

PROOF. (i): by setting $\kappa = \lambda$ in 1.11.11(i).

(ii) Since τ is a one-one finite transformation, it is easily seen that τ^{-1} is also a one-one finite transformation and $\tau \circ \tau^{-1} = \tau^{-1} \circ \tau = \alpha 1 Id$. Hence, by (i) and 1.11.11(iv), $\mathsf{s}_\tau \circ \mathsf{s}_{\tau^{-1}} = \mathsf{s}_{\tau^{-1}} \circ \mathsf{s}_\tau = A 1 Id$. This clearly implies that s_τ is one-one and $Rg\,\mathsf{s}_\tau = A$; the conclusion now follows by 1.11.11(iii).

(iii) τ can be written as a composition of transpositions and replacements: $\tau = \sigma_0 \circ \ldots \circ \sigma_{\nu-1}$. For each $\mu < \nu$, if σ_μ is a replacement, then s_{σ_μ} is a complete endomorphism by 1.11.11(i) and 1.5.3 (cf. the remark following 1.5.3). If σ_μ is a transposition, then s_{σ_μ} is a complete endomorphism by (ii). (An automorphism of a BA is automatically a complete automorphism; cf. the remarks beginning Section 0.2.)

(iv) By a simple inductive argument it is seen that we may restrict ourselves to the case $|\Gamma| = 1$, say $\Gamma = \{\kappa\}$. There is a $\mu \in \alpha \sim (\Delta x \cup \{\kappa\})$. Then

$$\begin{aligned}
s_\sigma c_\kappa x &= s_\sigma s^\kappa_\mu c_\kappa x && \text{by 1.5.8(i)}\\
&= s_{\sigma\circ[\kappa/\mu]} c_\kappa x && \text{by 1.11.11(i), 1.11.11(iv)}\\
&= s_{\tau\circ[\kappa/\mu]} c_\kappa x\\
&= s_\tau s^\kappa_\mu c_\kappa x\\
&= s_\tau c_\kappa x.
\end{aligned}$$

(v) Let $\Gamma = \{\kappa : \sigma\kappa \neq \tau\kappa\}$. Then $|\Gamma| < \omega$, $\Gamma \subseteq \alpha \sim \Delta x$, and hence $c_{(\Gamma)}x = x$ and $(\alpha \sim \Gamma)1\sigma = (\alpha \sim \Gamma)1\tau$. Thus (iv) yields (v).

(vi) Again a simple inductive argument shows that it is sufficient to treat the case in which $|\Gamma| = 1$, say $\Gamma = \{\kappa\}$. Two cases may then occur. First, we may have $\Delta = 0$. Then, for some $\mu \in \alpha \sim \{\kappa\}$ we have

$$c_\kappa s_\tau x = c_\kappa s_{[\kappa/\mu]\circ\tau} x = c_\kappa s^\kappa_\mu s_\tau x = s^\kappa_\mu s_\tau x = s_\tau x.$$

Secondly, we have $\Delta = \{\lambda\}$ for some λ. Then $(([\kappa/\lambda, \lambda/\kappa]\circ\tau)^{-1})^\star\lambda = \{\lambda\}$, and hence $[\kappa/\lambda, \lambda/\kappa]\circ\tau$ can be written as a product of transpositions and replacements ρ such that $(\rho^{-1})^\star\lambda = \{\lambda\}$. It is easily seen using 1.5.8(ii) that for any such transformation ρ we have $s_\rho c_\lambda y = c_\lambda s_\rho y$ for every $y \in A$, and hence $s_{[\kappa/\lambda,\lambda/\kappa]\circ\tau} c_\lambda x = c_\lambda s_{[\kappa/\lambda,\lambda/\kappa]\circ\tau} x$. Applying $s_{[\kappa/\lambda,\lambda/\kappa]}$ to both sides of this equation we get

$$\begin{aligned}
s_\tau c_\lambda x &= s_{[\kappa/\lambda,\lambda/\kappa]} c_\lambda s_{[\kappa/\lambda,\lambda/\kappa]\circ\tau} x\\
&= c_\kappa s_{[\kappa/\lambda,\lambda/\kappa]} s_{[\kappa/\lambda,\lambda/\kappa]\circ\tau} x && \text{by 1.5.21}\\
&= c_\kappa s_\tau x,
\end{aligned}$$

as desired.

(vii) By an easy induction using 1.8.4 we may restrict ourselves to the case $|\Gamma| = 2$, say $\Gamma = \{\kappa, \lambda\}$ with $\kappa \neq \lambda$. By 1.5.4 the desired equation $s_\tau d_{\kappa\lambda} = d_{\tau\kappa,\tau\lambda}$ clearly holds whenever τ is a replacement or a transposition, and hence it holds for any finite transformation τ.

(viii) By (vii), $s_\tau \bar{d}R \leq s_\tau - d_{\kappa\lambda} = -d_{\tau\kappa,\tau\lambda} = 0$.

(ix): obvious.

(x) Suppose $\kappa \notin \tau^*\Delta x$. Let $\Gamma = (\tau^{-1})^\star\kappa$, and let $\lambda \in \alpha \sim \{\kappa\}$. Choose σ such that $(\alpha \sim \Gamma)1\sigma = (\alpha \sim \Gamma)1\tau$ while $\sigma^*\Gamma \subseteq \{\lambda\}$. Now $\Delta x 1\tau = \Delta x 1\sigma$ since $\Delta x \cap \Gamma = 0$. Hence by (v) and (vi) we have

$$c_\kappa s_\tau x = c_\kappa s_\sigma x = s_\sigma c_{(0)} x = s_\tau x.$$

Thus $\kappa \notin \Delta(s_\tau x)$ as desired. This completes the proof of the theorem.

We now restrict ourselves to locally finite CA_α's and define for them the generalized substitution operator applicable to arbitrary, not only finite, transformations of α. Also the notion of generalized cylindrification discussed in Section 1.7 undergoes a further generalization and becomes relativized to an arbitrary, and not only finite, subset of α. The reader will readily notice, however, that both "generalizations" have a purely formal character: the new operations trivially reduce to the old ones simply using the fact that all the elements in the algebras discussed have finite dimension sets.

DEFINITION 1.11.13. *Let $\mathfrak{A}$ be a Lf_α with $\alpha \geqq \omega$. By the* ***generalized substitution operator*** s^+ *we understand the function which correlates with every transformation τ of α a unary operation s^+_τ on A defined by the formula*

$$\mathsf{s}^+_\tau x = \mathsf{s}_{\Delta x 1 \tau \cup (\alpha \sim \Delta x) 1 Id}\, x$$

for every $x \in A$. Also for an arbitrary (not necessarily finite) subset Γ of α we set

$$\mathsf{c}^+_{(\Gamma)} x = \mathsf{c}_{(\Gamma \cap \Delta x)} x.$$

THEOREM 1.11.14. *Let $\mathfrak{A} \in \mathsf{Lf}_\alpha$, $\alpha \geqq \omega$, and let $x, y \in A$. Let $\Gamma, \Delta \subseteq \alpha$ (both possibly infinite), and let $\sigma, \tau \in {}^\alpha\alpha$. Then:*

(i) $\mathsf{c}^+_{(\Gamma)} x = \mathsf{c}_{(\Gamma)} x$ *if Γ is finite;*
(ii) $\mathsf{c}^+_{(\Gamma)} 0 = 0$;
(iii) $x \leqq \mathsf{c}^+_{(\Gamma)} x$;
(iv) $\mathsf{c}^+_{(\Gamma)}(x \cdot \mathsf{c}^+_{(\Gamma)} y) = \mathsf{c}^+_{(\Gamma)} x \cdot \mathsf{c}^+_{(\Gamma)} y$;
(v) $\mathsf{c}^+_{(\Gamma)} \mathsf{c}^+_{(\Delta)} x = \mathsf{c}^+_{(\Gamma \cup \Delta)} x$;
(vi) $\mathsf{s}^+_\tau x = \mathsf{s}_\tau x$ *if τ is a finite transformation of α;*
(vii) s^+_τ *is an endomorphism of $\mathfrak{Bl}\mathfrak{A}$;*
(viii) $\mathsf{s}^+_{\alpha 1 Id} = A 1 Id$;
(ix) $\mathsf{s}^+_{\sigma \circ \tau} = \mathsf{s}^+_\sigma \circ \mathsf{s}^+_\tau$;
(x) *if $(\alpha \sim \Gamma) 1 \sigma = (\alpha \sim \Gamma) 1 \tau$, then $\mathsf{s}^+_\sigma \mathsf{c}^+_{(\Gamma)} x = \mathsf{s}^+_\tau \mathsf{c}^+_{(\Gamma)} x$;*
(xi) *if $\Delta x 1 \sigma = \Delta x 1 \tau$, then $\mathsf{s}^+_\sigma x = \mathsf{s}^+_\tau x$;*
(xii) *if $(\tau^{-1})^* \Gamma = \Theta$ and $\Theta 1 \tau$ is one-one, then $\mathsf{c}^+_{(\Gamma)} \mathsf{s}^+_\tau x = \mathsf{s}^+_\tau \mathsf{c}^+_{(\Theta)} x$;*
(xiii) $\mathsf{s}^+_\tau \mathsf{d}_{\kappa\lambda} = \mathsf{d}_{\tau\kappa, \tau\lambda}$ *for all $\kappa, \lambda < \alpha$;*
(xiv) $\mathsf{d}_{\kappa\lambda} \cdot x \leqq \mathsf{s}^+_{[\kappa/\lambda]} x$.[1)]

PROOF. Conditions (ii), (iii), (vi), (viii), (xi), and (xiii) are obvious. As to (i), we have, by 1.7.3,

$$\mathsf{c}_{(\Gamma)} x = \mathsf{c}_{(\Gamma \cap \Delta x)} \mathsf{c}_{(\Gamma \sim \Delta x)} x = \mathsf{c}_{(\Gamma \cap \Delta x)} x = \mathsf{c}^+_{(\Gamma)} x.$$

To prove (iv), note that $\mathsf{c}^+_{(\Gamma)} y = \mathsf{c}_{(\Gamma \cap \Delta y)} y$, and hence, by 1.7.7(i),

1) Various parts of 1.11.14 were used in Halmos [57], [57a] to construct postulate systems for arbitrary polyadic algebras with equality.

$\Delta(x \cdot \mathsf{c}^+_{(\Gamma)}y) \subseteq \Delta x \cup \Delta y$. Let $\Theta = \Gamma \cap (\Delta x \cup \Delta y)$. Then

$$\begin{aligned} \mathsf{c}^+_{(\Gamma)}x \cdot \mathsf{c}^+_{(\Gamma)}y &= \mathsf{c}_{(\Gamma \cap \Delta x)}x \cdot \mathsf{c}_{(\Gamma \cap \Delta y)}y \\ &= \mathsf{c}_{(\Theta)}x \cdot \mathsf{c}_{(\Theta)}y \\ &= \mathsf{c}_{(\Theta)}(x \cdot \mathsf{c}_{(\Theta)}y) \\ &= \mathsf{c}_{(\Gamma \cap \Delta(x \cdot \mathsf{c}^+{}_{(\Gamma)}y))}(x \cdot \mathsf{c}_{(\Gamma \cap \Delta y)}y) \\ &= \mathsf{c}^+_{(\Gamma)}(x \cdot \mathsf{c}^+_{(\Gamma)}y). \end{aligned}$$

Condition (v) is proved analogously.

Taking (vii) next, it is clear that $\mathsf{s}^+_\tau - z = -\mathsf{s}^+_\tau z$ for any $z \in A$. If $u, v \in A$, let $\rho = \Delta u 1 \tau \cup (\alpha \sim \Delta u) 1 Id$, $\rho' = \Delta v 1 \tau \cup (\alpha \sim \Delta v) 1 Id$, $\rho'' = \Delta(u+v) 1 \tau \cup (\alpha \sim \Delta(u+v)) 1 Id$, and $\rho''' = (\Delta u \cup \Delta v) 1 \tau \cup (\alpha \sim (\Delta u \cup \Delta v)) 1 Id$. Since $\Delta(u+v) \subseteq \Delta u \cup \Delta v$, we have $\Delta u 1 \rho''' = \Delta u 1 \rho$, $\Delta v 1 \rho''' = \Delta v 1 \rho'$, and $\Delta(u+v) 1 \rho''' = \Delta(u+v) 1 \rho''$. Hence

$$\begin{aligned} \mathsf{s}^+_\tau(u+v) &= \mathsf{s}_{\rho''}(u+v) & \\ &= \mathsf{s}_{\rho'''}(u+v) & \text{by 1.11.12(v)} \\ &= \mathsf{s}_{\rho'''}u + \mathsf{s}_{\rho'''}v & \text{by 1.11.12(iii)} \\ &= \mathsf{s}_{\rho}u + \mathsf{s}_{\rho'}v & \text{by 1.11.12(v)} \\ &= \mathsf{s}^+_\tau u + \mathsf{s}^+_\tau v. & \end{aligned}$$

To prove (ix), let $\tau' = \Delta x 1 \tau \cup (\alpha \sim \Delta x) 1 Id$, $\sigma' = \Delta(\mathsf{s}^+_\tau x) 1 \sigma \cup (\alpha \sim (\Delta \mathsf{s}^+_\tau x)) 1 Id$, $\rho = \Delta x 1 (\sigma \circ \tau) \cup (\alpha \sim \Delta x) 1 Id$, and $\sigma'' = \tau^* \Delta x 1 \sigma \cup (\alpha \sim \tau^* \Delta x) 1 Id$. Then $\Delta x 1 (\sigma'' \circ \tau') = \Delta x 1 \rho$, and

$$\begin{aligned} \mathsf{s}^+_\sigma \mathsf{s}^+_\tau x &= \mathsf{s}_{\sigma'} \mathsf{s}_{\tau'} x & \\ &= \mathsf{s}_{\sigma''} \mathsf{s}_{\tau'} x & \text{by 1.11.12(v),(x)} \\ &= \mathsf{s}_{\sigma'' \circ \tau'} x & \text{by 1.11.11(iv)} \\ &= \mathsf{s}_\rho x & \text{by 1.11.12(v)} \\ &= \mathsf{s}^+_{\sigma \circ \tau} x. & \end{aligned}$$

For (x), note that $\Delta \mathsf{c}^+_{(\Gamma)} x = \Delta \mathsf{c}_{(\Gamma \cap \Delta x)} x \subseteq \Delta x \sim \Gamma$ (by 1.7.7). Hence $\Delta \mathsf{c}^+_{(\Gamma)} x 1 \sigma = \Delta \mathsf{c}^+_{(\Gamma)} x 1 \tau$, and the result follows by (xi).

Considering (xii) we let $\Phi = \Gamma \cap \tau^* \Delta x$, let $\lambda \in \alpha \sim \Phi$, and for each $\kappa < \alpha$ let

$$\rho\kappa = \begin{cases} \tau\kappa \ \textit{if} \ \kappa \in \Delta x, \\ \lambda \ \textit{if} \ \kappa \in \Phi \sim \Delta x, \\ \kappa \ \textit{otherwise}. \end{cases}$$

Then $(\rho^{-1})^*\Phi = \Theta \cap \Delta x$, so $(\rho^{-1})^*\Phi 1 \rho$ is one-one, and

$$\begin{aligned} c^+_{(\Gamma)} s^+_\tau x &= c_{(\Phi)} s_\rho x \\ &= s_\rho c_{(\rho^{-1})^*\Phi} x && \text{by 1.11.12(vi)} \\ &= s_\rho c_{(\Theta \cap \Delta x)} x \\ &= s^+_\tau c^+_{(\Theta)} x. \end{aligned}$$

Finally, with regard to (xiv) we note that by (vi) and 1.11.11(i) we have $s^+_{[\kappa/\lambda]} x = s^\kappa_\lambda x$. The conclusion of (xiv) now follows immediately. The theorem is thus proved.

In contrast to 1.11.12(iii), it may be noted that s^+_σ is not always a complete endomorphism of $\mathfrak{B} \upharpoonright \mathfrak{A}$. For example, let $\mathfrak{B}$ be the α-dimensional full cylindric set algebra with base $2 = \{0, 1\}$ $(\alpha \geqq \omega)$, and let $\mathfrak{A}$ be the subalgebra of $\mathfrak{B}$ whose universe consists of all sets $X \subseteq {}^\alpha 2$ such that $|\Delta^{(\mathfrak{B})} X| < \omega$. It is easily seen that $\prod^{(\mathfrak{A})}_{\kappa,\lambda<\alpha} d_{\kappa\lambda} = 0$. If we let $\sigma = \alpha \times 1$, i.e., $\sigma\kappa = 0$ for all $\kappa < \alpha$, then $s^+_\sigma d_{\kappa\lambda} = 1$ for all $\kappa, \lambda < \alpha$, and so $\prod^{(\mathfrak{A})}_{\kappa,\lambda<\alpha} s^+_\sigma d_{\kappa\lambda} = 1 \neq 0$.

REMARK 1.11.15. From 1.11.14(ii)–(v) it follows that for Lf_α's Theorem 1.7.5 applies to operations $c^+_{(\Gamma)}$ (where Γ need not be finite).

The operators s^+_σ and $c^+_{(\Gamma)}$ will be discussed further in Part II. We may mention that, with the aid of 1.11.6, d_Γ and $\bar{d}R$ can be similarly generalized to infinite Γ and R if the CA discussed is dimension-complemented. All these generalizations clearly require some assumptions concerning the existence of certain infinite sums and products. For work related to this problem see Mangani [66b] and Lucas [68].

PROBLEMS

PROBLEM 1.1. *Boolean algebras can be treated as algebraic structures* $\mathfrak{A} = \langle A, +, - \rangle$ *where* $+$ *and* $-$ *have the usual meaning. Clearly, the class of* BA's *so treated is definitionally — and, in fact, polynomially — equivalent with the class of* BA's *described at the beginning of Section 1.1. Is it true that* BA's *under the new treatment can be characterized as algebras* $\langle A, +, - \rangle$ *satisfying the following postulates for any* $x, y, z \in A$:

(R_1) $x+y = y+x$,

(R_2) $x+(y+z) = (x+y)+z$,

(R_3) $-[-(x+y) + -(x+ -y)] = x$?

This problem originates with Herbert Robbins. It is related to an old result in Huntington [33*] and [33a*] from which it follows that Postulates (R_1) and (R_2) together with a postulate very similar to (R_3) do characterize BA's. It is known that every finite algebra satisfying (R_1)–(R_3) is a BA.

PROBLEM 1.2. *Is there a finite set of identities* (B_0)–(B_μ) *with the following properties:* (i) *the identities* (B_0)–(B_μ) *contain only variables and symbols for fundamental operations and distinguished elements of Boolean algebras;* (ii) *the identities* (B_0)–(B_μ) *hold in all Boolean algebras but do not form an adequate postulate system for Boolean algebras;* (iii) *the identities* (B_0)–(B_μ) *and* (C_2)–(C_7) *jointly form an adequate postulate system for cylindric algebras of dimension* $\alpha \geq 2$?

Notice that postulate (C_1) is not listed in condition (iii); for an explanation see the remark following 1.3.2.

PROBLEM 1.3. *In* 1.5.23 *an equation with* 3 *different variables is given which does not involve diagonal elements and is identically satisfied in all cylindric algebras of dimension* 4 *(or more), but not in all diagonal-free cylindric algebras of dimension* 4. *Is there an equation with the same properties in which only one variable or only two different variables occur?*

PROBLEM 1.4. *Is there an equation which does not involve diagonal elements and is identically satisfied in all cylindric algebras of dimension* 3 *but not in all diagonal-free cylindric algebras of the same dimension*?

Concerning analogous problems for cylindric algebras of dimension less than 3 see 1.5.24.

CHAPTER 2

GENERAL ALGEBRAIC NOTIONS APPLIED TO CYLINDRIC ALGEBRAS

2. GENERAL ALGEBRAIC NOTIONS APPLIED TO CYLINDRIC ALGEBRAS

In this chapter we discuss the application to cylindric algebras of the general algebraic notions defined in Chapter 0 for arbitrary algebras. We shall see that in many cases the original definitions of these notions can be modified using specific properties of CA's, so as to make them more adaptable for the purposes of our work. We shall pay particular attention to those properties of the notions discussed which, while formulated in general algebraic terms, do not apply to all algebras, but are applicable either to all CA's or at least to all algebras in some special classes of CA's; cf. 2.1.1, 2.1.14, and 2.3.22. Some further results which conceptually belong to this discussion will be found only in the second part of this work (since their proofs depend on the representation theory which is developed there). Sections 2.2 and 2.7 differ somewhat in their character from the remaining sections. In 2.2 we extend the notion of a relativized algebra from the theory of BA's to that of CA's. The discussion in 2.7 is related to the fact that cylindric algebras can be subsumed under a comprehensive class of algebras known in the literature as Boolean algebras with operators; without presupposing any knowledge of the general theory of those algebras, we re-formulate and re-establish some general results concerning Boolean algebras with operators in their application to the special class of cylindric algebras.

The algebraic notions involved in our discussion will be applied not only to the cylindric algebra $\mathfrak{A}$, but also to the Boolean algebra $\mathfrak{Bl}\,\mathfrak{A}$, the Boolean part of $\mathfrak{A}$. When applying a term from the general theory of algebras, not to a CA $\mathfrak{A}$ but to its Boolean part $\mathfrak{Bl}\,\mathfrak{A}$, we shall provide this term with the adjective "Boolean" and then refer it directly to $\mathfrak{A}$. Thus we shall speak of subalgebras of $\mathfrak{A}$, using this term in the sense of 0.1.8, but we shall also speak of *Boolean subalgebras* of $\mathfrak{A}$, meaning subalgebras of $\mathfrak{Bl}\,\mathfrak{A}$. This convention extends the related stipulations made in Chapter 1 (see the remarks following 1.1.1). Exceptions to this convention will be pointed out explicitly.

2.1. SUBALGEBRAS

We shall be concerned here with the notion of subuniverse and several other notions defined in its terms. Definitions of these notions were given in Section 0.1.

A subuniverse of an algebra $\mathfrak{A}$ is by definition any subset of its universe A which is closed under all fundamental operations of $\mathfrak{A}$. Among fundamental operations of CA's we find some operations in the usual sense — Boolean operations $+$, $\cdot$, $-$ and cylindrifications c_κ — as well as certain distinguished elements, namely 0, 1, and the diagonal elements $\mathsf{d}_{\kappa\lambda}$. In the general theory of algebras distinguished elements are identified with constant operations of rank 1, and therefore they belong to every non-empty subuniverse. Consequently, a subuniverse of a CA $\mathfrak{A}$ can be defined as any subset of A which is either empty or else contains the elements 0, 1, $\mathsf{d}_{\kappa\lambda}$ and is closed under Boolean operations and cylindrifications. As is well known from the theory of Boolean algebras, we may omit the elements 0, 1 and either one of the operations $+$, $\cdot$ from this list since, e.g., $0 = -(x+-x)$, $1 = x+-x$, and $x\cdot y = -(-x+-y)$. The resulting definition of subuniverse is more suitable for the discussion of CA's; the modification extends, of course, to such derivative notions as those of subalgebra and the subuniverse (or subalgebra) generated by a given set of elements.

THEOREM 2.1.1. $\mathsf{SCA}_\alpha = \mathsf{CA}_\alpha$; *that is, the class* CA_α *is closed under the formation of subalgebras.*

REMARK 2.1.2. A proof of 2.1.1 based directly upon the definitions of the notions involved is entirely routine. On the other hand, we can prove 2.1.1 by means of a very concise model-theoretic argument: we use the simple observation that by the very form of its definition CA_α is an equational class of algebras, and then we apply the general theorem 0.4.63 from which it follows that $\mathsf{SK} = \mathsf{K}$ holds for every equational class K. Similar remarks apply to the analogous theorems 2.3.1 and 2.4.1 below.

In connection with the next theorem see 1.6.1 and 1.11.1 for the definitions of Δ, Lf_α, and Dc_α. Recall that, by 0.1.8(i), the symbol "$\subseteq$" is used to denote, not only the relation between a subset and a set, but also that between a subalgebra and an algebra.

THEOREM 2.1.3. (i) *If* $\mathfrak{A} \subseteq \mathfrak{B} \in \mathsf{CA}_\alpha$ *and* $x \in A$, *then* $\Delta^{(\mathfrak{A})}x = \Delta^{(\mathfrak{B})}x$.
(ii) $\mathsf{SLf}_\alpha = \mathsf{Lf}_\alpha$ *and* $\mathsf{SDc}_\alpha = \mathsf{Dc}_\alpha$.

PROOF. (i) follows directly from 1.6.1, and (ii) follows from 2.1.1 and (i).

LEMMA 2.1.4. *If β is an infinite cardinal, $\Gamma \subseteq \alpha$, $\mathfrak{A} \in \mathsf{CA}_\alpha$, $X \subseteq A$, and $|\Delta x \cap \Gamma| < \beta$ for each $x \in X$, then $|\Delta x \cap \Gamma| < \beta$ for each $x \in SgX$.*

PROOF: by 1.6.4, 1.6.5, 1.6.7, and 1.6.8.

In connection with the statement of 2.1.4 recall that SgX is the subuniverse of $\mathfrak{A}$ generated by X, i.e., $SgX = 0$ if $X = 0$, and if $X \neq 0$, then SgX is the smallest subset B of A such that $X \subseteq B$, $\mathsf{d}_{\kappa\lambda} \in B$ for all $\kappa, \lambda < \alpha$, and B is closed under Boolean operations and cylindrifications.

THEOREM 2.1.5. *Assume $\alpha \geqq \omega$, $\mathfrak{A} \in \mathsf{CA}_\alpha$, and $0 \neq X \subseteq A$.*

(i) *If $|\Delta x| < |\alpha|$ for each $x \in X$, then $\mathfrak{Sg}X \in \mathsf{Dc}_\alpha$; if $|\Delta x| < \omega$ for each $x \in X$, then $\mathfrak{Sg}X \in \mathsf{Lf}_\alpha$.*

(ii) *If $\Gamma \subseteq \alpha$, $|\alpha \sim \Gamma| \geqq \omega$, and $\Delta x \subseteq \Gamma$ for each $x \in X$, then $\mathfrak{Sg}X \in \mathsf{Dc}_\alpha$; if in addition $|\Gamma| < \omega$, then $\mathfrak{Sg}X \in \mathsf{Lf}_\alpha$.*

PROOF. Apply 2.1.4, taking α for Γ to obtain (i) and $\alpha \sim \Gamma$ for Γ to obtain (ii).

COROLLARY 2.1.6. *If $\mathfrak{A} \in \mathsf{CA}_\alpha$ and $B = \{x : x \in A, |\Delta x| < \omega\}$, then $B \in Su\mathfrak{A}$ and the subalgebra $\mathfrak{B}$ of $\mathfrak{A}$ with universe B is the largest subalgebra of $\mathfrak{A}$ which is an Lf_α.*

Recall that $Su\mathfrak{A}$ is the class of all subuniverses of $\mathfrak{A}$.

THEOREM 2.1.7. *Assume $\alpha \geqq \omega$, $\mathfrak{A} \in \mathsf{CA}_\alpha$, and $0 \neq X \subseteq A$. Then the following two conditions are equivalent:*

(i) $\mathfrak{Sg}X \in \mathsf{Dc}_\alpha$;

(ii) $|\alpha \sim \bigcup_{y \in Y} \Delta y| \geqq \omega$ *for every finite $Y \subseteq X$.*

PROOF. (i) implies (ii) by 1.11.4, and from 0.1.17(ix) and 2.1.5(ii) we obtain the implication in the opposite direction.

THEOREM 2.1.8. *For a CA_α $\mathfrak{A}$ to be an Lf_α it is necessary and sufficient that there exist a set X such that $\mathfrak{A} = \mathfrak{Sg}X$ and Δx is a finite ordinal for every $x \in X$.*

PROOF. The sufficiency of the condition follows immediately from 2.1.5(i). Set

$$B = Sg\{x : x \in A, \Delta x = |\Delta x|\};$$

to show the necessity of the condition we shall prove by induction on $||\Delta x| \sim \Delta x|$ that $x \in B$ for every $x \in A$. If $||\Delta x| \sim \Delta x| = 0$, then $\Delta x = |\Delta x|$ (since Δx is finite) and hence $x \in B$ trivially. Assume now that $||\Delta x| \sim \Delta x| > 0$, and consider any

(1) $$\lambda \in |\Delta x| \sim \Delta x.$$

Then, because $|\Delta x| < \omega$, there exists a

(2) $$\kappa \in \Delta x \sim |\Delta x|.$$

By (1), (2), and 1.6.15 we have

$$\Delta(\mathsf{s}_\lambda^\kappa x) = (\Delta x \sim \{\kappa\}) \cup \{\lambda\},$$

and hence, again using (1) and (2),

$$|\Delta(\mathsf{s}_\lambda^\kappa x)| \sim \Delta(\mathsf{s}_\lambda^\kappa x) = |\Delta x| \sim \Delta(\mathsf{s}_\lambda^\kappa x) = (|\Delta x| \sim \Delta x) \sim \{\lambda\}.$$

This along with (1) and the induction hypothesis gives $\mathsf{s}_\lambda^\kappa x \in B$, and, because $x = \mathsf{s}_\kappa^\lambda \mathsf{s}_\lambda^\kappa x$ by (1), 1.5.8(i), and 1.5.10(v), we obtain $x \in B$. Hence $B = A$ and the proof is complete.

THEOREM 2.1.9. *Let* $\mathfrak{A} \in \mathsf{CA}_\alpha$ *and* $\mathfrak{A} = \mathfrak{Sg}X$. *We then have:*

(i) $|X| \leqq |A| \leqq |\alpha| \cup |X| \cup \omega$;

(ii) $|A| \geqq |\alpha| \cup |X|$ *in case* $\mathfrak{A}$ *is not discrete;*

(iii) $|A| = |\alpha| \cup |X|$ *in case* $\mathfrak{A}$ *is not discrete and either* $\alpha \geqq \omega$ *or* $|X| \geqq \omega$.

PROOF: (i) by 0.1.20; (ii) by (i) and 1.3.15; (iii) by (i) and (ii).

The cardinality of discrete CA's is essentially a problem of the theory of Boolean algebras. In connection with 2.1.9 compare 2.4.72 and 2.4.73 below.

THEOREM 2.1.10. *All finitely generated* CA_0*'s and* CA_1*'s are finite.*

PROOF. This is obvious for CA_0's since it is well known that every finitely-generated BA is finite. Let $\mathfrak{A} \in \mathsf{CA}_1$ and $\mathfrak{A} = \mathfrak{Sg}X$ with $|X| < \omega$. Let $Y = \mathfrak{Sg}^{(\mathfrak{Bl}\mathfrak{A})}X$; thus $|Y| < \omega$. Now $Y \cup \mathsf{c}_0{}^*Y$ is closed under c_0, and hence, as is easily seen, so is $A = Sg^{(\mathfrak{Bl}\mathfrak{A})}(Y \cup \mathsf{c}_0{}^*Y)$. Consequently, $|A| < \omega$.

The least upper bound for the cardinality of a finitely generated CA_1 (dependent on the number of its generators) will be established in 2.5.62.

Theorem 2.1.10 does not extend to CA_α's with $\alpha \geqq 2$. In fact we have

THEOREM 2.1.11. (i) *If* $\alpha \geqq 2$, *there is a* CA_α $\mathfrak{A}$ *with one generator such that* $|A| = |\alpha| \cup \omega$ *and hence* $\mathfrak{A}$ *is infinite.*[1)]

(ii) *If* $2 \leqq \alpha < \omega$ *and* $\nu < \omega$, *then there is a* CA_α $\mathfrak{A}$ *with one generator such that* $\nu < |A| < \omega$.

(iii) *For any distinct* $\kappa, \lambda < \alpha$ *the algebra* $\mathfrak{A}$ *in either* (*i*) *or* (*ii*) *can be chosen so that there exists an* $a \in A$ *such that* $\mathfrak{A} = \mathfrak{Sg}\{a\}$ *and* $\Delta a = \{\kappa, \lambda\}$.

PROOF. Suppose $\alpha \geqq 2$ and consider any fixed ξ such that $1 \leqq \xi \leqq \omega$.

1) This theorem was proved in Thompson [52], p. 51, Theorem 3, by a different method.

Let $\mathfrak{B}$ be the full cylindric set algebra with base ξ, and dimension α and let $\mathfrak{A} = \mathfrak{Sg}\{X\}$ where $X = \{\mu : \mu \in {}^{\alpha}\xi,\ \mu_0 < \mu_1\}$. We shall define a sequence $Y \in {}^{\omega}B$ by recursion: $Y_0 = {}^{\alpha}\xi$, $Y_1 = \mathbf{C}_0 X$, and $Y_{\lambda+1} = \mathbf{C}_0(\mathbf{C}_1(Y_\lambda \sim X) \cap X)$ if $0 < \lambda < \omega$. It is clear that $RgY \subseteq A$. Also, as is easily seen by induction,

$$Y_\lambda = \{\mu : \mu \in {}^{\alpha}\xi,\ \lambda \leqq \mu_1\}$$

for all $\lambda < \omega$. If $\xi = \omega$, then Y is an infinite sequence without repeating terms. Hence the algebra $\mathfrak{A}$ is infinite and (i) follows by 2.1.9 ($\mathfrak{A}$ is obviously non-discrete). In case $\xi = \nu < \omega$, clearly $|RgY| = \nu+1$. On the other hand, in this case $\mathfrak{A}$ is obviously finite if $\alpha < \omega$. Thus (ii) holds. In order to prove (iii) observe that, with obvious modifications, the above proof goes through when we take $X = \{\mu : \mu \in {}^{\alpha}\xi,\ \mu_\kappa < \mu_\lambda\}$.

If we carry the above proof further, we quickly arrive at the following interesting result:

If $2 \leqq \alpha < \omega$ and $1 \leqq \kappa < \omega$, then the cylindric set algebra (with dimension α) of all subsets of ${}^{\alpha}\kappa$ is generated by a single element.

THEOREM 2.1.12. *Suppose $\mathfrak{A} \in \mathsf{Dc}_\alpha$ with $\alpha \geqq \omega$ and $\mathfrak{A} = \mathfrak{Sg}X$. Let $Y = Sg^{(\mathfrak{Bl}\,\mathfrak{A})}(\{\mathsf{d}_{\kappa\lambda} : \kappa, \lambda < \alpha\} \cup \{\mathsf{s}_\sigma x : x \in X,\ \sigma$ a finite transformation of $\alpha\})$. Then A coincides with the unique subset of A which includes Y and is closed under $-$ and all the operations c_κ or, equivalently, is closed under all the operations c_κ and $\mathsf{c}^\partial_\kappa$ $(\kappa < \alpha)$.*

PROOF. Let K be the set of all B such that $Y \subseteq B \subseteq A$ and B is closed under $-$ and c_κ for each $\kappa < \alpha$. We want to prove that K contains exactly one set, namely A. In order to do this we shall show that $\bigcap K = A$. Since $X \subseteq Y$, it is clear that $X \subseteq \bigcap K$. Also, obviously $\mathsf{d}_{\kappa\lambda} \in \bigcap K$ for all $\kappa, \lambda < \alpha$, and $\bigcap K$ is closed under $-$ and c_κ for each $\kappa < \alpha$. It remains to show that $\bigcap K$ is closed under $+$. This will allow us to conclude that $Sg X \subseteq \bigcap K$, and hence $A \subseteq \bigcap K$ as desired.

For any given $y \in \bigcap K$, let B_y be the set of all $z \in \bigcap K$ such that $\mathsf{s}_\sigma z + y$ and $-\mathsf{s}_\sigma z + y \in \bigcap K$ for every finite transformation σ of α. Clearly B_y is closed under $-$. Suppose $\kappa < \alpha$, $z \in B_y$, and σ is a finite transformation of α. Let Γ be a finite subset of α such that $(\alpha \sim \Gamma)1\sigma = (\alpha \sim \Gamma)1 Id$ and $\sigma^*\Gamma \subseteq \Gamma$. Then there is a $\lambda \in \alpha \sim (\Delta z \cup \Delta y \cup \Gamma)$, and $\mathsf{s}_\sigma \mathsf{c}_\kappa z + y = \mathsf{s}_\sigma \mathsf{c}_\lambda \mathsf{s}^\kappa_\lambda z + y = \mathsf{c}_\lambda(\mathsf{s}_\sigma \mathsf{s}^\kappa_\lambda z + y)$ by 1.5.8(i), 1.5.9(i), and 1.11.12(vi). Since $z \in B_y$, we have $\mathsf{s}_\sigma \mathsf{s}^\kappa_\lambda z + y \in \bigcap K$, and hence $\mathsf{s}_\sigma \mathsf{c}_\kappa z + y = \mathsf{c}_\lambda(\mathsf{s}_\sigma \mathsf{s}^\kappa_\lambda z + y) \in \bigcap K$. Similarly, $-\mathsf{s}_\sigma \mathsf{c}_\kappa z + y = -\mathsf{s}_\sigma \mathsf{c}_\lambda \mathsf{s}^\kappa_\lambda z + y = -\mathsf{c}_\lambda(\mathsf{s}_\sigma \mathsf{s}^\kappa_\lambda z \cdot -y)$, and we infer that $-\mathsf{s}_\sigma \mathsf{c}_\kappa z + y \in \bigcap K$. Thus we have shown that B_y is closed under an arbitrary cylindrification c_κ. If $y \in Y$, then clearly

$Y \subseteq B_y$, and hence $B_y \in K$, $\bigcap K \subseteq B_y$. In particular, $z+y \in \bigcap K$ whenever $z \in \bigcap K$. This being true for any $y \in Y$, we infer that $Y \subseteq B_w$, and hence $B_w \in K$, $\bigcap K \subseteq B_w$, for each $w \in \bigcap K$. Thus $\bigcap K$ is closed under $+$.

Now let L be the set of all C such that $Y \subseteq C \subseteq A$ and C is closed under c_κ and c_κ^∂ for each $\kappa < \alpha$. To complete the proof of the theorem we want to show that $\bigcap L = A$. Since $\bigcap L$ includes Y and is closed under all the c_κ's, it suffices by the first part of the proof to prove that $\bigcap L$ is closed under $-$. To this end let D be the set of all $y \in \bigcap L$ such that $-y \in \bigcap L$. Since Y is closed under $-$, we have $Y \subseteq D$. Also, if $y \in D$ and $\kappa < \alpha$, then $-c_\kappa y = c_\kappa^\partial - y \in \bigcap L$ and $-c_\kappa^\partial y = c_\kappa - y \in \bigcap L$; hence $c_\kappa y$ and $c_\kappa^\partial y \in D$. This shows that D is closed under c_κ and c_κ^∂ for each $\kappa < \alpha$. Therefore $D = \bigcap L$, i.e., $\bigcap L$ is closed under $-$. The proof is finished.

Theorem 2.1.12 is an algebraic analogue of the well-known metalogical theorem according to which any formula of first-order logic can be put in prenex normal form.

THEOREM 2.1.13. *If* K *is a subset of* CA_α, Lf_α, *or* Dc_α *directed by* $\subseteq$, *then* $\bigcup K \in \mathsf{CA}_\alpha$, $\bigcup K \in \mathsf{Lf}_\alpha$, *or* $\bigcup K \in \mathsf{Dc}_\alpha$, *respectively.*

PROOF. Under the hypothesis $\mathsf{K} \subseteq \mathsf{CA}_\alpha$ compare Remark 2.1.2 and use information in 0.3.84. Under the remaining two hypotheses a direct argument based on 1.11.1 and 2.1.3(i) leads quickly to the indicated conclusions.

In the last part of this section we deal with minimal CA_α's. We recall Definition 0.1.8(ii) which introduces this notion in a general algebraic setting.

THEOREM 2.1.14. *Every* CA_α *has exactly one subalgebra which is minimal.*

DEFINITION 2.1.15. *We let* Mn_α *be the class of all minimal cylindric algebras of dimension* α.

Clearly, Mn_0 consists of all one- and two-element Boolean algebras, and Mn_1 consists of all one- and two-element CA_1's; note that all two-element CA_1's are isomorphic. Only in Mn_2 do we meet non-trivial algebras.

Recalling the remarks at the beginning of this section and comparing them with 0.1.23 we see that a CA $\mathfrak{A}$ is minimal iff the only subset of its universe A which contains all the elements $d_{\kappa\lambda}$ and is closed under Boolean operations and cylindrifications is A itself. Speaking more loosely, $\mathfrak{A}$ is minimal iff all its elements can be obtained from the diagonal elements by a succession of Boolean operations and cylindrifications. This leads to a simple interpretation of minimal CA's in the metalogical setting: a cylindric algebra of formulas associated with a set of sentences in a language Λ of predicate logic is minimal iff all its elements are equivalence classes correlated with those formulas of Λ which contain equations $\mathbf{v}_\kappa = \mathbf{v}_\lambda$ as their only atomic parts (and therefore which do not contain any non-logical constants). As a consequence, when discussing minimal CA's of formulas we can always assume, without any loss of generality, that the underlying language Λ itself does not contain any non-logical constants. Conversely, all CA's of formulas constructed in such a language are minimal.

THEOREM 2.1.16. $\mathsf{Mn}_\alpha \subseteq \mathsf{Lf}_\alpha$.
PROOF: by 2.1.5(i).

In connection with the next theorem, 2.1.17 (a particular case of Theorem 12 in Monk [64a]), recall that $Cl_\Gamma\mathfrak{A}$ is the set of (Γ)-closed elements of $\mathfrak{A}$ and $\mathfrak{Zd}\,\mathfrak{A}$ is the BA of zero-dimensional elements of $\mathfrak{A}$; cf. 1.6.18. See also 1.7.1 and 1.9.1 for the meaning of $\mathsf{c}_{(\kappa)}\bar{\mathsf{d}}(\kappa\times\kappa)$.

THEOREM 2.1.17. *Let* $\mathfrak{A} \in \mathsf{Mn}_\alpha$, $a_\kappa = \mathsf{c}_{(\kappa)}\bar{\mathsf{d}}(\kappa\times\kappa)$ *for every* $\kappa < (\alpha+1)\cap\omega$ (*whence, in particular,* $a_0 = 1$, *and* $a_1 = 1$ *if* $\alpha > 0$), *and*

$$B = \{\mathsf{d}_{\kappa\lambda} : \kappa, \lambda < \alpha\} \cup \{a_\kappa : \kappa < (\alpha+1)\cap\omega\}.$$

We then have:
(i) $A = Sg^{(\mathfrak{Bl}\,\mathfrak{A})}B$;
more generally, for each $\Gamma \subseteq \alpha$,
(ii) $Cl_\Gamma\mathfrak{A} = Sg^{(\mathfrak{Bl}\,\mathfrak{A})}(B \cap Cl_\Gamma\mathfrak{A})$
and
$B \cap Cl_\Gamma\mathfrak{A} = \{\mathsf{d}_{\kappa\lambda} : \kappa, \lambda \in \alpha \sim \Gamma\} \cup \{a_\kappa : \kappa < (\alpha+1)\cap\omega\}$;
(iii) $Zd\mathfrak{A} = Sg^{(\mathfrak{Bl}\,\mathfrak{A})}\{a_\kappa : \kappa < (\alpha+1)\cap\omega\}$.
PROOF. For each $\Delta \subseteq \alpha$, we let

$$B_\Delta = \{\mathsf{d}_{\kappa\lambda} : \kappa, \lambda \in \Delta\} \cup \{a_\kappa : \kappa < (\alpha+1)\cap\omega\};$$

observe that $B = B_\alpha$ and $B \cap Cl_\Gamma\mathfrak{A} = B_{\alpha\sim\Gamma}$. We begin by proving the following lemma:

(1) if $\Delta, \Theta \subseteq \alpha$ and $|\Theta| < \omega$, then $\mathsf{c}_{(\Theta)}^{*}Sg^{(\mathfrak{Bl}\,\mathfrak{A})}B_\Delta \subseteq Sg^{(\mathfrak{Bl}\,\mathfrak{A})}B_{\Delta\sim\Theta}$.

It clearly suffices to consider only the case where $\Theta = \{\kappa\}$. Let C and D be the set of all finite sums of constituents associated with B_Δ and $B_{\Delta\sim\{\kappa\}}$, respectively (cf. the remarks preceding 1.9.6). We must show that $\mathsf{c}_\kappa x \in D$ for each $x \in C$. In fact, since c_κ is additive by 1.2.6, it is enough to consider the case when $\kappa \in \Delta$ and x is simply a constituent of B_Δ, specifically, $x =$

$\prod X - \sum Y$ where X and Y consist exclusively of diagonal elements $\mathsf{d}_{\kappa\lambda}$ with $\lambda \in \Delta \sim \{\kappa\}$; indeed, it is clear that all the elements of B_Δ not of this form are contained in $B_{\Delta\sim\{\kappa\}}$ and this in turn implies that these elements are also κ-closed (this last implication is an instance of the last formula of (ii) which obviously holds). The proof of the lemma is completed by observing that if $\Lambda, \Omega \subseteq \alpha$ and $|\Lambda|, |\Omega| < \omega$, then by 1.9.11(ii) and 1.9.13, $\mathsf{c}_\kappa(\prod_{\nu\in\Lambda} - \mathsf{d}_{\kappa\nu}) \in D$ if $\kappa \notin \Lambda$, while by 1.5.1, 1.5.3, and 1.5.4 we have

$$\mathsf{c}_\kappa(\mathsf{d}_{\kappa\lambda} \cdot \prod_{\mu\in\Omega} \mathsf{d}_{\kappa\mu} \cdot \prod_{\nu\in\Lambda} - \mathsf{d}_{\kappa\nu}) = \prod_{\mu\in\Omega} \mathsf{d}_{\lambda\mu} \cdot \prod_{\nu\in\Lambda} - \mathsf{d}_{\lambda\nu}$$

provided $\kappa \neq \lambda$ and $\kappa \notin \Omega \cup \Lambda$.

An immediate consequence of (1) is that $Sg^{(\mathfrak{Bl}\mathfrak{A})}B$ is closed under c_κ for each $\kappa < \alpha$, and from this (i) follows at once.

In order to prove the first formula of (ii) it suffices to show that

$$(2) \qquad x \in Sg^{(\mathfrak{Bl}\mathfrak{A})}B_{\Delta x} \text{ for all } x \in A.$$

To see that it does suffice notice that for any $\Gamma \subseteq \alpha$ we have, by (2) and 0.1.17(ii),

$$Cl_\Gamma \mathfrak{A} \subseteq \bigcup\{Sg^{(\mathfrak{Bl}\mathfrak{A})}B_{\Delta x} : x \in A,\ \Delta x \cap \Gamma = 0\} \subseteq Sg^{(\mathfrak{Bl}\mathfrak{A})}(B \cap Cl_\Gamma \mathfrak{A}),$$

while the inclusion in the opposite direction follows immediately from 1.6.19. Now in order to prove (2) we first observe that for any $x \in A$ there exists, in view of (i) and 0.1.17(ix), a finite $\Delta \subseteq \alpha$ such that $x \in Sg^{(\mathfrak{Bl}\mathfrak{A})}B_\Delta$. Hence, letting $\Theta = \Delta \cap (\alpha \sim \Delta x)$, we conclude from (1) that

$$x = \mathsf{c}_{(\Theta)}x \in Sg^{(\mathfrak{Bl}\mathfrak{A})}B_{(\Delta\sim\Theta)} \subseteq Sg^{(\mathfrak{Bl}\mathfrak{A})}B_{\Delta x}.$$

This completes the proof of (ii) (it has previously been observed that the second formula of (ii) obviously holds). Finally, (iii) follows directly from (ii) and 1.6.18(ii).

REMARKS 2.1.18. To appreciate what has been achieved in Theorem 2.1.17, recall that all elements of a $\mathsf{Mn}_\alpha\ \mathfrak{A}$ are obtained from the diagonal elements $\mathsf{d}_{\kappa\lambda}$ by performing Boolean operations and cylindrifications arbitrarily many times and in an arbitrary order. Cylindrifications are operations of a more complicated nature and less easy to handle than Boolean operations. From 2.1.17(i) it appears that a process is available which considerably restricts, and in a sense eliminates, the use of cylindrifications in constructing elements of Mn_α's. To this end, for any given $\mathsf{Mn}_\alpha\ \mathfrak{A}$, we form a set B of its elements by adjoining to the $\mathsf{d}_{\kappa\lambda}$'s some additional elements a_κ whose construction involves cylindrifications but is otherwise quite simple, namely, $a_\kappa = \mathsf{c}_{(\kappa)}\bar{\mathsf{d}}(\kappa \times \kappa)$. It turns out now that all the elements of $\mathfrak{A}$ can be constructed from elements of B by using exclusively Boolean operations and hence can be represented, e.g., as finite sums of constituents of B. This is expressed concisely by the formula 2.1.17(i). By 2.1.17(ii) every element $x \in A$ can be obtained by means of Boolean operations from those elements in B whose dimension sets are included in Δx; in particular, by 2.1.17(iii), every zero-dimensional element can be so obtained from the elements a_κ alone. The set B can be called a *Boolean base* of the Mn_α $\mathfrak{A}$, and the process just described can be referred to as the process of *eliminating cylindrifications* for Mn_α's.

This leads to a simple metalogical interpretation of the results. In this interpretation the process of eliminating cylindrifications goes over into the well-known process of eliminating quantifiers, applied in the present case to the so-called theory of identity, i.e., the predicate logic with identity symbol but without any non-logical constants. To be more specific, we restrict ourselves to interpreting 2.1.17(iii). Let $\mathfrak{A}$ be the CA of formulas associated with the theory of identity, i.e., with the empty set of sentences in a language Λ without non-logical constants; thus $\mathfrak{A}$ is minimal. The zero-dimensional elements are the equivalence classes of formulas of Λ which contain at least one sentence. The elements a_κ are correlated with certain sentences σ_κ which express in a natural way the fact that the universe contains at least κ distinct elements ($\kappa < \omega$). The sentences obtained by combining σ_κ's by means of sentential connectives are usually referred to as numerical sentences. Theorem 2.1.17(iii) appears now as an algebraic translation of the well-known metalogical result by which every sentence in Λ is logically equivalent to a numerical sentence[1].

In Section 2.2 the method of eliminating cylindrifications will be developed for a class of CA's more comprehensive then that of minimal algebras, namely, for the so-called *monadic-generated* CA's; cf. 2.2.24. It should be emphasized, however, that this method can be fruitfully applied only to rather special and restricted classes of cylindric algebras. If we consider, in particular, cylindric algebras of formulas, the method proves applicable primarily to those algebras which are associated with decidable theories. In fact, in its metalogical form, it is actually one of the most important means of obtaining affirmative solutions of decision problems. These questions will be further discussed in Part II of this work.

When the process of eliminating cylindrifications has been successfully worked out for a given class K of cylindric algebras (or for a particular CA), it usually gives us much insight into the structure of the algebras involved. This

[1]) See Hilbert-Bernays [34*], pp. 164 ff. A footnote on p. 200 op. cit. gives references to earlier papers in this direction.

is especially true in those cases in which the construction of the Boolean base B is uniform for all algebras $\mathfrak{A}$ belonging to K and is simple enough to provide us with additional information about elements of B and Boolean relations holding among them. Since every element of $\mathfrak{A}$ is a finite sum of mutually disjoint constituents of B, it is especially important for us to learn which constituents are different from zero. This information helps us indeed in establishing a unique canonical form for elements of $\mathfrak{A}$. In case B is finite we may restrict our attention to minimal constituents; in fact, the set of non-zero minimal constituents of B coincides then with the set of all atoms of $\mathfrak{A}$, and its cardinality determines completely the structure of the Boolean part of $\mathfrak{A}$.

In the case of minimal CA's the results in 2.1.17 actually provide a uniform and simple construction of a Boolean base B for every Mn $\mathfrak{A}$. Using these results we shall establish in the next few theorems several basic properties of minimal CA's. Further applications of 2.1.17 will be found in Section 2.4 and 2.5 as well as in Part II. Of special importance is Theorem 2.5.25 in which a simple criterion for isomorphism of two Mn_α's will be established, which will lead to an exhaustive description of all isomorphism types of Mn_α's. In Part II we shall establish the fact that every Mn_α is representable, i.e., isomorphic to a subdirect product of cylindric set algebras (cf. Remark 1.1.13). We shall also see there that the decision problem for the set of identities holding in all Mn_α's admits an affirmative solution. Altogether these are indeed the best results that one can hope to get from the application of our method.

THEOREM 2.1.19. *Every* Mn_α *with* $\alpha < \omega$ *is finite.*

PROOF: by 2.1.17(i) and the theory of BA's.

The cardinality of finite-dimensional Mn_α's will be determined exactly in 2.4.69.

THEOREM 2.1.20. *Let* $\mathfrak{A} \in \mathsf{Mn}_\alpha$ *with* $\alpha \geqq \omega$. *We then have:*

(i) *either* $\mathsf{c}_0^\partial \mathsf{d}_{01} = 1$, $\mathfrak{A}$ *is discrete, and* $|A| \leqq 2$, *or else* $\mathsf{c}_0^\partial \mathsf{d}_{01} \neq 1$, $\mathfrak{A}$ *is not discrete, and* $|A| = |\alpha|$;

(ii) *either* $\mathsf{c}_0^\partial \mathsf{d}_{01} = 0$ *and* $\mathfrak{A}$ *is atomless, or else* $\mathsf{c}_0^\partial \mathsf{d}_{01} \neq 0$ *and* $\mathsf{c}_0^\partial \mathsf{d}_{01}$ *is the only atom of* $\mathfrak{A}$.

PROOF. By $1.2.1^\partial$ and 1.3.12(i),(iii) we have $\mathfrak{A}$ discrete iff $\mathsf{c}_0^\partial \mathsf{d}_{01} = 1$ (cf. Remark 1.4.5). Hence (i) follows from 2.1.9(iii) and the well-known fact that every BA with more than one element includes a two-element subalgebra.

If a_κ for $\kappa < \omega$ are defined as in Theorem 2.1.17, we see by 1.9.10 that $a_\kappa \cdot \mathsf{c}_0^\partial \mathsf{d}_{01} = 0$ for every $\kappa \geqq 2$ and thus $\mathsf{c}_0^\partial \mathsf{d}_{01}$ is disjoint from any a_κ different from 1. Using this fact we conclude from 2.1.17(iii) that $\mathsf{c}_0^\partial \mathsf{d}_{01}$ includes no zero-dimensional element other than itself and 0. Hence, by $1.6.20^{\partial}$,

(1) $$\text{either } \mathsf{c}_0^{\partial}\mathsf{d}_{01} = 0 \text{ or } \mathsf{c}_0^{\partial}\mathsf{d}_{01} \text{ is an atom of } \mathfrak{A}.$$

On the other hand, by 1.10.5(ii), 1.11.3(i), 1.11.8(i), and 2.1.16 we see that every atom of $\mathfrak{A}$ is included in $\mathsf{c}_0^{\partial}\mathsf{d}_{01}$. This fact along with (1) implies (ii).

REMARK 2.1.21. By combining 2.1.20 with certain known facts from the theory of BA's we conclude that the Boolean parts of Mn_α's with $\alpha = \omega$ are of four different isomorphism types; these are the types of one-element BA's, two-element BA's, denumerable atomless BA's, and denumerable BA's with a single atom.

THEOREM 2.1.22. *If* $\mathfrak{A} \in \mathsf{Mn}_\alpha$ *and* $x \in A$, *then* $|\Delta x| \neq 1$.

PROOF. By 2.1.17(ii),(iii) we have $Cl_{\alpha\sim\{\kappa\}}\mathfrak{A} = Zd\mathfrak{A}$ for every $\kappa < \alpha$.

A consequence of this theorem is the necessity of the condition $|\alpha \sim \Gamma| \geqq 2$ in the statement of Theorem 1.10.5(i). Indeed, if $\mathfrak{A}$ is the minimal subalgebra of any cylindric set algebra in the space ${}^{\alpha}2$ with $2 \leqq \alpha < \omega$, and Γ is any subset of α such that $|\alpha \sim \Gamma| = 1$, then using 2.1.17(iii) and 2.1.22 it can easily be shown that $Zd\mathfrak{A} = \{0, {}^{\alpha}2\}$ and thus that ${}^{\alpha}2$ is a Γ-atom. On the other hand, the formula ${}^{\alpha}2 \subseteq \mathbf{C}_0^{\partial}\mathbf{D}_{01}$ does not hold since $\mathbf{C}_0^{\partial}\mathbf{D}_{01}$ is empty. Hence we see that the second part of 1.10.5(i) may fail if the premiss $|\alpha \sim \Gamma| \geqq 2$ is omitted or replaced by $\alpha \sim \Gamma \neq 0$. Analogous remarks apply to 1.10.5(iii).

THEOREM 2.1.23. *Let* $\mathfrak{A} \in \mathsf{Mn}_\alpha$, *and let* a_κ *be as in Theorem* 2.1.17 *for every* $\kappa < (\alpha+1) \cap \omega$; *in addition, set* $b = \prod_{\lambda<(\alpha+1)\cap\omega} a_\lambda$ *if* $|\{a_\lambda : \lambda < (\alpha+1)\cap\omega\}| < \omega$, *and* $b = 0$ *otherwise. Then we have*:

(i) $\mathfrak{Z}\mathfrak{d}\,\mathfrak{A}$ *is atomic and* $At\,\mathfrak{Z}\mathfrak{d}\,\mathfrak{A} = (\{a_\kappa - a_{\kappa+1} : \kappa < \alpha \cap \omega\} \cup \{b\}) \sim \{0\}$;

(ii) *either* $\mathfrak{Z}\mathfrak{d}\,\mathfrak{A}$ *is finite or else it is denumerable and, in fact, isomorphic to the Boolean set algebra of all finite subsets of* ω *and their complements.*

PROOF. Set $B = \{a_\kappa : \kappa < (\alpha+1) \cap \omega\}$ and notice that, by 1.9.10,

(1) $$a_\kappa \geqq a_{\kappa+1} \text{ for every } \kappa < \alpha \cap \omega.$$

Therefore, since $a_0 = a_1 = 1$, we see that every constituent of B can be represented either as an element of B or as the difference of two elements of B. Thus from 2.1.17(iii) we conclude that every element of $\mathfrak{Z}\mathfrak{d}\,\mathfrak{A}$ can be represented as a finite sum of elements of $\{a_\kappa - a_{\kappa+1} : \kappa < \alpha \cap \omega\} \cup B$. From this fact together with (1) part (i) follows at once; in addition it follows that every element of $\mathfrak{Z}\mathfrak{d}\,\mathfrak{A}$ is the sum of a finite number of atoms, and by the theory of BA's this leads easily to part (ii).

REMARKS 2.1.24. Theorem 2.1.23(i) admits a simple metalogical interpretation. Let Λ and $\mathfrak{A}$, just as in 2.1.18, be respectively a language of predicate logic without non-logical constants and the CA of formulas associated with the theory of identity. Given any sentence φ in Λ, let $\Theta[\varphi]$, the theory generated by φ, be the set of all sentences which are consequences of φ. Notice that the elements $a_\kappa - a_{\kappa+1}$ correspond to the sentences $\tau_\kappa = \sigma_\kappa \wedge \neg \sigma_{\kappa+1}$ expressing the fact that the universe has exactly κ elements. The metalogical result to the effect that every sentence in Λ is equivalent to a numerical sentence (cf. 2.1.18) implies as a corollary that the theories $\Theta[\tau_\kappa]$ are complete and consistent, and that every sentence φ in Λ for which $\Theta[\varphi]$ is complete and consistent is equivalent to one of the sentences τ_κ. It is easily seen that the equivalence classes of sentences φ with this property are just the atoms of the algebra of sentences associated with the theory of identity, i.e., the Boolean algebra $\mathfrak{Zd}\,\mathfrak{A}$ of zero-dimensional elements, and hence 2.1.23(i) is simply an algebraic translation of our corollary.

Several equivalent formulations of the second part of the disjunction 2.1.23(ii) are known from the theory of Boolean algebras. One of them is: $\mathfrak{Zd}\,\mathfrak{A}$ has an ordered base of type ω; another is: $\mathfrak{A}$ has exactly one proper maximal ideal which is not principal (compare in this connection the discussion in Remark 2.5.56 below).

We shall resume the discussion of minimal algebras in the last part of Section 2.4 (beginning with 2.4.61).

2.2. RELATIVIZATION OF CYLINDRIC ALGEBRAS

The notion of a relativized algebra which will be discussed in this section does not have a general algebraic character. It has been introduced in the theory of Boolean algebras and will be extended here to arbitrary cylindric algebras. In opposition to BA's, the algebras obtained by relativizing CA's are not necessarily CA's themselves.

Since BA's coincide with CA_0's, the following definition comprehends that of a relativized BA as a special case.

DEFINITION 2.2.1. *Let* $\mathfrak{A} = \langle A, +, \cdot, -, 0, 1, c_\kappa, d_{\kappa\lambda}\rangle_{\kappa,\lambda<\alpha}$ *be a* CA_α, *and suppose* $b \in A$. *Let* $Rl_b\mathfrak{A} = \{x\cdot b : x \in A\}$, *and for all* $x; y \in Rl_b\mathfrak{A}$ *and all* $\kappa, \lambda < \alpha$, *let* $x+'y = x+y$, $x\cdot'y = x\cdot y$, $-'x = b\cdot -x$, $0' = 0$, $1' = b$, $c'_\kappa x = c_\kappa x\cdot b$, *and* $d'_{\kappa\lambda} = d_{\kappa\lambda}\cdot b$. *Let* $\mathfrak{Rl}_b\mathfrak{A} = \langle Rl_b\mathfrak{A}, +', \cdot', -', 0', 1', c'_\kappa, d'_{\kappa\lambda}\rangle_{\kappa,\lambda<\alpha}$. *We refer to* $\mathfrak{Rl}_b\mathfrak{A}$ *as the algebra obtained by* ***relativizing*** *the* CA_α $\mathfrak{A}$ *to* b, *or simply as a* ***cylindric-relativized algebra of dimension*** α. *The class of all cylindric-relativized algebras of dimension* α *is denoted by* Cr_α.

We came across a particular case of cylindric-relativized algebras as far back as in Section 1.1. In fact, as is easily seen, the algebras $\mathfrak{Rl}_V\mathfrak{A}$ where $\mathfrak{A}$ is the full cylindric set algebra in a space ${}^\alpha U$ and V is a subset of ${}^\alpha U$ of the form $\bigcup_{i\in I}{}^\alpha W_i$ with W_i's pairwise disjoint, are among the generalized cylindric set algebras defined in 1.1.13; actually generalized cylindric set algebras coincide with subalgebras of the relativized algebras just described.

In the discussion of Cr's we shall be particularly interested here in conditions under which a cylindric-relativized algebra is a cylindric algebra. We begin the discussion by noticing two almost obvious facts. The first is that an isomorphic image of a Cr_α is again a Cr_α. The second is stated here formally:

THEOREM 2.2.2. $CA_\alpha \subseteq Cr_\alpha$.

PROOF. $\mathfrak{A} = \mathfrak{Rl}_1\mathfrak{A}$ for any CA_α $\mathfrak{A}$.

THEOREM 2.2.3. *If* $\mathfrak{A} \in CA_\alpha$ *and* $b \in A$, *then* $\mathfrak{Rl}_b\mathfrak{A}$ *satisfies all of the axioms of 1.1.1 except perhaps* (C_4) *and* (C_6). $\mathfrak{Rl}_b\mathfrak{A}$ *is a* CA_α *iff the following two conditions are satisfied (in* $\mathfrak{A}$*) for all* $\kappa, \lambda < \alpha$ *and for all* $x \in Rl_b\mathfrak{A}$:

(i) $c_\kappa(c_\lambda x \cdot b) \cdot b \leqq c_\lambda(c_\kappa x \cdot b)$;

(ii) $b \leqq c_\kappa(d_{\kappa\lambda} \cdot b)$.

PROOF. Let $\mathfrak{Rl}_b\mathfrak{A} = \langle Rl_b\mathfrak{A}, +', \cdot', -', 0', 1', c'_\kappa, d'_{\kappa\lambda}\rangle_{\kappa,\lambda<\alpha}$. That condition (C_0) holds is known (cf. the opening comments of this section), and it is obvious that (C_1), (C_2), and (C_5) hold. To verify (C_3), notice that, if $\kappa < \alpha$ and $x, y \in Rl_b\mathfrak{A}$, then

$$\begin{aligned} c'_\kappa(x \cdot c'_\kappa y) &= c_\kappa(x \cdot c_\kappa y \cdot b) \cdot b \\ &= c_\kappa(x \cdot c_\kappa y) \cdot b \\ &= c_\kappa x \cdot c_\kappa y \cdot b && \text{by } (C_3) \text{ for } \mathfrak{A} \\ &= c'_\kappa x \cdot c'_\kappa y. \end{aligned}$$

As to (C_7), if $\kappa, \lambda < \alpha$, $\kappa \neq \lambda$, and $x \in Rl_b\mathfrak{A}$, then

$$\begin{aligned} c'_\kappa(d'_{\kappa\lambda} \cdot x) \cdot c'_\kappa(d'_{\kappa\lambda} \cdot -'x) &= c_\kappa(d_{\kappa\lambda} \cdot x \cdot b) \cdot c_\kappa(d_{\kappa\lambda} \cdot -x \cdot b) \cdot b \\ &\leqq c_\kappa(d_{\kappa\lambda} \cdot x) \cdot c_\kappa(d_{\kappa\lambda} \cdot -x) \\ &= 0. \end{aligned}$$

Thus $\mathfrak{Rl}_b\mathfrak{A}$ satisfies all the axioms of 1.1.1 except possibly (C_4) and (C_6).

Clearly (i) holds in $\mathfrak{A}$ for all $\kappa, \lambda < \alpha$ and every $x \in Rl_b\mathfrak{A}$ iff (C_4) holds in $\mathfrak{Rl}_b\mathfrak{A}$. If $\mathfrak{Rl}_b\mathfrak{A} \in \mathsf{CA}_\alpha$, then $1' = c'_\kappa d'_{\kappa\lambda}$, and hence $b \leqq c_\kappa(d_{\kappa\lambda} \cdot b)$, i.e., (ii) holds. Now suppose (ii) holds and $\kappa \neq \lambda, \mu$. Then

$$\begin{aligned} c'_\kappa(d'_{\lambda\kappa} \cdot d'_{\kappa\mu}) &= c_\kappa(d_{\lambda\kappa} \cdot d_{\kappa\mu} \cdot b) \cdot b \\ &= s^\kappa_\lambda(d_{\kappa\mu} \cdot b) \cdot b \\ &= d_{\lambda\mu} \cdot s^\kappa_\lambda b \cdot b \\ &= d_{\lambda\mu} \cdot b && \text{by (ii)} \\ &= d'_{\lambda\mu}. \end{aligned}$$

Hence (C_6) holds in $\mathfrak{Rl}_b\mathfrak{A}$. This completes the proof of 2.2.3.

THEOREM 2.2.4. (i) $\mathsf{CA}_0 = \mathsf{Cr}_0$ *and* $\mathsf{CA}_1 = \mathsf{Cr}_1$.

(ii) $\mathsf{CA}_\alpha \subset \mathsf{Cr}_\alpha$ *for* $\alpha > 1$.

PROOF. (i) is an obvious consequence of 2.2.3. In view of 2.2.2, for (ii) we need only show that $\mathsf{Cr}_\alpha \nsubseteq \mathsf{CA}_\alpha$ for $\alpha \geqq 2$. Let $\mathfrak{A}$ be the full cylindric set algebra in the space ${}^\alpha U$ where $|U| > 2$. Let $b = -\mathbf{D}_{01}$. Then $\mathfrak{Rl}_b\mathfrak{A} = \langle Rl_b\mathfrak{A}, +', \cdot', -', 0', 1', c'_\kappa, d'_{\kappa\lambda}\rangle_{\kappa,\lambda<\alpha}$ fails to satisfy the axioms (C_4) and (C_6) for cylindric algebras. For, $b = d'_{11}$, but $c'_0(d'_{10} \cdot d'_{01}) = 0 \neq b$, and (C_6) fails. Also, if $f, g \in {}^\alpha U$, $(\alpha \sim 2) \upharpoonright f = (\alpha \sim 2) \upharpoonright g$, and $g1 = f0 \neq f1 \neq g0 \neq g1$, then $f \in c'_0 c'_1\{g\} \sim c'_1 c'_0\{g\}$, and (C_4) fails.

In connection with 2.2.3 and the proof of 2.2.4(ii), we may mention that there are cylindric-relativized algebras in which (C_4) holds and (C_6) fails, and others in which (C_6) holds and (C_4) fails.

THEOREM 2.2.5. *Let* $\mathfrak{B} \in \mathsf{Cr}_\alpha$, *say* $\mathfrak{B} = \mathfrak{Rl}_b\mathfrak{A} = \langle Rl_b\mathfrak{A}, +', \cdot', -', 0', 1', \mathsf{c}'_\kappa, \mathsf{d}'_{\kappa\lambda}\rangle_{\kappa,\lambda<\alpha}$ *where* $\mathfrak{A} \in \mathsf{CA}_\alpha$ *and* $b \in A$. *Let* $x \in B$ *and* $\kappa, \lambda, \mu < \alpha$. *We then have*:

(i) $\mathsf{d}'_{\kappa\lambda} = \mathsf{d}'_{\lambda\kappa}$;

(ii) $\mathsf{c}'_\kappa \mathsf{d}'_{\lambda\mu} = \mathsf{d}'_{\lambda\mu}$ *if* $\kappa \neq \lambda, \mu$;

(iii) $\mathsf{c}'_\kappa(\mathsf{d}'_{\lambda\kappa} \cdot \mathsf{d}'_{\kappa\mu}) \leqq \mathsf{d}'_{\lambda\mu}$ *if* $\kappa \neq \lambda, \mu$;

(iv) $\mathsf{c}'_\kappa\mathsf{c}'_\lambda\mathsf{c}'_\kappa(x \cdot \mathsf{d}'_{\kappa\mu}) \cdot \mathsf{d}'_{\kappa\mu} \leqq \mathsf{c}'_\lambda x$ *if* $\kappa \neq \lambda \neq \mu \neq \kappa$;

(v) $\mathsf{c}'_\mu\mathsf{c}'_\lambda\mathsf{c}'_\mu(\mathsf{c}'_\lambda\mathsf{c}'_\kappa\mathsf{c}'_\lambda(\mathsf{c}'_\kappa\mathsf{c}'_\mu\mathsf{c}'_\kappa(x \cdot \mathsf{d}'_{\kappa\lambda}) \cdot \mathsf{d}'_{\lambda\mu}) \cdot \mathsf{d}'_{\mu\kappa}) \cdot \mathsf{d}'_{\kappa\lambda} \leqq \mathsf{c}'_\mu x$ *if* $\kappa \neq \lambda \neq \mu \neq \kappa$.

PROOF. Clearly (i) and (ii) hold in $\mathfrak{B}$; cf. (C_3), 1.3.1, and 1.3.3. As to (iii), by (C_6) we have

$$\mathsf{c}'_\kappa(\mathsf{d}'_{\lambda\kappa} \cdot \mathsf{d}'_{\kappa\mu}) = \mathsf{c}_\kappa(\mathsf{d}_{\lambda\kappa} \cdot \mathsf{d}_{\kappa\mu} \cdot b) \cdot b \leqq \mathsf{c}_\kappa(\mathsf{d}_{\lambda\kappa} \cdot \mathsf{d}_{\kappa\mu}) \cdot b = \mathsf{d}'_{\lambda\mu}.$$

Thus (iii) holds. To prove (iv) and (v), assume that $\kappa \neq \lambda \neq \mu \neq \kappa$ and $x \in B$. Then

$$\begin{aligned}
\mathsf{c}'_\kappa\mathsf{c}'_\lambda\mathsf{c}'_\kappa(x \cdot \mathsf{d}'_{\kappa\mu}) \cdot \mathsf{d}'_{\kappa\mu} &= \mathsf{c}_\kappa(\mathsf{c}_\lambda(\mathsf{c}_\kappa(x \cdot \mathsf{d}_{\kappa\mu}) \cdot b) \cdot b) \cdot \mathsf{d}_{\kappa\mu} \cdot b \\
&\leqq \mathsf{c}_\kappa(\mathsf{c}_\lambda\mathsf{c}_\kappa(x \cdot \mathsf{d}_{\kappa\mu}) \cdot b) \cdot \mathsf{d}_{\kappa\mu} \cdot b \\
&= \mathsf{c}_\lambda\mathsf{c}_\kappa(x \cdot \mathsf{d}_{\kappa\mu}) \cdot \mathsf{d}_{\kappa\mu} \cdot b \qquad \text{by } (C_2), (C_3), (C_4) \\
&= \mathsf{c}_\lambda(x \cdot \mathsf{d}_{\kappa\mu}) \cdot b \qquad \text{by } (C_3), 1.3.3, 1.3.9 \\
&\leqq \mathsf{c}'_\lambda x,
\end{aligned}$$

and (iv) holds.

To aid in the proof of (v) set

(1) $$y = \mathsf{c}_\mu x \cdot \mathsf{d}_{\kappa\lambda} \cdot \mathsf{d}_{\lambda\mu}$$

and observe that by 1.3.7 and 1.3.9 we have

(2) $$\mathsf{c}_\kappa y \cdot \mathsf{d}_{\mu\kappa} = y \text{ and } \mathsf{c}_\lambda y \cdot \mathsf{d}_{\kappa\lambda} = y.$$

Computing we get

$$\begin{aligned}
\mathsf{c}'_\kappa\mathsf{c}'_\mu\mathsf{c}'_\kappa(x \cdot \mathsf{d}'_{\kappa\lambda}) \cdot \mathsf{d}'_{\lambda\mu} &= \mathsf{c}_\kappa(\mathsf{c}_\mu(\mathsf{c}_\kappa(x \cdot \mathsf{d}_{\kappa\lambda}) \cdot b) \cdot b) \cdot \mathsf{d}_{\lambda\mu} \cdot b \\
&\leqq \mathsf{c}_\kappa\mathsf{c}_\mu\mathsf{c}_\kappa(x \cdot \mathsf{d}_{\kappa\lambda}) \cdot \mathsf{d}_{\lambda\mu} \cdot b \\
&= \mathsf{c}_\mu\mathsf{c}_\kappa(x \cdot \mathsf{d}_{\kappa\lambda}) \cdot \mathsf{d}_{\lambda\mu} \cdot b \qquad \text{by } (C_4), 1.2.3 \\
&= \mathsf{c}_\kappa(\mathsf{c}_\mu x \cdot \mathsf{d}_{\kappa\lambda} \cdot \mathsf{d}_{\lambda\mu}) \cdot b \qquad \text{by } (C_3), (C_4), 1.3.3.
\end{aligned}$$

Thus, by (1),

(3) $$\mathsf{c}'_\kappa\mathsf{c}'_\mu\mathsf{c}'_\kappa(x \cdot \mathsf{d}'_{\kappa\lambda}) \cdot \mathsf{d}'_{\lambda\mu} \leqq \mathsf{c}_\kappa y \cdot b.$$

Continuing the computation,

$$\begin{aligned} \mathsf{c}'_\lambda \mathsf{c}'_\kappa \mathsf{c}'_\lambda(\mathsf{c}_\kappa y \cdot b) \cdot \mathsf{d}'_{\mu\kappa} &= \mathsf{c}_\lambda(\mathsf{c}_\kappa(\mathsf{c}_\lambda(\mathsf{c}_\kappa y \cdot b) \cdot b) \cdot \mathsf{d}_{\mu\kappa}) \cdot b \\ &\leqq \mathsf{c}_\lambda \mathsf{c}_\kappa \mathsf{c}_\lambda \mathsf{c}_\kappa y \cdot \mathsf{d}_{\mu\kappa} \cdot b \\ &= \mathsf{c}_\lambda \mathsf{c}_\kappa y \cdot \mathsf{d}_{\mu\kappa} \cdot b && \text{by } (C_4), 1.2.3 \\ &= \mathsf{c}_\lambda(\mathsf{c}_\kappa y \cdot \mathsf{d}_{\mu\kappa}) \cdot b && \text{by } (C_3), 1.3.3. \end{aligned}$$

Hence from (2) we conclude that

(4) $$\mathsf{c}'_\lambda \mathsf{c}'_\kappa \mathsf{c}'_\lambda(\mathsf{c}_\kappa y \cdot b) \cdot \mathsf{d}'_{\mu\kappa} \leqq \mathsf{c}_\lambda y \cdot b.$$

Repeating exactly the proof of (4) except that κ, λ and μ are now replaced respectively by λ, μ, and κ we obtain

$$\mathsf{c}'_\mu \mathsf{c}'_\lambda \mathsf{c}'_\mu(\mathsf{c}_\lambda y \cdot b) \cdot \mathsf{d}'_{\kappa\lambda} \leqq \mathsf{c}_\mu y \cdot b;$$

then (1) gives

(5) $$\mathsf{c}'_\mu \mathsf{c}'_\lambda \mathsf{c}'_\mu(\mathsf{c}_\lambda y \cdot b) \cdot \mathsf{d}'_{\kappa\lambda} \leqq \mathsf{c}'_\mu x.$$

The inclusion (v) follows directly from (3), (4), and (5).

REMARK 2.2.6. In 2.2.5(iv) the sequence $\mathsf{c}'_\kappa \mathsf{c}'_\lambda \mathsf{c}'_\kappa$ of cylindrifications can be replaced by any finite sequence of cylindrifications c'_κ and c'_λ. Analogously, the three sequences $\mathsf{c}'_\mu \mathsf{c}'_\lambda \mathsf{c}'_\mu$, $\mathsf{c}'_\lambda \mathsf{c}'_\kappa \mathsf{c}'_\lambda$, and $\mathsf{c}'_\kappa \mathsf{c}'_\mu \mathsf{c}'_\kappa$ in 2.2.5(v) can be replaced, respectively, by any three sequences of cylindrifications with indices μ and λ in the first sequence, λ and κ in the second, and κ and μ in the third.

In view of 2.2.3 and 2.2.5 the question naturally arises whether the class Cr_α can be characterized as the class of all algebras satisfying postulates (C_0)–(C_3), (C_5), and (C_7) together with identities 2.2.5(i)–(iii) and all those identities obtained from 2.2.5(iv),(v) in the way just indicated. By 2.2.4(i) the answer to this question is trivially affirmative in case $\alpha = 0, 1$. By a result of Henkin it is also affirmative for $\alpha = 2$; in this case all the equations of 2.2.5 reduce to the one equation $\mathsf{d}'_{01} = \mathsf{d}'_{10}$. On the other hand, the answer is negative for every $\alpha > 2$ since it has turned out that for no such α is Cr_α an equational class. Recall that by 0.4.63 a class of algebras is equational iff it is closed under the formation of subalgebras, homomorphic images, and direct products. However, the class Cr_α is not closed under the formation of subalgebras for any $\alpha > 2$; this has been shown by Diane Resek for $\alpha \geqq 4$, and it has been observed by Henkin that with small modification the same method can be used for $\alpha = 3$. It may be noticed that by Theorem 2.2.8(iii) below the class Cr_α is closed under the formation of direct products; the problem whether it is closed under the formation of homomorphic images is still open. It is also not known whether Cr_α with $\alpha \geqq 3$ is an elementary class.

THEOREM 2.2.7. *Let* $\mathfrak{A} \in \mathsf{CA}_\alpha$ *and* $b \in A$. *Let* c'_κ, $\mathsf{d}'_{\kappa\lambda}$ *be defined as in 2.2.1, and* $\bar{\mathsf{s}}^\kappa_\lambda$, ${}_\mu\bar{\mathsf{s}}(\kappa, \lambda)$ *be the unary operations on* $Rl_b\mathfrak{A}$ *determined by the formulas*

$$\bar{\mathsf{s}}^\kappa_\lambda x = \begin{cases} x & \text{if } \kappa = \lambda, \\ \mathsf{c}'_\kappa(\mathsf{d}'_{\kappa\lambda} \cdot x) & \text{if } \kappa \neq \lambda, \end{cases} \qquad {}_\mu\bar{\mathsf{s}}(\kappa, \lambda)x = \bar{\mathsf{s}}^\mu_\kappa \bar{\mathsf{s}}^\kappa_\lambda \bar{\mathsf{s}}^\lambda_\mu x,$$

for every $x \leqq b$. *We then have*:

(i) $\bar{\mathsf{s}}^\kappa_\lambda x = \mathsf{s}^\kappa_\lambda x \cdot b$ *for every* $x \leqq b$;

(ii) *if* $\mathfrak{Rl}_b\mathfrak{A} \in \mathsf{CA}_\alpha$, *then* ${}_\mu\bar{\mathsf{s}}(\kappa, \lambda)x = {}_\mu\mathsf{s}(\kappa, \lambda)x \cdot b$ *for every* $x \leqq b$;

(iii) *if* $\mathfrak{Rl}_b\mathfrak{A} \in \mathsf{CA}_\alpha$ *and* ${}_\mu\mathsf{s}(\kappa, \lambda) = {}_\mu\mathsf{s}(\lambda, \kappa)$, *then* ${}_\mu\bar{\mathsf{s}}(\kappa, \lambda) = {}_\mu\bar{\mathsf{s}}(\lambda, \kappa)$.

PROOF. (i) is obvious. To prove (ii) we may assume that $\mathfrak{Rl}_b\mathfrak{A} \in \mathsf{CA}_\alpha$; we then have

$$\begin{aligned} {}_\mu\bar{\mathsf{s}}(\kappa, \lambda)x &= \bar{\mathsf{s}}^\mu_\kappa \bar{\mathsf{s}}^\kappa_\lambda \bar{\mathsf{s}}^\lambda_\mu x \\ &= \bar{\mathsf{s}}^\mu_\kappa \bar{\mathsf{s}}^\kappa_\lambda (\mathsf{s}^\lambda_\mu x \cdot b) && \text{by (i)} \\ &= \bar{\mathsf{s}}^\mu_\kappa (\mathsf{s}^\kappa_\lambda (\mathsf{s}^\lambda_\mu x \cdot b) \cdot b) && \text{by (i)} \\ &= \bar{\mathsf{s}}^\mu_\kappa (\mathsf{s}^\kappa_\lambda \mathsf{s}^\lambda_\mu x \cdot b) && \text{by 1.5.3, 2.2.3(ii)} \\ &= \mathsf{s}^\mu_\kappa (\mathsf{s}^\kappa_\lambda \mathsf{s}^\lambda_\mu x \cdot b) \cdot b && \text{by (i)} \\ &= \mathsf{s}^\mu_\kappa \mathsf{s}^\kappa_\lambda \mathsf{s}^\lambda_\mu x \cdot b && \text{by 1.5.3, 2.2.3(ii)} \\ &= {}_\mu\mathsf{s}(\kappa, \lambda)x \cdot b. \end{aligned}$$

(iii) is an immediate consequence of (ii). Note that A is the domain of both ${}_\mu\mathsf{s}(\kappa, \lambda)$ and ${}_\mu\mathsf{s}(\lambda, \kappa)$ while $Rl_b\mathfrak{A}$ is the domain of both ${}_\mu\bar{\mathsf{s}}(\kappa, \lambda)$ and ${}_\mu\bar{\mathsf{s}}(\lambda, \kappa)$.

THEOREM 2.2.8. (i) *If* $\langle \mathfrak{A}_i : i \in I \rangle \in {}^I\mathsf{CA}_\alpha$ *and* $b \in \mathsf{P}_{i \in I} A_i$, *then*:

(i′) $\mathsf{P}_{i \in I} \mathfrak{Rl}_{b_i} \mathfrak{A}_i = \mathfrak{Rl}_b(\mathsf{P}_{i \in I} \mathfrak{A}_i)$;

(i″) *if* F *is a filter on* I, *then* $(\mathsf{P}_{i \in I} \mathfrak{Rl}_{b_i} \mathfrak{A}_i)/\bar{F} \cong \mathfrak{Rl}_{b/\bar{F}}(\mathsf{P}_{i \in I} \mathfrak{A}_i/\bar{F})$.

(ii) *If* $\langle \mathfrak{B}_i : i \in I \rangle \in {}^I\mathsf{Cr}_\alpha$ *and* F *is a filter on* I, *then* $\mathsf{P}_{i \in I} \mathfrak{B}_i/\bar{F} \in \mathsf{Cr}_\alpha$.

(iii) $\mathsf{PCr}_\alpha = \mathsf{Cr}_\alpha$.

(iv) $\mathsf{Up}\,\mathsf{Cr}_\alpha = \mathsf{Cr}_\alpha$.

(v) $\mathsf{UpSCr}_\alpha = \mathsf{SCr}_\alpha$.

PROOF. (i′) Let $\mathfrak{B}_i = \mathfrak{Rl}_{b_i}\mathfrak{A}_i$ for each $i \in I$, $\mathfrak{C} = \mathsf{P}_{i \in I} \mathfrak{A}_i$, and $\mathfrak{D} = \mathsf{P}_{i \in I} \mathfrak{B}_i$. Clearly $Rl_b\mathfrak{C} = D$, $+^{(\mathfrak{Rl}_b\mathfrak{C})}$ and $\cdot^{(\mathfrak{Rl}_b\mathfrak{C})}$ coincide respectively with $+^{(\mathfrak{D})}$ and $\cdot^{(\mathfrak{D})}$, $0^{(\mathfrak{Rl}_b\mathfrak{C})} = 0^{(\mathfrak{D})}$, and $1^{(\mathfrak{Rl}_b\mathfrak{C})} = 1^{(\mathfrak{D})}$. If $x \in \mathsf{P}_{i \in I} B_i$, then

$$-^{(\mathfrak{Rl}_b\mathfrak{C})}x = b \cdot -^{(\mathfrak{C})}x = \langle b_i \cdot -^{(\mathfrak{A}_i)}x_i : i \in I \rangle = \langle -^{(\mathfrak{B}_i)}x_i : i \in I \rangle = -^{(\mathfrak{D})}x;$$

$$\mathsf{c}_\kappa^{(\mathfrak{Rl}_b\mathfrak{C})}x = b \cdot \mathsf{c}_\kappa^{(\mathfrak{C})}x = \langle b_i \cdot \mathsf{c}_\kappa^{(\mathfrak{A}_i)}x_i : i \in I \rangle = \langle \mathsf{c}_\kappa^{(\mathfrak{B}_i)}x_i : i \in I \rangle = \mathsf{c}_\kappa^{(\mathfrak{D})}x;$$

and

$$\mathsf{d}_{\kappa\lambda}^{(\mathfrak{Rl}_b\mathfrak{C})} = b \cdot \mathsf{d}_{\kappa\lambda}^{(\mathfrak{C})} = \langle b_i \cdot \mathsf{d}_{\kappa\lambda}^{(\mathfrak{A}_i)} : i \in I \rangle = \langle \mathsf{d}_{\kappa\lambda}^{(\mathfrak{B}_i)} : i \in I \rangle = \mathsf{d}_{\kappa\lambda}^{(\mathfrak{D})}.$$

Thus $\mathfrak{Rl}_b\mathfrak{C} = \mathfrak{D}$, and (i′) holds.

The isomorphism needed in (i″) is the function G determined by the conditions

$$DoG = \mathsf{P}B \text{ and } G(x/\bar{F}^{(\mathfrak{B})}) = x/\bar{F}^{(\mathfrak{A})}$$

for every $x \in \mathsf{P}B$, where $B = \langle B_i : i \in I \rangle$.

Conditions (ii), (iii), (iv) easily follow from (i). (v) is a consequence of (iv) and 0.3.69(ii).

REMARK 2.2.9. From 2.2.8(v) it follows, in view of 0.3.83, that SCr_α is a universal class. Actually, however, it is known that SCr_α is an equational class for every α. For $\alpha = 0, 1, 2$ this trivially follows from 2.2.4(i) and from Henkin's result concerning Cr_2 which was stated in 2.2.6. For all finite $\alpha \geqq 3$ the result was established by Diane Resek with the help of Don Pigozzi. Her proof provides actually an infinite system of equations characterizing each CA_α. In case $\alpha = 3$, this system (in a simplified form actually due to Henkin) consists of postulates (C_0)–(C_3), (C_5), and (C_7), equations 2.2.5(i)–(iii), and all the equations obtained from 2.2.5(iv),(v) in the way indicated in 2.2.6. The equations characterizing SCr_α for $\alpha > 3$ are similar in structure but become more complicated as α increases. Pigozzi and Resek, following a suggestion of Henkin, extended the result from finite to infinite α's.[1)] For each $\alpha \geqq 3$ there is no finite system of equations characterizing the class SCr_α.

The proofs of some of the results mentioned in 2.2.6 and 2.2.9 will be found in Part II.

THEOREM 2.2.10. *Let* $\mathfrak{A} \in \mathsf{CA}_\alpha$ *and* $b \in A$. *If* $\mathsf{c}_\kappa b \cdot \mathsf{c}_\lambda b = b \leqq \mathsf{s}^\kappa_\lambda b$ *for any two distinct* $\kappa, \lambda < \alpha$ *or, equivalently,* $\mathsf{s}^\kappa_\lambda b \cdot \mathsf{s}^\lambda_\kappa b = b$ *for any* $\kappa, \lambda < \alpha$, *then* $\mathfrak{Rl}_b\mathfrak{A} \in \mathsf{CA}_\alpha$.

PROOF. Let κ, λ be any two distinct ordinals $< \alpha$. Then either of the two conditions imposed on b in the hypothesis implies $b \leqq \mathsf{s}^\kappa_\lambda b$ and hence $\mathsf{c}_\kappa b \leqq \mathsf{s}^\kappa_\lambda b$ by 1.5.9(ii). Since the inclusion in the opposite direction always holds, we conclude that $\mathsf{c}_\kappa b = \mathsf{s}^\kappa_\lambda b$ for every pair of distinct $\kappa, \lambda < \alpha$, whenever one of the two conditions mentioned above holds. The equivalence of these two conditions is now evident, and we shall prove that either of them gives $\mathfrak{Rl}_b\mathfrak{A} \in \mathsf{CA}_\alpha$.

For this purpose it suffices by 2.2.3 to show that

(1) $$\mathsf{c}_\kappa(\mathsf{c}_\lambda x \cdot b) \cdot b \leqq \mathsf{c}_\lambda(\mathsf{c}_\kappa x \cdot b)$$

for any $x \leqq b$ and any two distinct $\kappa, \lambda < \alpha$. Under these assumptions we

1) Pigozzi has also found another, simpler proof of the result discussed which applies at once to all α's without distinction, but does not provide a system of equations characterizing SCr_α.

obtain, using (C_3) several times,

$$\begin{aligned} c_\kappa(c_\lambda x \cdot b) \cdot b &= c_\kappa(c_\lambda(x \cdot b) \cdot b) \cdot b \\ &= c_\kappa(c_\lambda(x \cdot c_\kappa b \cdot c_\lambda b) \cdot c_\kappa b \cdot c_\lambda b) \cdot b \\ &= c_\kappa(c_\lambda(x \cdot c_\kappa b) \cdot c_\kappa b \cdot c_\lambda b) \cdot b \\ &= c_\kappa(c_\lambda(x \cdot c_\kappa b) \cdot c_\lambda b) \cdot b \\ &= c_\kappa c_\lambda(x \cdot c_\kappa b \cdot c_\lambda b) \cdot b \\ &= c_\kappa c_\lambda(x \cdot b) \cdot b. \end{aligned}$$

Similarly, $c_\lambda(c_\kappa x \cdot b) \cdot b = c_\lambda c_\kappa(x \cdot b) \cdot b$, and (1) follows.

REMARK 2.2.11. The condition imposed on an element b in the hypothesis of Theorem 2.2.10 acquires an interesting meaning when referred to set algebras.

Consider, in fact, a Cartesian space ${}^\alpha U$ and a point p in this space. The set ${}^\alpha U^{(p)}$ of all those points $x \in {}^\alpha U$ which differ from p in only finitely many coordinates, i.e., for which the set $\{\xi : \xi < \alpha, x_\xi \neq p_\xi\}$ is finite, can be called the *weak Cartesian space* with base U and dimension α, determined by p. (Another name for ${}^\alpha U^{(p)}$ is the αth weak Cartesian power of U relative to p; cf. the Preliminaries.) In case $\alpha < \omega$ we obviously have ${}^\alpha U^{(p)} = {}^\alpha U$ for every $p \in {}^\alpha U$, i.e., weak Cartesian spaces coincide then with ordinary Cartesian spaces.

Let now $\mathfrak{A}$ be any cylindric set algebra in the space ${}^\alpha U$. Let $F\mathfrak{A}$ be the family of all sets $X \in A$ which satisfy the hypothesis of 2.2.10 (with b replaced by X), i.e., such that

$$s^\kappa_\lambda X \cap s^\lambda_\kappa X = X$$

for any two $\kappa, \lambda < \alpha$. It turns out (rather easily, though perhaps unexpectedly) that for $\alpha \geqq 3$ the family $F\mathfrak{A}$ consists of those sets $X \in A$ which are unions of pairwise-disjoint weak Cartesian subspaces of ${}^\alpha U$:

$$X = \bigcup_{i \in I} {}^\alpha Y_i^{(p_i)}$$

where $p_i \in {}^\alpha Y_i$ and $Y_i \subseteq U$ for every $i \in I$, and ${}^\alpha Y_i^{(p_i)} \cap {}^\alpha Y_j^{(p_j)} = 0$ for any two distinct $i, j \in I$. For $\alpha = 2$ the family proves to consist simply of all Cartesian subspaces ${}^\alpha Y$ of ${}^\alpha U$ which belong to A.

By 2.2.10 all the relativized algebras $\mathfrak{Rl}_V\mathfrak{A}$ with $V \in F\mathfrak{A}$ are CA's. If, in particular, V is a weak Cartesian space, a subalgebra of $\mathfrak{Rl}_V\mathfrak{A}$ can be referred to as a *weak cylindric set algebra*. As will be shown in Part II, weak Cartesian spaces and weak set algebras can be used in characterizing representable CA's instead of ordinary Cartesian spaces and set algebras. In order to be more specific, consider a fixed but arbitrary α and let K be the class of isomorphic images of subdirect products of weak cylindric set algebras of dimension α; equivalently, K can be characterized as the class of all algebras which are isomorphic to a subalgebra of $\mathfrak{Rl}_V\mathfrak{A}$ where $\mathfrak{A}$ is any ordinary cylindric set algebra in a space ${}^{\alpha}U$ and V is a union of pairwise disjoint weak Cartesian subspaces of ${}^{\alpha}U$ ($\mathfrak{A}$ can be assumed here to be the full cylindric set algebra in the space ${}^{\alpha}U$). It is an easy matter to show that, if W is a union of pairwise disjoint ordinary subspaces of ${}^{\alpha}U$, i.e., if W is of the form $\bigcup_{i\in I}{}^{\alpha}Y_i$ where $Y_i \subseteq U$ for every $i \in I$, and $Y_i \cap Y_j = 0$ for any distinct $i, j \in I$, then W is also the union of a system of pairwise disjoint weak subspaces of ${}^{\alpha}U$. Thus we see that K includes all generalized cylindric set algebras in the sense of 1.1.13, and it then follows from the discussion in 1.1.13 that K contains every representable CA_α. It will turn out that K actually coincides with the class of representable CA_α's. From this statement and the result on the family $F\mathfrak{A}$ previously stated we obtain the following corollary: a CA_α with $\alpha \geqq 3$ is representable iff it is isomorphic to a subalgebra of an algebra $\mathfrak{Rl}_V\mathfrak{A}$ where $\mathfrak{A}$ is a cylindric set algebra and $V \in F\mathfrak{A}$. A further consequence of these results is: if $\mathfrak{A}$ is a representable CA_α with $\alpha \geqq 2$ and b is any of the elements described in 2.2.10, then $\mathfrak{Rl}_b A$ is also a representable CA_α. All the notions and results mentioned in the present remarks will be discussed in Part II.

THEOREM 2.2.12. *Let* $\mathfrak{A} \in \mathsf{CA}_\alpha$ *and* $b \in A$. *If* $b \in Zd\mathfrak{A}$, *then* $\mathfrak{Rl}_b\mathfrak{A} \in \mathsf{CA}_\alpha$ *and* $\mathfrak{Zd}\,\mathfrak{Rl}_b\mathfrak{A} = \mathfrak{Rl}_b\mathfrak{Zd}\,\mathfrak{A}$.

PROOF. It follows directly from 2.2.10 that $\mathfrak{Rl}_b\mathfrak{A}$ is a CA_α. If $x \in Rl_b\mathfrak{Zd}\,\mathfrak{A}$, then $x \leqq b$ and $\Delta^{(\mathfrak{A})}x = 0$; hence, for any $\kappa < \alpha$, $c_\kappa^{(\mathfrak{Rl}_b\mathfrak{A})}x = c_\kappa^{(\mathfrak{A})}x \cdot b = x \cdot b = x$, so that $\Delta^{(\mathfrak{Rl}_b\mathfrak{A})}x = 0$ and $x \in Zd\mathfrak{Rl}_b\mathfrak{A}$. If $x \in Zd\mathfrak{Rl}_b\mathfrak{A}$, then $x \leqq b$ and $\Delta^{(\mathfrak{Rl}_b\mathfrak{A})}x = 0$; hence $c_\kappa^{(\mathfrak{A})}x \leqq c_\kappa^{(\mathfrak{A})}b = b$, $x = c_\kappa^{(\mathfrak{Rl}_b\mathfrak{A})}x = c_\kappa^{(\mathfrak{A})}x \cdot b = c_\kappa^{(\mathfrak{A})}x$ for any $\kappa < \alpha$, and so $x \in Rl_b\mathfrak{Zd}\,\mathfrak{A}$. Thus $Zd\mathfrak{Rl}_b\mathfrak{A} = Rl_b\mathfrak{Zd}\,\mathfrak{A}$. It then follows easily that $\mathfrak{Zd}\,\mathfrak{Rl}_b\mathfrak{A} = \mathfrak{Rl}_b\mathfrak{Zd}\,\mathfrak{A}$.

Unlike the conditions stated in 2.2.3, the condition on b given in 2.2.10, or *a fortiori*, in 2.2.12, which is sufficient for $\mathfrak{Rl}_b\mathfrak{A}$ to be a CA_α, is by no means necessary. In practice, however, when constructing in this part of our work a relativized cylindric algebra $\mathfrak{Rl}_b\mathfrak{A}$ from a given cylindric algebra $\mathfrak{A}$, we restrict ourselves mostly to the case considered in 2.2.12, i.e., we assume that the element b is zero-dimensional; the results stated in Corollary 2.3.27 and Theorem 2.4.7 in the following two sections account for the importance of this particular case. Relativization to elements which are not zero-dimensional is used primarily for constructing counterexamples. However, the special form of relativization described in the next theorem presents an exception to the last remark. The theorem is rather restricted in scope but, as we shall see below (cf. 2.2.23), it underlies a general method of extending identities established for CA's.

THEOREM 2.2.13. *Let* $\mathfrak{A} \in \mathsf{CA}_\alpha$, $a \in A$, *and* $\Delta a \subseteq \{\kappa\} \subseteq \alpha$; *assume that* $\prod_{\lambda<\alpha} \mathsf{s}^\kappa_\lambda a$ *exists and set* $b = \prod_{\lambda<\alpha} \mathsf{s}^\kappa_\lambda a$. *Then* $\mathfrak{Rl}_b \mathfrak{A} \in \mathsf{CA}_\alpha$.

PROOF. By 1.6.13 we have

(1) $$\Delta \mathsf{s}^\kappa_\lambda a \subseteq \{\lambda\} \text{ for every } \lambda < \alpha.$$

Consider any $\mu, \nu < \alpha$ such that $\mu \neq \nu$. Then

$$\begin{aligned} \mathsf{s}^\mu_\nu b &= \textstyle\prod_{\lambda<\alpha} \mathsf{s}^\mu_\nu \mathsf{s}^\kappa_\lambda a && \text{by 1.5.3} \\ &= \mathsf{s}^\mu_\nu \mathsf{s}^\kappa_\mu a \cdot \textstyle\prod_{\lambda \in \alpha \sim \{\mu\}} \mathsf{s}^\mu_\nu \mathsf{s}^\kappa_\lambda a \\ &= \mathsf{s}^\kappa_\nu a \cdot \textstyle\prod_{\lambda \in \alpha \sim \{\mu\}} \mathsf{s}^\kappa_\lambda a && \text{by (1), 1.5.8(i), 1.5.10(ii)} \\ &= \textstyle\prod_{\lambda \in \alpha \sim \{\mu\}} \mathsf{s}^\kappa_\lambda a. \end{aligned}$$

Similarly, $\mathsf{s}^\nu_\mu b = \prod_{\lambda \in \alpha \sim \{\nu\}} \mathsf{s}^\kappa_\lambda a$, and thus

$$\mathsf{s}^\mu_\nu b \cdot \mathsf{s}^\nu_\mu b = b.$$

Since this equality also holds (trivially) in case $\mu = \nu$, we may apply 2.2.10 to obtain the desired conclusion.

This theorem clearly contains 2.2.12 (excluding the second part of its conclusion) as a special case.

THEOREM 2.2.14. *If* $\mathfrak{A} \in \mathsf{CA}_\alpha$ *and* $b \in Zd\,\mathfrak{A}$, *then for any given* $\kappa < \alpha$ *the following three conditions are equivalent*:

(i) $\mathfrak{Rl}_b \mathfrak{A}$ *is a discrete* CA_α;

(ii) *for all* $x \leqq b$, $\mathsf{c}_\kappa x = x$;

(iii) *for all* $x \leqq b$, $x \in Zd\,\mathfrak{A}$.

Moreover, for any given $\lambda < \alpha$, $\lambda \neq \kappa$, *(i) is also equivalent to either of the two conditions*:

(iv) $b \leqq \mathsf{d}_{\kappa\lambda}$;

(v) $b \cdot \mathsf{c}_\kappa - \mathsf{d}_{\kappa\lambda} = 0$.

PROOF. The equivalence of (i), (ii), and (iii) follows immediately from 1.3.12, 1.6.3, and 2.2.12. By 1.3.12 and 2.2.1, (i) and (iv) are easily seen to imply one another; to see that (iv) and (v) are equivalent use 1.2.5.

Theorem 2.2.14 is closely related to some results in Chapter 1 and can be regarded simply as a restatement of Theorems 1.3.12 and 1.6.3. It may be noticed that 1.3.19 and 1.6.20 can be derived from 2.2.14 as immediate corollaries.

The relativization process of 2.2.1 can be applied to any $\mathfrak{A}$ similar to CA_α's. If we relativize a Cr_α, we clearly obtain a new Cr_α. In particular we have the following

THEOREM 2.2.15. *Let* $\mathfrak{A} \in \mathsf{CA}_\alpha$, $b \in A$, *and* $c \in Rl_b\mathfrak{A}$. *Let* $C = \{x \cdot c : x \in Rl_b\mathfrak{A}\}$ *and* $\mathfrak{C} = \langle C, +', \cdot', -', 0', 1', \mathsf{c}'_\kappa, \mathsf{d}'_{\kappa\lambda}\rangle_{\kappa,\lambda<\alpha}$ *where* $+'$, $\cdot'$, *etc. are defined as in 2.2.1 with* $\mathfrak{Rl}_b\mathfrak{A}$ *and* $\mathfrak{C}$ *in place of* $\mathfrak{A}$ *and* $\mathfrak{Rl}_b\mathfrak{A}$, *respectively. Then*

$$\mathfrak{C} = \mathfrak{Rl}_c\mathfrak{A}.$$

PROOF. Clearly $C = Rl_c\mathfrak{A}$, and $+'$, $\cdot'$, $-'$, $0'$, and $1'$ equal the corresponding fundamental operations of $\mathfrak{Rl}_c\mathfrak{A}$. If $x \in C$ and $\kappa, \lambda < \alpha$, then, using the obvious fact that $c \leqq b$, we have

$$\mathsf{c}'_\kappa x = \mathsf{c}_\kappa^{(\mathfrak{Rl}_b\mathfrak{A})} x \cdot c = \mathsf{c}_\kappa x \cdot b \cdot c = \mathsf{c}_\kappa x \cdot c = \mathsf{c}_\kappa^{(\mathfrak{Rl}_c\mathfrak{A})} x,$$

$$\mathsf{d}'_{\kappa\lambda} = \mathsf{d}_{\kappa\lambda}^{(\mathfrak{Rl}_b\mathfrak{A})} \cdot c = \mathsf{d}_{\kappa\lambda} \cdot c \cdot b = \mathsf{d}_{\kappa\lambda} \cdot c = \mathsf{d}_{\kappa\lambda}^{(\mathfrak{Rl}_c\mathfrak{A})}.$$

It is also true that whenever $\mathfrak{A} \in \mathbf{S}\mathsf{Cr}_\alpha$ and $b \in A$ then $\mathfrak{Rl}_b\mathfrak{A} \in \mathbf{S}\mathsf{Cr}_\alpha$. For the case $\alpha = 3$ compare 2.2.9 and 2.2.5.

We shall now apply some properties of relativized algebras to the discussion of a new class of cylindric algebras defined as follows:

DEFINITION 2.2.16. $\mathfrak{A}$ *is called a* ***monadic-generated*** CA_α *if* $\mathfrak{A} \in \mathsf{CA}_\alpha$ *and there is a set* X *such that* $\mathfrak{A} = \mathfrak{Sg}\, X$, *and* $|\Delta x| \leqq 1$ *for every* $x \in X$.

COROLLARY 2.2.17. (i) *All* CA_0's *and* CA_1's *are monadic-generated.*

(ii) *For every* α, *the class of monadic-generated* CA_α's *includes* Mn_α *and is included in* Lf_α.

PROOF: (i) is obvious; (ii) by 0.1.23 and 2.1.5(i).

REMARK 2.2.18. By 2.2.17, and in view of 1.11.2, every monadic-generated CA_α with $\alpha \geqq \omega$ can be isomorphically represented as an algebra of formulas associated with a set of sentences in a language Λ of predicate logic. From 2.2.16 it is seen that for Λ we can always choose a language in which all non-logical predicates are of rank 0 or 1 (and no operation symbols occur). Such a language is usually referred to as a language of *monadic predicate logic*. If, conversely, Λ is a language of monadic predicate logic, then every algebra of formulas associated with Λ is monadic-generated.

THEOREM 2.2.19. *For a* CA_α $\mathfrak{A}$ *to be monadic-generated it is necessary and sufficient that there exist a set* X *such that* $\mathfrak{A} = \mathfrak{Sg}\, X$ *and, for every* $x \in X$, $\Delta x = 0$ *or* $\Delta x = 1$.

PROOF. The condition is sufficient by 2.2.16. In order to prove necessity, choose $Y \subseteq A$ such that $\mathfrak{A} = \mathfrak{S}\mathfrak{g}Y$ and $|\Delta y| \leqq 1$ for every $y \in Y$. For each $y \in Y$ let $y' = \mathsf{s}_0^\xi y$ if $|\Delta y| = 1$ and $\Delta y = \{\xi\}$; let $y' = y$ if $|\Delta y| = 0$. Then it is readily seen from 1.5.8(i), 1.5.10(v) and 1.6.13 that $X = \{y' : y \in Y\}$ has the desired properties.

We now want to develop the process of eliminating cylindrifications for the class of monadic-generated CA_α's, thus extending the corresponding result for minimal CA_α's, Theorem 2.1.17. To this end, which will be achieved in Theorem 2.2.24, we shall need three lemmas, 2.2.20–2.2.22, which improve some earlier theorems, 1.9.10, 1.9.11(ii), and 1.9.13. In establishing the last two of these lemmas the concept of relativized algebras will be applied.[1)]

LEMMA 2.2.20. *Let $\mathfrak{A} \in \mathsf{CA}_\alpha$, $x \in A$, and $\Delta x \subseteq 1$. If $\Gamma \subseteq \Delta \subseteq \alpha$ and $|\Delta| < \omega$, then*

$$\mathsf{c}_{(\Delta)}(\bar{\mathsf{d}}(\Delta \times \Delta) \cdot \prod_{\mu \in \Delta} \mathsf{s}_\mu^0 x) \leqq \mathsf{c}_{(\Gamma)}(\bar{\mathsf{d}}(\Gamma \times \Gamma) \cdot \prod_{\mu \in \Gamma} \mathsf{s}_\mu^0 x).$$

PROOF. We immediately have

$$\bar{\mathsf{d}}(\Delta \times \Delta) \cdot \prod_{\mu \in \Delta} \mathsf{s}_\mu^0 x \leqq \bar{\mathsf{d}}(\Gamma \times \Gamma) \cdot \prod_{\mu \in \Gamma} \mathsf{s}_\mu^0 x.$$

Hence,

$$\begin{aligned} \mathsf{c}_{(\Delta)}(\bar{\mathsf{d}}(\Delta \times \Delta) \cdot \prod_{\mu \in \Delta} \mathsf{s}_\mu^0 x) &\leqq \mathsf{c}_{(\Delta)}(\bar{\mathsf{d}}(\Gamma \times \Gamma) \cdot \prod_{\mu \in \Gamma} \mathsf{s}_\mu^0 x) && \text{by 1.7.5} \\ &= \mathsf{c}_{(\Gamma)}\mathsf{c}_{(\Delta \sim \Gamma)}(\bar{\mathsf{d}}(\Gamma \times \Gamma) \cdot \prod_{\mu \in \Gamma} \mathsf{s}_\mu^0 x) && \text{by 1.7.3} \\ &= \mathsf{c}_{(\Gamma)}(\bar{\mathsf{d}}(\Gamma \times \Gamma) \cdot \prod_{\mu \in \Gamma} \mathsf{s}_\mu^0 x) && \text{by 1.6.13, 1.7.5, 1.9.3.} \end{aligned}$$

Corollary 1.9.10 is clearly a particular case of Lemma 2.2.20, obtained by setting $x = 1$.

LEMMA 2.2.21. *Let $\mathfrak{A} \in \mathsf{CA}_\alpha$, $x \in \mathfrak{A}$, and $\Delta x \subseteq 1$. If $\Gamma, \Delta \subseteq \alpha$ and $|\Gamma| = |\Delta| < \omega$, then*

$$\mathsf{c}_{(\Gamma)}(\bar{\mathsf{d}}(\Gamma \times \Gamma) \cdot \prod_{\mu \in \Gamma} \mathsf{s}_\mu^0 x) = \mathsf{c}_{(\Delta)}(\bar{\mathsf{d}}(\Delta \times \Delta) \cdot \prod_{\mu \in \Delta} \mathsf{s}_\mu^0 x).$$

PROOF. In this proof we make an essential use of 1.9.11(ii). (Again, the latter theorem is but a particular case of the present lemma.)

It is clear that we may assume $\Gamma, \Delta \neq 0$. For each $\Theta \subseteq \alpha$ with $|\Theta| < \omega$ we let $x_\Theta = \prod_{\mu \in \Theta} \mathsf{s}_\mu^0 x$; notice that $x_0 = 1$ and,

(1) for any finite $\Theta, \Theta' \subseteq \alpha$, $x_\Theta \cdot x_{\Theta'} = x_{\Theta \cup \Theta'}$ and $\Delta x_\Theta \subseteq \Theta$;

[1)] The basic results on monadic-generated CA's presented here originate with Monk. In particular, 2.2.21, 2.2.22, and 2.2.24 correspond respectively to Theorems 8, 10, and 12 of Monk [64a]. However, the actual proofs of 2.2.21 and 2.2.22 given here use the method of relativization which is based on 2.2.13 and is due to Pigozzi and Tarski.

the latter inclusion follows by 1.6.13. By 1.5.8(i), 1.5.9(i), and the premiss $\Delta x \subseteq \{0\}$ we have

(2) $$\mathsf{c}_\lambda \mathsf{s}_\lambda^0 x = \mathsf{c}_0 \mathsf{s}_0^\lambda x = \mathsf{c}_0 x \text{ for any } \lambda < \alpha,$$

and hence

(3) $$x_\Theta \leqq \mathsf{c}_0 x \text{ for any } \Theta \subseteq \alpha,\ 0 < |\Theta| < \omega.$$

Consider any finite $\Phi \subseteq \alpha$ and $\kappa \in \Phi$. Then, by (2),

$$\mathsf{c}_\kappa x_\Phi = x_{\Phi \sim \{\kappa\}} \cdot \mathsf{c}_\kappa \mathsf{s}_\kappa^0 x = x_{\Phi \sim \{\kappa\}} \cdot \mathsf{c}_0 x,$$

and thus, by (2) and (3), $\mathsf{c}_\kappa x_\Phi = x_{\Phi \sim \{\kappa\}}$ if $|\Phi| \geqq 2$ and $\mathsf{c}_\kappa x_\Phi = \mathsf{c}_0 x$ if $\Phi = \{\kappa\}$. From this result we obtain by a simple inductive argument:

(4) $$\mathsf{c}_{(\Theta)} x_\Theta = \mathsf{c}_0 x \text{ for any } \Theta \subseteq \alpha,\ 0 < |\Theta| < \omega.$$

Keeping (4) in mind, we prove the lemma under the assumption $\alpha < \omega$. Let

(5) $$\mathfrak{B} = \mathfrak{Rl}_{x_\alpha} \mathfrak{A} = \langle B, +', \cdot', -', 0', 1', \mathsf{c}'_\kappa, \mathsf{d}'_{\kappa\lambda} \rangle_{\kappa,\lambda<\alpha}.$$

Then $\mathfrak{B} \in \mathsf{CA}_\alpha$ by 2.2.13, and hence from 1.9.11(ii) we obtain

(6) $$\mathsf{c}'_{(\Gamma)} \bar{\mathsf{d}}'(\Gamma \times \Gamma) = \mathsf{c}'_{(\Delta)} \bar{\mathsf{d}}'(\Delta \times \Delta).$$

Now

$$\begin{aligned} \bar{\mathsf{d}}'(\Gamma \times \Gamma) &= \prod\{-'\mathsf{d}'_{\kappa\lambda} : \langle \kappa, \lambda \rangle \in (\Gamma \times \Gamma) \sim Id\} \\ &= \prod\{x_\alpha - (\mathsf{d}_{\kappa\lambda} \cdot x_\alpha) : \langle \kappa, \lambda \rangle \in (\Gamma \times \Gamma) \sim Id\} \\ &= x_\alpha \cdot \prod\{-\mathsf{d}_{\kappa\lambda} : \langle \kappa, \lambda \rangle \in (\Gamma \times \Gamma) \sim Id\} \\ &= x_\alpha \cdot \bar{\mathsf{d}}(\Gamma \times \Gamma). \end{aligned}$$

Using this result and arguing by induction on $|\Gamma|$ we have

$$\mathsf{c}'_{(\Gamma)} \bar{\mathsf{d}}'(\Gamma \times \Gamma) = x_\alpha \cdot \mathsf{c}_{(\Gamma)}(\bar{\mathsf{d}}(\Gamma \times \Gamma) \cdot x_\Gamma).$$

The latter equation when combined with a similar one for Δ enables us to conclude from (6) that

(7) $$x_\alpha \cdot \mathsf{c}_{(\Gamma)}(\bar{\mathsf{d}}(\Gamma \times \Gamma) \cdot x_\Gamma) = x_\alpha \cdot \mathsf{c}_{(\Delta)}(\bar{\mathsf{d}}(\Delta \times \Delta) \cdot x_\Delta).$$

Now (1) implies that

$$\Delta(\mathsf{c}_{(\Gamma)}(\bar{\mathsf{d}}(\Gamma \times \Gamma) \cdot x_\Gamma)) = \Delta(\mathsf{c}_{(\Delta)}(\bar{\mathsf{d}}(\Delta \times \Delta) \cdot x_\Delta)) = 0;$$

therefore, applying $\mathsf{c}_{(\alpha)}$ to both members of (7) and using (4), we obtain

(8) $$\mathsf{c}_0 x \cdot \mathsf{c}_{(\Gamma)}(\bar{\mathsf{d}}(\Gamma \times \Gamma) \cdot x_\Gamma) = \mathsf{c}_0 x \cdot \mathsf{c}_{(\Delta)}(\bar{\mathsf{d}}(\Delta \times \Delta) \cdot x_\Delta).$$

From (3) and the assumption $\Gamma, \Delta \neq 0$ we infer that $x_\Gamma, x_\Delta \leqq \mathsf{c}_0 x$ and hence that the initial factor $\mathsf{c}_0 x$ of both members of the equation (8) may be removed; the resulting equality is the conclusion of the lemma.

We now take up the case $\alpha \geqq \omega$. To begin with notice that both $\mathsf{c}_{(\Gamma)}(\bar{\mathsf{d}}(\Gamma \times \Gamma) \cdot x_\Gamma)$ and $\mathsf{c}_{(\Delta)}(\bar{\mathsf{d}}(\Delta \times \Delta) \cdot x_\Delta)$ belong to $Sg^{(\mathfrak{A})}\{x\}$. Thus we may assume without loss of generality that $\mathfrak{A}$ is generated by x and consequently, in view of 2.1.5(i), that $\mathfrak{A} \in \mathsf{Lf}_\alpha$. Let σ be any finite permutation of α such that $\sigma^*(\Gamma \cup \Delta) \subseteq \omega$ and let β be the least ordinal including $\sigma^*(\Gamma \cup \Delta)$. Let

(9) $$\mathfrak{C} = \langle A, +, \cdot, -, 0, 1, \mathsf{c}_\kappa, \mathsf{d}_{\kappa\lambda} \rangle_{\kappa,\lambda < \beta}.$$

Clearly $\mathfrak{C} \in \mathsf{CA}_\beta$. (By (9), $\mathfrak{C}$ is a reduct of the CA_α $\mathfrak{A}$ in the sense of 0.5.1, and actually a reduct in the more special sense in which this notion will be introduced and discussed in Section 2.6. However, the only property of reducts which we use in the present proof is that a reduct of a CA in the sense of 2.6.1 is always a CA itself.) Since $\beta < \omega$, we can, because of (9) and the first part of our proof, apply the present lemma with $\mathfrak{C}$, $\sigma^*\Gamma$, and $\sigma^*\Delta$ in place of $\mathfrak{A}$, Γ, and Δ, respectively. This gives

(10) $$\mathsf{c}_{(\sigma^*\Gamma)}(\bar{\mathsf{d}}(\sigma^*\Gamma \times \sigma^*\Gamma) \cdot x_{\sigma^*\Gamma}) = \mathsf{c}_{(\sigma^*\Delta)}(\bar{\mathsf{d}}(\sigma^*\Delta \times \sigma^*\Delta) \cdot x_{\sigma^*\Delta}).$$

We have by 1.5.11(i) and 1.11.11(i),(iv) that

(11) $$\mathsf{s}_{\sigma^{-1}} x_{\sigma^*\Gamma} = x_\Gamma \text{ and } \mathsf{s}_{\sigma^{-1}} x_{\sigma^*\Delta} = x_\Delta.$$

Applying $\mathsf{s}_{\sigma^{-1}}$ to both members of the equation (10) and using (11) and 1.11.12(vi),(ix), the conclusion of the lemma is easily obtained.

LEMMA 2.2.22. *Let $\mathfrak{A} \in \mathsf{CA}_\alpha$, $x \in A$, and $\Delta x \subseteq 1$. Assume $\Gamma \subseteq \alpha$, $|\Gamma| < \omega$, and $\kappa \in \alpha \sim \Gamma$. Then, setting $\Delta^+ = \Delta \cup \{\kappa\}$ for every $\Delta \subseteq \Gamma$, we have*

$$\mathsf{c}_\kappa(\bar{\mathsf{d}}(\{\kappa\} \times \Gamma) \cdot \mathsf{s}^0_\kappa x) =$$
$$\prod_{\Delta \subseteq \Gamma}[-(\bar{\mathsf{d}}(\Delta \times \Delta) \cdot \prod_{\mu \in \Delta} \mathsf{s}^0_\mu x) + \mathsf{c}_{(\Delta^+)}(\bar{\mathsf{d}}(\Delta^+ \times \Delta^+) \cdot \prod_{\mu \in \Delta^+} \mathsf{s}^0_\mu x)].$$

PROOF. For each finite $\Theta \subseteq \alpha$ we let $x_\Theta = \prod_{\mu \in \Theta} \mathsf{s}^0_\mu x$, as in the proof of 2.2.21; similarly we let $(-x)_\Theta = \prod_{\mu \in \Theta} -\mathsf{s}^0_\mu x$. Moreover, for every finite $\Theta \subseteq \alpha$ we set

$$h_\Theta = \bar{\mathsf{d}}(\Theta \times \Theta) \cdot x_\Theta.$$

We will first prove the lemma under the assumption $\alpha < \omega$. This proof falls into two distinct parts. In the first part we follow closely the lines of the argument in that part of the proof of 2.2.21 which leads to the formula (7); we use, however, 1.9.13 instead of 1.9.11(ii). In this way we arrive at the following result:

(1) If $\Phi \subseteq \alpha$, $|\Phi| < \omega$, and $\lambda \in \alpha \sim \Phi$, then

$$x_{\Phi \cup \{\lambda\}} \cdot c_\lambda(\bar{d}(\{\lambda\} \times \Phi) \cdot x_{\{\lambda\}}) = x_{\Phi \cup \{\lambda\}} \cdot \prod_{\Omega \subseteq \Phi}[-\bar{d}(\Omega \times \Omega) + c_{(\Omega \cup \{\lambda\})} h_{\Omega \cup \{\lambda\}}].$$

From here on we proceed as follows. Clearly, for every $\Delta \subseteq \Gamma$ we have

$$\bar{d}(\{\kappa\} \times \Gamma) \cdot x_{\{\kappa\}} \cdot h_\Delta \leqq h_{\Delta^+} \leqq c_{(\Delta^+)} h_{\Delta^+}.$$

Hence, since $\kappa \notin \Delta h_\Delta \cup \Delta c_{(\Delta^+)} h_{\Delta^+}$, we have

(2) $$c_\kappa(\bar{d}(\{\kappa\} \times \Gamma) \cdot x_{\{\kappa\}}) \leqq \prod_{\Delta \subseteq \Gamma}(-h_\Delta + c_{(\Delta^+)} h_{\Delta^+}).$$

On the other hand, for each $\Theta \subseteq \Gamma$,

$$\begin{aligned} & x_{\Theta^+} \cdot (-x)_{\Gamma \sim \Theta} \cdot \prod_{\Delta \subseteq \Gamma}(-h_\Delta + c_{(\Delta^+)} h_{\Delta^+}) \\ & \leqq x_{\Theta^+} \cdot (-x)_{\Gamma \sim \Theta} \cdot \prod_{\Delta \subseteq \Theta}(-h_\Delta + c_{(\Delta^+)} h_{\Delta^+}) \\ & = x_{\Theta^+} \cdot (-x)_{\Gamma \sim \Theta} \cdot \prod_{\Delta \subseteq \Theta}(-\bar{d}(\Delta \times \Delta) + -x_\Delta \cdot x_\Theta + c_{(\Delta^+)} h_{\Delta^+}) \\ & = x_{\Theta^+} \cdot (-x)_{\Gamma \sim \Theta} \cdot \prod_{\Delta \subseteq \Theta}(-\bar{d}(\Delta \times \Delta) + c_{(\Delta^+)} h_{\Delta^+}) \\ & = x_{\Theta^+} \cdot (-x)_{\Gamma \sim \Theta} \cdot c_\kappa(\bar{d}(\{\kappa\} \times \Theta) \cdot x_{\{\kappa\}}); \end{aligned}$$

this last equality is obtained by observing that $\kappa \notin \Theta$ and applying (1). Therefore, since $(-x)_{\Gamma \sim \Theta} \cdot x_{\{\kappa\}} \leqq \bar{d}(\{\kappa\} \times (\Gamma \sim \Theta))$ by 1.5.6, we may conclude that

(3) $$\begin{aligned} & x_{\Theta^+} \cdot (-x)_{\Gamma \sim \Theta} \cdot \prod_{\Delta \subseteq \Gamma}(-h_\Delta + c_{(\Delta^+)} h_{\Delta^+}) \leqq \\ & \qquad x_{\Theta^+} \cdot (-x)_{\Gamma \sim \Theta} \cdot c_\kappa(\bar{d}(\{\kappa\} \times \Gamma) \cdot x_{\{\kappa\}}). \end{aligned}$$

Now $\sum_{\Theta \subseteq \Gamma}(x_\Theta \cdot (-x)_{\Gamma \sim \Theta})$ is easily seen to be the sum of all minimal constituents of the set $\{s^0_\mu x : \mu \in \Gamma\}$, and hence equals 1; therefore, $x_{\{\kappa\}} = \sum_{\Theta \subseteq \Gamma}(x_{\Theta^+} \cdot (-x)_{\Gamma \sim \Theta})$. Making use of this latter equation we easily obtain from (3)

$$x_{\{\kappa\}} \cdot \prod_{\Delta \subseteq \Gamma}(-h_\Delta + c_{(\Delta^+)} h_{\Delta^+}) \leqq x_{\{\kappa\}} \cdot c_\kappa(\bar{d}(\{\kappa\} \times \Gamma) \cdot x_{\{\kappa\}}).$$

Applying c_κ to both sides of this inequality we get

(4) $$c_0 x \cdot \prod_{\Delta \subseteq \Gamma}(-h_\Delta + c_{(\Delta^+)} h_{\Delta^+}) \leqq c_0 x \cdot c_\kappa(\bar{d}(\{\kappa\} \times \Gamma) \cdot x_{\{\kappa\}}).$$

Furthermore, since $x_{\{\kappa\}} \leqq c_0 x$ and $-h_0 + c_{(0^+)} h_{0^+} = c_0 x$, (4) also holds if the initial factor $c_0 x$ is removed. The resulting equation together with (2) gives the conclusion of the lemma.

The argument by which the lemma for the case $\alpha \geqq \omega$ is obtained from the case $\alpha < \omega$ is entirely analogous to the argument given in the last paragraph of the proof of 2.2.21. The details are left to the reader.

REMARK 2.2.23. In proving Lemmas 2.2.21 and 2.2.22, and more specifically, in deriving these lemmas from the corresponding but more special results in Section 1.9, we have used a general method based upon Theorem 2.2.13. (The same method could be used, by the way, to derive 2.2.20 from 1.9.10.) The method can be described as an application of a general metamathematical theorem concerning identities in cylindric algebras. The formulation of this theorem, which will be discussed in Part II of our work, is somewhat involved and will not be given. We should like only to mention that it is closely related to a well-known metalogical theorem by which a valid sentence of predicate logic continues to be logically valid if all its quantifiers are relativized to a unary predicate not occurring in the sentence; cf., for instance, Tarski-Mostowski-Robinson [53*], pp. 24 ff.

THEOREM 2.2.24. *Let* $\mathfrak{A}$ *be a monadic-generated* CA_α, *and more specifically let* $\mathfrak{A} = \mathfrak{Sg}X$ *where* $\Delta x \subseteq 1$ *for every* $x \in X$. *Let*

$$C = \{a_\kappa(Y, Z) : Y \cup Z \subseteq X, |Y \cup Z| < \omega, Y \cap Z = 0, \kappa < (\alpha+1) \cap \omega\}$$

where

$$a_\kappa(Y, Z) = \mathsf{c}_{(\kappa)}[\bar{\mathsf{d}}(\kappa \times \kappa) \cdot \textstyle\prod_{\lambda<\kappa}(\prod_{y \in Y} \mathsf{s}^0_\lambda y - \sum_{z \in Z} \mathsf{s}^0_\lambda z)],$$

and let

$$B = \{\mathsf{d}_{\kappa\lambda} : \kappa, \lambda < \alpha\} \cup \{\mathsf{s}^0_\xi x : x \in X, \xi < \alpha\} \cup C.$$

We then have

(i) $$A = Sg^{(\mathfrak{Bl}\,\mathfrak{A})}B;$$

more generally, for each $\Gamma \subseteq \alpha$,

(ii) $$Cl_\Gamma \mathfrak{A} = Sg^{(\mathfrak{Bl}\,\mathfrak{A})}(B \cap Cl_\Gamma \mathfrak{A})$$

and

$$B \cap Cl_\Gamma \mathfrak{A} = \{\mathsf{d}_{\kappa\lambda} : \kappa, \lambda \in \alpha \sim \Gamma\} \cup \{\mathsf{s}^0_\xi x : x \in X, \xi \in \alpha \sim \Gamma\} \cup C;$$

(iii) $$Zd\,\mathfrak{A} = Sg^{(\mathfrak{Bl}\,\mathfrak{A})}C.$$

PROOF. For each $\Delta \subseteq \alpha$, we let

$$B_\Delta = \{\mathsf{d}_{\kappa\lambda} : \kappa, \lambda \in \Delta\} \cup \{\mathsf{s}^0_\xi x : x \in X, \xi \in \Delta\} \cup C;$$

observe that $B = B_\alpha$ and $B \cap Cl_\Gamma \mathfrak{A} = B_{\alpha \sim \Gamma}$. We begin by proving the following lemma:

(1) if $\Delta, \Theta \subseteq \alpha$ and $|\Theta| < \omega$, then $\mathsf{c}_{(\Theta)}^* Sg^{(\mathfrak{Bl}\,\mathfrak{A})} B_\Delta \subseteq Sg^{(\mathfrak{Bl}\,\mathfrak{A})} B_{\Delta \sim \Theta}$.

It clearly suffices to consider only the case where $\Theta = \{\kappa\}$. Let D and E be the set of all finite sums of constituents associated with B_Δ and $B_{\Delta \sim \{\kappa\}}$, respectively. We must show that $\mathsf{c}_\kappa w \in E$ for each $w \in D$. In fact, since c_κ is additive by 1.2.6, it is enough to consider the case when $\kappa \in \Delta$ and w is simply a constituent of B_Δ, specifically, $w = \prod U - \sum V$ where U and V consist exclusively of diagonal elements $\mathsf{d}_{\kappa\lambda}$ with $\lambda \in \Delta \sim \{\kappa\}$ and the elements $\mathsf{s}^0_\kappa x$ with $x \in X$; indeed, it is clear that all the elements of B_Δ not of this form are contained in $B_{\Delta \sim \{\kappa\}}$, and this in turn implies that these elements are also κ-closed. (This last implication is an instance of the last formula of (ii) which obviously holds in light of 1.6.15.) The proof of the lemma is completed by observing first of all that, by 2.2.21 and 2.2.22,

$$\mathsf{c}_\kappa(\prod_{\nu \in \Lambda} - \mathsf{d}_{\kappa\nu} \cdot \prod_{y \in Y} \mathsf{s}^0_\kappa y \cdot \prod_{z \in Z} - \mathsf{s}^0_\kappa z) \in E$$

if $\kappa \notin \Lambda$ and $Y, Z \subseteq X$ with $|Y \cup Z| < \omega$ and $Y \cap Z = 0$ (in applying 2.2.22 we take $\prod Y - \sum Z$ for x); while on the other hand, by 1.5.1, 1.5.3, 1.5.4, and 1.5.11(i) we have

$$\mathsf{c}_\kappa(\mathsf{d}_{\kappa\lambda} \cdot \prod_{\mu \in \Omega} \mathsf{d}_{\kappa\mu} \cdot \prod_{\nu \in \Lambda} - \mathsf{d}_{\kappa\nu} \cdot \prod_{y \in Y} \mathsf{s}^0_\kappa y \cdot \prod_{z \in Z} - \mathsf{s}^0_\kappa z) =$$
$$\prod_{\mu \in \Omega} \mathsf{d}_{\lambda\mu} \cdot \prod_{\nu \in \Lambda} - \mathsf{d}_{\lambda\nu} \cdot \prod_{y \in Y} \mathsf{s}^0_\lambda y \cdot \prod_{z \in Z} - \mathsf{s}^0_\lambda z$$

provided $\kappa \neq \lambda$, $\kappa \notin \Omega \cup \Lambda$, and $Y, Z \subseteq X$ with $|Y \cup Z| < \omega$ and $Y \cap Z = 0$.

To complete the argument, turn back to the proof of 2.1.17 and follow exactly the way in which that theorem was derived from lemma (1) stated there.

The specific assumption in 2.2.24 to the effect that $\Delta x \subseteq 1$ for every $x \in X$ does not imply any loss of generality in view of 2.2.19.

Corollary 2.2.25. *If the hypothesis of 2.2.24 is supplemented by the premiss $|X| < \omega$, and if, in formulas for C and $a_\kappa(Y, Z)$, Z is replaced by $X \sim Y$, then the conclusions 2.2.24(i)–(iii) continue to hold.*

Proof. In view of 1.2.6 and the remarks preceding 1.9.6, it is easily seen that each element $a_\kappa(Y, Z)$ with $Y \cup Z \subseteq X$, $|Y \cup Z| < \omega$, $Y \cap Z = 0$, and $\kappa < (\alpha+1) \cap \omega$ can be represented as a finite sum of elements of the form $a_\kappa(Y, X \sim Y)$ where $Y \subseteq X$ and $|Y| < \omega$. Hence, the corollary follows immediately from 2.2.24.

Corollary 2.2.26. *Every finitely generated* CA_α *with* $\alpha < \omega$ *which is monadic-generated is finite.*

Proof: by 2.2.24 and the theory of BA's.

From 2.2.24 we can actually derive a stronger conclusion: for fixed finite α and β, the cardinalities of all monadic-generated CA_α's with β generators are bounded above (by a finite cardinal). In Section 2.5 the least upper bound for these cardinalities will be established; cf. 2.5.60 and 2.5.61.

In view of 2.2.17, Corollary 2.2.26 comprehends as particular cases both Theorem 2.1.10 concerning CA_1's and Theorem 2.1.19 applying to Mn_α's. With the help of 2.2.24 and 2.2.25 we will establish in Section 2.5 some further properties of monadic-generated algebras analogous to those stated for minimal algebras in 2.1.23; cf. 2.5.65 and 2.5.66.

2.2.26 implies directly that every finitely generated monadic-generated CA of finite dimension is atomic. From 2.2.25 we can easily obtain a description of all atoms of such an algebra in terms of the set X of its generators (with the property that $\Delta x \subseteq 1$ for every $x \in X$). We shall state here explicitly only a rather special result in this direction which will be needed for our further discussion; namely, using 2.2.25, we shall give a description of all zero-dimensional atoms of a finitely generated CA_1:

THEOREM 2.2.27. *Assume $\mathfrak{A} \in \mathsf{CA}_1$, X is a finite subset of A, and $\mathfrak{A} = \mathfrak{S}\mathfrak{g} X$. Then*

$$At\mathfrak{A} \cap Zd\mathfrak{A} = \{\textstyle\prod_{Y \neq W \subseteq X} - \mathsf{c}_0(\prod W - \sum(X \sim W)) : Y \subseteq X\} \sim \{0\},$$

and hence

$$|At\mathfrak{A} \cap Zd\mathfrak{A}| \leqq 2^{|X|}.$$

PROOF. Let

$$h_Y = \textstyle\prod Y - \sum(X \sim Y)$$

for each $Y \subseteq X$. Notice that the h_Y's are exactly the minimal constituents of the set X and that distinct $Y, Z \subseteq X$ correspond to distinct constituents. Some simple consequences of these facts, which we will find useful in the present proof, are the following:

(1) $$\mathsf{c}_0^\partial h_Y \cdot \mathsf{c}_0 h_Z = 0 \text{ for any } Y, Z \subseteq X \text{ such that } Y \neq Z;$$

(2) $$\textstyle\prod_{W \subseteq X} - \mathsf{c}_0 h_W = 0;$$

(3) $$\textstyle\prod_{Y \neq W \subseteq X} - \mathsf{c}_0 h_W \leqq h_Y \text{ for every } Y \subseteq X.$$

In order to see (1) note that

$$\mathsf{c}_0^\partial h_Y \cdot \mathsf{c}_0 h_Z = \mathsf{c}_0(\mathsf{c}_0^\partial h_Y \cdot h_Z) \leqq \mathsf{c}_0(h_Y \cdot h_Z);$$

both (2) and (3) are simple consequences of the fact that $\sum_{W \subseteq X} h_W = 1$.

From 2.2.25 we have that the atoms of $\mathfrak{A}$ are just the non-zero minimal constituents of $B = X \cup \{\mathsf{c}_0 h_Y : Y \subseteq X\}$. Let

$$y = h_Y \cdot \textstyle\prod_{Z \in K} \mathsf{c}_0 h_Z \cdot \prod_{W \in SbX \sim K} - \mathsf{c}_0 h_W,$$

where $Y \subseteq X$ and $K \subseteq Sb\,X$, be an arbitrary minimal constituent of B, and assume, in addition, that y is zero-dimensional and different from 0. Then

$$y = \mathsf{c}_0^\partial y = \mathsf{c}_0^\partial h_Y \cdot \prod_{Z \in K} \mathsf{c}_0 h_Z \cdot \prod_{W \in Sb\,X \sim K} -\mathsf{c}_0 h_W.$$

Hence by (1) we have $K \subseteq \{Y\}$ since $y \neq 0$; however, because of (2), K cannot be empty for the same reason. Therefore, we have $K = \{Y\}$ and

$$y = h_Y \cdot \mathsf{c}_0 h_Y \cdot \prod_{Y \neq W \subseteq X} -\mathsf{c}_0 h_W.$$

From this last formula and (3) we conclude that

$$y = \prod_{Y \neq W \subseteq X} -\mathsf{c}_0 h_W.$$

This completes the proof.

In connection with this theorem compare 2.3.32.

Further applications of the method of eliminating cylindrifications for monadic-generated algebras will be found in Section 2.5, and in Part II where in particular it will be established that every monadic-generated algebra is representable (cf. 1.1.13).

2.3. HOMOMORPHISMS, ISOMORPHISMS, AND IDEALS

In this section we discuss for cylindric algebras the notions of homomorphism and isomorphism, as well as certain derivative notions, in the first place that of ideal. These are notions which for arbitrary algebraic structures were discussed in Section 0.2. As usual, the general definitions of these notions can be modified by using specific properties of CA's. In particular, a homomorphism f from a CA $\mathfrak{A}$ onto a CA $\mathfrak{B}$ (i.e., a member f of $Ho(\mathfrak{A}, \mathfrak{B})$) is defined to be a function with domain A and range B which preserves the fundamental operations and distinguished elements, in the sense that $f(x+^{(\mathfrak{A})}y) = fx+^{(\mathfrak{B})}fy, \ldots, f\mathsf{d}_{\kappa\lambda}^{(\mathfrak{A})} = \mathsf{d}_{\kappa\lambda}^{(\mathfrak{B})}$; but in listing these conditions we can omit those involving the distinguished elements 0 and 1 and any one of the three Boolean operations $+$, $\cdot$, $-$. Analogously, in defining an isomorphism from $\mathfrak{A}$ onto $\mathfrak{B}$ we can omit conditions involving two of the Boolean operations, either $+$ and $-$, or $\cdot$ and $-$. With respect to some derivative notions, in fact, those of ideal and simple algebra, the modifications which can be made in the general definitions are more essential; they will be described explicitly in Theorems 2.3.7 and 2.3.14. It will be seen from these theorems that in the case of CA's these two notions acquire a more elementary character, and can be defined in terms of simple conditions imposed on the elements of a CA.

THEOREM 2.3.1. $\mathsf{HCA}_\alpha = \mathsf{CA}_\alpha$; *that is, a homomorphic image of a* CA_α *is a* CA_α.

Concerning the proof of 2.3.1 cf. 2.1.2.

THEOREM 2.3.2. *If* $\mathfrak{A} \in \mathsf{CA}_\alpha$, $f \in Ho(\mathfrak{A}, \mathfrak{B})$, *and* $x \in A$, *then* $\Delta^{(\mathfrak{B})}fx \subseteq \Delta^{(\mathfrak{A})}x$.

THEOREM 2.3.3. $\mathsf{HLf}_\alpha = \mathsf{Lf}_\alpha$, $\mathsf{HDc}_\alpha = \mathsf{Dc}_\alpha$, *and* $\mathsf{HMn}_\alpha = \mathsf{Mn}_\alpha$.
PROOF: by 0.2.14(ii) and 2.3.2.

It goes without saying that CA_α, Lf_α, Dc_α, and Mn_α are closed under the formation of isomorphisms as well. Further observations of this kind will not be stated explicitly; cf. 0.2.7.

We now turn to the discussion of ideals in cylindric algebras. We assume as known the notion of an ideal in a Boolean algebra $\mathfrak{A}$. This notion can be subsumed under the general notion of a z-ideal defined in 0.2.39. Indeed, since the operation of symmetric difference, $\oplus$, is a group-forming polynomial (i.e., a binary polynomial operation such that $\langle A, \oplus \rangle$ is a group), we may apply 0.4.17 to infer that for each z the z-ideals are properly functioning, and that consequently they are in a natural one-one correspondence with the congruence relations of $\mathfrak{A}$.

In the theory of BA's we fix z in either one of two different ways, choosing $z = 0$ or $z = 1$; we agree to refer to 0-ideals simply as *ideals*, and to 1-ideals as *filters* of $\mathfrak{A}$. It can easily be shown that these definitions of an ideal and a filter in a BA are equivalent to the usual ones. (There is a third method of introducing filters for a BA $\mathfrak{A}$, namely, as ideals of its dual $\mathfrak{A}^{\partial}$.)

We shall now apply an analogous procedure to cylindric algebras except that we restrict ourselves to 0-ideals without extending the use of the term "filter".

THEOREM 2.3.4. *For any* $\mathfrak{A} \in \mathsf{CA}_\alpha$,

(i) $\oplus$ *is a group-forming polynomial over* $\mathfrak{A}$,

(ii) *the z-ideals of* $\mathfrak{A}$ *function properly for every* $z \in A$.

PROOF: (i) by the theory of BA's; (ii) by (i) and 0.4.17(iii).

DEFINITION 2.3.5. *By an* ***ideal*** *of a* CA_α $\mathfrak{A}$ *we mean a* 0-*ideal of* $\mathfrak{A}$, *i.e., any set* $0/R$ *where* R *is a congruence relation over* $\mathfrak{A}$.

REMARKS 2.3.6. The terminology developed in Section 0.2 for z-ideals will be applied to Boolean and cylindric algebras without the use of the subscript z ($z = 0$ in our case). Thus $Il\mathfrak{A}$ (the set of ideals of $\mathfrak{A}$), $Ig^{(\mathfrak{A})}X$ (the ideal of $\mathfrak{A}$ generated by X), etc. will denote Boolean or cylindric notions, dependent on whether $\mathfrak{A} \in \mathsf{BA}$ or $\mathfrak{A} \in \mathsf{CA}$. By a *Boolean ideal* of a CA $\mathfrak{A}$ we mean, of course, an ideal of $\mathfrak{B}\mathfrak{l}\,\mathfrak{A}$.

By 2.3.4 and 0.2.49, ideals of a CA $\mathfrak{A}$ are in natural one-one correspondence with congruence relations on $\mathfrak{A}$. Because of this fact we can eliminate congruence relations entirely from the discussion of CA's and can consistently use ideals instead; the advantages of this procedure were explained in 0.2.48.

Since ideals replace congruence relations, with every ideal I of a CA $\mathfrak{A}$ we can correlate a quotient algebra $\mathfrak{A}/I$. The significance of ideals for the theory of CA's results primarily from the fact that these quotient algebras provide a standard isomorphic representation for arbitrary homomorphic images. This representation is based upon the following two statements:

(I) *If* $\mathfrak{A} \in \mathsf{CA}_\alpha$ *and* $I \in Il\mathfrak{A}$, *then* $\mathfrak{A} \succcurlyeq \mathfrak{A}/I$. *In fact* $\langle x/I : x \in A\rangle \in Ho(\mathfrak{A}, \mathfrak{A}/I)$, *where* x/I *is determined for all* $x \in I$ *by the formula* $x/I = \{y : y \in A,\ x \oplus y \in I\}$.

(II) *If* $\mathfrak{A} \in \mathsf{CA}_\alpha$ *and* $\mathfrak{A} \succcurlyeq \mathfrak{B}$, *then there is an* $I \in Il\mathfrak{A}$ *such that* $\mathfrak{B} \cong \mathfrak{A}/I$. *In fact, if* $h \in Ho(\mathfrak{A}, \mathfrak{B})$, *we can let* $I = (h^{-1})^\star\, 0^{(\mathfrak{B})}$ *and we obtain* $h^{-1}|\langle x/I : x \in A\rangle \in Is(\mathfrak{B}, \mathfrak{A}/I)$.

In view of 2.3.4, (I) and (II) are direct applications of the general theory of algebras. Cf. 0.2.21, 0.2.23, and 0.2.52; also, use 0.4.17(i) to derive the specific formula for x/I in (I).

Using 0.2.49 and 2.3.4 we can automatically express various other results for CA_α's concerning congruence relations in the terminology of ideals. Except for a few cases where these results play a special role in the further development (cf., e.g., 2.4.40 below), we shall not explicitly formulate them. In what follows we shall often refer, during the course of an argument, to a definition or theorem concerning congruence relations when, actually, the corresponding result for ideals is needed. It will be left to the reader to supply the proper translation. An elementary characterization of ideals in CA's independent of the notion of congruence relations is given in

THEOREM 2.3.7. *A set* I *is an ideal of a* CA_α $\mathfrak{A}$ *iff the following conditions hold*:

(i) *I is a Boolean ideal of* $\mathfrak{A}$, *i.e.*,

(i′) $0 \in I \subseteq A$,

and, for all $x, y \in A$,

(i″) *if* $x, y \in I$, *then* $x + y \in I$,

(i‴) *if* $x \in I$ *and* $y \leqq x$, *then* $y \in I$.

(ii) *For any* $x \in I$ *and* $\kappa < \alpha$, $\mathsf{c}_\kappa x \in I$.[1]

PROOF. First suppose that I is an ideal of $\mathfrak{A}$, i.e., a 0-ideal of $\mathfrak{A}$ in the sense of 0.2.39. A direct proof that I satisfies (i) and (ii) is not difficult; because of 2.3.4, we may if we wish apply 0.4.17(ii) instead, with $\cdot$ replaced by $\oplus$ and z by 0. Condition (i′) coincides with 0.4.17(ii′). Conditions (i″), (i‴), (ii) can all be obtained by suitably applying 0.4.17(ii‴); we omit details.

Assume now, conversely, that (i) and (ii) hold. To show that I is an ideal of $\mathfrak{A}$, it suffices by 0.4.17(ii) to establish the following conditions:

(1) $$0 \in I \subseteq A.$$

[1]) In Tarski-Thompson [52] Theorem 2.3.7 is taken for the definition of ideals in CA's. It is also stated there that the relations between ideals and homomorphisms is exactly the same in cylindric algebras as in Boolean algebras; this amounts to saying in our present terminology that in both classes of algebras the ideals function properly.

(2) If $x \oplus y$, $x' \oplus y' \in I$, then $(x+x') \oplus (y+y') \in I$, $(x \cdot x') \oplus (y \cdot y') \in I$, and $-x \oplus -y \in I$ for all $x, y, x', y' \in A$.
(3) If $x \oplus y \in I$, then $c_\kappa x \oplus c_\kappa y \in I$ for all $x, y \in A$ and $\kappa < \alpha$.

(1) and (2) follow from (i) and the theory of Boolean ideals. If $x \oplus y \in I$, then $c_\kappa x \oplus c_\kappa y \leqq c_\kappa(x \oplus y) \in I$ by 1.2.8(ii) and (ii), and hence $c_\kappa x \oplus c_\kappa y \in I$ by (i). Thus (3) holds, and the proof is complete by 0.4.17(ii).

As a direct consequence of 2.3.7 we obtain the dual theorem $2.3.7^\partial$ which gives a simple characterization of dual ideals in a CA_α $\mathfrak{A}$, i.e., ideals in $\mathfrak{A}^\partial$ (cf. the remark following 1.4.3); as in the case of BA's the dual ideals prove to coincide with 1-ideals in $\mathfrak{A}$ itself. For further reference we state here explicitly $2.3.7^\partial$ in a somewhat simplified form in which conditions corresponding to (i″) and (i‴) are replaced by one condition:

A set I is a dual ideal of a CA_α $\mathfrak{A}$ if the following conditions hold:
(i) *I is a filter of $\mathfrak{A}$, i.e.,*
 (i′) $1 \in I \subseteq A$;
 (i″) *if $y \in A$ and $x, -x+y \in I$, then $y \in I$.*
(ii) *If $x \in I$ and $\kappa < \alpha$, then $c^\partial_\kappa x \in I$.*

Of course the original theorem can be simplified in a similar manner.

THEOREM 2.3.8. *If $\mathfrak{A} \in \mathsf{CA}_\alpha$ and $X \subseteq A$, then $\mathfrak{Ig}\, X$ is the set of all $y \in A$ such that $y \leqq c_{(\Gamma)}(x_0 + \ldots + x_{\kappa-1})$ for some $x \in {}^\kappa X$ with $\kappa < \omega$ and some $\Gamma \subseteq \alpha$ with $|\Gamma| < \omega$.*

PROOF. Let I be the set of all $y \in A$ such that $y \leqq c_{(\Gamma)}(x_0 + \ldots + x_{\kappa-1})$ for some $x \in {}^\kappa X$ with $\kappa < \omega$ and some $\Gamma \subseteq \alpha$ with $|\Gamma| < \omega$. Clearly $I \subseteq \mathfrak{Ig}\, X$. To prove the inclusion in the opposite direction, it suffices to show that $X \subseteq I \in Il\mathfrak{A}$. Obviously, $X \subseteq I$, $0 \in I \subseteq A$, $y \in I$ whenever $y \leqq x \in I$, and $c_\kappa x \in I$ whenever $x \in I$ and $\kappa < \alpha$. Thus, in view of 2.3.7, it only remains to prove that $y+z \in I$ when $y, z \in I$. Say

$$y \leqq c_{(\Gamma)}(x_0 + \ldots + x_{\kappa-1}) \text{ and } z \leqq c_{(\Delta)}(w_0 + \ldots + w_{\lambda-1})$$

where $\kappa, \lambda < \omega$, $x \in {}^\kappa X$, $w \in {}^\lambda X$, $\Gamma, \Delta \subseteq \alpha$, and $|\Gamma|, |\Delta| < \omega$. Then both y and z are $\leqq c_{(\Gamma \cup \Delta)}(x_0 + \ldots + x_{\kappa-1} + w_0 + \ldots + w_{\lambda-1})$, so that $y+z \leqq c_{(\Gamma \cup \Delta)}(x_0 + \ldots + x_{\kappa-1} + w_0 + \ldots + w_{\lambda-1})$ and $y+z \in I$.

We shall see from 2.3.9 and 2.3.10 that the construction given in 2.3.8 of the ideal generated by a set X can be simplified in various ways under special assumptions on X.

COROLLARY 2.3.9. *Let $\mathfrak{A} \in \mathsf{CA}_\alpha$.*

(i) *If* $X \subseteq A$ *and* $|\Delta x| < \omega$ *for each* $x \in X$, *then*
$IgX = \{y : y \in A, y \leqq \mathsf{c}_{(\Delta x_0)}x_0 + \ldots + \mathsf{c}_{(\Delta x_{\kappa-1})}x_{\kappa-1}$ *for some* $\kappa < \omega$ *and* $x \in {}^{\kappa}X\}$.

(ii) *If* $X \subseteq Zd\mathfrak{A}$, *then*
$Ig^{(\mathfrak{A})}X = Ig^{(\mathfrak{B}l\mathfrak{A})}X = \{y : y \in A, y \leqq x_0 + \ldots + x_{\kappa-1}$ *for some* $\kappa < \omega$ *and* $x \in {}^{\kappa}X\}$.

(iii) *If* $K \subseteq Il\mathfrak{A}$, *then*
$Ig^{(\mathfrak{A})}\bigcup K = Ig^{(\mathfrak{B}l\mathfrak{A})}\bigcup K = \{x_0 + \ldots + x_{\kappa-1} : \kappa < \omega, \{x_0, \ldots, x_\kappa\} \subseteq \bigcup K\}$.

COROLLARY 2.3.10. *Let* $\mathfrak{A} \in \mathsf{CA}_\alpha$.

(i) *If* $x \in A$, *then the principal ideal* $Ig\{x\}$ *is determined by the formula*

$$Ig\{x\} = \{y : y \in A, y \leqq \mathsf{c}_{(\Gamma)}x \text{ for some finite } \Gamma \subseteq \alpha\}.$$

(ii) *If* $x \in A$ *and* $|\Delta x| < \omega$, *then*

$$Ig\{x\} = Ig\{\mathsf{c}_{(\Delta x)}x\} = \{y \in A, y \leqq \mathsf{c}_{(\Delta x)}x\} = Rl_{\mathsf{c}_{(\Delta x)}x}\mathfrak{A}.$$

(iii) *If* $x \in Zd\mathfrak{A}$, *then*

$$Ig\{x\} = Ig^{(\mathfrak{B}l\mathfrak{A})}\{x\} = \{y : y \in A, y \leqq x\} = Rl_x\mathfrak{A}.$$

(iv) *If* $X \subseteq A$ *and* $|X| < \omega$, *then* $IgX = Ig\{\sum X\}$.

We know from 0.2.41 (and the subsequent remark) that in every algebra $\mathfrak{A}$ the z-ideals form a complete lattice under appropriate operations of join and meet. This general result can of course be applied to ideals in CA's. As we shall see from the next theorem, the resulting lattice proves to be distributive and actually exhibits a stronger property: it satisfies the distributive law for meets over arbitrary (infinite) joins. This is a generalization of a known result from the theory of Boolean algebras. Complete lattices which satisfy this particular distributive law coincide with *complete Brouwerian lattices*; cf. Birkhoff [67*], p. 128, Theorem 24. They can be characterized as duals of complete Brouwerian algebras in the sense of McKinsey-Tarski [46*] (which are also known as complete Brouwerian logics).

THEOREM 2.3.11. *Let* $\mathfrak{A} \in \mathsf{CA}_\alpha$; *let* $I + J = Ig(I \cup J)$ *for any* $I, J \in Il\mathfrak{A}$, *and*

$$\textstyle\sum_{k\in K} J_k = Ig(\bigcup_{k\in K} J_k)$$

for any set K *and any* $J \in {}^{K}Il\mathfrak{A}$. *We have then*:

(i) $\langle Il\mathfrak{A}, +, \cap\rangle$ *is a complete lattice with the zero element* $\{0\}$ *and the unit element* A. *In this lattice* $\sum_{k\in K} J_k$ *and* $\bigcap_{k\in K} J_k$ *are respectively the least upper bound and greatest lower bound of ideals* J_k *with* $k \in K$.

(ii) *If* $I \in Il\mathfrak{A}$ *and* $J \in {}^{K}Il\mathfrak{A}$, *then*

$$I \cap \textstyle\sum_{k\in K} J_k = \sum_{k\in K}(I \cap J_k).$$

Hence, in particular, the lattice $\langle Il\mathfrak{A}, +, \cap\rangle$ *is distributive.*

PROOF. (i) follows immediately from 0.1.11 in the light of the remark following 0.2.41. Now in order to prove

(1) $$I \cap \sum_{k\in K} J_k = \sum_{k\in K}(I \cap J_k)$$

it suffices to show $I \cap \sum_{k\in K} J_k \subseteq \sum_{k\in K}(I \cap J_k)$ since the inclusion in the opposite direction is obvious (and, incidently, holds in every lattice). Assume that $a \in I \cap \sum_{k\in K} J_k$; then, by 2.3.9(iii), $a = x_0 + \ldots + x_{\kappa-1}$ for some $\kappa < \omega$ and $\{x_0, \ldots, x_{\kappa-1}\} \subseteq \bigcup_{k\in K} J_k$. Furthermore, we have by 2.3.7(i‴) that $\{x_0, \ldots, x_{\kappa-1}\} \subseteq I$ since, for each $\lambda < \kappa$, $x_\lambda \leqq a$ while $a \in I$. Thus $\{x_0, \ldots, x_{\kappa-1}\} \subseteq \bigcup_{k\in K}(I \cap J_k)$ and consequently $a \in \sum_{k\in K}(I \cap J_k)$ by 2.3.9(iii).

The ordinary distributive law,

$$I \cap (J + J') = (I \cap J) + (I \cap J')$$

for any $I, J, J' \in Il\mathfrak{A}$, follows as a particular case of (1); and, by applying this law several times, we obtain the second distributive law,

$$I + (J \cap J') = (I + J) \cap (I + J')$$

for any $I, J, J' \in Il\mathfrak{A}$.

Our formulation of 2.3.11 assumes that $\bigcap_{k\in K} J_k = A$ in case $K = 0$ (as in the Preliminaries).

It is known that the distributive law for joins over infinite meets,

$$I + \bigcap_{k\in K} J_k = \bigcap_{k\in K}(I + J_k),$$

fails in the lattice of ideals of infinite BA's; hence it does not hold in general for the ideals of CA's.

An intrinsic characterization of lattices of ideals of BA's is known from the literature; cf. Tarski [37*], p. 188. From 1.3.11 and our next theorem it follows that this characterization applies to lattices of ideals of Lf_α's as well. For $\alpha \geqq \omega$, no intrinsic characterization is known for the lattices of ideals of arbitrary CA_α's.

In reference to the following theorem, recall that $Ism(\mathfrak{A}, \mathfrak{B})$ is the set of all isomorphisms from $\mathfrak{A}$ into $\mathfrak{B}$ (cf. 0.2.5(ii)).

THEOREM 2.3.12. *Let $\mathfrak{A} \in \mathsf{CA}_\alpha$; let F and G be the functions defined respectively for $I \in Il\mathfrak{A}$ and $J \in Il\mathfrak{Zd}\mathfrak{A}$ by the formulas*

$$FI = I \cap Zd\mathfrak{A}, \quad GJ = Ig^{(\mathfrak{A})}J;$$

finally, let $\mathfrak{L} = \langle Il\mathfrak{A}, +, \cap \rangle$ be the lattice of ideals of $\mathfrak{A}$ described in 2.3.11 and let $\mathfrak{L}' = \langle Il\mathfrak{Zd}\mathfrak{A}, +', \cap \rangle$ be the analogously defined lattice of ideals of *the Boolean algebra* $\mathfrak{Zd}\,\mathfrak{A}$. *Under these assumptions we have*:

(i) $GFI \subseteq I$ *for every* $I \in Il\mathfrak{A}$ *and* $FGJ = J$ *for every* $J \in Il\mathfrak{Zd}\,\mathfrak{A}$.

(ii) F *maps* $Il\mathfrak{A}$ *onto* $Il\mathfrak{Zd}\,\mathfrak{A}$ *and* G *is one-one; in fact,* $G \in Ism(\mathfrak{L}', \mathfrak{L})$.

If, moreover, $\mathfrak{A} \in \mathsf{Lf}_\alpha$, *we have*:

(iii) $F \in Is(\mathfrak{L}, \mathfrak{L}')$, $G \in Is(\mathfrak{L}', \mathfrak{L})$, *and* $F^{-1} = G$.

(iv) $FI = \{\mathsf{c}_{(\Delta x)}x : x \in I\}$ *for every* $I \in Il\mathfrak{A}$, *and*

$$GJ = \{x : x \in A,\ \mathsf{c}_{(\Delta x)}x \in J\}$$

for every $J \in Il\mathfrak{Zd}\,\mathfrak{A}$.

PROOF. We first observe that $RgG \subseteq Il\mathfrak{A}$ and $RgF \subseteq Il\mathfrak{Zd}\,\mathfrak{A}$ both hold; the first inclusion is obvious while the second is implied by 0.2.42(ii) and 2.3.7(i). The first part of (i) then follows from the definition of F and G, and the second part if an easy consequence of the fact that, by 2.3.9(ii),

(1) $$GJ = \{x : x \in A,\ x \leqq y \text{ for some } y \in J\}$$

for each $J \in Il\mathfrak{Zd}\,\mathfrak{A}$.

For the proof of (ii), first note that F maps $Il\mathfrak{A}$ onto $Il\mathfrak{Zd}\,\mathfrak{A}$ and G is one-one; both these facts follow immediately from the second part of (i), i.e., that $F \circ G = Il\mathfrak{Zd}\,\mathfrak{A} \upharpoonright Id$. In order to prove that $G \in Ism(\mathfrak{L}', \mathfrak{L})$, i.e., that G maps $\mathfrak{L}'$ isomorphically into $\mathfrak{L}$, it remains only to show that, for any $J, J' \in Il\mathfrak{Zd}\,\mathfrak{A}$, $G(J +' J') = GJ + GJ'$ and $G(J \cap J') = GJ \cap GJ'$. By (1) and 2.3.9(iii) it is sufficient for the proof of the first of these equations to show that, for every $x \in A$, $x \leqq a +^{(\mathfrak{A})} a'$ for some $a \in J$ and $a' \in J'$ iff $x = b +^{(\mathfrak{A})} b'$ for some $b \in GJ$, $b' \in GJ'$; but given a, a' we let $b = x \cdot a$ and $b' = x \cdot a'$ while, if b, b' are given, by (1) we may choose a, a' such that $b \leqq a \in J$ and $b' \leqq a' \in J'$. $G(J \cap J') = GJ \cap GJ'$ is proven analogously.

To prove (iii) and (iv) we observe first of all that, if $I \in Il\mathfrak{A}$ where $\mathfrak{A}$ is an arbitrary CA_α, we have by 1.7.11 and 2.3.7 that $x \in I$ iff $\mathsf{c}_{(\Delta x)}x \in I \cap Zd\mathfrak{A}$, i.e.,

(2) $$x \in I \text{ iff } \mathsf{c}_{(\Delta x)}x \in FI,$$

for all $x \in A$ such that $|\Delta x| < \omega$. From this it follows that the inclusion $I \subseteq GFI$ holds in case $\mathfrak{A} \in \mathsf{Lf}_\alpha$, since then we have $x \leqq \mathsf{c}_{(\Delta x)}x \in FI$ and thus $x \in GFI$, for each $x \in I$. This along with the first part of (i) shows that $G \circ F = Il\mathfrak{A} \upharpoonright Id$, and the latter fact combined with (ii) gives (iii). The inclusion $FI \supseteq \{\mathsf{c}_{(\Delta x)}x : x \in I\}$ follows from (2), while the inclusion in the opposite direction is a simple consequence of the fact that $\mathsf{c}_{(\Delta x)}x = x$ for every $x \in Zd\mathfrak{A}$. The second equality of (iv) is gotten immediately upon substituting GJ for I in (2) and recalling that, by (ii), $FGJ = J$. This completes the proof of the theorem.

Part (ii) of 2.3.12 cannot be improved by saying that $F \in Ho(\mathfrak{L}, \mathfrak{L}')$, because it is easy to construct a CA with ideals I and I' such that $FI +' FI' \subset F(I+I')$. From 2.3.12(iii) it is seen that, in case $\mathfrak{A} \in \mathsf{Lf}_\alpha$, the theory of ideals in $\mathfrak{A}$ reduces to that of ideals in $\mathfrak{Zd}\,\mathfrak{A}$. If $\mathfrak{Zd}\,\mathfrak{A}$ has a simple structure, its ideals and those of $\mathfrak{A}$ can be easily determined. For example, Theorem 2.1.23 can be used to describe the lattice of ideals of any Mn_α.

From 2.3.12 we see that every ideal I of an Lf_α has a set of zero-dimensional generators; in fact, by 2.3.12(iii) we have $I = IgFI$. Actually, using 2.3.9(i) we can say even more: if I is generated by $X \subseteq A$, then there is a set Y of zero-dimensional generators of I such that $|Y| \leqq |X|$; we can take for Y the set $\{\mathsf{c}_{(\Delta x)}x : x \in X\}$. This last observation leads us to the following interesting property of Lf_α's: Every ideal I of an $\mathsf{Lf}_\alpha\mathfrak{A}$ which is generated by a countable set has a (countable) irredundant base; i.e., there is $Y \subseteq A$ such that $I = IgY$ and $I \supset IgY'$ for every $Y' \subset Y$. In fact, if $I = Ig\{x_\kappa : \kappa < \xi\}$ where $\xi \leqq \omega$, we can set

$$Y = \{\mathsf{c}_{(\Delta x_\kappa)}x_\kappa - \textstyle\sum_{\lambda<\kappa}\mathsf{c}_{(\Delta x_\lambda)}x_\lambda : \kappa < \xi\} \sim \{0\}.$$

REMARKS 2.3.13. Theorem 2.3.12 leads to a clear and relatively simple metalogical interpretation of ideals in cylindric algebras. Actually, it is more convenient to interpret the dual ideals and relate them, with the help of $2.3.12^\partial$, to filters in Boolean algebras of zero-dimensional elements.

We consider a theory Θ in a language Λ, and let $\mathfrak{A}$ be the associated CA_α of formulas; then $\mathfrak{Zd}\,\mathfrak{A}$ is the BA of sentences associated with Θ (cf. 1.1.16 and the remark immediately following 1.6.18). We assume $\alpha \geqq \omega$. Each formula φ in Λ determines one of the elements of $\mathfrak{A}$, namely, the equivalence class $\varphi/\equiv_\Theta$. The elements of $\mathfrak{Zd}\,\mathfrak{A}$ are those elements $\sigma/\equiv_\Theta$ of $\mathfrak{A}$ determined by a sentence σ in Λ. Thus, corresponding to any dual ideal I of $\mathfrak{A}$ we may consider the correlated set Φ of formulas where $\Phi = \bigcup I$; and corresponding to any filter J of $\mathfrak{Zd}\,\mathfrak{A}$ we may consider the correlated set Σ of all those sentences which are in $\bigcup J$.

To characterize all sets Φ and Σ obtained in this way we recall the simple characterization of dual ideals in CA's and filters of BA's given in connection $2.3.7^\partial$ (see the remark immediately following 2.3.7). We easily see that the sets Σ are characterized by the following conditions: Σ contains all sentences valid in Θ; if σ and $\sigma \rightarrow \zeta$ belong to Σ, then ζ also belongs to Σ. From this we conclude at once that the sets Σ correlated with filters in $\mathfrak{Zd}\,\mathfrak{A}$ are simply theories which include Θ; such theories are usually called extensions of Θ. (The completeness theorem of predicate logic is essentially involved in obtaining this characterization of the sets Σ, and the same applies, not only to some later parts of these remarks, but also to certain remarks made further on in the chapter; compare here 1.11.2.) On the other hand, the sets Φ are characterized by the three conditions: they contain all formulas φ valid in Θ (i.e., whose closures $[\varphi]$ belong to Θ); if $\varphi, \varphi \to \psi \in \Phi$, then $\psi \in \Phi$; if $\varphi \in \Phi$, then, for every $\kappa < \alpha$, $\forall_{v_\kappa} \varphi \in \Phi$ (and hence also $[\varphi] \in \Phi$). This description of the sets Φ is simple enough. A still simpler characterization can be obtained, however, from part (iv) of 2.3.12$^\partial$: a set Φ of formulas is correlated with a dual ideal in $\mathfrak{A}$ iff there is an extension Θ' of the theory Θ such that Φ consists of just those formulas whose closures are valid in Θ'. This theory Θ' is simply the set of all those sentences which belong to Φ.

We wish to discuss briefly one further metalogical interpretation, namely, that of a quotient algebra $\mathfrak{A}/I$. Thus let $\mathfrak{A}$ be, as before, the algebra of formulas associated with a theory of Θ. Along with an ideal I we consider the corresponding dual ideal I^∂ given by the formula $I^\partial = \{-x : x \in I\}$. Although we have decided to use the ordinary ideals (i.e., 0-ideals) in the construction of quotient algebras, we could equally well use dual ideals (i.e., 1-ideals) for the same purpose; and indeed, by returning to congruence relations, we can easily see that actually $\mathfrak{A}/I$ and $\mathfrak{A}/I^\partial$ are the same algebra (cf. 0.2.47). Now the dual ideal I^∂ is a set of equivalence classes of formulas, and its union $\bigcup I^\partial$ is a set of formulas. As we know from the preceding remarks, there is a well determined extension Θ' of the theory Θ such that $\bigcup I^\partial$ consists of just those formulas whose closures belong to Θ'. Let $\mathfrak{A}'$ be the algebra of formulas associated with Θ'. A moment's reflection suffices to perceive that, although the algebra $\mathfrak{A}/I^\partial$ is not itself a cylindric algebra of formulas, it is isomorphic to $\mathfrak{A}'$; the argument is based upon the second isomorphism theorem, 0.2.27(iv).

A consequence of these observations is: if $\mathfrak{A}$ is the algebra of formulas associated with the theory Θ, then every homomorphic image of $\mathfrak{A}$ is isomorphic to an algebra of formulas $\mathfrak{A}'$ associated with some theory Θ' which is an extension of Θ. It is easily seen, again using 0.2.27(iv), that the converse of this statement also holds: every algebra of formulas $\mathfrak{A}'$ associated with an extension Θ' of Θ is a homomorphic image of $\mathfrak{A}$. Assume in particular that Θ_0 is the logic of the given language Λ (cf. the Preliminaries), so that every theory in Λ is an extension of Θ_0. Let $\mathfrak{A}_0$ be the algebra of formulas associated with the theory Θ_0. We see then that the class of all homomorphic images of $\mathfrak{A}_0$ coincides with the class of all cylindric algebras of formulas in the language Λ and their isomorphic images.

In the next few theorems we are concerned with simple CA's and maximal ideals. Notice first that, by results from the general theory of algebras, a CA $\mathfrak{A}$

is simple iff it has just two ideals (or, equivalently, if $Il\mathfrak{A} = \{\{0\}, A\}$ and $0 \neq 1$); also, for any given ideal I of $\mathfrak{A}$, the algebra $\mathfrak{A}/I$ is simple iff I is a maximal proper ideal.

THEOREM 2.3.14. *For any* CA_α $\mathfrak{A}$ *the following conditions are equivalent:*

(i) $\mathfrak{A}$ *is simple;*

(ii) $|A| > 1$, *and for every non-zero* $x \in A$ *there is a finite subset* Γ *of* α *such that* $\mathsf{c}_{(\Gamma)}x = 1$.

Each of these conditions implies:

(iii) $|Zd\mathfrak{A}| = 2$ *or, equivalently,* $|A| > 1$ *and* $Zd\mathfrak{A} = \{0, 1\}$.

In case $\mathfrak{A} \in \mathsf{Lf}_\alpha$ *the three conditions (i)–(iii) are equivalent to each other and also equivalent to*

(iv) $|A| > 1$, *and* $\mathsf{c}_{(\Delta x)}x = 1$ *for every non-zero* $x \in A$.[1)]

PROOF: by 0.2.34, 1.7.11(i), and 2.3.10(i),(ii).

It will be seen in Section 2.4 that for $\alpha \geqq \omega$ conditions (i) and (iii) are not in general equivalent; compare 2.4.14 and 2.4.50.

REMARKS 2.3.15. Examples of simple CA_α's are numerous. In case $\alpha = 0$ we know from the theory of BA's that the only simple algebras are those with two elements. In case $\alpha = 1$ an elementary construction of all simple CA_1's (and Df_1's) is fully described in 1.2.14. From this construction it is seen that there is a one-one correspondence between BA's and simple CA_1's and that actually the two classes of algebras are first-order definitionally equivalent (cf. 0.1.6). For $\alpha \geqq 2$ (as well as for $\alpha = 1$) trivial examples of simple algebras are provided by two-element algebras; as is seen from 2.3.14 and 1.6.3 these are the only simple CA_α's which are discrete. Non-discrete simple algebras can easily be found among the special algebras of Section 1.1. Thus a CA_α $\mathfrak{A}$ (with $\alpha \geqq \omega$) of formulas associated with a complete consistent theory Θ is always simple because it is locally finite and contains just two zero-dimensional elements (compare the remarks following 1.6.19); moreover, $\mathfrak{A}$ has more than two elements provided the sentence $\exists_{xy} x \neq y$ is valid in Θ. Conversely, the discussion in 1.11.2 and 2.3.13 shows that, for $\alpha \geqq \omega$, every simple Lf_α with more than two elements is isomorphic with an algebra of formulas associated with a complete consistent theory.

If $1 \leqq \alpha < \omega$ and $|U| \geqq 2$, then the full cylindric set algebra of dimension α and with base U is again a simple algebra with more than two elements. In case $\alpha \geqq \omega$ and $|U| \geqq 2$ the full cylindric set algebra in the space ${}^\alpha U$ is

1) The essential parts of this theorem were stated in Tarski-Thompson [52].

not simple. However, we shall now describe a class of simple set algebras which proves to be quite comprehensive.

Let $\alpha \geqq \omega$ and U be a non-empty set, and for each finite subset Γ of α define R_Γ to be the equivalence relation which holds between any two elements x and y of ${}^\alpha U$ iff $\Gamma 1 x = \Gamma 1 y$. If X is a non-empty subset of ${}^\alpha U$ for which there exists a finite $\Gamma \subseteq \alpha$ such that

$$\text{(I)} \qquad R_\Gamma {}^* X = X,$$

then, as is easily seen, $\mathsf{C}_{(\Gamma)} X = {}^\alpha U$; moreover, the set of all $X \subseteq {}^\alpha U$ which satisfy (I) for some finite $\Gamma \subseteq \alpha$ is clearly a subuniverse of the full set algebra in the space ${}^\alpha U$. Therefore, every set algebra in the space ${}^\alpha U$ which is generated by an arbitrary set of elements $X \subseteq {}^\alpha U$ satisfying the condition (I) for some finite $\Gamma \subseteq \alpha$ is an example of a simple set algebra of dimension $\alpha \geqq \omega$. Actually, if we denote by K_α the class of all algebras defined in this manner, we can easily convince ourselves that for each algebra $\mathfrak{A}$ the following three conditions are equivalent: $\mathfrak{A}$ is a simple Lf_α, $\mathfrak{A}$ is isomorphic with an algebra in K_α, and $\mathfrak{A}$ is isomorphic to an algebra of formulas associated with a complete and consistent theory.

In spite of certain similarities between 2.3.14(ii) and the condition used in defining Lf_α's, there are simple CA's which are not locally finite; cf. Theorem 2.5.24 below.

As an immediate consequence of 2.3.14 we have

COROLLARY 2.3.16. (i) *If a* CA_α $\mathfrak{A}$ *is simple, then every subalgebra of* $\mathfrak{A}$ *is simple.*

(ii) *If a* CA_α $\mathfrak{A}$ *is not simple, then there is a subalgebra of* $\mathfrak{A}$ *generated by one element which is not simple and which is actually not a subalgebra of any simple subalgebra of* $\mathfrak{A}$.

Neither part of 2.3.16 extends to arbitrary algebras.

THEOREM 2.3.17. *If* $\mathfrak{A} \in \mathsf{CA}_\alpha$, *then for every proper* $I \in Il\mathfrak{A}$ *there is a maximal proper* $J \in Il\mathfrak{A}$ *such that* $J \supseteq I$.

PROOF: by 0.2.46(iii), since $A = Ig\{1\}$ by 2.3.10.

THEOREM 2.3.18. *If* $\mathfrak{A} \in \mathsf{CA}_\alpha$ *and* $I \in Il\mathfrak{A}$, *then the following conditions are equivalent:*

(i) *I is a maximal proper ideal of* $\mathfrak{A}$;

(ii) *for each* $x \in A$, *either* $x \in I$ *or else* $-\mathsf{c}_{(\Gamma)} x \in I$ *for some finite subset* Γ *of* α, *but not both.*

In case $\mathfrak{A} \in \mathsf{Lf}_\alpha$ *each of these two conditions is equivalent to the condition:*

(iii) *for all* $x \in A$, *either* $x \in I$ *or* $-\mathsf{c}_{(\Delta x)} x \in I$, *but not both.*

PROOF: by 0.2.38 and 2.3.14.

From 2.3.18 and 0.2.42(ii) we further obtain

COROLLARY 2.3.19. *If* $\mathfrak{A} \subseteq \mathfrak{B} \in \mathsf{CA}_\alpha$ *and* I *is a maximal proper ideal of* $\mathfrak{B}$, *then* $A \cap I$ *is a maximal proper ideal of* $\mathfrak{A}$.

THEOREM 2.3.20. *For any* CA_α $\mathfrak{A}$ *the following conditions are equivalent:*

(i) $\mathfrak{A}$ *is simple*;

(ii) $|A| > 1$, *and* $\mathfrak{A} \cong \mathfrak{B}$ *whenever* $\mathfrak{A} \succcurlyeq \mathfrak{B}$ *and* $|B| > 1$.

PROOF. That (i) implies (ii) is obvious (cf. 0.2.36). If (ii) holds, let I be a maximal proper ideal of $\mathfrak{A}$ (by 2.3.17). Then $\mathfrak{A} \succcurlyeq \mathfrak{A}/I$ and $|A/I| > 1$, so $\mathfrak{A} \cong \mathfrak{A}/I$. Since $\mathfrak{A}/I$ is simple by 0.2.38, $\mathfrak{A}$ is simple as well.

REMARK 2.3.21. From 2.3.20 it is seen that for cylindric algebras the notions of simplicity and pseudo-simplicity coincide; cf. 0.2.37.

THEOREM 2.3.22. *If* $\mathfrak{A}$ *is a finitely generated* Dc_α *with* $\alpha \geqq \omega$ *which either is simple or satisfies the condition* $\mathsf{c}_0 - \mathsf{d}_{01} = 1$, *then* $\mathfrak{A}$ *is generated by a single element.*

PROOF. We can restrict ourselves to the case when $\mathsf{c}_0 - \mathsf{d}_{01} = 1$. In fact, every simple CA_α satisfies this condition (unless $|A| = 2$, in which case the conclusion of the theorem trivially holds). Further, it suffices to assume that $\mathfrak{A}$ is generated by two elements x, y, and show that it is generated by just one element. By 1.11.4 we may choose distinct $\kappa, \lambda \in \alpha \sim (\Delta x \cup \Delta y)$. Let $z = x \cdot \mathsf{d}_{\kappa\lambda} + y \cdot -\mathsf{d}_{\kappa\lambda}$. Then $\mathfrak{A} = \mathfrak{Sg}\{z\}$. To prove this, note that $z \cdot \mathsf{d}_{\kappa\lambda} = x \cdot \mathsf{d}_{\kappa\lambda}$ and $z \cdot -\mathsf{d}_{\kappa\lambda} = y \cdot -\mathsf{d}_{\kappa\lambda}$. Thus

$$\mathsf{c}_\kappa(z \cdot \mathsf{d}_{\kappa\lambda}) = \mathsf{c}_\kappa(x \cdot \mathsf{d}_{\kappa\lambda}) = x \cdot \mathsf{c}_\kappa \mathsf{d}_{\kappa\lambda} = x.$$

Hence $x \in Sg\{z\}$. Similarly,

$$\mathsf{c}_\kappa(z \cdot -\mathsf{d}_{\kappa\lambda}) = \mathsf{c}_\kappa(y \cdot -\mathsf{d}_{\kappa\lambda}) = y \cdot \mathsf{c}_\kappa - \mathsf{d}_{\kappa\lambda} = y \cdot \mathsf{c}_0 - \mathsf{d}_{01} = y.$$

Hence also $y \in Sg\{z\}$. Consequently, $\mathfrak{A} = \mathfrak{Sg}\{z\}$, as desired.

REMARKS 2.3.23. In its metalogical interpretation Theorem 2.3.22 (due in the present form to Henkin) is closely related to the following elementary fact: for any two finitary relations R and S there is a relation T such that T is definable in terms of R and S (in a language of predicate logic provided with infinitely many variables) and, conversely, both R and S are definable in terms of T. The relation T can be constructed in several different ways. Cf. Goodman [43*].

For each $\alpha \neq 0$, every finite discrete CA_α $\mathfrak{A}$ with more than two atoms is an example of a non-simple finitely generated Dc_α which is not generated by one element; for $\alpha \geqq \omega$ we can also easily obtain a non-discrete CA_α $\mathfrak{A}'$ with these properties by setting $\mathfrak{A}' = \mathfrak{A} \times \mathfrak{B}$ where $\mathfrak{A}$ is as above and $\mathfrak{B}$ is any finitely generated non-discrete Lf_α. Thus the hypothesis of simplicity in 2.3.22 cannot be entirely omitted. On the other hand, the premiss $\alpha \geqq \omega$ in 2.3.22 is not essential: the conclusion trivially holds in case $2 \leqq \alpha < \omega$.

It will be shown in Section 2.6, Remark 2.6.25, that for every finite $\alpha > 0$ there is a simple finite CA_α which is not generated by any single element. (However, we may recall that any full cylindric set algebra of dimension α, with $1 < \alpha < \omega$, and with finite base is generated by a single element; cf. the remark following 2.1.11.) The problem is open whether, in case $\alpha \geqq \omega$, Theorem 2.3.22 extends from Dc_α's to arbitrary finitely generated simple CA_α's.

THEOREM 2.3.24. *If* $\mathfrak{A}$ *is a simple* CA_α *and* $|A| > |\alpha| \cup \omega$, *then there is a* $\mathfrak{B} \subseteq \mathfrak{A}$ *such that* $|B| = |\alpha| \cup \omega$, $\mathfrak{B}$ *is simple, and* $\mathfrak{B}$ *is not finitely generated.*[1]

PROOF. Clearly there is a sequence $\langle \mathfrak{C}_\kappa : \kappa < \omega \rangle \in {}^\omega\mathbf{S}\mathfrak{A}$ such that $|C_\kappa| = |\alpha| \cup \omega$ and $C_\kappa \subset C_{\kappa+1}$, for all $\kappa < \omega$. By 2.3.16(i) the CA_α $\mathfrak{B} = \bigcup\{\mathfrak{C}_\kappa : \kappa < \omega\}$ has the properties indicated in the conclusion of the theorem.

REMARK 2.3.25. From 2.3.24 we can infer that for every α there is a simple Lf_α of power $|\alpha| \cup \omega$ which is not finitely generated. For, let $\beta = (|\alpha| \cup \omega)^+$, and let A be the family of all $X \subseteq {}^\alpha\beta$ with the following property: there is a finite $\Gamma \subseteq \alpha$ such that whenever $\sigma \in X$ we have also $\tau \in X$ for every sequence $\tau \in {}^\alpha\beta$ which coincides with σ on Γ. It is easily seen that A is a cylindric field of subsets of ${}^\alpha\beta$, and that the associated CA_α $\mathfrak{A}$ with universe A is simple, locally finite, and such that $|A| > |\alpha| \cup \omega$. Hence the CA_α $\mathfrak{B}$ given by 2.3.24 is simple, locally finite, not finitely generated, and satisfies $|B| = |\alpha| \cup \omega$.

As was pointed out in 2.3.6, a homomorphic image of a CA $\mathfrak{A}$ can always be represented isomorphically as a quotient algebra $\mathfrak{A}/I$ over an ideal I, and thus as an algebra whose elements are certain subsets of the universe A. It will be seen from the next theorem that in certain special though important cases the representation of a homomorphic image undergoes a further simplification: the quotient algebra $\mathfrak{A}/I$ can be transformed isomorphically into a relativized algebra $\mathfrak{Rl}_b\mathfrak{A}$, hence into an algebra whose elements are among those of A, and whose operations are very closely related to those of $\mathfrak{A}$.

THEOREM 2.3.26. *For every* $\mathfrak{A} \in \mathsf{CA}_\alpha$ *and* $b \in Zd\mathfrak{A}$ *we have:*

(i) *if* $h = \langle x \cdot b : x \in A \rangle$, *then* $h \in Ho(\mathfrak{A}, \mathfrak{Rl}_b\mathfrak{A})$ *and* $(h^{-1})^\star 0 = Ig\{-b\}$;

(ii) $\mathfrak{A} \succcurlyeq \mathfrak{Rl}_b\mathfrak{A}$.

[1] Theorem 2.3.24 is a very special case of a general model-theoretical result in Hanf [62a*], p. 38.

PROOF. (i) is easily checked by means of 2.2.1 and 2.3.10(iii). (ii) is an immediate consequence of (i).

COROLLARY 2.3.27. *For every* $\mathfrak{A} \in \mathsf{CA}_\alpha$ *we have:*

(i) *if* $b \in Zd\mathfrak{A}$, *then* $\mathfrak{A}/Ig\{b\} \cong \mathfrak{Rl}_{-b}\mathfrak{A}$;

(ii) *more generally, if* $b \in A$ *and* $|\Delta b| < \omega$, *then* $\mathfrak{A}/Ig\{b\} \cong \mathfrak{Rl}_{-\mathsf{c}_{(\Delta b)}b}\mathfrak{A}$.

PROOF. (i) follows easily from 2.3.26(i) (cf. 2.3.6, in particular statement (II)). (ii) follows from (i), 1.7.11(i), and 2.3.10(ii).

We see that, if $b \in Zd\mathfrak{A}$, the relativized algebra $\mathfrak{Rl}_b\mathfrak{A}$ is not only a CA, but is actually a homomorphic image of the original CA $\mathfrak{A}$; this supplements Theorem 2.2.12. Obviously, every algebra $\mathfrak{B}$ isomorphic with $\mathfrak{Rl}_b\mathfrak{A}$ is also a homomorphic image of $\mathfrak{A}$. We shall now show that, under certain rather restrictive assumptions concerning $\mathfrak{A}$ and $\mathfrak{B}$, the converse of this last statement is likewise true.

THEOREM 2.3.28. *Assume* $\alpha < \omega$, $\mathfrak{A}$ *is a finitely generated* CA_α, *and* $\mathfrak{B}$ *is a finite* CA_α.

(i) *If* $h \in Ho(\mathfrak{A}, \mathfrak{B})$, *then there is a* $b \in Zd\mathfrak{A}$ *such that* $(h^{-1})^\star 0^{(\mathfrak{B})} = Ig\{b\}$.

(ii) *If* $\mathfrak{A} \succcurlyeq \mathfrak{B}$, *then there is a* $c \in Zd\mathfrak{A}$ *such that* $\mathfrak{B} \cong \mathfrak{Rl}_c\mathfrak{A}$.

PROOF. It seems to be more convenient in this proof to deal with congruence relations rather than ideals. By 2.3.10(ii),(iv) we see that in order to prove (i) it is sufficient to prove that $(h^{-1})^\star 0^{(\mathfrak{B})}$ is a finitely generated ideal; hence, in view of 0.2.50(ii) and 2.3.4(ii), we need only show that $h|h^{-1}$ is a finitely generated congruence relation on A.

Because of the hypothesis of the theorem, there exists a finite subset X of A such that X generates $\mathfrak{A}$ and, in addition,

(1) $$h^*X = B.$$

Let

$$X' = X \cup \{x+y : x, y \in X\} \cup \{-x : x \in X\} \cup \bigcup_{\kappa<\alpha} \mathsf{c}_\kappa^* X \cup \{\mathsf{d}_{\kappa\lambda} : \kappa, \lambda < \alpha\}$$

and set

(2) $$R = Cg^{(\mathfrak{A})}(h|h^{-1} \cap (X \times X')).$$

Then, since X and hence also X' is finite, we have that R is finitely generated. Clearly $R \subseteq h|h^{-1}$. Consequently, to complete the proof of (i) it remains only to show $h|h^{-1} \subseteq R$, and for this purpose we shall first prove that every element of A is in the relation R to some element of X, i.e., that

(3) $$R^*X = A.$$

Suppose now that $a, b \in R^*X$, say xRa and yRb where $x, y \in X$; then

(4) $$(x+y)R(a+b).$$

However, from (1) we see that there is a $z \in X$ such that $hz = h(x+y)$; thus $zR(x+y)$ by (2). Combining this with (4) we get $zR(a+b)$, i.e., $a+b \in R^*X$. Therefore, R^*X is closed under the operation $+$, and in a similar way using (1) and (2) we can show that R^*X is closed under $-$ and all the cylindrifications c_κ. It follows immediately from (1) and (2) that R^*X contains all the diagonal elements $\mathsf{d}_{\kappa\lambda}$; thus R^*X is a subuniverse of $\mathfrak{A}$. Since the inclusion $X \subseteq R^*X$ is obvious, we obtain (3).

Now consider any $a, b \in A$ such that $ha = hb$. By (3) there are $x, y \in X$ such that

(5) $$xRa \text{ and } yRb.$$

Then, since $R \subseteq h|h^{-1}$, we have

$$hx = ha = hb = hy,$$

and hence xRy by (2); this combines with (5) to give aRb. Therefore, we have shown $h|h^{-1} \subseteq R$. This completes the proof of (i), and (ii) is an immediate consequence of (i), 2.3.6(II), and 2.3.27(i).

THEOREM 2.3.29. *Let $\mathfrak{A}$ be a finitely generated CA_α with $\alpha < \omega$.*

(i) *If $a \in A$, $h \in Ho\mathfrak{A}$, $|h^*A| < \omega$, and $ha \neq 0^{(h^*\mathfrak{A})}$, then there is an $x \in At\mathfrak{A}$ such that $x \leqq a$.*

(ii) *If for every non-zero $x \in A$ there is an $h \in Ho\mathfrak{A}$ such that $|h^*A| < \omega$ and $hx \neq 0^{(h^*\mathfrak{A})}$, then $\mathfrak{A}$ is atomic.*

PROOF. By 2.3.28(i) and the hypothesis of (i) we have $(h^{-1})^\star 0^{(\mathfrak{B})} = Ig\{b\}$ for some $b \in Zd\mathfrak{A}$; furthermore, the premiss $ha \neq 0^{(\mathfrak{B})}$ implies that we do not have $a \leqq b$, i.e., that

(1) $$a \cdot -b \neq 0.$$

By 2.3.6(II) and 2.3.27(i) we obtain

$$h^*\mathfrak{A} \cong \mathfrak{A}/Ig\{b\} \cong \mathfrak{Rl}_{-b}\mathfrak{A};$$

hence $\mathfrak{Rl}_{-b}\mathfrak{A}$ is finite and thus atomic. From this latter fact and (1) we conclude that there is an $x \in At\mathfrak{Rl}_{-b}\mathfrak{A}$ such that $x \leqq a$. The conclusion of (i) now follows since it is clear that $At\mathfrak{Rl}_{-b}\mathfrak{A} \subseteq At\mathfrak{A}$. (ii) follows immediately from (i).

The two immediately preceding theorems, due to Tarski, will find some interesting applications in Section 2.5; cf. 2.5.7 and 2.5.9.

THEOREM 2.3.30. *If* $\mathfrak{A} \in \mathsf{CA}_\alpha$, *with* $|A| > 2$ *and* $\alpha \geqq \omega$, *and* $b \in At\,\mathfrak{A}$, *then* $Ig\{b\}$ *is a proper ideal.*

PROOF. If $Ig\{b\}$ were not a proper ideal, then by 2.3.10(i) we would have $\mathsf{c}_{(\Gamma)}b = 1$ for some finite $\Gamma \subseteq \alpha$. In view of 1.10.5(ii),(iii) this means that $\Delta b = 0$ and thus $b = 1$. However, this is impossible since b is an atom and $|A| > 2$.

As a consequence of this theorem we have that every simple CA_α with $\alpha \geqq \omega$ and with more than two elements is atomless. This result will be generalized in 2.4.54 below.

THEOREM 2.3.31. *For any* CA_α $\mathfrak{A}$ *with* $\alpha \geqq 2$ *we have:*

(i) *If* $\mathfrak{A} = \mathfrak{Sg}\,X$ *and* $|X| < \omega$, *then*

$$At\,\mathfrak{A} \cap Zd\,\mathfrak{A} = \{(\prod Y - \sum(X \sim Y)) \cdot \mathsf{c}_0^\partial \mathsf{d}_{01} : Y \subseteq X\} \sim \{0\}.$$

(ii) *If* $\mathfrak{A}$ *is finitely generated, then*

$$\mathsf{c}_0^\partial \mathsf{d}_{01} = \sum(At\,\mathfrak{A} \cap Zd\,\mathfrak{A}),$$

and thus $\mathsf{c}_0^\partial \mathsf{d}_{01} = 0$ *whenever* $\mathfrak{A}$ *is atomless.*

PROOF. Let X be as in the hypothesis of (i) and let $\mathfrak{B} = \mathfrak{Rl}_{\mathsf{c}_0^\partial \mathsf{d}_{01}}\mathfrak{A}$. By 0.2.18(i) and 2.3.26(i) we have $B = Sg\{x \cdot \mathsf{c}_0^\partial \mathsf{d}_{01} : x \in X\}$. Furthermore, since $\mathfrak{B}$ is discrete by 2.2.14, we see that $B = Sg^{(\mathfrak{Bl}\,\mathfrak{B})}\{x \cdot \mathsf{c}_0^\partial \mathsf{d}_{01} : x \in X\}$. Therefore, from 2.2.1 and the theory of BA's we may conclude that

(1) $$At\,\mathfrak{B} = \{(\prod Y - \sum(X \sim Y)) \cdot \mathsf{c}_0^\partial \mathsf{d}_{01} : Y \subseteq X\} \sim \{0\}$$

and

(2) $$\mathsf{c}_0^\partial \mathsf{d}_{01} = \sum At\,\mathfrak{B}.$$

However, $At\,\mathfrak{B} = At\,\mathfrak{A} \cap Zd\,\mathfrak{A}$ by 1.10.5(ii); hence (i) and (ii) follow respectively from (1) and (2).

It may be interesting to compare 2.3.31 with 2.2.27; we notice that the descriptions of zero-dimensional atoms in finitely generated CA_α's are very different in the two cases $\alpha = 1$ and $\alpha > 1$. Nevertheless, we shall see in the following corollary that both descriptions yield the same upper bounds for the cardinalities of the sets of these atoms, and it will be seen from Theorem 2.5.11 below that this upper bound is actually the least upper bound.

COROLLARY 2.3.32. *For every* CA_α $\mathfrak{A}$, *if* $\mathfrak{A} = \mathfrak{Sg}\,X$ *and* $|X| < \omega$, *then*

$$|At\,\mathfrak{A} \cap Zd\,\mathfrak{A}| \leqq 2^{|X|}.$$

PROOF: for $\alpha = 0$ by the theory of BA's, for $\alpha = 1$ by 2.2.27, and for $\alpha \geqq 2$ by 2.3.31(i).

Don Pigozzi has observed that there is a simple direct proof of 2.3.32 which does not involve diagonal elements so that it applies also to diagonal-free algebras. Thus 2.3.32 holds for these algebras as well.

COROLLARY 2.3.33. *For every* $\mathsf{Dc}_\alpha\ \mathfrak{A}$ *we have:*

(i) *If* $\mathfrak{A} = \mathfrak{S}\mathfrak{g}\, X$ *and* $|X| < \omega$, *then* $|At\mathfrak{A}| \leqq 2^{|X|}$.

(ii) *If* $\mathfrak{A}$ *is finitely generated and* $\alpha \geqq 2$, *then* $\mathsf{c}_0^\partial \mathsf{d}_{01} = \sum At\mathfrak{A}$, *and thus* $\mathfrak{A}$ *is atomless iff* $\mathsf{c}_0^\partial \mathsf{d}_{01} = 0$.

PROOF. By 1.11.8(i), $At\mathfrak{A} \subseteq Zd\mathfrak{A}$. Thus (i) follows from 2.3.32 and (ii) from 2.3.31(ii).

Compare 2.3.33(ii) with 1.11.8(ii).

2.4. DIRECT PRODUCTS AND RELATED NOTIONS

We begin this section by discussing the closure properties of various classes of CA's under the formation of direct products. Direct decompositions of CA's and their zero-dimensional parts make up our next topic. This leads naturally to the discussion of the refinement and remainder properties; it turns out that these two properties hold for a comprehensive class of cylindric algebras. Now as we know from Section 0.3, several interesting and rather deep theorems involving direct products can be established for all algebras which possess the refinement and remainder properties. Since the results in their general algebraic form were stated in Section 0.3 without proof, in the present section we shall give detailed proofs of these results as applied to cylindric algebras. In the last part of the section we discuss subdirect and reduced products of CA's.

THEOREM 2.4.1. $\mathsf{PCA}_\alpha = \mathsf{CA}_\alpha$; *that is, the class* CA_α *is closed under the formation of direct products.*

Again this theorem can be proved directly, or one can appeal to the equational character of the class CA_α (cf. 2.1.2). From 2.1.1, 2.3.1, and 2.4.1 we see that $\mathbf{HSPCA}_\alpha = \mathsf{CA}_\alpha$, so that CA_α is an algebraically closed class (cf. 0.3.14).

THEOREM 2.4.2. *If* $\mathfrak{B} \in {}^I\mathsf{CA}_\alpha$ *and* $f \in \mathsf{P}_{i\in I} B_i$, *then* $\Delta f = \bigcup_{i\in I} \Delta f_i$.

PROOF. The condition $\kappa \in \Delta f$ is equivalent to $\mathsf{c}_\kappa f \neq f$, which in turn is equivalent to the existence of an $i \in I$ such that $\mathsf{c}_\kappa f_i \neq f_i$, i.e., to $\kappa \in \bigcup_{i\in I} \Delta f_i$.

COROLLARY 2.4.3. $\mathfrak{Zd}(\mathsf{P}_{i\in I}\mathfrak{B}_i) = \mathsf{P}_{i\in I}\mathfrak{Zd}\,\mathfrak{B}_i$.

COROLLARY 2.4.4. *Let* $\mathfrak{B} \in {}^I\mathsf{Lf}_\alpha$.

(i) *If* $\alpha < \omega$, *then* $\mathsf{P}_{i\in I}\mathfrak{B}_i \in \mathsf{Lf}_\alpha$.

(ii) *If* $\alpha \geqq \omega$, *then* $\mathsf{P}_{i\in I}\mathfrak{B}_i \in \mathsf{Lf}_\alpha$ *iff* $|\{i : i \in I,\ \mathfrak{B}_i \text{ is non-discrete}\}| < \omega$. *Thus we always have* $\mathsf{P}_{i\in I}\mathfrak{B}_i \in \mathsf{Lf}_\alpha$ *in case* $|I| < \omega$.

PROOF. (i) is obvious since $\mathsf{Lf}_\alpha = \mathsf{CA}_\alpha$ for $\alpha < \omega$, and (ii) follows readily from 2.4.2 with the help of 1.3.12(i),(ii) and 1.6.3.

COROLLARY 2.4.5. (i) *If* $0 < \lambda < \omega$, *then* ${}^{\lambda}\mathfrak{A} \in \mathsf{Dc}_\alpha$ *iff* $\mathfrak{A} \in \mathsf{Dc}_\alpha$.
(ii) *If* $\lambda \geqq |\alpha| > 0$, *then* ${}^{\lambda}\mathfrak{A} \in \mathsf{Dc}_\alpha$ *iff* $\mathfrak{A}$ *is a discrete* CA_α.

PROOF. In case $\alpha < \omega$, (i) and (ii) are both trivial since Dc_α consists just of the discrete CA_α's if $\alpha > 0$. Assume now that $\alpha \geqq \omega$. Then (i) follows from 1.11.4 and 2.4.2, and the only part of (ii) that requires proof is that ${}^{\lambda}\mathfrak{A} \notin \mathsf{Dc}_\alpha$ whenever $\mathfrak{A}$ is non-discrete and $\lambda \geqq |\alpha|$. This, however, is an easy consequence of 1.3.12(i),(ii) along with 2.4.2.

REMARKS 2.4.6. In connection with 2.4.5 the problem arises whether the classes Dc_α are closed under the formation of direct products of systems indexed by sets of cardinality $<|\alpha|$. Clearly, in case $0 < \alpha < \omega$, Dc_α is closed under the formation of arbitrary direct products, for Dc_α coincides then with the class of discrete CA_α's. The following example shows, however, that for $\alpha \geqq \omega$ none of the classes Dc_α is closed even under the formation of finite direct products. Let $\mathfrak{C}$ be the full cylindric set algebra of all subsets of ${}^{\alpha}2$. Let $X = \{f : f \in {}^{\alpha}2,\ f\kappa = 0 \text{ for all even } \kappa < \alpha\}$ and $Y = \{f : f \in {}^{\alpha}2,\ f\kappa = 0$ for all odd $\kappa < \alpha\}$. Let $\mathfrak{A} = \mathfrak{Sg}^{(\mathfrak{C})}\{X\}$ and $\mathfrak{B} = \mathfrak{Sg}^{(\mathfrak{C})}\{Y\}$. By 2.1.5(ii), $\mathfrak{A}, \mathfrak{B} \in \mathsf{Dc}_\alpha$, but $\mathfrak{A} \times \mathfrak{B} \notin \mathsf{Dc}_\alpha$ since $\Delta\langle X, Y\rangle = \alpha$ by 2.4.2.

From 2.4.4 and 2.4.5 we see that, in case $\alpha \geqq \omega$, neither of the classes Lf_α or Dc_α is closed under the formation of direct powers, although both are closed under the formation of finite direct powers. Hence, by a theorem of Mostowski [52*], p. 25, Theorem 5.32, the classes Lf_α and Dc_α are not elementary for $\alpha \geqq \omega$ (although they clearly are elementary for $\alpha < \omega$).

If $\mathfrak{A} \in \mathsf{Mn}_\alpha$ (with α arbitrary) and $|A| > 1$, then clearly $\mathfrak{A} \times \mathfrak{A} \notin \mathsf{Mn}_\alpha$. Thus 2.4.4 and even 2.4.5 cannot be extended to Mn_α's.

The next theorem can be called the fundamental theorem on direct decomposition of CA's. We know from 0.3.22 that every direct decomposition of an algebra $\mathfrak{A}$,

$$\mathfrak{A} \cong \mathsf{P}_{i \in I}\, \mathfrak{B}_i,$$

can be characterized by means of an appropriate system of congruence relations, and that the $\mathfrak{B}_i$ can be isomorphically represented by the corresponding quotient algebras; from remarks in 2.3.6 it follows that in the case of CA's the congruence relations can be replaced by ideals. It turns out that the ideals involved are of a very special kind, namely, principal ideals generated by zero-dimensional elements. Hence, a further simplification is possible: a direct decomposition of a CA can be characterized in terms of a system of zero-dimensional elements. (The elements forming such a system turn out to be the complements of those generating the above mentioned principal ideals.) Thus, in view of Theorem 2.3.27, $\mathfrak{A}$ can be isomorphically represented as the product of the relativized algebras correlated with these elements. This generalization of a well-known result from the theory of Boolean algebras is precisely the content of the next theorem.[1]

THEOREM 2.4.7. *Let* $\mathfrak{A} \in \mathsf{CA}_\alpha$. *For* $\mathfrak{A} \cong \mathsf{P}_{i\in I}\mathfrak{B}_i$ *it is necessary and sufficient that there exists a system* $\langle x_i : i \in I\rangle$ *of elements of* A *satisfying the following conditions*:

(i) $x_i \in Zd\mathfrak{A}$ *for every* $i \in I$;
(ii) $x_i \cdot x_j = 0$ *if* $i, j \in I$ *and* $i \neq j$;
(iii) $\sum_{i\in I} x_i = 1$;
(iv) $\mathfrak{Rl}_{x_i}\mathfrak{A} \cong \mathfrak{B}_i$ *for every* $i \in I$;
(v) *for all* $y \in {}^I A$ *the sum* $\sum_{i\in I}(y_i \cdot x_i)$ *exists.*

PROOF. To prove the necessity of (i)–(v) we may assume that $\mathfrak{A} = \mathsf{P}_{i\in I}\mathfrak{B}_i$. For each $i \in I$ let $x_i = \langle 0^{(\mathfrak{B}_j)} : j \in I \sim \{i\}\rangle \cup \{\langle i, 1^{(\mathfrak{B}_i)}\rangle\}$. Conditions (i)–(iii) are then obvious. For any $i \in I$ the function $\langle f_i : f \in Rl_{x_i}\mathfrak{A}\rangle$ is an isomorphism of $\mathfrak{Rl}_{x_i}\mathfrak{A}$ onto $\mathfrak{B}_i$; thus (iv) holds. Finally, if $y \in {}^I A$, then clearly $\sum_{i\in I}(y_i \cdot x_i) = \langle y_i(i) : i \in I\rangle$, and (v) holds.

In order to show that (i)–(v) are sufficient take $J_i = Ig\{-x_i\}$ for each $i \in I$. Suppose $z \in \bigcap_{i\in I} J_i$. Then by (i) and 2.3.10(iii) we have $z \leqq -x_i$ for each $i \in I$, and hence $z \leqq \prod_{i\in I} -x_i = 0$ by (iii). Thus

$$(1) \qquad \bigcap_{i\in I} J_i = \{0\}.$$

Now consider any $y \in {}^I A$. Setting $z = \sum_{i\in I}(y_i \cdot x_i)$, which exists by (v), we see from (ii) that

$$y_i \cdot x_i = z \cdot x_i \text{ for each } i \in I.$$

In view of (i) and 2.3.26(i) this implies that $z \in \bigcap_{i\in I}(y_i/J_i)$, and hence we have shown that

$$(2) \qquad \bigcap_{i\in I}(y_i/J_i) \neq 0 \text{ for every } y \in {}^I A.$$

Finally, by (i), (iv), and 2.3.27(i) we have

$$(3) \qquad \mathfrak{A}/J_i \cong \mathfrak{B}_i \text{ for each } i \in I.$$

From (1), (2), and (3) we may conclude by 0.2.49 and 0.3.22 that $\mathfrak{A} \cong \mathsf{P}_{i\in I}\mathfrak{B}_i$.

Note that, by 2.2.12, 2.4.7(i) implies that $\mathfrak{Rl}_{x_i}\mathfrak{A}$ is a CA_α; 2.2.12 will be used below in several instances without citation.

[1]) The subsequent part of the present section beginning with 2.4.7 and ending with 2.4.32 is essentially due to Tarski except as otherwise noted.

COROLLARY 2.4.8. *For every* CA_α $\mathfrak{A}$ *the following two conditions are equivalent:*

(i) $\mathfrak{B} \mid \mathfrak{A}$,

(ii) $\mathfrak{B} \cong \mathfrak{Rl}_b\mathfrak{A}$ *for some* $b \in Zd\mathfrak{A}$.

PROOF. The fact that (i) implies (ii) follows immediately from 2.4.7. To obtain the implication in the opposite direction we apply 2.4.7 again, letting $I = 2$, $x_0 = b$, and $x_1 = -b$.

From the theory of BA's it is well known that every BA $\mathfrak{B}$ with $|B| > 1$ is homomorphic to a two-element algebra and hence $\mathfrak{B} \succcurlyeq\mid\subseteq \mathfrak{C}$ for every BA $\mathfrak{C}$ ($\mathfrak{B} \succcurlyeq\mid\subseteq \mathfrak{C}$ trivially if $|C| = 1$). Another known result from the theory of BA's is: for every BA $\mathfrak{A}$ and every non-zero element $b \in A$ we have $\mathfrak{Rl}_b\mathfrak{A} \cong\mid\subseteq \mathfrak{A}$ (cf., for instance, Keisler-Tarski [64*], p. 292, Theorem 4.20). The two results are essentially equivalent by 0.3.8 and 2.4.8 (with $\alpha = 0$). The situation changes when we turn to CA_α's for any $\alpha \neq 0$. We can then easily construct two algebras $\mathfrak{B}$ and $\mathfrak{C}$ such that $|B| > 1$ but the formula $\mathfrak{B} \succcurlyeq\mid\subseteq \mathfrak{C}$ does not hold; hence, again by 0.3.8 and 2.4.8, there is a CA_α $\mathfrak{A}$ and a non-zero element $b \in A$ with $\Delta b = 0$ such that $\mathfrak{Rl}_b\mathfrak{A} \cong\mid\subseteq \mathfrak{A}$ fails.

A system $x \in {}^IA$ satisfying conditions (i)–(iii) and (v) of 2.4.7 may be called a *decomposition system* for the CA_α $\mathfrak{A}$; in analogy to the notation for congruence relations it could also be referred to as a *system of complementary elements*. Condition (v) in 2.4.7 and in the definition of decomposition system can obviously be omitted in case I is finite or, more generally, in case $|I| < \beta$ and $\mathfrak{A}$ is assumed to be β-complete (i.e., the sum $\sum_{j\in J} x_j$ is assumed to exist whenever $x \in {}^JA$ and $|J| < \beta$). Actually a weaker assumption suffices for this purpose. To formulate it conveniently we define:

DEFINITION 2.4.9. *Elements* x, y *of a* CA_α $\mathfrak{A}$ *are said to be* ***separable*** *if there is an element* $z \in Zd\mathfrak{A}$ *such that* $x \leqq z$ *and* $y \leqq -z$. $\mathfrak{A}$ *is called* ***separably** β**-complete*** *(β a cardinal) if* $\sum_{i\in I} x_i$ *exists for any system* $\langle x_i : i \in I\rangle$ *of pairwise separable elements of A for which* $|I| < \beta$; *it is called* ***separably complete*** *if it is separably β-complete for every cardinal β. A separably ω_1-complete* CA_α *is also called* ***separably countably complete.***

Obviously every CA which is β-complete in the ordinary sense is separably β-complete (but not conversely, as is easily seen). If $\beta \leqq \omega$, then every CA is separably β-complete. Furthermore, if $\mathfrak{A}$ is separably β-complete, then $\mathfrak{Zd}\mathfrak{A}$ is β-complete; this follows from a known theorem by which, for any given BA $\mathfrak{B}$, the existence of all sums $\sum_{i\in I} x_i$ of pairwise disjoint elements of $\mathfrak{B}$ with $|I| < \beta$ implies that $\mathfrak{B}$ is β-complete (cf. Smith-Tarski [57*], p. 249,

Theorem 3.5). Notice the following, seemingly weaker but actually equivalent, definition of separable β-completeness:

A $\mathsf{CA}_\alpha\ \mathfrak{A}$ *is separably β-complete iff* $\sum_{i\in I}(y_i\cdot z_i)$ *exists for every system* $y\in{}^IZd\mathfrak{A}$ *of pairwise disjoint elements and every system* $z\in{}^IA$ *with* $|I|<\beta$.

The proof of equivalence is based entirely on the general theory of Boolean algebras.

COROLLARY 2.4.10. *Assume that* $|I|<\omega$ *or, more generally, that for some cardinal* β, $|I|<\beta$ *and* $\mathfrak{A}$ *is a separably β-complete* CA_α. *For* $\mathfrak{A}\cong\mathsf{P}_{i\in I}\mathfrak{B}_i$ *it is necessary and sufficient that there exist a system* $\langle x_i:i\in I\rangle$ *of elements of* A *such that the conditions 2.4.7(i)–(iv) hold.*

Just as in the case of ordinary β-completeness we have

THEOREM 2.4.11. *For any system of* CA_α*'s* $\langle\mathfrak{B}_i:i\in I\rangle$ *and any cardinal* β, $\mathsf{P}_{i\in I}\mathfrak{B}_i$ *is separably β-complete iff* $\mathfrak{B}_i$ *is separably β-complete for each* $i\in I$.

PROOF. By 2.4.2, elements $f, g\in\mathsf{P}_{i\in I}B_i$ are separable iff f_i and g_i are separable for each $i\in I$. Hence the equivalence stated in the theorem follows from the theory of BA's.

When applying this theorem in our further discussion we shall generally omit explicit reference to it.

THEOREM 2.4.12. *If* $\mathfrak{A}$ *is a separably β-complete* CA_α, $\mathfrak{C}\in{}^I\mathsf{BA}$, *and* $|I|<\beta$, *then the following two conditions are equivalent*:

(i) *There is a system* $\mathfrak{B}\in{}^I\mathsf{CA}_\alpha$ *such that* $\mathfrak{A}\cong\mathsf{P}_{i\in I}\mathfrak{B}_i$ *and* $\mathfrak{Zd}\,\mathfrak{B}_i=\mathfrak{C}_i$ *for each* $i\in I$.

(ii) $\mathfrak{Zd}\,\mathfrak{A}\cong\mathsf{P}_{i\in I}\mathfrak{C}_i$.

PROOF: by 2.2.12, 2.4.3, and 2.4.10.

Recall the definition 0.3.34 of algebras with the refinement property.

THEOREM 2.4.13. *Every* CA_α *has the refinement property.*

PROOF. Assume that $\mathfrak{A}\cong\mathsf{P}_{i\in I}\mathfrak{B}_i\cong\mathsf{P}_{j\in J}\mathfrak{C}_j$. By this assumption and 2.4.7 we may choose two decomposition systems (cf. the remarks following 2.4.8) $\langle x_i:i\in I\rangle$ and $\langle y_j:j\in J\rangle$ for $\mathfrak{A}$ such that $\mathfrak{Rl}_{x_i}\mathfrak{A}\cong\mathfrak{B}_i$ for all $i\in I$ and $\mathfrak{Rl}_{y_j}\mathfrak{A}\cong\mathfrak{C}_j$ for all $j\in J$. Let $z=\langle x_i\cdot y_j:i\in I, j\in J\rangle$. Then it can easily be checked that, for each $i\in I$, $\langle z_{ij}:j\in J\rangle$ is a decomposition system for $\mathfrak{Rl}_{x_i}\mathfrak{A}$, and similarly, for each $j\in J$, $\langle z_{ij}:i\in I\rangle$ is a decomposition system for $\mathfrak{Rl}_{y_j}\mathfrak{A}$. Hence, by 2.4.7, $\mathfrak{B}_i\cong\mathsf{P}_{j\in J}\mathfrak{Rl}_{z_{ij}}\mathfrak{A}$ for each $i\in I$, and $\mathfrak{C}_j\cong\mathsf{P}_{i\in I}\mathfrak{Rl}_{z_{ij}}\mathfrak{A}$ for each $j\in J$, as desired.

As we know from the remarks following 0.3.35, the principal significance of the refinement property for the theory of direct decompositions is that every algebra which has this property and is totally decomposable also has the unique decomposition property. Our next task, therefore, is to describe those CA's which are totally decomposable; this will be done in 2.4.18. To this end, however, we first have to turn our attention to directly indecomposable CA's.

THEOREM 2.4.14. *For any* CA_α $\mathfrak{A}$ *the following conditions are equivalent*:

(i) $\mathfrak{A}$ *is directly indecomposable*;

(ii) $\mathfrak{Zd}\,\mathfrak{A}$ *is directly indecomposable*;

(iii) $|Zd\,\mathfrak{A}| = 2$, *or, equivalently*, $|A| > 1$ *and* $Zd\,\mathfrak{A} = \{0, 1\}$.[1]

PROOF. (i) and (ii) are equivalent by 2.4.12 (with $\beta = \omega$), and the equivalence of (ii) and (iii) is a well-known fact from the theory of Boolean algebras.

By comparing this theorem with 2.3.14 we conclude that directly indecomposable Lf_α's coincide with simple Lf_α's. This fact, along with some supplementary information, will be stated formally in Theorem 2.4.43 below.

COROLLARY 2.4.15. *If* $\mathfrak{A}$ *is directly indecomposable, then so is every subalgebra of* $\mathfrak{A}$.

THEOREM 2.4.16. (i) *If a* CA_α $\mathfrak{A}$ *is not directly indecomposable, then there is a subalgebra of* $\mathfrak{A}$, *generated by a single element, which is not directly indecomposable and, in fact, is not a subalgebra of any directly indecomposable subalgebra of* $\mathfrak{A}$.

(ii) *If* K *is a non-empty set of directly indecomposable* CA_α*'s directed by the relation* $\subseteq$, *then* UK *is directly indecomposable.*

PROOF: (i) by 2.4.14; (ii) by (i) and 0.1.28.

THEOREM 2.4.17. (i) *Every finite directly indecomposable* CA_α *is simple.*

(ii) *Every finite* CA_α *is isomorphic to a direct product of simple* CA_α*'s.*

PROOF: (i) by 1.3.15, 2.3.14, and 2.4.14 (cf. the remark after 2.4.14); (ii) by (i) and 0.3.31.

THEOREM 2.4.18. *For any* CA_α $\mathfrak{A}$ *the following four conditions are equivalent*:

(i) $\mathfrak{A}$ *is separably complete and* $\mathfrak{Zd}\,\mathfrak{A}$ *is atomic*;

(ii) $\mathfrak{A}$ *is totally decomposable*;

(iii) $\mathfrak{A}$ *is uniquely totally decomposable*;

(iv) $\mathfrak{A}$ *and all of its direct factors are uniquely totally decomposable.*

All these conditions are satisfied if, in particular, $\mathfrak{A}$ *is finite.*

[1]) This theorem and the essential parts of two related results, Theorems 2.4.43 and 2.4.52, were stated in Tarski-Thompson [52].

PROOF. Obviously (iv) implies (iii), and (iii) implies (ii).

Assume now that (ii) holds, i.e., $\mathfrak{A}$ is totally decomposable; then clearly $\mathfrak{Zd}\,\mathfrak{A}$ is atomic. Moreover, there is a system $\langle x_i : i \in I \rangle$ of elements of A satisfying 2.4.7(i)–(iii),(v) with $\mathfrak{Rl}_{x_i}\mathfrak{A}$ directly indecomposable for each $i \in I$, i.e., x_i is an atom of $\mathfrak{Zd}\,\mathfrak{A}$ for each $i \in I$. Now let $\langle y_j : j \in J \rangle$ be any system of pairwise separable elements of A. Then for any $i \in I$ there is at most one $j \in J$ such that $x_i \cdot y_j \neq 0$. For assume that $x_i \cdot y_j \neq 0$ and consider any $k \neq j$. Then there is a $z \in Zd\mathfrak{A}$ such that $y_j \leqq z$ and $y_k \leqq -z$, and thus $x_i \leqq z$ since x_i is an atom of $\mathfrak{Zd}\,\mathfrak{A}$ and $x_i \cdot z \neq 0$; hence $x_i \cdot y_k \leqq z \cdot -z = 0$. Therefore, the sum $\sum_{j \in J}(x_i \cdot y_j)$ exists for each $i \in I$, and we set

$$z = \langle \textstyle\sum_{j \in J}(x_i \cdot y_j) : i \in I \rangle;$$

then, since $\langle x_i : i \in I \rangle$ satisfies 2.4.7(v), it follows that $\sum_{i \in I} z_i$ exists. It is easily checked that $\sum_{i \in I} z_i$ is the least upper bound of $\{y_j : j \in J\}$. Thus (ii) implies (i).

Finally, if (i) holds, then, by 2.4.11, $\mathfrak{A}$ and all its direct factors are separably complete; furthermore, their zero-dimensional parts are both complete and atomic. From 2.4.12 it follows then that $\mathfrak{A}$, as well as every direct factor of $\mathfrak{A}$, is totally decomposable, and the refinement property implies that the decomposition is unique (cf. 0.3.35 and 2.4.13). Thus conditions (i)–(iv) are actually equivalent. The last part of the theorem follows by 0.3.31(i).

REMARK 2.4.19. We know from Chapter 0 (cf. 0.3.33) that condition 2.4.18(iv) has far-reaching implications. Many basic problems concerning direct products admit positive solutions for algebras which satisfy this condition, i.e., which hereditarily have the unique decomposition property, while the discussion of these problems for arbitrary algebras presents considerable difficulties. To concentrate on the same problems as in Chapter 0 we recall, for instance, that every algebra $\mathfrak{A}$ satisfying 2.4.18(iv) has the following two properties:

(α) *For any algebra* $\mathfrak{B}$, *if* $\mathfrak{A}|\mathfrak{B}|\mathfrak{A}$, *then* $\mathfrak{A} \cong \mathfrak{B}$.

(β) *For any algebras* $\mathfrak{B}$ *and* $\mathfrak{C}$, *if* $\mathfrak{A} \cong \mathfrak{B} \times \mathfrak{B} \cong \mathfrak{C} \times \mathfrak{C}$, *then* $\mathfrak{B} \cong \mathfrak{C}$.

Unfortunately, as is seen from 2.4.18(i), the class of CA's which hereditarily have the unique decomposition property is rather restricted. As was mentioned in 0.3.33, the properties (α) and (β) may fail for Boolean algebras, i.e., for CA_0's, and hence also for discrete CA_α's of arbitrary dimension α. Thus the problem arises of finding a naturally defined class K of algebras which is essentially more comprehensive than the class described in 2.4.18(i) and for which (α), (β) and various analogous properties can be established. We shall present an affirmative solution of this problem here by showing that the class of all separably countably complete CA_α's can serve as K.

Our discussion will be based exclusively upon certain properties of the class of separably countably complete CA's listed in the next theorem. The discussion will thus have a general algebraic character, in that it will apply to any class of algebras having these properties.

THEOREM 2.4.20. *The class* K *of all separably countably complete* CA_α's *satisfies the following conditions*:

(i) *if* $\mathfrak{A}, \mathfrak{B} \in \mathsf{K}$, *then* $\mathfrak{A} \times \mathfrak{B} \in \mathsf{K}$;

(ii) *if* $\mathfrak{B} | \mathfrak{A} \in \mathsf{K}$, *then* $\mathfrak{B} \in \mathsf{K}$;

(iii) *every algebra* $\mathfrak{A} \in \mathsf{K}$ *has the finite refinement property*;

(iv) *every algebra* $\mathfrak{A} \in \mathsf{K}$ *has the remainder property.*[1]

PROOF. (i)–(iii) follow immediately from 2.4.11 and 2.4.13. To prove (iv), let $\langle \mathfrak{A}_\kappa : \kappa < \omega \rangle$ and $\langle \mathfrak{B}_\kappa : \kappa < \omega \rangle$ be systems of CA_α's such that $\mathfrak{A}_0$ is separably countably complete, and such that $\mathfrak{A}_\kappa \cong \mathfrak{A}_{\kappa+1} \times \mathfrak{B}_\kappa$ for each $\kappa < \omega$. Using 2.4.7 one sees that there is a sequence $a \in {}^\omega A_0$ such that the following conditions hold for each $\kappa < \omega$:

(1) $$\Delta a_\kappa = 0;$$

(2) $$a_\kappa \geqq a_{\kappa+1};$$

(3) $$\mathfrak{Rl}_{a_\kappa} \mathfrak{A}_0 \cong \mathfrak{A}_\kappa;$$

(4) $$\mathfrak{Rl}_{a_\kappa - a_{\kappa+1}} \mathfrak{A}_0 \cong \mathfrak{B}_\kappa.$$

Let $c = \prod_{\kappa < \omega} a_\kappa$ and $\mathfrak{C} = \mathfrak{Rl}_c \mathfrak{A}_0$ (cf. the remarks preceding 2.4.10). From (1), using 1.2.10 and 1.2.11, we obtain

(5) $$\Delta c = 0;$$

(6) $$\Delta(a_\kappa - a_{\kappa+1}) = 0 \text{ for each } \kappa < \omega.$$

Furthermore, from (2) and the elementary theory of BA's we obtain

(7) $$c + \sum_{\lambda < \omega} (a_{\kappa+\lambda} - a_{\kappa+\lambda+1}) = a_\kappa \text{ for each } \kappa < \omega.$$

By (4)–(7) and 2.4.10 the desired conclusion of the theorem follows.

[1]) It is known from the literature that the countably complete BA's form a class K satisfying conditions 2.4.20(i)–(iv) and hence also all the consequences of these conditions (cf. Tarski [49*], pp. 200–214). This result is, of course, a particular case of 2.4.20. On the other hand, a generalization of 2.4.20 is also known; it concerns algebras which, like CA_α's, have BA's as reducts (cf. Tarski [66]).

In connection with 2.4.20(iii) we note that by 2.4.13 all algebras $\mathfrak{A} \in \mathsf{K}$ actually have the full refinement property; however only the finite refinement property will be used in the subsequent discussion.

The fact that every separably countably complete CA_α has the properties (α) and (β) stated above will be expressed in Theorems 2.4.24 and 2.4.31 (the latter in a more general form). To prove these results we shall need a series of auxiliary theorems, some of which, e.g., 2.4.29, are of interest in their own right.

If T is the class of all isomorphism types correlated with algebras $\mathfrak{A}$ of any class K of algebras satisfying 2.4.20(i)–(iv), then the structure $\langle T, \times, \mathsf{P}\rangle$ (cf. 0.3.11) proves to be a finitely closed generalized cardinal algebra in the sense of Tarski [49*], p. 69, Definition 5.1, except that only the finite refinement property is assumed to hold. The proofs of Theorems 2.4.21 to 2.4.31 below have been obtained essentially by paraphrasing those of corresponding results in the general theory of cardinal algebras which can be found in op. cit., Sections 1 and 2.

We begin with a simple consequence of the finite refinement property (which actually applies to arbitrary cylindric algebras, and not only to separably countably complete CA's).

THEOREM 2.4.21. *If $\mathfrak{A}$, $\mathfrak{B}$, and $\mathfrak{C}$ are separably countably complete CA_α's and $\mathfrak{A} \times \mathfrak{B} \cong \mathfrak{C} \times \mathfrak{C}$, then there exist $\mathfrak{A}'$, $\mathfrak{B}'$, $\mathfrak{C}'$ such that $\mathfrak{A} \cong \mathfrak{A}' \times \mathfrak{A}' \times \mathfrak{C}'$, $\mathfrak{B} \cong \mathfrak{B}' \times \mathfrak{B}' \times \mathfrak{C}'$, and $\mathfrak{C} \cong \mathfrak{A}' \times \mathfrak{B}' \times \mathfrak{C}'$.*

PROOF. By applying 2.4.20(iii) we obtain $\mathfrak{D} \in {}^4\mathsf{CA}_\alpha$ such that $\mathfrak{A} \cong \mathfrak{D}_0 \times \mathfrak{D}_1$, $\mathfrak{B} \cong \mathfrak{D}_2 \times \mathfrak{D}_3$, and $\mathfrak{C} \cong \mathfrak{D}_0 \times \mathfrak{D}_2 \cong \mathfrak{D}_1 \times \mathfrak{D}_3$. Applying 2.4.20(iii) again to this last formula we obtain $\mathfrak{E} \in {}^4\mathsf{CA}_\alpha$ such that $\mathfrak{D}_0 \cong \mathfrak{E}_0 \times \mathfrak{E}_1$, $\mathfrak{D}_2 \cong \mathfrak{E}_2 \times \mathfrak{E}_3$, $\mathfrak{D}_1 \cong \mathfrak{E}_0 \times \mathfrak{E}_2$, and $\mathfrak{D}_3 \cong \mathfrak{E}_1 \times \mathfrak{E}_3$. Letting $\mathfrak{A}' = \mathfrak{E}_0$, $\mathfrak{B}' = \mathfrak{E}_3$, and $\mathfrak{C}' = \mathfrak{E}_1 \times \mathfrak{E}_2$, the desired conclusions are easily verified.

THEOREM 2.4.22. *If the CA_α $\mathfrak{A}$ is separably countably complete, then the following conditions are equivalent:*

$$\text{(i) } \mathfrak{A} \cong \mathfrak{A} \times \mathfrak{B}, \quad \text{(ii) } {}^\omega\mathfrak{B} \mid \mathfrak{A}, \quad \text{(iii) } \mathfrak{A} \cong \mathfrak{A} \times {}^\omega\mathfrak{B}.$$

PROOF. (i) implies (ii) by 2.4.20(iv). If $\mathfrak{A} \cong \mathfrak{C} \times {}^\omega\mathfrak{B}$, then $\mathfrak{A} \times {}^\omega\mathfrak{B} \cong \mathfrak{C} \times {}^\omega\mathfrak{B} \times {}^\omega\mathfrak{B} \cong \mathfrak{C} \times {}^\omega\mathfrak{B} \cong \mathfrak{A}$. Thus (ii) implies (iii). Finally, if $\mathfrak{A} \cong \mathfrak{A} \times {}^\omega\mathfrak{B}$, then $\mathfrak{A} \times \mathfrak{B} \cong \mathfrak{A} \times {}^\omega\mathfrak{B} \times \mathfrak{B} \cong \mathfrak{A} \times {}^\omega\mathfrak{B} \cong \mathfrak{A}$, and hence (iii) implies (i).

THEOREM 2.4.23. *If $\mathfrak{A}$ is a separably countably complete CA_α and $\mathfrak{A} \cong \mathfrak{A} \times \mathfrak{B}_\kappa$ for each $\kappa < \omega$, then $\mathfrak{A} \cong \mathfrak{A} \times \mathsf{P}_{\kappa<\omega}\mathfrak{B}_\kappa$.*

PROOF. By 2.4.22 we have $\mathfrak{A} \cong \mathfrak{A} \times {}^\omega\mathfrak{B}_\kappa$ for each $\kappa < \omega$. Hence by 2.4.20(iv) there is a $\mathfrak{C} \in \mathsf{CA}_\alpha$ such that $\mathfrak{A} \cong \mathfrak{C} \times \mathsf{P}_{\kappa<\omega}{}^\omega\mathfrak{B}_\kappa$. Therefore

$$\mathfrak{A} \times \mathsf{P}_{\kappa<\omega}\mathfrak{B}_\kappa \cong \mathfrak{C} \times \mathsf{P}_{\kappa<\omega}{}^\omega\mathfrak{B}_\kappa \times \mathsf{P}_{\kappa<\omega}\mathfrak{B}_\kappa \cong \mathfrak{C} \times \mathsf{P}_{\kappa<\omega}{}^\omega\mathfrak{B}_\kappa \cong \mathfrak{A}.$$

THEOREM 2.4.24. *If $\mathfrak{A}$ and $\mathfrak{B}$ are separably countably complete CA_α's, $\mathfrak{A} \mid \mathfrak{B}$, and $\mathfrak{B} \mid \mathfrak{A}$, then $\mathfrak{A} \cong \mathfrak{B}$.*

PROOF. Say $\mathfrak{B} \cong \mathfrak{A} \times \mathfrak{C}$ and $\mathfrak{A} \cong \mathfrak{B} \times \mathfrak{D}$. Then $\mathfrak{A} \cong \mathfrak{A} \times \mathfrak{C} \times \mathfrak{D}$. Hence, by 2.4.22, $\mathfrak{A} \cong \mathfrak{A} \times {}^\omega\mathfrak{C} \times {}^\omega\mathfrak{D}$, and so $\mathfrak{A} \cong \mathfrak{A} \times {}^\omega\mathfrak{C} \times {}^\omega\mathfrak{D} \times \mathfrak{C} \cong \mathfrak{A} \times \mathfrak{C} \cong \mathfrak{B}$.

Theorem 2.4.24 may be called the Cantor-Bernstein Theorem for direct products.

THEOREM 2.4.25. *If $\mathfrak{A}$, $\mathfrak{B}$, and $\mathfrak{C}$ are separably countably complete CA_α's and $\mathfrak{A} \times \mathfrak{C} \cong \mathfrak{B} \times \mathfrak{C}$, then there are $\mathfrak{A}'$, $\mathfrak{B}'$, and $\mathfrak{D}$ such that $\mathfrak{A} \cong \mathfrak{A}' \times \mathfrak{D}$, $\mathfrak{B} \cong \mathfrak{B}' \times \mathfrak{D}$, and $\mathfrak{C} \cong \mathfrak{A}' \times \mathfrak{C} \cong \mathfrak{B}' \times \mathfrak{C}$.*

PROOF. Let K be the set of all quadruples $\mathfrak{L} = \langle \mathfrak{L}_0, \mathfrak{L}_1, \mathfrak{L}_2, \mathfrak{L}_3 \rangle$ of separably countably complete CA_α's such that $\mathfrak{L}_0 \times \mathfrak{L}_2 \cong \mathfrak{L}_1 \times \mathfrak{L}_2$, and let R be the binary relation which holds between two members $\mathfrak{M}, \mathfrak{N} \in \mathsf{K}$ iff $\mathfrak{M}_0 \cong \mathfrak{N}_0 \times \mathfrak{N}_3$, $\mathfrak{M}_1 \cong \mathfrak{N}_1 \times \mathfrak{N}_3$, and $\mathfrak{M}_2 \cong \mathfrak{N}_1 \times \mathfrak{N}_2$. If $\mathfrak{M} \in \mathsf{K}$, then by the finite refinement property we obtain an $\mathfrak{N} \in \mathsf{K}$ such that $\mathfrak{M}\mathsf{R}\mathfrak{N}$. Since, moreover, $\langle \mathfrak{A}, \mathfrak{B}, \mathfrak{C}, \mathfrak{A} \rangle \in \mathsf{K}$ by hypothesis, we apply the principle of dependent choices and obtain an infinite sequence of quadruples $\mathfrak{L}^{(0)}, \mathfrak{L}^{(1)}, \mathfrak{L}^{(2)}, \ldots \in \mathsf{K}$ such that $\mathfrak{L}^{(0)} = \langle \mathfrak{A}, \mathfrak{B}, \mathfrak{C}, \mathfrak{A} \rangle$ and $\mathfrak{L}^{(0)} \mathsf{R}\, \mathfrak{L}^{(1)} \mathsf{R}\, \mathfrak{L}^{(2)} \ldots$. In other words, we obtain four sequences $\mathfrak{E}, \mathfrak{F}, \mathfrak{G}, \mathfrak{H} \in {}^\omega\mathsf{CA}_\alpha$ satisfying the following conditions (where κ ranges over all finite ordinals):

(1) $$\mathfrak{E}_0 = \mathfrak{A},\ \mathfrak{F}_0 = \mathfrak{B},\ \mathfrak{G}_0 = \mathfrak{C}, \text{ and } \mathfrak{H}_0 = \mathfrak{A};$$

(2) $$\mathfrak{E}_\kappa \times \mathfrak{G}_\kappa \cong \mathfrak{F}_\kappa \times \mathfrak{G}_\kappa;$$

(3) $$\mathfrak{E}_\kappa \cong \mathfrak{E}_{\kappa+1} \times \mathfrak{H}_{\kappa+1};$$

(4) $$\mathfrak{F}_\kappa \cong \mathfrak{F}_{\kappa+1} \times \mathfrak{H}_{\kappa+1};$$

(5) $$\mathfrak{G}_\kappa \cong \mathfrak{F}_{\kappa+1} \times \mathfrak{G}_{\kappa+1};$$

(6) $\mathfrak{E}_\kappa, \mathfrak{F}_\kappa, \mathfrak{G}_\kappa, \mathfrak{H}_\kappa$ are separably countably complete CA_α's.

Applying 2.4.20(iv) to (3) and (4) we obtain CA_α's $\mathfrak{A}'$ and $\mathfrak{B}'$ such that, for each $\kappa < \omega$,

(7) $$\mathfrak{E}_\kappa \cong \mathfrak{A}' \times \mathsf{P}_{\lambda<\omega}\mathfrak{H}_{\kappa+\lambda+1} \text{ and } \mathfrak{F}_\kappa \cong \mathfrak{B}' \times \mathsf{P}_{\lambda<\omega}\mathfrak{H}_{\kappa+\lambda+1}.$$

Let $\mathfrak{D} = \mathsf{P}_{\kappa<\omega}\mathfrak{H}_{\kappa+1}$. Then from (7) with $\kappa = 0$ and (1) we get

(8) $$\mathfrak{A} \cong \mathfrak{A}' \times \mathfrak{D} \text{ and } \mathfrak{B} \cong \mathfrak{B}' \times \mathfrak{D}.$$

Now by (7) we see that ${}^{\omega}\mathfrak{A}' \mid \mathsf{P}_{\kappa<\omega}\mathfrak{E}_{\kappa+1}$, while from (1), (2), (5), and 2.4.20(iv) we have $\mathsf{P}_{\kappa<\omega}\mathfrak{E}_{\kappa+1} \mid \mathfrak{G}_0 = \mathfrak{C}$. Thus ${}^{\omega}\mathfrak{A}' \mid \mathfrak{C}$, and hence by 2.4.22 $\mathfrak{C} \cong \mathfrak{A}' \times \mathfrak{C}$. Similarly, $\mathfrak{C} \cong \mathfrak{B}' \times \mathfrak{C}$. Together with (8) this completes the proof of the theorem.

THEOREM 2.4.26. *If $\mathfrak{A}$, $\mathfrak{B}$ and $\mathfrak{C}$ are separably countably complete CA_α's and $\mathfrak{A}\times\mathfrak{C} \cong \mathfrak{B}\times\mathfrak{C}$, or $\mathfrak{A}\times\mathfrak{C} \mid \mathfrak{B}\times\mathfrak{C}$, then there is a $\mathfrak{C}' \in \mathsf{CA}_\alpha$ such that $\mathfrak{A}\times\mathfrak{C}' \cong \mathfrak{B}\times\mathfrak{C}'$, or $\mathfrak{A}\times\mathfrak{C}' \mid \mathfrak{B}\times\mathfrak{C}'$, respectively, and $\mathfrak{C} \cong \mathfrak{C}'\times\mathfrak{C}$.*

PROOF. First suppose $\mathfrak{A}\times\mathfrak{C} \cong \mathfrak{B}\times\mathfrak{C}$. Let $\mathfrak{A}'$, $\mathfrak{B}'$, $\mathfrak{D}$ satisfy the conclusion of 2.4.25. Let $\mathfrak{C}' = {}^{\omega}\mathfrak{A}'\times{}^{\omega}\mathfrak{B}'$. Then $\mathfrak{A}'\times\mathfrak{C}' \cong \mathfrak{C}' \cong \mathfrak{B}'\times\mathfrak{C}'$, and hence

$$\mathfrak{A}\times\mathfrak{C}' \cong \mathfrak{D}\times\mathfrak{A}'\times\mathfrak{C}' \cong \mathfrak{D}\times\mathfrak{B}'\times\mathfrak{C}' \cong \mathfrak{B}\times\mathfrak{C}'.$$

Finally, by 2.4.22 we have $\mathfrak{C} \cong \mathfrak{C}\times{}^{\omega}\mathfrak{A}' \cong \mathfrak{C}\times{}^{\omega}\mathfrak{B}'$, and hence $\mathfrak{C} \cong \mathfrak{C}'\times\mathfrak{C}$.

The second part of the theorem easily follows from the first.

THEOREM 2.4.27. *If CA_α's $\mathfrak{B}$ and $\mathfrak{C}$ are separably countably complete, $\kappa < \omega$, and $\mathfrak{A}\times{}^{\kappa}\mathfrak{C} \mid \mathfrak{B}\times{}^{\kappa+1}\mathfrak{C}$, then $\mathfrak{A} \mid \mathfrak{B}\times\mathfrak{C}$.*

PROOF. The conclusion is obvious for $\kappa = 0$. For $\kappa = 1$ we apply 2.4.26 to get a CA_α $\mathfrak{C}'$ such that $\mathfrak{A}\times\mathfrak{C}' \mid \mathfrak{B}\times\mathfrak{C}\times\mathfrak{C}' \cong \mathfrak{B}\times\mathfrak{C}$; hence $\mathfrak{A} \mid \mathfrak{B}\times\mathfrak{C}$. The general case is obtained from the case $\kappa = 1$ by induction.

THEOREM 2.4.28. *If $\mathfrak{A}$, $\mathfrak{B}$, $\mathfrak{C}$ are separably countably complete CA_α's, $\kappa < \omega$, and $\mathfrak{A}\times{}^{\kappa}\mathfrak{C} \cong \mathfrak{B}\times{}^{\kappa}\mathfrak{C}$, then $\mathfrak{A}\times\mathfrak{C} \cong \mathfrak{B}\times\mathfrak{C}$.*

PROOF. By 2.4.27 we have $\mathfrak{A}\times\mathfrak{C} \mid \mathfrak{B}\times\mathfrak{C}$, and similarly $\mathfrak{B}\times\mathfrak{C} \mid \mathfrak{A}\times\mathfrak{C}$. Hence, by 2.4.24, $\mathfrak{A}\times\mathfrak{C} \cong \mathfrak{B}\times\mathfrak{C}$.

THEOREM 2.4.29. *If the CA_α $\mathfrak{B}$ is separably countably complete, $\langle\mathfrak{A}_\kappa : \kappa < \omega\rangle$ is a system of CA_α's, and $\mathsf{P}_{\kappa<\lambda}\mathfrak{A}_\kappa \mid \mathfrak{B}$ for each $\lambda < \omega$, then $\mathsf{P}_{\kappa<\omega}\mathfrak{A}_\kappa \mid \mathfrak{B}$.*

PROOF. By hypothesis, for each $\lambda < \omega$ choose $\mathfrak{C}_\lambda$ such that $\mathfrak{B} \cong \mathfrak{C}_\lambda\times\mathsf{P}_{\kappa<\lambda}\mathfrak{A}_\kappa$. Then for every $\lambda < \omega$ we have

$$\mathfrak{C}_\lambda\times\mathsf{P}_{\kappa<\lambda}\mathfrak{A}_\kappa \cong \mathfrak{C}_{\lambda+1}\times\mathfrak{A}_\lambda\times\mathsf{P}_{\kappa<\lambda}\mathfrak{A}_\kappa.$$

Hence we may apply 2.4.25 to obtain $\mathfrak{A}', \mathfrak{B}', \mathfrak{D} \in {}^{\omega}\mathsf{CA}_\alpha$ such that, for every $\lambda < \omega$,

$$\mathfrak{C}_\lambda \cong \mathfrak{A}'_\lambda\times\mathfrak{D}_\lambda,\ \mathfrak{C}_{\lambda+1}\times\mathfrak{A}_\lambda \cong \mathfrak{B}'_\lambda\times\mathfrak{D}_\lambda,$$

and

$$\mathsf{P}_{\kappa<\lambda}\mathfrak{A}_\kappa \cong \mathfrak{A}'_\lambda\times\mathsf{P}_{\kappa<\lambda}\mathfrak{A}_\kappa \cong \mathfrak{B}'_\lambda\times\mathsf{P}_{\kappa<\lambda}\mathfrak{A}_\kappa.$$

Hence $\mathfrak{B}'_\lambda\times\mathfrak{B} \cong \mathfrak{B}'_\lambda\times\mathfrak{C}_\lambda\times\mathsf{P}_{\kappa<\lambda}\mathfrak{A}_\kappa \cong \mathfrak{C}_\lambda\times\mathsf{P}_{\kappa<\lambda}\mathfrak{A}_\kappa \cong \mathfrak{B}$. Since this holds for every $\lambda < \omega$, we infer from 2.4.23 that $\mathfrak{B} \cong \mathfrak{B}\times\mathsf{P}_{\kappa<\omega}\mathfrak{B}'_\kappa$.

Now for each $\lambda < \omega$ let $\mathfrak{B}''_\lambda = \mathfrak{C}_\lambda \times \mathsf{P}_{\kappa<\lambda}\mathfrak{A}'_\kappa \times \mathsf{P}_{\kappa<\omega}\mathfrak{B}'_{\lambda+\kappa}$. Then $\mathfrak{B}''_0 = \mathfrak{C}_0 \times \mathsf{P}_{\kappa<\omega}\mathfrak{B}'_\kappa \cong \mathfrak{B} \times \mathsf{P}_{\kappa<\omega}\mathfrak{B}'_\kappa \cong \mathfrak{B}$. Furthermore, for each $\lambda < \omega$ we have

$$\begin{aligned}\mathfrak{B}''_{\lambda+1} \times \mathfrak{A}_\lambda &= \mathfrak{C}_{\lambda+1} \times \mathsf{P}_{\kappa<\lambda+1}\mathfrak{A}'_\kappa \times \mathsf{P}_{\kappa<\omega}\mathfrak{B}'_{\lambda+\kappa+1} \times \mathfrak{A}_\lambda \\ &\cong \mathsf{P}_{\kappa<\lambda+1}\mathfrak{A}'_\kappa \times \mathsf{P}_{\kappa<\omega}\mathfrak{B}'_{\lambda+\kappa} \times \mathfrak{D}_\lambda \\ &\cong \mathfrak{C}_\lambda \times \mathsf{P}_{\kappa<\lambda}\mathfrak{A}'_\kappa \times \mathsf{P}_{\kappa<\omega}\mathfrak{B}'_{\lambda+\kappa} \\ &= \mathfrak{B}''_\lambda.\end{aligned}$$

Hence, applying 2.4.20(iv), we see that $\mathsf{P}_{\kappa<\omega}\mathfrak{A}_\kappa \mid \mathfrak{B}$, as desired.

THEOREM 2.4.30. *If* $\mathfrak{A}$, $\mathfrak{B}$, $\mathfrak{C}$ *are separably countably complete* CA_α*'s,* $0 < \kappa < \omega$, *and* ${}^\kappa\mathfrak{A} \times \mathfrak{C} \mid {}^\kappa\mathfrak{B} \times \mathfrak{C}$, *then* $\mathfrak{A} \times \mathfrak{C} \mid \mathfrak{B} \times \mathfrak{C}$.

PROOF.[1] The case $\kappa = 1$ is trivial. We now assume that $\kappa > 1$, and make the following inductive assumption:

(1) If $0 < \lambda < \kappa$, $\mathfrak{A}''$, $\mathfrak{B}''$, $\mathfrak{C}''$ are separably countably complete, and ${}^\lambda\mathfrak{A}'' \times \mathfrak{C}'' \mid {}^\lambda\mathfrak{B}'' \times \mathfrak{C}''$, then $\mathfrak{A}'' \times \mathfrak{C}'' \mid \mathfrak{B}'' \times \mathfrak{C}''$.

We assume that ${}^\kappa\mathfrak{A} \times \mathfrak{C} \mid {}^\kappa\mathfrak{B} \times \mathfrak{C}$. Let K be the set of all triples $\mathfrak{L} = \langle \mathfrak{L}_0, \mathfrak{L}_1, \mathfrak{L}_2 \rangle$ of separably countably complete CA_α's such that ${}^\kappa\mathfrak{L}_0 \times \mathfrak{L}_2 \mid {}^\kappa\mathfrak{L}_1 \times \mathfrak{L}_2$, and let R be the binary relation holding between two triples $\mathfrak{G}, \mathfrak{H} \in \mathsf{K}$ iff

(2) $\mathfrak{G}_0 \times \mathfrak{G}_2 \cong {}^2\mathfrak{H}_0 \times \mathfrak{H}_2$, $\mathfrak{G}_2 \mid {}^2\mathfrak{H}_1 \times \mathfrak{H}_2$, and $\mathfrak{G}_1 \times \mathfrak{G}_2 \cong \mathfrak{H}_0 \times \mathfrak{H}_1 \times \mathfrak{H}_2$.

Clearly $\langle \mathfrak{A}, \mathfrak{B}, \mathfrak{C} \rangle \in \mathsf{K}$. In order to apply the principle of dependent choices, we have to show that, for any $\mathfrak{G} \in \mathsf{K}$, there is an $\mathfrak{H} \in \mathsf{K}$ such that $\mathfrak{G}\mathsf{R}\mathfrak{H}$. Since $\mathfrak{G} \in \mathsf{K}$ there is a $\mathfrak{D}$ for which

(3) $${}^\kappa\mathfrak{G}_0 \times \mathfrak{G}_2 \times \mathfrak{D} \cong {}^\kappa\mathfrak{G}_1 \times \mathfrak{G}_2.$$

Let $\lambda < \omega$ be such that $2^\lambda < \kappa \leqq 2^{\lambda+1}$. Then

$$({}^{2^\lambda}\mathfrak{G}_0 \times \mathfrak{G}_2) \mid ({}^\kappa\mathfrak{G}_0 \times \mathfrak{G}_2) \mid ({}^\kappa\mathfrak{G}_1 \times \mathfrak{G}_2) \mid ({}^{2^\lambda}({}^2\mathfrak{G}_1) \times \mathfrak{G}_2),$$

so $({}^{2^\lambda}\mathfrak{G}_0 \times \mathfrak{G}_2) \mid ({}^{2^\lambda}({}^2\mathfrak{G}_1) \times \mathfrak{G}_2)$. Hence by (1) we have $\mathfrak{G}_0 \times \mathfrak{G}_2 \mid {}^2\mathfrak{G}_1 \times \mathfrak{G}_2$, say

(4) $$\mathfrak{G}_0 \times \mathfrak{G}_2 \times \mathfrak{E} \cong {}^2\mathfrak{G}_1 \times \mathfrak{G}_2.$$

[1]) The proof of 2.4.30 given here is a rearrangement and slight amplification of a known proof of the corresponding result in the general theory of cardinal algebras; cf. Tarski [49*], p. 30, Theorem 2.31. This simplification (due to Monk) can also be carried through in the general case.

Consequently, we have

$$
\begin{aligned}
{}^{\kappa-1}(\mathfrak{G}_2\times\mathfrak{E})\times{}^{\kappa-1}\mathfrak{G}_0 &\cong {}^{\kappa-1}(\mathfrak{G}_0\times\mathfrak{G}_2\times\mathfrak{E})\\
&\cong {}^{\kappa-1}({}^2\mathfrak{G}_1\times\mathfrak{G}_2) && \text{by (4)}\\
&\cong {}^{\kappa-2}\mathfrak{G}_1\times{}^{\kappa-2}\mathfrak{G}_2\times{}^{\kappa}\mathfrak{G}_1\times\mathfrak{G}_2\\
&\cong {}^{\kappa-2}\mathfrak{G}_1\times{}^{\kappa-2}\mathfrak{G}_2\times{}^{\kappa}\mathfrak{G}_0\times\mathfrak{G}_2\times\mathfrak{D} && \text{by (3)}\\
&\cong ({}^{\kappa-2}\mathfrak{G}_1\times{}^{\kappa-1}\mathfrak{G}_2\times\mathfrak{D})\times{}^{\kappa}\mathfrak{G}_0.
\end{aligned}
$$

Therefore, applying 2.4.27 we obtain ${}^{\kappa-1}(\mathfrak{G}_2\times\mathfrak{E})\,|\,({}^{\kappa-2}\mathfrak{G}_1\times{}^{\kappa-1}\mathfrak{G}_2\times\mathfrak{D}\times\mathfrak{G}_0)$, say

(5) $${}^{\kappa-2}\mathfrak{G}_1\times{}^{\kappa-1}\mathfrak{G}_2\times\mathfrak{D}\times\mathfrak{G}_0\cong{}^{\kappa-1}(\mathfrak{G}_2\times\mathfrak{E})\times\mathfrak{F}.$$

Hence

$$
\begin{aligned}
{}^{2\kappa-2}\mathfrak{G}_1\times{}^{\kappa-1}\mathfrak{G}_2\times\mathfrak{F} &\cong {}^{\kappa-1}({}^2\mathfrak{G}_1\times\mathfrak{G}_2)\times\mathfrak{F}\\
&\cong {}^{\kappa-1}\mathfrak{G}_0\times{}^{\kappa-1}(\mathfrak{G}_2\times\mathfrak{E})\times\mathfrak{F} && \text{by (4)}\\
&\cong {}^{\kappa}\mathfrak{G}_0\times\mathfrak{G}_2\times\mathfrak{D}\times{}^{\kappa-2}\mathfrak{G}_1\times{}^{\kappa-2}\mathfrak{G}_2 && \text{by (5)}\\
&\cong {}^{2\kappa-2}\mathfrak{G}_1\times{}^{\kappa-1}\mathfrak{G}_2 && \text{by (3).}
\end{aligned}
$$

Thus by two applications of 2.4.28 we obtain

(6) $$\mathfrak{G}_1\times\mathfrak{G}_2\times\mathfrak{F}\cong\mathfrak{G}_1\times\mathfrak{G}_2.$$

Hence we have

$$
\begin{aligned}
(\mathfrak{G}_0\times\mathfrak{G}_2)\times(\mathfrak{G}_2\times\mathfrak{E}\times\mathfrak{F}) &\cong (\mathfrak{G}_0\times\mathfrak{G}_2\times\mathfrak{E})\times(\mathfrak{G}_2\times\mathfrak{F})\\
&\cong {}^2\mathfrak{G}_1\times\mathfrak{G}_2\times\mathfrak{G}_2\times\mathfrak{F} && \text{by (4)}\\
&\cong {}^2(\mathfrak{G}_1\times\mathfrak{G}_2) && \text{by (6).}
\end{aligned}
$$

Now applying 2.4.21 we obtain $\mathfrak{H}_0$, $\mathfrak{H}_1$, $\mathfrak{H}_2$ such that

(7) $$\mathfrak{G}_0\times\mathfrak{G}_2\cong{}^2\mathfrak{H}_0\times\mathfrak{H}_2,\quad \mathfrak{G}_2\times\mathfrak{E}\times\mathfrak{F}\cong{}^2\mathfrak{H}_1\times\mathfrak{H}_2,$$
$$\text{and }\mathfrak{G}_1\times\mathfrak{G}_2\cong\mathfrak{H}_0\times\mathfrak{H}_1\times\mathfrak{H}_2.$$

Thus

$$
\begin{aligned}
{}^{\kappa}\mathfrak{H}_0\times\mathfrak{H}_2\times{}^{\kappa-2}(\mathfrak{H}_1\times\mathfrak{H}_2) &\cong {}^2\mathfrak{H}_0\times\mathfrak{H}_2\times{}^{\kappa-2}(\mathfrak{H}_0\times\mathfrak{H}_1\times\mathfrak{H}_2)\\
&\cong \mathfrak{G}_0\times\mathfrak{G}_2\times{}^{\kappa-2}(\mathfrak{G}_1\times\mathfrak{G}_2) && \text{by (7)}\\
&\cong \mathfrak{G}_0\times\mathfrak{G}_2\times{}^{\kappa-2}(\mathfrak{G}_1\times\mathfrak{G}_2\times\mathfrak{F}) && \text{by (6)}\\
&\cong ({}^{\kappa-2}\mathfrak{G}_1\times{}^{\kappa-1}\mathfrak{G}_2\times\mathfrak{G}_0)\times{}^{\kappa-2}\mathfrak{F}\\
&\quad |\ {}^{\kappa-1}(\mathfrak{G}_2\times\mathfrak{E}\times\mathfrak{F}) && \text{by (5)}\\
&\cong {}^{\kappa-1}({}^2\mathfrak{H}_1\times\mathfrak{H}_2) && \text{by (7)}\\
&\cong {}^{\kappa-1}\mathfrak{H}_1\times{}^{\kappa-1}(\mathfrak{H}_1\times\mathfrak{H}_2).
\end{aligned}
$$

Hence by 2.4.27 we have ${}^{\kappa}\mathfrak{H}_0 \times \mathfrak{H}_2 \mid {}^{\kappa}\mathfrak{H}_1 \times \mathfrak{H}_2$. This shows that $\mathfrak{H} \in \mathsf{K}$ and, by virtue of (7), that $\mathfrak{G}\mathsf{R}\mathfrak{H}$.

We can now apply the principle of dependent choices (just as in the proof of 2.4.25). We conclude that there exist three sequences $\mathfrak{A}', \mathfrak{B}', \mathfrak{C}' \in {}^{\omega}\mathsf{CA}_\alpha$ satisfying the following conditions for all $\mu < \omega$:

(8) $\mathfrak{A}'_\mu$, $\mathfrak{B}'_\mu$, and $\mathfrak{C}'_\mu$ are separably countably complete CA_α's;

(9) $$\mathfrak{A}'_0 = \mathfrak{A},\ \mathfrak{B}'_0 = \mathfrak{B},\ \text{and}\ \mathfrak{C}'_0 = \mathfrak{C};$$

(10) $${}^{\kappa}\mathfrak{A}'_\mu \times \mathfrak{C}'_\mu \mid {}^{\kappa}\mathfrak{B}'_\mu \times \mathfrak{C}'_\mu;$$

(11) $$\mathfrak{A}'_\mu \times \mathfrak{C}'_\mu \cong {}^{2}\mathfrak{A}'_{\mu+1} \times \mathfrak{C}'_{\mu+1};$$

(12) $$\mathfrak{C}'_\mu \mid {}^{2}\mathfrak{B}'_{\mu+1} \times \mathfrak{C}'_{\mu+1};$$

(13) $$\mathfrak{B}'_\mu \times \mathfrak{C}'_\mu \cong \mathfrak{A}'_{\mu+1} \times \mathfrak{B}'_{\mu+1} \times \mathfrak{C}'_{\mu+1}.$$

Applying 2.4.20(iv) to (11) we obtain a CA_α $\mathfrak{J}$ such that, for every $\mu < \omega$,

(14) $$\mathfrak{A}'_\mu \times \mathfrak{C}'_\mu \cong \mathfrak{J} \times \mathsf{P}_{\nu<\omega} \mathfrak{A}'_{\mu+\nu+1}.$$

From (13) we get by induction

(15) $$\mathfrak{B}'_0 \times \mathfrak{C}'_0 \cong \mathfrak{B}'_\mu \times \mathfrak{C}'_\mu \times \mathsf{P}_{\nu<\mu} \mathfrak{A}'_{\nu+1}$$

for every $\mu < \omega$. Now by (13) and (14) we have $\mathfrak{J} \mid \mathfrak{A}'_{\mu+1} \times \mathfrak{C}'_{\mu+1} \mid \mathfrak{B}'_\mu \times \mathfrak{C}'_\mu$; thus, by (15), $\mathfrak{J} \times \mathsf{P}_{\nu<\mu} \mathfrak{A}'_{\nu+1} \mid \mathfrak{B}'_0 \times \mathfrak{C}'_0$ for each $\mu < \omega$. Hence, by 2.4.29, $\mathfrak{J} \times \mathsf{P}_{\nu<\omega} \mathfrak{A}'_{\nu+1} \mid \mathfrak{B}'_0 \times \mathfrak{C}'_0$. From (14), with $\mu = 0$, we then infer that $\mathfrak{A}'_0 \times \mathfrak{C}'_0 \mid \mathfrak{B}'_0 \times \mathfrak{C}'_0$, i.e., $\mathfrak{A} \times \mathfrak{C} \mid \mathfrak{B} \times \mathfrak{C}$, and the inductive proof of the theorem is complete.

THEOREM 2.4.31. *If two* CA_α*'s* $\mathfrak{A}$ *and* $\mathfrak{B}$ *are separably countably complete,* $0 < \kappa < \omega$, *and* ${}^{\kappa}\mathfrak{A} \mid {}^{\kappa}\mathfrak{B}$, *or* ${}^{\kappa}\mathfrak{A} \cong {}^{\kappa}\mathfrak{B}$, *then* $\mathfrak{A} \mid \mathfrak{B}$, *or* $\mathfrak{A} \cong \mathfrak{B}$, *respectively.*

PROOF: by 2.4.24 and 2.4.30.

We wish to state here without proof two further consequences of 2.4.20 which deserve interest:

(γ) *If all* $\mathfrak{B}_\lambda$*'s with* $\lambda < \omega$ *are separably countably complete* CA_α*'s, and* $\mathfrak{A}_\kappa \mid \mathfrak{B}_\lambda$ *for all* $\kappa, \lambda < \omega$, *then there is a* $\mathfrak{C}$ *such that* $\mathfrak{A}_\kappa \mid \mathfrak{C} \mid \mathfrak{B}_\lambda$ *for all* $\kappa, \lambda < \omega$.

(δ) *If* $\mathfrak{A}$ *is a separably countably complete* CA_α, $0 < \kappa$, $\lambda < \omega$, κ *and* λ *are relatively prime, and* ${}^{\kappa}\mathfrak{A} \cong {}^{\lambda}\mathfrak{B}$, *then there is a* $\mathfrak{C}$ *such that* $\mathfrak{A} \cong {}^{\lambda}\mathfrak{C}$ *and* $\mathfrak{B} \cong {}^{\kappa}\mathfrak{C}$.

(γ) is called the *interpolation theorem* and (δ) *Euclid's Theorem*. The derivation of these results from 2.4.20 can again be found in Tarski [49*], pp. 27 ff. and 33 ff.

REMARKS 2.4.32. As was pointed out at an earlier place, the results stated in 2.4.21–2.4.31 extend to members of an arbitrary class K of algebras satisfying conditions (i)–(iv) of 2.4.20. In particular, we can apply these results to locally finite cylindric algebras taking for K the class of separably countably complete Lf_α's. In case $\alpha \geqq \omega$ this class K, however, does not present much interest since it turns out to be very restricted: K consists then just of those algebras which can be isomorphically represented as products $\mathfrak{A} \times \mathsf{P}_{\kappa<\nu} \mathfrak{B}_\kappa$, where $\mathfrak{A}$ is a countably complete discrete CA_α and $\langle \mathfrak{B}_\kappa : \kappa < \nu \rangle$ is a finite sequence of directly indecomposable Lf_α's. Even more restricted, by the way, is the class of all countably complete Lf_α's with $\alpha \geqq \omega$, which simply coincides with that of countably complete discrete CA_α's. (The last two observations were made by Monk.)

There is a method which permits us to extend most of the results presented in 2.4.21–2.4.31 to a much wider and more interesting class of Lf's of infinite dimension. For any given $\alpha \geqq \omega$ let L_α be the class of all those Lf_α's $\mathfrak{A}$ which are separably countably complete in the following weak sense: the sum $\sum_{i \in I} x_i$ exists for every countable system x of pairwise separable elements of $\mathfrak{A}$ such that $\bigcup_{i \in I} \Delta x_i$ is finite. A subclass of L_α is the class L_α' of all those Lf_α's $\mathfrak{A}$ in which $\sum_{i \in I} x_i$ exists for every countable $x \in {}^I A$ with $|\bigcup_{i \in I} \Delta x_i| < \omega$. As will be seen in Part II, the algebras of L_α' appear in a natural way in the discussion of a class of infinitary languages related to those mentioned in 1.11.2.

To extend the results discussed to algebras in L_α we proceed as follows. Given any system $\mathfrak{A} = \langle \mathfrak{A}_i : i \in I \rangle$ of Lf_α's, let $\mathfrak{B}$ be the largest locally finite subalgebra of $\mathsf{P}_{i \in I} \mathfrak{A}_i$; the existence of such a subalgebra is secured by 2.1.6. We refer to $\mathfrak{B}$ as the Lf_α-direct product or, more briefly, the Lf-product of the system $\mathfrak{A}$, and we denote it by ${}_{\mathsf{l}}\mathsf{P}\mathfrak{A}$ or ${}_{\mathsf{l}}\mathsf{P}_{i \in I} \mathfrak{A}_i$. (The use of this notation does not extend beyond the present remarks.) As is easily seen, ${}_{\mathsf{l}}\mathsf{P}_{i \in I} \mathfrak{A}_i$ is a subdirect product of Lf_α's $\mathfrak{A}_i$, $i \in I$, in the sense of 0.3.40. It also proves to be a particular case of the general notion of the K-direct product (where K is any class of similar algebras) which originates in the theory of categories; cf. Preller [68*]. In general $\mathsf{P}\mathfrak{A}$ and ${}_{\mathsf{l}}\mathsf{P}\mathfrak{A}$ do not coincide simply because the ordinary direct product of Lf_α's is not always an Lf_α. Nevertheless, $\mathsf{P}\mathfrak{A}$ and ${}_{\mathsf{l}}\mathsf{P}\mathfrak{A}$ have many properties in common. In particular, the general commutative and associative laws formulated in 0.3.3 continue to hold if P is replaced by ${}_{\mathsf{l}}\mathsf{P}$ (and the algebras involved are assumed to be locally finite), and the same applies to 0.3.4. Furthermore, Theorem 2.4.20 remains valid when we take K to be L_α, if the notions occurring in this theorem, and in particular that of the remainder property in 2.4.20(iv), are understood to be defined in terms of Lf-products rather than ordinary direct products; the proof requires but little change in the original argument. On the other hand, the only properties of direct products used in the proofs of the results stated in 2.4.21–2.4.31 are various special commutative and associative laws and the properties listed in 2.4.20. Hence, by repeating almost literally the original proofs, we can conclude that results entirely analogous to those in 2.4.21–2.4.31 hold for Lf-products of arbitrary algebras in L_α. We now make the following two observations: (I) in view of 2.4.4(ii), $\mathsf{P}\mathfrak{A}$ and ${}_{\mathrm{l}}\mathsf{P}\mathfrak{A}$ coincide for every finite system $\mathfrak{A}$ of Lf_α's; (II) with the exception of 2.4.22, 2.4.23, and 2.4.29 all the results in 2.4.21–2.4.31 involve exclusively finite products. Our final conclusion is that, with only the exceptions mentioned, all the results stated in 2.4.21–2.4.31 for direct products of separably countably complete CA_α's continue to hold for (ordinary) direct products of arbitrary algebras in L_α.

REMARKS 2.4.33. An interesting general result which encompasses most of the theorems 2.4.21–2.4.31 can be derived (as a rather special consequence) from the work of Bradford [65*]. Essentially the result has a purely mathematical character, but it will be more convenient here to put it in metamathematical terms. To shorten the formulation we introduce some special stipulations. We notice that in most of the theorems 2.4.21 to 2.4.31 we were concerned with formulas of the type

(E) $$^{\kappa_0}\mathfrak{A}_0 \times \ldots \times {}^{\kappa_{\mu-1}}\mathfrak{A}_{\mu-1} \cong {}^{\lambda_0}\mathfrak{A}_0 \times \ldots \times {}^{\lambda_{\mu-1}}\mathfrak{A}_{\mu-1}$$

where the variables $\mathfrak{A}_0, \ldots, \mathfrak{A}_{\mu-1}$ range over algebras and $\kappa_0, \ldots, \kappa_{\mu-1}$, $\lambda_0, \ldots, \lambda_{\mu-1}$ stand for constants representing finite cardinals or the cardinal ω. (Two such formulas with the same μ and $\mathfrak{A}_0, \ldots, \mathfrak{A}_{\mu-1}$ may still differ, of course, in the exponents $\kappa_0, \ldots, \lambda_0, \ldots$.) With every formula of type (E) we correlate the following formula:

(E′) $$\kappa_0 \cdot \alpha_0 + \ldots + \kappa_{\mu-1} \cdot \alpha_{\mu-1} = \lambda_0 \cdot \alpha_0 + \ldots + \lambda_{\mu-1} \cdot \alpha_{\mu-1}$$

in which the variables $\alpha_0, \ldots, \alpha_{\mu-1}$ range over cardinals, $\kappa_0, \ldots, \kappa_{\mu-1}$, $\lambda_0, \ldots, \lambda_{\mu-1}$ stand for the same constants as before, and $+$ and $\cdot$ denote, respectively, cardinal addition and cardinal multiplication. We notice that all the theorems 2.4.21 through 2.4.31 with the exception of 2.4.23 and 2.4.29 can be put in the following schematic form:

(Σ) *If* $\mathfrak{A}_0, \ldots, \mathfrak{A}_{\nu-1}$ *are any algebras in* K *which satisfy a given finite system of formulas of type* (E) *with* $\mu = \nu$, *then there are algebras* $\mathfrak{A}_\nu, \ldots, \mathfrak{A}_{\nu+\pi-1}$ *in* K *such that* $\mathfrak{A}_0, \ldots, \mathfrak{A}_{\nu+\pi-1}$ *satisfy another finite system of formulas of type* (E) *with* $\mu = \nu + \pi$.

Here we do not exclude the case $\pi = 0$. With every statement of form (Σ) we correlate a statement of form (Σ') obtained by referring to cardinals

$\alpha_0, \ldots, \alpha_{\nu+\pi-1}$ less than or equal to ω instead of algebras in K, and by replacing formulas of (E) by the correlated formulas of type (E′). The general result in which we are interested can now be formulated as follows:

For every statement (Σ) *the following three conditions are equivalent*:

(I) *The statement* (Σ) *holds for every class* K *satisfying 2.4.20(i)–(iv) with the word "finite" in (iii) omitted.*

(II) *The statement* (Σ) *holds for the class* C_α *of all separably countably complete* CA_α*'s.*

(III) *The correlated statement* (Σ') *is true.*

It may be noticed that condition (III) can be replaced by an obviously equivalent condition (III′) formulated in purely algebraic terms:

(III′) *The statement* (Σ) *holds for the class of all complete and atomic Boolean algebras with at most* ω *atoms.*

Actually, the equivalence of (III) and (III′) is used in proving that (II) implies (III). Note that in (I) we require each $\mathfrak{A} \in K$ to have the refinement property; it is not known whether the finite refinement property suffices for this purpose, but it is known that the denumerable refinement property (i.e., the property formulated in 0.3.34(i) with the restrictions $|I| \leqq \omega$, $|J| \leqq \omega$) does suffice.

The proof of this general result is long and difficult; it uses in particular the fact that Theorem 2.4.31 holds in every class K satisfying conditions 2.4.20(i)–(iv). From the proof of the result it is seen that, not only are its three parts (I)–(III) equivalent, but their equivalence can be established on the basis of the system of set theory underlying our whole work. (Concerning the underlying system see the beginning of the Preliminaries.) The general result has many important consequences. It clearly implies as particular cases all theorems 2.4.21 to 2.4.31 except 2.4.23 and 2.4.29; it also implies the statement (δ) formulated above. Some further interesting implications of the general result are of metamathematical character. To obtain them we use the following known facts (essentially due to Presburger [30*]): there is an automatic method for deciding whether or not any given statement of form (Σ') holds in our system of set theory, and in each particular case either the statement or its negation is actually provable in this system. By combining these facts with the general result we arrive at the following conclusions: there is an automatic method for deciding whether or not any given statement of form (Σ) holds for every class K satisfying conditions 2.4.20(i)–(iv) — or, equivalently, whether or not such a statement holds for the class C_α; given a statement of form (Σ), either it is provable (in set theory) that the statement holds for C_α, or else it is provable that its negation holds for C_α.

We may point out a possible slight improvement of the preceding development. Instead of basing the discussion on the result that the class K of all separably countably complete CA_α's satisfies 2.4.20(i)–(iv), we can dispense with K altogether and use merely the fact that each individual separably countably complete CA_α $\mathfrak{D}$ hereditarily has the refinement property (possibly restricted to finite or denumerable decompositions) and also has the remainder property; the results of the discussion will still have a general algebraic character and will apply to all algebras with these properties. By analyzing the proofs of Theorems 2.4.21–2.4.31 as well as of the general result stated above, we can convince ourselves that with very slight changes in the formulations of the results the proofs remain valid in the new setting. To illustrate the way in which the results have to be re-formulated, we restrict our attention to 2.4.31; it now assumes the form:

If $\mathfrak{D}$ *is any separably countably complete* CA_α, $0 < \kappa < \omega$, *and* ${}^\kappa\mathfrak{A} \mid {}^\kappa\mathfrak{B} \mid \mathfrak{D}$, *or* ${}^\kappa\mathfrak{A} \cong {}^\kappa\mathfrak{B} \mid \mathfrak{D}$, *then* $\mathfrak{A} \mid \mathfrak{B}$, *or* $\mathfrak{A} \cong \mathfrak{B}$, *respectively.*

The whole improvement just described may seem very insignificant and is undoubtedly of no relevance whatsoever for the theory of CA's. It has nevertheless some importance from a general algebraic point of view since it permits one to apply the results of our discussion to any algebra $\mathfrak{D}$ which has the properties mentioned above, but which is not a member of any class K satisfying 2.4.20(i)–(iv). Examples of such algebras are known.

REMARK 2.4.34. The problem naturally arises whether the assumption of separable countable completeness in the preceding theorems (2.4.21–2.4.31) is essential, and whether, by means of some other method, these theorems could be extended to arbitrary CA's. We shall concentrate our attention upon Theorems 2.4.24 and 2.4.31. It is known that these two theorems do not apply to all BA's; counterexamples can be found in Kinoshita [53*] and Hanf [57*]. Hence, automatically, the theorems do not apply to all discrete CA_α's. Actually, it is easily seen that they fail for some non-discrete CA_α's as well. In fact, let $\mathfrak{A}'$ and $\mathfrak{B}'$ be two discrete CA_α's for which 2.4.24, or 2.4.31, fails, i.e., for which $\mathfrak{A}' \mid \mathfrak{B}' \mid \mathfrak{A}'$, or $\mathfrak{A}' \times \mathfrak{A}' \cong \mathfrak{B}' \times \mathfrak{B}'$, holds but $\mathfrak{A}' \cong \mathfrak{B}'$ does not hold. Further, let $\mathfrak{C}$ be any non-discrete directly indecomposable CA_α (cf. 2.3.15). From 0.3.36(ii) and 2.4.13 it follows that the algebras $\mathfrak{A} = \mathfrak{A}' \times \mathfrak{C}$ and $\mathfrak{B} = \mathfrak{B}' \times \mathfrak{C}$ provide the desired counterexample.

We may still be dissatisfied with this result and ask the further question: do the theorems discussed apply at least to those CA_α's which are *hereditarily non-discrete*, in the sense that they have no discrete factors with more than one element? (The use of the term "hereditarily non-discrete" in this sense is not in full agreement with the general stipulation made at the beginning of 0.3.32, since every CA_α has a one-element direct factor, which is trivially discrete.) In Theorem 2.4.36 we shall give a negative answer to this question.

The proof of 2.4.36 will be based upon a lemma, 2.4.35. This lemma is in a sense a converse of 2.4.12. By 2.4.12 the study of direct decompositions of a separably complete CA_α $\mathfrak{A}$ reduces to the study of direct decompositions of the correlated BA $\mathfrak{Zd}\,\mathfrak{A}$. By 2.4.35, corresponding to each BA $\mathfrak{B}$ there is an hereditarily non-discrete CA_α $\mathfrak{A}$ such that $\mathfrak{Zd}\,\mathfrak{A} = \mathfrak{B}$, and isomorphisms between factors of $\mathfrak{B}$ induce isomorphisms between factors of $\mathfrak{A}$.

LEMMA 2.4.35. *If* $\mathfrak{B}$ *is a* BA, *then there is a* CA_α $\mathfrak{A}$ *satisfying the following conditions*:

(i) $\mathfrak{B} = \mathfrak{Zd}\,\mathfrak{A}$;

(ii) *for all* $a, b \in B$, $\mathfrak{Rl}_a\mathfrak{B} \cong \mathfrak{Rl}_b\mathfrak{B}$ *iff* $\mathfrak{Rl}_a\mathfrak{A} \cong \mathfrak{Rl}_b\mathfrak{A}$.

In addition, $\mathfrak{A}$ *can be chosen so that*

(iii) $\mathfrak{A} \in \mathsf{Lf}_\alpha$,

and

(iv) $\mathfrak{A}$ *has no discrete factor with more than one element.*[1)]

PROOF. It is well known that $\mathfrak{B}$ is isomorphic to a Boolean set algebra $\mathfrak{D} = \langle D, \cup, \cap, {}_U\sim, 0, U\rangle$, whose elements are subsets of some set U satisfying the following conditions:

(1) if X is an arbitrary system of members of D indexed by a set I and $\bigcup_{i\in I} X_i = U$, then $\bigcup_{i\in J} X_i = U$ for some finite $J \subseteq I$;

(2) for every $x, y \in U$ with $x \neq y$, there is an $X \in D$ such that $x \in X$ while $y \notin X$.

(The existence of such a set algebra $\mathfrak{D}$ is an easy consequence of the well-known theorem of Stone [34*] by which every Boolean algebra is isomorphic to the Boolean set algebra of all closed and open sets of a totally disconnected compact space.) Without loss of generality, we may assume that $\mathfrak{B} = \mathfrak{D}$, i.e., that $\mathfrak{B}$ is a Boolean algebra of subsets of U satisfying (1) and (2).

Choose a directly indecomposable CA_α $\mathfrak{C}$ and consider the cylindric algebra ${}^U\mathfrak{C}$. For each $f \in {}^UC$, let F_f be the function on C determined by the formula $F_f c = (f^{-1})^\star c$ for every $c \in C$. Furthermore, let $A' = \{f : f \in {}^UC,\ Rg F_f \subseteq B\}$.

[1)] Lemma 2.4.35 follows from an analysis of the proof of a general result in Jónsson [57*] concerning a comprehensive class of algebras, the so-called centerless algebras.

Then

(3) each $f \in A'$ has a finite range.

In order to prove this it suffices to show that, for each $f \in A'$, $F_f c$ is empty for all but a finite number of $c \in C$. This, however, follows immediately from (1) and the fact that $F_f c \cap F_f c'$ is empty whenever $c \neq c'$. We also have

(4) $$A' \in Su\,{}^U\mathfrak{C}.$$

In fact we clearly have $0, 1, \mathsf{d}_{\kappa\lambda} \in A'$ for all $\kappa, \lambda < \alpha$. If $f \in A'$ and $c \in C$, then $F_{-f}c = F_f - c \in B$, and thus $-f \in A'$. If $f, g \in A'$ and $c \in C$, then

$$F_{f+g}c = \bigcup_{d+e=c}(F_f d \cap F_g e);$$

since, by (3), F_f and F_g take the empty set as value for all but a finite number of elements of C, it follows that $F_{f+g}c \in B$. Hence $f+g \in A'$. Similarly, it is seen that A' is closed under c_κ for each $\kappa < \alpha$, and (4) is established.

Let $\mathfrak{A}'$ be the subalgebra of ${}^U\mathfrak{C}$ with universe A'. By 0.3.58 and 2.3.15, $\mathfrak{C}$ can be chosen so that it is a non-discrete Lf_α in addition to being directly indecomposable. Hence, by (3) we have

(5) $$\mathfrak{A}' \in \mathsf{Lf}_\alpha.$$

We also get:

(6) $\mathfrak{A}'$ has no discrete factor with more than one element.

To show this, consider any non-zero $f \in A'$. Then $fx \neq 0^{(\mathfrak{C})}$ for some $x \in U$. Since $\mathfrak{C}$ is directly indecomposable and non-discrete, it must have more than two elements, and all elements different from $0^{(\mathfrak{C})}$ and $1^{(\mathfrak{C})}$ must have non-empty dimension sets; hence there is a $c \in C$ such that $c \leqq^{(\mathfrak{C})} fx$ and $\Delta^{(\mathfrak{C})}c$ is non-empty. Let

$$g = (F_f fx \times \{c\}) \cup (U \sim F_f fx) 1 f.$$

Clearly, $g \in A'$, $g \leqq f$, and Δg is non-empty by 2.4.2. Therefore, using 2.2.14 and 2.4.8 we see that $\mathfrak{A}'$ has no non-trivial discrete factors, and (6) is established.

Now consider the function $G \in {}^B A'$ such that, for each $X \in B$,

$$GX = \langle 1^{(\mathfrak{C})} : x \in X \rangle \cup \langle 0^{(\mathfrak{C})} : x \in U \sim X \rangle.$$

Clearly $G \in Ism(\mathfrak{B}, \mathfrak{Zd}\,\mathfrak{A}')$. Suppose $f \in Zd\mathfrak{A}'$. Since Δf is empty, we have $\Delta^{(\mathfrak{C})}fx$ empty for each $x \in X$ by 2.4.2. Since $\mathfrak{C}$ is directly indecomposable, it follows by 2.4.14 that $fx = 0^{(\mathfrak{C})}$ or $fx = 1^{(\mathfrak{C})}$ for each $x \in U$. Since $f \in A'$, we have that $F_f 1^{(\mathfrak{C})} \in B$, and thus $GF_f 1^{(\mathfrak{C})} = f$. This shows that $RgG = Zd\mathfrak{A}'$, and hence we obtain

(7) $$G \in Is(\mathfrak{B}, \mathfrak{Zd}\,\mathfrak{A}').$$

Suppose $X, Y \in B$ and $\mathfrak{R}\mathfrak{l}_{GX}\mathfrak{A}' \cong \mathfrak{R}\mathfrak{l}_{GY}\mathfrak{A}'$. Clearly $(Sb\,X \cap B)1\,G \in Is(\mathfrak{R}\mathfrak{l}_X\mathfrak{B}, \mathfrak{Z}\mathfrak{d}\,\mathfrak{R}\mathfrak{l}_{GX}\mathfrak{A}')$ and $(Sb\,Y \cap B)1\,G \in Is(\mathfrak{R}\mathfrak{l}_Y\mathfrak{B}, \mathfrak{Z}\mathfrak{d}\,\mathfrak{R}\mathfrak{l}_{GY}\mathfrak{A}')$, so that $\mathfrak{R}\mathfrak{l}_X\mathfrak{B} \cong \mathfrak{R}\mathfrak{l}_Y\mathfrak{B}$.

Suppose now, conversely, that $\mathfrak{R}\mathfrak{l}_X\mathfrak{B} \cong \mathfrak{R}\mathfrak{l}_Y\mathfrak{B}$, and let $H \in Is(\mathfrak{R}\mathfrak{l}_X\mathfrak{B}, \mathfrak{R}\mathfrak{l}_Y\mathfrak{B})$. Consider any $y \in Y$. Notice that, by (1), all the sets $H^{-1}Z$ where $y \in Z \in B \cap Sb\,Y$ have at least one element in common, and that from (2) we can conclude without difficulty that they have at most one element in common. Hence, there is a uniquely determined function h on Y to X such that, for each $y \in Y$,

$$\{hy\} = \bigcap\{H^{-1}Z : y \in Z \in B \cap Sb\,Y\}.$$

By a similar argument, which also uses (1) and (2), we can show that f actually establishes a one-one correspondence between the elements of Y and X, and, furthermore, that

$$HW = (h^{-1})^*W$$

for each $W \in B \cap Sb\,X$.

Now for each $f \in Rl_{GX}\mathfrak{A}'$ let

$$\bar{H}f = \langle fhy : y \in Y\rangle \cup \langle 0^{(\mathfrak{C})} : y \in U \sim Y\rangle.$$

It will be shown that $\bar{H} \in Is(\mathfrak{R}\mathfrak{l}_{GX}\mathfrak{A}', \mathfrak{R}\mathfrak{l}_{GY}\mathfrak{A}')$. If $0^{(\mathfrak{C})} \neq c \in C$, then we have $F_{\bar{H}f}c = (h^{-1})^*F_fc = HF_fc \in B$. Also, $F_{\bar{H}f}0^{(\mathfrak{C})} = (h^{-1})^*F_f0^{(\mathfrak{C})} \cup (U \sim Y) \in B$. Hence $\bar{H}f \in Rl_{GY}\mathfrak{A}'$, and $Rg\bar{H} \subseteq Rl_{GY}\mathfrak{A}'$. $\bar{H}$ is clearly seen to be a homomorphism from $\mathfrak{R}\mathfrak{l}_{GX}\mathfrak{A}'$ into $\mathfrak{R}\mathfrak{l}_{GY}\mathfrak{A}'$. If $0 \neq f \in Rl_{GX}\mathfrak{A}'$, choose $x \in X$ such that $fx \neq 0^{(\mathfrak{C})}$; then $(\bar{H}f)h^{-1}x \neq 0^{(\mathfrak{C})}$, and thus $\bar{H}f \neq 0$. Therefore, $\bar{H}$ is one-one. If $g \in Rl_{GY}\mathfrak{A}'$, let

$$f = \langle gh^{-1}x : x \in X\rangle \cup \langle 0^{(\mathfrak{C})} : x \in U \sim X\rangle;$$

then $f \in Rl_{GX}\mathfrak{A}'$ and $\bar{H}f = g$. Hence $Rg\bar{H} = Rl_{GY}\mathfrak{A}'$, and it has been shown that $\bar{H} \in Is(\mathfrak{R}\mathfrak{l}_{GX}\mathfrak{A}', \mathfrak{R}\mathfrak{l}_{GY}\mathfrak{A}')$.

In this way the following equivalence has been established:

(8) for all $X, Y \in B$, $\mathfrak{R}\mathfrak{l}_X\mathfrak{B} \cong \mathfrak{R}\mathfrak{l}_Y\mathfrak{B}$ iff $\mathfrak{R}\mathfrak{l}_{GX}\mathfrak{A}' \cong \mathfrak{R}\mathfrak{l}_{GY}\mathfrak{A}'$.

We have thus constructed a CA_α $\mathfrak{A}'$ which, by (5) and (6), satisfies conditions (iii) and (iv) of the conclusion of the lemma (with $\mathfrak{A}$ replaced by $\mathfrak{A}'$), and, moreover, which satisfies conditions (7) and (8) for an appropriately chosen G. It remains to be shown that $\mathfrak{A}'$ can be replaced by an isomorphic CA_α $\mathfrak{A}$ in such a way that the identity function can be taken for G, and consequently conditions (7) and (8) will go over into parts (i) and (ii) of the conclusion. To achieve this we simply apply the method of the general theory of algebras (called sometimes the "exchange method") which was used in the proof of 0.2.15.

THEOREM 2.4.36. *For any given* $\alpha \geqq 1$ *there are* Lf_α*'s* $\mathfrak{A}$, $\mathfrak{B}$, *and* $\mathfrak{D}$ *satisfying the following conditions*:

(i) *none of the algebras* $\mathfrak{A}$, $\mathfrak{B}$, *or* $\mathfrak{D}$ *has a discrete direct factor with more than one element*;

(ii) $\mathfrak{A} \cong \mathfrak{A} \times \mathfrak{D} \times \mathfrak{D}$ *holds while* $\mathfrak{A} \cong \mathfrak{A} \times \mathfrak{D}$ *fails*;

(iii) $\mathfrak{A} \mid \mathfrak{B} \mid \mathfrak{A}$ *and* $\mathfrak{A} \times \mathfrak{A} \cong \mathfrak{B} \times \mathfrak{B}$ *hold while* $\mathfrak{A} \cong \mathfrak{B}$ *fails.*

PROOF. By Hanf [57*] there are BA's $\mathfrak{A}'$ and $\mathfrak{D}'$ such that $\mathfrak{A}' \cong \mathfrak{A}' \times \mathfrak{D}' \times \mathfrak{D}'$ holds while $\mathfrak{A}' \cong \mathfrak{A}' \times \mathfrak{D}'$ fails. Let $\mathfrak{A}$ be a CA_α which satisfies conditions (i)–(iv) of 2.4.35 with $\mathfrak{B}$ replaced by $\mathfrak{A}'$. From 2.4.35(i),(ii) we easily conclude that there is a CA_α $\mathfrak{D}$ such that $\mathfrak{A} \cong \mathfrak{A} \times \mathfrak{D} \times \mathfrak{D}$ holds while $\mathfrak{A} \cong \mathfrak{A} \times \mathfrak{D}$ fails. By 2.3.3 and 2.4.35(iii), $\mathfrak{A}, \mathfrak{D}, \mathfrak{A} \times \mathfrak{D} \in \mathsf{Lf}_\alpha$. Furthermore, 2.4.35(iv) shows that $\mathfrak{A}$, and hence also $\mathfrak{D}$ and $\mathfrak{A} \times \mathfrak{D}$, has no discrete factor with more than one element. Taking $\mathfrak{B} = \mathfrak{A} \times \mathfrak{D}$ we see then that conditions (i)–(iii) hold.

We can use Lemma 2.4.35 for still another purpose. When applied to CA's the three notions of pseudo-direct indecomposability characterized by the conditions (ii′), (ii″), and (iii‴) in 0.3.29 prove to be distinct from each other and from the notion of direct indecomposability. This can easily be seen by considering discrete CA_α's corresponding to the following BA's: (1) a denumerable atomless BA, (2) the Boolean set algebra of all finite subsets of ω and their complements, and (3) the BA ${}^\omega\mathfrak{A}$ where $\mathfrak{A}$ is a two-element BA. However, with the help of 2.4.35 we can also obtain CA_α's $\mathfrak{A}$ and $\mathfrak{B}$ with $\alpha \geqq 1$ which are hereditarily non-discrete in addition to being locally finite, and such that $\mathfrak{A}$ satisfies (ii′) but is not directly indecomposable, while $\mathfrak{B}$ satisfies (iii″) but not (ii′). As regards the relation between (ii″) and (iii‴), if $\mathfrak{A}$ is any directly indecomposable CA_α and $\beta \geq \omega \cup |A|$, then ${}^\beta\mathfrak{A}$ satisfies the first of these conditions but not the second. However, from 2.4.4(ii) it is easily seen that for all non-discrete Lf_α's with $\alpha \geqq \omega$ these two conditions coincide.

In passing we may mention that all the results established so far in this section extend with practically no changes to arbitrary Df's.

The notion of hereditarily non-discrete CA_α's, which played an essential part in the last portion of our discussion, has some intrinsic importance; therefore we wish to discuss it a little further in two theorems of more general character, 2.4.37 and 2.4.38. These two theorems are closely related to 1.6.20 and, like the latter, were proved by Don Pigozzi; also the remarks following 2.4.38 are due to him.

THEOREM 2.4.37. *For any* $\mathfrak{A}, \mathfrak{B}, \mathfrak{C} \in \mathsf{CA}_\alpha$ *with* $\alpha \geqq 2$ *the following two conpitions are equivalent*:

(i) $\mathfrak{A} \cong \mathfrak{B} \times \mathfrak{C}$, $\mathfrak{B}$ *is discrete, and* $\mathfrak{C}$ *has no discrete factor with more than one element*;

(ii) $\mathfrak{B} \cong \mathfrak{Rl}_{c_0 \cdot d_{01}} \mathfrak{A}$ *and* $\mathfrak{C} \cong \mathfrak{Rl}_{c_0 - d_{01}} \mathfrak{A}$.

PROOF. Consider any $a \in Zd\mathfrak{A}$ such that $\mathfrak{Rl}_{-a}\mathfrak{A}$ is discrete and $\mathfrak{Rl}_a\mathfrak{A}$ has no discrete factor with more than one element. Then, by 2.2.14, $-a \cdot c_0 - d_{01} = 0$ (cf. the remark after 1.3.18). Also by 2.2.14 we have, for all non-zero $x \in A$, that $x \leqq a$ implies $x \cdot c_0 - d_{01} \neq 0$, and this just means that $a \leqq c_0 - d_{01}$. Hence, $a = c_0 - d_{01}$, and in light of 2.4.7 we see that (i) implies (ii). The implication in the opposite direction is proved similarly.

COROLLARY 2.4.38. *A* CA_α $\mathfrak{A}$ *with* $\alpha \geqq 2$ *has no discrete direct factor with more than one element iff* $c_0 - d_{01} = 1$.

From 2.4.37 it follows immediately that every CA_α with $\alpha \geqq 2$ has a direct decomposition, unique up to isomorphism, into two factors one of which is discrete and the other hereditarily non-discrete. The proof of this result makes essential use of diagonal elements and as a consequence it does not extend to CA_1's (nor, by the way, to arbitrary Df_α's with $\alpha \geqq 1$). To obtain a simple example of a CA_1 $\mathfrak{A}$ without the desired decomposition we consider a two-element CA_1 $\mathfrak{B}$ and a four-element simple CA_1 $\mathfrak{C}$ (cf. 2.3.15), and we take for $\mathfrak{A}$ the subalgebra of ${}^{\omega}\mathfrak{B} \times {}^{\omega}\mathfrak{C}$ generated by the set of all atoms.

From 2.4.38 it is seen that, for any given $\alpha \geqq 2$, the class of hereditarily non-discrete CA_α's is equational, and hence is closed under the formation of subalgebras, homomorphic images, and direct products. It can easily be shown that, in case $\alpha = 1$, this class is not equational and is actually not closed under the formation of either subalgebras or homomorphic images. It is, however, closed under the formation of direct products, and even under subdirect products (a fact which was implicitly used in the proof of 2.4.35).

It should be pointed out that in some of our earlier results we dealt with hereditary non-discreteness without using the term explicitly. In particular, by 1.11.8(ii), every hereditarily non-discrete Dc_α with $\alpha \geqq 2$ is atomless while, by 2.3.31(ii), every finitely generated atomless CA_α with $\alpha \geqq 2$ is hereditarily non-discrete; see also 2.1.20(ii).

We now turn to the discussion of subdirect products and related notions. In the second part of this work the notion of subdirect product will play a large role because, as we have seen, the class of representable CA's is defined in its terms; cf. Remarks 1.1.13.

In the next few theorems we shall be concerned with subdirect decompositions of CA's given by formulas of the type

$$\mathfrak{A} \cong | \subseteq_d \mathsf{P}_{i \in I} \mathfrak{B}_i.$$

The first two of these theorems are simple adaptations of certain results from the general theory of algebras which make them useful tools in our further discussion. We know that every subdirect decomposition of an algebra $\mathfrak{A}$ can be characterized in terms of a system of congruence relations of $\mathfrak{A}$ and that, in case $\mathfrak{A}$ is a CA, congruence relations can be replaced by ideals (cf. 2.3.6). In 2.4.40 we formulate explicitly the final outcome of this transformation. In 2.4.39 we state a simple condition, formulated in terms of homomorphisms, which is necessary and sufficient for a given CA $\mathfrak{A}$ to be representable isomorphically as a subdirect product of some CA's belonging to a given class K. It will be seen from the proof of 2.4.39 that this condition has been obtained by simplifying a closely related but somewhat more complicated condition which performs the same function for an arbitrary algebra $\mathfrak{A}$ and a class K of arbitrary algebras (similar to $\mathfrak{A}$); it will also be seen that in this simplification only Boolean properties are used.

THEOREM 2.4.39. *For any* $\mathsf{K} \subseteq \mathsf{CA}_\alpha$ *and* $\mathfrak{A} \in \mathsf{CA}_\alpha$ *the following conditions are equivalent*:

(i) $\mathfrak{A} \cong | \subseteq_d \mathsf{P}\mathfrak{B}$ *for some system* $\mathfrak{B}$ *of algebras in* K;

(ii) *for every non-zero* $x \in A$ *there is a* $\mathfrak{B} \in \mathsf{K}$ *and an* $f \in Ho(\mathfrak{A}, \mathfrak{B})$ *such that* $fx \neq 0^{(\mathfrak{B})}$.

PROOF. With the help of 0.3.46 we easily show that, for algebras and classes of algebras of arbitrary character, condition (i) is equivalent to:

(ii′) *for any two distinct elements* $x, y \in A$, *there is a* $\mathfrak{B} \in \mathsf{K}$ *and an* $f \in Ho(\mathfrak{A}, \mathfrak{B})$ *such that* $fx \neq fy$.

In the domain of CA's, (ii′) obviously implies (ii). To obtain the implication in the opposite direction consider any $x, y \in A$ with $x \neq y$. Then $x \oplus^{(\mathfrak{A})} y \neq 0^{(\mathfrak{A})}$ whence, by (ii), $f(x \oplus^{(\mathfrak{A})} y) \neq f0^{(\mathfrak{A})}$; therefore, $fx \oplus^{(\mathfrak{B})} fy \neq 0^{(\mathfrak{B})}$, and finally $fx \neq fy$. This completes the proof.

THEOREM 2.4.40. *For any* $\mathfrak{A} \in \mathsf{CA}_\alpha$ *and any system* $\mathfrak{B} \in {}^I\mathsf{CA}_\alpha$, *the following conditions are equivalent*:

(i) $\mathfrak{A} \cong | \subseteq_d \mathsf{P}\mathfrak{B}$;

(ii) *there is a system* $J \in {}^I Il\mathfrak{A}$ *such that* $\bigcap_{i \in I} J_i = \{0\}$, *and* $\mathfrak{A}/J_i \cong \mathfrak{B}_i$ *for each* $i \in I$.

PROOF: by 0.2.49, 0.3.46, and 2.3.4.

It should be pointed out that, in contrast to our findings on direct decompositions, it is not possible in general to characterize a subdirect decomposition of a CA $\mathfrak{A}$ in terms of a system of its elements such that the decomposition can be represented isomorphically by means of the corresponding relativized algebras (cf. the remarks preceding 2.4.7). This can, of course, be done (using 2.3.26) in case all ideals of $\mathfrak{A}$ are principal and are generated by zero-dimensional elements. However, the class of CA's satisfying these conditions proves to consist just of those algebras which are direct products of finitely many simple algebras. While it thus contains all finite CA's, by 2.4.17(ii), it is otherwise rather narrow.

The following two theorems are formal analogues of results on direct decompositions, but are restricted to the case of finitely many factors. The first states that CA's have the finite refinement property for subdirect decompositions; by the second theorem, an irredundant subdirect decomposition of a CA into finitely many subdirectly indecomposable factors, if it exists, is uniquely determined up to isomorphism. (Notice that, by 0.3.54 and 2.3.1, every CA_α can be subdirectly decomposed into subdirectly indecomposable CA_α's, but the number of factors in this decomposition is not in general finite.) Both theorems discussed are consequences of the distributivity of the lattice of ideals in every CA.

THEOREM 2.4.41. *Let* $\mathfrak{A} \in \mathsf{CA}_\alpha$. *If* $\mu, \nu < \omega$, $\mathfrak{A} \cong | \subseteq_{\mathsf{d}} \mathsf{P}_{\kappa<\mu} \mathfrak{B}_\kappa$, *and* $\mathfrak{A} \cong | \subseteq_{\mathsf{d}} \mathsf{P}_{\lambda<\nu} \mathfrak{C}_\lambda$, *then there is a system* $\langle \mathfrak{D}_{\kappa\lambda} : \kappa < \mu, \lambda < \nu \rangle$ *of* CA_α*'s such that* $\mathfrak{B}_\kappa \cong | \subseteq_{\mathsf{d}} \mathsf{P}_{\lambda<\nu} \mathfrak{D}_{\kappa\lambda}$ *for each* $\kappa < \mu$, *and* $\mathfrak{C}_\lambda \cong | \subseteq_{\mathsf{d}} \mathsf{P}_{\kappa<\mu} \mathfrak{D}_{\kappa\lambda}$ *for each* $\lambda < \nu$.

PROOF. In accordance with 2.4.40, choose sequences $K \in {}^\mu Il\mathfrak{A}$ and $L \in {}^\nu Il\mathfrak{A}$ such that $\bigcap_{\kappa<\mu} K_\kappa = \bigcap_{\lambda<\nu} L_\lambda = \{0\}$, $\mathfrak{A}/K_\kappa \cong \mathfrak{B}_\kappa$ for each $\kappa < \mu$, and $\mathfrak{A}/L_\lambda \cong \mathfrak{C}_\lambda$ for each $\lambda < \nu$. By 2.3.11(ii) we have, for any $\kappa < \mu$ and $\lambda, \lambda' < \nu$,

$$Ig(K_\kappa \cup L_\lambda) \cap Ig(K_\kappa \cup L_{\lambda'}) = Ig(K_\kappa \cup (L_\lambda \cap L_{\lambda'})).$$

Hence, by an easy inductive generalization,

$$(1) \qquad \bigcap_{\lambda<\nu} Ig(K_\kappa \cup L_\lambda) = Ig(K_\kappa \cup \bigcap_{\lambda<\nu} L_\lambda) = Ig(K_\kappa \cup \{0\}) = K_\kappa$$

for every $\kappa < \mu$. Similarly, we have

$$(2) \qquad \bigcap_{\kappa<\mu} Ig(K_\kappa \cup L_\lambda) = L_\lambda$$

for each $\lambda < \nu$. For $\kappa < \mu$ and $\lambda < \nu$ let $\mathfrak{D}_{\kappa\lambda} = \mathfrak{A}/Ig(K_\kappa \cup L_\lambda)$. The desired conclusions of the theorem follow by (1), (2), 0.2.27, and 2.4.40.

It can be shown by means of an example that Theorem 2.4.41 does not extend to infinite compositions.

THEOREM 2.4.42. *Let* $\mathfrak{A} \in \mathsf{CA}_\alpha$. *Assume that* $\mu, \nu < \omega$, $\mathfrak{A} \cong|\subseteq_{\mathsf{d}} \mathsf{P}_{\kappa<\mu}\mathfrak{B}_\kappa$, $\mathfrak{A} \cong|\subseteq_{\mathsf{d}} \mathsf{P}_{\lambda<\nu}\mathfrak{C}_\lambda$, *and all the algebras* $\mathfrak{B}_\kappa$ *with* $\kappa < \mu$ *and* $\mathfrak{C}_\lambda$ *with* $\lambda < \nu$ *are subdirectly indecomposable. In addition assume that there is no set* $\Gamma \subset \mu$ *for which* $\mathfrak{A} \cong|\subseteq_{\mathsf{d}} \mathsf{P}_{\kappa\in\Gamma}\mathfrak{B}_\kappa$, *and no set* $\Delta \subset \nu$ *for which* $\mathfrak{A} \cong|\subseteq_{\mathsf{d}} \mathsf{P}_{\lambda\in\Delta}\mathfrak{C}_\lambda$. *Under these assumptions we have* $\mu = \nu$, *and there is a permutation* φ *of* μ *such that* $\mathfrak{B}_\kappa \cong \mathfrak{C}_{\varphi\kappa}$ *for each* $\kappa < \mu$.

PROOF. By 2.4.40 and the irredundancy of the two decompositions, there are sequences $K \in {}^\mu Il\mathfrak{A}$ and $L \in {}^\nu Il\mathfrak{A}$ satisfying the following conditions:

(1) $$\bigcap_{\kappa<\mu} K_\kappa = \bigcap_{\lambda<\nu} L_\lambda = \{0\};$$

(2) $$\mathfrak{A}/K_\kappa \cong \mathfrak{B}_\kappa \text{ for each } \kappa < \mu, \text{ and}$$
$$\mathfrak{A}/L_\lambda \cong \mathfrak{C}_\lambda \text{ for each } \lambda < \nu;$$

(3) $$K_\kappa \nsubseteq K_{\kappa'} \text{ for } \kappa, \kappa' < \mu, \quad \kappa \neq \kappa', \text{ and}$$
$$L_\lambda \nsubseteq L_{\lambda'} \text{ for } \lambda, \lambda' < \nu, \quad \lambda \neq \lambda'.$$

It was seen in the proof of 2.4.41 that (1) and 2.3.11(ii) imply

$$K_\kappa = \bigcap_{\lambda<\nu} Ig(K_\kappa \cup L_\lambda)$$

for each $\kappa < \mu$. Therefore, it follows from (2), 0.3.53, and the subdirect indecomposability of $\mathfrak{B}_\kappa$ that, for some $\lambda < \nu$, we have $Ig(K_\kappa \cup L_\lambda) = K_\kappa$, and thus $L_\lambda \subseteq K_\kappa$. Similarly, for each $\lambda < \nu$ there is a $\kappa < \mu$ such that $K_\kappa \subseteq L_\lambda$. Consequently, (3) implies that $\mu = \nu$ and that there exists a permutation φ of μ such that $K_\kappa = L_{\varphi\kappa}$ for all $\kappa < \mu$. The desired conclusion now follows by (2).

It should be pointed out that 2.4.41 and 2.4.42 do not seem to have implications comparable in strength with the consequences of their analogues in the theory of direct decompositions. In particular we see no way of deriving 2.4.42 directly from 2.4.41; indeed, the significance of 2.4.41 for the theory of subdirect decompositions does not appear great.

THEOREM 2.4.43. *If* $\mathfrak{A} \in \mathsf{Lf}_\alpha$ *(in particular, if* $\mathfrak{A}$ *is any* CA_α *with* $\alpha < \omega$*), then the following conditions are equivalent*:

(i) $\mathfrak{A}$ *is simple*;

(ii) $\mathfrak{A}$ *is subdirectly indecomposable*;

(iii) $\mathfrak{A}$ *is weakly subdirectly indecomposable*;

(iv) $\mathfrak{A}$ *is directly indecomposable.*

PROOF: by 0.3.58, 2.3.14, and 2.4.14.

THEOREM 2.4.44. *For any* $\mathfrak{A} \in \mathsf{CA}_\alpha$ *the following conditions are equivalent.*

(i) $\mathfrak{A}$ *is subdirectly indecomposable.*

(ii) *There is an* $x \in A$, $x \neq 0$, *such that for every* $y \in A$ *with* $y \neq 0$ *there is a finite subset* Γ *of* α *for which* $x \leqq \mathsf{c}_{(\Gamma)}y$.

PROOF. By 0.3.52, (i) holds iff the set $\{Ig\{x\}: x \in A \sim \{0\}\}$ of proper principal ideals of $\mathfrak{A}$ has a least member (under the relation $\subseteq$). The equivalence of (i) and (ii) now follows by 2.3.10(i).

COROLLARY 2.4.45. (i) *If* $\mathfrak{A}$ *is a subdirectly indecomposable* CA_α, *then there is a subalgebra* $\mathfrak{B}$ *of* $\mathfrak{A}$ *generated by one element which is subdirectly indecomposable, and actually such that every algebra* $\mathfrak{C}$ *with* $\mathfrak{B} \subseteq \mathfrak{C} \subseteq \mathfrak{A}$ *is subdirectly indecomposable.*

(ii) *If* K *is a non-empty set of* CA_α*'s directed by the relation* $\subseteq$ *and no member of* K *is subdirectly indecomposable, then* UK *is not subdirectly indecomposable.*

THEOREM 2.4.46. *For any* $\mathfrak{A} \in \mathsf{CA}_\alpha$ *the following conditions are equivalent:*

(i) $\mathfrak{A}$ *is weakly subdirectly indecomposable.*

(ii) $|A| > 1$ *and for every non-zero element* x *of* A *we have*

$$\sum\{\mathsf{c}_{(\Gamma)}x : \Gamma \subseteq \alpha,\ |\Gamma| < \omega\} = 1.$$

PROOF. Suppose $\mathfrak{A}$ is not weakly subdirectly indecomposable. Then, by 0.3.50, let I and J be ideals of $\mathfrak{A}$ such that $I, J \neq \{0\}$ and $I \cap J = \{0\}$. Choose $x \in I \sim \{0\}$ and $y \in J \sim \{0\}$. Then for every finite $\Gamma \subseteq \alpha$ we have $\mathsf{c}_{(\Gamma)}x \cdot y \in I \cap J$ and hence $\mathsf{c}_{(\Gamma)}x \cdot y = 0$, or $\mathsf{c}_{(\Gamma)}x \leqq -y$. It follows that $\sum\{\mathsf{c}_{(\Gamma)}x : \Gamma \subseteq \alpha, |\Gamma| < \omega\}$ either does not exist, or exists and is different from 1.

The converse is proved similarly.

COROLLARY 2.4.47. (i) *A subalgebra of a weakly subdirectly indecomposable* CA_α *is weakly subdirectly indecomposable.*

(ii) *If a* CA_α $\mathfrak{A}$ *is not weakly subdirectly indecomposable, then there is a subalgebra of* $\mathfrak{A}$ *generated by two elements which is not weakly subdirectly indecomposable and, in fact, is not a subalgebra of any weakly subdirectly indecomposable subalgebra of* $\mathfrak{A}$.

PROOF. (i) obviously follows by 2.4.46. To obtain (ii) assume that $\mathfrak{A}$ is not weakly subdirectly indecomposable. Then by 2.4.46 there exist two elements $x, y \in A$ such that $x \neq 0$, $y \neq 1$, and $\mathsf{c}_{(\Gamma)}x \leqq y$ for every finite $\Gamma \subseteq \alpha$. The subalgebra $\mathfrak{Sg}\{x, y\}$ clearly has the desired properties.

REMARKS 2.4.48. We have previously established two results, 2.3.16(i) and 2.4.15, entirely analogous to 2.4.47(i). The three results can be restated by saying that the classes of simple, directly indecomposable, and weakly subdirectly indecomposable CA_α's are closed under the formation of subalgebras.

These three classes are also closed under the formation of unions of sets of algebras directed by inclusion, i.e., they are local in the sense that they satisfy 0.1.28(i). For the class of directly indecomposable CA_α's this was explicitly established in 2.4.16(ii). For the other two classes it follows from the corresponding results in Chapter 0, 0.2.35(ii) and 0.3.51(ii), concerning arbitrary algebras. In addition we have two results specific for CA_α's which, in view of 0.1.28, immediately imply the local character of the two classes discussed. These are 2.3.16(ii) and 2.4.47(ii); they are somewhat stronger than the corresponding results in Chapter 0, 0.2.35(i) and 0.3.51(i).

On the other hand, it will be seen at the end of 2.4.50 that the class of subdirectly indecomposable CA_α's with $\alpha \geqq \omega$ is neither local nor closed under the formation of subalgebras. Somewhat related properties of this class, which will not be discussed here in detail, were stated in 2.4.45.

THEOREM 2.4.49. (i) *If $\mathfrak{A}$ is a CA_α with at least one atom, then $\mathfrak{A}$ is weakly subdirectly indecomposable iff it is subdirectly indecomposable.*

(ii) *If $\mathfrak{A}$ is an $|\alpha|^+$-complete CA_α, then $\mathfrak{A}$ is weakly subdirectly indecomposable iff it is directly indecomposable.*

PROOF. (i) By 0.3.58 we know that $\mathfrak{A}$ is weakly subdirectly indecomposable if $\mathfrak{A}$ is subdirectly indecomposable. Now suppose $\mathfrak{A}$ is weakly subdirectly indecomposable. Let x be an atom of $\mathfrak{A}$. If y is a non-zero element of A, by 2.4.46 we see that there is a finite $\Gamma \subseteq \alpha$ such that $x \cdot \mathsf{c}_{(\Gamma)} y \neq 0$. Hence, $x \leqq \mathsf{c}_{(\Gamma)} y$. By 2.4.44, $\mathfrak{A}$ is subdirectly indecomposable.

(ii) By 0.3.58, $\mathfrak{A}$ is directly indecomposable if $\mathfrak{A}$ is weakly subdirectly indecomposable. Now suppose $\mathfrak{A}$ is directly indecomposable. If x is a non-zero element of A, it is easily seen, using 1.2.6, that $\Delta(\sum\{\mathsf{c}_{(\Gamma)} x : \Gamma \subseteq \alpha, |\Gamma| < \omega\}) = 0$. Hence $\sum\{\mathsf{c}_{(\Gamma)} x : \Gamma \subseteq \alpha, |\Gamma| < \omega\} = 1$ by 2.4.14. Thus $\mathfrak{A}$ is weakly subdirectly indecomposable by 2.4.46.

REMARKS 2.4.50. By 2.4.43 the notion of simplicity and the various notions of indecomposability discussed in this section coincide when restricted to Lf_α's; by 2.4.48 and 2.4.49 some of these notions coincide also under some other assumptions. From the following examples it will be seen, however, that for the whole class CA_α with $\alpha \geqq \omega$ no two of these notions coincide.

Example (I). We shall obtain a CA_α $\mathfrak{B}$ which is subdirectly indecomposable but not simple by considering the full cylindric set algebra $\mathfrak{A}$ in the space ${}^\alpha 2$ and letting X_0 be the weak Cartesian subspace of ${}^\alpha 2$ determined by $\alpha \times \{0\}$, i.e., the set of all $f \in {}^\alpha 2$ such that $\{\kappa : \kappa < \alpha, f\kappa = 1\}$ is finite. With the help of 2.4.44 and 2.3.14 it is easily seen that the weak cylindric set algebra

$\mathfrak{B} = \mathfrak{Rl}_{X_0}\mathfrak{A}$ has the desired properties; for x in both 2.4.44(ii) and 2.3.14(ii) we take $\{f\}$ where $f\kappa = 0$ for all $\kappa < \alpha$.

Example (II). To obtain a CA_α $\mathfrak{C}$ which is weakly subdirectly indecomposable but not subdirectly indecomposable, we consider $\mathfrak{A}$, X_0, and $\mathfrak{B}$ as defined in (I). Let $\{\Delta_\kappa : \kappa < \omega\}$ be a partition of α into infinite pairwise disjoint sets, and, for each $\kappa < \omega$, let

$$Y_\kappa = \{f : f \in X_0, f\mu = 0 \text{ for all } \mu \in \bigcup_{\lambda<\kappa} \Delta_\lambda\}.$$

Finally set $\mathfrak{C} = \mathfrak{Sg}^{(\mathfrak{B})}\{Y_\kappa : \kappa < \omega\}$. It is easily seen that for every $Z \in C$ there is a $\kappa < \omega$ such that $\Delta Z \subseteq \bigcup_{\lambda<\kappa}\Delta_\lambda$, and hence, for this κ, $Z \subseteq \mathsf{c}_{(\Gamma)}Y_\kappa$ does not hold for any finite $\Gamma \subseteq \alpha$ unless $Z = 0$. This shows $\mathfrak{C}$ not to be subdirectly indecomposable by 2.4.44, while, on the other hand $\mathfrak{C}$ is easily seen to be weakly subdirectly indecomposable by (I) and 2.4.47(i).

Example (III). Finally we outline the construction of a CA_α $\mathfrak{D}$ which is directly indecomposable without being weakly subdirectly indecomposable. Let $\mathfrak{A}$ and X_0 be as in (I), and let X_1 be the weak Cartesian subspace of ${}^\alpha 2$ determined by $\alpha \times \{1\}$; set $X = X_0 \cup X_1$. Finally, denote the set of all finite subsets of X by K, and take $\mathfrak{D} = \mathfrak{Sg}^{(\mathfrak{Rl}_X\mathfrak{A})}K$. $\mathfrak{D}$ is easily seen not to be weakly subdirectly indecomposable.

It is, however, more difficult to show that $\mathfrak{D}$ is directly indecomposable. For each $\Gamma \subseteq \alpha$ and finite $\Delta \subseteq \alpha$ we have

(1) $\quad \mathsf{c}_{(\Delta)}{}^*Sg^{(\mathfrak{B}\upharpoonright\mathfrak{D})}(\{\mathsf{d}_{\kappa\lambda} : \kappa, \lambda \in \Gamma\} \cup K) \subseteq Sg^{(\mathfrak{B}\upharpoonright\mathfrak{D})}(\{\mathsf{d}_{\kappa\lambda} : \kappa, \lambda \in \Gamma \sim \Delta\} \cup K).$

The critical step in the proof of (1) consists in first observing that, for every $\mu < \alpha$ and all $Z, W \in D$, $\mathsf{c}_\mu Z \cdot \mathsf{c}_\mu^\partial W \subseteq \mathsf{c}_\mu(Z \cdot W)$ and hence

$$\mathsf{c}_\mu(Z \cdot W) = \mathsf{c}_\mu Z \cdot [\mathsf{c}_\mu^\partial W + \mathsf{c}_\mu(Z \cdot W)],$$

and then applying the latter formula to the case where $Z \in Sg^{(\mathfrak{B}\upharpoonright\mathfrak{D})}\{\mathsf{d}_{\kappa\lambda} : \kappa, \lambda \in \Gamma\}$ and $X \sim W \in K$ (cf. 2.1.17(i)). We now use (1) to show that $Z \in K$ or $X \sim Z \in K$ whenever $Z \in Zd\mathfrak{D}$ (compare here the proof of 2.1.17). Finally, by applying 2.4.14, we conclude that $\mathfrak{D}$ is directly indecomposable.

Example (II) is due to Stephen Comer. He also observed that this example can be used to show that the class of subdirectly indecomposable CA_α's with $\alpha \geqq \omega$ is not closed under the formation of subalgebras. As was noticed by Don Pigozzi, by means of the same example one can show that this class is not local. In fact, the algebra $\mathfrak{C}$, which is not subdirectly indecomposable, is the union of its directly indecomposable subalgebras $\mathfrak{E}_\lambda = \mathfrak{Sg}^{(\mathfrak{C})}\{Y_\kappa : \kappa < \lambda\}$ for $\lambda < \omega$. Pigozzi also pointed out that the algebra $\mathfrak{C}$ is a Dc_α and that the constructions in Examples (I) and (III) can be modified so as to yield Dc_α's; thus no part of 2.4.43 extends from Lf_α's to Dc_α's.

Recall that, by 0.3.48, an algebra is semisimple iff it is isomorphic to a subdirect product of simple algebras.

DEFINITION 2.4.51. *The class of all semisimple* CA_α*'s will be denoted by* Ss_α.

THEOREM 2.4.52. *Given a* CA_α $\mathfrak{A}$, *consider the three following conditions:*

(i) $\mathfrak{A} \in \mathsf{Lf}_\alpha$;

(ii) *every proper ideal of* $\mathfrak{A}$ *is the intersection of all maximal proper ideals which include it*;

(iii) $\mathfrak{A} \in \mathsf{Ss}_\alpha$.

Condition (*i*) *implies* (*ii*), *and* (*ii*) *implies* (*iii*).

PROOF. It follows directly from 0.2.27(i),(iii) and 0.3.49 (cf. also 2.3.6) that (ii) is equivalent to

(ii′) $\mathbf{H}\mathfrak{A} \subseteq \mathsf{Ss}_\alpha$.

Thus (ii) obviously implies (iii). $\mathsf{Lf}_\alpha \subseteq \mathsf{Ss}_\alpha$ by 0.3.54 and 2.4.43, and hence (i) implies (ii′) by 2.3.3; thus (i) implies (ii).

In case α is finite, $\mathsf{Lf}_\alpha = \mathsf{CA}_\alpha$ and hence conditions (i)–(iii) of 2.4.52 are trivially equivalent. For $\alpha \geqq \omega$ we will see in 2.5.24 that (ii) does not imply (i). Also in this case (iii) does not imply (ii) because it will be shown in 2.4.59 that $\mathbf{H}\mathsf{Ss}_\alpha \neq \mathsf{Ss}_\alpha$ for $\alpha \geqq \omega$.

THEOREM 2.4.53. $\mathsf{Ss}_\alpha = \mathbf{S}\mathsf{Ss}_\alpha = \mathbf{P}\mathsf{Ss}_\alpha$. *If* $\alpha < \omega$, *then* $\mathsf{CA}_\alpha = \mathsf{Ss}_\alpha = \mathbf{H}\mathsf{Ss}_\alpha$.

PROOF: by 0.3.42(iii), 0.3.43, 2.3.1, 2.3.16(i), and 2.4.52.

THEOREM 2.4.54. *Suppose* $\alpha \geqq \omega$ *and* $\mathfrak{A} \in \mathsf{Ss}_\alpha$.

(i) *If* $x \in At\mathfrak{A}$, *then* $\Delta x = 0$.

(ii) *If* $\mathsf{c}_0^\partial \mathsf{d}_{01} = 0$, *then* $\mathfrak{A}$ *is atomless.*

PROOF. If $\mathfrak{A}$ is simple and $x \in At\mathfrak{A}$, then by 2.3.30 we must have $|A| = 2$ and $x = 1$. Thus both (i) and (ii) hold in case $\mathfrak{A}$ is simple. Turning to the general case we let $x \in At\mathfrak{A}$. By 0.3.48 there is a system $\mathfrak{B} = \langle \mathfrak{B}_i : i \in I \rangle$ of simple CA_α's and an $h \in Ism(\mathfrak{A}, \mathbf{P}\mathfrak{B})$ such that

$$(1) \qquad (\mathsf{pj}_i \circ h)^* A = B_i \text{ for every } i \in I.$$

Let j be an arbitrary element of I such that $(hx)_j \neq 0^{(\mathfrak{B}_j)}$, and consider any $y \in A$ such that

$$(2) \qquad (hy)_j <^{(\mathfrak{B}_j)} (hx)_j.$$

Then $x \cdot y = 0$ since x is an atom and y cannot include x because of (2). Thus, again using (2), we get

$$(hy)_j = (hx)_j \cdot^{(\mathfrak{B}_j)} (hy)_j = h(x \cdot y)_j = 0^{(\mathfrak{B}_j)};$$

in view of (1) this shows that $(hx)_j$ is an atom. Therefore, we have proved that

$$(hx)_i \in At\,\mathfrak{B}_i \cup \{0^{(\mathfrak{B}_i)}\} \text{ for every } i \in I.$$

Now this result together with the fact that the theorem holds for simple algebras gives $\Delta^{(\mathfrak{B}_i)}(hx)_i = 0$ for each $i \in I$. Hence $\Delta x = 0$ by 2.4.2, and (i) holds. Part (ii) follows from (i) and 1.10.5(ii).

This theorem as well as some results upon which its proof rests (2.3.30 and part (iii) of 1.10.5) are due to Don Pigozzi.

By 2.4.54(ii) any simple CA_α $\mathfrak{A}$ with $\alpha \geqq \omega$ and with more than two elements is atomless. From 1.11.8(ii) we see that, if $\mathfrak{A}$ is dimension-complemented, then the same conclusion holds when the premiss that $\mathfrak{A}$ is simple is weakened to require only that $\mathfrak{A}$ be directly indecomposable. It would be interesting to find a natural common generalization of 2.4.54 and 1.11.8.

For the next theorem recall the definitions of reduced product and ultraproduct given in 0.3.62.

THEOREM 2.4.55. *Every reduced product and, in particular, every ultraproduct of a system of* CA_α*'s is a* CA_α.

PROOF: by 2.3.1 and 2.4.1.

REMARK 2.4.56. In addition to the class CA_α, the class of discrete CA_α's and that of hereditarily non-discrete CA_α's for $\alpha \geqq 2$ are equational and hence also closed under the formation of reduced products and ultraproducts; cf. 0.4.63, 1.3.10, and the remarks following 2.4.38.

From 2.3.14 it is easily seen that the class of simple CA_α's with $\alpha < \omega$ is elementary, and therefore by 0.3.77 we have

THEOREM 2.4.57. *For $\alpha < \omega$ every ultraproduct of a system of simple* CA_α*'s is a simple* CA_α.

REMARKS 2.4.58. For each $\alpha < \omega$ the class of simple CA_α's is elementary in the narrower sense (i.e., it can be characterized by a single first order sentence). Hence its complement, i.e., the class of CA_α's which are not simple, is also elementary and therefore is closed under the formation of ultraproducts. Of course, in view of 2.4.43, we can replace "simple" by "subdirectly indecomposable", etc., in this remark and in 2.4.57.

Turning now to the case $\alpha \geqq \omega$, we are confronted with four distinct classes — the classes of CA_α's which are simple, or subdirectly indecomposable, or weakly subdirectly indecomposable, or directly indecomposable. It can be shown by means of an example that none of these classes is closed under the formation of ultraproducts or even ultrapowers, and hence none is elementary.

In view of 0.3.58 it suffices for this purpose to exhibit a simple CA_α $\mathfrak{A}$ and an ultrafilter F on ω such that ${}^\omega\mathfrak{A}/\bar{F}$ is not directly indecomposable. In fact it suffices to take for $\mathfrak{A}$ any simple non-discrete CA_α (cf. 2.3.15) and for F any non-principal ultrafilter on ω. As regards the complements of these classes, they are not elementary either, but, with the exception of the class of subdirectly indecomposable CA_α's, they are closed under the formation of ultraproducts. This follows from some simple observations in the general theory of algebras, which, by the way, are not dependent on the assumption $\alpha \geqq \omega$; cf. 0.5.22.

REMARK 2.4.59. We want still to discuss the problem whether the classes Lf_α, Dc_α, and Ss_α are closed under ultraproducts and ultrapowers. For $\alpha < \omega$ the problem is trivial in view of 1.11.3, 2.4.53, and 2.4.57. The following example will show that none of the three classes has these closure properties when $\alpha \geqq \omega$. Let $\mathfrak{A}$ be any simple, non-discrete Lf_α (cf. 2.3.15). Let $I = \{\Gamma : \Gamma \subseteq \alpha, |\Gamma| < \omega\}$, and let F be any ultrafilter on I such that $\{\Delta : \Gamma \subseteq \Delta \in I\} \in F$ for every $\Gamma \in I$. We claim that the algebra $\mathfrak{B} = {}^I\mathfrak{A}/\bar{F}$ is not semisimple. Suppose, on the contrary, that $\mathfrak{B}$ were semisimple. Since $\mathfrak{B}$ is hereditarily non-discrete, it would be isomorphic to a subdirect product of a system $\langle \mathfrak{C}_j : j \in J \rangle$ of simple, non-discrete CA_α's (cf. the remarks following 2.4.38). Now for each $\mathfrak{C}_j$ we have

$$\prod_{\Gamma \in I} \mathsf{d}_\Gamma^{(\mathfrak{C}_j)} = 0^{(\mathfrak{C}_j)}. \tag{1}$$

In order to prove (1) suppose that $x \in \mathfrak{C}_j$ and $x \leqq \mathsf{d}_\Gamma^{(\mathfrak{C}_j)}$ for every $\Gamma \in I$, and consider any $\Delta \in I$. Choosing distinct $\kappa, \lambda \in \alpha \sim \Delta$ we have, by 1.8.6 and the fact that $\mathfrak{C}_j$ is non-discrete, $\mathsf{c}_{(\Delta)}^{(\mathfrak{C}_j)} x \leqq \mathsf{d}_{\kappa\lambda}^{(\mathfrak{C}_j)} < 1^{(\mathfrak{C}_j)}$. Hence $\mathsf{c}_{(\Gamma)}^{(\mathfrak{C}_j)} x \neq 1^{(\mathfrak{C}_j)}$ for every $\Gamma \in I$, and by 2.3.14 this implies $x = 0^{(\mathfrak{C}_j)}$ since $\mathfrak{C}_j$ is simple.

Because (1) holds for each $\mathfrak{C}_j$ we have also that (1) holds when $\mathfrak{C}_j$ is replaced by $\mathfrak{B}$. This, however, is impossible because it is clear that $\langle \mathsf{d}_\Delta^{(\mathfrak{A})} : \Delta \in I \rangle / \bar{F} \neq 0^{(\mathfrak{B})}$, while

$$\langle \mathsf{d}_\Delta^{(\mathfrak{A})} : \Delta \in I \rangle / \bar{F} \leqq \mathsf{d}_\Gamma^{(\mathfrak{B})}$$

for every $\Gamma \in I$. Thus $\mathfrak{B}$ is not semisimple. It can also easily be shown that $\mathfrak{B}$ is not a Dc_α.

The above example implies, of course, that the classes Lf_α, Dc_α, and Ss_α, with $\alpha \geqq \omega$, are not elementary. For the first two of these classes this result was previously obtained by a different method; cf. 2.4.6. Using 2.4.53 we can further conclude from this result that $\mathbf{H}\mathsf{Ss}_\alpha \neq \mathsf{Ss}_\alpha$[1] (in opposition to the

[1]) This observation was first made (using a different argument) and communicated to the authors by Jerzy Łoś.

formulas stated in 2.4.53). Taking $\mathfrak{A}$ in the same example to be minimal, we also arrive at the conclusion that the class Mn_α with $\alpha \geqq \omega$ is not closed under the formation of ultraproducts and ultrapowers.

REMARK 2.4.60. In 2.4.58 and 2.4.59 we have come across various classes of CA_α's which are not elementary and hence not universal, but which have many properties in common with universal classes: they are local and closed under the formation of subalgebras and isomorphic images; equivalently, they satisfy the condition (II) formulated in 0.3.85. The following classes belong here: Lf_α, Dc_α, and the classes of simple, directly indecomposable, and weakly subdirectly indecomposable CA_α's, all with $\alpha \geqq \omega$ (cf. 2.4.48).

To conclude this section we take up again the discussion of minimal algebras. We begin with a classification of all those CA's in which the minimal subalgebras are simple or, what amounts to the same by 2.4.43, directly indecomposable:

DEFINITION 2.4.61. *A* CA_α $\mathfrak{A}$ *is said to be of* ***characteristic*** κ *if the minimal subalgebra of* $\mathfrak{A}$ *is simple, and either* (i) $\kappa = 0$, *and* $\mathsf{c}_{(\lambda+1)}\bar{\mathsf{d}}((\lambda+1)\times(\lambda+1)) \neq 0$ *for every* $\lambda < \alpha \cap \omega$, *or else* (ii) κ *is the least* $\lambda < \alpha \cap \omega$ *such that*

$$\mathsf{c}_{(\lambda+1)}\bar{\mathsf{d}}((\lambda+1)\times(\lambda+1)) = 0.$$

COROLLARY 2.4.62. *For any* CA_α $\mathfrak{A}$ *with a simple minimal subalgebra there is exactly one* κ *such that* $\mathfrak{A}$ *is of characteristic* κ; *this* κ *is finite and* $< \alpha$.

To obtain examples of CA's with various characteristics, consider a cylindric set algebra $\mathfrak{A}$ of dimension α and base $U \neq 0$; as is easily seen from 2.4.61, $\mathfrak{A}$ is of characteristic 0 in case $|U| \geqq \alpha \cap \omega$, and of characteristic $|U|$ otherwise. In metalogical interpretation, the characteristic of a CA_α associated with a (first order) theory Θ is 0 iff Θ is consistent and all its models are of cardinality $\geqq \alpha \cap \omega$; it is of characteristic λ where $0 < \lambda < \alpha \cap \omega$ iff Θ is again consistent and all its models are of cardinality λ.

In the next theorem we give a modification of Definition 2.4.61 in which the notion of simplicity is not explicitly involved.

THEOREM 2.4.63. (i) *A* CA_α $\mathfrak{A}$ *is of characteristic* 0 *iff* $0 \neq 1$ *and* $\mathsf{c}_{(\lambda)}\bar{\mathsf{d}}(\lambda\times\lambda) = 1$ *for every* $\lambda < (\alpha+1)\cap\omega$.

(ii) *In case* $\kappa > 0$, *a* CA_α $\mathfrak{A}$ *is of characteristic* κ *iff* $\kappa < \alpha\cap\omega$, $0 \neq 1$, $\mathsf{c}_{(\kappa)}\bar{\mathsf{d}}(\kappa\times\kappa) = 1$, *and* $\mathsf{c}_{(\kappa+1)}\bar{\mathsf{d}}((\kappa+1)\times(\kappa+1)) = 0$.

PROOF. By 2.1.17(iii) and 2.3.14 the minimal subalgebra of $\mathfrak{A}$ is simple iff $0 \neq 1$ and $\mathsf{c}_{(\lambda)}\bar{\mathsf{d}}(\lambda\times\lambda) \in \{0, 1\}$ for every $\lambda < (\alpha+1)\cap\omega$. (i) and (ii) now follow from 1.9.10 and 2.4.61.

REMARK 2.4.64. From 2.4.63 it is seen that, for any $\kappa < \alpha \cap \omega$, the class of CA_α's of characteristic κ becomes equational when one includes in it all one-element algebras.

COROLLARY 2.4.65. (i) *Every* CA_α *with more than one element and with* $\alpha \leqq 1$ *is of characteristic* 0.

(ii) *A* CA_α $\mathfrak{A}$ *is of characteristic* 1 *iff* $\mathfrak{A}$ *is discrete, A has more than one element, and* $\alpha > 1$.

No characteristic is ascribed to a cylindric algebra whose minimal subalgebra is not simple, i.e., consists of just one element or contains more than two zero-dimensional elements. In Remark 2.5.29 we shall discuss informally a more general notion, that of *characteristic set*, which refers to arbitrary CA's. We shall see right now, however, that our present notion of characteristic provides by itself a classification of all those CA's whose minimal subalgebra contains only a finite number of zero-dimensional elements; this is a rather comprehensive class of CA's which includes, in particular, all finite-dimensional CA's and also all finite CA's. A classification of algebras of this class in terms of characteristic results from the next two theorems, by which every such algebra can be directly decomposed into finitely many algebras with definite characteristics, and this decomposition is unique up to isomorphism.

THEOREM 2.4.66. *Let* $\mathfrak{A}$ *be a* CA_α *with* $\alpha < \omega$ *or, more generally, any* CA_α *in which the set B of all zero-dimensional elements of its minimal subalgebra is finite. Let B′ be the subset of B consisting of all elements* $\mathsf{c}_{(\lambda)}\bar{\mathsf{d}}(\lambda \times \lambda)$ *with* $\lambda < (\alpha+1) \cap \omega$. *Set* $b_0 = \prod B'$, $b_\kappa = \mathsf{c}_{(\kappa)}\bar{\mathsf{d}}(\kappa \times \kappa) - \mathsf{c}_{(\kappa+1)}\bar{\mathsf{d}}((\kappa+1) \times (\kappa+1))$ *when* $0 < \kappa < \alpha \cap \omega$, *and* $\Gamma = \{\kappa : \kappa < \alpha \cap \omega, b_\kappa \neq 0\}$.

Under these assumptions the set Γ is finite, each of the algebras $\mathfrak{Rl}_{b_\kappa}\mathfrak{A}$ *is a* CA_α *of characteristic* κ, *and*

$$\mathfrak{A} \cong \mathsf{P}_{\kappa \in \Gamma} \mathfrak{Rl}_{b_\kappa} \mathfrak{A}.^{1)}$$

PROOF. By 2.1.23(i) we have that $\{b_\kappa : \kappa < \alpha \cap \omega\} \sim \{0\}$ is the set of atoms of the zero-dimensional part of the minimal subalgebra of $\mathfrak{A}$. Consequently, the hypothesis assures us that Γ is finite. Thus, since conditions (i)–(iii) of 2.4.7 obviously hold when Γ is taken for I and $\langle b_\kappa : \kappa \in \Gamma \rangle$ is taken for $\langle x_i : i \in I \rangle$, we have by 2.4.10 that

$$\mathfrak{A} \cong \mathsf{P}_{\kappa \in \Gamma} \mathfrak{Rl}_{b_\kappa} \mathfrak{A}.$$

[1]) Theorem 2.4.66 is analogous to Theorems 4.35 and 4.36 of Jónsson-Tarski [52] concerning relation algebras.

It can easily be checked, with the aid of 2.4.63, that $\mathfrak{Rl}_{b_\kappa}\mathfrak{A}$ is of characteristic κ for each $\kappa \in \Gamma$.

THEOREM 2.4.67. *Let $\mathfrak{B}$ and $\mathfrak{B}'$ be two systems of* CA_α*'s indexed respectively by two finite subsets Γ and Γ' of ω. If $\mathfrak{B}_\kappa$ is of characteristic κ for each $\kappa \in \Gamma$, $\mathfrak{B}'_\lambda$ is of characteristic λ for each $\lambda \in \Gamma'$, and*

$$\mathsf{P}_{\kappa\in\Gamma}\mathfrak{B}_\kappa \cong \mathsf{P}_{\lambda\in\Gamma'}\mathfrak{B}'_\lambda,$$

then $\Gamma = \Gamma'$, and $\mathfrak{B}_\kappa \cong \mathfrak{B}'_\kappa$ for each $\kappa \in \Gamma$.

PROOF. By 2.4.13 we see that there are algebras $\mathfrak{C}_{\kappa\lambda}$ with $\langle\kappa, \lambda\rangle \in \Gamma\times\Gamma'$ such that $\mathfrak{B}_\kappa \cong \mathsf{P}_{\lambda\in\Gamma'}\mathfrak{C}_{\kappa\lambda}$ for all $\kappa \in \Gamma$ and $\mathfrak{B}'_\lambda \cong \mathsf{P}_{\kappa\in\Gamma}\mathfrak{C}_{\kappa\lambda}$ for all $\lambda \in \Gamma'$. For any fixed $\langle\kappa, \lambda\rangle \in \Gamma\times\Gamma'$ we have $\mathfrak{C}_{\kappa\lambda} \mid \mathfrak{B}_\kappa$ and $\mathfrak{C}_{\kappa\lambda} \mid \mathfrak{B}'_\lambda$, and hence by 2.4.64 we may conclude that either $|C_{\kappa\lambda}| = 1$ or $\mathfrak{C}_{\kappa\lambda}$ is at the same time of characteristic κ and λ. Thus, by 2.4.62, $|C_{\kappa\lambda}| = 1$ whenever $\kappa \neq \lambda$. The desired conclusion now follows at once.

In the theorems which follow we shall apply the notion of characteristic (and in particular the decomposition theorem 2.4.66) to the problem of computing cardinalities of Mn_α's. In view of Theorem 2.1.20(i) we restrict ourselves to the case of finite-dimensional algebras.

THEOREM 2.4.68. *Let $\mathfrak{A}$ be a* Mn_α *of characteristic κ. For each finite $\Delta \subseteq \alpha$ let $\pi_\kappa\Delta$ be the total number of partitions of Δ in case $\kappa = 0$, and the number of partitions of Δ into at most κ subsets in case $\kappa > 0$. We then have:*

(i) $|A| = 2^{\pi_\kappa\alpha}$ *for any* $\alpha < \omega$

and, more generally,

(ii) $|Cl_{\alpha\sim\Delta}\mathfrak{A}| = 2^{\pi_\kappa\Delta}$ *for an arbitrary α and any finite $\Delta \subseteq \alpha$.*

PROOF. It suffices to prove (ii) since (i) is an obvious special case of (ii). By 2.1.17(ii) and 2.3.14 we have $Cl_{\alpha\sim\Delta}\mathfrak{A} = Sg^{(\mathfrak{Bl}\mathfrak{A})}(\{\mathsf{d}_{\kappa\lambda} : \kappa, \lambda \in \Delta\})$. Since $|\Delta| < \omega$, $\mathfrak{Cl}_{\alpha\sim\Delta}\mathfrak{A}$ is finite, and the atoms of $\mathfrak{Cl}_{\alpha\sim\Delta}\mathfrak{A}$ are exactly the non-zero elements of the form $\prod_{\Omega\in P}\mathsf{d}_\Omega \cdot \bar{\mathsf{d}}(\Theta\times\Theta)$ where P is a partition of Δ and $|\Theta \cap \Omega| = 1$ for each $\Omega \in P$ (cf. 1.9.7 and 1.9.8). Now by 1.8.3(i), 1.8.6, and 1.9.4(i) we have

$$\mathsf{c}_{(\Delta\sim\Theta)}[\prod_{\Omega\in P}\mathsf{d}_\Omega \cdot \bar{\mathsf{d}}(\Theta\times\Theta)] = \bar{\mathsf{d}}(\Theta\times\Theta).$$

Thus $\prod_{\Omega\in P}\mathsf{d}_\Omega \cdot \bar{\mathsf{d}}(\Theta\times\Theta) = 0$ iff κ is finite and P is a partition of Δ into more than κ subsets. The desired conclusion readily follows.

By 2.4.68(i) and 2.1.20(i) any two Mn_α's of the same characteristic have the same cardinality. For instance, $|A| = 2$ for every Mn_α $\mathfrak{A}$ of characteristic 1, and $|A| = 2^{2^{\alpha+1}}$ for every finite-dimensional Mn_α of characteristic 2.

We shall see in the next section (Theorem 2.5.30) that any two Mn_α's of the same characteristic are actually isomorphic.

THEOREM 2.4.69. *Let $\mathfrak{A}$ be an Mn_α with $\alpha < \omega$. Let Γ and $\pi_\kappa\Delta$ be defined as in 2.4.66 and 2.4.68, respectively. Then*

$$|A| = \prod_{\kappa \in \Gamma} 2^{\pi_\kappa \alpha}.$$

PROOF: by 2.3.3, 2.4.66, and 2.4.68(i).

Theorem 2.4.71 below can be regarded as a kind of converse of the theorem just proved. In establishing 2.4.71 we shall make use of the following

THEOREM 2.4.70. *If $\mathfrak{A}$ and $\mathfrak{B}$ are Mn_α's of characteristic κ and λ, respectively, and $\kappa \neq \lambda$, then $\mathfrak{A} \times \mathfrak{B} \in \mathsf{Mn}_\alpha$.*

PROOF. Let $\mathfrak{C}$ be the minimal subalgebra of $\mathfrak{A} \times \mathfrak{B}$. We may suppose, without loss of generality, that $\kappa < \lambda$, and we consider first the case $\kappa \neq 0$. Then $\mathsf{c}_{(\lambda)}^{(\mathfrak{A})}\bar{\mathsf{d}}(\lambda \times \lambda)^{(\mathfrak{A})} = 0^{(\mathfrak{A})}$ while $\mathsf{c}_{(\lambda)}^{(\mathfrak{B})}\bar{\mathsf{d}}(\lambda \times \lambda)^{(\mathfrak{B})} = 1^{(\mathfrak{B})}$. Hence, $\langle 0^{(\mathfrak{A})}, 1^{(\mathfrak{B})} \rangle \in C$, and it easily follows that $\langle 0^{(\mathfrak{A})}, b \rangle \in C$ for every $b \in B$. By complementation we also find $\langle 1^{(\mathfrak{A})}, 0^{(\mathfrak{B})} \rangle \in C$, and therefore, by symmetry, $\langle a, 0^{(\mathfrak{B})} \rangle \in C$ for every $a \in A$. Hence, as is readily seen, $\mathfrak{C} = \mathfrak{A} \times \mathfrak{B}$. The same conclusion is obtained by a similar argument in case $\kappa = 0$.

Notice that the algebra $\mathfrak{A} \times \mathfrak{B}$ of 2.4.70 is not simple and hence has no characteristic. On the other hand, if $\kappa = \lambda$, then $\mathfrak{A} \times \mathfrak{B}$ is of characteristic κ (by 2.4.64), but is not minimal.

THEOREM 2.4.71. *Suppose $0 < \alpha < \omega$, and let $\pi_\kappa\Delta$ be defined as in 2.4.68. If Γ is any non-empty finite subset of α, then there is an $\mathfrak{A} \in \mathsf{Mn}_\alpha$ such that*

$$|A| = \prod_{\kappa \in \Gamma} 2^{\pi_\kappa \alpha}.$$

PROOF. Choose $\mathfrak{B}_\kappa$ to be a Mn_α of characteristic κ for each $\kappa \in \Gamma$. (From the remark following 2.4.62 we see that such a $\mathfrak{B}_\kappa$ can always be chosen.) Then, by 2.4.70 and an easy inductive argument, we obtain $\mathsf{P}_{\kappa \in \Gamma}\mathfrak{B}_\kappa \in \mathsf{Mn}_\alpha$. The desired conclusion follows by 2.4.68(i).

Using the properties of minimal algebras we shall establish now the greatest lower bound for the cardinalities of all non-discrete CA_α's for fixed α such that $2 \leqq \alpha < \omega$.

THEOREM 2.4.72. *Assume $2 \leqq \alpha < \omega$ and let $\mathfrak{A}$ be a non-discrete CA_α. Then*

(i) $|A| \geqq 2^{2^{\alpha-1}}$,

(ii) $|A| = 2^{2^{\alpha-1}}$ *iff $\alpha = 2$ and $\mathfrak{A}$ is an Mn_2 of characteristic 0, or $\alpha > 2$ and $\mathfrak{A}$ is an Mn_α of characteristic 2.*

PROOF. Let b_κ and Γ be defined as in Theorem 2.4.66. Then by 2.4.66 we have

(1) $$\mathfrak{A} \cong \mathsf{P}_{\kappa\in\Gamma}\mathfrak{Rl}_{b_\kappa}\mathfrak{A}.$$

Since $\mathfrak{A}$ is assumed to be non-discrete, we see from 2.4.65(ii) that

(2) $$\Gamma \sim \{1\} \neq 0.$$

For each $\kappa \in \Gamma \sim \{1\}$ let $\mathfrak{B}_\kappa$ be the minimal subalgebra of $\mathfrak{Rl}_{b_\kappa}\mathfrak{A}$. Then, clearly,

(3) $$\mathfrak{B}_\kappa \text{ is a } \mathsf{Mn}_\alpha \text{ of characteristic } \kappa \text{ for each } \kappa \in \Gamma \sim \{1\},$$

and from (1) we obtain

(4) $$|A| \geqq |B_\kappa|, \text{ and } |A| = |B_\kappa| \text{ iff } \mathfrak{A} = \mathfrak{B}_\kappa, \text{ for each } \kappa \in \Gamma \sim \{1\}.$$

If $\alpha = 2$, then by (2) we have $\Gamma \sim \{1\} = \{0\}$. Thus both (i) and (ii) for the case $\alpha = 2$ follow immediately from (3), (4), and the fact that by 2.4.68(i) the cardinality of any Mn_2 of characteristic 0 is 4.

Assume now that $\alpha > 2$. Then by (3) and 2.4.68(i) we have $|B_\kappa| > 2^{2^{\alpha-1}}$ for every $\kappa \in \Gamma \sim \{1, 2\}$, and $|B_2| = 2^{2^{\alpha-1}}$ if $2 \in \Gamma$. (i) and (ii) then follow easily from (2) and (4).

REMARK 2.4.73. For completeness we want to make some comments about two cases not covered by the hypothesis of 2.4.72. The case $\alpha = 0$ is trivial since there are no non-discrete CA_0's. As regards the case $\alpha = 1$, notice that every Mn_1 has at most two elements and hence is discrete. Thus, if $\mathfrak{A}$ is a CA_1 with the property that $|A|$ is the greatest lower bound for the cardinalities of all non-discrete CA_1's, then $\mathfrak{A}$ cannot be a minimal algebra. Indeed, it is easily seen that for $\mathfrak{A}$ to have this property it is necessary and sufficient that it be a simple CA_1 with exactly four elements (cf. 2.3.15).

2.5. FREE ALGEBRAS

The two principal notions which were introduced and studied in the second half of Section 0.4 are those of $\mathfrak{Fr}_\beta \mathsf{K}$, the free algebra with β generators over a class K, and of a set which K-freely generates a given algebra; cf. 0.4.19(ii) and 0.4.23. In the present section we shall be interested in applications of these notions to classes K consisting of algebras similar to CA_α's; more specifically, we shall discuss free algebras over the class CA_α and some of its subclasses.

Throughout this section β will represent an arbitrary cardinal $\neq 0$. We recall that $\mathfrak{Fr}_\beta \mathsf{K}$ is generated by the set of all elements $\xi/Cr_\beta\mathsf{K}$ with $\xi < \beta$. (To avoid confusion notice that "0" in "$0/Cr_\beta\mathsf{K}$" and "1" in "$1/Cr_\beta\mathsf{K}$" always denote the ordinals zero and one, and not the elements of a CA.) The precise definition of the congruence relation $Cr_\beta\mathsf{K}$ and the generators $\xi/Cr_\beta\mathsf{K}$ (given in 0.4.19(ii)) is largely irrelevant for our purpose. Two consequences of this definition are, however, essential for what follows: (I) The set $\{\xi/Cr_\beta\mathsf{K} : \xi < \beta\}$, which is always of cardinality β unless K consists exclusively of one-element algebras, K-freely generates the algebra $\mathfrak{Fr}_\beta\mathsf{K}$ in the sense of 0.4.23 (cf. 0.4.27). (II) We have $\mathfrak{Fr}_\beta\mathsf{K} \in \mathbf{SP}\mathsf{K}$, and hence $\mathfrak{Fr}_\beta\mathsf{K} \in \mathsf{K}$ whenever $\mathsf{K} = \mathbf{SP}\mathsf{K}$ or, in particular, when $\mathsf{K} = \mathbf{HSP}\mathsf{K}$, i.e., when K is equational (cf. 0.4.22). In view of 2.1.1 and 2.4.1, (II) implies

THEOREM 2.5.1. $\mathfrak{Fr}_\beta\mathsf{CA}_\alpha \in \mathsf{CA}_\alpha$. *More generally, if* $0 \neq \mathsf{K} \subseteq \mathsf{CA}_\alpha$, *then* $\mathfrak{Fr}_\beta\mathsf{K} \in \mathsf{CA}_\alpha$.

REMARK 2.5.2. Some further properties of free CA's can also be derived directly or almost directly from general theorems of Section 0.4. For instance, by 0.4.40(ii) and 0.4.63, $\mathfrak{Fr}_\beta\mathsf{CA}_\alpha$ with $\beta \geqq \omega$ does not belong to any proper equational subclass of CA_α. Thus, in particular, for $\alpha \geqq 2$ we have that $\mathfrak{Fr}_\beta\mathsf{CA}_\alpha$ is neither discrete nor hereditarily non-discrete and does not have any definite characteristic (cf. 2.4.64 and the remarks following 2.4.38); this is easily seen to be true even if $\beta < \omega$.

The structures of free CA_α's are very involved, except for the lowest values of α, $\alpha = 0$, $\alpha = 1$, and perhaps $\alpha = 2$. Even simple problems concerning these algebras may present considerable difficulty, and many of them still remain open. We are going to discuss here a number of problems referring to the cardinality of free algebras, their atomic character, and the dimensionality of their elements.

THEOREM 2.5.3. (i) *If* $\alpha < 2$ *and* $\beta < \omega$, *then* $\mathfrak{Fr}_\beta \mathsf{CA}_\alpha$ *is finite.*

(ii) *If* $\alpha \geqq 2$ *or* $\beta \geqq \omega$, *then* $\mathfrak{Fr}_\beta \mathsf{CA}_\alpha$ *is infinite and, in fact,* $|Fr_\beta \mathsf{CA}_\alpha| = |\alpha| \cup \beta \cup \omega$.

PROOF. (i) is an immediate consequence of 2.1.10. $|Fr_\beta \mathsf{CA}_\alpha| \leqq |\alpha| \cup \beta \cup \omega$ for any α and β by 2.1.9(i). If $\alpha \geqq 2$, then $|Fr_\beta \mathsf{CA}_\alpha| \geqq |\alpha| \cup \beta \cup \omega$ by 0.4.33(ii), 0.4.39, 2.1.9(ii), and 2.1.11(i). If $\beta \geqq \omega$, then $|Fr_\beta \mathsf{CA}_\alpha| \geqq |\alpha| \cup \beta = |\alpha| \cup \beta \cup \omega$ by 2.1.9(ii).

A precise evaluation of the cardinality of $\mathfrak{Fr}_\beta \mathsf{CA}_\alpha$ in case $\alpha < 2$ and $\beta < \omega$ will be given in 2.5.62.

The discussion of the atomic character of free CA's will be based upon several auxiliary theorems.[1)]

LEMMA 2.5.4. *For every simple* $\mathfrak{A} \in \mathsf{CA}_2$ *and every finite* $X \subseteq A$ *there is a* $\mathfrak{B} \in \mathsf{CA}_2$ *satisfying the conditions*:

(i) $\mathfrak{B}$ *is finite and, in fact,* $|B| \leqq 2^{2^{9 \cdot |X|+1}}$;

(ii) $X \subseteq B$ *and* $\mathfrak{Bl}\,\mathfrak{B} \subseteq \mathfrak{Bl}\,\mathfrak{A}$;

(iii) $\mathsf{d}_{01}^{(\mathfrak{A})} = \mathsf{d}_{01}^{(\mathfrak{B})}$ *and, if* $x \in X$, *then* $\mathsf{c}_\kappa^{(\mathfrak{A})} x = \mathsf{c}_\kappa^{(\mathfrak{B})} x \in B$ *for* $\kappa = 0, 1$.

PROOF. In all cases throughout this proof κ and λ are intended to represent both 0 and 1. Let

$$(1) \qquad X' = X \cup \mathsf{c}_0^* X \cup \mathsf{c}_1^* X \cup \{\mathsf{d}_{01}\}, \ \mathfrak{B} = \mathfrak{Sg}^{(\mathfrak{Bl}\,\mathfrak{A})}(X' \cup (\mathsf{s}_1^0)^* X' \cup (\mathsf{s}_0^1)^* X').$$

Consider $c_\kappa \in {}^A A$ where

$$(2) \qquad c_\kappa x = \prod \{y : x \leqq y \in \mathsf{c}_\kappa^* A \cap B\}$$

for every $x \in A$. From (1) we immediately conclude that

$$(3) \qquad |B| \leqq 2^{2^{9 \cdot |X|+1}}$$

[1)] These auxiliary theorems, 2.5.4–2.5.6, are due to Henkin; they were originally obtained as lemmas for certain metalogical results which will be presented in Part II. Among proper theorems on the atomic structure of free CA's, 2.5.7 was established by Tarski using 2.5.6 and one of his earlier results, 2.3.29. Also 2.5.9 is due to him. The original proof of 2.5.9 was not deep but depended heavily on metamathematical intuition. The possibility of deriving 2.5.9 from 2.3.28 in a purely algebraic manner was indicated by Don Pigozzi, by whom 2.5.11 and 2.5.13 were found as well.

and

(4) $$X \subseteq B,\ \langle B, +, \cdot, -, 0, 1\rangle \subseteq \mathfrak{Bl}\,\mathfrak{A}.$$

Clearly, from (2) and 1.2.9 one sees that, if $\mathsf{c}_\kappa x \in \mathsf{c}^*_\kappa A \cap B$, i.e., $\mathsf{c}_\kappa x \in B$, then $c_\kappa x = \mathsf{c}_\kappa x$. Thus

(5) $$(\mathsf{c}_\kappa^{-1})^* B \upharpoonright \mathsf{c}_\kappa = (\mathsf{c}_\kappa^{-1})^* B \upharpoonright c_\kappa,$$

and hence, in view of (1),

(6) $$\mathsf{c}_\kappa x = c_\kappa x \in B \text{ for all } x \in X.$$

We now set

(7) $$\mathfrak{B} = \langle B, +, \cdot, -, 0, 1, B \upharpoonright c_\kappa, \mathsf{d}_{\kappa\lambda}\rangle_{\kappa,\lambda<2},$$

and proceed to prove that $\mathfrak{B} \in \mathsf{CA}_2$. This we shall do by showing axioms (C_0)–(C_7) of Definition 1.1.1 all hold in $\mathfrak{B}$.

(C_0) obviously holds in $\mathfrak{B}$, while (C_1) and (C_2) are immediate consequences of (2); another simple consequence of (2) is:

(8) $$\text{if } x, y \in A \text{ and } x \leqq y, \text{ then } c_\kappa x \leqq c_\kappa y.$$

We now consider (C_3). From (2) we obtain $c^*_\kappa A \subseteq \mathsf{c}^*_\kappa A \cap B$ and $(\mathsf{c}^*_\kappa A \cap B) \upharpoonright c_\kappa \subseteq Id$. Therefore, since it is clear that $\mathsf{c}^*_\kappa A \cap B \in Su\,\mathfrak{Bl}\,\mathfrak{A}$, we conclude that

(9) $$(Sg^{(\mathfrak{Bl}\,\mathfrak{A})} c^*_\kappa A) \upharpoonright c_\kappa \subseteq Id.$$

Consider any $x, y \in B$. We have $x \cdot c_\kappa y \leqq c_\kappa x \cdot c_\kappa y$ by (C_2), whence $c_\kappa(x \cdot c_\kappa y) \leqq c_\kappa x \cdot c_\kappa y$ by (8) and (9). On the other hand, $x \cdot c_\kappa y \leqq c_\kappa(x \cdot c_\kappa y)$ by (C_2), and thus $x \leqq c_\kappa(x \cdot c_\kappa y) + -c_\kappa y$. We now use (8) and (9) to get $c_\kappa x \leqq c_\kappa(x \cdot c_\kappa y) + -c_\kappa y$ and then $c_\kappa x \cdot c_\kappa y \leqq c_\kappa(x \cdot c_\kappa y)$. Therefore, (C_3) holds in $\mathfrak{B}$.

For any $x \in A$ we see by 1.2.9 that $c_\kappa x \geqq \mathsf{c}_\kappa x$; hence $c_\kappa \mathsf{d}_{\kappa\lambda} = 1$ by 1.3.2, and, using in addition the premiss that $\mathfrak{A}$ is simple, $c_0 c_1 x = 1 = c_1 c_0 x$ in case $x \neq 0$, while, by (2), $c_0 c_1 0 = 0 = c_1 c_0 0$. Thus (C_6) and (C_4) hold in $\mathfrak{B}$. Of course (C_5) holds. Finally we turn to (C_7). By (1), 1.5.8(i), and 1.5.10(v) it is easily seen that

$$(\mathsf{s}^\kappa_\lambda)^*[X' \cup (\mathsf{s}^0_1)^* X' \cup (\mathsf{s}^1_0)^* X'] \subseteq X' \cup (\mathsf{s}^0_1)^* X' \cup (\mathsf{s}^1_0)^* X'.$$

Hence, since $\mathsf{s}^\kappa_\lambda$ is a Boolean endomorphism of $\mathfrak{A}$ (cf. 1.5.3) we have $(\mathsf{s}^\kappa_\lambda)^* B \subseteq B$. Using this result and (5) we conclude that, if $\kappa \neq \lambda$, then $\mathsf{c}_\kappa(\mathsf{d}_{\kappa\lambda} \cdot x) = c_\kappa(\mathsf{d}_{\kappa\lambda} \cdot x)$ for every $x \in B$. Therefore,

$$c_\kappa(\mathsf{d}_{\kappa\lambda} \cdot x) \cdot c_\kappa(\mathsf{d}_{\kappa\lambda} \cdot -x) = \mathsf{c}_\kappa(\mathsf{d}_{\kappa\lambda} \cdot x) \cdot \mathsf{c}_\kappa(\mathsf{d}_{\kappa\lambda} \cdot -x) = 0$$

for all $x \in B$, and (C_7) holds in $\mathfrak{B}$. This completes the proof that $\mathfrak{B} \in \mathsf{CA}_2$, and now (3), (4), (6), and (7) are used to establish conditions (i)–(iii) of the conclusion of the lemma.

THEOREM 2.5.5. *If* $\mathfrak{A} = \mathfrak{Fr}_\beta \mathsf{CA}_2$ *and* $a \in A \sim \{0^{(\mathfrak{A})}\}$, *then there is an* $h \in Ho\,\mathfrak{A}$ *such that* $|h^*A| < \omega$ *and* $ha \neq 0^{(h^*\mathfrak{A})}$.

PROOF. Let $g_\xi = \xi / Cr_\beta \mathsf{CA}_\alpha$ for each $\xi < \beta$. By 0.1.17(ix) we may choose a finite $\Gamma \subseteq \beta$ such that $a \in Sg\{g_\xi : \xi \in \Gamma\}$. Let $A_0 = \{g_\xi : \xi \in \Gamma\} \cup \{\mathsf{d}_{01}\}$ and

$$A_{\mu+1} = A_\mu \cup \{-x : x \in A_\mu\} \cup \{x \cdot y : x, y \in A_\mu\} \cup \{\mathsf{c}_\kappa x : x \in A_\mu, \kappa < 2\}$$

for each $\mu < \omega$. Of course $|A_\mu| < \omega$ for every $\mu < \omega$. It is easily seen that $Sg\{g_\xi : \xi \in \Gamma\} = \bigcup_{\mu<\omega} A_\mu$; so there is a $\lambda < \omega$ such that

(1) $$a \in A_\lambda.$$

By 1.11.3(ii), 2.4.39, and 2.4.52 there exists a simple CA_2 $\mathfrak{C}$ and a function f such that

(2) $$f \in Ho(\mathfrak{A}, \mathfrak{C}) \text{ and } fa \neq 0^{(\mathfrak{C})}.$$

Taking $\mathfrak{C}$ and f^*A_λ respectively for $\mathfrak{A}$ and X in Lemma 2.5.4 we conclude that there is a $\mathfrak{B} \in \mathsf{CA}_2$ satisfying the following conditions:

(3) $$f^*A_\lambda \subseteq B \text{ and } B \text{ is finite;}$$

(4) $$\mathfrak{Bl}\,\mathfrak{B} \subseteq \mathfrak{Bl}\,\mathfrak{C};$$

(5) $\mathsf{d}_{01}^{(\mathfrak{B})} = \mathsf{d}_{01}^{(\mathfrak{C})}$ and, if $x \in f^*A_\lambda$, then $\mathsf{c}_\kappa^{(\mathfrak{B})} x = \mathsf{c}_\kappa^{(\mathfrak{C})} x$ for $\kappa = 0, 1$.

By (3) and 0.4.27(ii) there is a function h such that

(6) $$h \in Hom(\mathfrak{A}, \mathfrak{B})$$

and

(7) $$hg_\xi = fg_\xi \text{ for every } \xi \in \Gamma.$$

We will now prove by induction on μ that $A_\mu 1 h = A_\mu 1 f$ for all $\mu \leqq \lambda$. By (2), (5), (6), and (7) this is true for $\mu = 0$. Now assume $\mu < \lambda$ and $A_\mu 1 h = A_\mu 1 f$, and consider any $x \in A_{\mu+1}$. If $x \in A_\mu \cup \{-y : y \in A_\mu\} \cup \{y \cdot z : y, z \in A_\mu\}$, then $hx = fx$ follows by the induction hypothesis along with (2), (4), and (6). If $x = \mathsf{c}_\kappa y$ with $y \in A_\mu$ and $\kappa = 0$ or $\kappa = 1$, then by the induction hypothesis along with (2), (5), and (6) we have

$$hx = \mathsf{c}_\kappa^{(\mathfrak{B})}(hy) = \mathsf{c}_\kappa^{(\mathfrak{B})}(fy) = \mathsf{c}_\kappa^{(\mathfrak{C})}(fy) = fx.$$

This completes the inductive proof; thus, in particular, $A_\lambda 1 h = A_\lambda 1 f$. Consequently, in view of (1), (2), and (4) we have $ha \neq 0^{(\mathfrak{B})}$. This together with (3) and (6) shows that h satisfies the conclusion of the theorem.

THEOREM 2.5.6. *If* $\alpha \leqq 2$ *and* K *is the class of finite* CA_α's, *then* $\mathbf{HSPK} = \mathsf{CA}_\alpha$.

PROOF. For $\alpha = 0$ and 1 this follows immediately from 0.3.18, 2.1.10, and the remark just after 2.4.1. By 2.4.39 and 2.5.5 we have $\mathsf{CA}_2 \subseteq \mathbf{SPK}$ and hence $\mathsf{CA}_2 \subseteq \mathbf{HSPK}$. The inclusion in the opposite direction follows from the remark after 2.4.1.

THEOREM 2.5.7. (i) *If* $\alpha, \beta < \omega$ *and* K *is any non-empty class of finite* CA_α's, *then* $\mathfrak{Fr}_\beta \mathsf{K}$ *is an atomic* CA_α.

(ii) *If* $\alpha \leqq 2$ *and* $\beta < \omega$, *then* $\mathfrak{Fr}_\beta \mathsf{CA}_\alpha$ *is atomic.*

(iii) *In both (i) and (ii) the zero-dimensional parts of the free algebras involved are also atomic.*

PROOF. By 0.3.43, 0.4.22, and 2.1.1 we see that there is a system $\langle A_i : i \in I \rangle$ of CA_α's in $\mathbf{S}\mathsf{K}$ such that $\mathfrak{Fr}_\beta \mathsf{K} \cong | \subseteq_{\mathsf{d}} \mathsf{P}_{i \in I} \mathfrak{A}_i$. Hence (i) follows by 2.3.29(ii) and 2.4.39. (ii) is an immediate consequence of (i), 0.4.36, and 2.5.6. (iii) follows easily from (i) and (ii) using 1.10.3(i).

REMARK 2.5.8. It will be shown in Part II of this work that 2.5.7(ii) cannot be extended to any α such that $2 < \alpha < \omega$. The proof depends on certain metamathematical results, and it would be difficult to carry it through at the present stage of our discussion. On the other hand, since, by 2.5.3(ii), $\mathfrak{Fr}_\beta \mathsf{CA}_2$ is infinite, 2.5.7(ii) implies that $\mathfrak{Fr}_\beta \mathsf{CA}_2$ with $\beta < \omega$ contains infinitely many atoms. We shall now show that this result can be easily extended to all $\mathfrak{Fr}_\beta \mathsf{CA}_\alpha$'s with $2 \leqq \alpha < \omega$ and $\beta < \omega$.

THEOREM 2.5.9. *If* $2 \leqq \alpha < \omega$ *and* $\beta < \omega$, *then* $\mathfrak{Fr}_\beta \mathsf{CA}_\alpha$ *has infinitely many atoms.*

PROOF. Let $\mathfrak{A} = \mathfrak{Fr}_\beta \mathsf{CA}_\alpha$ for α, β such that $2 \leqq \alpha < \omega$ and $\beta < \omega$. It suffices to show that, for any $\kappa < \omega$, $\mathfrak{A}$ contains at least κ atoms. By 2.1.11(ii) we can find a finite CA_α $\mathfrak{B}$, having at least κ atoms, which is generated by a single element. Then by 0.4.39 and 2.3.28(ii) there is a $b \in Zd\mathfrak{A}$ such that $\mathfrak{B} \cong \mathfrak{Rl}_b \mathfrak{A}$. Since $At\mathfrak{Rl}_b\mathfrak{A} \subseteq At\mathfrak{A}$ for any $b \in Zd\mathfrak{A}$, we obtain the desired result.

REMARK 2.5.10. The results stated in 2.5.7–2.5.9 do not give us much insight into the intrinsic structure of the algebras $\mathfrak{Fr}_\beta \mathsf{CA}_\alpha$ with $2 < \alpha < \omega$ and $\beta < \omega$, or even into the structure of their Boolean parts. For instance, the following simple problem concerning these algebras is open: Does the sum of all atoms in any such algebra exist? In other words, can the Boolean part of $\mathfrak{Fr}_\beta\mathsf{CA}_\alpha$ be represented as the direct product of an atomic and an atomless algebra for some, or for all, α and β with $2 < \alpha < \omega$ and $\beta < \omega$? We know even less about the atoms of $\mathfrak{Fr}_\beta\mathsf{CA}_\alpha$ in case $\alpha \geqq \omega$ and $\beta < \omega$. All that is known on this subject is given in the following

THEOREM 2.5.11. *If $\beta < \omega$ and α is an arbitrary ordinal, then $\mathfrak{Fr}_\beta\mathsf{CA}_\alpha$ has exactly 2^β zero-dimensional atoms.*

PROOF. Let $\mathfrak{A} = \mathfrak{Fr}_\beta\mathsf{CA}_\alpha$ and for each $\xi < \beta$ let $g_\xi = \xi/Cr_\beta\mathsf{CA}_\alpha$. The theorem follows from the theory of BA's if $\alpha = 0$. We now consider the case when $\alpha = 1$. By 2.2.27 we have

$$At\,\mathfrak{A} \cap Zd\,\mathfrak{A} = \{\textstyle\prod_{\Gamma \neq \Theta \subseteq \beta} - \mathsf{c}_0 h_\Theta : \Gamma \subseteq \beta\} \sim \{0\}$$

where $h_\Theta = \prod_{\xi\in\Theta} g_\xi \cdot \prod_{\eta\in\beta\sim\Theta} - g_\eta$ for every $\Theta \subseteq \beta$. It is easy to convince oneself that $\prod_{\Gamma\neq\Theta\subseteq\beta} - \mathsf{c}_0 h_\Theta$ and $\prod_{\Delta\neq\Theta\subseteq\beta} - \mathsf{c}_0 h_\Theta$ represent distinct atoms provided that they are non-zero and that $\Gamma \neq \Delta$ (compare in this regard formula (2) of the proof of 2.2.27). Therefore, it remains only to show that

(1) $$\textstyle\prod_{\Gamma\neq\Theta\subseteq\beta} - \mathsf{c}_0 h_\Theta \neq 0 \text{ for every } \Gamma \subseteq \beta.$$

Given such a Γ there exists by 0.4.27(ii) and 0.4.33(ii) an $f \in Hom(\mathfrak{A}, \mathfrak{A})$ such that $fg_\xi = 1$ if $\xi \in \Gamma$, and $fg_\xi = 0$ if $\xi \in \beta \sim \Gamma$. Then $fh_\Theta = 0$ if $\Gamma \neq \Theta \subseteq \beta$, and hence $f(\prod_{\Gamma\neq\Theta\subseteq\beta} - \mathsf{c}_0 h_\Theta) = 1 \neq 0$ (since $\mathfrak{A}$ is clearly not a one-element algebra). This gives (1) and thus also the theorem in case $\alpha = 1$.

We now assume that $\alpha \geqq 2$. In view of 2.3.31 the only thing we have to show is that there are at least 2^β zero-dimensional atoms, and by 1.10.5(ii) this amounts to showing that

(2) $$\mathfrak{Rl}_{\mathsf{c}_0^\partial \mathsf{d}_{01}}\mathfrak{A} \text{ has at least } 2^\beta \text{ atoms.}$$

Let $\mathfrak{B} = \mathfrak{Rl}_{\mathsf{c}_0^\partial \mathsf{d}_{01}}\mathfrak{A}$. By 0.2.18(i), 2.2.14, and 2.3.26(i) we have

(3) $$\mathfrak{B} = \mathfrak{Sg}^{(\mathfrak{Bl}\mathfrak{B})}\{g_\xi \cdot \mathsf{c}_0^\partial \mathsf{d}_{01} : \xi < \beta\}.$$

For each $\Gamma \subseteq \beta$ let

$$x_\Gamma = \textstyle\prod_{\xi\in\Gamma}(g_\xi \cdot \mathsf{c}_0^\partial \mathsf{d}_{01}) \cdot \prod_{\xi\in\beta\sim\Gamma} -^{(\mathfrak{B})}(g_\xi \cdot \mathsf{c}_0^\partial \mathsf{d}_{01});$$

we see that $\{x_\Gamma : \Gamma \subseteq \beta\}$ is just the set of minimal constituents of $\{g_\xi \cdot \mathsf{c}_0^\partial \mathsf{d}_{01} : \xi < \beta\}$ (cf. the remarks preceding 1.9.6). Let $\mathfrak{C}$ be a two-element CA_α. Then for each $\Gamma \subseteq \beta$ there is by 0.4.27(ii) and 0.4.33(ii) an $f \in Hom(\mathfrak{A}, \mathfrak{C})$ such that $fg_\xi = 1^{(\mathfrak{C})}$ iff $\xi \in \Gamma$; notice that $f(\mathsf{c}_0^\partial \mathsf{d}_{01}) = 1^{(\mathfrak{C})}$ since $\mathfrak{C}$ is discrete. Hence, $fx_\Gamma = 1^{(\mathfrak{C})}$. This shows that $x_\Gamma \neq 0$ for every $\Gamma \subseteq \beta$, while, on the other hand, it is easily seen that x_Γ and x_Δ are distinct if $\Gamma, \Delta \subseteq \beta$, and $\Gamma \neq \Delta$. Therefore, there are exactly 2^β non-zero constituents of $\{g_\xi \cdot \mathsf{c}_0^\partial \mathsf{d}_{01} : \xi < \beta\}$; this fact when combined with (3) gives (2). The proof is complete.

REMARK 2.5.12. It seems plausible that, in case $\alpha \geqq \omega$ and $\beta < \omega$, all the atoms of the algebra $\mathfrak{A} = \mathfrak{Fr}_\beta \mathsf{CA}_\alpha$ are zero-dimensional and hence that $\mathfrak{A}$ has a direct decomposition into two factors,

$$\mathfrak{A} = \mathfrak{Rl}_{\mathsf{c}^\partial \mathsf{d}_{01}} \mathfrak{A} \times \mathfrak{Rl}_{\mathsf{c}_0 - \mathsf{d}_{01}} \mathfrak{A},$$

the first of which is finite with exactly 2^{2^β} elements, atomic, and discrete, while the second is infinite, atomless, and hereditarily non-discrete (cf. 2.4.37). This conjecture has not been confirmed.

The discussion of the atomic character of free CA_α's which are not finitely generated turns out to be much simpler:

THEOREM 2.5.13. *If $\beta \geqq \omega$, then $\mathfrak{Fr}_\beta \mathsf{CA}_\alpha$ is atomless.*

PROOF. Let $\mathfrak{A} = \mathfrak{Fr}_\beta \mathsf{CA}_\alpha$ and for each $\xi < \beta$ let $g_\xi = \xi / Cr_\beta \mathsf{CA}_\alpha$. Consider any non-zero $x \in A$. By 0.1.17(ix) there is a finite $\Gamma \subseteq \beta$ such that

(1) $$x \in Sg^{(\mathfrak{A})}\{g_\xi : \xi \in \Gamma\}.$$

Choose $\eta \in \beta \sim \Gamma$; then by 0.4.27(ii) and 0.4.33(ii) there are $f, h \in Hom(\mathfrak{A}, \mathfrak{A})$ such that

(2) $$fg_\xi = hg_\xi = g_\xi \text{ for all } \xi \in \Gamma$$

while

(3) $$fg_\eta = 1 \text{ and } hg_\eta = 0.$$

From (1), (2), and 0.2.14(iii) we have $fx = gx = x$. Therefore, by (3) it may be concluded that $f(x \cdot g_\eta) = h(x \cdot -g_\eta) = x \neq 0$, so $x \cdot g_\eta \neq 0$ and $x \cdot -g_\eta \neq 0$. Thus x is not an atom, and the theorem is proved.

REMARK 2.5.14. The last theorem implies that, in case $\alpha \leqq \omega$ and $\beta = \omega$, the Boolean part of $\mathfrak{Fr}_\beta \mathsf{CA}_\alpha$ is denumerable and atomless. Hence, as is well known from the theory of BA's, it is a free Boolean algebra with ω generators.

In the following theorem we shall determine the dimension set of generators of free CA_α's. An important supplement to this result will be Theorem 2.6.23 in the next section, in which the dimensionality of arbitrary elements of free CA's will be discussed.

THEOREM 2.5.15. *If $\mathsf{K} = \mathsf{CA}_\alpha$ or, more generally, K is any subclass of CA_α which contains some non-discrete algebra, then in the algebra $\mathfrak{Fr}_\beta \mathsf{K}$ we have $\Delta(\xi / Cr_\beta \mathsf{K}) = \alpha$ for every $\xi < \beta$.*

PROOF: by 0.4.27(ii), 1.3.12, 1.6.4, and 2.3.2.

COROLLARY 2.5.16. *If* K *is as in 2.5.15, then* $\mathfrak{Fr}_\beta\mathsf{K} \notin \mathsf{Dc}_\alpha$.

In the next few theorems we deal with free algebras over various special classes of CA's.

THEOREM 2.5.17. $\mathfrak{Fr}_\beta\mathsf{Lf}_\alpha$, $\mathfrak{Fr}_\beta\mathsf{Ss}_\alpha$, $\mathfrak{Fr}_\beta\mathsf{Mn}_\alpha \in \mathsf{Ss}_\alpha$.

PROOF: by 0.4.22, 2.1.16, 2.4.52, and 2.4.53.

REMARK 2.5.18. In addition to the free algebras listed in 2.5.17, the algebra $\mathfrak{Fr}_\beta\mathsf{Dc}_\alpha$ is also semisimple. For $\alpha < \omega$ this follows from 1.11.3(iii); for $\alpha \geqq \omega$ this is a direct consequence of Theorem 2.6.52 in the next section, by which $\mathfrak{Fr}_\beta\mathsf{Lf}_\alpha = \mathfrak{Fr}_\beta\mathsf{Dc}_\alpha = \mathfrak{Fr}_\beta\mathsf{Ss}_\alpha$. On the other hand, it will be also seen in the next section that $\mathfrak{Fr}_\beta\mathsf{CA}_\alpha$ is not semisimple (cf. 2.6.53).

Besides CA_α and Ss_α various other subclasses K of CA_α are known for which $\mathfrak{Fr}_\beta\mathsf{K} \in \mathsf{K}$. This applies, for instance, to all the equational classes of CA_α's mentioned in 2.5.2.

From 2.5.15 it follows that, for $\alpha \geqq \omega$, neither $\mathfrak{Fr}_\beta\mathsf{Dc}_\alpha$ nor any of the free algebras listed in 2.5.17 is a Dc_α or *a fortiori* an Lf_α; thus, in particular, $\mathfrak{Fr}_\beta\mathsf{Lf}_\alpha \notin \mathsf{Lf}_\alpha$. Analogously, we have

THEOREM 2.5.19. $\mathfrak{Fr}_\beta\mathsf{Mn}_\alpha \notin \mathsf{Mn}_\alpha$.

PROOF. Let $a = 0/Cr_\beta\mathsf{Mn}_\alpha$ (recall that in this expression "0" denotes the ordinal zero), and let h be an endomorphism of $\mathfrak{Fr}_\beta\mathsf{Mn}_\alpha$ such that $ha = 0$. Each member of the minimal subalgebra of $\mathfrak{Fr}_\beta\mathsf{Mn}_\alpha$ is left fixed by h, but $ha \neq a$ since clearly $a \neq 0$. Hence $\mathfrak{Fr}_\beta\mathsf{Mn}_\alpha \notin \mathsf{Mn}_\alpha$.

From a result in the general theory of algebraic structures, namely 0.4.54, we know that under a certain weak assumption on the class K the algebras $\mathfrak{Fr}_\beta\mathsf{K}$ have a number of interesting properties listed as conditions (i)–(v) in 0.4.54; the most interesting of them is perhaps condition (i) to the effect that $\mathfrak{Fr}_\beta\mathsf{K}$ is not generated by any set of cardinality $< \beta$. The assumption in question is that K should contain at least one finite algebra with more than one element; even this assumption is superfluous in case $\beta \geqq \omega$. In any case the condition is obviously satisfied by CA_α and various subclasses of CA_α, and hence we obtain

THEOREM 2.5.20. *Let* K *be one of the following classes*: CA_α, Lf_α, Ss_α, *or* Mn_α, *all with* α *arbitrary, or* Dc_α *with* $\alpha > 0$. *We then have*:

(i) *if* $Fr_\beta\mathsf{K} = SgX$, *then* $|X| \geqq \beta$;

(ii) *if* X K*-freely generates* $\mathfrak{Fr}_\beta\mathsf{K}$, *then* $|X| = \beta$.

Furthermore, for any cardinal $\gamma \neq 0$ *we have*:

(iii) *if* $\mathfrak{Fr}_\beta\mathsf{K} \succcurlyeq \mathfrak{Fr}_\gamma\mathsf{K}$, *then* $\beta \geqq \gamma$;

(iv) *if* $\mathfrak{Fr}_\beta\mathsf{K} \cong \mathfrak{Fr}_\gamma\mathsf{K}$, *then* $\beta = \gamma$.

PROOF. It is easily seen that each of the classes K listed in the theorem contains, for every value of α, a two-element algebra. Hence the conclusion follows by 0.4.54. (It may be noted that for $\mathsf{K} = \mathsf{CA}_\alpha$ and $\beta < \omega$ our theorem is an immediate consequence of 2.3.31 and 2.5.11.)

REMARK 2.5.21. Since Dc_0 is empty, it follows from 0.4.19(ii) that $\mathfrak{Fr}_\beta\mathsf{Dc}_0$ is always a one-element CA_α, and therefore the class Dc_0 satisfies none of the conditions 0.4.54(i)–(v).

The problem of extending 2.5.20 to other classes of CA's may present considerable difficulty. Consider, for instance, the class of all hereditarily non-discrete CA_α's and denote it by H_α. If $\alpha < \omega$, the class H_α is easily seen to contain a finite algebra with more than one element, and hence to satisfy the conclusions of 0.4.54. On the other hand, H_α with $\alpha \geqq \omega$ does not contain any non-trivial finite algebra. Nevertheless, we shall be able to show in Part II that $\mathfrak{Fr}_\beta\mathsf{H}_\alpha$ satisfies the conclusions of 0.4.54 in this case as well; the proof will be based upon 0.5.17 and will use some methods which are not available at this stage of the discussion. The same remarks apply to all classes $\mathsf{K} \cap \mathsf{H}_\alpha$ where K is any one of the classes listed in 2.5.20, with the exception of Dc_α for $\alpha < \omega$.

REMARK 2.5.22. Another property of free algebras which has been discussed in the general theory of algebras and is closely related to those stated in 0.4.54 can be formulated as follows: for $\beta < \omega$, if $\mathfrak{Fr}_\beta\mathsf{K}$ is generated by a set $X \subseteq \mathfrak{Fr}_\beta\mathsf{K}$ of cardinality $|X| = \beta$, then $\mathfrak{Fr}_\beta\mathsf{K}$ is K-freely generated by X. This property is known to apply to many familiar classes of algebras. In 0.4.55 it has been established under a certain assumption on K which is essentially stronger than the one occurring in 0.4.54: it is the assumption to the effect that every identity which holds in all finite algebras of K holds also in all algebras of K. In application to the classes CA_α we have succeeded in establishing the property discussed only for the lowest values of α. In fact, we have,

THEOREM 2.5.23. *If* $\alpha \leqq 2$, $|X| = \beta < \omega$, *and* $\mathfrak{Fr}_\beta\mathsf{CA}_\alpha = \mathfrak{Sg}\,X$, *then* $\mathfrak{Fr}_\beta\mathsf{CA}_\alpha$ *is* CA_α*-freely generated by* X.

PROOF: by 0.4.55 and 2.5.6.

The problem whether the conclusion of 2.5.23 holds for any class CA_α with $\alpha \geqq 3$ is still open. It may be noticed that, even if the solution of this problem is affirmative, it cannot be obtained by a direct application of 0.4.55. In fact, it easily follows from 1.3.15 that for $\alpha \geqq \omega$ there are identities which hold in all finite CA_α's but not in all CA_α's. In Part II we shall see that this is also true in case $3 \leqq \alpha < \omega$.

Properties of free algebras can sometimes be fruitfully applied to the discussion of problems in which the notion of free algebras is not directly involved. Rather interesting examples of this kind will be found in the next two theorems, the first of which is due to Monk.

THEOREM 2.5.24. *For every* $\alpha \geqq \omega$ *there is a simple* CA_α *which is not a* Dc_α.

PROOF. Let $\mathfrak{A} = \mathfrak{Fr}_1\mathsf{CA}_\alpha$ and let

$$I = Ig^{(\mathfrak{A})}(\{-\mathsf{c}_\kappa a : \kappa \in \alpha \sim 1\} \cup \{\mathsf{c}_0^\partial a\})$$

where $a = 0/Cr_1\mathsf{CA}_\alpha$. We claim that I is proper. For, if $1 \in I$, then by 2.3.8 there exists a finite subset Γ of α, a $\lambda < \omega$, and a $\kappa \in {}^\lambda(\alpha \sim 1)$ such that

$$(1) \qquad 1 \leqq \mathsf{c}_{(\Gamma)}(\mathsf{c}_0^\partial a + -\mathsf{c}_{\kappa_0} a + \ldots + -\mathsf{c}_{\kappa_{\lambda-1}} a).$$

Let $\mathfrak{C}$ be the cylindric set algebra of all subsets of ${}^\alpha\omega$, and let

$$Y = \{\varphi : \varphi \in {}^\alpha\omega,\ \varphi_0 \leqq \varphi_{\kappa_\mu} \text{ for some } \mu < \lambda\}.$$

By 0.4.27(ii) there exists a homomorphism h from $\mathfrak{A}$ into $\mathfrak{C}$ such that $ha = Y$. Notice that $\mathbf{C}_{\kappa_\mu} Y = {}^\alpha\omega$ for all $\mu < \lambda$, and also $\mathbf{C}_0^\partial Y = 0$ (see 1.1.5 for notation). Hence, by (1) we have

$$\begin{aligned} {}^\alpha\omega = h1 &\leqq h[\mathsf{c}_{(\Gamma)}(\mathsf{c}_0^\partial a + -\mathsf{c}_{\kappa_0} a + \ldots + -\mathsf{c}_{\kappa_{\lambda-1}} a)] \\ &= \mathbf{C}_{(\Gamma)}(\mathbf{C}_0^\partial Y \cup ({}^\alpha\omega \sim \mathbf{C}_{\kappa_0} Y) \cup \ldots \cup ({}^\alpha\omega \sim \mathbf{C}_{\kappa_{\lambda-1}} Y)) \\ &= 0, \end{aligned}$$

i.e., ${}^\alpha\omega = 0$, which is false.

Thus I is proper. Let J be a maximal proper ideal of $\mathfrak{A}$ such that $I \subseteq J$; $\mathfrak{B} = \mathfrak{A}/J$ is then simple. Now $a/J \neq 1^{(\mathfrak{B})}$, for otherwise we would have $-a/J = 0^{(\mathfrak{B})}$, $-a \in J$, and hence $-\mathsf{c}_0^{\partial(\mathfrak{A})} a = \mathsf{c}_0^{(\mathfrak{A})} -a \in J$; however, this is impossible since J is proper and $\mathsf{c}_0^{\partial(\mathfrak{A})} a \in J$. On the other hand, $\mathsf{c}_\kappa^{(\mathfrak{B})}(a/J) = 1^{(\mathfrak{B})}$ for all $\kappa \in \alpha \sim 1$. Therefore, $\Delta^{(\mathfrak{B})}(a/J) \supseteq \alpha \sim 1$, and hence $\mathfrak{B} \notin \mathsf{Dc}_\alpha$ by 1.11.4.

THEOREM 2.5.25. *Given any* CA_α $\mathfrak{A}$, *let* $a_\kappa^{(\mathfrak{A})} = \mathsf{c}_{(\kappa)}\bar{\mathsf{d}}(\kappa \times \kappa)$ *for every* $\kappa < (\alpha+1) \cap \omega$, *and let* $\nabla\mathfrak{A} = \{\kappa : 1 \leqq \kappa < \alpha \cap \omega,\ a_\kappa^{(\mathfrak{A})} - a_{\kappa+1}^{(\mathfrak{A})} \neq 0^{(\mathfrak{A})}\}$. *In order that two* Mn_α*'s* $\mathfrak{A}$ *and* $\mathfrak{A}'$ *be isomorphic it is necessary and sufficient that they satisfy the following two conditions*:

(i) $\nabla\mathfrak{A} = \nabla\mathfrak{A}'$

(ii) $a_\kappa^{(\mathfrak{A})} = 0^{(\mathfrak{A})}$ *for some* $\kappa < (\alpha+1) \cap \omega$ *iff* $a_{\kappa'}^{(\mathfrak{A}')} = 0^{(\mathfrak{A}')}$ *for some* $\kappa' < (\alpha+1) \cap \omega$.

In case the set $\nabla\mathfrak{A}$ (or $\nabla\mathfrak{A}'$) *is infinite, we have* $a_\kappa^{(\mathfrak{A})} \neq 0^{(\mathfrak{A})}$ (or $a_\kappa^{(\mathfrak{A}')} \neq 0^{(\mathfrak{A}')}$) *for every* $\kappa < (\alpha+1) \cap \omega$, *and hence condition* (*i*) *alone is necessary and sufficient for* $\mathfrak{A} \cong \mathfrak{A}'$.

PROOF. For brevity we denote $a_\kappa^{(\mathfrak{Fr}_1\mathsf{CA}_\alpha)}$ simply by a_κ, for every $\kappa < (\alpha+1)\cap\omega$. Let $\mathfrak{C}$ be the minimal subalgebra of $\mathfrak{Fr}_1\mathsf{CA}_\alpha$ and let $B = \{a_\kappa : \kappa < (\alpha+1)\cap\omega\}$. Then by 2.1.17(iii) we have that $\mathfrak{Zd}\,\mathfrak{C}$ is generated by B. Because $a_\kappa \geqq a_{\kappa+1}$ for every $\kappa < \alpha\cap\omega$ (cf. 1.9.10) and $a_0 = a_1 = 1$, we readily see that every constituent of B can be represented either as an element of B or as the difference of two elements of B. Therefore,

(1) every non-zero element of $\mathfrak{Zd}\,\mathfrak{C}$ can be expressed as a finite sum of elements of $\{a_\kappa - a_{\kappa+1} : 1 \leqq \kappa < \alpha\cap\omega\} \cup B$.

Let $\mathfrak{A}$ and $\mathfrak{A}'$ be Mn_α's. Conditions (i) and (ii) are clearly necessary for $\mathfrak{A}$ and $\mathfrak{A}'$ to be isomorphic. In order to show they are also sufficient we assume that $\mathfrak{A}$ and $\mathfrak{A}'$ satisfy (i) and (ii). In view of the fact that

(2) $$a_\kappa^{(\mathfrak{A})} \geqq a_{\kappa+1}^{(\mathfrak{A})} \text{ and } a_\kappa^{(\mathfrak{A}')} \geqq a_{\kappa+1}^{(\mathfrak{A}')} \text{ for every } \kappa < \alpha\cap\omega,$$

it follows from (i) and (ii) that,

(3) $$\text{for every } \kappa < (\alpha+1)\cap\omega,\ a_\kappa^{(\mathfrak{A})} = 0^{(\mathfrak{A})} \text{ iff } a_\kappa^{(\mathfrak{A}')} = 0^{(\mathfrak{A}')}.$$

By 0.4.39 there exist $h \in Ho(\mathfrak{Fr}_1\mathsf{CA}_\alpha, \mathfrak{A})$ and $h' \in Ho(\mathfrak{Fr}_1\mathsf{CA}_\alpha, \mathfrak{A}')$. Hence $h \in Ho(\mathfrak{C}, \mathfrak{A})$ and $h' \in Ho(\mathfrak{C}, \mathfrak{A}')$ ($C1h$ and $C1h'$ are both uniquely determined; in fact, it is easily seen that for any Mn_α $\mathfrak{D}$ we have $|Ho(\mathfrak{D}, \mathfrak{E})| \leqq 1$ for every $\mathfrak{E} \in \mathsf{CA}_\alpha$). Also $ha_\kappa = a_\kappa^{(\mathfrak{A})}$ and $h'a_\kappa = a_\kappa^{(\mathfrak{A}')}$, for every $\kappa < (\alpha+1)\cap\omega$. Therefore from condition (i) and (3) we see that, for each $x \in \{a_\kappa - a_{\kappa+1} : \kappa < \alpha\cap\omega\} \cup B$, $hx = 0^{(\mathfrak{A})}$ iff $h'x = 0^{(\mathfrak{A}')}$; from this fact and (1) we infer:

$$(h^{-1})^\star 0^{(\mathfrak{A})} \cap Zd\mathfrak{C} = (h'^{-1})^\star 0^{(\mathfrak{A}')} \cap Zd\mathfrak{C}.$$

Hence, by 2.1.16 and 2.3.12(iii),

$$(h^{-1})^\star 0^{(\mathfrak{A})} \cap C = (h'^{-1})^\star 0^{(\mathfrak{A}')} \cap C.$$

$\mathfrak{A} \cong \mathfrak{A}'$ now follows by 2.3.6(II), so that conditions (i) and (ii) are sufficient as well as necessary.

The remaining parts of the theorem are immediate consequences of (2).

THEOREM 2.5.26. *Let $a_\kappa^{(\mathfrak{A})}$ and $\nabla\mathfrak{A}$ be defined as in 2.5.25. Let $\Gamma \subseteq (\alpha\cap\omega) \sim \{0\}$. Then there is an Mn_α $\mathfrak{A}$ such that $\nabla\mathfrak{A} = \Gamma$. If, moreover, the set Γ is finite, $\mathfrak{A}$ can be chosen to satisfy the condition $a_\kappa^{(\mathfrak{A})} = 0^{(\mathfrak{A})}$ for any $\kappa < (\alpha+1)\cap\omega$ such that $\kappa > \lambda$ for all $\lambda \in \Gamma$; it can also be chosen to satisfy the condition $a_\kappa^{(\mathfrak{A})} \neq 0^{(\mathfrak{A})}$ for every $\kappa < (\alpha+1)\cap\omega$.*

PROOF. For every $\mu < \alpha\cap\omega$ choose $\mathfrak{B}_\mu$ to be a minimal algebra of characteristic μ; by the remark following 2.4.62 we see that such a $\mathfrak{B}_\mu$ can always be found. Therefore from 2.4.63 we have:

(1) $a_\kappa^{(\mathfrak{B}_0)} - a_{\kappa+1}^{(\mathfrak{B}_0)} = 0^{(\mathfrak{B}_0)}$ for every $\kappa < \alpha \cap \omega$, and
$a_\kappa^{(\mathfrak{B}_0)} \neq 0^{(\mathfrak{B}_0)}$ for every $\kappa < (\alpha+1) \cap \omega$;

(2) if $1 \leqq \mu < \alpha \cap \omega$, then $a_\kappa^{(\mathfrak{B}_\mu)} - a_{\kappa+1}^{(\mathfrak{B}_\mu)} \neq 0^{(\mathfrak{B}_\mu)}$ iff $\kappa = \mu$, and
$a_\lambda^{(\mathfrak{B}_\mu)} \neq 0^{(\mathfrak{B}_\mu)}$ iff $\lambda \leqq \mu$.

Let $\mathfrak{C}$ be the minimal subalgebra of $\mathsf{P}_{\mu \in \Gamma} \mathfrak{B}_\mu$ and recall that $a_\kappa^{(\mathfrak{C})} = \langle a_\kappa^{(\mathfrak{B}_\mu)} : \mu \in \Gamma \rangle$ for every $\kappa < (\alpha+1) \cap \omega$.

Taking $\mathfrak{A} = \mathfrak{C}$ we have by (2) that $\nabla \mathfrak{A} = \Gamma$; in addition, if Γ is finite, then we see from (2) again that $a_\kappa^{(\mathfrak{A})} = 0^{(\mathfrak{A})}$ for some $\kappa < (\alpha+1) \cap \omega$. On the other hand, taking $\mathfrak{A} = \mathfrak{C} \times \mathfrak{B}_0$ we have by (1) that $\nabla \mathfrak{A} = \nabla \mathfrak{C}$, and hence $\nabla \mathfrak{A} = \Gamma$ in this case also; however, again using (1) we see that now $a_\kappa^{(\mathfrak{A})} \neq 0^{(\mathfrak{A})}$ for every $\kappa < (\alpha+1) \cap \omega$ independent of the cardinality of Γ. This completes the proof.

Remark 2.5.27. Theorems 2.5.25 and 2.5.26 provide an exhaustive description of the isomorphism types of Mn_α's. We see from 2.5.25 that with each isomorphism type τ of Mn_α's a uniquely determined set $\Xi\tau$ of positive integers $< \alpha$ is correlated; in fact, $\Xi\tau = \nabla \mathfrak{A}$ (in the sense of 2.5.25) for every Mn_α $\mathfrak{A}$ such that the isomorphism type of $\mathfrak{A}$ is τ. In case $\Xi\tau$ is infinite, the correlation is one-one. If, however, $\Xi\tau$ is finite, the correlation is two-one, i.e., for every finite set Γ of positive integers there are just two isomorphism types τ with $\Xi\tau = \Gamma$. One of these types corresponds to those Mn_α's $\mathfrak{A}$ (with $\nabla \mathfrak{A} = \Gamma$) in which $a_\kappa^{(\mathfrak{A})} = 0^{(\mathfrak{A})}$ for some $\kappa < (\alpha+1) \cap \omega$, and the other to all the remaining Mn_α's $\mathfrak{A}$ (with the same $\nabla \mathfrak{A}$). By 2.5.26, for any given α, the range of Ξ consists of all sets of positive integers $< \alpha$.

From these remarks it is seen that the variety of the isomorphism types of Mn_α's is rather restricted. In particular, we obviously have

Corollary 2.5.28. *If $\alpha \neq 0$, the class of all isomorphism types of Mn_α's has the cardinality $2^{\alpha \cap \omega}$. In case $\alpha = 0$ this class has the cardinality* 2.

Remark 2.5.29. In the light of Remark 2.5.27 the set $\Xi\tau$ can be referred to as the *characteristic set* of the isomorphism type τ of Mn_α's. This terminology can be extended in a natural way from isomorphism types to Mn_α's themselves and even to arbitrary CA_α's; by the characteristic set of a given CA_α $\mathfrak{A}$ we understand of course the characteristic set of its unique minimal subalgebra $\mathfrak{B}$. In metalogical interpretation, the characteristic set of a CA_α of formulas associated with a theory Θ consists of all those $\lambda < \alpha \cap \omega$ which are cardinalities of models of Θ. Thus, in case $\alpha \geqq \omega$, the characteristic set coincides with what is called the spectrum of Θ; cf. Scholz [52*].

It is easy to realize that a positive integer λ is the characteristic of a CA_α $\mathfrak{A}$ in the sense of 2.4.61 iff the singleton $\{\lambda\}$ is the characteristic set of $\mathfrak{A}$ in the sense just mentioned, and, moreover, $a_\kappa^{(\mathfrak{A})} = 0^{(\mathfrak{A})}$ for some $\kappa < (\alpha+1)\cap\omega$; similarly, 0 is the characteristic of $\mathfrak{A}$ iff the characteristic set of $\mathfrak{A}$ is empty, and $a_\kappa^{(\mathfrak{A})} \neq 0^{(\mathfrak{A})}$ for every $\kappa < (\alpha+1)\cap\omega$. Hence as an immediate consequence of 2.5.25 we obtain

COROLLARY 2.5.30. *Any two* Mn_α*'s of the same characteristic are isomorphic.*

In Remarks 0.4.65 we discussed concisely the notion of a free algebra with defining relations, i.e., with certain conditions imposed on generators. A special case of this notion proves to be of significance for the theory of cylindric algebras and will be discussed here with more details.

As was shown in 2.5.15, all the generators of a free CA_α have the largest possible dimension set, i.e., α. We wish to construct algebras which differ from free CA_α's in one respect only: the dimension sets of the generators may be arbitrary subsets of α. In other words, we are interested in free algebras over CA_α, or any subclass K of CA_α, with a set of conditions of the form $\mathsf{c}_\kappa g_\xi = g_\xi$ imposed on each generator g_ξ; the sets of conditions for different generators do not have to be identical. As was indicated in 0.4.65, to define these algebras we can use a certain variant of our original definition 0.4.19(ii) of free algebras. In this variant ordered pairs of words (elements of the absolutely free algebra) serve as mathematical substitutes for equations which express the conditions imposed on generators. In the present case, however, the conditions are so simple that we can use a less sophisticated device: each of the algebras we wish to construct is determined by a function Δ which correlates with each ordinal $\xi < \beta$ a set $\Delta\xi \subseteq \alpha$, namely, the set of all those ordinals $\lambda < \alpha$ which are not excluded as possible elements of the dimension set of the generator g_ξ. Thus the conditions imposed on g_ξ are expressed by all the equations $\mathsf{c}_\kappa g_\xi = g_\xi$ where κ is any ordinal in $\alpha \sim \Delta\xi$.[1] As an immediate consequence of these conditions, the dimension set Δg_ξ of g_ξ is a subset of $\Delta\xi$; actually it will turn out that the dimension set equals $\Delta\xi$. The definition can as easily be applied to arbitrary classes of algebras similar to CA_α's; it assumes the following form:

DEFINITION 2.5.31. *Given any class* K *of algebras similar to* CA_α*'s and any function* $\Delta \in {}^\beta Sb\alpha$*, we let*

$$Cr_\beta^{(\Delta)}\mathsf{K} = \bigcap\{R : R \in Co\mathfrak{Fr}_\beta,\ \mathfrak{Fr}_\beta/R \in \mathbf{IS}\mathsf{K},\ (\mathsf{c}_\kappa^{(\mathfrak{Fr}_\beta)}\xi)/R = \xi/R\}$$

for each $\xi < \beta$ *and each* $\kappa \in \alpha \sim \Delta\xi$*;*

[1]) Henkin-Tarski [61], p. 94, called attention to these algebras for the first time.

we also let

$$Fr_\beta^{(\Delta)}\mathsf{K} = Fr_\beta / Cr_\beta^{(\Delta)}\mathsf{K} \text{ and } \mathfrak{Fr}_\beta^{(\Delta)}\mathsf{K} = \mathfrak{Fr}_\beta / Cr_\beta^{(\Delta)}\mathsf{K}.$$

The algebra $\mathfrak{Fr}_\beta^{(\Delta)}\mathsf{K}$ is called a ***dimension-restricted free algebra over*** K ***with*** β ***generators;*** *more specifically, we say that $\mathfrak{Fr}_\beta^{(\Delta)}\mathsf{K}$ is* ***dimension-restricted by the function*** Δ, *or, simply, that it is* ***Δ-dimension-restricted.***

In what follows we concentrate on the class $\mathsf{K} = \mathsf{CA}_\alpha$.

COROLLARY 2.5.32. *If $\Delta = \beta \times \{\alpha\}$, then $\mathfrak{Fr}_\beta^{(\Delta)}\mathsf{CA}_\alpha = \mathfrak{Fr}_\beta \mathsf{CA}_\alpha$.*
PROOF: by 0.4.19(ii) and 2.5.31.

THEOREM 2.5.33. *For every function $\Delta \in {}^\beta Sb\alpha$ we have:*
(i) $\mathfrak{Fr}_\beta^{(\Delta)}\mathsf{CA}_\alpha \in \mathsf{CA}_\alpha$;
(ii) $\Delta(\xi / Cr_\beta^{(\Delta)}\mathsf{CA}_\alpha) = \Delta\xi$ *for every* $\xi < \beta$.
PROOF. Let $\mathfrak{A} = \mathfrak{Fr}_\beta^{(\Delta)}\mathsf{CA}_\alpha$, and $g_\xi = \xi / Cr_\beta^{(\Delta)}\mathsf{CA}_\alpha$ for every $\xi < \beta$. Using 0.3.46 (and the immediately following remark) together with 2.4.2, we can infer from 2.5.31 the existence of a system $\mathfrak{B} \in {}^I\mathsf{CA}_\alpha$ and a function $f \in Is\mathfrak{A}$ such that

(1) $$f^*\mathfrak{A} \subseteq_{\mathsf{d}} \mathsf{P}_{i \in I}\mathfrak{B}_i,$$

and

(2) $$\Delta(fg_\xi) \subseteq \Delta\xi \text{ for each } \xi < \beta.$$

Furthermore, in light of 0.2.23(i) and 0.4.27(i) the system $\mathfrak{B}$ can be chosen so that,

(3) for every CA_α $\mathfrak{C}$ and $y = \langle y_\xi : \xi < \beta \rangle \in {}^\beta C$ with $\Delta^{(\mathfrak{C})}y_\xi \subseteq \Delta\xi$, there is an $i \in I$ and $h \in Ism(\mathfrak{B}_i, \mathfrak{C})$ such that $h([fg_\xi]_i) = y_\xi$ for every $\xi < \beta$.

(i) follows immediately from (1). In the full cylindric set algebra in the space ${}^\alpha 2$ we can find a system of sets $\langle X_\xi : \xi < \beta \rangle$ such that $\Delta X_\xi = \Delta\xi$ for every $\xi < \beta$. Hence, from (3) we see that there is an $i \in I$ for which

(4) $$\Delta[fg_\xi]_i = \Delta\xi \text{ for every } \xi < \beta.$$

(2) and (4) lead directly to (ii).

In agreement with Remarks 0.4.65 many results concerning ordinary free CA's can be extended with appropriate changes to dimension-restricted free CA_α's. This applies to most of the results of general algebraic character established in Section 0.4 and to some specific results stated in this section; e.g., Theorem 2.5.33(i),(ii) is an extension of 2.5.1 and partly of 2.5.15. In particular, as was indicated in 0.4.65, we can extend the notion of a K-freely generating set and formulate with its help a generalization of 0.4.29, i.e., the fundamental theorem on free algebras. This will be done in the next definition and in Theorem 2.5.35 below.

DEFINITION 2.5.34. *Assume that* $\mathfrak{A} \in \mathsf{CA}_\alpha$, $x = \langle x_\xi : \xi < \beta \rangle \in {}^\beta A$, *and* $\Delta \in {}^\beta Sb\alpha$. *We say that the sequence* x CA_α***-freely generates*** $\mathfrak{A}$ ***under the dimension-restricting function*** Δ *(or, simply,* x ***freely generates*** $\mathfrak{A}$ ***under*** Δ*) if the following two conditions are satisfied:*

(i) $\mathfrak{A} = \mathfrak{Sg}^{(\mathfrak{A})} Rg x$, *and* $\Delta^{(\mathfrak{A})} x_\xi \subseteq \Delta\xi$ *for every* $\xi < \beta$;

(ii) *whenever* $\mathfrak{B} \in \mathsf{CA}_\alpha$, $y = \langle y_\xi : \xi < \beta \rangle \in {}^\beta B$, *and* $\Delta^{(\mathfrak{B})} y_\xi \subseteq \Delta\xi$ *for every* $\xi < \beta$, *there is a (unique)* $h \in Hom(\mathfrak{A}, \mathfrak{B})$ *such that* $h \circ x = y$.

In order to understand why it is necessary in this definition to deal with the notion of a sequence x, rather than a set X, freely generating $\mathfrak{A}$ under Δ, the reader is referred to the discussion in 0.4.65.

By extending 2.5.34 in an obvious way we could define the general notion of a sequence x K-freely generating an algebra $\mathfrak{A}$ under a dimension-restricting function Δ (where $\mathfrak{A}$ is any algebra similar to CA_α's and K is any class of such algebras).

COROLLARY 2.5.35. *Under the assumptions of 2.5.34, if* x CA_α*-freely generates* $\mathfrak{A}$ *under* Δ, *then* $\Delta^{(\mathfrak{A})} x_\xi = \Delta\xi$ *for every* $\xi < \beta$.

Compare the proof of 2.5.15.

THEOREM 2.5.36. *If* $\mathfrak{A} = \mathfrak{Fr}_\beta^{(\Delta)} \mathsf{CA}_\alpha$, *and* $x = \langle \xi / Cr_\beta^{(\Delta)} \mathsf{CA}_\alpha : \xi < \beta \rangle$ *where* $\Delta \in {}^\beta Sb\alpha$, *then* x CA_α*-freely generates* $\mathfrak{A}$ *under* Δ.

PROOF. Condition (i) of 2.5.34 is obviously satisfied. In order to show that condition (ii) also holds we consider any $\mathfrak{B} \in \mathsf{CA}_\alpha$ and $y \in {}^\beta B$ where $\Delta^{(\mathfrak{B})} y_\xi \subseteq \Delta\xi$ for every $\xi < \beta$. By 0.4.27(i) there is an $f \in Hom(\mathfrak{Fr}_\beta, \mathfrak{B})$ such that $f\xi = y_\xi$ for every $\xi < \beta$. Hence, because of 0.2.23(i), we have $\mathfrak{Fr}_\beta/(f|f^{-1}) \cong \mathfrak{Sg}^{(\mathfrak{B})} Rg y \in \mathsf{CA}_\alpha$ and $\Delta(\xi/(f|f^{-1})) = \Delta^{(\mathfrak{B})} y_\xi \subseteq \Delta\xi$ for every $\xi < \beta$; this shows that $Cr_\beta^{(\Delta)} \mathsf{CA}_\alpha \subseteq (f|f^{-1})$. Therefore, applying 0.2.23(iii) we obtain an $h \in Hom(\mathfrak{A}, \mathfrak{B})$ such that $h \circ x = y$. Condition (ii) of 2.5.34 is thus satisfied.

THEOREM 2.5.37. *For any* CA_α $\mathfrak{A}$ *and any* $\Delta \in {}^\beta Sb\alpha$, *in order that* $\mathfrak{A} \cong \mathfrak{Fr}_\beta^{(\Delta)} \mathsf{CA}_\alpha$ *it is necessary and sufficient that there exists a sequence* $x \in {}^\beta A$ *which* CA_α*-freely generates* $\mathfrak{A}$ *under* Δ.

PROOF. For $\mathfrak{A}$ to be isomorphic to $\mathfrak{Fr}_\beta^{(\Delta)} \mathsf{CA}_\alpha$ it is clearly necessary by 2.5.36 that $\mathfrak{A}$ be freely generated under Δ by some sequence $x \in {}^\beta A$. In order to show this condition is also sufficient, assume that it holds and choose

$f \in Hom(\mathfrak{A}, \mathfrak{Fr}_\beta^{(\Delta)}\mathsf{CA}_\alpha)$ such that

(1) $$fx_\xi = \xi/Cr_\beta^{(\Delta)}\mathsf{CA}_\alpha \text{ for every } \xi < \beta;$$

notice that by 0.2.18(i) we actually have

(2) $$f \in Ho(\mathfrak{A}, \mathfrak{Fr}_\beta^{(\Delta)}\mathsf{CA}_\alpha).$$

By 2.5.36 we can also find an $h \in Hom(\mathfrak{Fr}_\beta^{(\Delta)}\mathsf{CA}_\alpha, \mathfrak{A})$ such that

(3) $$h(\xi/Cr_\beta^{(\Delta)}\mathsf{CA}_\alpha) = x_\xi \text{ for every } \xi < \beta.$$

Using 0.2.18(iii) we conclude from (1), (2), and (3) that $\mathfrak{A} \cong \mathfrak{Fr}_\beta^{(\Delta)}\mathsf{CA}_\alpha$.

THEOREM 2.5.38. *Assume* $\Delta \in {}^\beta Sb\alpha$.

(i) *The function* $\langle \xi/Cr_\beta^{(\Delta)}\mathsf{CA}_\alpha : \xi < \beta \rangle$ *is one-one.*

(ii) *If either* $\alpha \geqq \omega$ *or* $\beta \geqq \omega$ *or, finally,* $|\Delta\eta| \geqq 2$ *for some* $\eta < \beta$, *then* $\mathfrak{Fr}_\beta^{(\Delta)}\mathsf{CA}_\alpha$ *is infinite and, in fact,* $|Fr_\beta^{(\Delta)}\mathsf{CA}_\alpha| = |\alpha| \cup \beta \cup \omega$.

PROOF. To prove (i) let $\mathfrak{B}$ be any discrete CA_α such that $|B| \geqq \beta$, and let y be a β-termed sequence without repetitions and with range included in B. Then by 2.5.36 there is an $h \in Hom(\mathfrak{Fr}_\beta^{(\Delta)}\mathsf{CA}_\alpha, \mathfrak{B})$ such that $f(\xi/Cr_\beta^{(\Delta)}\mathsf{CA}_\alpha) = y_\xi$. Thus $\langle \xi/Cr_\beta^{(\Delta)}\mathsf{CA}_\alpha : \xi < \beta \rangle$ must also be a sequence without repetitions. Turning to (ii) we have $|Fr_\beta^{(\Delta)}\mathsf{CA}_\alpha| \leqq |\alpha| \cup \beta \cup \omega$ for any α, β, and Δ by 2.1.9(i). If $\alpha \geqq \omega$ or $\beta \geqq \omega$, then $|Fr_\beta^{(\Delta)}\mathsf{CA}_\alpha| \geqq |\alpha| \cup \beta = |\alpha| \cup \beta \cup \omega$ by (i) and 2.1.9(ii). If $|\Delta\eta| \geqq 2$ for some $\eta < \beta$, then $|Fr_\beta^{(\Delta)}\mathsf{CA}_\alpha| \geqq |\alpha| \cup \beta \cup \omega$ by (i), 2.1.9(ii), 2.1.11(iii), and 2.5.36.

The case when $\alpha, \beta < \omega$ and $|\Delta\xi| < 2$ for every $\xi < \beta$, which is not covered by the hypothesis of 2.5.38(ii), will be dealt with in Theorem 2.5.59 below.

The discussion of atoms of dimension-restricted free CA's does not lead to any problems or results not already encountered in discussing ordinary free CA's. We may only point out that, in case $\beta < \omega$, $\mathfrak{Fr}_\beta^{(\Delta)}\mathsf{CA}_2$ is atomic and $\mathfrak{Fr}_\beta^{(\Delta)}\mathsf{CA}_\alpha$ for an arbitrary α has exactly 2^β zero-dimensional atoms.

In 2.5.20 we saw that the general properties of free algebras established under certain assumptions in 0.4.54 hold in particular for the free CA's. We shall now show that some of these properties when appropriately modified apply also to dimension-restricted free CA's.

THEOREM 2.5.39. *Let* $\Delta \in {}^\beta Sb\alpha$ *and* $X \subseteq Fr_\beta^{(\Delta)}\mathsf{CA}_\alpha$. *Then*

(i) $|X| \geqq \beta$ *whenever* $\mathfrak{Fr}_\beta^{(\Delta)}\mathsf{CA}_\alpha = \mathfrak{Sg}\, X$.

Moreover, for any β' *and* $\Delta' \in {}^{\beta'}Sb\alpha$, *we have:*

(ii) $\beta \geqq \beta'$ *whenever* $\mathfrak{Fr}_\beta^{(\Delta)}\mathsf{CA}_\alpha \succcurlyeq \mathfrak{Fr}_{\beta'}^{(\Delta')}\mathsf{CA}_\alpha$;

(iii) $\beta = \beta'$ *whenever* $\mathfrak{Fr}_\beta^{(\Delta)}\mathsf{CA}_\alpha \cong \mathfrak{Fr}_{\beta'}^{(\Delta')}\mathsf{CA}_\alpha$.

PROOF. Let K be the class of all discrete CA_α's. K is equational by 1.3.12 and hence $\mathfrak{Fr}_\beta\mathsf{K} \in \mathsf{K}$ by 0.4.22 and 0.4.63. Thus, in view of 2.5.36 there exists an $h \in Ho(\mathfrak{Fr}_\beta^{(\Delta)}\mathsf{CA}_\alpha, \mathfrak{Fr}_\beta\mathsf{K})$ such that $h(\xi/Cr_\beta^{(\Delta)}\mathsf{CA}_\alpha) = \xi/Cr_\beta\mathsf{K}$ for every $\xi < \beta$, and applying 0.2.18(i) and 0.4.54(i) we get

$$|X| \geqq |h^*X| \geqq \beta.$$

This proves (i); (ii) and (iii) are immediate consequences of (i).

The next two theorems are due to Don Pigozzi; the idea used in the proof of the first of them was suggested by an unpublished argument of William Hanf.

THEOREM 2.5.40. *Assume* $\Delta \in {}^\beta Sb\alpha$, *and let* φ *be any one-one function from* $|\beta+1|$ *onto* $\beta+1$. *Then*

$$\mathfrak{Fr}_\beta^{(\Delta)}\mathsf{CA}_\alpha \times \mathfrak{Fr}_\beta^{(\Delta)}\mathsf{CA}_\alpha \cong \mathfrak{Fr}_{|\beta+1|}^{(\Delta')}\mathsf{CA}_\alpha$$

where $\Delta' = (\Delta \cup \{\langle\beta, 0\rangle\}) \circ \varphi$. *If in particular* $\beta < \omega$, *we obtain*

$$\mathfrak{Fr}_\beta^{(\Delta)}\mathsf{CA}_\alpha \times \mathfrak{Fr}_\beta^{(\Delta)}\mathsf{CA}_\alpha \cong \mathfrak{Fr}_{\beta+1}^{(\Delta\cup\{\langle\beta,0\rangle\})}\mathsf{CA}_\alpha.$$

PROOF. Let $\mathfrak{A} = \mathfrak{Fr}_{|\beta+1|}^{(\Delta')}\mathsf{CA}_\alpha$ and $g_\xi = \varphi^{-1}\xi/(Cr_{|\beta+1|}^{(\Delta')}\mathsf{CA}_\alpha)$ for every $\xi < \beta+1$. By 2.5.33 we have

(1) $$\Delta g_\beta = 0.$$

We shall prove that

(2) $\mathfrak{Rl}_{g_\beta}\mathfrak{A}$ is CA_α-freely generated by $\langle g_\xi \cdot g_\beta : \xi < \beta\rangle$ under Δ.

From 0.2.18(ii), 2.3.2, 2.3.26(i), and 2.5.33 we see that condition (i) of 2.5.34 holds when $\mathfrak{Rl}_{g_\beta}\mathfrak{A}$ and $\langle g_\xi \cdot g_\beta : \xi < \beta\rangle$ are taken for $\mathfrak{A}$ and x, respectively.

Consider any $\mathfrak{B} \in \mathsf{CA}_\alpha$ and $y \in {}^\beta B$ such that $\Delta^{(\mathfrak{B})}y_\xi \subseteq \Delta\xi$ for every $\xi < \beta$. By 2.5.36 there is an $f \in Hom(\mathfrak{A}, \mathfrak{B})$ such that $fg_\xi = y_\xi$ for every $\xi < \beta$, and $fg_\beta = 1^{(\mathfrak{B})}$. Set $\mathfrak{Rl}_{g_\beta}\mathfrak{A} = \langle Rl_{g_\beta}\mathfrak{A}, +', \cdot', -', 0', 1', \mathsf{c}'_\kappa, \mathsf{d}'_{\kappa\lambda}\rangle_{\kappa,\lambda<\alpha}$ and let $x, z \in Rl_{g_\beta}\mathfrak{A}$. Clearly $f(x+'z) = fx+^{(\mathfrak{B})}fz$, $f(x\cdot'z) = fx\cdot^{(\mathfrak{B})}fz$, and $f0' = 0^{(\mathfrak{B})}$; also,

$$f1' = fg_\beta = 1^{(\mathfrak{B})},\quad f-'x = f(g_\beta - x) = 1^{(\mathfrak{B})} -^{(\mathfrak{B})} x = -^{(\mathfrak{B})}x,$$

and in a similar manner we can show $f\mathsf{c}'_\kappa x = \mathsf{c}_\kappa^{(\mathfrak{B})}fx$ and $f\mathsf{d}'_{\kappa\lambda} = \mathsf{d}_{\kappa\lambda}^{(\mathfrak{B})}$, for any $\kappa, \lambda < \alpha$. This proves that $f \in Hom(\mathfrak{Rl}_{g_\beta}\mathfrak{A}, \mathfrak{B})$, and in the same way we get $f(g_\xi \cdot g_\beta) = y_\xi$ for every $\xi < \beta$. Hence condition (ii) of 2.5.34 also holds when $\mathfrak{A}$ is replaced by $\mathfrak{Rl}_{g_\beta}\mathfrak{A}$ and x by $\langle g_\xi \cdot g_\beta : \xi < \beta\rangle$. Thus (2) holds, and in an entirely analogous manner we obtain:

(3) $\mathfrak{Rl}_{-g_\beta}\mathfrak{A}$ is CA_α-freely generated by $\langle g_\xi \cdot -g_\beta : \xi < \beta\rangle$ under Δ.

Together, (1), (2), and (3), when combined with 2.4.7 and 2.5.37, give the conclusion of the theorem.

THEOREM 2.5.41. *If $\beta < \omega$ and $\kappa, \lambda < \omega$, then the formula*

$$^{\kappa}\mathfrak{Fr}_{\beta}^{(\Delta)}\mathsf{CA}_{\alpha} \cong {}^{\lambda}\mathfrak{Fr}_{\beta}^{(\Delta)}\mathsf{CA}_{\alpha}$$

for any $\Delta \in {}^{\beta}Sb\alpha$, and, in particular, the formula

$$^{\kappa}\mathfrak{Fr}_{\beta}\mathsf{CA}_{\alpha} \cong {}^{\lambda}\mathfrak{Fr}_{\beta}\mathsf{CA}_{\alpha},$$

implies $\kappa = \lambda$.

PROOF. By means of a simple inductive argument 2.5.39 and 2.5.40 jointly lead to the following:

(1) if $\mu, \nu < \omega$, then $^{2^{\mu}}\mathfrak{Fr}_{\beta}^{(\Delta)}\mathsf{CA}_{\alpha} \cong {}^{2^{\nu}}\mathfrak{Fr}_{\beta}^{(\Delta)}\mathsf{CA}_{\alpha}$ implies $\mu = \nu$.

To complete the proof we assume by way of contradiction that

(2) $$^{\kappa}\mathfrak{Fr}_{\beta}^{(\Delta)}\mathsf{CA}_{\alpha} \cong {}^{\lambda}\mathfrak{Fr}_{\beta}^{(\Delta)}\mathsf{CA}_{\alpha}$$

for some $\kappa, \lambda < \omega$ with $\kappa < \lambda$. Expressing $\lambda - \kappa$ in the form $2^{\pi} \cdot \lambda'$ with λ' odd, we obtain from (2), by a simple inductive argument, that

(3) $$^{\kappa}\mathfrak{Fr}_{\beta}^{(\Delta)}\mathsf{CA}_{\alpha} \cong {}^{\kappa + 2^{\pi}\cdot\lambda'\cdot\sigma}\mathfrak{Fr}_{\beta}^{(\Delta)}\mathsf{CA}_{\alpha}$$

for every natural number σ. Then, choosing μ so that 2^{μ} exceeds both κ and 2^{π}, forming the direct products of both sides of (3) with $^{2^{\mu}-\kappa}\mathfrak{Fr}_{\beta}^{(\Delta)}\mathsf{CA}_{\alpha}$, and considering values of σ of the form $2^{\mu-\pi}\cdot\rho$, we see that

(4) $$^{2^{\mu}}\mathfrak{Fr}_{\beta}^{(\Delta)}\mathsf{CA}_{\alpha} \cong {}^{2^{\mu}\cdot(1+\rho\cdot\lambda')}\mathfrak{Fr}_{\beta}^{(\Delta)}\mathsf{CA}_{\alpha}$$

for every $\rho < \omega$. Since λ' is odd, it follows from the Euler-Fermat theorem of elementary number theory (see, for instance, Sierpiński [64*], p. 243, Theorem 8) that there exists a ρ such that $1+\rho\cdot\lambda'$ is a positive power of 2; (4) with this choice of ρ then contradicts (1). This completes the proof.

The problem to what extent the condition $\beta < \omega$ in 2.5.41 is necessary will be discussed briefly in 2.5.54.

In the next theorem and its corollary we concern ourselves with characteristic sets and minimal subalgebras of ordinary and dimension-restricted free algebras.

THEOREM 2.5.42. *Let $\mathfrak{A} = \mathfrak{Fr}_{\beta}^{(\Delta)}\mathsf{CA}_{\alpha}$ with $\Delta \in {}^{\beta}Sb\alpha$ (or, in particular, let $\mathfrak{A} = \mathfrak{Fr}_{\beta}\mathsf{CA}_{\alpha}$). Let $a_{\kappa}^{(\mathfrak{A})}$ and $\nabla\mathfrak{A}$ be defined as in 2.5.25. Then $a_{\kappa}^{(\mathfrak{A})} < a_{\lambda}^{(\mathfrak{A})}$ for any κ, λ with $1 \leq \lambda < \kappa < (\alpha+1) \cap \omega$, or, equivalently, $\nabla\mathfrak{A} = (\alpha \cap \omega) \sim \{0\}$; moreover, $a_{\alpha}^{(\mathfrak{A})} \neq 0$ in case $\alpha < \omega$.*

PROOF. Because of 2.5.32 we consider only the case $\mathfrak{A} = \mathfrak{Fr}_\beta^{(\Delta)}\mathsf{CA}_\alpha$. For every $\mathfrak{B} \in \mathsf{CA}_\alpha$ there exists by 2.5.36 at least one homomorphism from $\mathfrak{A}$ into $\mathfrak{B}$ (since $Zd\,\mathfrak{B}$ is clearly always non-empty). Thus for any κ, λ with $1 \leqq \lambda < \kappa < (\alpha+1) \cap \omega$ we have $a_\kappa^{(\mathfrak{A})} < a_\lambda^{(\mathfrak{A})}$ if $a_\kappa^{(\mathfrak{B})} < a_\lambda^{(\mathfrak{B})}$ for some $\mathfrak{B} \in \mathsf{CA}_\alpha$; in case $\alpha < \omega$ we also have $a_\alpha^{(\mathfrak{A})} \neq 0^{(\mathfrak{A})}$ if $a_\alpha^{(\mathfrak{C})} \neq 0^{(\mathfrak{C})}$ for some $\mathfrak{C} \in \mathsf{CA}_\alpha$. The desired conclusion now follows from 2.5.26.

From this theorem we see that the characteristic set (in the sense of 2.5.29) of a free CA_α, whether dimension-restricted or not, is uniquely determined by α and is actually the largest possible set which can occur as the characteristic set of a CA_α. By comparing 2.5.42 with 2.5.25 we obtain at once

COROLLARY 2.5.43. *If* $\mathfrak{A} = \mathfrak{Fr}_\beta^{(\Delta)}\mathsf{CA}_\alpha$ *with* $\Delta \in {}^\beta Sb\alpha$ *(or* $\mathfrak{A} = \mathfrak{Fr}_\beta\mathsf{CA}_\alpha$*), and* $\mathfrak{A}' = \mathfrak{Fr}_{\beta'}^{(\Delta')}\mathsf{CA}_\alpha$ *with* $\Delta' \in {}^{\beta'}Sb\alpha$ *(or* $\mathfrak{A} = \mathfrak{Fr}_{\beta'}\mathsf{CA}_\alpha$*), then the minimal subalgebras of* $\mathfrak{A}$ *and* $\mathfrak{A}'$ *are isomorphic.*

Our discussion in the remaining part of this section centers around those dimension-restricted free algebras which are finite-dimensional or, more generally, locally finite-dimensional.

THEOREM 2.5.44. *For any* $\Delta \in {}^\beta Sb\alpha$, $\mathfrak{Fr}_\beta^{(\Delta)}\mathsf{CA}_\alpha \in \mathsf{Lf}_\alpha$ *iff* $|\Delta\xi| < \omega$ *for every* $\xi < \beta$.

PROOF: by 2.5.1(i) and 2.5.33.

THEOREM 2.5.45. *The following two conditions are equivalent:*

(i) $\mathfrak{A} \in \mathsf{Lf}_\alpha$,

(ii) *there exists a* $\beta \leqq |A|$ *and a* $\Delta \in {}^\beta Sb\alpha$ *with* $|\Delta\xi| < \omega$ *for each* $\xi < \beta$ *such that* $\mathfrak{Fr}_\beta^{(\Delta)}\mathsf{CA}_\alpha \succcurlyeq \mathfrak{A}$.

The theorem remains true if the symbols "$\leqq |A|$*" in (ii) are replaced by "*$= |A|$*" or else if they are simply omitted.*

PROOF. Let $\beta = |A|$ and $a \in {}^\beta A$ be such that $Rg\,a = A$; obviously $|\Delta a_\xi| < \omega$ for each $\xi < \beta$ by (i). If $\Delta = \langle \Delta a_\xi : \xi < \beta \rangle$, then $\mathfrak{Fr}_\beta^{(\Delta)}\mathsf{CA}_\alpha \succcurlyeq A$ by 2.5.36. This shows that (i) implies (ii). The implication in the opposite direction follows immediately from 2.3.3 and 2.5.44. The last part of the theorem is obvious.

REMARK 2.5.46. The importance of locally finite dimension-restricted free CA's is due primarily to their metalogical representation.

Let $\mathfrak{A} = \mathfrak{Fr}_\beta^{(\Delta)}\mathsf{CA}_\alpha$ where $\alpha \geqq \omega$ and Δ is finite-valued (i.e., $\Delta\xi$ is a finite subset of α for every $\xi < \beta$). To simplify the matter assume that $\alpha = \omega$, and assume also that, for every $\xi < \beta$, $\Delta\xi$ is actually a finite ordinal and not merely a finite set of ordinals. (From Corollary 2.5.52 below it will be seen that this latter assumption does not restrict the generality of the observations we are going to make.) We correlate with $\mathfrak{A}$ a well-determined language Λ. All the non-logical constants of this language are predicates. The variables are arranged in a sequence of length ω and the predicates in a sequence of length β; the rank of the ξth predicate is $\Delta\xi$. By comparing the content of 2.5.45 with Remark 1.11.2 we conclude that every homomorphic image of the algebra $\mathfrak{A} = \mathfrak{Fr}_\beta^{(\Delta)}\mathsf{CA}_\alpha$ is isomorphic to an algebra of formulas in the language Λ and that, conversely, every such algebra of formulas is a homomorphic image of $\mathfrak{A}$. In Remarks 2.3.13 we have come across another algebra with the same property, namely, the algebra of formulas $\mathfrak{A}_0$ associated with the logic of the language Λ or, what amounts to the same, with the empty set of sentences. Hence it follows immediately that each one of the two algebras $\mathfrak{A}$ and $\mathfrak{A}_0$ is homomorphic to the other. Indeed, one can easily show (with the help of 0.2.18(iii)) that these two algebras are isomorphic. These results are rather simple consequences of the completeness theorem for predicate logic and will be established precisely in Part II. As a consequence, the zero-dimensional part of $\mathfrak{A}$ is isomorphic with the algebra of sentences associated with the logic of the language Λ.

REMARKS 2.5.47. Several deep results are known concerning locally finite free algebras $\mathfrak{Fr}_\beta^{(\Delta)}\mathsf{CA}_\alpha$ of infinite dimension and their zero-dimensional parts. All of these results were originally formulated and established, not for abstract free algebras, but for their metalogical representations, i.e., for algebras of formulas and algebras of sentences associated with logics. For many of these results none of the available proofs are algebraic "in spirit" — in spite of the fact that the results are primarily algebraic in character or at least have a clear algebraic content; the known algebraic proofs are rather clumsy translations of the original metalogical arguments. For the remaining results the known algebraic proofs depend essentially on the representation theory. For these reasons we restrict ourselves to stating results and postpone a detailed presentation to Part II.

We begin with a result which concerns the isomorphism of algebras $\mathfrak{Zd}\,\mathfrak{Fr}_\beta^{(\Delta)}\mathsf{CA}_\alpha$ with varying α, β, and Δ.

(I) *Let* $\alpha \geqq \omega$, $\beta < \omega$, $\Delta \in {}^\beta Sb\alpha$, $|\Delta\xi| < \omega$ *for every* $\xi < \beta$, *and* $|\Delta\eta| \geqq 2$ *for some* $\eta < \beta$. *Assume that analogous conditions hold for* α', β', *and* Δ'. *Then* $\mathfrak{Zd}\,\mathfrak{Fr}_\beta^{(\Delta)}\mathsf{CA}_\alpha \cong \mathfrak{Zd}\,\mathfrak{Fr}_{\beta'}^{(\Delta')}\mathsf{CA}_{\alpha}$.

In metalogical form (and with inessential restriction to the case $\alpha = \alpha' = \omega$) Theorem (I) is stated without proof in Hanf [62*]. The result facilitates considerably the study of intrinsic properties of the algebras involved: to show that such a property applies to every algebra $\mathfrak{Zd}\,\mathfrak{Fr}_\beta^{(\Delta)}\mathsf{CA}_\alpha$ satisfying the assumption of (I), it suffices to establish the property for any particular algebra of this kind (e.g., for the algebra $\mathfrak{Zd}\,\mathfrak{Fr}_1^{(\Delta)}\mathsf{CA}_\omega$ with $\Delta = \{\langle 0, 2\rangle\}$). However, as we have seen in 2.5.39(iii) and as we shall see even more strongly in 2.5.53, the result in no way extends from the zero-dimensional parts of dimension-restricted free algebras to the algebras themselves.

In Theorem (I) each of the premisses $\beta < \omega$, $|\Delta\xi| < \omega$ for every $\xi < \beta$, and $|\Delta\eta| \geqq 2$ for some $\eta < \beta$ is known to be essential. In 2.5.66 below the situation which arises when the last premiss fails is considered. In case $\beta \geq \omega$, the algebras $\mathfrak{Zd}\,\mathfrak{Fr}_\beta^{(\Delta)}\mathsf{CA}_\alpha$ and $\mathfrak{Zd}\,\mathfrak{Fr}_{\beta'}^{(\Delta')}\mathsf{CA}_\alpha$ are isomorphic only if $\beta = \beta'$; if, in particular, $\beta = \beta' = \omega$, then both algebras are denumerable atomless BA's and hence they are indeed isomorphic to each other and to the algebra $\mathfrak{Fr}_\omega\mathsf{BA}$. (The last remarks hold for any finite-valued Δ, independent of whether $|\Delta\eta| \geqq 2$ for some $\eta < \beta$.) Compare Remark 2.5.14.

In Hanf [62*] two further, more special results are stated (also without proofs). They concern direct decompositions of algebras discussed in (I):

(II) *Assume α, β, and Δ satisfy the premisses of (I), and let $\mathfrak{A} = \mathfrak{Zd}\,\mathfrak{Fr}_\beta^{(\Delta)}\mathsf{CA}_\alpha$. Then $\mathfrak{A} \cong \mathfrak{A} \times \mathfrak{A}$; in other words, there is an $a \in A$ such that $\mathfrak{A} \cong \mathfrak{Rl}_a\mathfrak{A} \cong \mathfrak{Rl}_{-a}\mathfrak{A}$.*

(III) *Again assume α, β, and Δ satisfy the premisses of (I), and let $\mathfrak{A} = \mathfrak{Zd}\,\mathfrak{Fr}_\beta^{(\Delta)}\mathsf{CA}_\alpha$. Then if $\mathfrak{A} = \mathfrak{B} \times \mathfrak{C}$, we have $\mathfrak{A} \cong \mathfrak{B}$ or $\mathfrak{A} \cong \mathfrak{C}$; in other words, $\mathfrak{A} \cong \mathfrak{Rl}_a\mathfrak{A}$ or $\mathfrak{A} \cong \mathfrak{Rl}_{-a}\mathfrak{A}$ for every $a \in A$.*

Hence it is seen that the algebra $\mathfrak{A}$ is not directly indecomposable, but is pseudo-directly indecomposable in the sense of 0.3.29(ii″). From (II) and (III) we can also draw some conclusions of a more specific nature. For instance, some known metalogical results imply that the algebra $\mathfrak{A}$ is neither atomic nor atomless, and that it actually contains infinitely many atoms. Hence we conclude by (III) that $\mathfrak{A}$ is not a direct product of an atomic and an atomless algebra, or, equivalently, that the sum of all atoms does not exist. Similarly we can show by means of (III) that sums (or products) of some other infinite subsets of $\mathfrak{A}$ do not exist. In general, the problem whether the sum of a given infinite set $B \subseteq A$ exists may present considerable difficulties.

As in the case of (I), the theorems (II) and (III) do not extend from the zero-dimensional parts of dimension-restricted free algebras to the algebras themselves. For (II) this follows directly from 2.5.41; for (III) this can easily be shown with the help of 2.5.42. Still another striking difference between algebras $\mathfrak{Fr}_\beta^{(\Delta)}\mathsf{CA}_\alpha$ satisfying the premisses of (I) and their zero-dimensional parts can be noted here: as was pointed out, $\mathfrak{Zd}\,\mathfrak{Fr}_\beta^{(\Delta)}\mathsf{CA}_\alpha$ has infinitely many atoms while, using 1.11.8(i) and arguing as in the proof of 2.5.11, it can easily be shown that $\mathfrak{Fr}_\beta^{(\Delta)}\mathsf{CA}_\alpha$ has exactly 2^β atoms.

The following theorem has both an intrinsic interest and a wide range of applications:

(IV) *Assume* $\alpha \geqq \omega$, $\Delta \in {}^\beta Sb\alpha$, *and* $|\Delta\xi| < \omega$ *for every* $\xi < \beta$. *Let* $\mathfrak{A} = \mathfrak{Fr}_\beta^{(\Delta)}\mathsf{CA}_\alpha$. *Under these assumptions, whenever* $X, Z \subseteq \{\xi/Cr_\beta^{(\Delta)}\mathsf{CA}_\alpha : \xi < \beta\}$, $x \in SgX$, $z \in SgZ$, *and* $x \leqq z$, *there exists a* $y \in Sg((X \cap Z) \cup \{0^{(\mathfrak{A})}\})$ *such that* $x \leqq y \leqq z$.

A metalogical version of (IV) (with $\alpha = \omega$) is stated and proved in Craig [57*], p. 267, Theorem 5, and is known as Craig's interpolation theorem.

Given any ordinal α, any function $\Delta \in {}^\beta Sb\alpha$, and any class $\mathsf{K} \subseteq \mathsf{CA}_\alpha$, we let $\mathfrak{A} = \mathfrak{Fr}_\beta^{(\Delta)}\mathsf{K}$ (see Definition 2.5.31), and we say that $\mathfrak{A}$ has the *interpolation property* if it satisfies the conclusion of (IV) (beginning with the word „whenever") with CA_α replaced by K. Using this terminology (IV) expresses the fact that $\mathfrak{Fr}_\beta^{(\Delta)}\mathsf{CA}_\alpha$ has the interpolation property provided $\alpha \geqq \omega$ and Δ is finite-valued.

The interpolation property of $\mathfrak{Fr}_\beta^{(\Delta)}\mathsf{K}$ turns out to be closely related to a certain property of K which is of a general algebraic nature and is known as the *amalgamation property*. A class K of arbitrary algebras is said to have this property if, for any $\mathfrak{A}, \mathfrak{B}, \mathfrak{C} \in \mathsf{K}$, any $f \in Ism(\mathfrak{A}, \mathfrak{B})$, and any $g \in Ism(\mathfrak{A}, \mathfrak{C})$, there is a $\mathfrak{D} \in \mathsf{K}$, an $h \in Ism(\mathfrak{B}, \mathfrak{D})$, and a $k \in Ism(\mathfrak{C}, \mathfrak{D})$ such that $h \circ f = k \circ g$. Only a few classes of algebras are known at present to have the amalgamation property; cf. Jónsson [65*].

In the realm of CA's it has been shown in Daigneault [64a], p. 97, Theorem 2.7, by means of an algebraic argument based upon the representation theory, that the class Lf_α, for any $\alpha \geqq \omega$, has the amalgamation property.[1] In Daigneault [64], Theorem (IV) is derived from this result by a purely algebraic method. Thus the two papers just quoted provide together a new, algebraic proof of Craig's interpolation theorem.

In the opposite direction Don Pigozzi has found the following two related theorems:

(V) *Let* α *be any ordinal, let* $\mathsf{K} \subseteq \mathsf{Lf}_\alpha$ *and* $\mathsf{K} = \mathbf{H}\mathsf{K}$. *If* $\mathfrak{Fr}_\beta^{(\Delta)}\mathsf{K}$ *belongs to* K *and has the interpolation property for every cardinal* $\beta > 0$ *and every finite-valued* $\Delta \in {}^\beta Sb\alpha$, *then* K *has the amalgamation property.*

[1]) Strictly speaking, Daigneault [64a] deals, not with Lf's, but with closely related locally finite-dimensional polyadic algebras. However, the argument used in the paper can easily be adapted to CA's.

(VI) *Let* α *be any ordinal and* K *be any subclass of* CA_α *which is algebraically closed in the sense of 0.3.14. If* $\mathfrak{Fr}_\beta\mathsf{K}$ *(i.e.,* $\mathfrak{Fr}_\beta^{(\Delta)}\mathsf{K}$ *with* $\Delta = \beta \times \{\alpha\}$*) has the interpolation property for every cardinal* $\beta > 0$*, then* K *has the amalgamation property.*

Using (V) we can derive the amalgamation property of Lf_α from the interpolation theorem (IV); it suffices to let $\mathsf{K} = \mathsf{Lf}_\alpha$ in (V) and to notice that $\mathfrak{Fr}_\beta^{(\Delta)}\mathsf{Lf}_\alpha = \mathfrak{Fr}_\beta^{(\Delta)}\mathsf{CA}_\alpha$ in case Δ is finite-valued. (VI) cannot be applied to $\mathsf{K} = \mathsf{Lf}_\alpha$ with $\alpha \geqq \omega$ (since Lf_α is not algebraically closed), but has another interesting implication. It is shown in Comer [69a] that the class CA_α with $2 \leqq \alpha < \omega$ does not have the amalgamation property and that the same applies to the class R_α of all representable CA_α's. Hence, using (VI), Pigozzi infers that, for all cardinals β greater than a fixed finite cardinal, neither $\mathfrak{Fr}_\beta\mathsf{CA}_\alpha$ nor $\mathfrak{Fr}_\beta\mathsf{R}_\alpha$ has the interpolation property; actually, the latter result had been found earlier to hold for all $\beta \geqq 2$, by a direct proof due independently to Henkin and Pigozzi. As we know this implies that the interpolation theorem fails in predicate logic provided with only finitely many variables.

An important metalogical consequence of (IV) is the well-known Beth's definability theorem. For a formulation and derivation of the theorem see Beth [53*] and Craig [57a*], p. 275, Theorem 1; we do not state this theorem explicitly since in all algebraic translations known at present it assumes a form without intrinsic algebraic interest. However, as shown in Daigneault [64], the theorem is closely related to a rather simple algebraic statement expressing a stronger form of the amalgamation property; see here also Daigneault [64a].

Another consequence of (IV), of an essentially algebraic character, is

(VII) *Assume* $\alpha \geqq \omega$, $\Delta \in {}^\beta Sb\alpha$, *and* $|\Delta\xi| < \omega$ *for every* $\xi < \beta$; *let* $\mathfrak{A} = \mathfrak{Fr}_\beta^{(\Delta)}\mathsf{CA}_\alpha$. *Then every ideal of* $\mathfrak{A}$ *has an irredundant base; in other words, for every* $I \in Il\mathfrak{A}$ *there is a* $G \subseteq A$ *such that* $IgG = I$, *while* $IgG' \subset I$ *whenever* $G' \subset G$.

In case $\beta \leqq \omega$ the proof of (VII) is simple. Actually (VII) follows then as a particular case of a more general theorem, not involving free algebras, by which every ideal of a Lf_α $\mathfrak{A}$ which has a countable base has also an irredundant countable base; cf. the remarks following 2.3.12. In its general form Theorem (VII) was first proved in Resnikoff [65*]. In metalogical interpretation (VII) means, of course, that every theory in a language of predicate logic (with arbitrarily many predicates) has an independent axiom system.

REMARK 2.5.48. In Remarks 2.5.46 and 2.5.47 we have disregarded almost entirely algebras $\mathfrak{Fr}_\beta^{(\Delta)}\mathsf{CA}_\alpha$ in which α is finite (and therefore Δ is automatically finite-valued). The problem of a metalogical representation of these algebras, analogous to that considered in 2.5.46, is involved and will not be discussed at this point. As regards the results stated in 2.5.47, we have pointed out that (IV) cannot be extended to the case $2 \leqq \alpha < \omega$. In (V) and (VI) no conditions on α are imposed. The problem whether the remaining results, i.e., (I), (II), (III), and (VII), hold for algebras $\mathfrak{Fr}_\beta^{(\Delta)}\mathsf{CA}_\alpha$ with $2 \leqq \alpha < \omega$ has not yet been studied.

The following simple theorem concerns algebras $\mathfrak{Fr}_\beta^{(\Delta)}\mathsf{CA}_\alpha$ in which both α and β are finite.

THEOREM 2.5.49. *If* $\alpha, \beta < \omega$ *and* $\Delta \in {}^\beta Sb\alpha$, *then* $\mathfrak{Fr}_\beta^{(\Delta)}\mathsf{CA}_\alpha | \mathfrak{Fr}_\beta\mathsf{CA}_\alpha$ *or, equivalently,* $\mathfrak{Fr}_\beta^{(\Delta)}\mathsf{CA}_\alpha \cong \mathfrak{Rl}_b\mathfrak{Fr}_\beta\mathsf{CA}_\alpha$ *for some* $b \in Zd\mathfrak{Fr}_\beta\mathsf{CA}_\alpha$.

PROOF. Let $g_\xi = \xi/Cr_\beta\mathsf{CA}_\alpha$ for every $\xi < \beta$ and let

(1) $$J = Ig^{(\mathfrak{Fr}_\beta\mathsf{CA}_\alpha)}\{\mathsf{c}_\kappa g_\xi \oplus g_\xi : \xi < \beta,\ \kappa \in \alpha \sim \Delta\xi\}.$$

By 0.2.27(ii)–(iv), 0.2.45, and 2.5.31 we have

(2) $$\mathfrak{Fr}_\beta^{(\Delta)}\mathsf{CA}_\alpha \cong \mathfrak{Fr}_\beta\mathsf{CA}_\alpha/J.$$

If $x = \sum\{\mathsf{c}_\kappa g_\xi \oplus g_\xi : \xi < \beta,\ \kappa \in \alpha \sim \Delta\xi\}$, then from (1) and 2.3.10(iv) we obtain $J = Ig^{(\mathfrak{Fr}_\beta\mathsf{CA}_\alpha)}\{x\}$. Therefore, (2), 2.3.27(ii), and 2.4.8 imply that the conclusion of the theorem holds when $b = -\mathsf{c}_{(\alpha)}x$.

REMARK 2.5.50. Theorem 2.5.49 can be somewhat improved by omitting the premiss $\beta < \omega$ and assuming instead that there are only finitely many $\xi < \beta$ such that $\Delta\xi \neq \alpha$. A further improvement of 2.5.49 is obtained by considering the general notion of a free CA_α with defining relations (cf. the remarks preceding 2.5.31). In fact, it can be shown that for any $\alpha < \omega$ the following two conditions are equivalent: (I) $\mathfrak{B}|\mathfrak{Fr}_\beta\mathsf{CA}_\alpha$; (II) there is a finite set $\mathfrak{S}$ of defining relations (or, equivalently, a set $\mathfrak{S}$ consisting of one defining relation) such that $\mathfrak{B}$ is isomorphic to $\mathfrak{Fr}_\beta^{(\mathfrak{S})}\mathsf{CA}_\alpha$ in the sense of 0.4.65. The proof is as simple as that of 2.5.49. From this result, by means of an additional argument of a general algebraic character, we can derive the following conclusion: if $\alpha, \beta < \omega$ and $\mathfrak{B}$ is a CA_α of cardinality $\leqq \beta$, then $\mathfrak{B}|\mathfrak{Fr}_\beta\mathsf{CA}_\alpha$. This same conclusion, however, can also be obtained as a corollary of 2.3.28.

The following theorem, 2.5.51, and its corollary, 2.5.53, were communicated to the authors by Don Pigozzi. The corollary provides a simple criterion for the isomorphism of two dimension-restricted free CA_α's which are both locally finite and finitely generated.

THEOREM 2.5.51. *Let $\Delta, \Delta' \in {}^{\beta}Sb\alpha$.*

(i) *In order that*

$$\mathfrak{Fr}_{\beta}^{(\Delta)}\mathsf{CA}_{\alpha} \cong \mathfrak{Fr}_{\beta}^{(\Delta')}\mathsf{CA}_{\alpha},$$

it is sufficient that there be a permutation π of β such that

$$|\Delta\xi \sim \Delta'(\pi\xi)| = |\Delta'(\pi\xi) \sim \Delta\xi| < \omega$$

for every $\xi < \beta$.

(ii) *In case $\beta < \omega$ and $|\Delta\xi| < \omega$ for every $\xi < \beta$, the condition of (i) is also necessary.*

PROOF. For any CA_{α} $\mathfrak{B}$ and any pair of finite, non-empty sets $\Theta, \Omega \subseteq \alpha$ with the same cardinality we set

$$(s_{\Omega}^{\Theta})^{(\mathfrak{B})} = (\mathsf{s}_{\eta_{\kappa}}^{\xi_{\kappa}})^{(\mathfrak{B})} \circ \ldots \circ (\mathsf{s}_{\eta_0}^{\xi_0})^{(\mathfrak{B})}$$

where $\kappa = |\Theta|-1$ and ξ, η are the (uniquely determined) strictly increasing sequences in ${}^{|\Theta|}\alpha$ for which $\Theta = Rg\xi$ and $\Omega = Rg\eta$; in addition we take $(s_0^0)^{(\mathfrak{B})} = B 1 Id$. The following lemma, which is a rather obvious generalization of certain results in Sections 1.5 and 1.6, plays an essential role in the proof of (i).

(1) Let $\mathfrak{B} \in \mathsf{CA}_{\alpha}$ and Θ, Ω be any pair of finite subsets of α such that $|\Theta| = |\Omega|$ and $\Theta \cap \Omega = 0$; let x be any element of B such that $\Delta x \cap \Omega = 0$. Then

$$\Delta[(s_{\Omega}^{\Theta})^{(\mathfrak{B})}x] \subseteq (\Delta x \sim \Theta) \cup \Omega$$

and

$$(s_{\Theta}^{\Omega})^{(\mathfrak{B})}(s_{\Omega}^{\Theta})^{(\mathfrak{B})}x = x.$$

(1) is proved by induction on $|\Theta| = |\Omega|$. If $|\Theta| = 0$, both conclusions of (1) hold trivially. Assume, therefore, that $|\Theta| > 0$ and let ξ and η be the largest ordinals in Θ and Ω, respectively. For convenience set $\Theta' = \Theta \sim \{\xi\}$ and $\Omega' = \Omega \sim \{\eta\}$. Then

$$\begin{aligned} \Delta(s_{\Omega}^{\Theta}x) &= \Delta(\mathsf{s}_{\eta}^{\xi}s_{\Omega'}^{\Theta'}x) \\ &\subseteq [\Delta(s_{\Omega'}^{\Theta'}x) \sim \{\xi\}] \cup \{\eta\} && \text{by 1.6.13} \\ &\subseteq (\Delta x \sim \Theta) \cup \Omega && \text{by the induction hypothesis.} \end{aligned}$$

Using 1.5.10(iii) and the premiss $\Theta \cap \Omega = 0$ we get (by a simple induction argument)

$$s_{\Theta}^{\Omega}s_{\Omega}^{\Theta}x = \mathsf{s}_{\eta}^{\xi}s_{\Theta'}^{\Omega'}\mathsf{s}_{\xi}^{\eta}s_{\Omega'}^{\Theta'}x = \mathsf{s}_{\eta}^{\xi}\mathsf{s}_{\xi}^{\eta}s_{\Theta'}^{\Omega'}s_{\Omega'}^{\Theta'}x.$$

Hence, from 1.5.8(i), 1.5.10(v), the premiss $\Delta x \cap \Omega = 0$, and the induction hypothesis, the second conclusion of (1) is obtained.

Turning to the proof of the theorem we set $\mathfrak{A} = \mathfrak{Fr}_\beta^{(\Delta)}\mathsf{CA}_\alpha$, $\mathfrak{A}' = \mathfrak{Fr}_\beta^{(\Delta')}\mathsf{CA}_\alpha$, and $g_\xi = \xi/Cr_\beta^{(\Delta)}\mathsf{CA}_\alpha$, $g'_\xi = \xi/Cr_\beta^{(\Delta')}\mathsf{CA}_\alpha$, for every $\xi < \beta$. It is clear that in the proof of (i) we may restrict our attention to the case when $\pi = \beta 1 Id$ and hence

$$|\Delta\xi \sim \Delta'\xi| = |\Delta'\xi \sim \Delta\xi| < \omega \text{ for every } \xi < \beta.$$

In order to prove (i) it suffices by 2.5.37 to show that

(2) $\langle (s^{\Delta'\xi\sim\Delta\xi}_{\Delta\xi\sim\Delta'\xi})^{(\mathfrak{A}')} g'_\xi : \xi < \beta \rangle$ CA_α-freely generates $\mathfrak{A}'$ under Δ.

To demonstrate (2) notice first of all that, for each $\xi < \beta$, all the conditions in the hypothesis of (1) hold when $\mathfrak{B}$, Θ, Ω, and x are respectively taken to be $\mathfrak{A}'$, $\Delta'\xi \sim \Delta\xi$, $\Delta\xi \sim \Delta'\xi$, and g'_ξ (since $\Delta^{(\mathfrak{A}')} g'_\xi \subseteq \Delta'\xi$ by 2.5.33(ii)). Hence, applying (1), we easily conclude that

$$\mathfrak{A}' = \mathfrak{Sg}^{(\mathfrak{A}')}\{(s^{\Delta'\xi\sim\Delta\xi}_{\Delta\xi\sim\Delta'\xi})^{(\mathfrak{A}')} g'_\xi : \xi < \beta\}$$

and

$$\Delta^{(\mathfrak{A}')}[(s^{\Delta'\xi\sim\Delta\xi}_{\Delta\xi\sim\Delta'\xi})^{(\mathfrak{A}')} g'_\xi] \subseteq \Delta\xi \text{ for every } \xi < \beta.$$

Consequently,

(3) condition (i) of 2.5.34 is satisfied if we take $\mathfrak{A}'$ for $\mathfrak{A}$ and $\langle (s^{\Delta'\xi\sim\Delta\xi}_{\Delta\xi\sim\Delta'\xi})^{(\mathfrak{A}')} g'_\xi : \xi < \beta \rangle$ for x.

Now consider any CA_α $\mathfrak{B}$ and any $y \in {}^\beta B$ such that

(4) $$\Delta^{(\mathfrak{B})} y_\xi \subseteq \Delta\xi \text{ for every } \xi < \beta.$$

Then, for every $\xi < \beta$, all the conditions in the hypothesis of (1) are satisfied when we take $\Theta = \Delta\xi \sim \Delta'\xi$, $\Omega = \Delta'\xi \sim \Delta\xi$, and $x = y_\xi$. Thus (1) and (4) together imply

(5) $$\Delta^{(\mathfrak{B})}[(s^{\Delta\xi\sim\Delta'\xi}_{\Delta'\xi\sim\Delta\xi})^{(\mathfrak{B})} y_\xi] \subseteq \Delta'\xi \text{ for every } \xi < \beta,$$

and

(6) $$(s^{\Delta'\xi\sim\Delta\xi}_{\Delta\xi\sim\Delta'\xi})^{(\mathfrak{B})}(s^{\Delta\xi\sim\Delta'\xi}_{\Delta'\xi\sim\Delta\xi})^{(\mathfrak{B})} y_\xi = y_\xi,$$

also for every $\xi < \beta$. Since $\langle g'_\xi : \xi < \beta \rangle$ CA_α-freely generates $\mathfrak{A}'$ by 2.5.36, (5) guarantees the existence of an $h \in Hom(\mathfrak{A}', \mathfrak{B})$ such that

$$hg'_\xi = (s^{\Delta\xi\sim\Delta'\xi}_{\Delta'\xi\sim\Delta\xi})^{(\mathfrak{B})} y_\xi$$

for every $\xi < \beta$. Thus, using (6) we have

$$\begin{aligned} h[(s^{\Delta'\xi\sim\Delta\xi}_{\Delta\xi\sim\Delta'\xi})^{(\mathfrak{A}')} g'_\xi] &= (s^{\Delta'\xi\sim\Delta\xi}_{\Delta\xi\sim\Delta'\xi})^{(\mathfrak{B})} hg'_\xi \\ &= (s^{\Delta'\xi\sim\Delta\xi}_{\Delta\xi\sim\Delta'\xi})^{(\mathfrak{B})}(s^{\Delta\xi\sim\Delta'\xi}_{\Delta'\xi\sim\Delta\xi})^{(\mathfrak{B})} y_\xi \\ &= y_\xi \end{aligned}$$

for every $\xi < \beta$. This shows that, when $\mathfrak{A}$ and x are taken to be as in (3), condition (ii) of 2.5.34 also holds. This fact along with (3) gives (2), and part (i) is proved.

We now take up the proof of part (ii); thus we assume

(7) $$\mathfrak{A} \cong \mathfrak{A}'.$$

In view of 2.5.44, the fact that Δ is assumed to be finite-valued implies that Δ' is also finite-valued. Let ρ and ρ' be permutations of β such that

$$|\Delta(\rho 0)| \geqq |\Delta(\rho 1)| \geqq \dots \geqq |\Delta(\rho(\beta-1))|$$

and

$$|\Delta'(\rho' 0)| \geqq |\Delta'(\rho' 1)| \geqq \dots \geqq |\Delta'(\rho'(\beta-1))|.$$

Then, because of part (i) and the assumption (7), we have

$$\mathfrak{Fr}_\beta^{\langle|\Delta(\rho\xi)|:\xi<\beta\rangle}\mathsf{CA}_\alpha \cong \mathfrak{Fr}_\beta^{\langle|\Delta'(\rho'\xi)|:\xi<\beta\rangle}\mathsf{CA}_\alpha,$$

while on the other hand, it is easily seen that the existence of a permutation π of β such that

$$|\Delta\xi \sim \Delta'(\pi\xi)| = |\Delta'(\pi\xi) \sim \Delta\xi| \text{ for every } \xi < \beta$$

is equivalent to the condition

$$|\Delta(\rho\xi)| = |\Delta'(\rho'\xi)| \text{ for every } \xi < \beta.$$

Thus in order to prove (ii) it suffices to assume $\Delta\xi$ and $\Delta'\xi$ to be finite ordinals for every $\xi < \beta$, to assume

(8) $$\Delta 0 \geqq \Delta 1 \geqq \dots \geqq \Delta(\beta-1),\ \Delta' 0 \geqq \Delta' 1 \geqq \dots \geqq \Delta'(\beta-1),$$

and to prove

(9) $$\Delta\xi = \Delta'\xi \text{ for every } \xi < \beta.$$

For each $\kappa < \alpha \cap \omega$ we let π_κ be defined as in 2.4.68. Choosing $\mathfrak{B}_\kappa$ to be a minimal algebra of characteristic κ for each $\kappa < \alpha \cap \omega$ (cf. the remark following 2.4.62), we let L_κ be the set of all sequences $x \in {}^\beta B_\kappa$ such that $\Delta^{(\mathfrak{B}_\kappa)} x_\xi \subseteq \Delta\xi$ for every $\xi < \beta$; similarly L'_κ is the set of all sequences $y \in {}^\beta B_\kappa$ such that $\Delta^{(\mathfrak{B}_\kappa)} y_\xi \subseteq \Delta'_\xi$ for every $\xi < \beta$. It follows from 2.4.68(ii) that, for each $\kappa < \alpha \cap \omega$,

(10) $$|L_\kappa| = 2^{\Sigma\{\pi_\kappa \Delta\xi:\xi<\beta\}} \text{ and } |L'_\kappa| = 2^{\Sigma\{\pi_\kappa \Delta'\xi:\xi<\beta\}}.$$

By 0.2.14(iii), 0.3.2, and 2.5.36 we see that for each $\kappa < \alpha \cap \omega$ there is a one-one correspondence between the elements of L_κ and the elements of $Hom(\mathfrak{A}, \mathfrak{B}_\kappa)$, and also one between the elements of L'_κ and the elements of $Hom(\mathfrak{A}', \mathfrak{B}_\kappa)$. Hence by (7) we have $|L_\kappa| = |L'_\kappa|$ for $\kappa < \alpha \cap \omega$, and thus from (10) we conclude that

(11) $$\sum_{\xi<\beta} \pi_\kappa(\Delta\xi) = \sum_{\xi<\beta} \pi_\kappa(\Delta'\xi) \text{ for every } \kappa < \alpha \cap \omega.$$

At this point we assume by way of contradiction that condition (9) fails to hold, i.e., that there exists an ordinal $\xi < \beta$ such that $\Delta\xi \neq \Delta'\xi$. Let λ be the least such ordinal ξ. Then

$$\sum_{\xi<\lambda} \pi_\kappa(\Delta\xi) = \sum_{\xi<\lambda} \pi_\kappa(\Delta'\xi)$$

and hence, by (11),

(12) $$\sum_{\lambda\leqq\xi<\beta} \pi_\kappa(\Delta\xi) = \sum_{\lambda\leqq\xi<\beta} \pi_\kappa(\Delta'\xi)$$

for every $\kappa < \alpha \cap \omega$. On the other hand, if we assume (obviously without loss of generality) that $\Delta\lambda > \Delta'\lambda$, then it is easily seen from (8) and the definition of $\pi_{\Delta\lambda}$ and $\pi_{\Delta\lambda-1}$ that

$$\sum_{\lambda\leqq\xi<\beta} \pi_{\Delta\lambda}(\Delta\xi) > \sum_{\lambda\leqq\xi<\beta} \pi_{\Delta\lambda-1}(\Delta\xi)$$

while

$$\sum_{\lambda\leqq\xi<\beta} \pi_{\Delta\lambda}(\Delta'\xi) = \sum_{\lambda\leqq\xi<\beta} \pi_{\Delta\lambda-1}(\Delta'\xi).$$

These two formulas together contradict the fact that (12) holds both with $\kappa = \Delta\lambda$ and $\kappa = \Delta\lambda - 1$. This completes the proof of the theorem.

COROLLARY 2.5.52. *Let* $\Delta \in {}^\beta Sb\alpha$ *and* $|\Delta\xi| < \omega$ *for every* $\xi < \beta$. *Set* $\Delta' = \langle |\Delta\xi| : \xi < \beta \rangle$. *Then*

$$\mathfrak{Fr}_\beta^{(\Delta)}\mathsf{CA}_\alpha \cong \mathfrak{Fr}_\beta^{(\Delta')}\mathsf{CA}_\alpha.$$

COROLLARY 2.5.53. *Let* $\Delta \in {}^\beta Sb\alpha$ *and* $\Delta' \in {}^{\beta'} Sb\alpha$; *let* $\beta < \omega$, *and* $|\Delta\xi| < \omega$ *for every* $\xi < \beta$. *In order that*

$$\mathfrak{Fr}_\beta^{(\Delta)}\mathsf{CA}_\alpha \cong \mathfrak{Fr}_{\beta'}^{(\Delta')}\mathsf{CA}_\alpha,$$

it is necessary and sufficient that $\beta = \beta'$ *and there be a permutation* π *of* β *such that*

$$|\Delta\xi| = |\Delta'(\pi\xi)|$$

for every $\xi < \beta$.

PROOF: by 2.5.39(iii) and 2.5.51.

REMARKS 2.5.54. The problem to what extent both finiteness conditions in 2.5.51(ii) and 2.5.53 are essential has not yet been cleared up, although some fragmentary results are known. Notice that under the assumption that Δ is finite-valued our problem can be formulated as follows. We consider the condition:

$$|\{\xi:\xi < \beta, |\Delta\xi| = \kappa\}| = |\{\xi:\xi < \beta', |\Delta'\xi| = \kappa\}| \text{ for every } \kappa < \omega.$$

From 2.5.51 it is easily seen that this condition is sufficient for the isomorphism of $\mathfrak{Fr}_\beta^{(\Delta)}\mathsf{CA}_\alpha$ and $\mathfrak{Fr}_{\beta'}^{(\Delta')}\mathsf{CA}_\alpha$, and that it is necessary in the case when β is finite; the question is open whether it is necessary in the general case.

Closely related to the problem just discussed is the analogous problem concerning Theorem 2.5.41: is the condition $\beta < \omega$ in this theorem essential? For instance, using 2.5.40 it is easily seen that, if the finiteness conditions in 2.5.53 can be omitted, the following statement holds: the algebras ${}^\kappa\mathfrak{Fr}_\beta^{(\Delta)}\mathsf{CA}_\alpha$ and ${}^\lambda\mathfrak{Fr}_\beta^{(\Delta)}\mathsf{CA}_\alpha$ with $\kappa < \lambda < \omega$ are isomorphic iff the set $\{\xi:\xi < \beta, \Delta\xi = 0\}$ is infinite.

REMARKS 2.5.55. In the course of the present section we have acquired some scattered information concerning intrinsic properties of free algebras $\mathfrak{Fr}_\beta\mathsf{CA}_\alpha$ and dimension-restricted free algebras $\mathfrak{Fr}_\beta^{(\Delta)}\mathsf{CA}_\alpha$. With the exception, however, of some particularly simple cases (such as those of $\mathfrak{Fr}_\beta\mathsf{CA}_0$ and $\mathfrak{Fr}_\beta\mathsf{CA}_1$) we are still very far from being able to give an exhaustive intrinsic characterization of any of these algebras, and we have no general methods which would permit us to solve various problems in this domain.

As we have seen, for instance, some very simple problems concerning atoms in free algebras remain open; cf. Remarks 2.5.10 and 2.5.12. Some open problems which concern the dimensionality of elements of free algebras are also known; cf. Remark 2.6.24.

We wish to discuss here in detail certain interesting and important open problems concerning primarily the zero-dimensional parts of locally finite algebras $\mathfrak{Fr}_\beta^{(\Delta)}\mathsf{CA}_\alpha$. They have a less elementary character than those just mentioned; their formulation and discussion involve some auxiliary notions and results from the general theory of Boolean algebras. Since these notions are not widely known and some of the results are not available in the literature, we shall provide the necessary information in the next remarks.

REMARKS 2.5.56. The main topic of these remarks is a division of all BA's into two mutually exclusive classes. For either of the two classes a number of interesting properties are known which are characteristic for all algebras of the class involved; there is a great conceptual diversity among properties characterizing the same class, and the fact that they are all mutually equivalent appears to be a rather unexpected and non-trivial result. As the basis for a formal definition of either class we can choose, of course, any one of its characteristic properties.

To fix ideas we agree to say that a BA $\mathfrak{B}$ is *of class* 1 iff $\mathfrak{B}$ together with all its homomorphic images is atomic; $\mathfrak{B}$ is *of class* 2 iff it has at least one non-atomic homomorphic image. Thus, for instance, every finite BA is of class 1, and so is, more generally, every BA generated by the set of its atoms; on the other hand, every infinite complete BA, whether atomic or not, is easily seen to be of class 2. We shall state here a number of basic results concerning this classification of BA's. In most cases we find it convenient to refer the results exclusively to algebras of class 1, omitting the dual results which concern algebras of class 2.

In our discussion we shall often deal with ordered sets of elements of a BA $\mathfrak{B}$, meaning by this the subsets of B which are simply ordered by the Boolean inclusion $\leqq$. Recall that a set X simply ordered by a relation R is called *dense* (or *densely ordered*) if $|X| \geqq 2$ and for any elements $x, z \in X$, the formulas xRz and $x \neq z$ imply the existence of an element $y \in X$ such that $xRyRz$ and $x \neq y \neq z$. X is called *scattered* if it does not have any densely ordered subset; thus, in particular, every well-ordered set is scattered.

To obtain a useful (and practically indispensable) tool for studying the two classes of algebras, we correlate, with every BA $\mathfrak{B}$, a transfinite sequence of ideals I_ξ and the sequence of corresponding quotient algebras $\mathfrak{D}_\xi$, where ξ ranges over arbitrary ordinals. The two sequences are defined recursively by the formulas:

$$I_\xi = I_\xi^{(\mathfrak{B})} = Ig^{(\mathfrak{B})} \bigcup \{\bigcup At\,\mathfrak{D}_\eta : \eta < \xi\}, \quad \mathfrak{D}_\xi = \mathfrak{D}_\xi^{(\mathfrak{B})} = \mathfrak{B}/I_\xi.$$

The algebra $\mathfrak{D}_\xi$ may be called the ξth *derivative* of the algebra $\mathfrak{B}$. (From the familiar results on a topological representation of BA's it is easily seen that the topological space correlated with $\mathfrak{D}_\xi$ is homeomorphic with the topological ξ^{th} derivative of the space correlated with $\mathfrak{B}$.)

There is a close connection between the sequences of ideals and derivative algebras thus defined and the division of BA's into classes 1 and 2. This is brought to light by the following argument. Obviously, for any given BA $\mathfrak{B}$ we have $I_\xi \subseteq I_\eta \subseteq B$ whenever $\xi \leqq \eta$. Hence there must be an ordinal ξ such that $I_\xi = I_\eta$ for every $\eta \geqq \xi$. Two cases are now possible: (1) $I_\xi = B$ and thus $\mathfrak{D}_\xi$ is a trivial (one-element) algebra. In this case $\mathfrak{B}$ can be shown to be of class 1. (2) $I_\xi = I_{\xi+1} \neq B$. It is then easily seen that the BA $\mathfrak{D}_\xi$, which is a homomorphic image of $\mathfrak{B}$, is atomless, so that $\mathfrak{B}$ is of class 2. This argument plays an essential role in the proof of the next theorem, where a number of conditions are stated each of which is characteristic for algebras of class 1.

(I) *For every* BA $\mathfrak{B}$ *the following conditions are equivalent:*

(i) $\mathfrak{B}$ *is of class* 1;
(ii) *no non-trivial homomorphic image of* $\mathfrak{B}$ *is atomless;*
(iii) *every subalgebra of* $\mathfrak{B}$ *is atomic;*
(iv) *no non-trivial subalgebra of* $\mathfrak{B}$ *is atomless;*
(v) *no ordered set* $X \subseteq B$ *is dense;*
(vi) *no countable ordered set* $X \subseteq B$ *is dense;*
(vii) *there is an ordinal* ξ *such that the algebra* $\mathfrak{D}_\xi$ *is trivial;*
(viii) *either* $\mathfrak{B}$ *is trivial or else there is an ordinal* ζ *such that the algebra* $\mathfrak{D}_\zeta$ *is finite and not trivial.*

A detailed proof of the equivalence of conditions (i)–(vii) can be found in Mostowski-Tarski [39*], Lemmas 3.4, 3.5, 3.10, and Theorem 3.12. If (vii) holds, there is a least ξ for which $\mathfrak{D}_\xi$ is trivial and hence $I_\xi = B$. If, in addition, $\mathfrak{B}$ is not trivial, then $\xi \neq 0$. Using the fact that the ideal I_ξ is principal, we easily show that ξ must be of the form $\xi = \zeta+1$. From $I_{\zeta+1} = B$ it follows that $\mathfrak{D}_\zeta$ is finite; since $\zeta < \xi$, $\mathfrak{D}_\zeta$ is not trivial. Thus, (vii) implies (viii); conversely, (viii) implies (vii), for $\mathfrak{D}_{\zeta+1}$ is trivial whenever $\mathfrak{D}_\zeta$ is finite.

In view of both the definition of class 1 and the condition (I)(iii), Boolean algebras of class 1 can be referred to as *hereditarily atomic* (using the term "hereditarily" in this context in a much stronger sense than the one specified in 0.3.32). The notion of hereditarily atomic BA's was introduced and studied in Mostowski-Tarski [39*]; some further results can be found in Day [67*].

By (I)(i),(viii), for every non-trivial $\mathfrak{B}$ of class 1 there is an ordinal ζ such that $\mathfrak{D}_\zeta$ is finite and not trivial. If ζ is any such ordinal, then $\mathfrak{D}_\eta$ is trivial for every $\eta > \zeta$, and $\mathfrak{D}_\eta$ is infinite for every $\eta < \zeta$. Thus the ordinal ζ is uniquely determined. We denote it by $\gamma\mathfrak{B}$, and we let in addition $\gamma\mathfrak{B} = 0$ in case $\mathfrak{B}$ is trivial. Furthermore, we set $\kappa\mathfrak{B} = |At\,\mathfrak{D}_{\gamma\mathfrak{B}}|$, so that $\kappa\mathfrak{B}$ is always finite, and $\kappa\mathfrak{B} = 0$ iff $\mathfrak{B}$ is trivial. The ordered pair $\langle\gamma\mathfrak{B}, \kappa\mathfrak{B}\rangle$ may be called the *characteristic pair* of the BA $\mathfrak{B}$ of class 1. As we shall see, the characteristic pair $\langle\gamma\mathfrak{B}, \kappa\mathfrak{B}\rangle$ plays an important role in the discussion of BA's of class 1. The following results are known in this connection:

(II) *If* $\mathfrak{B}$ *is a* BA *of class* 1 *and* $\mathfrak{C} \preccurlyeq \mathfrak{B}$, *then* $\mathfrak{C}$ *is also of class* 1 *and we have either* $\gamma\mathfrak{C} < \gamma\mathfrak{B}$, *or else* $\gamma\mathfrak{C} = \gamma\mathfrak{B}$ *and* $\kappa\mathfrak{C} \leqq \kappa\mathfrak{B}$.

(III) *If* $\mathfrak{B}$ *is a* BA *of class* 1 *and* $\mathfrak{C} \subseteq \mathfrak{B}$, *then* $\mathfrak{C}$ *is also of class* 1 *and we have* $\gamma\mathfrak{C} < \gamma\mathfrak{B}$, *or else* $\gamma\mathfrak{C} = \gamma\mathfrak{B}$ *and* $\kappa\mathfrak{C} \leqq \kappa\mathfrak{B}$.

(IV) *If* $\mathfrak{C}, \mathfrak{D}$ *are* BA's *of class* 1 *and* $\mathfrak{B} \cong \mathfrak{C} \times \mathfrak{D}$, *then* $\mathfrak{B}$ *is of class* 1 *and*

$\gamma\mathfrak{B} = \gamma\mathfrak{C} \cup \gamma\mathfrak{D}$, *while* $\kappa\mathfrak{B}$ *equals* $\kappa\mathfrak{C}$, $\kappa\mathfrak{D}$, *or* $\kappa\mathfrak{C}+\kappa\mathfrak{D}$ *dependent on whether* $\gamma\mathfrak{C} > \gamma\mathfrak{D}$, $\gamma\mathfrak{C} < \gamma\mathfrak{D}$, *or* $\gamma\mathfrak{C} = \gamma\mathfrak{D}$.

(V) *If* $\mathfrak{B}$ *is a* BA, $\mathfrak{C}, \mathfrak{D}$ *are* BA's *of class* 1, $\mathfrak{C}, \mathfrak{D} \subseteq \mathfrak{B}$, $\mathfrak{B} = \mathfrak{Sg}(C \cup D)$, *and* $x \cdot y \neq 0$ *whenever* $x \in C$, $y \in D$, *and* $x, y \neq 0$, *then* $\mathfrak{B}$ *is of class* 1, $\gamma\mathfrak{B} = \gamma\mathfrak{C} \# \gamma\mathfrak{D}$ *(the natural sum of* $\gamma\mathfrak{C}$ *and* $\gamma\mathfrak{D}$*), and* $\kappa\mathfrak{B} = \kappa\mathfrak{C} \cdot \kappa\mathfrak{D}$.

For the notion of the natural sum of two ordinals (due to Gerhard Hessenberg) which occurs in (V) see, e.g., Sierpiński [65*], pp. 366 f.

The proofs of (II)–(V) are not available in the literature. They are based on (I) and on the following four lemmas derived by transfinite induction from the definition of the ideals I_ξ:

(II′) *If* $\mathfrak{B}$ *is a* BA *and* $J \in Il\,\mathfrak{B}$, *then* $I_\xi^{(\mathfrak{B})}/J \subseteq I_\xi^{(\mathfrak{B}/J)}$ *(for every ordinal* ξ*).*

(III′) *If* $\mathfrak{B}, \mathfrak{C}$ *are* BA's *and* $\mathfrak{C} \subseteq \mathfrak{B}$, *then* $C \cap I_\xi^{(\mathfrak{B})} \subseteq I_\xi^{(\mathfrak{C})}$.

(IV′) *If* $\mathfrak{C}, \mathfrak{D}$ *are* BA's, *then* $I_\xi^{(\mathfrak{C}\times\mathfrak{D})} = I_\xi^{(\mathfrak{C})} \times I_\xi^{(\mathfrak{D})}$.

(V′) *Under the hypothesis of* (V) *we have:*

$I_\xi^{(\mathfrak{B})} = Ig \bigcup \{I_\eta^{(\mathfrak{C})} \cdot I_\zeta^{(\mathfrak{D})} : \eta \# \zeta \leqq \xi + 1\}$ *where* $I_\eta^{(\mathfrak{C})} \cdot I_\zeta^{(\mathfrak{D})} = \{x \cdot y : x \in I_\eta^{(\mathfrak{C})}, y \in I_\zeta^{(\mathfrak{D})}\}$.

Notice the following more concise, but essentially equivalent, formulation of the hypothesis of (V) and (V′): $\mathfrak{B}$ *is isomorphic to the* BA-*free product of* BA's $\mathfrak{C}$ *and* $\mathfrak{D}$ *of class* 1. For the equivalence of the two formulations see Sikorski [64*], pp. 39 ff. The general algebraic notion of a free product was mentioned in 0.4.65.[1)]

(IV) and (V) do not extend to infinite systems of algebras; see Day [67*], p. 381.

A simple and rather general method of constructing BA's of class 1, or class 2, with some further prescribed properties, e.g., BA's of class 1 with a given characteristic pair $\langle\gamma, \kappa\rangle$, is provided by the discussion of the so-called BA's with ordered bases. By an *ordered base* of a BA $\mathfrak{B}$ we understand any ordered set $X \subseteq B$ (i.e., X is simply ordered by $\leqq$) which generates $\mathfrak{B}$. We find it convenient, however, to impose on X two additional conditions, $0 \in X$

[1)] Theorems (II)–(IV) were originally established by Mostowski and Tarski for arbitrary BA's with well-ordered bases (which will be discussed below); compare the next footnote in this section. Theorem (V) is an easy consequence of topological results in Telgársky [68*]. When formulated in terms of free products, it improves a result in Mayer-Pierce [60*], Theorem 5.9, which applies only to countable BA's with well-ordered bases. The present forms of Theorems (III)–(V) are due to Don Pigozzi. He also supplied a purely algebraic proof of (V) using Lemma (V′).

and $1 \notin X$; since, under these additional conditions, a trivial BA would have no ordered base, we stipulate that the empty set will be considered as an ordered base of a trivial BA. For a detailed discussion of BA's with ordered bases see Mostowski-Tarski [39*] and Mayer-Pierce [60*].[1)]

In connection with BA's with ordered bases two notions very rarely involved in the theory of CA's will occur in these remarks: the notion of the order type of an arbitrary simply ordered set and that of the sum $\gamma+\delta$ of two order types. In agreement with the Preliminaries we do not distinguish between ordinals and order types of well-ordered sets.

The applications of BA's with ordered bases to the main topic of the present remarks rests upon two results, (VI) and (VII). The first of them concerns arbitrary BA's with ordered bases; in the second we characterize those BA's with ordered basis which are of class 1.

(VI) (i) *For every order type* β *there is a* BA $\mathfrak{B}$ *with an ordered base of type* $1+\beta$.

(ii) *Let* $\mathfrak{B}$ *and* $\mathfrak{C}$ *be two* BA*'s with ordered bases. For* $\mathfrak{B} \cong \mathfrak{C}$ *it is sufficient (and also necessary) that* $\mathfrak{B}$ *and* $\mathfrak{C}$ *have ordered bases of the same order type.*

(VII) *For every* BA $\mathfrak{B}$ *the following conditions are equivalent*:

(i) $\mathfrak{B}$ *has an ordered base and is of class* 1;
(ii) $\mathfrak{B}$ *has a scattered ordered base*;
(iii) $\mathfrak{B}$ *has an ordered base, and every ordered base of* $\mathfrak{B}$ *is scattered.*

A proof of (VI) and (VII) can be found in Mostowski-Tarski [39*], Theorems 1.7, 2.1, and 3.15.

As a consequence of (VII) the class of BA's of class 1 with ordered bases contains, in particular, all BA's with well-ordered bases. The study of these algebras is simpler and leads to more complete results than that of BA's with arbitrary ordered bases or even of BA's with scattered ordered bases. In general, an infinite BA with an ordered base has ordered bases of many different order types, and no simple characterization of the set of all these types has yet

[1)] As a sequel to Mostowski-Tarski [39*] the same authors prepared in 1939 a paper which was to appear in Fundamenta Mathematicae, Volume 33. However, the type-setting was destroyed during the Second World War and the paper was never reconstructed. The paper is mentioned in Horn-Tarski [48*], p. 484, footnote 15 and Hanf [64*], p. 327, footnote 3. It contained a number of results on BA's with scattered and well-ordered bases and their applications to countable BA's; as examples we may mention Theorems (II)–(IV) above restricted to BA's with well-ordered bases, Theorems (VIII) and (XI)–(XIII) below, the construction of a BA with a well-ordered base which has a subalgebra without any ordered base, and the construction of a BA with a scattered ordered base but with no well-ordered base.

been found. Such a characterization, however, can easily be provided if we restrict our attention to well-ordered bases:

(VIII) *If* $\mathfrak{B}$ *is a non-trivial* BA *with a well-ordered base, and* Γ *is the set of all order types of well-ordered bases of* $\mathfrak{B}$, *then* $\mathfrak{B}$ *is of class* 1, *and*

$$\Gamma = \{\xi : \omega^{\gamma\mathfrak{B}} \cdot \kappa\mathfrak{B} \leqq \xi < \omega^{\gamma\mathfrak{B}} \cdot (\kappa\mathfrak{B}+1)\}.$$ [1)]

To prove (VIII) we first notice that, if a BA has an ordered base of type $1+\xi+1+\eta$, it has also an ordered base of type $1+\eta+1+\xi$. We next show by transfinite induction that, if $\mathfrak{B}$ has a well-ordered base of type $\omega^{\gamma}\cdot\kappa$ where γ is any ordinal and $0 < \kappa < \omega$, then $\gamma = \gamma\mathfrak{B}$ and $\kappa = \boldsymbol{\kappa}\mathfrak{B}$. (Here and until the end of this section the symbols of the form ω^{ξ} denote, of course, ordinal powers of ω.) From these two facts we easily derive the conclusion of (VIII) using some familiar results of the arithmetic of transfinite ordinals.

Using (VI)(i) and (VIII), for any given ordinal γ and finite ordinal $\kappa \neq 0$ we can construct a BA of class 1 with the characteristic pair $\langle\gamma, \kappa\rangle$; in fact, we can take for $\mathfrak{B}$ a BA with a well-ordered base of type $\omega^{\gamma}\cdot\kappa$. A simple corollary from (VI)(ii) and (VIII) is:

(IX) *Let* $\mathfrak{B}$ *and* $\mathfrak{C}$ *be two* BA's *with well-ordered bases. For* $\mathfrak{B} \cong \mathfrak{C}$ *it is sufficient (as well as necessary) that* $\gamma\mathfrak{B} = \gamma\mathfrak{C}$ *and* $\boldsymbol{\kappa}\mathfrak{B} = \boldsymbol{\kappa}\mathfrak{C}$.

(VIII) and (IX) provide jointly a description of all isomorphism types of BA's with well-ordered bases. We see in particular that, for any given infinite cardinal β, there are just β^+ isomorphism types of BA's of cardinality β which have well-ordered bases.

We now correlate two ordinals, $\boldsymbol{\delta}\mathfrak{B}$ and $\boldsymbol{\varepsilon}\mathfrak{B}$, with any given BA $\mathfrak{B}$. $\boldsymbol{\delta}\mathfrak{B}$ is defined as the union (the least upper bound) of all ordinals ξ which are order types of well-ordered sets $X \subseteq B$; by $\boldsymbol{\varepsilon}\mathfrak{B}$ we denote the union of all ordinals $\boldsymbol{\delta}\mathfrak{Rl}_x\mathfrak{B}$ where x is any element of B such that $\mathfrak{Rl}_x\mathfrak{B}$ is of class 1. The ordinals $\boldsymbol{\delta}\mathfrak{B}$ and $\boldsymbol{\varepsilon}\mathfrak{B}$ will be seen to play an especially important role in the discussion of countable BA's.

(X) *Let* $\mathfrak{B}$ *be a* BA; *let* $\alpha = |At\,\mathfrak{B}|+1$ *if* $\mathfrak{B}$ *is finite and* $\alpha = |B|^+$ *if* $\mathfrak{B}$ *is infinite. We then have*:

(i) $\boldsymbol{\varepsilon}\mathfrak{B} \leqq \boldsymbol{\delta}\mathfrak{B} \leqq \alpha$; $\boldsymbol{\varepsilon}\mathfrak{B} < \alpha$ *if* $|B| \geqq \omega$;

(ii) $\boldsymbol{\delta}\mathfrak{B}$ *and* $\boldsymbol{\varepsilon}\mathfrak{B}$ *are ordinals of the form* $\omega^{\gamma}\cdot\kappa$ *with* $0 < \kappa < \omega$;

(iii) *if* $\mathfrak{B}$ *has a well-ordered base, then*

$$\boldsymbol{\delta}\mathfrak{B} = \boldsymbol{\varepsilon}\mathfrak{B} = \omega^{\gamma\mathfrak{B}}\cdot(\boldsymbol{\kappa}\mathfrak{B}+1).$$

[1)] Concerning this theorem compare the footnote on the preceding page. The theorem is related to Theorem 4.6 in Mayer-Pierce [60*].

(X) is easily established with the help of (III) and (VIII).

In connection with (X)(i) it will be seen from (XIII)(i),(iv) that there are infinite BA's $\mathfrak{B}$ for which $\delta\mathfrak{B} = \alpha$; such BA's may even be atomless, so that $\varepsilon\mathfrak{B} = 0$. In connection with (X)(iii) it is known that there are BA's of class 1, and indeed BA's with scattered ordered bases, such that $\delta\mathfrak{B} < \omega^{\gamma\mathfrak{B}} \cdot (\kappa\mathfrak{B}+1)$.

From (IX) and (X)(iii) it follows that the ordinal $\delta\mathfrak{B}$ alone determines uniquely the isomorphism type of a BA $\mathfrak{B}$ with a well-ordered base.

We turn now to the results concerning the division of countable BA's into classes 1 and 2. The discussion of these algebras is greatly influenced by the following two fundamental results:

(XI) (i) *Every countable* BA *has an ordered base.*

(ii) *Every countable* BA *which has a scattered ordered base has also a well-ordered base.*

(i) is a consequence of the well-known fact that a countable BA with a dense ordered base is isomorphic to $\mathfrak{Fr}_\omega$BA. (ii) follows easily from familiar theorems on the topological representation of BA's combined with results in Mazurkiewicz-Sierpiński [20*].

It is known that neither of the two parts of (XI) extends to uncountable BA's.[1]

With the essential help of (XI) we can now supplement (I) by stating a series of new conditions which are necessary and sufficient for a countable BA to be of class 1; this is done in (XII). Some further conditions, which assume more natural forms when referred to countable BA's of class 2, will be formulated separately in (XIII).

(XII) *For every countable* BA $\mathfrak{B}$ *the following conditions are equivalent*:

(i) $\mathfrak{B}$ *is of class* 1;

(ii) $\mathfrak{B}$ *has a scattered ordered base*;

(iii) $\mathfrak{B}$ *has a well-ordered base*;

(iv) $\delta\mathfrak{B} < \omega_1$; *in other words, there is a countable ordinal* ξ *which is not the order type of any well-ordered set* $X \subseteq B$;

(v) *the set of all maximal proper ideals of* $\mathfrak{B}$ *is countable.*

(XIII) *For every countable* BA $\mathfrak{B}$ *the following conditions are equivalent*:

(i) $\mathfrak{B}$ *is of class* 2;

(ii) *every countable* BA *is a homomorphic image of* $\mathfrak{B}$;

[1]) See Mayer-Pierce [60*], Theorem 2.10 and Lemma 3.4. Compare also footnote [1]) on page 367.

(iii) *every countable* BA *is isomorphic to a subalgebra of* $\mathfrak{B}$;

(iv) $\delta\mathfrak{B} = \omega_1$;

(v) *the set of all maximal proper ideals of* $\mathfrak{B}$ *has the cardinality of the continuum.*

It is convenient to prove (XII) and (XIII) simultaneously. The equivalence of conditions (i)–(iii) of (XII) is a direct consequence of (VII) and (XI). Using (X)(i),(iii) and (I), as well as the well-known theorem of Cantor by which a countable densely ordered set includes subsets of each countable order type, we show that (XII)(iii) implies (XII)(iv), and that (XIII)(i) implies (XIII)(iv); hence the implications in the opposite directions follow automatically. Each of the conditions (ii) and (iii) of (XIII) implies (XIII)(i) by (II) and (III); the implications in the opposite directions are obtained with the help of the same theorem of Cantor and Theorem 2.2 of Mostowski-Tarski [39*]. Finally, the fact that conditions (v) in (XII) and (XIII) are respectively equivalent to the remaining conditions of these two theorems is an easy consequence of results in Mostowski-Tarski [39*], Corollaries 4.2 and 4.3.

It can be shown that conditions (i) and (iii) of (XIII) are equivalent for all BA's, whether countable or not. On the other hand, from Theorem 4.4 in Smith-Tarski [57*] it is seen that there are BA's $\mathfrak{B}$ of class 2 which do not satisfy (XIII)(ii).

The fact that every countable BA of class 1 satisfies condition (XII)(v) extends to uncountable BA's in the following sense:

(XIV) *If* $\mathfrak{B}$ *is an infinite* BA *of class* 1, *then the set of all maximal proper ideals of* $\mathfrak{B}$ *has the same cardinality as* $\mathfrak{B}$.

This is an unpublished result of Tarski. He also noticed that for every BA $\mathfrak{B}$ of class 1 one can prove *without the axiom of choice* that every proper ideal is included in a maximal proper ideal.

From the above results it appears clearly that the evaluation of $\delta\mathfrak{B}$ and $\varepsilon\mathfrak{B}$ is of paramount importance for the study of any given countable $\mathfrak{B}$. By (X)(i) we have $\delta\mathfrak{B} \leq \omega_1$. If $\delta\mathfrak{B}$ proves to be less than ω_1, in which case $\delta\mathfrak{B} = \varepsilon\mathfrak{B}$, then, by (XII), $\mathfrak{B}$ is of class 1 and has a well-ordered base; hence, by (X)(iii), $\delta\mathfrak{B} = \omega^{\gamma\mathfrak{B}}\cdot(\kappa\mathfrak{B}+1)$ and, by (VIII), $\mathfrak{B}$ has a well-ordered base of type $\omega^{\gamma\mathfrak{B}}\cdot\kappa\mathfrak{B}$. Thus, in this case the isomorphism type of $\mathfrak{B}$ is uniquely determined; if we know the exact value of $\delta\mathfrak{B}$, and hence of $\gamma\mathfrak{B}$ and $\kappa\mathfrak{B}$, we get a perfect insight into the structure of $\mathfrak{B}$. We obtain also an exhaustive description of all isomorphism types of countable BA's of class 1, and we see that the variety of all these types is of cardinality ω_1.

If, on the other hand, $\delta\mathfrak{B} = \omega_1$, then, by (XII), $\mathfrak{B}$ has no well-ordered base. Still, by (XI), it has an ordered base, but each of its ordered bases has a densely ordered subset. However, the isomorphism type of $\mathfrak{B}$ is now by no means uniquely determined by the value of $\delta\mathfrak{B}$. On the contrary, it is known that the variety of all isomorphism types of countable BA's $\mathfrak{B}$ without a well-ordered base, hence with $\delta\mathfrak{B} = \omega_1$, has the cardinality of the continuum; no exhaustive description of all these types is available. In this case the evaluation of $\varepsilon\mathfrak{B}$ is especially important. We know from (X) that $\varepsilon\mathfrak{B}$ is of the form $\omega^{\gamma}\cdot\kappa$ with $\gamma < \omega_1$ and $\kappa < \omega$; for any $\gamma < \omega_1$ and $\kappa < \omega$ there is a countable BA $\mathfrak{B}$ without a well-ordered base such that $\varepsilon\mathfrak{B} = \omega^{\gamma}\cdot\kappa$. The value of $\varepsilon\mathfrak{B}$ still does not determine uniquely the isomorphism type of $\mathfrak{B}$ itself, but it determines at least the variety of all isomorphism types of BA's with well-ordered bases which can be obtained from $\mathfrak{B}$ by relativization (or, in other words, which are direct factors of $\mathfrak{B}$). This throws much light on the structure of $\mathfrak{B}$ and in particular on the order types of all its ordered bases; however, we shall not here go into details.

Remarks 2.5.57. The problems discussed in Remarks 2.5.56 (especially in their last portion) extend in a natural way to the theory of cylindric algebras. Given a (countable) CA $\mathfrak{A}$, we can take for $\mathfrak{B}$ the Boolean part of $\mathfrak{A}$, $\mathfrak{Bl}\,\mathfrak{A}$, or the zero-dimensional part of $\mathfrak{A}$, $\mathfrak{Zd}\,\mathfrak{A}$; we can then try to evaluate $\delta\mathfrak{B}$ and $\varepsilon\mathfrak{B}$, and to draw from this evaluation further conclusions concerning the structure of $\mathfrak{B}$.

The problems thus arising for $\mathfrak{B} = \mathfrak{Bl}\,\mathfrak{A}$ are sometimes quite difficult, even though they may not present special interest. Let, for instance, $\mathfrak{A} = \mathfrak{Fr}_\beta\mathsf{CA}_\alpha$ with $\alpha, \beta \leqq \omega$. If $\alpha \leqq 1$ and $\beta < \omega$, then $\mathfrak{B}$ is finite by 2.5.3(i) and hence of class 1; if $\beta = \omega$, then $\mathfrak{B}$ is atomless by 2.5.13 and hence of class 2; in both cases the computation of $\delta\mathfrak{B}$ and $\varepsilon\mathfrak{B}$ is trivial. In the remaining cases it is easily seen that $\mathfrak{B}$ has non-atomic homomorphic images and hence is again of class 2. However, the value of $\varepsilon\mathfrak{B}$ is not known. This problem is closely related to the results and problems mentioned in 2.5.7–2.5.12. For instance, in case $\alpha = \omega$ (and $\beta < \omega$) it seems likely that $\varepsilon\mathfrak{B} = 2^{\beta}+1$ (see 2.5.12); the problem of determining $\varepsilon\mathfrak{B}$ in case $\alpha = 2$ should not present much difficulty. However, we would not risk any conjecture concerning the case $2 < \alpha < \omega$. The general situation is similar if we consider $\mathfrak{Fr}_\beta^{(\Delta)}\mathsf{CA}_\alpha$ instead of $\mathfrak{Fr}_\beta\mathsf{CA}_\alpha$.

We arrive at genuinely interesting problems when we take for $\mathfrak{B}$ the algebra $\mathfrak{Zd}\,\mathfrak{A}$. This applies particularly to the case when $\mathfrak{A}$ is locally finite and of infinite dimension. As we know, $\mathfrak{A}$ can then be isomorphically represented as the algebra of formulas and $\mathfrak{B}$ as the algebra of sentences associated with a theory Θ in a language of predicate logic. By 2.5.56(XII),(XIII) $\mathfrak{B}$ is of class 1 or 2 dependent on, e.g., whether Θ has countably or continuum many complete and consistent extensions. The evaluation of $\boldsymbol{\delta}\mathfrak{B}$ and $\boldsymbol{\varepsilon}\mathfrak{B}$ may depend on difficult metalogical results concerning the decidability of the theory Θ and the variety of its extensions; the solution of the problem may throw much light on the structure of Θ. Let, in particular, $\mathfrak{A}$ be any algebra $\mathfrak{Fr}_{\beta}^{(\Delta)}\mathsf{CA}_{\alpha}$ with $\omega_1 > \alpha \geqq \omega$, β finite, Δ finite-valued, and $|\Delta\xi| \geqq 2$ for some $\xi < \beta$. As we know, $\mathfrak{A}$ can be represented as the algebra of formulas associated with the logic Θ of a language Λ containing at least one predicate of rank $\geqq 2$ (and in which the sequence of variables is of length α). It is known that every algebra $\mathfrak{B} = \mathfrak{Zd}\,\mathfrak{A}$ with α, β, and Δ satisfying the above conditions, is a countable BA of class 2; this follows, e.g., from the fact that $\mathfrak{B}$ is not atomic (cf. remarks following 2.5.47(III)). Also, Theorem (I) in 2.5.47 implies that $\boldsymbol{\varepsilon}\mathfrak{B}$ is the same for all such algebras $\mathfrak{B}$. However, the evaluation of $\boldsymbol{\varepsilon}\mathfrak{B}$ in this case is an outstanding open problem (originating with Tarski). We know from (X)(ii) in 2.5.56 that $\boldsymbol{\varepsilon}\mathfrak{B}$ is always an ordinal of the form $\omega^{\gamma}\cdot\kappa$; from 2.5.47(II),(III) we can easily conclude that in this particular case $\kappa = 1$ so that $\boldsymbol{\varepsilon}\mathfrak{B} = \omega^{\gamma}$. As was noticed by William Hanf, $\boldsymbol{\varepsilon}\mathfrak{B} \geqq \omega^{\omega}$; it is still not known, however, whether or not $\boldsymbol{\varepsilon}\mathfrak{B} > \omega^{\omega}$. In metalogical terms this can be expressed as follows. We do not know whether there is a sentence σ in the language Λ satisfying the conditions: (1) there is a sequence φ of sentences, of length ω^{ω}, such that $\varphi_0 = \sigma$, $\varphi_{\xi} \rightarrow \varphi_{\eta}$ is logically valid, and $\varphi_{\eta} \rightarrow \varphi_{\xi}$ is not logically valid whenever $\eta < \xi < \omega^{\omega}$ (thus, φ is a sequence of sentences strictly increasing in their logical strength); (2) there is no such sequence of length $\omega^{\omega}\cdot 2$ with these properties. The problem may seem rather special. The fact, however, that it is not yet solved gives strong evidence of our inadequate insight into the algebraic structure of predicate logic.

Notice that in case $\mathfrak{A} = \mathfrak{Fr}_{\beta}^{(\Delta)}\mathsf{CA}_{\alpha}$ where $2 \leqq \alpha < \omega$, and β, Δ satisfy the same conditions as before, it is still true that $\mathfrak{B} = \mathfrak{Zd}\,\mathfrak{A}$ is a countable BA of class 2, while the problem of evaluating $\boldsymbol{\varepsilon}\mathfrak{B}$ has the same status as before. In all cases the most important problem is the one of providing an intrinsic characterization for the algebra $\mathfrak{B}$, or, in other words, the problem of determining the isomorphism type of $\mathfrak{B}$. Since $\mathfrak{B}$ is a countable BA and has therefore an ordered base, the problem reduces to that of describing the order type of some ordered base of $\mathfrak{B}$. It seems very likely that the evaluation of $\boldsymbol{\varepsilon}\mathfrak{B}$ would be an important step leading to such a description.

The solution of the problems discussed above turns out to be easy only in application to those dimension-restricted free cylindric algebras which are monadic-generated (cf. Theorem 2.5.68 below). It is just with this class of algebras that we deal in the last portion of the section.

THEOREM 2.5.58. *For any* $\Delta \in {}^{\beta}Sb\alpha$, $\mathfrak{Fr}_{\beta}^{(\Delta)}\mathsf{CA}_{\alpha}$ *is monadic-generated iff* $|\Delta\xi| \leqq 1$ *for every* $\xi < \beta$.

PROOF. We shall only prove that, if $\mathfrak{Fr}_{\beta}^{(\Delta)}\mathsf{CA}_{\alpha}$ is monadic-generated, then $|\Delta\xi| \leqq 1$ for every $\xi < \beta$ (the implication in the opposite direction being trivial).

To arrive at a contradiction set $\mathfrak{A} = \mathfrak{Fr}_{\beta}^{(\Delta)}\mathsf{CA}_{\alpha}$ and assume that $\mathfrak{A}$ is monadic-generated while $|\Delta\eta| \geqq 2$ for a certain $\eta < \beta$. Let

(1) $$\kappa, \lambda \in \Delta\eta \text{ where } \kappa \neq \lambda.$$

Apply 0.5.11(ii) and use the assumption that $\mathfrak{A}$ is monadic-generated to get $Y \subseteq A$ and $\Gamma \subseteq \alpha$ such that

(2) $$|Y|, |\Gamma| < \omega,$$

(3) $$|\Delta y| \leqq 1 \text{ for every } y \in Y,$$

and

(4) $$\eta/Cr_{\beta}^{(\Delta)}\mathsf{CA}_{\alpha} \in C$$

where C is the intersection of all subsets U of A satisfying the following conditions: $Y \subseteq U$, U is a Boolean subuniverse of $\mathfrak{A}$, and $\mathsf{c}_{\mu}^{*}U \subseteq U$ and $\mathsf{d}_{\mu\nu} \in U$ for every $\mu, \nu \in \Gamma$. Moreover, it is clear that we can choose Γ such that

(5) $$\kappa, \lambda \in \Gamma.$$

Now C is easily seen to be the universe of a cylindric algebra $\mathfrak{C}$ of dimension $|\Gamma|$ which is generated by Y (compare the remarks immediately following statement (9) in the proof of 2.2.21); in particular, in view of (5) we have:

(6) C contains $0, 1, \mathsf{d}_{\kappa\lambda}$ and is closed under the operations $+, \cdot, -, \mathsf{c}_{\kappa}$, and c_{λ}.

Furthermore, because of (2) and (3), it is easily seen that $\mathfrak{C}$ is a finitely generated CA of finite dimension which is monadic-generated. Therefore, by 2.2.26,

(7) $$|C| < \omega.$$

Let $\mathfrak{B}$ be the full cylindric set algebra in the space ${}^{\alpha}\omega$ and let

$$X = \{\mu : \mu \in {}^{\alpha}\omega, \ \mu_{\kappa} < \mu_{\lambda}\}.$$

Clearly, $\Delta^{(\mathfrak{B})}X = \{\kappa, \lambda\}$; hence, by (1) and 2.5.36, there is an $h \in Hom(\mathfrak{A}, \mathfrak{B})$ such that $h(\eta/Cr_{\beta}^{(\Delta)}\mathsf{CA}_{\alpha}) = X$. From (4) and (6) we infer that $h^{*}C$ contains $X, 0, {}^{\alpha}\omega, \mathbf{D}_{\kappa\lambda}$ and is closed under the operations $\cup, \cap, \sim, \mathbf{C}_{\kappa}$, and $\mathbf{C}_{\lambda}$. Hence, the argument of 2.1.11 shows that h^*C is infinite. This contradicts (7), and the proof is complete.

REMARK 2.5.59. From 2.2.16 and 2.5.46 it is seen that every monadic-generated $\mathfrak{Fr}_\beta^{(\Delta)}\mathsf{CA}_\alpha$ with $\alpha \geqq \omega$ can be isomorphically represented as an algebra of formulas associated with a monadic logic, i.e., with the logic of a language Λ which contains β predicates of rank 0 or 1 and no other non-logical constants.

THEOREM 2.5.60. *The following two conditions are equivalent:*

(i) $\mathfrak{A}$ *is a monadic-generated* CA_α,

(ii) *there exists a* $\beta \leqq |A|$ *and* $\Delta \in {}^\beta Sb\alpha$ *with* $|\Delta\xi| \leqq 1$ *for each* $\xi < \beta$ *such that* $\mathfrak{Fr}_\beta^{(\Delta)}\mathsf{CA}_\alpha \succcurlyeq \mathfrak{A}$.

The theorem remains true if the symbols "$\leqq |A|$*" in (ii) are replaced by "*$= |A|$*" or else if they are omitted.*

The proof is analogous to the proof of 2.5.45.

In 2.2.24 we have developed the method of eliminating cylindrifications for arbitrary monadic-generated CA's. We shall give some applications of this method to the discussion of free monadic-generated CA's. In the first place we shall use it to compute the cardinalities of finite algebras of this kind. The result of this computation (in a form due to Tarski) is given in

THEOREM 2.5.61. *Assume that* $\Delta \in {}^\beta Sb\alpha$, $\alpha, \beta < \omega$, *and* $|\Delta\xi| \leqq 1$ *for each* $\xi < \beta$. *Then* $\mathfrak{Fr}_\beta^{(\Delta)}\mathsf{CA}_\alpha$ *is finite.*

More specifically, set $\Gamma = \{\xi : \xi < \beta, |\Delta\xi| = 1\}$, *and let K be the set of all functions F from* $Sb\Gamma$ *to* $Sb(Sb\alpha \sim \{0\})$ *such that*

(i) $\{\bigcup F_\Theta : \Theta \subseteq \Gamma\} \sim \{0\}$ *is a partition of* α, *and* $\bigcup F_\Theta \neq \bigcup F_{\Theta'}$, *whenever* $\Theta, \Theta' \subseteq \Gamma$ *and* $\Theta \neq \Theta'$;

(ii) *for each* $\Theta \subseteq \Gamma$, F_Θ *is a partition of* $\bigcup F_\Theta$.

Then $|Fr_\beta^{(\Delta)}\mathsf{CA}_\alpha| = 2^\delta$, *where*

$$\delta = |At\mathfrak{Fr}_\beta^{(\Delta)}\mathsf{CA}_\alpha| = 2^{|\beta\sim\Gamma|}\cdot\textstyle\sum_{F\in K}\prod_{\Theta\subseteq\Gamma}(\alpha+1-|F_\Theta|).$$

PROOF. Let $\mathfrak{A} = \mathfrak{Fr}_\beta^{(\Delta)}\mathsf{CA}_\alpha$ and $g_\xi = \xi/Cr_\beta^{(\Delta)}\mathsf{CA}_\alpha$ for every $\xi < \beta$. In view of 2.5.52 we may assume without losing any generality that $\Delta\xi \subseteq 1$ for every $\xi < \beta$. Furthermore, we will suppose for the time being that $\Delta\xi = 1$ for every $\xi < \beta$, i.e., we suppose $\Gamma = \beta$. For each $\Theta \subseteq \beta$ and $\Omega \subseteq \alpha$ we let

$$h_{\Theta,\Omega} = \textstyle\prod_{\mu\in\Omega}(\prod_{\xi\in\Theta}\mathsf{s}^0_\mu g_\xi - \sum_{\eta\in\beta\sim\Theta}\mathsf{s}^0_\mu g_\eta).$$

If

$$B = \{\mathsf{d}_{\kappa\lambda} : \kappa, \lambda < \alpha\} \cup \{\mathsf{s}^0_\mu g_\xi : \xi < \beta, \mu < \alpha\} \cup$$
$$\{\mathsf{c}_{(\kappa)}(\bar{\mathsf{d}}(\kappa\times\kappa)\cdot h_{\Theta,\kappa}) : \Theta \subseteq \beta, \kappa \leqq \alpha\},$$

then by 2.2.25 we have

$$\mathfrak{A} = \mathfrak{Sg}^{(\mathfrak{B}\mathfrak{l}\,\mathfrak{A})}B.$$

Hence, by the theory of BA's, $\mathfrak{A}$ is finite, and the atoms of $\mathfrak{A}$ are just the minimal constituents of B which are non-zero. We shall now discuss some special relations which hold among the elements of B and which enable us to give a more useful characterization of the atoms of $\mathfrak{A}$.

First of all, from 1.5.6 we easily obtain $\mathsf{s}^0_\kappa g_\xi \cdot -\mathsf{s}^0_\lambda g_\xi \leqq -\mathsf{d}_{\kappa\lambda}$ for every $\xi < \beta$ and any $\kappa, \lambda < \alpha$. Hence,

(1) $h_{\Theta,\Omega} \cdot h_{\Theta',\Omega'} \leqq \bar{\mathsf{d}}(\Omega \times \Omega')$ whenever $\Theta, \Theta' \subseteq \beta$, $\Theta \neq \Theta'$, and $\Omega, \Omega' \subseteq \alpha$.

Secondly, from Lemma 2.2.20 we have (taking $\prod_{\xi\in\Theta} g_\xi - \sum_{\eta\in\beta\sim\Theta} g_\eta$ for x)

(2) $\mathsf{c}_{(\kappa)}(\bar{\mathsf{d}}(\kappa\times\kappa)\cdot h_{\Theta,\kappa}) \geqq \mathsf{c}_{(\kappa+1)}(\bar{\mathsf{d}}((\kappa+1)\times(\kappa+1))\cdot h_{\Theta,\kappa+1})$ whenever $\Theta \subseteq \beta$ and $\kappa \leqq \alpha$;

here we adopt the *ad hoc* convention that $\mathsf{c}_{(\alpha+1)}(\bar{\mathsf{d}}((\alpha+1)\times(\alpha+1))\cdot h_{\Theta,\alpha+1}) = 0$ for every $\Theta \subseteq \beta$. Finally, using Lemmas 2.2.20 and 2.2.21 one can easily establish the following:

(3) for any $\Theta \subseteq \beta$, $\Omega \subseteq \alpha$, and partition P of Ω, if $\kappa \leqq |P|$, then

$$\prod_{\Lambda\in P}(\mathsf{d}_\Lambda \cdot \bar{\mathsf{d}}(\Lambda\times(\Omega\sim\Lambda)))\cdot h_{\Theta,\Omega} \leqq \mathsf{c}_{(\kappa)}(\bar{\mathsf{d}}(\kappa\times\kappa)\cdot h_{\Theta,\kappa}).$$

We are now able to give the promised characterization of the atoms of $\mathfrak{A}$. Let L be the set of all ordered pairs $\langle F, \tau\rangle$ such that $F \in K$, $\tau \in {}^{Sb\beta}(\alpha+1)$, and $\tau_\Theta \geqq |F_\Theta|$ for every $\Theta \subseteq \beta$. For each $\langle F, \tau\rangle \in L$ we let

$$x_{F,\tau} = \prod_{\Theta\subseteq\beta}[\prod_{\Lambda\in F_\Theta}(\mathsf{d}_\Lambda \cdot \bar{\mathsf{d}}(\Lambda\times(\bigcup F_\Theta\sim\Lambda)))\cdot h_{\Theta\cup F_\Theta}\cdot$$
$$\mathsf{c}_{(\tau_\Theta)}(\bar{\mathsf{d}}(\tau_\Theta\times\tau_\Theta)\cdot h_{\Theta,\tau_\Theta}) - \mathsf{c}_{(\tau_\Theta+1)}(\bar{\mathsf{d}}((\tau_\Theta+1)\times(\tau_\Theta+1))\cdot h_{\Theta,\tau_\Theta+1})].$$

With the aid of (1), (2), and (3) it can be shown without too much difficulty that

$$At\,\mathfrak{A} = \{x_{F,\tau} : \langle F, \tau\rangle \in L\} \sim \{0\};$$

furthermore, it is clear that $x_{F,\tau}$ and $x_{F',\tau'}$ are disjoint whenever $\langle F, \tau\rangle \neq \langle F', \tau'\rangle$. Therefore, since an easy computation gives

$$|L| = \sum_{F\in K}\prod_{\Theta\subseteq\beta}(\alpha+1-|F_\Theta|),$$

the proof will be complete (at least for the case $\Gamma = \beta$) as soon as we show that $x_{F,\tau} \neq 0$ for every $\langle F, \tau\rangle \in L$.

Let $\langle F, \tau\rangle$ be a fixed element of L. We will describe a cylindric set algebra $\mathfrak{C}$ of dimension α and a homomorphism f from $\mathfrak{A}$ into $\mathfrak{C}$ such that $fx_{F,\tau}$ is non-empty. Let $U = \{\langle\Theta, \lambda\rangle : \Theta \subseteq \beta, \lambda < \tau_\Theta\}$, and let $\mathfrak{C}$ be the full cylindric set algebra in the space ${}^\alpha U$. For each $\xi < \beta$ we take

$$G_\xi = \{y : y \in {}^\alpha U, \xi \in \mathsf{pj}_0 y_0\}.$$

By 2.5.36 there exists a homomorphism f from $\mathfrak{A}$ onto $\mathfrak{C}$ such that $fg_\xi = G_\xi$ for every $\xi < \beta$. It can easily be shown that

$$fh_{\Theta,\Omega} = \{y : y \in {}^\alpha U, \Omega 1 \mathsf{pj}_0 \circ y = \Omega \times \{\Theta\}\}$$

for every $\Theta \subseteq \beta$ and $\Omega \subseteq \alpha$. From this it follows almost immediately that $fx_{F,\tau} \neq 0^{(\mathfrak{C})}$.

We turn finally to the general case where $\beta \sim \Gamma$ may not be empty. Using 2.5.40 and arguing inductively we can prove

$$|A| = |Sg^{(\mathfrak{A})}\{g_\xi : \xi \in \Gamma\}|^{2^{|\beta \sim \Gamma|}}.$$

The desired conclusion now follows by the first part of the proof.

A moment's reflection on the proof just given brings to light the following interesting fact: for any non-zero x of a monadic-generated $\mathfrak{Fr}_\beta^{(\Delta)}\mathsf{CA}_\alpha$ there is a cylindric set algebra $\mathfrak{B}$ and an $f \in Ho(\mathfrak{Fr}_\beta^{(\Delta)}\mathsf{CA}_\alpha, \mathfrak{B})$ such that $fx \neq 0^{(\mathfrak{B})}$; furthermore, this result can easily be shown to hold for arbitrary β and α. Thus by 2.4.39 we see that every monadic-generated $\mathfrak{Fr}_\beta^{(\Delta)}\mathsf{CA}_\alpha$ is representable in the sense of 1.1.13. Using now the fact (to be established in Part II) that the class of representable CA_α's is closed under the formation of homomorphic images, and applying 2.6.50, we can conclude that every monadic-generated CA_α is representable. If $\alpha \geqq \omega$, this is of course but a particular case of a general result mentioned in 1.11.2. A more detailed discussion of the representability of monadic-generated CA's will be found in Part II; in this connection see also 2.6.56 below and compare it with 2.6.26.

COROLLARY 2.5.62. *If $\beta < \omega$, then*

$$|Fr_\beta\mathsf{CA}_0| = 2^{2^\beta} \text{ and } |Fr_\beta\mathsf{CA}_1| = 2^{2^{\beta+2^\beta-1}}.$$ [1]

PROOF. The determination of $|Fr_\beta\mathsf{CA}_0|$ is trivial. (We can either use 2.5.61 or appeal directly to the theory of BA's.) In order to apply 2.5.61 to compute $|Fr_\beta\mathsf{CA}_1|$, observe that, for every $F \in K$, $\mathsf{U}F_\Theta$ is non-empty for exactly one $\Theta \subseteq \beta$, and for that Θ we have $|F_\Theta| = 1$.

[1]) The first formula is well known from the theory of BA's. The second formula (in a metalogical setting) was first obtained by Carnap [46*]; also a particular case of 2.5.65 with $\Gamma = \beta = \gamma$ can be found there.

REMARK 2.5.63. As further particular cases of Theorem 2.5.61, assuming the hypothesis of this theorem and setting $\gamma = |\Gamma|$, we obtain

(I) $$|At\mathfrak{Fr}_\beta^{(\Delta)}\mathsf{CA}_1| = 2^{\beta+2^\gamma-1},$$

(II) $$|At\mathfrak{Fr}_\beta^{(\Delta)}\mathsf{CA}_2| = (5\cdot 2^\beta+4\cdot 2^{\beta+\gamma})\cdot 3^{2^\gamma-2},$$

(III) $$|At\mathfrak{Fr}_\beta^{(\Delta)}\mathsf{CA}_3| = (34\cdot 2^\beta+99\cdot 2^{\beta+\gamma}+27\cdot 2^{\beta+2\cdot\gamma})\cdot 4^{2^\gamma-3}.$$

We shall outline the computation leading to (III). Clearly we may assume $\Gamma = \gamma$, i.e., Γ is an ordinal.

It proves convenient to partition K into six subsets which we shall represent respectively by L_1, L_2, L_3, M_1, M_2, and N. They are defined in the following manner: For $\kappa = 1, 2, 3$ we have $F \in L_\kappa$ iff $F \in K$ and there is a $\Theta \subseteq \gamma$ such that $|\bigcup F_\Theta| = 3$ and $|F_\Theta| = \kappa$. If $\lambda = 1, 2$, then $F \in M_\lambda$ iff $F \in K$ and there is a $\Theta \subseteq \gamma$ such that $|\bigcup F_\Theta| = 2$ and $|F_\Theta| = \lambda$. Finally, $F \in N$ iff $|\bigcup F_\Theta| \leqq 1$ for every $\Theta \subseteq \gamma$.

Let $\pi F = \prod_{\Theta\subseteq\gamma}(4-|F_\Theta|)$ for every $F \in K$. Then a simple computation gives $|L_1| = |L_3| = 2^\gamma$, $|L_2| = 3\cdot 2^\gamma$, and $\pi F = 4^{2^\gamma-1}\cdot(4-\kappa)$ for every $F \in L_\kappa$ and $\kappa = 1, 2, 3$. Thus

(1) $$\sum_{F\in L_1\cup L_2\cup L_3}\pi F = 10\cdot 2^\gamma\cdot 4^{2^\gamma-1}.$$

For $\lambda = 1, 2$ we get $|M_\lambda| = 3\cdot 2^\gamma\cdot(2^\gamma-1)$ and $\pi F = 4^{2^\gamma-2}\cdot 3\cdot(4-\lambda)$ for each $F \in M_\lambda$. Hence

(2) $$\sum_{F\in M_1\cup M_2}\pi F = 45\cdot 2^\gamma\cdot(2^\gamma-1)\cdot 4^{2^\gamma-2}.$$

Finally, $|N| = 2^\gamma\cdot(2^\gamma-1)\cdot(2^\gamma-2)$ and $\pi F = 4^{2^\gamma-3}\cdot 3^3$ for every $F \in N$. Thus

(3) $$\sum_{F\in N}\pi F = 27\cdot 2^\gamma\cdot(2^\gamma-1)(2^\gamma-2)\cdot 4^{2^\gamma-3}.$$

The formulas (1), (2), and (3) lead easily to a numerical expression for $\sum_{F\in K}\prod_{\Theta\subseteq\gamma}(4-|F_\Theta|)$; on simplifying this expression and applying 2.5.61 we arrive at (III).

It might be interesting to find a formula for the cardinality of $\mathfrak{Fr}_\beta^{(\Delta)}\mathsf{CA}_\alpha$ which would be more arithmetical in character than the one given in 2.5.61, and from which all the special formulas stated in 2.5.62 and 2.5.63 could be obtained by a simple computation.

As will be seen in the next theorem, 2.5.65, a simple numerical formula can be obtained for the cardinality, not of the algebra $\mathfrak{Fr}_\beta^{(\Delta)}\mathsf{CA}_\alpha$, but of its zero-dimensional part. We begin with a lemma (which will find some application in our later discussion as well).

LEMMA 2.5.64. *Assume that* $\Delta \in {}^{\beta}Sb\alpha$ *with* $\Delta\xi = 1$ *for every* $\xi < \beta$ *and consider the algebra* $\mathfrak{Fr}_{\beta}^{(\Delta)}\mathsf{CA}_{\alpha}$. *For each* $\xi < \beta$ *set* $g_\xi = \xi/Cr_{\beta}^{(\Delta)}\mathsf{CA}_{\alpha}$. *Let* Λ *be any finite subset of* β, *and for each* $\Theta \subseteq \Lambda$ *and* $\kappa < (\alpha+1) \cap \omega$ *set*

$$a_\kappa\Theta = \mathsf{c}_{(\kappa)}[\bar{\mathsf{d}}(\kappa\times\kappa)\cdot\prod_{\mu<\kappa}(\prod_{\xi\in\Theta}\mathsf{s}_\mu^0 g_\xi - \sum_{\eta\in\Lambda\sim\Theta}\mathsf{s}_\mu^0 g_\eta)];$$

if $\alpha < \omega$, *set* $a_{\alpha+1}\Theta = 0$. *Finally, let* $\tau \in {}^{Sb\Lambda}((\alpha+1)\cap\omega)$. *We then have*

$$\prod_{\Theta\subseteq\Lambda}(a_{\tau_\Theta}\Theta - a_{\tau_\Theta+1}\Theta) \neq 0.$$

PROOF. Let $U = \{\langle\Theta, \lambda\rangle : \Theta \subseteq \Lambda, \lambda < \tau_\Theta\}$, and let $\mathfrak{C}$ be the full cylindric set algebra in the space ${}^{\alpha}U$. For each $\xi < \beta$ we take

$$G_\xi = \{y : y \in {}^{\alpha}U, \xi \in \mathsf{pj}_0 y_0\}.$$

By 2.5.36 there exists an $f \in Hom(\mathfrak{Fr}_{\beta}^{(\Delta)}\mathsf{CA}_{\alpha}, \mathfrak{C})$ such that $fg_\xi = G_\xi$ for each $\xi < \beta$, and it can easily be checked that $f(\prod_{\Theta\subseteq\Lambda}(a_{\tau_\Theta}\Theta - a_{\tau_\Theta+1}\Theta)) = {}^{\alpha}U$. Compare the proof of 2.5.61.

THEOREM 2.5.65. *Let* $\mathfrak{A} = \mathfrak{Fr}_{\beta}^{(\Delta)}\mathsf{CA}_{\alpha}$ *where* $\Delta \in {}^{\beta}Sb\alpha$, $\alpha, \beta < \omega$, *and* $|\Delta\xi| \leqq 1$ *for each* $\xi < \beta$. *Then* $\mathfrak{Zd}\,\mathfrak{A}$ *is finite. In fact, setting* $\Gamma = \{\xi : \xi < \beta, |\Delta\xi| = 1\}$ *and* $\gamma = |\Gamma|$, *we have*

$$|At\,\mathfrak{Zd}\,\mathfrak{A}| = (\alpha+1)^{2^\gamma}\cdot 2^{\beta-\gamma} \textit{ and } |Zd\,\mathfrak{A}| = 2^{(\alpha+1)^{2^\gamma}\cdot 2^{\beta-\gamma}}.$$

PROOF. Set $g_\xi = \xi/Cr_{\beta}^{(\Delta)}\mathsf{CA}_{\alpha}$ for every $\xi < \beta$. As usual we may assume, because of 2.5.52, that $\Delta\xi \subseteq 1$ for every $\xi < \beta$. Furthermore, we consider first of all the case when $\Delta\xi = 1$ for every $\xi < \beta$, i.e., we suppose $\Gamma = \beta$. For each $\kappa \leqq \alpha$ and $\Theta \subseteq \beta$ let $a_\kappa\Theta$ be defined as in 2.5.64 with $\Lambda = \beta$. Then by Lemma 2.2.20 (taking $\prod_{\xi\in\Theta}g_\xi - \sum_{\eta\in\beta\sim\Theta}g_\eta$ for x) we have

(1) $\qquad a_\kappa\Theta \leqq a_\lambda\Theta$ whenever $\lambda \leqq \kappa \leqq \alpha$ and $\Theta \subseteq \beta$.

Set $\mathfrak{B} = \mathfrak{Zd}\,\mathfrak{A}$. Then by 2.2.25 we have

$$B = Sg\{a_\kappa\Theta : \kappa \leqq \alpha, \Theta \subseteq \beta\},$$

and the atoms of $\mathfrak{B}$ are just the non-zero minimal constituents of $\{\alpha_\kappa\Theta : \kappa \leqq \alpha, \Theta \subseteq \beta\}$. In light of (1), every minimal constituent can be represented in the form

(2) $$\prod_{\Theta\subseteq\beta}(a_{\tau_\Theta}\Theta - a_{\tau_\Theta+1}\Theta)$$

for some $\tau \in {}^{Sb\beta}(\alpha+1)$; for each such τ we infer from 2.5.64 that the element represented by (2) is non-zero. Therefore, since any two constituents represented by (2) for distinct τ's are disjoint and thus themselves distinct, we see that

$\mathfrak{B}$ has exactly $|{}^{Sb\beta}(\alpha+1)| = (\alpha+1)^{2^\beta}$ atoms. We now use 2.5.40 to obtain the conclusion of the theorem.

2.5.38 and 2.5.61 show that, among the algebras $\mathfrak{Fr}_\beta^{(\Delta)}\mathsf{CA}_\alpha$, the only ones which are finite are those in which α, β, and Δ satisfy the assumptions of 2.5.61; from 2.5.65 and the next theorem we shall see that they are also the only ones whose zero-dimensional parts are finite. In connection with the next theorem recall the notion of a well-ordered base introduced in 2.5.56.

THEOREM 2.5.66. *Assume* $\mathfrak{A} = \mathfrak{Fr}_\beta^{(\Delta)}\mathsf{CA}_\alpha$ *where* $\Delta \in {}^\beta Sb\alpha$, α *is arbitrary,* $\beta < \omega$, *and* $|\Delta\xi| \leqq 1$ *for each* $\xi < \beta$. *Set* $\Gamma = \{\xi : \xi < \beta, |\Delta\xi| = 1\}$ *and* $\gamma = |\Gamma|$. *Then* $\mathfrak{Zd}\,\mathfrak{A}$ *has a well-ordered base of type* $((\alpha+1) \cap \omega)^{2^\gamma} \cdot 2^{\beta-\gamma}$.

PROOF. In case $\alpha < \omega$ the conclusion follows immediately from 2.5.65. Thus we may assume $\alpha \geqq \omega$. Also, in view of 2.5.52, we may assume that $\Delta\xi \subseteq 1$ for every $\xi < \beta$; finally, we assume for the time being that $\Delta\xi = 1$ for every $\xi < \beta$, i.e., we suppose $\Gamma = \beta = \gamma$.

Set $g_\xi = \xi/Cr_\beta^{(\Delta)}\mathsf{CA}_\alpha$ for each $\xi < \beta$, and set $\mathfrak{B} = \mathfrak{Zd}\,\mathfrak{A}$. For each $\kappa < \omega$ and $\Theta \subseteq \Lambda$ let $a_\kappa\Theta$ be defined as in 2.5.64 with $\beta = \Lambda$. Then by Lemma 2.2.20 we have

(1) $$a_\kappa\Theta \leqq a_\lambda\Theta \text{ whenever } \lambda \leqq \kappa < \omega \text{ and } \Theta \subseteq \beta.$$

For each $\Theta \subseteq \beta$ set

(2) $$\mathfrak{C}_\Theta = \mathfrak{Sg}^{(\mathfrak{B})}\{a_\kappa\Theta : \kappa < \omega\};$$

then

(3) $$\mathfrak{C}_\Theta \subseteq \mathfrak{B} \text{ for every } \Theta \subseteq \beta.$$

By 2.2.25 we have

(4) $$\mathfrak{B} = \mathfrak{Sg}(\textstyle\bigcup_{\Theta \subseteq \beta} C_\Theta).$$

Now consider any $c \in \mathsf{P}_{\Theta \subseteq \beta} C_\Theta$. From (1) and (2) it follows that, for each $\Theta \subseteq \beta$, c_Θ can be represented as the sum of a finite number of elements of the form $a_\kappa\Theta - a_{\kappa+1}\Theta$ with $\kappa < \omega$. Thus the element $\prod_{\Theta \subseteq \beta} c_\Theta$ can be written as the sum of a finite number of elements of the form

$$\textstyle\prod_{\Theta \subseteq \beta}(a_{\tau_\Theta}\Theta - a_{\tau_\Theta+1}\Theta)$$

where $\tau \in {}^{Sb\beta}\omega$. Therefore from 2.5.64 we conclude that $\prod_{\Theta \subseteq \beta} c_\Theta = 0$ iff $c_\Omega = 0$ for some $\Omega \subseteq \beta$, whence

(5) $$\textstyle\prod_{\Theta \subseteq \beta} c_\Theta \neq 0 \text{ for every } c \in \mathsf{P}_{\Theta \subseteq \beta}(C_\Theta \sim \{0\}).$$

It is clearly seen from (1), (2), and 2.5.64 that, for each $\Theta \subseteq \beta$, $\mathfrak{C}_\Theta$ has a well-ordered base of type ω. Therefore, in view of (3)–(5) and 2.5.56(V),(VIII), (XII), we conclude by simple induction that $\mathfrak{B}$ has a well-ordered base of type ω^{2^β}.

We have thus completed the proof in the special case $\Gamma = \beta = \gamma$. Hence the conclusion in the general case is obtained by induction using 2.4.3, 2.5.40, and 2.5.56(IV),(IX),(XII).

COROLLARY 2.5.67. *Under the hypothesis of 2.5.66 we have:*

(i) *every homomorphic image of* $\mathfrak{Zd}\,\mathfrak{A}$ *is atomic;*

(ii) *every subalgebra of* $\mathfrak{Zd}\,\mathfrak{A}$ *is atomic;*

(iii) *no set* $X \subseteq Zd\mathfrak{A}$ *is densely ordered;*

(iv) *the set of all maximal proper ideals of* $\mathfrak{Zd}\,\mathfrak{A}$ *is countable.*

PROOF: by 2.5.56(I),(XII) and 2.5.66.

COROLLARY 2.5.68. *Under the hypothesis of 2.5.66 we have:*

(i) *if* $\alpha \geqq \omega$, *then*

$$\gamma\mathfrak{Zd}\,\mathfrak{A} = 2^\gamma,\ \kappa\mathfrak{Zd}\,\mathfrak{A} = 2^{\beta-\gamma},\ \delta\mathfrak{Zd}\,\mathfrak{A} = \varepsilon\mathfrak{Zd}\,\mathfrak{A} = \omega^{2^\gamma}\cdot(2^{\beta-\gamma}+1);$$

(ii) *if* $\alpha < \omega$, *then*

$$\gamma\mathfrak{Zd}\,\mathfrak{A} = 0,\ \kappa\mathfrak{Zd}\,\mathfrak{A} = (\alpha+1)^{2^\gamma}\cdot 2^{\beta-\gamma},\ \delta\mathfrak{Zd}\,\mathfrak{A} = \varepsilon\mathfrak{Zd}\,\mathfrak{A} = (\alpha+1)^{2^\gamma}\cdot 2^{\beta-\gamma}+1.$$

PROOF: by 2.5.66 and (VIII), (X)(iii) of 2.5.56; recall the definitions of $\gamma\mathfrak{B}$, $\kappa\mathfrak{B}$, $\delta\mathfrak{B}$, and $\varepsilon\mathfrak{B}$ given in 2.5.56.

REMARKS 2.5.69. As another easy consequence of 2.5.66 combined with 2.5.47(I) we can obtain the formula $\varepsilon\mathfrak{Zd}\,\mathfrak{A} \geqq \omega^\omega$ where $\mathfrak{A}$ is any dimension-restricted finitely generated free CA_α, with $\alpha \geqq \omega$, which is locally finite but not monadic-generated. As was mentioned in 2.5.57, it is still an open problem whether in this formula (due to William Hanf) the symbol "$\geqq$" can be replaced by "$=$".

As a final remark we call the readers attention to the possibility of characterizing the isomorphism types of monadic-generated CA_α's with β generators for any fixed α and $\beta < \omega$. This characterization is a natural generalization of Theorems 2.5.25 and 2.5.26, which concern minimal algebras, and it makes essential use of the method of eliminating cylindrifications for arbitrary monadic-generated CA's. We leave the formulation of this characterization as an exercise.

2.6. REDUCTS

The general notion of a reduct was introduced and discussed in Section 0.5. We recall that a reduct of an algebra $\mathfrak{A}$ is a new algebra obtained by leaving the universe of $\mathfrak{A}$ unchanged and keeping some but discarding others from the list of the fundamental operations of $\mathfrak{A}$. Only special cases of this notion are involved in the theory of CA's. Thus, the reduct of a CA_α obtained by removing all the diagonal elements is a Df_α, a diagonal-free cylindric algebra of dimension α as defined in 1.1.2. (It was pointed out however, in 1.5.24, that not every Df_α can be obtained in this way.)

Especially important for our purposes are those reducts of CA_α's which are themselves CA_β's for some ordinal β usually different from α. They are obtained from CA_α's by removing some cylindrifications together with the corresponding diagonal elements. Only these special reducts will be referred to as reducts of CA's in the subsequent discussion. In the immediately following Definition 2.6.1 we introduce an appropriate notation for these reducts. We leave it to the reader to convince himself that actually 2.6.1 can be construed as a special case of 0.5.1.

For several reasons the notion of a reduct is of considerable significance for the theory of cylindric algebras. One of them, and perhaps the main one, will be discussed in Remarks 2.6.26. Another reason is that the discussion of an infinite-dimensional CA reduces in certain situations to the discussion of some finite-dimensional algebras, and in fact, the finite reducts of $\mathfrak{A}$; the main result which makes this reduction possible is Theorem 2.6.4 below.

DEFINITION 2.6.1. *Let* $\mathfrak{B} = \langle B, +, \cdot, -, 0, 1, \mathsf{c}_\kappa, \mathsf{d}_{\kappa\lambda}\rangle_{\kappa,\lambda<\beta}$ *be a* CA_β, α *an ordinal, and* ρ *a sequence of length* α *without repeating terms and with* $Rg\rho \subseteq \beta$. *Then the algebra* $\mathfrak{A} = \langle B, +, \cdot, -, 0, 1, \mathsf{c}_{\rho\kappa}, \mathsf{d}_{\rho\kappa,\rho\lambda}\rangle_{\kappa,\lambda<\alpha}$ *is called the* ***ρ-reduct*** *of* $\mathfrak{B}$ *and is denoted by* $\mathfrak{Rd}_\alpha^{(\rho)}\mathfrak{B}$ *or, simply, by* $\mathfrak{Rd}^{(\rho)}\mathfrak{B}$. *If* $\alpha < \omega$, *then* $\mathfrak{A}$ *is referred to as a* ***finite reduct*** *of* $\mathfrak{B}$. *In case* $\rho = \alpha 1 Id$ (*and thus* $\alpha \leqq \beta$), *we call* $\mathfrak{A}$ *the* ***α-reduct*** *of* $\mathfrak{B}$ *and denote it by* $\mathfrak{Rd}_\alpha\mathfrak{B}$.

For any class K *of* CA_β*'s we set*

$$\mathsf{Rd}_\alpha^{(\rho)}\mathsf{K} = \{\mathfrak{Rd}_\alpha^{(\rho)}\mathfrak{B} : \mathfrak{B} \in \mathsf{K}\}$$

and, if $\alpha \leqq \beta$,

$$\mathsf{Rd}_\alpha \mathsf{K} = \{\mathfrak{Rd}_\alpha \mathfrak{B} : \mathfrak{B} \in \mathsf{K}\}.$$

Finally, a CA $\mathfrak{C}$ *is called a* ***subreduct*** *of* $\mathfrak{B}$, *in symbols* $\mathfrak{C} \subseteq^{\mathrm{r}} \mathfrak{B}$, *if* $\mathfrak{C} \subseteq \mathfrak{Rd}_\gamma \mathfrak{B}$ *for some* $\gamma \leqq \beta$.

THEOREM 2.6.2. (i) *If* α *and* β *are ordinals and* ρ *is a sequence in* ${}^\alpha\beta$ *without repetitions, then* $\mathsf{Rd}_\alpha^{(\rho)} \mathsf{CA}_\beta \subseteq \mathsf{CA}_\alpha$ *and* $\mathsf{Rd}_\alpha^{(\rho)} \mathsf{Lf}_\beta \subseteq \mathsf{Lf}_\alpha$.

(ii) $\mathfrak{A} = \mathfrak{Rd}_\alpha \mathfrak{A}$ *for each* $\mathfrak{A} \in \mathsf{CA}_\alpha$, *and* $\mathsf{Rd}_\alpha \mathsf{CA}_\alpha = \mathsf{CA}_\alpha$.

(iii) $\mathfrak{Bl}\,\mathfrak{B} = \mathfrak{Rd}_0 \mathfrak{B}$ *for every* $\mathfrak{B} \in \mathsf{CA}_\beta$, *and* $\mathsf{Rd}_0 \mathsf{CA}_\beta = \mathsf{BA}$.

PROOF. (i), (ii), and the first part of (iii) are all obvious. The last part of (iii) follows from 1.3.11.

REMARK 2.6.3. It is easily seen that Theorem 2.6.2(i) cannot be extended to various other classes of CA's, e.g., to the class of dimension-complemented CA's. By an argument similar to the proof of 2.5.24 we can show that the same applies also to the class of semisimple CA's. Furthermore, the inclusions in 2.6.2(i) cannot be replaced by equalities. If, e.g., $\mathfrak{A}$ is a non-discrete CA_α with $\alpha > 0$, then, by 1.3.15, $\mathfrak{A} \notin \mathsf{Rd}_\alpha \mathsf{CA}_\beta$ for any β such that $|\beta| > |A|$, and similarly for Lf_α's.

In reading the next theorem recall that, for any class K of algebras and any cardinal $\delta \geqq 1$, $\mathbf{S}_\delta \mathsf{K}$ is the class of all subalgebras with fewer than δ generators of algebras of K; cf. 0.1.15(ii).

THEOREM 2.6.4. *Let* $\mathsf{K} \subseteq \mathsf{CA}_\alpha$ *and* R *be the set of all finite sequences without repeating terms and with range included in* α. *Then each of the following conditions is sufficient for* $\mathfrak{A} \in \mathbf{SUp}\,\mathsf{K}$:

(i) $\mathfrak{Rd}^{(\rho)}\mathfrak{A} \in \mathsf{Rd}^{(\rho)}\mathsf{K}$ *for every* $\rho \in R$;
(ii) $\mathfrak{Rd}^{(\rho)}\mathfrak{A} \in \mathbf{S}\mathsf{Rd}^{(\rho)}\mathsf{K}$ *for every* $\rho \in R$;
(iii) $\mathsf{Rd}^{(\rho)}\mathbf{S}_\omega\mathfrak{A} \subseteq \mathsf{Rd}^{(\rho)}\mathsf{K}$ *for every* $\rho \in R$;
(iv) $\mathsf{Rd}^{(\rho)}\mathbf{S}_\omega\mathfrak{A} \subseteq \mathbf{S}\mathsf{Rd}^{(\rho)}\mathsf{K}$ *for every* $\rho \in R$;
(v) $\mathbf{S}_\omega\mathfrak{Rd}^{(\rho)}\mathfrak{A} \subseteq \mathsf{Rd}^{(\rho)}\mathsf{K}$ *for every* $\rho \in R$;
(vi) $\mathbf{S}_\omega\mathfrak{Rd}^{(\rho)}\mathfrak{A} \subseteq \mathbf{S}\mathsf{Rd}^{(\rho)}\mathsf{K}$ *for every* $\rho \in R$.

Moreover, if $\mathsf{K} = \mathbf{S}\mathsf{K} = \mathbf{Up}\,\mathsf{K}$, *then each of these conditions except* (*v*) *is equivalent to the condition*

(vii) $\mathfrak{A} \in \mathsf{K}$.

PROOF. Theorem 2.6.4 is essentially a particular case of a result from Chapter 0, in fact Theorem 0.5.15(ii). Since the latter was not proved explicitly and the present theorem is important for our discussion, we outline a direct proof of it independent of 0.5.15(ii).

It is easily seen that each of the conditions (i)–(v) implies (v). We wish now to show that (vi) implies $\mathfrak{A} \in \mathbf{SUp}\,\mathsf{K}$. Let

$$I = \{F : F \subseteq A,\ |F| < \omega\} \times R.$$

For $\langle F, \rho\rangle \in I$, let

$$M_{F\rho} = \{\langle G, \sigma\rangle : \langle G, \sigma\rangle \in I,\ F \subseteq G,\ Rg\rho \subseteq Rg\sigma\}.$$

Then there is an ultrafilter U on I such that $M_{F\rho} \in U$ for every $\langle F, \rho\rangle \in I$. By (vi) we obtain for each $\langle F, \rho\rangle \in I$ a $\mathfrak{B}_{F\rho} \in \mathsf{K}$ such that

(1) $$\mathfrak{Sg}^{(\mathfrak{Rd}^{(\rho)}\mathfrak{A})}F \subseteq \mathfrak{Rd}^{(\rho)}\mathfrak{B}_{F\rho},$$

and we form the ultraproduct

$$\mathfrak{C} = \mathsf{P}_{\langle F,\rho\rangle\in I}\mathfrak{B}_{F\rho}/\bar{U} \in \mathsf{Up}\,\mathsf{K}.$$

By (1) there is a function f from A into $\mathsf{P}_{\langle F,\rho\rangle\in I}B_{F\rho}$ such that $(fa)_{\langle F,\rho\rangle} = a$ whenever $a \in F$; then, using (1), it is a routine matter to check that $\langle fa/\bar{U} : a \in A\rangle$ is an isomorphism from $\mathfrak{A}$ into $\mathfrak{C}$. Hence $\mathfrak{A} \in \mathbf{SUp}\,\mathsf{K}$ as desired.

This completes the proof of the theorem except for its last part which involves condition (vii). This last part however follows trivially from what we have shown so far.

THEOREM 2.6.5. *Under the hypothesis of 2.6.4, if* $\mathsf{K} = \mathsf{CA}_\alpha$, *then the conditions (i)–(iv), (vi), and (vii) are mutually equivalent and also equivalent to each of the following conditions:*

(viii) $\mathfrak{Rd}^{(\rho)}\mathfrak{A} \in \mathsf{CA}_{Do\rho}$ *for every* $\rho \in R$;

(ix) $\mathsf{Rd}^{(\rho)}\mathbf{S}_\omega\mathfrak{A} \subseteq \mathsf{CA}_{Do\rho}$ *for every* $\rho \in R$;

(x) $\mathbf{S}_\omega\mathfrak{Rd}^{(\rho)}\mathfrak{A} \subseteq \mathsf{CA}_{Do\rho}$ *for every* $\rho \in R$.

PROOF. The mutual equivalence of (i)–(iv),(vi) and (vii) follows immediately from 2.6.4. The equivalence of (vii) to each of the conditions (viii)–(x) is an easy consequence of 2.6.2(i) along with the definition of CA_α.

As opposed to 2.6.4, Theorem 2.6.5 does not involve the notion of ultraproduct. However, 2.6.5 was derived here with the help of 2.6.4, and hence ultraproducts play an essential role in its proof. On the other hand, 2.6.5 can easily be established by a metamathematical method not involving ultraproducts. Actually the original proof of 2.6.5 had a metamathematical character, and the argument outlined here is a mathematical transformation of the original proof; compare the remarks at the bottom of p. 105. Entirely analogous observations apply to Theorems 2.6.7 and 2.6.34 below.

For the following corollary recall the definition of reduct union given in 0.5.8.

COROLLARY 2.6.6. *If* L *is a class of* CA*'s directed by the relation* $\subseteq^{\mathrm{r}}$, *then* $\mathsf{U}^{\mathrm{r}}\mathsf{L}$ *is a* CA.

THEOREM 2.6.7. *If* $\alpha \leqq \beta$ *and* $0 < \gamma = \bigcup\gamma$, *i.e.,* γ *is a non-zero limit ordinal, then*

$$\bigcap_{\xi<\gamma} \mathsf{SRd}_\alpha \mathsf{CA}_{\beta+\xi} = \mathsf{SRd}_\alpha \mathsf{CA}_{\beta+\gamma};$$

in particular,

$$\bigcap_{\kappa<\omega} \mathsf{SRd}_\alpha \mathsf{CA}_{\alpha+\kappa} = \mathsf{SRd}_\alpha \mathsf{CA}_{\alpha+\omega}.$$

PROOF. The inclusion $\bigcap_{\xi<\gamma} \mathsf{SRd}_\alpha \mathsf{CA}_{\beta+\xi} \supseteq \mathsf{SRd}_\alpha \mathsf{CA}_{\beta+\gamma}$ follows from 0.5.5 and 2.6.2(i). Now suppose $\mathfrak{A} \in \bigcap_{\xi<\gamma} \mathsf{SRd}_\alpha \mathsf{CA}_{\beta+\xi}$; for every $\xi < \gamma$ let

$$(1) \qquad \mathfrak{A} \subseteq \mathfrak{Rd}_\alpha \mathfrak{B}_\xi$$

where $\mathfrak{B}_\xi \in \mathsf{CA}_{\beta+\xi}$. For each $\xi < \gamma$ we choose $\mathfrak{C}_\xi$ to be an algebra similar to $\mathsf{CA}_{\beta+\gamma}$'s such that

$$(2) \qquad \mathfrak{B}_\xi = \mathfrak{Rd}_{\beta+\xi} \mathfrak{C}_\xi.$$

(Even though $\mathfrak{C}_\xi$ is not necessarily a $\mathsf{CA}_{\beta+\gamma}$, the notation $\mathfrak{Rd}_{\beta+\xi}\mathfrak{C}_\xi$ can be given a clear meaning along the lines of 2.6.1.) Let F be any ultrafilter on γ such that $\gamma \sim \xi \in F$ for every $\xi < \gamma$, and let

$$(3) \qquad \mathfrak{D} = \mathsf{P}_{\xi<\gamma} \mathfrak{C}_\xi / \overline{F}.$$

For any $\eta < \gamma$ we have

$$\begin{aligned} \mathfrak{Rd}_{\beta+\eta}\mathfrak{D} &= \mathsf{P}_{\xi<\gamma} \mathfrak{Rd}_{\beta+\eta} \mathfrak{C}_\xi / \overline{F} && \text{by 0.5.13(vi)} \\ &\cong \mathsf{P}_{\eta\leqq\xi<\gamma} \mathfrak{Rd}_{\beta+\eta} \mathfrak{C}_\xi / \overline{G}_\eta && \text{by 0.3.65(ii)} \\ &= \mathsf{P}_{\eta\leqq\xi<\gamma} \mathfrak{Rd}_{\beta+\eta} \mathfrak{B}_\xi / \overline{G}_\eta && \text{by (2), 0.5.5} \end{aligned}$$

where $G_\eta = \{\Gamma \cap (\gamma \sim \eta) : \Gamma \in F\}$. Therefore, for every $\eta < \gamma$, $\mathfrak{Rd}_{\beta+\eta}\mathfrak{D} \in \mathsf{CA}_{\beta+\eta}$ by 2.6.2(i), and thus from 2.6.6 and the premiss $0 < \gamma = \bigcup\gamma$ we conclude that

$$(4) \qquad \mathfrak{D} \in \mathsf{CA}_{\beta+\gamma}.$$

Using (1)–(3) we easily establish that $h = \langle\langle x : \xi < \gamma\rangle / \overline{F} : x \in A\rangle$ is an isomorphism from $\mathfrak{A}$ into $\mathfrak{Rd}_\alpha\mathfrak{D}$. Hence, (4) gives us $\mathfrak{A} \in \mathsf{SRd}_\alpha \mathsf{CA}_{\beta+\gamma}$; this finishes the proof.

In the next few theorems we discuss, for any two different ordinals α, β with $\alpha \leqq \beta$, the relation between arbitrary CA_α's and the α-dimensional subreducts of CA_β's. We disregard the trivial case $\alpha = \beta$ which was dealt with in 2.6.2(ii).

THEOREM 2.6.8. *If* $\alpha \leqq 1$, *then*

$$\mathsf{CA}_\alpha = \mathbf{HSP}\mathfrak{Rd}_\alpha\mathsf{CA}_\beta = \mathbf{S}\mathfrak{Rd}_\alpha\mathsf{CA}_\beta$$

for every $\beta > \alpha$.

PROOF. By 2.6.2(i) we have

$$\mathsf{CA}_\alpha \supseteq \mathbf{HSP}\mathfrak{Rd}_\alpha\mathsf{CA}_\beta \supseteq \mathbf{S}\mathfrak{Rd}_\alpha\mathsf{CA}_\beta.$$

Hence it suffices to show that

(1) $$\mathsf{CA}_\alpha \subseteq \mathbf{S}\mathfrak{Rd}_\alpha\mathsf{CA}_\beta.$$

In case $\alpha = 0$, (1) follows from 2.6.2(iii) and the theory of BA's. Suppose now that $\mathfrak{A}$ is any simple CA_1. Let U be the set of all maximal proper ideals of $\mathfrak{Bl}\mathfrak{A}$, and let $\mathfrak{B}$ be the full cylindric set algebra of dimension β and with base U. Making use of the well-known result due to Stone [34*], and the construction of all simple CA_1's described in 1.2.14, one can easily show that $\langle\{F : F \in {}^\beta U, x \notin F_0\} : x \in A\rangle$ is an isomorphism from $\mathfrak{A}$ into $\mathfrak{Rd}_1\mathfrak{B}$. Thus, if K is the class of all simple CA_1's, we have $\mathsf{K} \subseteq \mathbf{S}\mathfrak{Rd}_1\mathsf{CA}_\beta$; applying 0.1.14(i), 0.3.12(ii), and 0.5.13(iii) we obtain

$$\mathbf{SP}\mathsf{K} \subseteq \mathbf{SPS}\mathfrak{Rd}_1\mathsf{CA}_\beta \subseteq \mathbf{S}\mathfrak{Rd}_1\mathsf{CA}_\beta.$$

However, by 2.4.52 we have $\mathsf{CA}_1 \subseteq \mathbf{SP}\mathsf{K}$, and thus (1) holds for $\alpha = 1$ also.

THEOREM 2.6.9. *If* $\alpha \geqq \omega$, *then*

$$\mathsf{CA}_\alpha = \mathbf{HSP}\mathfrak{Rd}_\alpha\mathsf{CA}_\beta = \mathbf{S}\mathfrak{Rd}_\alpha\mathsf{CA}_\beta$$

for every $\beta > \alpha$.[1)]

PROOF. As in the proof of 2.6.8 it suffices to show that

(1) $$\mathsf{CA}_\alpha \subseteq \mathbf{S}\mathfrak{Rd}_\alpha\mathsf{CA}_\beta.$$

Consider first the case $\beta = \alpha+1$. By 0.1.14(i), 0.3.69(ii), and 0.5.13(viii) we obtain

(2) $$\mathbf{S}\mathfrak{Rd}_\alpha\mathsf{CA}_{\alpha+1} = \mathbf{SS}\mathfrak{Rd}_\alpha\mathsf{CA}_{\alpha+1} = \mathsf{Up}\mathbf{S}\mathfrak{Rd}_\alpha\mathsf{CA}_{\alpha+1}.$$

Now suppose $\mathfrak{A} \in \mathsf{CA}_\alpha$ and ρ is a finite sequence without repetitions and with range included in α. Let σ be a one-one function from $\alpha+1$ onto α such

[1)] This is a result of Monk first announced in Henkin-Tarski [61], p. 99; a proof is given in Monk [61], p. 1448, Theorem 1.

that $Rg\rho 1 \sigma \subseteq Id$. Then $\mathfrak{Rd}^{(\rho)}\mathfrak{A} = \mathfrak{Rd}^{(\rho)}\mathfrak{Rd}_\alpha \mathfrak{Rd}_{\alpha+1}{}^{(\sigma)}\mathfrak{A}$ by 0.5.5, and consequently, in view of 2.6.2(i) we get

$$\mathfrak{Rd}^{(\rho)}\mathfrak{A} \in \mathsf{Rd}^{(\rho)}\mathbf{S}\mathsf{Rd}_\alpha \mathsf{CA}_{\alpha+1}.$$

Hence 2.6.4(i) holds for $\mathsf{K} = \mathbf{S}\mathsf{Rd}_\alpha \mathsf{CA}_{\alpha+1}$, and so, by (2) and 2.6.4, $\mathfrak{A} \in \mathbf{S}\mathsf{Rd}_\alpha \mathsf{CA}_{\alpha+1}$. This shows that the inclusion (1) holds and hence that the theorem is true in the special case $\beta = \alpha+1$.

To prove (1) holds for arbitrary $\beta > \alpha$ we proceed by transfinite induction on β. Suppose $\beta = \beta'+1$ for some β'. Then by the induction hypothesis, or by 2.6.2(ii) if $\beta' = \alpha$, we have $\mathsf{CA}_\alpha \subseteq \mathbf{S}\mathsf{Rd}_\alpha \mathsf{CA}_{\beta'}$. Thus applying the special case of the theorem just proved, along with 0.1.14(i), 0.5.5, and 0.5.4(iii), one gets

$$\mathsf{CA}_\alpha \subseteq \mathbf{S}\mathsf{Rd}_\alpha \mathbf{S}\mathsf{Rd}_{\beta'} \mathsf{CA}_\beta \subseteq \mathbf{SS}\mathsf{Rd}_\alpha \mathsf{Rd}_{\beta'} \mathsf{CA}_\beta = \mathbf{S}\mathsf{Rd}_\alpha \mathsf{CA}_\beta.$$

If $\beta = \bigcup\beta$, i.e., if β is a limit ordinal, then the inclusion (1) follows directly from 2.6.7 and the induction hypothesis. This completes the proof.

THEOREM 2.6.10. *If* $\beta \geqq \alpha \cup \omega$, *then* $\mathbf{S}\mathsf{Rd}_\alpha \mathsf{CA}_\beta = \mathbf{S}\mathsf{Rd}_\alpha \mathsf{CA}_{\alpha\cup\omega}$.

PROOF. By 0.5.5 and 2.6.2(i) we obtain

$$\mathbf{S}\mathsf{Rd}_\alpha \mathsf{CA}_\beta = \mathbf{S}\mathsf{Rd}_\alpha \mathsf{Rd}_{\alpha\cup\omega} \mathsf{CA}_\beta \subseteq \mathbf{S}\mathsf{Rd}_\alpha \mathsf{CA}_{\alpha\cup\omega}.$$

On the other hand,

$$\begin{aligned} \mathbf{S}\mathsf{Rd}_\alpha \mathsf{CA}_{\alpha\cup\omega} &= \mathbf{S}\mathsf{Rd}_\alpha \mathbf{S}\mathsf{Rd}_{\alpha\cup\omega} \mathsf{CA}_\beta && \text{by 2.6.9} \\ &\subseteq \mathbf{SS}\mathsf{Rd}_\alpha \mathsf{Rd}_{\alpha\cup\omega} \mathsf{CA}_\beta && \text{by 0.5.4(iii)} \\ &= \mathbf{S}\mathsf{Rd}_\alpha \mathsf{CA}_\beta && \text{by 0.1.14(i), 0.5.5.} \end{aligned}$$

In Theorem 2.6.14 we shall show that 2.6.8 and 2.6.9 do not extend to the classes CA_α with $2 \leqq \alpha < \omega$. The proof will be based on several lemmas.

LEMMA 2.6.11. *If* $\mathfrak{A} \in \mathsf{CA}_\alpha$ *and* $1 \leqq \kappa < \alpha \cap \omega$, *then*

$$\mathsf{c}_0(x_0 \cdot x_\kappa) \cdot \mathsf{c}_0(x_0 \cdot -x_\kappa) \cdot [\textstyle\prod_{\lambda<\kappa} -\mathsf{c}_0\mathsf{c}_1(\mathsf{c}_1 x_\lambda \cdot \mathsf{s}_1^0 \mathsf{c}_1 x_\lambda \cdot -\mathsf{d}_{01})] \cdot -\mathsf{c}_0 \prod_{\lambda<\kappa} -\mathsf{c}_1 x_\lambda = 0$$

holds for every $x \in {}^{\kappa+1}A$.[1])

PROOF. The proof will take the form of a series of more or less involved computations.

[1]) Theorems 2.6.11, 2.6.13, 2.6.14, and 2.6.16 were obtained by Henkin. The proof of the last three results are based on the method of splitting elements which also originates with him. However, the general and purely algebraic form of this method as described in 2.6.12 is based on an idea of Monk.

For any $\mu \leqq \kappa$ we have

$$\begin{aligned} -\mathsf{c}_0\prod_{\lambda<\kappa} -\mathsf{c}_1 x_\lambda &= \mathsf{c}_0^\partial \sum_{\lambda<\kappa} \mathsf{c}_1 x_\lambda \\ &\leqq \mathsf{s}_\mu^0 \sum_{\lambda<\kappa} \mathsf{c}_1 x_\lambda \qquad \text{by 1.2.7}^\partial\text{, 1.4.4(vi)} \\ &= \sum_{\lambda<\kappa} \mathsf{s}_\mu^0 \mathsf{c}_1 x_\lambda . \end{aligned}$$

Hence

$$(1) \qquad -\mathsf{c}_0\prod_{\lambda<\kappa} -\mathsf{c}_1 x_\lambda \leqq \prod_{\mu\leqq\kappa}\sum_{\lambda<\kappa} \mathsf{s}_\mu^0 \mathsf{c}_1 x_\lambda = \sum_{\varphi\in{}^{\kappa+1}\kappa}\prod_{\mu\leqq\kappa} \mathsf{s}_\mu^0 \mathsf{c}_1 x_{\varphi\mu}.$$

If $\mu < \nu < \alpha \cap \omega$ and $\lambda < \kappa$, then

$$\begin{aligned} -\mathsf{c}_0\mathsf{c}_1(\mathsf{c}_1 x_\lambda \cdot \mathsf{s}_1^0 \mathsf{c}_1 x_\lambda \cdot -\mathsf{d}_{01}) &= \mathsf{c}_0^\partial \mathsf{c}_1^\partial -(\mathsf{c}_1 x_\lambda \cdot \mathsf{s}_1^0 \mathsf{c}_1 x_\lambda \cdot -\mathsf{d}_{01}) \\ &\leqq \mathsf{s}_\mu^0 \mathsf{s}_\nu^1 -(\mathsf{c}_1 x_\lambda \cdot \mathsf{s}_1^0 \mathsf{c}_1 x_\lambda \cdot -\mathsf{d}_{01}) \quad \text{by 1.2.7}^\partial\text{, 1.4.4(vi)} \\ &= \mathsf{s}_\mu^0 -(\mathsf{c}_1 x_\lambda \cdot \mathsf{s}_\nu^1 \mathsf{s}_1^0 \mathsf{c}_1 x_\lambda \cdot -\mathsf{d}_{0\nu}) \quad \text{by 1.5.4(i), 1.5.8(i)} \\ &= -\mathsf{s}_\mu^0 (\mathsf{c}_1 x_\lambda \cdot \mathsf{s}_\nu^0 \mathsf{c}_1 x_\lambda \cdot -\mathsf{d}_{0\nu}) \quad \text{by 1.5.11(i)} \\ &= -(\mathsf{s}_\mu^0 \mathsf{c}_1 x_\lambda \cdot \mathsf{s}_\nu^0 \mathsf{c}_1 x_\lambda \cdot -\mathsf{d}_{\mu\nu}) \quad \text{by 1.5.4(i), 1.5.10(i).} \end{aligned}$$

Thus we have shown that, for any μ, ν, λ such that $\mu < \nu < \alpha \cap \omega$ and $\lambda < \kappa$, we have

$$(2) \qquad -\mathsf{c}_0\mathsf{c}_1(\mathsf{c}_1 x_\lambda \cdot \mathsf{s}_1^0 \mathsf{c}_1 x_\lambda \cdot -\mathsf{d}_{01}) \leqq -(\mathsf{s}_\mu^0 \mathsf{c}_1 x_\lambda \cdot \mathsf{s}_\nu^0 \mathsf{c}_1 x_\lambda \cdot -\mathsf{d}_{\mu\nu}),$$

and consequently we obtain

$$(3) \quad \bar{\mathsf{d}}((\kappa+1)\times(\kappa+1)) \cdot \prod_{\lambda<\kappa} -\mathsf{c}_0\mathsf{c}_1(\mathsf{c}_1 x_\lambda \cdot \mathsf{s}_1^0 \mathsf{c}_1 x_\lambda \cdot -\mathsf{d}_{01}) \leqq \prod_{\lambda<\kappa} -(\mathsf{s}_\mu^0 \mathsf{c}_1 x_\lambda \cdot \mathsf{s}_\nu^0 \mathsf{c}_1 x_\lambda) \quad \text{whenever } \mu < \nu \leqq \kappa.$$

For any $\varphi \in {}^{\kappa+1}\kappa$ there exists π, ρ such that $\pi < \rho \leqq \kappa$ and $\varphi\pi = \varphi\rho$. Thus from (3) we have

$$\begin{aligned} \bar{\mathsf{d}}((\kappa+1)\times(\kappa+1)) \cdot \prod_{\lambda<\kappa} -\mathsf{c}_0\mathsf{c}_1(\mathsf{c}_1 x_\lambda \cdot \mathsf{s}_1^0 \mathsf{c}_1 x_\lambda \cdot -\mathsf{d}_{01}) &\leqq -(\mathsf{s}_\pi^0 \mathsf{c}_1 x_{\varphi\pi} \cdot \mathsf{s}_\rho^0 \mathsf{c}_1 x_{\varphi\rho}) \\ &\leqq -\prod_{\mu\leqq\kappa} \mathsf{s}_\mu^0 \mathsf{c}_1 x_{\varphi\mu}, \end{aligned}$$

and hence

$$(4) \quad \bar{\mathsf{d}}((\kappa+1)\times(\kappa+1)) \cdot \prod_{\lambda<\kappa} -\mathsf{c}_0\mathsf{c}_1(\mathsf{c}_1 x_\lambda \cdot \mathsf{s}_1^0 \mathsf{c}_1 x_\lambda \cdot -\mathsf{d}_{01}) \leqq -\sum_{\varphi\in{}^{\kappa+1}\kappa}\prod_{\mu\leqq\kappa} \mathsf{s}_\mu^0 \mathsf{c}_1 x_{\varphi\mu}.$$

Combining (1) and (4) we get

$$[\prod_{\lambda<\kappa} -\mathsf{c}_0\mathsf{c}_1(\mathsf{c}_1 x_\lambda \cdot \mathsf{s}_1^0 \mathsf{c}_1 x_\lambda \cdot -\mathsf{d}_{01})] \cdot -\mathsf{c}_0\prod_{\lambda<\kappa} -\mathsf{c}_1 x_\lambda \leqq -\bar{\mathsf{d}}((\kappa+1)\times(\kappa+1));$$

applying c_0^∂ to both members of this inequality and using 1.4.4(iii) we obtain finally

$$(5)\quad [\prod_{\lambda<\kappa} -c_0c_1(c_1x_\lambda \cdot s_1^0 c_1 x_\lambda \cdot -d_{01})] \cdot -c_0 \prod_{\lambda<\kappa} -c_1 x_\lambda \leqq -c_0\bar{d}((\kappa+1)\times(\kappa+1)).$$

From now until the end of the proof we shall assume that μ and ν are any ordinals such that $1 \leqq \mu < \nu \leqq \kappa$; we also assume for the time being that $\kappa \geqq 2$. If $\mu > 1$, then

$$\begin{aligned} s_\nu^\mu -c_0c_1(c_1x_0\cdot s_1^0c_1x_0\cdot -d_{01}) &= -c_0c_1s_\nu^\mu(c_1x_0\cdot s_1^0c_1x_0\cdot -d_{01}) && \text{by 1.5.8(ii)}\\ &= -c_0c_1(s_\nu^\mu c_1x_0\cdot s_1^0 s_\nu^\mu c_1x_0\cdot -d_{01}) && \text{by 1.5.10(iii)}\\ &= s_\mu^1 -c_0c_1(s_\nu^\mu c_1x_0\cdot s_1^0 s_\nu^\mu c_1x_0\cdot -d_{01}) && \text{by 1.5.8(i)}\\ &\leqq -s_\nu^\mu c_1x_0 + -s_\mu^1 s_1^0 s_\nu^\mu c_1x_0 + d_{0\mu} && \text{by 1.5.4(i), 1.5.8(i),(ii)}\\ &= -s_\nu^\mu c_1x_0 + -s_\mu^0 s_\nu^\mu c_1x_0 + d_{0\mu} && \text{by 1.5.11(i);} \end{aligned}$$

on the other hand, if $\mu = 1$, then

$$\begin{aligned} s_\nu^\mu -c_0c_1(c_1x_0\cdot s_1^0c_1x_0\cdot -d_{01}) &= -c_0c_1(c_1x_0\cdot s_1^0c_1x_0\cdot -d_{01}) && \text{by 1.5.8(i)}\\ &\leqq -c_1x_0 + -s_1^0c_1x_0 + d_{01}\\ &= -s_\nu^\mu c_1x_0 + -s_\mu^0 s_\nu^\mu c_1x_0 + d_{0\mu} && \text{by 1.5.8(i).} \end{aligned}$$

Hence, for all μ, ν we have

$$(6)\quad s_\nu^\mu -c_0c_1(c_1x_0\cdot s_1^0c_1x_0\cdot -d_{01}) \leqq -s_\nu^\mu c_1x_0 + -s_\mu^0 s_\nu^\mu c_1x_0 + d_{0\mu}.$$

Continuing the computation,

$$\begin{aligned} s_\nu^\mu[c_0(x_0\cdot x_\kappa)\cdot c_0(x_0\cdot -x_\kappa)] &= c_0s_\nu^\mu(x_0\cdot x_\kappa)\cdot c_0s_\nu^\mu(x_0\cdot -x_\kappa) && \text{by 1.5.8(ii)}\\ &= c_\mu s_\mu^0 s_\nu^\mu(x_0\cdot x_\kappa)\cdot c_0s_\nu^\mu(x_0\cdot -x_\kappa) && \text{by 1.5.9(i), 1.5.10(i)}\\ &= c_\mu c_0[s_\mu^0 s_\nu^\mu(x_0\cdot x_\kappa)\cdot s_\nu^\mu(x_0\cdot -x_\kappa)] && \text{by 1.5.9(ii).} \end{aligned}$$

Applying this last result we get

$$\begin{aligned}
&s^{\mu}_{\nu}[-c_0c_1(c_1x_0\cdot s^0_1c_1x_0\cdot -d_{01})\cdot c_0(x_0\cdot x_\kappa)\cdot c_0(x_0\cdot -x_\kappa)]\\
&\quad = s^{\mu}_{\nu}-c_0c_1(c_1x_0\cdot s^0_1c_1x_0\cdot -d_{01})\cdot c_\mu c_0[s^0_\mu s^\mu_\nu(x_0\cdot x_\kappa)\cdot s^\mu_\nu(x_0\cdot -x_\kappa)]\\
&\quad = c_\mu c_0[s^{\mu}_{\nu}-c_0c_1(c_1x_0\cdot s^0_1c_1x_0\cdot -d_{01})\cdot s^0_\mu s^\mu_\nu(x_0\cdot x_\kappa)\cdot s^\mu_\nu(x_0\cdot -x_\kappa)] &&\text{by 1.5.9(ii)}\\
&\quad \leqq c_\mu c_0[(-s^\mu_\nu c_1x_0+-s^0_\mu s^\mu_\nu c_1x_0+d_{0\mu})\cdot s^0_\mu s^\mu_\nu(x_0\cdot x_\kappa)\cdot s^\mu_\nu(x_0\cdot -x_\kappa)] &&\text{by (6)}\\
&\quad = c_\mu c_0[d_{0\mu}\cdot s^0_\mu s^\mu_\nu(x_0\cdot x_\kappa)\cdot s^\mu_\nu(x_0\cdot -x_\kappa)]\\
&\quad = c_\mu c_0[d_{0\mu}\cdot s^\mu_\nu(x_0\cdot x_\kappa)\cdot s^\mu_\nu(x_0\cdot -x_\kappa)] &&\text{by 1.5.5}\\
&\quad = 0.
\end{aligned}$$

Therefore, in view of 1.5.5,

(7) $d_{\mu\nu}\cdot -c_0c_1(c_1x_0\cdot s^0_1c_1x_0\cdot -d_{01})\cdot c_0(x_0\cdot x_\kappa)\cdot c_0(x_0\cdot -x_\kappa)=0.$

We now return to (2) and apply it with $\lambda=0$:

$$\begin{aligned}
-c_0c_1(c_1x_0\cdot s^0_1c_1x_0\cdot -d_{01}) &\leqq -[s^0_\mu c_1x_0\cdot s^0_\nu c_1x_0\cdot -d_{\mu\nu}]\\
&\leqq -[s^0_\mu(x_0\cdot x_\kappa)\cdot s^0_\nu(x_0\cdot -x_\kappa)\cdot -d_{\mu\nu}]\\
&= -[s^0_\mu(x_0\cdot x_\kappa)\cdot s^0_\nu(x_0\cdot -x_\kappa)] &&\text{by 1.5.6;}
\end{aligned}$$

similarly, we get

$$-c_0c_1(c_1x_0\cdot s^0_1c_1x_0\cdot -d_{01})\leqq -[s^0_\mu(x_0\cdot -x_\kappa)\cdot s^0_\nu(x_0\cdot x_\kappa)].$$

Hence one sees that

(8) $-c_0c_1(c_1x_0\cdot s^0_1c_1x_0\cdot -d_{01})\leqq$
$$-\textstyle\sum_{1\leqq\mu<\nu\leqq\kappa}[s^0_\mu(x_0\cdot x_\kappa)\cdot s^0_\nu(x_0\cdot -x_\kappa)+s^0_\mu(x_0\cdot -x_\kappa)\cdot s^0_\nu(x_0\cdot x_\kappa)].$$

Now set $\Gamma=(\kappa+1)\sim\{0\}$. We then have

$$\begin{aligned}
&\bar{d}(\Gamma\times\Gamma)\cdot -c_0\bar{d}((\kappa+1)\times(\kappa+1))\cdot c_0(x_0\cdot x_\kappa)\\
&\quad = c_0[\bar{d}(\Gamma\times\Gamma)\cdot -c_0\bar{d}((\kappa+1)\times(\kappa+1))\cdot x_0\cdot x_\kappa]\\
&\quad \leqq c_0[\bar{d}(\Gamma\times\Gamma)\cdot \textstyle\sum_{0\leqq\pi<\mu\leqq\kappa}d_{\pi\mu})\cdot x_0\cdot x_\kappa]\\
&\quad \leqq c_0\textstyle\sum_{1\leqq\mu\leqq\kappa}(d_{0\mu}\cdot x_0\cdot x_\kappa)\\
&\quad = \textstyle\sum_{1\leqq\mu\leqq\kappa}s^0_\mu(x_0\cdot x_\kappa).
\end{aligned}$$

Combining this result with the one obtained from it by replacing x_κ by $-x_\kappa$ we get

$$\bar{d}(\Gamma\times\Gamma)\cdot -c_0\bar{d}((\kappa+1)\times(\kappa+1))\cdot c_0(x_0\cdot x_\kappa)\cdot c_0(x_0\cdot -x_\kappa)\leqq$$
$$\textstyle\sum_{1\leqq\mu<\nu\leqq\kappa}[s^0_\mu(x_0\cdot x_\kappa)\cdot s^0_\nu(x_0\cdot -x_\kappa)+s^0_\mu(x_0\cdot -x_\kappa)\cdot s^0_\nu(x_0\cdot x_\kappa)].$$

This last inequality together with (8) gives

(9) $\bar{d}(\Gamma\times\Gamma)\cdot -c_0\bar{d}((\kappa+1)\times(\kappa+1))\cdot c_0(x_0\cdot x_\kappa)\cdot c_0(x_0\cdot -x_\kappa)\cdot$

$$-c_0c_1(c_1x_0\cdot s_1^0c_1x_0\cdot -d_{01}) = 0.$$

We easily see that

$$-c_0\bar{d}((\kappa+1)\times(\kappa+1))\cdot c_0(x_0\cdot x_\kappa)\cdot c_0(x_0\cdot -x_\kappa)\cdot -c_0c_1(c_1x_0\cdot s_1^0c_1x_0\cdot -d_{01}) =$$

$$(\bar{d}(\Gamma\times\Gamma)+\textstyle\sum_{1\leqq\mu<\nu\leqq\kappa}d_{\mu\nu})\cdot -c_0\bar{d}((\kappa+1)\times(\kappa+1))\cdot c_0(x_0\cdot x_\kappa)\cdot c_0(x_0\cdot -x_\kappa)\cdot$$

$$-c_0c_1(c_1x_0\cdot s_1^0c_1x_0\cdot -d_{01}).$$

Hence, since the right side of this equation equals 0 by (7) and (9), we have

(10) $c_0(x_0\cdot x_\kappa)\cdot c_0(x_0\cdot -x_\kappa)\cdot -c_0c_1(c_1x_0\cdot s_1^0c_1x_0\cdot -d_{01}) \leqq$

$$c_0\bar{d}((\kappa+1)\times(\kappa+1)).$$

Notice that by 1.3.20 this inclusion also holds in case $\kappa = 1$; the lemma now follows from (5) and (10).

In the next lemma a new method of constructing CA's is described. In view of its intuitive content it could be called the *method of splitting elements.* The method has a rather special character but turns out to be a useful tool in constructing counterexamples.

LEMMA 2.6.12. *Assume that* $\mathfrak{A} = \langle A, +, \cdot, -, 0, 1, c_\kappa, d_{\kappa\lambda}\rangle_{\kappa,\lambda<\alpha}$ *is a* CA_α *with* $\alpha \geqq 2$, *and a is any element of* $\mathfrak{A}$ *such that* $a \leqq -d_{\kappa\lambda}$ *whenever* $\kappa<\lambda<\alpha$. *Let* $\mathfrak{B} = \langle B, +', \cdot', -', 0', 1', c'_\kappa, d'_{\kappa\lambda}\rangle_{\kappa,\lambda<\alpha}$ *be the algebra similar to* $\mathfrak{A}$ *determined by the following stipulations*:

(i) $\langle B, +', \cdot', -', 0', 1'\rangle = \mathfrak{Bl}\mathfrak{A}\times\mathfrak{Rl}_a\mathfrak{Bl}\mathfrak{A}$;

(ii) $d'_{\kappa\lambda} = \langle d_{\kappa\lambda}, a\cdot d_{\kappa\lambda}\rangle$ *for all* $\kappa,\lambda<\alpha$;

(iii) $c'_\kappa\langle x, y\rangle = \langle c_\kappa(x+y), a\cdot c_\kappa(x+y)\rangle$ *for all* $x, y\in\mathfrak{A}$ *with* $y\leqq a$.

Under these assumptions $\mathfrak{B}\in CA_\alpha$.

Furthermore, $\mathfrak{A}\cong\mathfrak{Rl}_{\langle 1,0\rangle}\mathfrak{B}$ *and for each* $x\in A$ *such that* $x\leqq a$ *we have disjoint elements* $\langle x, 0\rangle$ *and* $\langle 0, x\rangle$ *in* $\mathfrak{B}$ *such that* $c'_\kappa\langle x, 0\rangle = c'_\kappa\langle 0, x\rangle$ *for every* $\kappa<\alpha$.

PROOF. The axioms (C_0), (C_1), (C_2), and (C_5) of cylindric algebras are obviously satisfied in $\mathfrak{B}$. We shall verify the remaining axioms by computation.

(C_3). Consider any $\langle x, y\rangle, \langle u, v\rangle\in B$ and $\kappa<\alpha$. Then

$$\begin{aligned}
\mathsf{c}'_\kappa(\langle x, y\rangle \cdot' \mathsf{c}'_\kappa\langle u, v\rangle) &= \mathsf{c}'_\kappa(\langle x, y\rangle \cdot' \langle \mathsf{c}_\kappa(u+v), a\cdot \mathsf{c}_\kappa(u+v)\rangle)\\
&= \mathsf{c}'_\kappa\langle x\cdot \mathsf{c}_\kappa(u+v), y\cdot \mathsf{c}_\kappa(u+v)\rangle && \text{since } y \leqq a\\
&= \langle \mathsf{c}_\kappa[x\cdot \mathsf{c}_\kappa(u+v) + y\cdot \mathsf{c}_\kappa(u+v)], a\cdot \mathsf{c}_\kappa[x\cdot \mathsf{c}_\kappa(u+v)+y\cdot \mathsf{c}_\kappa(u+v)]\rangle\\
&= \langle \mathsf{c}_\kappa(x+y)\cdot \mathsf{c}_\kappa(u+v), a\cdot \mathsf{c}_\kappa(x+y)\cdot \mathsf{c}_\kappa(u+v)\rangle && \text{by } (C_3) \text{ for } \mathfrak{A}\\
&= \mathsf{c}'_\kappa\langle x, y\rangle \cdot' \mathsf{c}'_\kappa\langle u, v\rangle.
\end{aligned}$$

(C_4). Let $\langle x, y\rangle \in B$ and $\kappa, \lambda < \alpha$. Then

$$\begin{aligned}
\mathsf{c}'_\kappa \mathsf{c}'_\lambda\langle x, y\rangle &= \mathsf{c}'_\kappa\langle \mathsf{c}_\lambda(x+y), a\cdot \mathsf{c}_\lambda(x+y)\rangle\\
&= \langle \mathsf{c}_\kappa[\mathsf{c}_\lambda(x+y) + a\cdot \mathsf{c}_\lambda(x+y)], a\cdot \mathsf{c}_\kappa[\mathsf{c}_\lambda(x+y)+a\cdot \mathsf{c}_\lambda(x+y)]\rangle\\
&= \langle \mathsf{c}_\kappa \mathsf{c}_\lambda(x+y), a\cdot \mathsf{c}_\kappa \mathsf{c}_\lambda(x+y)\rangle;
\end{aligned}$$

similarly,

$$\mathsf{c}'_\lambda \mathsf{c}'_\kappa\langle x, y\rangle = \langle \mathsf{c}_\lambda \mathsf{c}_\kappa(x+y), a\cdot \mathsf{c}_\lambda \mathsf{c}_\kappa(x+y)\rangle.$$

Thus $\mathsf{c}'_\kappa \mathsf{c}'_\lambda\langle x, y\rangle = \mathsf{c}'_\lambda \mathsf{c}'_\kappa\langle x, y\rangle$ by (C_4) for $\mathfrak{A}$. This shows that (C_4) holds for $\mathfrak{B}$.

(C_6). Assume $\mu \neq \kappa, \lambda$. Then

$$\begin{aligned}
\mathsf{c}'_\mu(\mathsf{d}'_{\kappa\mu} \cdot' \mathsf{d}'_{\mu\lambda}) &= \mathsf{c}'_\mu(\langle \mathsf{d}_{\kappa\mu}, a\cdot \mathsf{d}_{\kappa\mu}\rangle \cdot' \langle \mathsf{d}_{\mu\lambda}, a\cdot \mathsf{d}_{\mu\lambda}\rangle)\\
&= \mathsf{c}'_\mu(\langle \mathsf{d}_{\kappa\mu}, 0\rangle \cdot' \langle \mathsf{d}_{\mu\lambda}, 0\rangle) && \text{since } a \leqq -\mathsf{d}_{\kappa\mu}, -\mathsf{d}_{\mu\lambda}\\
&= \langle \mathsf{c}_\mu(\mathsf{d}_{\kappa\mu}\cdot \mathsf{d}_{\mu\lambda}), a\cdot \mathsf{c}_\mu(\mathsf{d}_{\kappa\mu}\cdot \mathsf{d}_{\mu\lambda})\rangle\\
&= \langle \mathsf{d}_{\kappa\lambda}, a\cdot \mathsf{d}_{\kappa\lambda}\rangle && \text{by } (C_6) \text{ for } \mathfrak{A}\\
&= \mathsf{d}'_{\kappa\lambda}.
\end{aligned}$$

(C_7). Assume $\kappa \neq \lambda$ and $\langle x, y\rangle \in B$. Then

$$\begin{aligned}
&\mathsf{c}'_\kappa(\mathsf{d}'_{\kappa\lambda} \cdot' \langle x, y\rangle) \cdot' \mathsf{c}'_\kappa(\mathsf{d}'_{\kappa\lambda} \cdot' -'\langle x, y\rangle)\\
&\quad= \mathsf{c}'_\kappa(\langle \mathsf{d}_{\kappa\lambda}, 0\rangle \cdot' \langle x, y\rangle) \cdot' \mathsf{c}'_\kappa(\langle \mathsf{d}_{\kappa\lambda}, 0\rangle \cdot' -'\langle x, y\rangle) && \text{since } a \leqq -\mathsf{d}_{\kappa\lambda}\\
&\quad= \mathsf{c}'_\kappa\langle \mathsf{d}_{\kappa\lambda}\cdot x, 0\rangle \cdot' \mathsf{c}'_\kappa\langle \mathsf{d}_{\kappa\lambda}\cdot -x, 0\rangle\\
&\quad= \langle \mathsf{c}_\kappa(\mathsf{d}_{\kappa\lambda}\cdot x), a\cdot \mathsf{c}_\kappa(\mathsf{d}_{\kappa\lambda}\cdot x)\rangle \cdot' \langle \mathsf{c}_\kappa(\mathsf{d}_{\kappa\lambda}\cdot -x), a\cdot \mathsf{c}_\kappa(\mathsf{d}_{\kappa\lambda}\cdot -x)\rangle\\
&\quad= \langle \mathsf{c}_\kappa(\mathsf{d}_{\kappa\lambda}\cdot x)\cdot \mathsf{c}_\kappa(\mathsf{d}_{\kappa\lambda}\cdot -x), a\cdot \mathsf{c}_\kappa(\mathsf{d}_{\kappa\lambda}\cdot x)\cdot \mathsf{c}_\kappa(\mathsf{d}_{\kappa\lambda}\cdot -x)\rangle\\
&\quad= \langle 0, 0\rangle && \text{by } (C_7) \text{ for } \mathfrak{A}\\
&\quad= 0'.
\end{aligned}$$

This completes the proof that $\mathfrak{B} \in \mathsf{CA}_\alpha$. In order to show that $\mathfrak{A} \cong \mathfrak{Rl}_{\langle 1,0\rangle}\mathfrak{B}$ we let h be the function from A into $Rl_{\langle 1,0\rangle}\mathfrak{B}$ such that $hx = \langle x, 0\rangle$ for every $x \in A$. Clearly $h \in Is(\mathfrak{B}\mathfrak{l}\mathfrak{A}, \mathfrak{B}\mathfrak{l}\mathfrak{R}\mathfrak{l}_{\langle 1,0\rangle}\mathfrak{B})$, and $h\mathsf{d}_{\kappa\lambda} = \mathsf{d}_{\kappa\lambda}^{(\mathfrak{R}\mathfrak{l}_{\langle 1,0\rangle}\mathfrak{B})}$ for all $\kappa, \lambda < \alpha$. Also, for any $x \in A$ and $\kappa < \alpha$ we have

$$\begin{aligned} h\mathsf{c}_\kappa x &= \langle \mathsf{c}_\kappa x, 0\rangle \\ &= \langle \mathsf{c}_\kappa x, a \cdot \mathsf{c}_\kappa x\rangle \cdot' \langle 1, 0\rangle \\ &= \mathsf{c}'_\kappa\langle x, 0\rangle \cdot' \langle 1, 0\rangle \\ &= \mathsf{c}_\kappa^{(\mathfrak{R}\mathfrak{l}_{\langle 1,0\rangle}\mathfrak{B})} hx. \end{aligned}$$

Thus $h \in Is(\mathfrak{A}, \mathfrak{R}\mathfrak{l}_{\langle 1,0\rangle}\mathfrak{B})$. Finally, to complete the proof of the lemma we observe that for each $x \in A$ with $x \leqq a$ one gets

$$\mathsf{c}'_\kappa\langle x, 0\rangle = \langle \mathsf{c}_\kappa x, a \cdot \mathsf{c}_\kappa x\rangle = \mathsf{c}'_\kappa\langle 0, x\rangle$$

for every $\kappa < \alpha$.

LEMMA 2.6.13. *For every β such that $2 \leqq \beta < \omega$ there is a CA_β which is generated by a singleton and in which the equation of 2.6.11 with $\kappa = \beta$ fails.*

Furthermore, for each α such that $2 \leqq \alpha < \beta$, there is a $\mathfrak{C} \in \mathsf{Rd}_\alpha\mathsf{CA}_\beta$ such that $\mathfrak{C}$ is generated by a set with cardinality β and the same equation fails.

PROOF. Let $\mathfrak{A} = \langle A, +, \cdot, -, 0, 1, \mathsf{c}_\kappa, \mathsf{d}_{\kappa\lambda}\rangle_{\kappa,\lambda<\beta}$ be the full cylindric set algebra of dimension β and with base $\beta = \{0, 1, \dots, \beta-1\}$; let $a = \{\beta 1 Id\}$. Then

(1) $$a \leqq -\mathsf{d}_{\kappa\lambda} \text{ whenever } \kappa < \lambda < \beta;$$

furthermore, one can easily check that

(2) $$\mathsf{c}_1 a \cdot \mathsf{s}_1^0 \mathsf{c}_1 a,\ \mathsf{c}_0 a \cdot \mathsf{s}_0^1 \mathsf{c}_0 a \leqq \mathsf{d}_{01}$$

and

(3) $$a + \textstyle\sum_{1 \leqq \lambda < \beta}(\mathsf{c}_0 a \cdot \mathsf{d}_{0\lambda}) = \mathsf{c}_0 a.$$

Let $\mathfrak{B} = \langle B, +', \cdot', 0', 1', \mathsf{c}'_\kappa, \mathsf{d}'_{\kappa\lambda}\rangle_{\kappa,\lambda<\beta}$ be the algebra which is defined in Lemma 2.6.12 when α is replaced everywhere by β. Then because of (1) we have by 2.6.12 that $\mathfrak{B} \in \mathsf{CA}_\beta$. Let $\mathfrak{D} = \mathfrak{S}\mathfrak{g}^{(\mathfrak{B})}\{\langle a, 0\rangle\}$ and set

$$\begin{gathered} x_0 = \mathsf{c}'_1\langle a, 0\rangle = \langle \mathsf{c}_1 a, a\rangle, \\ x_\lambda = \mathsf{c}'_0\langle a, 0\rangle \cdot' \mathsf{d}'_{0\lambda} = \langle \mathsf{c}_0 a \cdot \mathsf{d}_{0\lambda}, 0\rangle \text{ when } 1 \leqq \lambda < \beta, \\ x_\beta = \langle a, 0\rangle. \end{gathered}$$

Then

(4) $\mathfrak{D}$ is a CA_β generated by a singleton, and $\{x_0, x_1, \dots, x_\beta\} \subseteq D$.

We now compute the element of $\mathfrak{D}$ represented by

(5) $\mathsf{c}'_0(x_0 \cdot' x_\beta) \cdot' \mathsf{c}'_0(x_0 \cdot' -'x_\beta) \cdot'$

$$[\prod'_{\lambda<\beta} -'\mathsf{c}'_0\mathsf{c}'_1(\mathsf{c}'_1 x_\lambda \cdot' \mathsf{s}_1^{0(\mathfrak{D})}\mathsf{c}'_1 x_\lambda \cdot' - \mathsf{d}'_{01}] \cdot' -'\mathsf{c}'_0 \prod'_{\lambda<\beta} -'\mathsf{c}'_1 x_\lambda.$$

Observe first of all that from (1) and (2) we have

(6) $$a \cdot \mathsf{s}_1^0 \mathsf{c}_1 a = a \cdot \mathsf{s}_0^1 \mathsf{c}_0 a = 0.$$

Then

$$\begin{aligned}
\mathsf{c}'_1 x_0 \cdot' \mathsf{s}_1^{0(\mathfrak{D})}\mathsf{c}'_1 x_0 \cdot' -'\mathsf{d}'_{01} &= \langle \mathsf{c}_1 a, a\rangle \cdot' \mathsf{s}_1^{0(\mathfrak{D})}\langle \mathsf{c}_1 a, a\rangle \cdot' \langle -\mathsf{d}_{01}, a\rangle \\
&= \langle \mathsf{c}_1 a, a\rangle \cdot' \langle \mathsf{s}_1^0 \mathsf{c}_1 a, 0\rangle \cdot' \langle -\mathsf{d}_{01}, a\rangle && \text{by (6)} \\
&= \langle \mathsf{c}_1 a \cdot \mathsf{s}_1^0 \mathsf{c}_1 a \cdot -\mathsf{d}_{01}, 0\rangle = \langle 0, 0\rangle && \text{by (2);}
\end{aligned}$$

$$\begin{aligned}
&\mathsf{c}'_1 x_1 \cdot' \mathsf{s}_1^{0(\mathfrak{D})}\mathsf{c}'_1 x_1 \cdot' -'\mathsf{d}'_{01} \\
&\quad = \mathsf{c}'_1\langle \mathsf{c}_0 a \cdot \mathsf{d}_{01}, 0\rangle \cdot' \mathsf{s}_1^{0(\mathfrak{D})}\mathsf{c}'_1\langle \mathsf{c}_0 a \cdot \mathsf{d}_{01}, 0\rangle \cdot' \langle -\mathsf{d}_{01}, a\rangle \\
&\quad = \langle \mathsf{s}_0^1 \mathsf{c}_0 a, 0\rangle \cdot' \mathsf{s}_1^{0(\mathfrak{D})}\langle \mathsf{s}_0^1 \mathsf{c}_0 a, 0\rangle \cdot' \langle -\mathsf{d}_{01}, a\rangle && \text{by (6)} \\
&\quad = \langle \mathsf{s}_0^1 \mathsf{c}_0 a, 0\rangle \cdot' \langle \mathsf{c}_0 a, a\rangle \cdot' \langle -\mathsf{d}_{01}, a\rangle && \text{by 1.5.11(i)} \\
&\quad = \langle \mathsf{s}_0^1 \mathsf{c}_0 a \cdot \mathsf{c}_0 a \cdot -\mathsf{d}_{01}, 0\rangle = \langle 0, 0\rangle && \text{by (2).}
\end{aligned}$$

For each λ such that $2 \leqq \lambda < \beta$ one obtains

$$\begin{aligned}
&\mathsf{c}'_1 x_\lambda \cdot' \mathsf{s}_1^{0(\mathfrak{D})}\mathsf{c}'_1 x_\lambda \cdot' -'\mathsf{d}'_{01} \\
&\quad = \mathsf{c}'_1\langle \mathsf{c}_0 a \cdot \mathsf{d}_{0\lambda}, 0\rangle \cdot' \mathsf{s}_1^{0(\mathfrak{D})}\mathsf{c}'_1\langle \mathsf{c}_0 a \cdot \mathsf{d}_{0\lambda}, 0\rangle \cdot' \langle -\mathsf{d}_{01}, a\rangle \\
&\quad = \langle \mathsf{c}_1\mathsf{c}_0 a \cdot \mathsf{d}_{0\lambda}, 0\rangle \cdot' \mathsf{s}_1^{0(\mathfrak{D})}\langle \mathsf{c}_1\mathsf{c}_0 a \cdot \mathsf{d}_{0\lambda}, 0\rangle \cdot' \langle -\mathsf{d}_{01}, a\rangle && \text{by (1)} \\
&\quad = \langle \mathsf{c}_1\mathsf{c}_0 a \cdot \mathsf{d}_{0\lambda}, 0\rangle \cdot' \langle \mathsf{c}_1\mathsf{c}_0 a \cdot \mathsf{d}_{1\lambda}, 0\rangle \cdot' \langle -\mathsf{d}_{01}, a\rangle \\
&\qquad && \text{by (1), 1.5.4(i), 1.5.8(i)} \\
&\quad = \langle \mathsf{c}_1\mathsf{c}_0 a \cdot \mathsf{d}_{0\lambda} \cdot \mathsf{d}_{1\lambda} \cdot -\mathsf{d}_{01}, 0\rangle = \langle 0, 0\rangle && \text{by 1.3.7.}
\end{aligned}$$

Hence

(7) $$\prod'_{\lambda<\beta} -'\mathsf{c}'_0\mathsf{c}'_1(\mathsf{c}'_1 x_\lambda \cdot' \mathsf{s}_1^{0(\mathfrak{D})}\mathsf{c}'_1 x_\lambda \cdot' -\mathsf{d}'_{01}) = \langle 1, a\rangle.$$

Furthermore, we have

$$\begin{aligned}
\sum'_{\lambda<\beta} \mathsf{c}'_1 x_\lambda &= \mathsf{c}'_1 \sum'_{\lambda<\beta} x_\lambda \\
&= \mathsf{c}'_1(\langle a, 0\rangle +' \sum'_{1\leqq\lambda<\beta}\langle \mathsf{c}_0 a \cdot \mathsf{d}_{0\lambda}, 0\rangle) \\
&= \mathsf{c}'_1\langle a + \sum_{1\leqq\lambda<\beta}(\mathsf{c}_0 a \cdot \mathsf{d}_{0\lambda}), 0\rangle \\
&= \mathsf{c}'_1\langle \mathsf{c}_0 a, 0\rangle && \text{by (3)} \\
&= \langle \mathsf{c}_1\mathsf{c}_0 a, a\rangle;
\end{aligned}$$

thus

$$(8)\qquad -'c'_0\Pi'_{\lambda<\beta}-'c'_1x_\lambda = -'c'_0-'\langle c_1c_0a, a\rangle = \langle c_1c_0a, a\rangle.$$

We also have

$$(9)\qquad c'_0(x_0\cdot' x_\beta) = c'_0(c'_1\langle a, 0\rangle\cdot'\langle a, 0\rangle) = c'_0\langle a, 0\rangle = \langle c_0a, a\rangle.$$

Finally,

$$\begin{aligned} c'_0(x_0\cdot' -'x_\beta) &= c'_0(c'_1\langle a, 0\rangle\cdot'\langle -a, a\rangle) \\ &= c'_0(\langle c_1a, a\rangle\cdot'\langle -a, a\rangle) \\ &= c'_0\langle c_1a\cdot -a, a\rangle \\ &= \langle c_0(c_1a\cdot -a) + c_0a, a\rangle \\ &= \langle c_0c_1a, a\rangle. \end{aligned}$$

Combining this last result with (7)–(9) we see that the element of $\mathfrak{D}$ represented by (5) is $\langle c_0a, a\rangle \neq 0$. This fact along with (4) gives the first part of the lemma.

It is easily seen from the above discussion that, for any α such that $2 \leqq \alpha < \beta$, the algebra $\mathfrak{C} = \mathfrak{Sg}^{(\mathfrak{Rd}_\alpha\mathfrak{D})}\{x_\lambda : 1 \leqq \lambda \leqq \beta\}$ satisfies the conclusion of the second part of the lemma.

THEOREM 2.6.14. *Let* $2 \leqq \alpha < \omega$ *and* $\beta > \alpha$. *Then*

(i) $\mathsf{CA}_\alpha \supset \mathbf{HSP}\mathsf{Rd}_\alpha\mathsf{CA}_\beta \supseteq \mathbf{S}\mathsf{Rd}_\alpha\mathsf{CA}_\beta$;

in fact we have

(ii) $\mathbf{S}_\delta\mathsf{CA}_\alpha \supset \mathbf{S}_\delta\mathbf{HSP}\mathsf{Rd}_\alpha\mathsf{CA}_\beta \supseteq \mathbf{S}_\delta\mathsf{Rd}_\alpha\mathsf{CA}_\beta$

for every cardinal $\delta > 1$.

PROOF. It clearly suffices to prove (ii). By 2.6.2(i) we have

$$\mathbf{S}_\delta\mathsf{CA}_\alpha \supseteq \mathbf{S}_\delta\mathbf{HSP}\mathsf{Rd}_\alpha\mathsf{CA}_\beta \supseteq \mathbf{S}\mathsf{Rd}_\alpha\mathsf{CA}_\beta.$$

From 2.6.11 we know that the equation stated in that lemma with $\kappa = \alpha$ holds identically in every CA_β and thus in every member of $\mathsf{Rd}_\alpha\mathsf{CA}_\beta$; hence by 0.4.63 it holds identically in every member of $\mathbf{HSP}\mathsf{Rd}_\alpha\mathsf{CA}_\beta$. On the other hand, applying the first part of 2.6.13 with α in place of β we obtain an $\mathfrak{A} \in \mathbf{S}_2\mathsf{CA}_\alpha$ in which the equation fails. Therefore, $\mathfrak{A} \in \mathbf{S}_\delta\mathsf{CA}_\alpha \sim \mathbf{HSP}\mathsf{Rd}_\alpha\mathsf{CA}_\beta$.

REMARK 2.6.15. From Theorems 2.6.8 and 2.6.9 we conclude that, in case $\alpha \geqq \omega$ or $\alpha \leqq 1$, (I) every CA_α is a subreduct of some CA_β for each $\beta > \alpha$. Using the terminology introduced in 0.3.14 and recalling the facts stated in 0.4.63 we draw from these theorems two further conclusions under the same assumptions on α: (II) the class CA_α is algebraically generated by the class of α-dimensional reducts of CA_β's (again for each $\beta > \alpha$), i.e., it is the least equational class containing all these reducts; (III) the class of all α-dimensional subreducts of CA_β's is algebraically closed and hence equational. By 2.6.14 neither conclusion (I) nor conclusion (II) holds in case $2 \leqq \alpha < \omega$, and actually, each of them fails for every $\beta > \alpha$. As regards conclusion (III) see Corollary 2.6.61 below. Since, by 0.3.12(ii) and 0.5.13(iii), the class $\mathbf{SRd}_\alpha\mathsf{CA}_\beta$ is closed under the formation of direct products, the problem whether (III) holds or fails amounts to the problem whether the class is closed under the formation of homomorphic images.

Notice the following consequence of Lemmas 2.6.11 and 2.6.13 (which partly improves 2.6.14):

THEOREM 2.6.16. *For any α and β with $2 \leqq \alpha < \beta < \omega$ we have*

(i) $\mathbf{SRd}_\alpha\mathsf{CA}_\beta \supset \mathbf{SRd}_\alpha\mathsf{CA}_{\beta+1}$;

moreover, we have

(ii) $\mathbf{S}_\delta\mathsf{Rd}_\alpha\mathsf{CA}_\beta \supset \mathbf{S}_\delta\mathsf{Rd}_\alpha\mathsf{CA}_{\beta+1}$ *and* $\mathbf{HSPS}_\delta\mathsf{Rd}_\alpha\mathsf{CA}_\beta \supset \mathbf{HSPS}_\delta\mathsf{Rd}_\alpha\mathsf{CA}_{\beta+1}$

for every cardinal $\delta > \beta$.

PROOF. As in 2.6.14 it suffices to prove (ii). By 0.5.5 and 2.6.2(i) we have

$$\mathsf{Rd}_\alpha\mathsf{CA}_\beta \supseteq \mathsf{Rd}_\alpha\mathsf{Rd}_\beta\mathsf{CA}_{\beta+1} = \mathsf{Rd}_\alpha\mathsf{CA}_{\beta+1};$$

hence

(1) $$\mathbf{S}_\delta\mathsf{Rd}_\alpha\mathsf{CA}_\beta \supseteq \mathbf{S}_\delta\mathsf{Rd}_\alpha\mathsf{CA}_{\beta+1},\quad \mathbf{HSPS}_\delta\mathsf{Rd}_\alpha\mathsf{CA}_\beta \supseteq \mathbf{HSPS}_\delta\mathsf{Rd}_\beta\mathsf{CA}_{\beta+1}.$$

By the last part of 2.6.13 there is a

(2) $$\mathfrak{C} \in \mathbf{S}_{\beta+1}\mathsf{Rd}_\alpha\mathsf{CA}_\beta \subseteq \mathbf{S}_\delta\mathsf{Rd}_\alpha\mathsf{CA}_\beta$$

in which the equation of 2.6.11 with $\kappa = \beta$ fails. Thus by 0.4.64 and 2.6.11 we have

(3) $$\mathfrak{C} \notin \mathbf{HSP}\mathsf{Rd}_\alpha\mathsf{CA}_{\beta+1} \supseteq \mathbf{HSPS}_\delta\mathsf{Rd}_\alpha\mathsf{CA}_{\beta+1}.$$

Statements (1)–(3) together imply part (ii) of the theorem.

REMARK 2.6.17. It should be pointed out that the negative results obtained in the preceding discussion, and in particular in Lemma 2.6.13, depend essentially on the properties of diagonal elements. As a consequence, the situation changes radically if instead of CA's we consider diagonal-free algebras. In fact, every $\mathfrak{A} \in \mathsf{Df}_\alpha$ is clearly a reduct of a $\mathfrak{B} \in \mathsf{Df}_\beta$ for every $\beta \geqq \alpha$: if $\mathfrak{A} = \langle A, +, \cdot, -, 0, 1, \mathsf{c}_\kappa \rangle_{\kappa<\alpha}$, we obtain $\mathfrak{B} = \langle A, +, \cdot, -, 0, 1, \mathsf{c}_\kappa \rangle_{\kappa<\beta}$ by setting $\mathsf{c}_\kappa = A 1 Id$ for every κ such that $\alpha \leqq \kappa < \beta$. (We assume here that the underlying definition of reducts of Df_β's is entirely analogous to 2.6.1.) Hence the formula

$$\mathsf{Df}_\alpha = \mathsf{Rd}_\alpha\mathsf{Df}_\beta = \mathbf{S}\mathsf{Rd}_\alpha\mathsf{Df}_\beta = \mathbf{HSP}\mathsf{Rd}_\alpha\mathsf{Df}_\beta$$

(with $\beta \geqq \alpha$) holds for every α without exception.

From 2.6.8–2.6.14 we can draw some conclusions connecting reducts and free algebras.

THEOREM 2.6.18. *Let* $\beta > \alpha$, $\mathfrak{B} \in \mathsf{CA}_\beta$, *and* $X \subseteq B$. *If* X CA_β*-freely generates* $\mathfrak{B}$, *then* X $\mathsf{Rd}_\alpha\mathsf{CA}_\beta$*-freely generates* $\mathfrak{Sg}^{(\mathfrak{Rd}_\alpha\mathfrak{B})}X$.

PROOF. Consider any $\mathfrak{A} \in \mathsf{Rd}_\alpha\mathsf{CA}_\beta$ and $f \in {}^XA$. Let $\mathfrak{C} \in \mathsf{CA}_\beta$ such that

$$(1) \qquad \mathfrak{A} = \mathfrak{Rd}_\alpha\mathfrak{C}.$$

Clearly $f \in {}^XC$, and thus by hypothesis there is an $h \in Hom(\mathfrak{B}, \mathfrak{C})$ such that $h \supseteq f$. Then $h \in Hom(\mathfrak{Rd}_\alpha\mathfrak{B}, \mathfrak{A})$ by (1), 0.5.4(ii) and 0.5.12(ii), and hence by 0.2.3 we conclude that

$$h \in Hom(\mathfrak{Sg}^{(\mathfrak{Rd}_\alpha\mathfrak{B})}X, \mathfrak{A}).$$

This shows that the conclusion of the theorem holds.

THEOREM 2.6.19. *For any* α *and* β *with* $\beta > \alpha$ *and any cardinal* $\delta \neq 0$ *the following three conditions are equivalent*:

(i) $\mathfrak{Fr}_\delta\mathsf{CA}_\alpha \cong \mathfrak{Fr}_\delta\mathsf{Rd}_\alpha\mathsf{CA}_\beta$;

(ii) *if* $\mathfrak{B} \in \mathsf{CA}_\beta$ *and* X CA_β*-freely generates* $\mathfrak{B}$, *then* X CA_α*-freely generates* $\mathfrak{Sg}^{(\mathfrak{Rd}_\alpha\mathfrak{B})}X$;

(iii) $\alpha \geqq \omega$ *or* $\alpha \leqq 1$.

PROOF. (iii) implies (ii) by 0.4.26(ii), 2.6.8, 2.6.9, and 2.6.18. To prove (ii) implies (i) observe that by 0.4.42 and 2.6.2(i) one immediately gets

$$(1) \qquad \mathfrak{Fr}_\delta\mathsf{CA}_\alpha \succcurlyeq \mathfrak{Fr}_\delta\mathsf{Rd}_\alpha\mathsf{CA}_\beta.$$

By 0.4.27(ii), $\delta/Cr_\delta\mathsf{CA}_\beta$ (i.e., $\{\xi/Cr_\delta\mathsf{CA}_\beta : \xi < \delta\}$) CA_β-freely generates $\mathfrak{Fr}_\delta\mathsf{CA}_\beta$. Hence, upon assuming (ii) we have that $\delta/Cr_\delta\mathsf{CA}_\beta$ CA_α-freely generates

$$\mathfrak{A} = \mathfrak{Sg}^{(\mathfrak{Rd}_\alpha\mathfrak{Fr}_\delta\mathsf{CA}_\beta)}(\delta/Cr_\delta\mathsf{CA}_\beta),$$

and this implies by 0.4.29 that

$$(2) \qquad \mathfrak{A} \cong \mathfrak{Fr}_\delta\mathsf{CA}_\alpha$$

(we use here the obvious fact that $|\delta/Cr_\delta\mathsf{CA}_\beta| = \delta = |\delta/Cr_\delta\mathsf{CA}_\alpha|$). On the other hand, since $\mathfrak{A} \in \mathbf{S}_{\delta+}\mathsf{Rd}_\alpha\mathsf{CA}_\beta$, we have $\mathfrak{Fr}_\delta\mathsf{Rd}_\alpha\mathsf{CA}_\beta \succcurlyeq \mathfrak{A}$ by 0.4.39. This fact together with (1), (2), and 0.4.43 give (ii). Hence (ii) implies (i).

Finally, we prove (i) implies (iii). Assume that (i) holds. Then by 0.4.22 and 0.4.39 we obtain

$$\mathbf{S}_{\delta^+}\mathsf{CA}_\alpha \subseteq \mathbf{H}\mathfrak{Fr}_\delta\mathsf{CA}_\alpha = \mathbf{H}\mathfrak{Fr}_\delta\mathsf{Rd}_\alpha\mathsf{CA}_\beta \subseteq \mathbf{HSP}\mathsf{Rd}_\alpha\mathsf{CA}_\beta.$$

Thus, from 2.6.14(ii) (with δ^+ in place of δ) one concludes that $\alpha \geqq \omega$ or $\alpha \leqq 1$, i.e., that (iii) holds. This finishes the proof.

THEOREM 2.6.20. *For any* β, β', *and* α *with* $\beta, \beta' > \alpha$ *and every cardinal* $\delta \neq 0$ *such that* $\delta \geqq \beta \cap \beta' \cap \omega$ *the following three conditions are equivalent*:

(i) $\mathbf{S}\mathsf{Rd}_\alpha\mathsf{CA}_\beta = \mathbf{S}\mathsf{Rd}_\alpha\mathsf{CA}_{\beta'}$;

(ii) $\mathfrak{Fr}_\delta\mathsf{Rd}_\alpha\mathsf{CA}_\beta \cong \mathfrak{Fr}_\delta\mathsf{Rd}_\alpha\mathsf{CA}_{\beta'}$;

(iii) $\beta = \beta'$ *or* $\beta, \beta' \geqq \omega$ *or, finally*, $\alpha \leqq 1$.

PROOF. (i) obviously holds if $\beta = \beta'$; it holds also in case $\beta, \beta' \geqq \omega$ or in case $\alpha \leqq 1$ since in the former case both members of (i) equal $\mathbf{S}\mathsf{Rd}_\alpha\mathsf{CA}_{\alpha\cup\omega}$ by 2.6.10 while in the latter both these members equal CA_α by 2.6.8. Hence (iii) implies (i), and (i) is seen in turn to imply (ii) by 1.4.43.

Now suppose that (ii) holds. Then by 0.4.40 we have

(1) $$\mathbf{HSPS}_{\delta^+}\mathsf{Rd}_\alpha\mathsf{CA}_\beta = \mathbf{HSPS}_{\delta^+}\mathsf{Rd}_\alpha\mathsf{CA}_{\beta'}.$$

In order to derive (iii) it is clear that we may assume $\beta \neq \beta'$ and, in particular, $\beta < \beta'$. Then using (1), 0.5.5, and 2.6.2(i) we get

(2) $$\mathbf{HSPS}_{\delta^+}\mathsf{Rd}_\alpha\mathsf{CA}_{\beta'} = \mathbf{HSPS}_{\delta^+}\mathsf{Rd}_\alpha\mathsf{Rd}_{\beta+1}\mathsf{CA}_{\beta'} \subseteq \mathbf{HSPS}_{\delta^+}\mathsf{Rd}_\alpha\mathsf{CA}_{\beta+1}.$$

If it is not the case that $\beta, \beta' \geqq \omega$, then $\beta < \omega$ and hence we conclude from (2), 2.6.16(ii) (with δ replaced by δ^+), and the premiss $\delta \geqq \beta \cap \beta' \cap \omega = \beta$ that $\alpha \leqq 1$. Therefore (iii) holds, and we have shown that it is implied by (ii). The proof is complete.

THEOREM 2.6.21. *For any ordinal* α, *any permutation* ρ *of* α, *and cardinal* $\delta \neq 0$, *we have*

$$\mathfrak{Rd}^{(\rho)}\mathfrak{Fr}_\delta\mathsf{CA}_\alpha \cong \mathfrak{Fr}_\delta\mathsf{CA}_\alpha.$$

PROOF. From 0.4.27(ii) we know that $\delta/Cr_\delta\mathsf{CA}_\alpha$ CA_α-freely generates $\mathfrak{Fr}_\delta\mathsf{CA}_\alpha$. Therefore, since $\mathfrak{Rd}^{(\rho)}\mathfrak{Fr}_\delta\mathsf{CA}_\alpha, \mathfrak{Rd}^{(\rho^{-1})}\mathfrak{Fr}_\delta\mathsf{CA}_\alpha \in \mathsf{CA}_\alpha$ by 2.6.2(i), there exist

(1) $$h \in Hom(\mathfrak{Fr}_\delta\mathsf{CA}_\alpha, \mathfrak{Rd}^{(\rho)}\mathfrak{Fr}_\delta\mathsf{CA}_\alpha)$$

and

(2) $$g \in Hom(\mathfrak{Fr}_\delta\mathsf{CA}_\alpha, \mathfrak{Rd}^{(\rho^{-1})}\mathfrak{Fr}_\delta\mathsf{CA}_\alpha)$$

such that

(3) $$\delta/Cr_\delta\mathsf{CA}_\alpha 1 h = \delta/Cr_\delta\mathsf{CA}_\alpha 1 g \subseteq Id.$$

By 0.5.5 one has

$$\mathfrak{Rd}^{(\rho)}\mathfrak{Rd}^{(\rho^{-1})}\mathfrak{Fr}_\delta\mathsf{CA}_\alpha = \mathfrak{Rd}^{(\rho^{-1}\circ\rho)}\mathfrak{Fr}_\delta\mathsf{CA}_\alpha = \mathfrak{Fr}_\delta\mathsf{CA}_\alpha,$$

and thus from (2), 0.2.15, 0.5.4(ii), and 0.5.12(ii) we obtain

(4) $$g \in Hom(\mathfrak{Rd}^{(\rho)}\mathfrak{Fr}_\delta\mathsf{CA}_\alpha, \mathfrak{Fr}_\delta\mathsf{CA}_\alpha).$$

From (1) and 0.2.10(i) we then infer that

(5) $$g \circ h \in Hom(\mathfrak{Fr}_\delta\mathsf{CA}_\alpha, \mathfrak{Fr}_\delta\mathsf{CA}_\alpha).$$

Since, by (3), $\delta/Cr_\delta\mathsf{CA}_\alpha 1 (g\circ h) = \delta/Cr_\delta\mathsf{CA} 1 Id$, and since $\delta/Cr_\delta\mathsf{CA}_\alpha$ generates $\mathfrak{Fr}_\delta\mathsf{CA}_\alpha$, it follows from (5) and 0.2.14(iii) that $g\circ h = Fr_\delta\mathsf{CA}_\alpha 1 Id$. In a similar way it is seen that $h\circ g = Fr_\delta\mathsf{CA}_\alpha 1 Id$. Hence by (4) g is an isomorphism from $\mathfrak{Rd}^{(\rho)}\mathfrak{Fr}_\delta\mathsf{CA}_\alpha$ onto $\mathfrak{Fr}_\delta\mathsf{CA}_\alpha$.

REMARK 2.6.22. For our further discussion we need a somewhat stronger form of Theorem 2.6.21: we shall use the fact (which clearly emerges from the proof of the theorem) that there is an $h \in Is(\mathfrak{Fr}_\delta\mathsf{CA}_\alpha, \mathfrak{Rd}^{(\rho)}\mathfrak{Fr}_\delta\mathsf{CA}_\alpha)$ such that $h(\xi/Cr_\delta\mathsf{CA}_\alpha) = \xi/Cr_\delta\mathsf{CA}_\alpha$ for each $\xi < \delta$.

Using this fact we can now establish a result (due to Tarski) concerning the dimensionality of elements of an arbitrary free CA_α.

THEOREM 2.6.23. *For any cardinal* $\delta \neq 0$, *if* $x \in Fr_\delta\mathsf{CA}_\alpha$, *then* $|\Delta x| < \omega$ *or else* $|\alpha \sim \Delta x| < \omega$.

PROOF. Clearly we may assume that $\alpha \geqq \omega$; we assume further that

(1) $$|\Delta x| \geqq \omega.$$

By 0.5.11(ii) there is a finite sequence σ without repeating terms and with $Rg\sigma \subseteq \alpha$ such that, setting

(2) $$\mathfrak{A} = \mathfrak{Rd}^{(\sigma)}\mathfrak{Fr}_\delta\mathsf{CA}_\alpha \text{ and } G = \{\xi/Cr_\delta\mathsf{CA}_\alpha : \xi < \delta\},$$

we have

(3) $$x \in Sg^{(\mathfrak{A})}G.$$

Because of (1) there is a $\kappa < \alpha$ such that

(4) $$\kappa \in \Delta x \sim Rg\sigma.$$

Let λ be any ordinal in $\alpha \sim Rg\sigma$, and let ρ be a permutation of α such that

(5) $$Rg\sigma 1 \rho \subseteq Id \text{ and } \rho\kappa = \lambda.$$

By 2.6.22 there is an

(6) $$h \in Is(\mathfrak{Fr}_\delta \mathsf{CA}_\alpha, \mathfrak{Rd}^{(\rho)}\mathfrak{Fr}_\delta \mathsf{CA}_\alpha)$$

such that, in light of (2),

(7) $$G 1 h \subseteq Id.$$

Observe that because of (5) and 0.5.5 we have

$$\mathfrak{Rd}^{(\sigma)}\mathfrak{Rd}^{(\rho)}\mathfrak{Fr}_\delta \mathsf{CA}_\alpha = \mathfrak{Rd}^{(\rho\circ\sigma)}\mathfrak{Fr}_\delta \mathsf{CA}_\alpha = \mathfrak{Rd}^{(\sigma)}\mathfrak{Fr}_\delta \mathsf{CA}_\alpha.$$

This result together with (2), (6), and 0.5.12(ii) gives $h \in Is(\mathfrak{A}, \mathfrak{A})$. Therefore, by (7), 0.2.3, and 0.2.18(i) we get

$$h \in Is(\mathfrak{Sg}^{(\mathfrak{A})}G, \mathfrak{Sg}^{(\mathfrak{A})}G),$$

and hence from (7) and 0.2.14(iii) we conclude that $Sg^{(\mathfrak{A})}G 1 h \subseteq Id$. In view of (3) this inclusion gives us

(8) $$hx = x.$$

From (4) one obtains $\mathsf{c}_\kappa x \neq x$, and hence from (6) and (8) we have, setting $\mathfrak{B} = \mathfrak{Rd}^{(\rho)}\mathfrak{Fr}_\delta \mathsf{CA}_\alpha$,

$$\mathsf{c}_\kappa^{(\mathfrak{B})} x \neq x;$$

however,

$$\mathsf{c}_\kappa^{(\mathfrak{B})} x = \mathsf{c}_{\rho\kappa} x = \mathsf{c}_\lambda x$$

by (5), and thus we conclude that $\lambda \in \Delta x$. Since λ was an arbitrary ordinal in $\alpha \sim Rg\sigma$, we have shown that, under the assumption of (1), $|\alpha \sim \Delta x| \leqq |Rg\sigma| < \omega$. This completes the proof.

REMARK 2.6.24. In connection with 2.6.23 the following has been conjectured: for every $\alpha \geqq \omega$ and every cardinal $\delta \neq 0$, all finite-dimensional elements of $\mathfrak{Fr}_\delta \mathsf{CA}_\alpha$ belong to the subuniverse generated by $\{x \cdot \mathsf{c}_0^\partial \mathsf{d}_{01} : x \in G\}$ where G is the set of standard generators of $\mathfrak{Fr}_\delta \mathsf{CA}_\alpha$ (i.e., $G = \delta / Cr_\delta \mathsf{CA}_\alpha$). We have been unable to verify this conjecture.

REMARK 2.6.25. Using the method of splitting elements described in 2.6.12 we can construct for each finite $\alpha > 0$ a finite simple CA_α $\mathfrak{C}$ which is not generated by any single element. (This construction is due to Henkin; for related problems see 2.3.23.) We shall briefly describe how this is done.

Consider any finite $\alpha > 0$ and let $\mathfrak{A}$ be the full cylindric set algebra in the space ${}^{\alpha}\alpha$; let $a = \{\alpha 1 Id\}$. It is then easily seen that the algebra $\mathfrak{B}$ constructed in Lemma 2.6.12 contains a pair of distinct atoms b, b' such that $b, b' \leqq -\mathsf{d}_{\kappa\lambda}^{(\mathfrak{B})}$ whenever $\kappa < \lambda < \alpha$, and $\mathsf{c}_{\kappa}^{(\mathfrak{B})}b = \mathsf{c}_{\kappa}^{(\mathfrak{B})}b'$ for every $\kappa < \alpha$. Furthermore, because $\mathfrak{A}$ is simple, $\mathfrak{B}$ is readily seen to be simple also. Now repeating this construction with $\mathfrak{B}$ and b in place of $\mathfrak{A}$ and a, respectively, we arrive at an algebra $\mathfrak{C}$ which can be shown with little difficulty to contain elements c, c', c'' satisfying the following conditions: (I) c, c', c'' are distinct atoms of $\mathfrak{C}$; (II) $c, c', c'' \leqq -\mathsf{d}_{\kappa\lambda}^{(\mathfrak{C})}$ whenever $\kappa < \lambda < \alpha$; (III) $\mathsf{c}_{\kappa}^{(\mathfrak{C})}c = \mathsf{c}_{\kappa}^{(\mathfrak{C})}c' = \mathsf{c}_{\kappa}^{(\mathfrak{C})}c''$ for every $\kappa < \alpha$. Of course, $\mathfrak{C}$ is both finite and simple.

Let $\mathfrak{D}$ be the Boolean algebra obtained by relativizing $\mathfrak{B}\mathfrak{l}\,\mathfrak{C}$ to the element $c+c'+c''$. By (I) we have $|D| = 8$, and using (I)–(III) we can easily show that

$$\{y\cdot(c+c'+c''):y \in Sg^{(\mathfrak{C})}\{x\}\} = Sg^{(\mathfrak{D})}\{x\cdot(c+c'+c'')\}$$

for every $x \in C$. Therefore, since the right hand member of this equality can never contain more than four elements, we see that $Sg^{(\mathfrak{C})}\{x\} \subset C$ for every $x \in C$. Thus, in $\mathfrak{C}$ we have an example of a finite simple CA_{α} which is not generated by any single element.

The method described above cannot be used to obtain a finitely-generated simple CA of infinite dimension which is not generated by a single element. The reason is that, as we shall see in Theorem 2.6.50 below, every non-zero element of a simple CA_{α} with $\alpha \geqq \omega$ meets at least one $\mathsf{d}_{\kappa\lambda}$ with $\kappa \neq \lambda$.

REMARKS 2.6.26. We now turn our attention to a special kind of subreducts referred to as neat reducts, and to the related notion of neat embeddings; see 2.6.28 for the definitions. We shall sometimes say that a CA_{α} $\mathfrak{C}$ has the *neat embedding property* if it is neatly embeddable in a CA_{β} with $\beta = \alpha+\omega$.

Neat reducts will play an important role in the representation theory of cylindric algebras. It will be shown in Part II that the class of representable CA's (in the sense of 1.1.13) simply coincides with the class of CA's having the neat embedding property. At the same time the notion of the neat embedding property appears to be more suitable for an abstract algebraic treatment than that of representability. This is the main reason why our discussion of neat reducts and their subalgebras in the present section will be comprehensive and detailed. In particular, we shall establish here several results which in their original formulation referred to representable algebras rather than to algebras with the neat embedding property.[1]

THEOREM 2.6.27. *If* $\mathfrak{B} \in \mathsf{CA}_\beta$ *and* $\alpha \leqq \beta$, *then* $Cl_{\beta\sim\alpha}\mathfrak{B}$ *is a non-empty subuniverse of* $\mathfrak{Rd}_\alpha\mathfrak{B}$.

This theorem justifies the following

DEFINITION 2.6.28. *Let* $\mathfrak{B}$ *be a* CA_β *and* α *any ordinal* $\leqq \beta$. *The* ***neat*** α***-reduct*** *of* $\mathfrak{B}$, *in symbols* $\mathfrak{Nr}_\alpha\mathfrak{B}$, *is the subalgebra* $\mathfrak{A}$ *of* $\mathfrak{Rd}_\alpha\mathfrak{B}$ *such that* $A = Cl_{\beta\sim\alpha}\mathfrak{B}$; *we also set* $Nr_\alpha\mathfrak{B} = Cl_{\beta\sim\alpha}\mathfrak{B}$. *For any class* K *of* CA_β*'s we set*

$$\mathsf{Nr}_\alpha\mathsf{K} = \{\mathfrak{Nr}_\alpha\mathfrak{A} : \mathfrak{A} \in \mathsf{K}\}.$$

A CA $\mathfrak{C}$ *is called a* ***neat subreduct*** *of* $\mathfrak{B}$, *or is said to be* ***neatly embedded*** *or* ***neatly embeddable*** *in* $\mathfrak{B}$, *if* $\mathfrak{C} \subseteq \mathfrak{Nr}_\gamma\mathfrak{B}$ *for some* $\gamma \leqq \beta$.

The remaining part of this section is devoted to the discussion of neat embeddings. Roughly speaking, after establishing in Theorems 2.6.29–2.6.31 some elementary and very general properties of neat reducts, the following topics are successively discussed: closure properties of classes $\mathsf{Nr}_\alpha\mathsf{CA}_\beta$; relationships between $\mathsf{Nr}_\alpha\mathsf{CA}_\beta$ and $\mathsf{Nr}_\alpha\mathsf{CA}_\gamma$ for different β and γ; conditions for CA's to be neatly embeddable in other CA's (in particular, conditions involving Lf's and Dc's); conditions for the uniqueness of a CA_β in which a given CA_α is neatly embeddable.

The reader will notice that some of the subsequent theorems are analogues of certain results previously stated for arbitrary reducts and classes of reducts. For instance, 2.6.30 is an exact analogue of 2.6.2, and 2.6.34 of 2.6.7. Theorem 2.6.10 does not extend in its full strength to neat reducts, but in 2.6.35 we find an analogue of this result under more restrictive assumptions. The analogues of 2.6.29 and 2.6.31 for arbitrary reducts have not been explicitly formulated since they are simply particular cases of general algebraic results given in Chapter 0, in fact, 0.5.4(ii),(iii) and 0.5.5.

THEOREM 2.6.29. *Let* $\alpha \leqq \beta$. *If* $\mathfrak{B} \subseteq \mathfrak{A} \in \mathsf{CA}_\beta$, *then* $\mathfrak{Nr}_\alpha\mathfrak{B} \subseteq \mathfrak{Nr}_\alpha\mathfrak{A}$. *If* $\mathsf{K} \subseteq \mathsf{CA}_\beta$, *then* $\mathsf{Nr}_\alpha\mathsf{SK} \subseteq \mathsf{SNr}_\alpha\mathsf{K}$.

THEOREM 2.6.30. (i) *If* $\alpha \leqq \beta$, *then* $\mathsf{Nr}_\alpha\mathsf{CA}_\beta \subseteq \mathsf{CA}_\alpha$ *and* $\mathsf{Nr}_\alpha\mathsf{Lf}_\beta \subseteq \mathsf{Lf}_\alpha$.

(ii) $\mathfrak{A} = \mathfrak{Nr}_\alpha\mathfrak{A}$ *for each* $\mathfrak{A} \in \mathsf{CA}_\alpha$, *and* $\mathsf{Nr}_\alpha\mathsf{CA}_\alpha = \mathsf{CA}_\alpha$.

(iii) $\mathfrak{Zd}\mathfrak{B} = \mathfrak{Nr}_0\mathfrak{B}$ *for every* $\mathfrak{B} \in \mathsf{CA}_\beta$, *and* $\mathsf{Nr}_0\mathsf{CA}_\beta = \mathsf{BA}$.

[1]) The notions of neat reduct and neat embedding were introduced in Henkin [55a], p. 40; also the result by which the CA's with the neat embedding property coincide with representable CA's was stated there.

PROOF. The last part of (iii) follows from 1.3.11. All the other parts of the theorem are obvious.

THEOREM 2.6.31. *If* $\alpha \leqq \beta \leqq \gamma$, *then* $\mathfrak{Nr}_\alpha\mathfrak{C} = \mathfrak{Nr}_\alpha\mathfrak{Nr}_\beta\mathfrak{C}$ *for every* $\mathfrak{C} \in \mathsf{CA}_\gamma$. *Hence*

$$\mathsf{Nr}_\alpha\mathsf{CA}_\beta \supseteq \mathsf{Nr}_\alpha\mathsf{Nr}_\beta\mathsf{CA}_\gamma = \mathsf{Nr}_\alpha\mathsf{CA}_\gamma.$$

THEOREM 2.6.32. *If* $\alpha \leqq \beta$, *then*

(i) $\mathbf{P}\mathsf{Nr}_\alpha\mathsf{CA}_\beta = \mathsf{Nr}_\alpha\mathsf{CA}_\beta$ *and* $\mathsf{UpNr}_\alpha\mathsf{CA}_\beta \subseteq \mathbf{S}\mathsf{Nr}_\alpha\mathsf{CA}_\beta$;

(ii) $\mathbf{HSP}\mathsf{Nr}_\alpha\mathsf{CA}_\beta = \mathbf{S}\mathsf{Nr}_\alpha\mathsf{CA}_\beta$.

PROOF. Let $\mathfrak{B} = \langle \mathfrak{B}_i : i \in I \rangle$ be any system of CA_β's. Then by 0.5.13(i) and 2.4.2 we get

$$\mathsf{P}_{i \in I}\mathfrak{Nr}_\alpha\mathfrak{B}_i = \mathfrak{Nr}_\alpha(\mathsf{P}_{i \in I}\mathfrak{B}_i).$$

Thus the first formula of (i) holds. In view of this fact it is sufficient in proving (ii) to show that $\mathbf{S}\mathsf{Nr}_\alpha\mathsf{CA}_\beta$ is closed under the formation of homomorphic images. For this purpose we assume $\mathfrak{B} \in \mathsf{CA}_\beta$, $\mathfrak{A} \subseteq \mathfrak{Nr}_\alpha\mathfrak{B}$, and $I \in Il\mathfrak{A}$, and derive $\mathfrak{A}/I \in \mathbf{S}\mathsf{Nr}_\alpha\mathsf{CA}_\beta$ from these assumptions. In fact let $J = Ig^{(\mathfrak{B})}I$. Using 2.3.8 it is easily verified that $J \cap A = I$. Then one checks that $\{\langle x/I, x/J \rangle : x \in A\}$ is an isomorphism from $\mathfrak{A}/I$ into $\mathfrak{Nr}_\alpha(\mathfrak{B}/J)$. Thus $\mathfrak{A}/I \in \mathbf{S}\mathsf{Nr}_\alpha\mathsf{CA}_\beta$, and (ii) is proved. The second formula of (i) is an immediate consequence of (ii) and 0.3.69(i).

REMARK 2.6.33. From 2.6.32(ii) we see that the class $\mathbf{S}\mathsf{Nr}_\alpha\mathsf{CA}_\beta$ with $\alpha \leqq \beta$ is algebraically closed and hence equational; in view of 2.6.26 the same applies to the class of all representable CA_α's.[1] As regards the class $\mathsf{Nr}_\alpha\mathsf{CA}_\beta$ we know only (from 2.6.32(i)) that it is closed under the formation of arbitrary direct products.

THEOREM 2.6.34. *If* $\alpha \leqq \beta$ *and* $0 < \gamma = \bigcup\gamma$, *then*

$$\bigcap_{\xi<\gamma}\mathbf{S}\mathsf{Nr}_\alpha\mathsf{CA}_{\beta+\xi} = \mathbf{S}\mathsf{Nr}_\alpha\mathsf{CA}_{\beta+\gamma};$$

in particular,

$$\bigcap_{\kappa<\omega}\mathbf{S}\mathsf{Nr}_\alpha\mathsf{CA}_{\alpha+\kappa} = \mathbf{S}\mathsf{Nr}_\alpha\mathsf{CA}_{\alpha+\omega}.$$

PROOF. The inclusion $\bigcap_{\xi<\gamma}\mathbf{S}\mathsf{Nr}_\alpha\mathsf{CA}_{\beta+\xi} \supseteq \mathbf{S}\mathsf{Nr}_\alpha\mathsf{CA}_{\beta+\gamma}$ follows from 2.6.31.

[1]) The theorem that the class of representable CA_α's is equational was first announced in Tarski [55*], p. 63, for $\alpha < \omega$ and in Henkin-Tarski [61], p. 103, Theorem 2.19, for an arbitrary α. The proof of the more general theorem concerning the class $\mathbf{S}\mathsf{Nr}_\beta\mathsf{CA}_\alpha$ for arbitrary β was given in Monk [61], p. 1452, Theorem 2′; this is essentially the proof outlined in the text.

Suppose $\mathfrak{A} \in \bigcap_{\xi<\gamma} \mathbf{S}\mathsf{Nr}_\alpha\mathsf{CA}_{\beta+\xi}$, and for each $\xi < \gamma$ let $\mathfrak{B}_\xi$ be a $\mathsf{CA}_{\beta+\xi}$ such that

(1) $$\mathfrak{A} \subseteq \mathfrak{Nr}_\alpha \mathfrak{B}_\xi.$$

We now proceed exactly as in the proof of 2.6.7 to obtain a $\mathfrak{D} \in \mathsf{CA}_{\beta+\gamma}$ and an $h \in Ism(\mathfrak{A}, \mathfrak{Rd}_\alpha \mathfrak{D})$. However, in this case it is easily checked that $h^*A \subseteq Nr_\alpha \mathfrak{D}$; hence $\mathfrak{A} \in \mathbf{S}\mathsf{Nr}_\alpha\mathsf{CA}_{\beta+\gamma}$ and the proof is finished.

THEOREM 2.6.35. *If* $\beta \geqq \omega$, *then* $\mathbf{S}\mathsf{Nr}_\alpha\mathsf{CA}_{\alpha+\beta} = \mathbf{S}\mathsf{Nr}_\alpha\mathsf{CA}_{\alpha+\omega}$.

PROOF. We immediately get from 2.6.31 that $\mathbf{S}\mathsf{Nr}_\alpha\mathsf{CA}_{\alpha+\beta} \subseteq \mathbf{S}\mathsf{Nr}_\alpha\mathsf{CA}_{\alpha+\omega}$. To obtain the inclusion in the opposite direction one needs only prove that

(1) $$\mathsf{Nr}_\alpha\mathsf{CA}_{\alpha+\xi} \subseteq \mathsf{Nr}_\alpha\mathsf{CA}_{\alpha+\xi+1} \text{ for every } \xi \geqq \omega,$$

since the desired inclusion follows from (1), 2.6.29, and 2.6.34 by a straightforward argument using transfinite induction (compare the proof of 2.6.9). Suppose then that $\mathfrak{A} \in \mathsf{Nr}_\alpha\mathsf{CA}_{\alpha+\xi}$, say $\mathfrak{A} = \mathfrak{Nr}_\alpha\mathfrak{B}$ where $\mathfrak{B} \in \mathsf{CA}_{\alpha+\xi}$. Let ρ be a one-one function in ${}^{\alpha+\xi+1}(\alpha+\xi)$ such that $\alpha 1 \rho \subseteq Id$. Then $\mathfrak{Rd}^{(\rho)}\mathfrak{B} \in \mathsf{CA}_{\alpha+\xi+1}$ by 2.6.2(i), while it is an easy matter to show that

$$\mathfrak{A} = \mathfrak{Nr}_\alpha\mathfrak{B} = \mathfrak{Nr}_\alpha\mathfrak{Rd}^{(\rho)}\mathfrak{B}.$$

This gives (1).

REMARK 2.6.36. By 2.6.26 and 2.6.31–2.6.35 the class $\mathbf{S}\mathsf{Nr}_\alpha\mathsf{CA}_{\alpha+\omega}$ has several strong and important properties, due to which it becomes the focal point in the theory of neat reducts.

REMARK 2.6.37. As was pointed out in 2.6.33, all the classes $\mathbf{S}\mathsf{Nr}_\alpha\mathsf{CA}_\beta$ with $\beta \geqq \alpha$ are algebraically closed. With the help of 2.6.34 and 2.6.35 we can establish another rather interesting closure property of these classes. A class K of CA_α's has this property if, together with any algebra $\mathfrak{A}$, it contains every algebra obtained by relativizing $\mathfrak{A}$ to any one of the elements b described in 2.2.10 (i.e., satisfying the formula $\mathsf{s}^\kappa_\lambda b \cdot \mathsf{s}^\lambda_\kappa b = b$ for all $\kappa, \lambda < \alpha$). By 2.2.10 the whole class CA_α has this closure property. From Remarks 2.2.11 it is seen that the same applies to the class of all representable CA_α's and hence, by 2.6.35, to every class $\mathbf{S}\mathsf{Nr}_\alpha\mathsf{CA}_\beta$ with $\beta \geqq \alpha+\omega$. Don Pigozzi has given another proof of this result which does not involve the representation theory, and moreover has improved the result by extending it to all classes $\mathbf{S}\mathsf{Nr}_\alpha\mathsf{CA}_\beta$ with $\beta \geqq \alpha$, thus establishing the following

THEOREM 2.6.38. *Let* $\mathfrak{A} \in \mathsf{CA}_\alpha$ *and consider* $a \in A$ *such that* $\mathsf{s}^\kappa_\lambda a \cdot \mathsf{s}^\lambda_\kappa a = a$ *for all* $\kappa, \lambda < \alpha$. *Then for every* $\beta \geqq \alpha$, *if* $\mathfrak{A} \in \mathbf{S}\mathsf{Nr}_\alpha\mathsf{CA}_\beta$, *then* $\mathfrak{Rl}_a\mathfrak{A} \in \mathbf{S}\mathsf{Nr}_\alpha\mathsf{CA}_\beta$.

PROOF. Assume for the time being that $\beta = \alpha+\mu$ where $\mu < \omega$. By hypothesis there is a $\mathfrak{B} \in \mathsf{CA}_{\alpha+\mu}$ such that

(1) $$\mathfrak{A} \subseteq \mathfrak{Nr}_\alpha \mathfrak{B}.$$

Set

(2) $$b = a \cdot \prod_{\nu<\mu} \mathsf{s}^0_{\alpha+\nu} a.$$

We will prove by induction on μ that

(3) $$\mathfrak{Rl}_a \mathfrak{A} \cong | \subseteq \mathfrak{Nr}_\alpha \mathfrak{Rl}_b \mathfrak{B};$$

it turns out to be useful to prove simultaneously that

(4) $$\mathsf{s}^\kappa_\lambda b \cdot \mathsf{s}^\lambda_\kappa b = b \text{ for every } \kappa, \lambda < \alpha+\mu.$$

If $\mu = 0$, then $b = a$, $\mathfrak{Nr}_\alpha \mathfrak{Rl}_b \mathfrak{B} = \mathfrak{Rl}_a \mathfrak{B}$ by 2.6.30(ii), and hence (3) and (4) are obvious.

Assume now that $\mu = 1$. Consider any $\kappa, \lambda < \alpha$ such that $\kappa \neq \lambda$. Then, by 1.5.10(i),

$$\mathsf{s}^\kappa_\alpha a = \mathsf{s}^\kappa_\alpha(\mathsf{s}^\kappa_\lambda a \cdot \mathsf{s}^\lambda_\kappa a) = \mathsf{s}^\kappa_\lambda a \cdot \mathsf{s}^\kappa_\alpha \mathsf{s}^\lambda_\kappa a;$$

thus, using the fact that $a \leqq \mathsf{s}^\kappa_\lambda a$ by hypothesis, and applying 1.5.10(ii), we get

$$a \cdot \mathsf{s}^\kappa_\alpha a = a \cdot \mathsf{s}^\kappa_\alpha \mathsf{s}^\lambda_\kappa a = a \cdot \mathsf{s}^\kappa_\alpha \mathsf{s}^\lambda_\alpha a.$$

We employ a similar argument and use in addition 1.5.10(iv) to obtain

$$a \cdot \mathsf{s}^\lambda_\alpha a = a \cdot \mathsf{s}^\lambda_\alpha \mathsf{s}^\kappa_\alpha a = a \cdot \mathsf{s}^\kappa_\alpha \mathsf{s}^\lambda_\alpha a.$$

Therefore, $a \cdot \mathsf{s}^\kappa_\alpha a = a \cdot \mathsf{s}^\lambda_\alpha a$ for all $\kappa, \lambda < \alpha$ and hence, in view of (2) and the assumption $\mu = 1$, we have

(5) $$b = a \cdot \prod_{\nu \in \Gamma} \mathsf{s}^\nu_\alpha a \text{ whenever } 0 \neq \Gamma \subseteq \alpha,\ |\Gamma| < \omega.$$

Now consider arbitrary $\kappa, \lambda < \alpha$. Then

$$\begin{aligned} \mathsf{s}^\kappa_\lambda b \cdot \mathsf{s}^\lambda_\kappa b &= \mathsf{s}^\kappa_\lambda(a \cdot \mathsf{s}^\kappa_\alpha a) \cdot \mathsf{s}^\lambda_\kappa(a \cdot \mathsf{s}^\lambda_\alpha a) && \text{by (5)} \\ &= \mathsf{s}^\kappa_\lambda a \cdot \mathsf{s}^\kappa_\alpha a \cdot \mathsf{s}^\lambda_\kappa a \cdot \mathsf{s}^\lambda_\alpha a && \text{by 1.5.10(i)} \\ &= a \cdot \mathsf{s}^\kappa_\alpha a \cdot \mathsf{s}^\lambda_\alpha a \\ &= b && \text{by (5),} \end{aligned}$$

and

$$\begin{aligned} \mathsf{s}^\kappa_\alpha b \cdot \mathsf{s}^\alpha_\kappa b &= \mathsf{s}^\kappa_\alpha(a \cdot \mathsf{s}^\kappa_\alpha a) \cdot \mathsf{s}^\alpha_\kappa(a \cdot \mathsf{s}^\kappa_\alpha a) && \text{by (5)} \\ &= \mathsf{s}^\kappa_\alpha a \cdot \mathsf{s}^\kappa_\alpha a \cdot a \cdot a && \text{by (1), 1.5.8(i), 1.5.10(v)} \\ &= b && \text{by (5).} \end{aligned}$$

Thus

(6) $$\mathsf{s}^\kappa_\lambda b \cdot \mathsf{s}^\lambda_\kappa b = b \text{ for all } \kappa, \lambda \leqq \alpha$$

and consequently

(7) $$\mathfrak{Rl}_b \mathfrak{B} \in \mathsf{CA}_{\alpha+1}$$

by 2.2.10.

Let $f = \langle x \cdot \mathsf{s}^0_\alpha a : x \in Rl_a \mathfrak{A} \rangle$ and set

(8) $$\mathfrak{A}' = \mathfrak{Rl}_a \mathfrak{A}, \ \mathfrak{B}' = \mathfrak{Rl}_b \mathfrak{B}.$$

Throughout the remainder of the proof x and y will represent arbitrary elements of A', i.e.,

(9) $$x, y \in A \text{ and } x, y \leqq a.$$

By (8) we have

$$f -^{(\mathfrak{A}')} x = f(-x \cdot a) = -x \cdot a \cdot \mathsf{s}^0_\alpha a = -x \cdot b = -^{(\mathfrak{B}')} x,$$

and hence it is easy to see that

(10) $$f \in Hom(\mathfrak{Bl}\mathfrak{A}', \mathfrak{Bl}\mathfrak{B}').$$

For any $\kappa < \alpha$ we compute:

$$\begin{aligned} f(\mathsf{c}^{(\mathfrak{A}')}_\kappa x) &= f(\mathsf{c}_\kappa x \cdot a) \\ &= \mathsf{c}_\kappa x \cdot a \cdot \mathsf{s}^0_\alpha a \\ &= \mathsf{c}_\kappa x \cdot b && \text{by (2)} \\ &= \mathsf{c}_\kappa x \cdot \mathsf{s}^\kappa_\alpha a \cdot b && \text{by (5)} \\ &= \mathsf{c}_\kappa (x \cdot \mathsf{s}^\kappa_\alpha a) \cdot b && \text{by 1.5.9(ii)} \\ &= \mathsf{c}_\kappa (x \cdot \mathsf{s}^0_\alpha a) \cdot b && \text{by (5), (9)} \\ &= \mathsf{c}_\kappa f x \cdot b \\ &= \mathsf{c}^{(\mathfrak{B}')}_\kappa f x. \end{aligned}$$

Therefore, since it is easily seen that $f\mathsf{d}_{\kappa\lambda}^{(\mathfrak{A}')} = \mathsf{d}_{\kappa\lambda}^{(\mathfrak{B}')}$ for all $\kappa, \lambda < \alpha$, in light of (10) we have

(11) $$f \in Hom(\mathfrak{A}', \mathfrak{Rd}_\alpha \mathfrak{B}').$$

On the other hand,

$$\begin{aligned} \mathsf{c}^{(\mathfrak{B}')}_\alpha f x &= \mathsf{c}^{(\mathfrak{B}')}_\alpha (x \cdot \mathsf{s}^0_\alpha a) \\ &= \mathsf{c}_\alpha (x \cdot \mathsf{s}^0_\alpha a) \cdot b && \text{by (8)} \\ &= x \cdot \mathsf{c}_\alpha \mathsf{s}^0_\alpha a \cdot b && \text{by (1), (9)} \\ &= x \cdot \mathsf{c}_0 a \cdot b && \text{by (1), 1.5.8(i), 1.5.9(i)} \\ &= f x && \text{by (5), (9),} \end{aligned}$$

whence

(12) $$f^*\mathfrak{A}' \subseteq \mathfrak{Nr}_\beta \mathfrak{B}'.$$

Finally, if $fx = fy$, then $x \cdot \mathsf{s}_\alpha^0 a = y \cdot \mathsf{s}_\alpha^0 a$; applying c_α to both members of this equation and using (1), (9), and 1.5.8(i) we get $x \cdot \mathsf{c}_\alpha \mathsf{s}_\alpha^0 a = y \cdot \mathsf{c}_\alpha \mathsf{s}_\alpha^0 a$. Consequently,

$$x = x \cdot \mathsf{c}_0 a = y \cdot \mathsf{c}_0 a = y$$

by (9), 1.5.8(i), and 1.5.9(i), and thus f is one-one. This result combined with (11) and (12) shows that f is an isomorphism from $\mathfrak{A}'$ into $\mathfrak{Nr}_\alpha \mathfrak{B}'$. In view of (8) we have thus shown that (3) holds if $\mu = 1$; (4) also holds in this case by (6). Observe that we have succeeded in establishing the theorem in the special case $\beta = \alpha + 1$; this fact will be needed below.

Assume now that $\mu \geqq 2$ and set

(13) $$b' = a \cdot \prod_{\nu < \mu - 1} \mathsf{s}_{\alpha+\nu}^0 a.$$

By (1) and 2.6.31 we have

$$\mathfrak{A} \subseteq \mathfrak{Nr}_\alpha \mathfrak{Nr}_{\alpha+\mu-1} \mathfrak{B}.$$

Therefore, applying the induction hypothesis we obtain from (13)

(14) $$\mathfrak{Rl}_a \mathfrak{A} \cong | \subseteq \mathfrak{Nr}_\alpha \mathfrak{Rl}_{b'} \mathfrak{Nr}_{\alpha+\mu-1} \mathfrak{B}$$

and

(15) $$\mathsf{s}_\lambda^\kappa b' \cdot \mathsf{s}_\kappa^\lambda b' = b' \text{ for all } \kappa, \lambda < \alpha + \mu - 1.$$

On the other hand, by 1.5.10(i) we have

$$b' \cdot \mathsf{s}_{\alpha+\mu-1}^0 b' = b' \cdot \mathsf{s}_{\alpha+\mu-1}^0 a \cdot \prod_{\nu < \mu - 1} \mathsf{s}_{\alpha+\nu}^0 a,$$

and hence by (2) it follows that

(16) $$b = b' \cdot \mathsf{s}_{\alpha+\mu-1}^0 b'.$$

Therefore, from (15), (16), and the special case $\beta = \alpha + 1$ of our theorem previously established we conclude that (3) holds with $\mathfrak{A}$, a, and α replaced respectively by $\mathfrak{Nr}_{\alpha+\mu-1} \mathfrak{B}$, b', and $\alpha + \mu - 1$, i.e., we have

(17) $$\mathfrak{Rl}_{b'} \mathfrak{Nr}_{\alpha+\mu-1} \mathfrak{B} \cong | \subseteq \mathfrak{Nr}_{\alpha+\mu-1} \mathfrak{Rl}_b \mathfrak{B}.$$

(3) now follows directly from (14), (17), and 2.6.31, while (4) follows from (15) and (16) in a manner identical to that in which (6) was obtained from (2). This completes the inductive proof of the fact that (3) and (4) hold for all $\mu < \omega$.

Thus we have proved the theorem in the case $\beta = \alpha + \mu$ where $\mu < \omega$.

The general case follows immediately from this particular one using 2.6.34 and 2.6.35.

THEOREM 2.6.39. *If* $\alpha \leqq 1$, *then* $\mathsf{CA}_\alpha = \mathsf{SNr}_\alpha \mathsf{CA}_\beta$ *for every* $\beta \geqq \alpha$.

PROOF. If $\beta > \alpha$ we repeat almost literally the proof of 2.6.8; if $\beta = \alpha$ we apply 2.6.30(ii).

REMARK 2.6.40. As particular cases of 2.6.39 we obtain the following formulas for $\alpha = 0, 1$:

(I) $\mathsf{CA}_\alpha = \mathsf{SNr}_\alpha \mathsf{CA}_{\alpha+1}$;

(II) $\mathsf{CA}_\alpha = \mathsf{SNr}_\alpha \mathsf{CA}_{\alpha+\omega}$;

(III) $\mathsf{SNr}_\alpha \mathsf{CA}_{\alpha+1} = \mathsf{SNr}_\alpha \mathsf{CA}_{\alpha+\omega}$.

The problem naturally arises whether these formulas hold for any $\alpha \geqq 2$. As regards (I) and (II) we can conclude immediately from 2.6.14(i) and 2.6.31 that they fail for every α such that $2 \leqq \alpha < \omega$; in Theorem 2.6.43 below we shall strengthen this observation by showing that (I) and (II) fail for $\alpha \geqq \omega$ as well. Concerning (III) it will be shown in Part II that (III) holds for $\alpha = 2$ and fails for $\alpha \geqq 3$.

In view of Remark 2.6.26, and as a consequence of the above remarks, we have that every CA_α with $\alpha \leqq 1$ is representable while for each $\alpha \geqq 2$ there is a CA_α which is not representable.

LEMMA 2.6.41. *If* $\mathfrak{A} \in \mathsf{SNr}_\alpha \mathsf{CA}_{\alpha+1}$ *for* $\alpha \geqq 2$, *then*

$$\mathsf{c}_1(x \cdot y \cdot \mathsf{c}_0(x \cdot -y)) \cdot -\mathsf{c}_0(\mathsf{c}_1 x \cdot -\mathsf{d}_{01}) = 0$$

for all $x, y \in A$.

PROOF. Let $\mathfrak{A} \subseteq \mathfrak{Nr}_\alpha \mathfrak{B}$ where $\mathfrak{B} \in \mathsf{CA}_{\alpha+1}$ and then repeat exactly the proof of 1.11.7 with λ, κ, and μ replaced respectively by 0, 1, and α.

LEMMA 2.6.42. *For every* $\alpha \geqq 2$ *there is a* CA_α *in which the equation*

$$\mathsf{c}_1(y \cdot \mathsf{c}_0(\mathsf{c}_1 y \cdot -y)) \cdot -\mathsf{c}_0(\mathsf{c}_1 y \cdot -\mathsf{d}_{01}) = 0$$

(*and hence a fortiori the equation stated in 2.6.41*) *fails.*

PROOF. Let $\mathfrak{A}$ be the full cylindric set algebra in the space ${}^\alpha\alpha$ and set $a = \{\alpha 1 Id\}$. Observe that

(1) $$a \leqq -\mathsf{d}_{\kappa\lambda} \text{ whenever } \kappa < \lambda < \alpha$$

and that

(2) $$\mathsf{c}_1 a \cdot -\mathsf{c}_0(\mathsf{c}_1 a \cdot -\mathsf{d}_{01}) = \{(\alpha \sim \{1\}) 1 Id \cup \{\langle 1, 0\rangle\}\} \neq 0.$$

Let $\mathfrak{B}$ be the algebra defined in the statement of Lemma 2.6.12. Then $\mathfrak{B} \in \mathsf{CA}_\alpha$, and taking $y = \langle a, 0\rangle$ we have, by a simple computation which essentially uses only (1) and 2.6.12(i)–(iii) (compare here the proof of 2.6.13),

$$\mathsf{c}'_1(y \cdot' \mathsf{c}'_0(\mathsf{c}'_1 y \cdot' -'y)) \cdot' -'\mathsf{c}'_0(\mathsf{c}'_1 y \cdot' -'\mathsf{d}'_{01}) = \langle \mathsf{c}_1 a \cdot -\mathsf{c}_0(\mathsf{c}_1 a \cdot -\mathsf{d}_{01}), 0\rangle.$$

Thus, in view of (2), the equation in question fails in $\mathfrak{B}$.

THEOREM 2.6.43. *If* $\alpha \geqq 2$, *then*

(i) $\mathsf{CA}_\alpha \supset \mathbf{S}\mathsf{Nr}_\alpha\mathsf{CA}_{\alpha+1} \supseteq \mathbf{S}\mathsf{Nr}_\alpha\mathsf{CA}_{\alpha+\omega}$;

more specifically, we have

(ii) $\mathbf{S}_\delta\mathsf{CA}_\alpha \supset \mathbf{S}_\delta\mathsf{Nr}_\alpha\mathsf{CA}_{\alpha+1} \supseteq \mathbf{S}_\delta\mathsf{Nr}_\alpha\mathsf{CA}_{\alpha+\omega}$

for every cardinal $\delta > 1$.

PROOF: by 2.6.31, 2.6.41, and 2.6.42.

REMARKS 2.6.44. For the purposes of proving 2.6.43 we have exhibited an equation,

(1) $$\mathsf{c}_1(x \cdot y \cdot \mathsf{c}_0(x \cdot -y)) \cdot -\mathsf{c}_0(\mathsf{c}_1 x \cdot -\mathsf{d}_{01}) = 0,$$

which, by 2.6.41 and 2.6.42, is identically satisfied in every CA_α (with $\alpha \geqq 2$) neatly embeddable in some $\mathsf{CA}_{\alpha+1}$, but not in all possible CA_α's. The same applies to the symmetric equation,

(2) $$\mathsf{c}_0(x \cdot y \cdot \mathsf{c}_1(x \cdot -y)) \cdot -\mathsf{c}_1(\mathsf{c}_0 x \cdot -\mathsf{d}_{01}) = 0.$$

If we agree to denote by L_α the class of CA_α's in which equations (1) and (2) are identically satisfied, we have the formula

$$\mathbf{S}\mathsf{Nr}_\alpha\mathsf{CA}_{\alpha+\omega} \subseteq \mathbf{S}\mathsf{Nr}_\alpha\mathsf{CA}_{\alpha+1} \subseteq \mathsf{L}_\alpha$$

for $\alpha \geqq 2$. It will be seen in Part II that for $\alpha = 2$ (but not for any $\alpha > 2$) the three classes of algebras involved in this formula coincide; in fact, it will be shown that L_2 coincides with the class of representable CA_2's.

Many other equations are known which, like (1) and (2) above, fail in certain CA_α's but are identically satisfied in every CA_α neatly embeddable in some $\mathsf{CA}_{\alpha+1}$. For instance, in Chapter 1 we came across the following equations (see the remarks following 1.5.14 and 1.5.18 as well as Remark 1.5.22):

$$\mathsf{s}^2_0\mathsf{s}^0_1\mathsf{s}^1_2\mathsf{c}_2 x = \mathsf{s}^2_1\mathsf{s}^1_0\mathsf{s}^0_2\mathsf{c}_2 x,$$

$$\mathsf{s}^2_0\mathsf{s}^0_1\mathsf{s}^1_2\mathsf{s}^2_0\mathsf{s}^0_1\mathsf{s}^1_2\mathsf{c}_2 x = \mathsf{c}_2 x,$$

$$\mathsf{c}_0 x \cdot \mathsf{c}_1 y \cdot \mathsf{c}_2 z - \mathsf{c}_0\mathsf{c}_1\mathsf{c}_2[\mathsf{c}_2(\mathsf{c}_1 x \cdot \mathsf{c}_0 y) \cdot \mathsf{c}_1(\mathsf{c}_2 x \cdot \mathsf{c}_0 z) \cdot \mathsf{c}_0(\mathsf{c}_2 y \cdot \mathsf{c}_1 z)] = 0.$$

For obvious reasons, in discussing CA's in which these equations hold or fail we have to restrict ourselves to CA_α's with $\alpha \geqq 3$. It is easily seen that each of these equations holds identically in every cylindric set algebra of dimension α, and hence also in every representable CA_α. Actually, as a direct consequence of 1.5.14, 1.5.17, and 1.5.23 we obtain a stronger result: the three equations hold in every CA_α which is neatly embeddable in some $\mathsf{CA}_{\alpha+1}$. On the other hand, we can construct particular CA_α's in which these equations fail. As opposed, however, to the case of equations (1) and (2), the method which was developed in 2.6.12 proves to be inadequate for such constructions.

From 2.6.39–2.6.43 we can draw some conclusions connecting neat reducts and dimension-restricted free algebras.

THEOREM 2.6.45. *Let* $\beta \geqq \alpha$, δ *be any cardinal* >0, *and* $\Delta \in {}^\delta Sb\alpha$. *Let* $\mathfrak{B} \in \mathsf{CA}_\beta$ *and* $x \in {}^\delta B$. *If the sequence* x CA_β*-freely generates* $\mathfrak{B}$ *under the dimension-restricting function* Δ, *then* x $\mathsf{Nr}_\alpha\mathsf{CA}_\beta$*-freely generates* $\mathfrak{Sg}^{(\mathfrak{Nr}_\alpha\mathfrak{B})}Rgx$ *under* Δ.

PROOF. Consider any $\mathfrak{A} \in \mathsf{Nr}_\alpha\mathsf{CA}_\beta$ and $y \in {}^\delta A$ such that $\Delta^{(\mathfrak{A})}y_\xi \subseteq \Delta\xi$ for every $\xi < \delta$. Let $\mathfrak{C} \in \mathsf{CA}_\beta$ such that

(1) $$\mathfrak{A} = \mathfrak{Nr}_\alpha\mathfrak{C}.$$

Clearly $y \in {}^\delta C$ and

(2) $$\Delta^{(\mathfrak{C})}y_\xi \subseteq \alpha \text{ for each } \xi < \delta.$$

Thus, by hypothesis, there is an $h \in Hom(\mathfrak{B}, \mathfrak{C})$ such that $h \circ x = y$. Therefore, $h \in Hom(\mathfrak{Rd}_\alpha\mathfrak{B}, \mathfrak{Rd}_\alpha\mathfrak{C})$ by 0.5.12(ii); using 0.2.3 and 0.2.18(i) we then conclude that

(3) $$h \in Hom(\mathfrak{Sg}^{(\mathfrak{Rd}_\alpha\mathfrak{B})}Rgx, \mathfrak{Sg}^{(\mathfrak{Rd}_\alpha\mathfrak{C})}h^*Rgx).$$

$Rgx \subseteq Nr_\alpha\mathfrak{B}$ by hypothesis and $h^*Rgx = Rgy \subseteq Nr_\alpha\mathfrak{C}$ by (2). Hence from (1) and (3) we get $h \in Hom(\mathfrak{Sg}^{(\mathfrak{Nr}_\alpha\mathfrak{B})}Rgx, \mathfrak{A})$. Since $\mathfrak{A} \in \mathsf{Nr}_\alpha\mathsf{CA}_\beta$ and $y \in {}^\delta A$ with $\Delta^{(\mathfrak{A})}y_\xi \subseteq \Delta\xi$ for all $\xi < \delta$ were arbitrarily chosen, it is seen that the conclusion of the theorem holds.

THEOREM 2.6.46. *For any* α *and* β *with* $\beta > \alpha$ *and any cardinal* $\delta \neq 0$ *the following three conditions are equivalent*:

(i) $\mathfrak{Fr}_\delta\mathsf{CA}_\alpha \cong \mathfrak{Fr}_\delta\mathsf{Nr}_\alpha\mathsf{CA}_\beta$;

(ii) *if* $\mathfrak{B} \in \mathsf{CA}_\beta$, $x \in {}^\delta B$, *and* x CA_β*-freely generates* $\mathfrak{B}$ *under* $\delta \times \{\alpha\}$, *then* Rgx CA_α*-freely generates* $\mathfrak{Sg}^{(\mathfrak{Nr}_\alpha\mathfrak{B})}Rgx$;

(iii) $\alpha \leqq 1$.

PROOF. (iii) implies (ii) by 0.4.26(ii), 2.6.39, and 2.6.45. To prove (ii) implies (i) observe that by 0.4.42 and 2.6.30(i) one immediately gets

(1) $$\mathfrak{Fr}_\delta\mathsf{CA}_\alpha \succcurlyeq \mathfrak{Fr}_\delta\mathsf{Nr}_\alpha\mathsf{CA}_\beta.$$

By 2.5.36, $\langle \xi/Cr_\delta^{(\delta\times\{\alpha\})}\mathsf{CA}_\beta : \xi < \delta\rangle$ CA_β-freely generates $\mathfrak{Fr}_\delta^{(\delta\times\{\alpha\})}\mathsf{CA}_\beta$ under $\delta\times\{\alpha\}$. Hence from (ii) and 2.5.33(i) we have that $\delta/Cr_\delta^{(\delta\times\{\alpha\})}\mathsf{CA}_\beta$ CA_α-freely generates the subalgebra $\mathfrak{A}$ of $\mathfrak{Nr}_\alpha\mathfrak{Fr}_\delta^{(\delta\times\{\alpha\})}\mathsf{CA}_\beta$ that it generates (cf. here the remark immediately following 2.5.34); this implies by 0.4.29 that

$$(2) \qquad \mathfrak{A} \cong \mathfrak{Fr}_\delta\mathsf{CA}_\alpha.$$

On the other hand, since $\mathfrak{A} \in \mathbf{S}_{\delta^+}\mathsf{Nr}_\alpha\mathsf{CA}_\beta$, we have $\mathfrak{Fr}_\delta\mathsf{Nr}_\alpha\mathsf{CA}_\beta \succcurlyeq \mathfrak{A}$ by 0.4.39. This fact together with (1), (2), and 0.4.43 gives (i). Hence (ii) implies (i).

Finally, we prove (i) implies (iii). Assume that (i) holds. Then

$$\begin{aligned} \mathbf{S}_{\delta^+}\mathsf{CA}_\alpha &\subseteq \mathbf{H}\mathfrak{Fr}_\delta\mathsf{CA}_\alpha && \text{by 0.4.39} \\ &= \mathbf{H}\mathfrak{Fr}_\delta\mathsf{Nr}_\alpha\mathsf{CA}_\beta && \text{by (i)} \\ &\subseteq \mathbf{HSPNr}_\alpha\mathsf{CA}_\beta && \text{by 0.4.22} \\ &\subseteq \mathbf{SNr}_\alpha\mathsf{CA}_{\alpha+1} && \text{by 2.6.31, 2.6.32(ii).} \end{aligned}$$

Thus, from 2.6.43(ii) (with δ^+ in place of δ) one concludes that $\alpha \leqq 1$, i.e., that (iii) holds. This finishes the proof.

The next theorem is an unpublished result of Henkin from 1960.

THEOREM 2.6.47. *For any $\alpha \geqq \omega$ and any $\mathfrak{A} \in \mathsf{CA}_\alpha$ the following conditions are equivalent*:

(i) $\mathfrak{A} \in \mathbf{SNr}_\alpha\mathsf{CA}_{\alpha+\omega}$;

(ii) *for all $\kappa, \lambda < \omega$, every sequence $\rho \in {}^\kappa\alpha$ without repeating terms, and every non-zero $x \in A$, there is a $\mathfrak{B} \in \mathsf{CA}_{\kappa+\lambda}$ and an $h \in Hom(\mathfrak{Rd}_\kappa^{(\rho)}\mathfrak{A}, \mathfrak{Nr}_\kappa\mathfrak{B})$ such that $hx \neq 0^{(\mathfrak{B})}$.*

PROOF. To prove (i) implies (ii), assume that $\mathfrak{A} \in \mathbf{SNr}_\alpha\mathsf{CA}_{\alpha+\omega}$, say $\mathfrak{A} \subseteq \mathfrak{Nr}_\alpha\mathfrak{C}$ with $\mathfrak{C} \in \mathsf{CA}_{\alpha+\omega}$. Let $\kappa, \lambda < \omega$, ρ be a one-one function in ${}^\kappa\alpha$, and $x \in A \sim \{0^{(\mathfrak{A})}\}$. Clearly there is a $\sigma \in {}^{\kappa+\lambda}(\alpha+\omega)$ such that σ is one-one, $\sigma \supseteq \rho$, and $\sigma^*((\kappa+\lambda) \sim \kappa) \subseteq (\alpha+\omega) \sim \alpha$. Then using 0.5.12(ii) we readily see that

$$A \upharpoonright Id \in Hom(\mathfrak{Rd}_\kappa^{(\rho)}\mathfrak{A}, \mathfrak{Nr}_\kappa\mathfrak{Rd}_{\kappa+\lambda}^{(\sigma)}\mathfrak{C}).$$

Thus, setting $\mathfrak{B} = \mathfrak{Rd}_{\kappa+\lambda}^{(\sigma)}\mathfrak{C}$ and referring to 2.6.2(i) we see that the conclusion of (ii) holds.

To prove (ii) implies (i), assume that $\kappa, \lambda < \omega$ and ρ is a one-one function in ${}^\kappa\alpha$. By (ii), 2.4.39, and 2.6.32(ii) we infer that $\mathfrak{Rd}_\kappa^{(\rho)}\mathfrak{A} \in \mathbf{SNr}_\kappa\mathsf{CA}_{\kappa+\lambda}$. Since λ is an arbitrary finite ordinal, we see by 2.6.34 and 2.6.35 that $\mathfrak{Rd}_\kappa^{(\rho)}\mathfrak{A} \in \mathbf{SNr}_\kappa\mathsf{CA}_{\alpha+\omega}$.

Thus $\mathfrak{Rd}_\kappa^{(\rho)}\mathfrak{A} \subseteq \mathfrak{Nr}_\kappa\mathfrak{B}$ for some $\mathfrak{B} \in \mathsf{CA}_{\alpha+\omega}$. Let σ be a permutation of $\alpha+\omega$ such that

$$(1) \qquad \kappa 1 \sigma = \rho$$

and $((\alpha+\omega) \sim \alpha) 1 \sigma \subseteq Id$. Then, by 0.5.5,

$$(2) \qquad \mathfrak{Rd}_\kappa^{(\rho)}\mathfrak{A} \subseteq \mathfrak{Nr}_\kappa\mathfrak{B} = \mathfrak{Nr}_\kappa\mathfrak{Rd}_{\alpha+\omega}^{(\sigma)}\mathfrak{Rd}_{\alpha+\omega}^{(\sigma^{-1})}\mathfrak{B}.$$

However, it is an easy matter to check that for any γ such that $\sigma^*\kappa \subseteq \gamma \leqq \alpha+\omega$ we have

$$\mathfrak{Nr}_\kappa\mathfrak{Rd}_{\alpha+\omega}^{(\sigma)}\mathfrak{Rd}_{\alpha+\omega}^{(\sigma^{-1})}\mathfrak{B} \subseteq \mathfrak{Rd}_\kappa^{(\kappa 1 \sigma)}\mathfrak{Nr}_\gamma\mathfrak{Rd}_{\alpha+\omega}^{(\sigma^{-1})}\mathfrak{B}.$$

Therefore, taking this result (with $\gamma = \alpha$) together with (1), (2), and 2.6.2(i), we conclude that

$$\mathfrak{Rd}_\kappa^{(\rho)}\mathfrak{A} \in \mathsf{SRd}_\kappa^{(\rho)}\mathsf{Nr}_\alpha\mathsf{CA}_{\alpha+\omega}.$$

Since ρ is an arbitrary finite sequence without repetitions and with $Rg\rho \subseteq \alpha$, we infer by 2.6.4 and 2.6.32(ii) that (i) holds. This completes the proof.

THEOREM 2.6.48. $\mathsf{SNr}_\alpha\mathsf{CA}_{\alpha+\omega} = \mathsf{SNr}_\alpha\mathsf{Dc}_{\alpha+\omega}$. *If* $\alpha < \omega$, *then* $\mathsf{SNr}_\alpha\mathsf{CA}_\omega = \mathsf{SNr}_\alpha\mathsf{Lf}_\omega$.

PROOF. To prove the first part of the theorem assume that $\mathfrak{A} \subseteq \mathfrak{Nr}_\alpha\mathfrak{B}$ and $\mathfrak{B} \in \mathsf{CA}_{\alpha+\omega}$. Set $\mathfrak{C} = \mathfrak{Sg}^{(\mathfrak{B})}\mathfrak{A}$. Clearly $\mathfrak{A} \subseteq \mathfrak{Nr}_\alpha\mathfrak{C}$ while, by 2.1.5(ii), $\mathfrak{C} \in \mathsf{Dc}_{\alpha+\omega}$. Thus $\mathsf{SNr}_\alpha\mathsf{CA}_{\alpha+\omega} \subseteq \mathsf{SNr}_\alpha\mathsf{Dc}_{\alpha+\omega}$; the inclusion in the opposite direction is obvious. The second part is proved analogously.

THEOREM 2.6.49. (i) $\mathsf{Dc}_\alpha \subseteq \mathsf{SNr}_\alpha\mathsf{CA}_{\alpha+\omega}$.
(ii) *If* $\alpha \geqq \omega$, *then* $\mathsf{Lf}_\alpha \subseteq \mathsf{SNr}_\alpha\mathsf{CA}_{\alpha+\omega}$.[1)]

PROOF. To understand the basic idea of the proof we first concentrate on the second part of the theorem. As we know from 1.11.2, every Lf_α with $\alpha \geqq \omega$ can be represented isomorphically as an algebra $\mathfrak{F}$ of formulas in a

[1)] Theorem 2.6.49(ii) and its proof given below were found by Tarski around 1952 (but were not published until now). The result was treated as an essential lemma for obtaining a purely algebraic demonstration of the representability of locally finite infinite-dimensional CA_α's. The desired conclusion was derived from the lemma by means of an additional argument, since the general result to the effect that $\mathsf{SNr}_\alpha\mathsf{CA}_{\alpha+\omega}$ coincides with the class of representable CA_α's was not available at that time. Another proof of 2.6.49(ii) is known which is based upon the theory of substitution operators developed in Section 1.11; the basic ideas of this proof appear in Daigneault-Monk [63], § 5. Monk noticed that Tarski's original proof of 2.6.49(ii), as opposed to the one just mentioned, extends with practically no changes to dimension-complemented CA_α's and thus leads to 2.6.49(i).

language Λ of predicate logic. In view of 2.6.28 the proof that $\mathfrak{F}$ has the neat embedding property reduces to showing that Λ can be extended to a new language Λ' satisfying the following conditions: the vocabulary of Λ' is obtained from that of Λ by adding ω new variables, all the formulas in Λ are also formulas in Λ', and the consequence relations which hold between the formulas in Λ remain unchanged. Intuitively, the possibility of extending Λ in this way seems a rather obvious and trivial matter, especially since the process may be broken, loosely speaking, into infinitely many steps each of which consists of enriching the vocabulary by means of one new variable. Essentially what we do in the proof below is to analyze precisely the process of enriching the language by new variables and to describe it in purely algebraic terms.

To prove the theorem it clearly suffices to prove only part (i), and this only for the case $\alpha \geqq \omega$. To this end we let $\mathfrak{A} \in \mathsf{Dc}_\alpha$ and proceed to show that $\mathfrak{A} \in \mathbf{SNr}_\alpha \mathsf{Dc}_{\alpha+1}$. Let

$$R = (\alpha \times A) 1 Id \cup \{\langle\langle\kappa, x\rangle, \langle\lambda, y\rangle\rangle : \kappa, \lambda < \alpha, \quad x, y \in A, \quad \lambda \notin \Delta x, \quad y = \mathsf{s}^\kappa_\lambda x\}.$$

It is easily seen using 1.5.11(i) and 1.6.13 that R is an equivalence relation on $\alpha \times A$. The following observation plays an important part in the proof.

(1) Let $\kappa < \omega$, $x \in {}^\kappa A$, $\lambda \in {}^\kappa\alpha$, and $\Gamma \subseteq \alpha$ such that $|\Gamma| < \omega$. Then there is a $\mu < \alpha$ and $y \in {}^\kappa A$ such that $\mu \notin \Gamma$ and $\langle\lambda_\nu, x_\nu\rangle R \langle\mu, y_\nu\rangle$ for every $\nu < \kappa$.

To show (1) choose any μ in the set $\alpha \sim (\Gamma \cup \bigcup_{\nu<\kappa} \Delta x_\nu)$ (which is non-empty by 1.11.4), and then set $y_\nu = \mathsf{s}^{\lambda_\nu}_\mu x_\nu$ for each $\nu < \kappa$. The conclusion of (1) is easily seen to hold.

It is now a simple matter to show that there exists an algebra

(2) $$\mathfrak{C} = \langle(\alpha \times A)/R, +', \cdot', -', 0', 1', \mathsf{c}'_\kappa, \mathsf{d}'_{\kappa\lambda}\rangle_{\kappa,\lambda<\alpha}$$

similar to CA_α's such that the following conditions hold:

(3) $\langle\mu, x\rangle/R +' \langle\mu, y\rangle/R = \langle\mu, x+y\rangle/R$ for all $\mu < \alpha, \quad x, y \in A$;

(4) $\langle\mu, x\rangle/R \cdot' \langle\mu, y\rangle/R = \langle\mu, x \cdot y\rangle/R$ for all $\mu < \alpha, \quad x, y \in A$;

(5) $$-'\langle\mu, x\rangle/R = \langle\mu, -x\rangle/R \text{ for all } \mu < \alpha, \quad x \in A;$$

(6) $$0' = \langle\mu, 0\rangle/R \text{ for every } \mu < \alpha;$$

(7) $$1' = \langle\mu, 1\rangle/R \text{ for every } \mu < \alpha;$$

(8) $$\mathsf{c}'_\kappa(\langle\mu, x\rangle/R) = \langle\mu, \mathsf{c}_\kappa x\rangle/R \text{ for } \kappa < \alpha, \quad x \in A, \quad \mu \in \alpha \sim \{\kappa\};$$

(9) $$\mathsf{d}'_{\kappa\lambda} = \langle\mu, \mathsf{d}_{\kappa\lambda}\rangle/R \text{ for all } \kappa, \lambda < \alpha, \quad \mu \in \alpha \sim \{\kappa, \lambda\}.$$

Using mainly (1) and the definition of R the reader may check that each of the conditions (3)–(9) in fact serves as a definition of the operation (or the element) involved in its formulation. With regard, e.g., to condition (3) this means precisely the following: We can show that for any elements $a, b \in (\alpha \times A)/R$ there is just one element $c \in (\alpha \times A)/R$ such that, whenever a and b are of the form $a = \langle \mu, x \rangle / R$ and $b = \langle \mu, y \rangle / R$ with $\mu < \alpha$ and $x, y \in A$, we have $c = \langle \mu, x + y \rangle / R$. The proof of the latter statement is left to the reader.

It is clear also how one proceeds in using (1)–(9) to show that $\mathfrak{C}$ satisfies axioms (C_0)–(C_7) for CA_α's; as an example which fully illustrates the method we shall prove that (C_7) holds. Consider any $\kappa, \lambda, \mu < \alpha$ with $\kappa \neq \lambda$ and any $x \in A$. Then by (1) there is a $\nu < \alpha$ and $y \in A$ such that $\nu \in \alpha \sim (\Delta x \cup \{\kappa, \lambda\})$ and $\langle \mu, x \rangle R \langle \nu, y \rangle$, and

$$\begin{aligned}
& \mathsf{c}'_\kappa(\mathsf{d}'_{\kappa\lambda} \cdot' \langle \mu, x \rangle / R) \cdot' \mathsf{c}'_\kappa(\mathsf{d}'_{\kappa\lambda} \cdot' -' \langle \mu, x \rangle / R) \\
&= \mathsf{c}'_\kappa(\mathsf{d}'_{\kappa\lambda} \cdot' \langle \nu, y \rangle / R) \cdot' \mathsf{c}'_\kappa(\mathsf{d}'_{\kappa\lambda} \cdot' -' \langle \nu, y \rangle / R) \\
&= \mathsf{c}'_\kappa(\mathsf{d}'_{\kappa\lambda} \cdot' \langle \nu, y \rangle / R) \cdot' \mathsf{c}'_\kappa(\mathsf{d}'_{\kappa\lambda} \cdot' \langle \nu, -y \rangle / R) && \text{by (5)} \\
&= \mathsf{c}'_\kappa(\langle \nu, \mathsf{d}_{\kappa\lambda} \rangle / R \cdot' \langle \nu, y \rangle / R) \cdot' \mathsf{c}'_\kappa(\langle \nu, \mathsf{d}_{\kappa\lambda} \rangle / R \cdot' \langle \nu, -y \rangle / R) && \text{by (9)} \\
&= \mathsf{c}'_\kappa(\langle \nu, \mathsf{d}_{\kappa\lambda} \cdot y \rangle / R) \cdot' \mathsf{c}'_\kappa(\langle \nu, \mathsf{d}_{\kappa\lambda} \cdot -y \rangle / R) && \text{by (4)} \\
&= \langle \nu, \mathsf{c}_\kappa(\mathsf{d}_{\kappa\lambda} \cdot y) \rangle / R \cdot' \langle \nu, \mathsf{c}_\kappa(\mathsf{d}_{\kappa\lambda} \cdot -y) \rangle / R && \text{by (8)} \\
&= \langle \nu, \mathsf{c}_\kappa(\mathsf{d}_{\kappa\lambda} \cdot y) \cdot \mathsf{c}_\kappa(\mathsf{d}_{\kappa\lambda} \cdot -y) \rangle / R && \text{by (4)} \\
&= \langle \nu, 0 \rangle / R && \text{by } (C_7) \\
&= 0' && \text{by (6).}
\end{aligned}$$

Thus (C_7) holds for $\mathfrak{C}$, and in a similar manner the other axioms are demonstrated. Hence $\mathfrak{C} \in \mathsf{CA}_\alpha$ and, moreover, from (8) it is easily seen that

(10) $$\mathfrak{C} \in \mathsf{Dc}_\alpha.$$

By an argument using (1)–(9) which is entirely analogous to the one just employed we can show that, if

(11) $$h = \{\langle x, \langle \mu, x \rangle / R \rangle : x \in A, \mu \in \alpha \sim \Delta x\},$$

then h is a one-one function from A into C, and in fact,

(12) $$h \in Ism(\mathfrak{A}, \mathfrak{C}).$$

To finish showing that $\mathfrak{A} \in \mathsf{SNr}_\alpha\mathsf{Dc}_{\alpha+1}$ we define a unary operation c'_α on $(\alpha \times A)/R$, and elements $\mathsf{d}'_{\kappa\alpha} = \mathsf{d}'_{\alpha\kappa} \in (\alpha \times A)/R$ for each $\kappa \leqq \alpha$, such that, if

(13) $$\mathfrak{B} = \langle (\alpha \times A)/R, +', \cdot', -', 0', \mathsf{c}'_\kappa, \mathsf{d}'_{\kappa\lambda} \rangle_{\kappa,\lambda<\alpha+1},$$

then

(14) $$\mathfrak{B} \in \mathsf{Dc}_{\alpha+1}$$

and

(15) $$h^*\mathfrak{A} \subseteq \mathfrak{Nr}_\alpha \mathfrak{B}.$$

Set

(16) $$\mathsf{c}'_\alpha = \{\langle \langle \mu, x \rangle/R, \langle \mu, \mathsf{c}_\mu x \rangle/R \rangle : \mu \in \alpha,\ x \in A\}.$$

If $\langle \mu, x \rangle R \langle \mu', x' \rangle$, then $\langle \mu, \mathsf{c}_\mu x \rangle R \langle \mu', \mathsf{c}_{\mu'} x' \rangle$ by 1.5.8(i) and 1.5.9(i), and hence c'_κ is a unary operation on $(\alpha \times A)/R$. Using 1.5.4(i) and 1.6.4 we can show that the conditions

(17) $$\mathsf{d}'_{\kappa\alpha} = \langle \mu, \mathsf{d}_{\kappa\mu} \rangle/R \quad \text{for every} \quad \mu \in \alpha \sim \{\kappa\}$$

uniquely defines an element of $(\alpha \times A)/R$ for each $\kappa < \alpha$. Finally, set

(18) $$\mathsf{d}'_{\alpha\kappa} = \mathsf{d}'_{\kappa\alpha} \text{ for each } \kappa < \alpha, \quad \mathsf{d}'_{\alpha\alpha} = 1'.$$

In view of (2), (10), and (13), in order to prove (14) it suffices to verify the axioms of $\mathsf{CA}_{\alpha+1}$'s just in those cases in which α appears as an index of a cylindrification symbol c_κ or of a diagonal symbol $\mathsf{d}_{\kappa\lambda}$. Thus (C_0) need not be considered, and (C_1), (C_2), and (C_5) are easily checked. Again as an example of the method employed in verifying the remaining axioms we shall prove that (C_7) holds. Consider any $\kappa, \mu < \alpha$ and $x \in A$. By (1) there is a $\nu < \alpha$ and $y \in A$ such that $\nu \in \alpha \sim (\Delta x \cup \{\kappa\})$ and $\langle \mu, x \rangle R \langle \nu, y \rangle$. Then

$$\begin{aligned}
&\mathsf{c}'_\kappa(\mathsf{d}'_{\kappa\alpha} \cdot' \langle \mu, x \rangle/R) \cdot' \mathsf{c}'_\kappa(\mathsf{d}'_{\kappa\alpha} \cdot' -'\langle \mu, x \rangle/R) \\
&\quad = \mathsf{c}'_\kappa(\mathsf{d}'_{\kappa\alpha} \cdot' \langle \nu, y \rangle/R) \cdot' \mathsf{c}'_\kappa(\mathsf{d}'_{\kappa\alpha} \cdot' \langle \nu, -y \rangle/R) && \text{by (5)} \\
&\quad = \mathsf{c}'_\kappa(\langle \nu, \mathsf{d}_{\kappa\nu} \rangle/R \cdot' \langle \nu, y \rangle/R) \cdot' \mathsf{c}'_\kappa(\langle \nu, \mathsf{d}_{\kappa\nu} \rangle/R \cdot' \langle \nu, -y \rangle/R) && \text{by (17)} \\
&\quad = 0' && \text{by (4), (6), (8), } (C_7).
\end{aligned}$$

$$\begin{aligned}
&\mathsf{c}'_\alpha(\mathsf{d}'_{\alpha\kappa} \cdot' \langle \mu, x \rangle/R) \cdot' \mathsf{c}'_\alpha(\mathsf{d}'_{\alpha\kappa} \cdot' -'\langle \mu, x \rangle/R) \\
&\quad = \mathsf{c}'_\alpha(\mathsf{d}'_{\alpha\kappa} \cdot' \langle \nu, y \rangle/R) \cdot' \mathsf{c}'_\alpha(\mathsf{d}'_{\alpha\kappa} \cdot' \langle \nu, -y \rangle/R) && \text{by (5)} \\
&\quad = \mathsf{c}'_\alpha(\langle \nu, \mathsf{d}_{\nu\kappa} \rangle/R \cdot' \langle \nu, y \rangle/R) \cdot' \mathsf{c}'_\alpha(\langle \nu, \mathsf{d}_{\nu\kappa} \rangle/R \cdot' \langle \nu, -y \rangle/R) && \text{by (17), (18)} \\
&\quad = \mathsf{c}'_\alpha(\langle \nu, \mathsf{d}_{\nu\kappa} \cdot y \rangle/R) \cdot' \mathsf{c}'_\alpha(\langle \nu, \mathsf{d}_{\nu\kappa} \cdot -y \rangle/R) && \text{by (4)} \\
&\quad = \langle \nu, \mathsf{c}_\nu(\mathsf{d}_{\nu\kappa} \cdot y) \rangle/R \cdot' \langle \nu, \mathsf{c}_\nu(\mathsf{d}_{\nu\kappa} \cdot -y) \rangle/R && \text{by (16)} \\
&\quad = 0' && \text{by (4), (6), } (C_7).
\end{aligned}$$

Thus (C_7) holds, and (C_3), (C_4), and (C_6) are verified similarly; this proves (14). Finally, for any $x \in A$ and $\mu \in \alpha \sim \Delta x$ we have by (16)

$$\mathsf{c}'_\alpha(\langle \mu, x \rangle / R) = \langle \mu, \mathsf{c}_\mu x \rangle / R = \langle \mu, x \rangle / R.$$

Hence, in view of (11), we see that (15) holds.

We have proved that $\mathsf{Dc}_\alpha \subseteq \mathbf{S}\mathsf{Nr}_\alpha \mathsf{Dc}_{\alpha+1}$ for every $\alpha \geqq \omega$; using this result one concludes by means of a simple inductive argument that, for every $\alpha \geqq \omega$ and $\kappa < \omega$,

$$\mathsf{Dc}_\alpha \subseteq \mathbf{S}\mathsf{Nr}_\alpha \mathsf{Dc}_{\alpha+\kappa}.$$

The theorem now follows by 2.6.34.

By comparing 2.6.49 with 2.6.26 we conclude that all dimension-complemented CA_α's, and hence all infinite-dimensional locally finite CA's, are representable. The condition $\alpha \geqq \omega$ in the second part 2.6.49. cannot be omitted in view of 1.11.3(ii) and 2.6.43.

THEOREM 2.6.50. *If* $\mathfrak{A} \in \mathsf{CA}_\alpha$ *and* $\alpha \geqq \omega$, *then, in the sequence of conditions (i)–(iv) below, each of the conditions (i)–(iii) implies the immediately following one*:

(i) $\mathfrak{A} \in \mathsf{Dc}_\alpha \cup \mathsf{Ss}_\alpha$;

(ii) *for every finite* $\Gamma \subseteq \alpha$ *and every non-zero* $x \in A$ *there exist distinct* $\kappa, \lambda \in \alpha \sim \Gamma$ *such that* $x \cdot \mathsf{d}_{\kappa\lambda} \neq 0$;

(iii) *for every finite sequence* ρ *without repeating terms and with range included in* α, *and for every non-zero* $x \in A$ *there is a function* h *and* $\kappa < \alpha$ *such that* h *is an endomorphism of* $\mathfrak{Rd}^{(\rho)}\mathfrak{A}$, $\kappa \in \alpha \sim Rg\rho$, $\mathsf{c}_\kappa^{(\mathfrak{A})} \circ h = h$, *and* $hx \neq 0$;

(iv) $\mathfrak{A} \in \mathbf{S}\mathsf{Nr}_\alpha \mathsf{CA}_{\alpha+\omega}$.

PROOF. To derive (ii) from (i) assume the hypothesis of (ii) holds. First suppose $\mathfrak{A} \in \mathsf{Dc}_\alpha$. Choose distinct $\kappa, \lambda \in \alpha \sim (\Gamma \cup \Delta x)$ (such κ, λ exist by 1.11.4). If $x \cdot \mathsf{d}_{\kappa\lambda} = 0$, then $x = \mathsf{c}_\kappa(x \cdot \mathsf{d}_{\kappa\lambda}) = 0$, which is impossible. Thus $x \cdot \mathsf{d}_{\kappa\lambda} \neq 0$. Next suppose $\mathfrak{A} \in \mathsf{Ss}_\alpha$. By 0.3.49 there is a maximal ideal I of $\mathfrak{A}$ such that $x \notin I$, and thus by 2.3.18 there is a finite subset Δ of α for which

$$-\mathsf{c}_{(\Delta)}x \in I. \tag{1}$$

Choose distinct $\kappa, \lambda \in \alpha \sim (\Gamma \cup \Delta)$. If now $x \cdot \mathsf{d}_{\kappa\lambda} = 0$, we obtain successively $\mathsf{c}_{(\Delta)}x \cdot \mathsf{d}_{\kappa\lambda} = 0$, $\mathsf{d}_{\kappa\lambda} \leqq -\mathsf{c}_{(\Delta)}x$, $\mathsf{d}_{\kappa\lambda} \in I$ by (1), and finally $1 = \mathsf{c}_\kappa \mathsf{d}_{\kappa\lambda} \in I$. The formula is impossible, however, since I is a proper ideal.

To derive (iii) from (ii) consider any ρ and x which satisfy the premisses of (iii), and choose distinct $\kappa, \lambda \in \alpha \sim Rg\rho$ such that $x \cdot \mathsf{d}_{\kappa\lambda} \neq 0$. Letting

$h = \mathsf{s}_\lambda^\kappa$ and applying 1.5.3, 1.5.4(ii), and 1.5.8(i),(ii), the conclusion of (iii) is easily obtained.

Finally we shall show that (iii) implies (iv). To this end we assume (iii) to hold and we prove by induction that the following condition holds for every $\lambda < \omega$:

(2) for every $\kappa < \omega$, every one-one $\rho \in {}^\kappa\alpha$, and every non-zero $x \in A$, there exist σ, h such that $\sigma \in {}^{\kappa+\lambda}\alpha$, σ is one-one, $\rho \subseteq \sigma$, h is an endomorphism of $\mathfrak{Rd}_\kappa^{(\rho)}\mathfrak{A}$, $\mathsf{c}_{\sigma\mu}^{(\mathfrak{A})} \circ h = h$ whenever $\kappa \leqq \mu < \kappa+\lambda$, and $hx \neq 0$.

(2) holds trivially for $\lambda = 0$ and follows directly from (iii) for $\lambda = 1$. Suppose now that (2) holds for any given $\lambda \geqq 1$; we want to show that (2) continues to hold when λ is replaced by $\lambda+1$. Thus consider κ, ρ, and x satisfying the premisses of (2). By the induction hypothesis there are σ, h such that $\sigma \in {}^{\kappa+\lambda}\alpha$, σ is one-one, $\rho \subseteq \sigma$, h is an endomorphism of $\mathfrak{Rd}_\kappa^{(\rho)}\mathfrak{A}$, $\mathsf{c}_{\sigma\mu}^{(\mathfrak{A})} \circ h = h$ whenever $\kappa \leqq \mu < \kappa+\lambda$, and $hx \neq 0$. By (iii) there exist k, v such that k is an endomorphism of $\mathfrak{Rd}_{\kappa+\lambda}^{(\sigma)}\mathfrak{A}$, $v \in \alpha \sim Rg\sigma$, $\mathsf{c}_v^{(\mathfrak{A})} \circ k = k$, and $khx \neq 0$. Let

$$\sigma' = \sigma \cup \{\langle \kappa+\lambda, v\rangle\} \text{ and } h' = k \circ h.$$

Then $\sigma' \in {}^{\kappa+\lambda+1}\alpha$, σ' is one-one, $\rho \subseteq \sigma'$, h' is an endomorphism of $\mathfrak{Rd}_\kappa^{(\rho)}\mathfrak{A}$ (here we use 0.5.5 and 0.5.12(ii)), $\mathsf{c}_{\sigma'\mu}^{(\mathfrak{A})} \circ h' = h'$ whenever $\kappa \leqq \mu < \kappa+\lambda+1$, and $h'x \neq 0$. Thus (2) holds with λ replaced by $\lambda+1$, and the induction is complete. Since (2) has been shown to hold for every $\lambda < \omega$, we infer immediately from 2.6.47 that $\mathfrak{A} \in \mathsf{SNr}_\alpha\mathsf{CA}_{\alpha+\omega}$. This completes the proof of 2.6.50.

Algebras satisfying condition 2.6.50(ii) were introduced in Monk [61] under the name *diagonal cylindric algebras.*[1)]

Theorem 2.6.50 clearly implies 2.6.49 and hence provides a new proof of that result. It may be noticed however that, while ultraproducts are not involved in the original proof of 2.6.49, a careful examination would show that they are involved in the proof of 2.6.50. For this reason the original proof of 2.6.49 has the advantage of being conceptually simpler.

REMARK 2.6.51. In 2.6.50 three implications are stated, and the problem naturally arises whether the implications in the opposite directions also hold. We shall show by examples that this is not true as far as the first two impli-

[1)] Theorem 2.6.50 without condition (iii) is due to Monk; the connections between (iii) and the other parts of the theorem were found by Henkin. Cf. Monk [61], in particular footnote 7, p. 1454.

cations are concerned. For the implication between conditions (iii) and (iv) of 2.6.50 the problem is open.

We assume that $\alpha \geqq \omega$. To show that 2.6.50(ii) does not imply 2.6.50(i) consider the full cylindric set algebra $\mathfrak{A}$ in the space ${}^{\alpha}2$. Condition (ii) is then clearly satisfied. Also $\mathfrak{A}$ is not a Dc_α since $\Delta a = \alpha$ for each atom $a \in \mathfrak{A}$. Finally $\mathfrak{A}$ is not semisimple. Indeed the element $X = \{\langle 0:\kappa < \alpha\rangle\}$ of $\mathfrak{A}$ belongs to every maximal proper ideal I of $\mathfrak{A}$. For, if we assume $X \notin I$, then by 2.3.18 there is a finite subset Γ of α such that $\sim\mathsf{C}^{(\Gamma)}X \in I$. Choose $\kappa \in \alpha \sim \Gamma$ and let

$$\varphi = \langle 0:\mu \in \alpha \sim \{\kappa\}\rangle \cup \{\langle \kappa, 1\rangle\}.$$

Then $\varphi \in \sim\mathsf{C}_{(\Gamma)}X$, so $\{\varphi\} \in I$. But $X \subseteq \mathsf{C}_\kappa\{\varphi\}$, so $X \in I$, which is impossible. Thus $\mathfrak{A}$ is not semisimple.

To show that 2.6.50(iii) does not imply 2.6.50(ii), let $\mathfrak{A}$ be the full cylindric set algebra in the space ${}^{\alpha}\alpha$. Obviously $\{\alpha 1 Id\} \cap \mathsf{D}_{\kappa\lambda} = 0$ in case $\kappa < \lambda < \alpha$, so $\mathfrak{A}$ does not satisfy 2.6.50(ii). To verify 2.6.50(iii), suppose that ρ is a finite sequence without repetitions with $Rg\rho \subseteq \alpha$ and $X \subseteq {}^{\alpha}\alpha$, $X \neq 0$. Let $\kappa \in \alpha \sim Rg\rho$ and choose $\tau \in {}^{\alpha}\alpha$ such that $\kappa \notin Rg\tau$, $Rg\rho 1 \tau \subseteq Id$, and τ is one-one. Let

$$H = \langle\{\varphi:\varphi \in {}^{\alpha}\alpha,\ \varphi\circ\tau \in Y\}:Y \subseteq {}^{\alpha}\alpha\rangle.$$

The conclusion of 2.4.50(iii) with $x = X$ and $h = H$ is then easily verified.[1)]

THEOREM 2.6.52. *If* $\alpha \geqq \omega$ *and* $\mathsf{Lf}_\alpha \subseteq \mathsf{K} \subseteq \mathsf{SNr}_\alpha\mathsf{CA}_{\alpha+\omega}$, *then*

$$\mathsf{SNr}_\alpha\mathsf{CA}_{\alpha+\omega} = \mathsf{SUpK} = \mathsf{HSPK},$$

and, for every cardinal $\delta \neq 0$,

$$\mathfrak{Fr}_\delta\mathsf{Nr}_\alpha\mathsf{CA}_{\alpha+\omega} = \mathfrak{Fr}_\delta\mathsf{K}.$$

In particular, the conclusions hold if we take Lf_α, Dc_α, *or* Ss_α *for the class* K.

PROOF. From 2.6.32(ii) we have

$$\mathsf{SUpLf}_\alpha \subseteq \mathsf{SUpK} \subseteq \mathsf{HSPK} \subseteq \mathsf{SNr}_\alpha\mathsf{CA}_{\alpha+\omega}.$$

Hence, in order to obtain the first formula of the conclusion it remains to show

$$(1) \qquad \mathsf{SNr}_\alpha\mathsf{CA}_{\alpha+\omega} \subseteq \mathsf{SUpLf}_\alpha.$$

1) Both counterexamples given in 2.6.51 are due to Monk. However, the fact that for every $\alpha \geqq \omega$ there are representable CA_α's which are not semisimple was first noticed and communicated to us by Jerzy Łoś; cf. footnote 1 on p. 328.

Let $\mathfrak{A} \in \mathbf{S}\mathsf{Nr}_\alpha\mathsf{CA}_{\alpha+\omega}$. By 2.6.35 we have $\mathbf{S}\mathsf{Nr}_\alpha\mathsf{CA}_{\alpha+\omega} = \mathbf{S}\mathsf{Nr}_\alpha\mathsf{CA}_{\alpha+\alpha}$; thus there is a $\mathfrak{B} \in \mathsf{CA}_{\alpha+\alpha}$ such that

(2) $$\mathfrak{A} \subseteq \mathfrak{Nr}_\alpha\mathfrak{B}.$$

Consider any finite sequence ρ without repeating terms and with $Rg\rho \subseteq \alpha$. Choose $\sigma \in {}^\alpha(\alpha+\alpha)$ such that σ is one-one,

(3) $$Rg\rho 1 \sigma \subseteq Id,$$

and

(4) $$\sigma^*(\alpha \sim Rg\rho) \subseteq (\alpha+\alpha) \sim \alpha.$$

Let $\mathfrak{C} = \mathfrak{Sg}^{(\mathfrak{Rd}^{(\sigma)}\mathfrak{B})}\mathfrak{A}$. Then by (2), (4), 2.1.5(ii), and 2.6.2(i) we have

(5) $$\mathfrak{C} \in \mathsf{Lf}_\alpha;$$

also, from (2), (3), 0.5.4(ii), and 0.5.5 we get $\mathfrak{Rd}^{(\rho)}\mathfrak{A} \subseteq \mathfrak{Rd}^{(\rho)}\mathfrak{B} = \mathfrak{Rd}^{(\rho)}\mathfrak{Rd}^{(\sigma)}\mathfrak{B}$, and hence

$$\mathfrak{Rd}^{(\rho)}\mathfrak{A} \subseteq \mathfrak{Rd}^{(\rho)}\mathfrak{C}.$$

Therefore, in view of (5) one obtains $\mathfrak{Rd}^{(\rho)}\mathfrak{A} \in \mathsf{Rd}^{(\rho)}\mathsf{Lf}_\alpha$. Since this holds for all finite sequences without repetitions and with range included in α, 2.6.4 implies that $\mathfrak{A} \in \mathbf{SUp}\,\mathsf{Lf}_\alpha$; consequently (1) holds.

The formula involving free algebras follows immediately from the first formula and 0.4.37(ii). The last part of the theorem is an easy consequence of the first part together with 1.11.3(i), 2.4.52, and 2.6.50.

REMARKS 2.6.53. By virtue of 0.3.70, 0.3.83, and 0.4.64, Theorem 2.6.52 implies that, for every class K including Lf_α and included in the class of all representable CA_α's, the latter is the least equational class and also the least universal class over K.[1] On the other hand, James S. Johnson has pointed out that for no $\alpha \geqq \omega$ is the class of representable CA_α's the least elementary class over Lf_α, or even over Dc_α (thus solving two problems stated in Monk [65]). This results from the following simple observation: the property that each atom of $\mathfrak{A}$ is zero-dimensional, which is an elementary property in view of 1.10.5(ii), holds for every $\mathsf{Dc}_\alpha\,\mathfrak{A}$ by 1.11.8(i), but obviously not for every representable CA_α. In this connection it would be interesting to find a simple intrinsic characterization of algebras in the least elementary class over Lf_α.

It may be noted that for $\alpha \geqq \omega$ the class K_α of all simple CA_α's also satisfies the conclusion (though not the hypothesis) of Theorem 2.6.52. In fact, as easy

[1]) The first of these results can essentially be found in Henkin-Tarski [61], Theorem 2.19, and the second in Monk [65], Theorem 1.11.

consequences of the definition of semisimplicity, we have the following formulas for every ordinal α (whether infinite or not) and every cardinal $\delta \neq 0$:

$$\mathbf{HSPK}_\alpha = \mathbf{HSs}_\alpha = \mathbf{HSPSs}_\alpha \text{ and } \mathfrak{Fr}_\delta \mathsf{K}_\alpha = \mathfrak{Fr}_\delta \mathsf{Ss}_\alpha.$$

On the other hand, taking $\mathsf{K} = \mathsf{CA}_\alpha$ for any $\alpha \geqq 2$ we see by 2.6.43 that the first formula of the conclusion of 2.6.52 is not satisfied, and hence, in view of 0.4.43 and 2.6.32(ii), that the second formula is not satisfied whenever $\delta \geqq \omega$. Actually, using 2.6.41 and 2.6.42 it is not difficult to show that, for every cardinal $\delta \neq 0$, $\mathfrak{Fr}_\delta \mathsf{Nr}_\alpha \mathsf{CA}_{\alpha+\omega} \neq \mathfrak{Fr}_\delta \mathsf{CA}_\alpha$; using in addition 2.6.50 we conclude for the same δ (and $\alpha \geqq \omega$) that $\mathfrak{Fr}_\delta \mathsf{CA}_\alpha \notin \mathsf{Ss}_\alpha$.

With regard to the next theorem recall the definition of characteristic given in 2.4.61.

THEOREM 2.6.54. *Assume that* $\mathfrak{A} \in \mathsf{CA}_\alpha$ *and* $\alpha \geqq \omega$. *Then each of the following conditions implies 2.6.50(ii) and hence also the formula* $\mathfrak{A} \in \mathbf{S}\mathsf{Nr}_\alpha \mathsf{CA}_{\alpha+\omega}$:

(i) *for some* $R \subseteq {}^2\alpha \sim Id$, $\prod_{\langle\kappa,\lambda\rangle \in R} -\mathsf{d}_{\kappa\lambda}$ *exists and equals* 0, *and* $|\alpha \sim FdR| \geqq \omega$;

(ii) $\mathfrak{A}$ *is of characteristic* $\kappa > 0$;

(iii) *for every* $\kappa < \alpha$ *and every* $x \in A$, $\mathsf{c}_\kappa x = \sum_{\lambda<\alpha} \mathsf{s}^\kappa_\lambda x$.

PROOF. Obviously (ii) implies (i). Now assume (i), and that the hypothesis of 2.6.50 (ii) holds, i.e., that Γ is a finite subset of α and x is a non-zero element of $\mathfrak{A}$. Clearly we may assume that Γ is non-empty. Let $\gamma, \delta \in {}^{|\Gamma|}\alpha$ such that γ, δ are one-one, $Rg\gamma = \Gamma$, and $Rg\delta \subseteq \alpha \sim (\Gamma \cup FdR)$. Let $f = \mathsf{s}^{\gamma_0}_{\delta_0} \circ \ldots \circ \mathsf{s}^{\gamma_\nu}_{\delta_\nu}$ where $\nu = |\Gamma|-1$. Then by (i) and 1.5.3 we have $\prod_{\langle\mu,\nu\rangle \in R} -f\mathsf{d}_{\mu\nu} = 0$ and hence

$$\sum_{\langle\mu,\nu\rangle \in R} f\mathsf{d}_{\mu\nu} = 1. \tag{1}$$

From 1.5.4(i) we see that, if $\langle\mu,\nu\rangle \in R$, then $f\mathsf{d}_{\mu\nu}$ has the form $\mathsf{d}_{\kappa\lambda}$ where $\kappa \neq \lambda$ and $\kappa, \lambda \in \alpha \sim \Gamma$. Thus from (1) we conclude that there exist $\kappa, \lambda \in \alpha \sim \Gamma$ such that $x \cdot \mathsf{d}_{\kappa\lambda} \neq 0$, i.e., that the conclusion of 2.6.50(ii) holds.

In turn, assume that (iii) holds. To derive 2.6.50(ii) we argue by contradiction, i.e., we assume there is a finite $\Gamma \subseteq \alpha$ and $x \in A$ such that $x \neq 0$ and

$$x \cdot \mathsf{d}_{\kappa\lambda} = 0 \text{ for all } \kappa, \lambda \in \alpha \sim \Gamma, \kappa \neq \lambda. \tag{2}$$

We now claim that there is a $\kappa \in \alpha \sim \Gamma$ such that

$$x \cdot \prod_{\lambda \in \Gamma} -\mathsf{d}_{\kappa\lambda} \neq 0. \tag{3}$$

Indeed, suppose otherwise. Then for each $\mu \in \alpha \sim \Gamma$ we have $x = \sum_{\lambda \in \Gamma}(x \cdot \mathsf{d}_{\mu\lambda})$.

Let $\Delta \subseteq \alpha \sim \Gamma$ with $|\Delta| = |\Gamma|+1$. Then

$$x = \prod_{\mu\in\Delta}\sum_{\lambda\in\Gamma}(x\cdot \mathsf{d}_{\mu\lambda}) = \sum_{\gamma\in{}^{\Delta}\Gamma}\prod_{\mu\in\Delta}(x\cdot \mathsf{d}_{\mu,\gamma\mu}).$$

Thus there is a $\gamma \in {}^{\Delta}\Gamma$ such that $\prod_{\mu\in\Delta}(x\cdot \mathsf{d}_{\mu,\gamma\mu}) \neq 0$. Since $|\Delta| > |\Gamma|$, there exist distinct $\mu, \nu \in \Delta$ for which $\gamma\mu = \gamma\nu$. We then have $x\cdot \mathsf{d}_{\mu,\gamma\mu}\cdot \mathsf{d}_{\nu,\gamma\nu} \neq 0$ and hence, in view of 1.3.7, $x\cdot \mathsf{d}_{\mu\nu} \neq 0$ which contradicts (2). Thus we have established that (3) holds for some $\kappa \in \alpha \sim \Gamma$. Using (iii) we also have

$$(4) \qquad \mathsf{c}_\kappa(x\cdot\prod_{\lambda\in\Gamma}-\mathsf{d}_{\kappa\lambda}) = \sum_{\mu<\alpha}\mathsf{s}^{\kappa}_{\mu}(x\cdot\prod_{\lambda\in\Gamma}-\mathsf{d}_{\kappa\lambda})$$
$$= \sum_{\mu\in\alpha\sim\Gamma}\mathsf{s}^{\kappa}_{\mu}(x\cdot\prod_{\lambda\in\Gamma}-\mathsf{d}_{\kappa\lambda}).$$

The second equality of (4) follows from the fact that, for each $\mu \in \Gamma$,

$$\mathsf{s}^{\kappa}_{\mu}(x\cdot\prod_{\lambda\in\Gamma}-\mathsf{d}_{\kappa\lambda}) = \mathsf{s}^{\kappa}_{\mu}x\cdot\prod_{\lambda\in\Gamma}-\mathsf{d}_{\mu\lambda} = 0,$$

by 1.5.4(i). If $\mu \in \alpha \sim \Gamma$ and $\mu \neq \kappa$, then $\mathsf{s}^{\kappa}_{\mu}x = \mathsf{c}_\kappa(\mathsf{d}_{\kappa\mu}\cdot x) = 0$ by (2). Thus we infer from (4) that

$$(5) \qquad \mathsf{c}_\kappa(x\cdot\prod_{\lambda\in\Gamma}-\mathsf{d}_{\kappa\lambda}) = x\cdot\prod_{\lambda\in\Gamma}-\mathsf{d}_{\kappa\lambda}.$$

Finally, for any $\mu \in \alpha \sim (\Gamma \cup \{\kappa\})$ we have

$$x\cdot\prod_{\lambda\in\Gamma}-\mathsf{d}_{\kappa\lambda} = \mathsf{c}_\kappa\mathsf{d}_{\kappa\mu}\cdot x\cdot\prod_{\lambda\in\Gamma}-\mathsf{d}_{\kappa\lambda} \qquad \text{by 1.3.2}$$
$$= \mathsf{c}_\kappa(\mathsf{d}_{\kappa\mu}\cdot x\cdot\prod_{\lambda\in\Gamma}-\mathsf{d}_{\kappa\lambda}) \qquad \text{by (5)}$$
$$= 0 \qquad \text{by (2).}$$

However, this is impossible because of (3). Thus the assumption that 2.6.50(ii) fails had led us to a contradiction. This finishes the proof.

REMARKS 2.6.55. It will be shown in Part II that all finite-dimensional CA's of positive characteristic are representable and hence have the neat embedding property. Thus condition 2.6.54(ii) implies $\mathfrak{A} \in \mathsf{SNr}_\alpha\mathsf{CA}_{\alpha+\omega}$, for all α. The conjecture that all infinite dimensional CA's satisfying (iii) are representable, which is confirmed by 2.6.54, was communicated to the authors by Jan Mycielski. Theorem 2.6.54 is due to Monk.

Only very few interesting classes of finite-dimensional CA's are known which consist exclusively of representable algebras, hence of algebras with the neat embedding property. As we have just mentioned the class of finite-dimensional CA's of positive characteristic is of this kind. Other examples are the class of all CA_1's (cf. 2.6.39) and the class of all CA_2's satisfying identically equations (1) and (2) of 2.6.44. Following Monk [64a] we shall now extend the result concerning CA_1's to arbitrary monadic-generated CA's; this will be another application of the method of eliminating cylindrifications (cf. 2.2.24).

THEOREM 2.6.56. *If* $\mathfrak{A}$ *is a monadic-generated* CA_α, *then* $\mathfrak{A} \in \mathbf{SNr}_\alpha\mathsf{CA}_{\alpha+\omega}$.

PROOF. By 2.5.60 there exists a non-zero cardinal $\delta \leqq |A|$ and a $\Delta \in {}^\delta Sb\alpha$ with $|\Delta\xi| \leqq 1$ for each $\xi < \delta$ such that $\mathfrak{Fr}_\delta^{(\Delta)}\mathsf{CA}_\alpha \succcurlyeq \mathfrak{A}$. Thus by 2.6.32 it suffices to prove

(1) $$\mathfrak{Fr}_\delta^{(\Delta)}\mathsf{CA}_\alpha \in \mathbf{SNr}_\alpha\mathsf{CA}_{\alpha+\omega}.$$

Set
$$\mathfrak{B} = \mathfrak{Fr}_\delta^{(\Delta)}\mathsf{CA}_\alpha \text{ and } \mathfrak{B}' = \mathfrak{Fr}_\delta^{(\Delta)}\mathsf{CA}_{\alpha+\omega}.$$

Let Λ be any finite subset of δ, and for each $\Theta \subseteq \Lambda$ and $\kappa < (\alpha+1)\cap\omega$ let $a_\kappa\Theta$ be the element of $\mathfrak{B}$ defined in 2.5.64 with δ in place of β; let $a'_\kappa\Theta$ be the similarly defined element of $\mathfrak{B}'$ with δ and $\alpha+\omega$ in place of β and α, respectively.

By 2.5.33(ii), 2.5.36 and 2.6.30(i) there is a homomorphism h from $\mathfrak{B}$ into $\mathfrak{Nr}_\alpha\mathfrak{B}'$ such that $h(\xi/Cr_\delta^{(\Delta)}\mathsf{CA}_\alpha) = \xi/Cr_\delta^{(\Delta)}\mathsf{CA}_{\alpha+\omega}$ for each $\xi < \delta$. It is easily seen that

(2) $$h(a_\kappa\Theta) = a'_\kappa\Theta \text{ for all } \Theta \subseteq \Lambda \text{ and } \kappa < (\alpha+1)\cap\omega.$$

Consider any $x \in Zd\,\mathfrak{B} \sim \{0^{(\mathfrak{B})}\}$. Then by 2.2.20 and 2.2.24(iii) there is a finite $\Lambda \subseteq \delta$ and a $\tau \in {}^{Sb\Lambda}((\alpha+1)\cap\omega)$ such that
$$\prod_{\Theta\subseteq\Lambda}(a_{\tau\Theta}\Theta - a_{\tau\Theta+1}\Theta) \leqq x$$
(compare the proof of 2.5.65). Then from (2) we obtain
$$\prod_{\Theta\subseteq\Lambda}(a'_{\tau\Theta}\Theta - a'_{\tau\Theta+1}\Theta) \leqq hx,$$
and hence $hx \neq 0$ by 2.5.64. This shows that $(h^{-1})^\star 0^{(\mathfrak{B}')} \cap Zd\,\mathfrak{B} = \{0^{(\mathfrak{B})}\}$ and thus, by 2.2.17(ii) and 2.3.12(iii), that $(h^{-1})^\star 0^{(\mathfrak{B}')} = \{0^{(\mathfrak{B})}\}$. Therefore, by 0.2.23(i) we conclude that $\mathfrak{B} \cong|\subseteq \mathfrak{Nr}_\alpha\mathfrak{B}'$ and hence that (1) holds. This completes the proof.

COROLLARY 2.6.57. $\mathsf{Mn}_\alpha \subseteq \mathbf{SNr}_\alpha\mathsf{CA}_{\alpha+\omega}$.

In Theorems 2.6.58 and 2.6.59 we discuss the relation between arbitrary subreducts and neat subreducts; in Corollary 2.6.61 we give an application of these results to a problem on arbitrary subreducts in whose formulation neat subreducts are not involved at all. 2.6.61 improves a result of Monk. Both 2.6.59 and 2.6.61 are due to Don Pigozzi.

THEOREM 2.6.58. *If* $2 \leqq \alpha < \beta, \gamma$, *then* $\mathbf{SRd}_\alpha\mathsf{CA}_\beta \sim \mathbf{SNr}_\alpha\mathsf{CA}_\gamma \neq 0$.

PROOF. By 2.6.42 there is an $\mathfrak{A} \in \mathsf{CA}_\beta$ in which the equation of 2.6.41 fails. Clearly it fails also in $\mathfrak{Rd}_\alpha\mathfrak{A}$ and hence, by 2.6.31 and 2.6.41,

$$\mathfrak{Rd}_\alpha\mathfrak{A} \in \mathsf{SRd}_\alpha\mathsf{CA}_\beta \sim \mathsf{SNr}_\alpha\mathsf{CA}_\gamma.$$

THEOREM 2.6.59. *Let* $\mathfrak{A}$ *be a simple* CA_α *and let* $\beta \geqq \alpha$. *If* $\mathfrak{A} \in \mathsf{SRd}_\alpha\mathsf{CA}_{\beta+1}$, *then* $\mathfrak{A} \in \mathsf{SNr}_\alpha\mathsf{CA}_\beta$.

PROOF. Consider first the case $\beta = \alpha+\kappa$ for some $\kappa < \omega$. The desired result follows immediately from 2.2.6(i) and 2.6.30(ii) if $\kappa = 0$, so we assume $\kappa > 0$. Let $\mathfrak{A} \subseteq \mathfrak{Rd}_\alpha\mathfrak{B}$ where $\mathfrak{B} \in \mathsf{CA}_{\alpha+\kappa+1}$, and let

$$h = \mathsf{s}^{\alpha}_{\alpha+\kappa} \circ \mathsf{s}^{\alpha+1}_{\alpha+\kappa} \circ \ldots \circ \mathsf{s}^{\alpha+\kappa-1}_{\alpha+\kappa}.$$

Then by 1.5.3, 1.5.4(ii), and 1.5.8(ii) we see that h is an endomorphism of $\mathfrak{Rd}_\alpha\mathfrak{B}$, and hence $h \in Hom(\mathfrak{A}, \mathfrak{Rd}_\alpha\mathfrak{B})$ by 0.2.3. Clearly h is not a constant function; thus, by the premiss that $\mathfrak{A}$ is simple and by 0.2.36,

$$h \in Ism(\mathfrak{A}, \mathfrak{Rd}_\alpha\mathfrak{B}). \tag{1}$$

By 0.5.5 we have $\mathfrak{Rd}_\alpha\mathfrak{B} = \mathfrak{Rd}_\alpha\mathfrak{Rd}_{\alpha+\kappa}\mathfrak{B}$, and from 1.5.9(ii) we easily conclude that $h^*\mathfrak{A} \subseteq \mathfrak{Nr}_\alpha\mathfrak{Rd}_{\alpha+\kappa}\mathfrak{B}$. Therefore, (1) gives us $\mathfrak{A} \cong|\subseteq \mathfrak{Nr}_\alpha\mathfrak{Rd}_{\alpha+\kappa}\mathfrak{B}$, and hence $\mathfrak{A} \in \mathsf{SNr}_\alpha\mathsf{CA}_{\alpha+\kappa}$ by 2.6.2(i).

Now consider the case $\beta \geqq \alpha+\omega$. Then by 2.6.2, 2.6.7, and 2.6.10 we have $\mathfrak{B} \in \bigcap_{\kappa<\omega}\mathsf{SRd}_\alpha\mathsf{CA}_{\alpha+\kappa+1}$, and hence $\mathfrak{B} \in \bigcap_{\kappa<\omega}\mathsf{SNr}_\alpha\mathsf{CA}_{\alpha+\kappa}$ by the first part of the proof. Then $\mathfrak{B} \in \mathsf{SNr}_\alpha\mathsf{CA}_\beta$ by 2.6.34 and 2.6.35. This finishes the proof.

REMARK 2.6.60. Notice that 2.6.59 has been proved without the help of 2.6.50. On the other hand, from 2.6.59, with the help of 2.6.9 and 2.6.32(ii), we easily derive the inclusion

$$\mathsf{Ss}_\alpha \subseteq \mathsf{SNr}_\alpha\mathsf{CA}_{\alpha+\omega} \text{ (for } \alpha \geqq \omega\text{)},$$

which is an essential part of 2.6.50.

COROLLARY 2.6.61. *If* $2 \leqq \kappa < \omega$ *and* $\kappa+2 \leqq \beta$, *then* $\mathsf{HSRd}_\kappa\mathsf{CA}_\beta \sim \mathsf{SRd}_\kappa\mathsf{CA}_\beta \neq 0$.

PROOF. Let $\gamma = \beta-1$ if β is finite and let $\gamma = \beta$ otherwise. To derive the conclusion, suppose, on the contrary, that $\mathsf{HSRd}_\kappa\mathsf{CA}_\beta \subseteq \mathsf{SRd}_\kappa\mathsf{CA}_\beta$. Then, by 0.3.40, 2.4.52, and 2.6.2(i), every $\mathfrak{A} \in \mathsf{SRd}_\kappa\mathsf{CA}_\beta$ is a subdirect product of simple algebras in $\mathsf{SRd}_\kappa\mathsf{CA}_\beta$. Consequently, employing 2.6.32(ii) and 2.6.59 (along with 2.6.10 in case $\beta \geqq \omega$) we obtain $\mathfrak{A} \in \mathsf{SNr}_\alpha\mathsf{CA}_\gamma$. We have shown that

$$\mathsf{SRd}_\kappa\mathsf{CA}_\beta \subseteq \mathsf{SNr}_\kappa\mathsf{CA}_\gamma;$$

however, since $2 \leqq \kappa < \beta, \gamma$ by hypothesis, this is impossible by 2.6.58.

We do not know whether the conclusion of 2.6.61 holds in case $\beta = \kappa+1$.

In the last portion of this section we discuss special kinds of neat subreducts introduced in the following

DEFINITION 2.6.62. *Given any* CA's $\mathfrak{A}$ *and* $\mathfrak{B}$, *we call* $\mathfrak{A}$ *a* ***generating neat subreduct*** *or, more briefly, a* ***generating subreduct*** *of* $\mathfrak{B}$, *if* $\mathfrak{A}$ *is a neat subreduct of* $\mathfrak{B}$ *and* $\mathfrak{B} = \mathfrak{Sg}^{(\mathfrak{B})}A$.[1]

THEOREM 2.6.63. *If a* CA $\mathfrak{A}$ *is a neat subreduct of a* CA $\mathfrak{B}$, *then it is a generating subreduct of* $\mathfrak{Sg}^{(\mathfrak{B})}A$.

PROOF: Obvious.

THEOREM 2.6.64. *Assume that a* CA $\mathfrak{A}$ *is a generating subreduct of a* CA $\mathfrak{B}$. *If* $\mathfrak{A}$ *is locally finite or dimension-complemented, then the same applies to* $\mathfrak{B}$.

PROOF. Because of 1.11.3, in proving (ii) we need only consider infinite-dimensional $\mathfrak{B}$. Assume first of all that $\mathfrak{A} \in \mathsf{Dc}_\alpha$ and $\mathfrak{B} \in \mathsf{CA}_\beta$. Consider any finite subset X of A and let $\Gamma = \beta \sim \bigcup_{x\in X} \Delta x$; note that

(1) $$|\Gamma| \geqq \omega,$$

by 1.11.4. From 2.1.4 we infer that $|\Delta y \cap \Gamma| < \omega$ for each $y \in Sg^{(\mathfrak{B})}X$, and hence, in view of (1), $\Delta y \neq \beta$ for each $y \in Sg^{(\mathfrak{B})}X$. Therefore, since $Sg^{(\mathfrak{B})}A = \bigcup\{Sg^{(\mathfrak{B})}X\colon X \subseteq A,\ |X| < \omega\}$ by 0.1.17(ix), we conclude that $\mathfrak{Sg}^{(\mathfrak{B})}A \in \mathsf{Dc}_\beta$. In view of 2.6.62, this proves (ii) in case $\mathfrak{A}$ is a Dc. If $\mathfrak{A}$ is an Lf, the desired result follows directly from 2.1.5(i).

LEMMA 2.6.65. *Assume* $\beta \geqq \alpha \geqq \omega$, $\mathfrak{A} \in \mathsf{Dc}_\alpha$, *and* $\mathfrak{A}$ *is a generating subreduct of a* Dc_β $\mathfrak{B}$. *If* σ *is a finite transformation of* β *such that* $\sigma^*\alpha \subseteq \alpha$, *then*

$$A \upharpoonright \mathsf{s}_\sigma^{(\mathfrak{B})} = \mathsf{s}_{\alpha\upharpoonright\sigma}^{(\mathfrak{A})},$$

and hence

$$(\mathsf{s}_\sigma^{(\mathfrak{B})})^*A \subseteq A.$$

PROOF. Because of 1.11.12(v) we may assume that $(\beta \sim \alpha)\upharpoonright\sigma \subseteq Id$. If $\sigma = [\kappa/\lambda]$ or $\sigma = [\kappa/\lambda, \lambda/\kappa]$, with $\kappa, \lambda < \alpha$, the conclusion follows from 1.11.11(i) or 1.11.11(ii); thus it holds in general by 1.11.11(iv) and the fact that σ is a composition of transpositions and replacements involving only elements of α (cf. the Preliminaries).

LEMMA 2.6.66. *Under the assumptions of 2.6.65, B coincides with the set*

[1]) For this notion (under a different name) and the essential ideas of the proofs of 2.6.63–2.6.74 see Daigneault-Monk [63] (where some further historical references can also be found).

of all elements $\mathsf{s}_\sigma^{(\mathfrak{B})}x$ *where* $x \in A$ *and* σ *is a finite transformation of* β *such that* $\alpha 1 \sigma$ *is one-one.*

PROOF. Let C be the set of all elements $\mathsf{s}_\sigma^{(\mathfrak{B})}x$ where $x \in A$ and σ is a finite transformation of β such that $\alpha 1 \sigma$ is one-one. To prove the theorem it suffices to show that

$$(1) \qquad A \subseteq C \text{ and } C \in Su\mathfrak{B}.$$

Since $\mathsf{s}_{\beta 1 Id}x = x$ for all $x \in A$ by 1.11.12(i), it is clear that $A \subseteq C$. If $y \in C$, then $-y \in C$ by 1.11.11(iii). Now suppose $y_0, y_1 \in C$, say

$$(2) \qquad y_0 = \mathsf{s}_{\sigma_0}x_0 \text{ and } y_1 = \mathsf{s}_{\sigma_1}x_1,$$

where $x_0, x_1 \in A$ and σ_0 and σ_1 are finite transformations of β which are one-one on α. Set

$$\Gamma = (\sigma_0^*\alpha \cup \sigma_1^*\alpha) \sim \alpha;$$

notice that Γ is finite since σ_0 and σ_1 are finite transformations. Choose $\tau \in {}^\beta\beta$ such that

$$(\beta \sim \Gamma) 1 \tau \subseteq Id \text{ and } \tau^*\Gamma \subseteq \alpha \sim (\sigma_0^*\Delta x_0 \cup \sigma_1^*\Delta x_1)$$

(we use here 1.11.4). Furthermore, we can take τ to be one-one on Γ, so that there is a $\pi \in {}^\beta\beta$ for which

$$(\beta \sim \tau^*\Gamma) 1 \pi \subseteq Id \text{ and } \Gamma 1 \pi \circ \tau \subseteq Id.$$

It is then easily seen that

$$(3) \qquad (\tau \circ \sigma_\kappa)^*\alpha \subseteq \alpha \text{ for } \kappa = 0, 1,$$

$$(4) \qquad \pi \text{ is one-one on } \alpha,$$

and

$$(5) \qquad \Delta x_\kappa 1 \sigma_\kappa \subseteq \pi \circ \tau \circ \sigma_\kappa \text{ for } \kappa = 0, 1.$$

Therefore,

$$\begin{aligned} y_0 + y_1 &= \mathsf{s}_{\sigma_0}x_0 + \mathsf{s}_{\sigma_1}x_1 \\ &= \mathsf{s}_{\pi\circ\tau\circ\sigma_0}x_0 + \mathsf{s}_{\pi\circ\tau\circ\sigma_1}x_1 && \text{by (5), 1.11.12(v)} \\ &= \mathsf{s}_\pi(\mathsf{s}_{\tau\circ\sigma_0}x_0 + \mathsf{s}_{\tau\circ\sigma_1}x_1) && \text{by 1.11.11(iii), (iv).} \end{aligned}$$

However, $\mathsf{s}_{\tau\circ\sigma_0}x_0 + \mathsf{s}_{\tau\circ\sigma_1}x_1 \in A$ by (3) and 2.6.65; hence $y_0 + y_1 \in C$ by (4).

If $\kappa, \lambda < \beta$, there clearly exist $\mu, \nu < \alpha$ and a finite transformation σ of β such that $\alpha 1 \sigma$ is one-one and $\sigma\mu = \kappa$, $\sigma\nu = \lambda$. Hence $\mathsf{d}_{\kappa\lambda} = \mathsf{s}_\sigma \mathsf{d}_{\mu\nu} \in C$. Finally, suppose $y \in C$ and $\kappa < \beta$; we will show that $\mathsf{c}_\kappa y \in C$. Let $y = \mathsf{s}_\sigma x$ where $x \in A$, σ is a finite transformation of β, and $\alpha 1 \sigma$ is one-one. Let τ be a finite transformation of β such that $\alpha 1 \sigma = \alpha 1 \tau$ and either $(\tau^{-1})^\bigstar \kappa = 0$ or $(\tau^{-1})^\bigstar \kappa = \{\lambda\}$ for some $\lambda < \alpha$. Then $y = \mathsf{s}_\sigma x = \mathsf{s}_\tau x$ by 1.11.12(v). Therefore, by 1.11.12(vi), either $\mathsf{c}_\kappa y = \mathsf{c}_\kappa \mathsf{s}_\tau x = \mathsf{s}_\tau x$ or $\mathsf{c}_\kappa y = \mathsf{c}_\kappa \mathsf{s}_\tau x = \mathsf{s}_\tau \mathsf{c}_\lambda x$; in either case we have $\mathsf{c}_\kappa y \in C$. Hence the second formula of (1) holds, and this completes the proof.

THEOREM 2.6.67. *Assume that $\mathfrak{A} \in \mathsf{CA}_\alpha$, $\mathfrak{B} \in \mathsf{CA}_\alpha$, and $\mathfrak{A}$ is a generating subreduct of $\mathfrak{B}$. Then we have:*

(i) *$\mathfrak{Nr}_\alpha \mathfrak{B}$ is also a generating subreduct of $\mathfrak{B}$, and the same applies to every CA_α $\mathfrak{C}$ such that $\mathfrak{A} \subseteq \mathfrak{C} \subseteq \mathfrak{Nr}_\alpha \mathfrak{B}$.*

(ii) *If, in addition, $\alpha \geqq \omega$ and $\mathfrak{A} \in \mathsf{Dc}_\alpha$, then $\mathfrak{A} = \mathfrak{Nr}_\alpha \mathfrak{B}$, and hence $\mathfrak{Nr}_\alpha \mathfrak{B}$ is then the unique α-dimensional generating subreduct of $\mathfrak{B}$.*

PROOF. (i) obviously follows from 2.6.62. To prove (ii), notice that $\mathfrak{B} \in \mathsf{Dc}_\beta$ by 2.6.64, and hence $\mathfrak{B}$ admits the substitution operator $\mathsf{s}^{(\mathfrak{B})}$.

We have to show that $\mathfrak{Nr}_\alpha \mathfrak{B} \subseteq \mathfrak{A}$. Suppose then that $x \in B$ and

(1) $$\Delta x \subseteq \alpha.$$

By 2.6.66 there exists a $y \in A$ and a finite transformation σ of β such that $\alpha 1 \sigma$ is one-one and

(2) $$x = \mathsf{s}_\sigma^{(\mathfrak{B})} y.$$

Let τ be a finite transformation of β such that

(3) $$\alpha 1 \tau = Id \text{ and } (\tau \circ \sigma)^* \alpha \subseteq \alpha.$$

Then we have

$$\begin{aligned} x &= \mathsf{s}_\tau^{(\mathfrak{B})} x && \text{by (1), (3), 1.11.12(i),(v)} \\ &= \mathsf{s}_\tau^{(\mathfrak{B})} \mathsf{s}_\sigma^{(\mathfrak{B})} y && \text{by (2)} \\ &= \mathsf{s}_{\tau\circ\sigma}^{(\mathfrak{B})} y && \text{by 1.11.11(iv)} \\ &= \mathsf{s}_{\alpha 1 (\tau\circ\sigma)}^{(\mathfrak{A})} y && \text{by (2), 2.6.65.} \end{aligned}$$

Thus $x \in A$ and the proof is complete.

COROLLARY 2.6.68. *Under the assumptions of 2.6.67(ii), $\mathfrak{Zd}\,\mathfrak{A} = \mathfrak{Zd}\,\mathfrak{B}$.*

PROOF: by 2.6.30(iii), 2.6.31, and 2.6.67(ii).

REMARK 2.6.69. Theorem 2.6.67 does not imply that $\mathfrak{Nr}_\alpha \mathfrak{B}$ is actually a generating subreduct of $\mathfrak{B}$, even under the assumptions: $\beta \geqq \alpha \geqq \omega$ and $\mathfrak{Nr}_\alpha \mathfrak{B}$ is dimension-complemented. For example, if $\beta \sim \alpha$ is infinite, and

$\mathfrak{B} = \mathfrak{Fr}_\gamma \mathsf{CA}_\beta$ where γ is any cardinal $\neq 0$, then $\mathfrak{Nr}_\alpha \mathfrak{B}$ is locally finite by 2.6.23; hence it is clearly not a generating subreduct of $\mathfrak{B}$.

It will be shown in Part II that for each α, β such that $\beta \geqq \alpha \geqq \omega$ there is a CA_α $\mathfrak{A}$ and a CA_β $\mathfrak{B}$ such that $\mathfrak{A}$ is a generating subreduct of $\mathfrak{B}$ different from $\mathfrak{Nr}_\alpha \mathfrak{B}$; in fact, both $\mathfrak{A}$ and $\mathfrak{B}$ can be taken to be representable. Thus Dc_α cannot simply be replaced by CA_α in Theorem 2.6.67(ii); it is known that this replacement also cannot be made in certain consequences of 2.6.67, namely 2.6.71 and 2.6.72.

THEOREM 2.6.70. *If $\beta \geqq \alpha \geqq \omega$ and $\mathfrak{B} \in \mathsf{Lf}_\beta$, then $\mathfrak{Nr}_\alpha \mathfrak{B}$ is the unique α-dimensional generating subreduct of $\mathfrak{B}$.*

PROOF. In view of 1.11.3(i), 2.1.3(ii), 2.6.30(i), and 2.6.67 we see that $\mathfrak{B}$ can have at most one generating subreduct of dimension α. Therefore we need only show that $\mathfrak{Nr}_\alpha \mathfrak{B}$ is such a subreduct, i.e., that

(1) $$\mathfrak{B} = \mathfrak{Sg}^{(\mathfrak{B})} Nr_\alpha \mathfrak{B}.$$

Let $x \in B$ and let σ be a finite transformation of β such that $\Delta x 1 \sigma$ is one-one and $\sigma^* \Delta x \subseteq \alpha$; then, by 1.11.12(x), $\Delta \mathsf{s}_\sigma x \subseteq \alpha$, and hence

(2) $$\mathsf{s}_\sigma x \in Nr_\alpha \mathfrak{B}.$$

Now let τ be a finite transformation of β such that $\Delta x 1 (\tau \circ \sigma) \subseteq Id$. Then by 1.11.11(iv), 1.11.12(i),(v) we have

(3) $$x = \mathsf{s}_{\tau \circ \sigma} x = \mathsf{s}_\tau \mathsf{s}_\sigma x.$$

It is clear from the definition of the substitution operator that any subuniverse of $\mathfrak{B}$ is closed under s_τ. Hence by (2) and (3) we have $x \in Sg^{(\mathfrak{B})} Nr_\alpha \mathfrak{B}$. Since x is an arbitrary element of $\mathfrak{B}$, (1) holds; this completes the proof.

THEOREM 2.6.71. *Assume $\alpha \geqq \omega$, $\mathfrak{A} \in \mathsf{Dc}_\alpha$, $\mathfrak{A}$ is a generating subreduct of a* CA $\mathfrak{B}$, *and $I \in Il\, \mathfrak{B}$. Then $I = Ig^{(\mathfrak{B})}(I \cap A)$.*

PROOF. From the premiss that $\mathfrak{A}$ is a generating subreduct of $\mathfrak{B}$ we infer by 2.1.5 that

(1) $$|\Delta x \sim \alpha| < \omega \text{ for every } x \in B,$$

and from the same premiss and the fact that $\mathfrak{A} \in \mathsf{Dc}_\alpha$ we infer by 2.6.67 that

(2) $$\mathfrak{A} = \mathfrak{Nr}_\alpha \mathfrak{B}.$$

Now consider any $x \in I$. Then by (1) and (2) we see that $\mathsf{c}_{(\Delta x \sim \alpha)} x$ exists and is contained in A and hence in $I \cap A$. Since $x \leqq \mathsf{c}_{(\Delta x \sim \alpha)} x$, we conclude that

$x \in Ig^{(\mathfrak{B})}(I \cap A)$. This shows that $I \subseteq Ig^{(\mathfrak{B})}(I \cap A)$. The inclusion in the opposite direction is obvious.

THEOREM 2.6.72. *Assume* $\beta \geqq \alpha \geqq \omega$, $\mathfrak{A} \in \mathsf{Dc}_\alpha$, $\mathfrak{B}, \mathfrak{B}' \in \mathsf{CA}_\beta$, *and* $\mathfrak{A}$ *is a generating subreduct of both* $\mathfrak{B}$ *and* $\mathfrak{B}'$. *Then there is an* $h \in Is(\mathfrak{B}, \mathfrak{B}')$ *such that* $A \upharpoonright h = A \upharpoonright Id$.

PROOF. Let x be any sequence such that

(1) $$Rgx = A$$

and the domain of x is a cardinal γ. Let

(2) $$\Delta = \langle \Delta^{(\mathfrak{A})} x_\xi : \xi < \gamma \rangle,$$

and set $\mathfrak{D} = \mathfrak{Fr}_\gamma^{(\Delta)} \mathsf{CA}_\beta$ and $g_\xi = \xi / Cr_\gamma^{(\Delta)} \mathsf{CA}_\beta$ for every $\xi < \gamma$; finally, set

(3) $$\mathfrak{C} = \mathfrak{Sg}^{(\mathfrak{Rd}_\alpha \mathfrak{D})}\{g_\xi : \xi < \gamma\}.$$

Clearly,

(4) $$\mathfrak{C} \text{ is a generating subreduct of } \mathfrak{D}.$$

Consider any finite subset Θ of γ and let $\Gamma = \alpha \sim \bigcup_{\xi \in \Theta} \Delta^{(\mathfrak{Rd}_\alpha \mathfrak{D})} g_\xi$. Then by (2), 1.11.4, 2.5.33(ii), and the premiss $\mathfrak{A} \in \mathsf{Dc}_\alpha$ we have

(5) $$|\Gamma| = |\alpha \sim \bigcup_{\xi \in \Theta} \Delta^{(\mathfrak{A})} x_\xi| \geqq \omega.$$

From 2.1.4 we infer that $|\Delta y \cap \Gamma| < \omega$ for each $y \in Sg^{(\mathfrak{Rd}_\alpha \mathfrak{D})}\{g_\xi : \xi \in \Theta\}$. Therefore, since $C = \bigcup\{Sg^{(\mathfrak{Rd}_\alpha \mathfrak{D})}\{g_\xi : \xi \in \Theta\} : \Theta \subseteq \gamma, |\Theta| < \omega\}$ by (3) and 0.1.17(ix), we conclude that

(6) $$\mathfrak{C} \in \mathsf{Dc}_\alpha.$$

By 2.5.36 there exist $f \in Hom(\mathfrak{D}, \mathfrak{B})$ and $f' \in Hom(\mathfrak{D}, \mathfrak{B}')$ such that $fg_\xi = f'g_\xi = x_\xi$ for every $\xi < \gamma$. Hence in view of (1) we have

(7) $$C \upharpoonright f = C \upharpoonright f'.$$

By hypothesis A generates both $\mathfrak{B}$ and $\mathfrak{B}'$; thus 0.2.18(i) implies that $f^*D = B$ and $f'^*D = B'$. From (7) we have

$$(f^{-1})^\star 0^{(\mathfrak{B})} \cap C = (f'^{-1})^\star 0^{(\mathfrak{B}')} \cap C.$$

Thus, applying (4), (6), and 2.6.71, we get

$$(f^{-1})^\star 0^{(\mathfrak{B})} = (f'^{-1})^\star 0^{(\mathfrak{B}')}.$$

Therefore we easily see from (7) and 0.2.23(i) that

$$h = \{\langle fx, f'x\rangle : x \in D\}$$

is the desired isomorphism.

REMARKS 2.6.73. Theorem 2.6.71 has still other interesting consequences. It is not hard to see that, if $\alpha \geqq \omega$, $\mathfrak{A} \in \mathsf{Dc}_\alpha$, and $\mathfrak{A}$ is a generating subreduct of $\mathsf{CA}\,\mathfrak{B}$, then the function $F = \langle I \cap A : I \in Il\mathfrak{B}\rangle$ is an isomorphism from the lattice of all ideals of $\mathfrak{B}$ onto the lattice of all ideals of $\mathfrak{A}$. More generally, one can show that, with obvious changes, all four parts of Theorem 2.3.12 remain valid when $\mathfrak{Zd}\,\mathfrak{A}$, $\mathfrak{A}$, and Δx are replaced respectively by $\mathfrak{A}$, $\mathfrak{B}$, and $\Delta x \sim \alpha$. Finally, we note that the above remarks apply whenever $\mathfrak{A} = \mathfrak{Nr}_\alpha\mathfrak{B}$ and $\mathfrak{A}$ is a generating subreduct of $\mathfrak{B}$, regardless of whether $\mathfrak{A}$ is a Dc_α; this is obvious from the proof of 2.6.71. These observations are due to Don Pigozzi.

In the last theorem of this section we give an application of the notion of generating subreducts to problems not involving this notion.

THEOREM 2.6.74. *If $\beta \geqq \alpha \geqq \omega$, then*
(i) $\mathsf{Dc}_\alpha \subseteq \mathsf{Nr}_\alpha\mathsf{Dc}_\beta$;
(ii) $\mathsf{Lf}_\alpha = \mathsf{Nr}_\alpha\mathsf{Lf}_\beta = \mathbf{S}\mathsf{Nr}_\alpha\mathsf{Lf}_\beta$.

PROOF. If $\mathfrak{A} \in \mathsf{Dc}_\alpha$, we have by 2.6.34, 2.6.35, and 2.6.49 that $\mathfrak{A} \in \mathbf{S}\mathsf{Nr}_\alpha\mathsf{CA}_\beta$; from 2.6.63 we further deduce that $\mathfrak{A} \in \mathbf{S}\mathsf{Nr}_\alpha\mathsf{Dc}_\beta$, and then an application of 2.6.67 shows that $\mathfrak{A} \in \mathsf{Nr}_\alpha\mathsf{Dc}_\beta$. This proves (i); (ii) is proven similarly, noting that $\mathbf{S}\mathsf{Nr}_\alpha\mathsf{Lf}_\beta \subseteq \mathsf{Lf}_\alpha$ by 2.6.30(i).

In view of 1.11.4 the inclusion 2.6.74(i) can be replaced by equality under the additional assumption $\beta < \alpha+\omega$. If, however, $\beta \geqq \alpha+\omega$, we can easily construct an $\mathfrak{A} \in \mathsf{Nr}_\alpha\mathsf{Dc}_\beta \sim \mathsf{Dc}_\alpha$.

2.7. CANONICAL EMBEDDING ALGEBRAS AND ATOM STRUCTURES

It is well known that the representation theorem and the embedding theorem are among the most basic results of the theory of Boolean algebras. By the first of these theorems every BA is representable in the sense that it can be isomorphically represented as a Boolean set algebra; by the second every BA can be embedded as a subalgebra in a complete and atomic BA (see Stone [34*]). The two results are closely related to each other, and either of them can be derived from the other by means of an elementary argument.

When we pass from Boolean algebras, i.e., cylindric algebras of dimension 0, to cylindric algebras of arbitrary dimension, the situation undergoes a substantial change. We know a notion of representability which is applicable to arbitrary CA's and appears as a natural extension of the notion of representability for BA's. In fact, a CA is called representable if it is isomorphic to a generalized cylindric set algebra (or, what amounts to the same, to a subdirect product of ordinary cylindric set algebras; cf. 1.1.13). However, we know also that, with respect to this notion, the representation theorem does not hold for arbitrary cylindric algebras: there are CA_α's with any given $\alpha \geqq 2$ which are not representable (cf. 2.6.44).

On the other hand, the general embedding theorem continues to hold in the theory of CA's: every CA can be embedded as a subalgebra of a complete and atomic CA. This is, in fact, the content of the main result of the present section, Theorem 2.7.20. Moreover, from the general embedding theorem we can derive in turn a kind of general representation theorem involving a different notion of representability; cf. 2.7.43. Roughly speaking, a CA is representable in the new sense if it is isomorphic to a CA whose Boolean part is a Boolean set algebra, and whose cylindrifications are derived from certain equivalence relations on the unit set of this algebra. A precise description of these representative algebras (including also a characterization of diagonal elements) will of course be given in the text below and will enable us to discuss in some detail, at the end of the section, the mutual relationship between the two notions of representability. We shall try to make it clear why the new notion presents much less interest than the original notion from the intuitive point of view, and why, therefore, the intuitive value of the general representation theorem for CA's established in this section does not equal that of the corresponding result for BA's.

The extension theorem for CA's and the representation theorem implied by it are special cases of results obtained in Jónsson-Tarski [51] for a much wider class of algebras, the so-called *Boolean algebras with operators.* These are algebras obtained by enriching BA's with operations of finite rank which are assumed only to be additive, i.e., distributive over Boolean addition; thus, in particular, all reducts of CA's obtained by discarding the operation $-$ are BA's with operators. In the present work we do not presuppose the general theory of Boolean algebras with operators, and we are interested only in consequences of this theory for cylindric algebras (and possibly some related classes of algebras). It turns out, however, that in order to establish these consequences in a relatively simple and natural way we need results of a much more general character concerning, if not all BA's with operators, at least all those which are similar to CA's. For these reasons we apply the following procedure. First we agree that by an α-dimensional Boolean algebra with operators, in symbols a Bo_α, we shall mean any BA with operators which is similar to a CA_α. We then deal with the general theory of BA's with operators in its restriction to Bo_α's for an arbitrary α, and we develop this theory to the extent needed for our purposes; since all extra-Boolean operations in Bo_α's are of rank 1, the restriction to these algebras carries with it a rather considerable simplification of proofs. From general results thus established we derive in turn the main applications to CA's, such as the embedding theorem. From then on most of the results are formulated exclusively for CA's, even though nearly all of them extend to arbitrary Bo_α's (and actually to all BA's with operators). Henceforth, when speaking of *Boolean algebras with operators* we shall mean exclusively Bo_α's with an arbitrary α.

DEFINITION 2.7.1. *By an **α-dimensional Boolean algebra with operators**, where α is any ordinal, we mean an algebraic structure*

$$\mathfrak{A} = \langle A, +, \cdot, -, 0, 1, \mathsf{c}_\kappa, \mathsf{d}_{\kappa\lambda} \rangle_{\kappa,\lambda<\alpha}$$

such that $\mathfrak{A}$ is of the same similarity type as CA_α's and the following conditions hold:

(i) $\langle A, +, \cdot, -, 0, 1 \rangle$ *is a Boolean algebra*;

(ii) $\mathsf{c}_\kappa(x+y) = \mathsf{c}_\kappa x + \mathsf{c}_\kappa y$ *for every* $\kappa < \alpha$ *and* $x, y \in A$.

The class of all Boolean algebras with operators is denoted by Bo, *and the class of all α-dimensional Boolean algebras with operators by* Bo_α.

The algebra $\mathfrak{A}$ *is said to be a* **complete** Bo_α *if it is a* Bo_α, $\langle A, +, \cdot, -, 0, 1\rangle$ *is a complete Boolean algebra, and*

(ii′) $\mathsf{c}_\kappa \sum X = \sum \mathsf{c}_\kappa^* X$ *for every* $\kappa < \alpha$ *and non-empty* $X \subseteq A$;

finally, $\mathfrak{A}$ *is said to be* **normal** *if*

(iii) $\mathsf{c}_\kappa 0 = 0$ *for every* $\kappa < \alpha$.

The algebra $\langle A, +, \cdot, -, 0, 1\rangle$ *is called the* **Boolean part** *of* $\mathfrak{A}$ *and is denoted by* $\mathfrak{Bl}\,\mathfrak{A}$.

In connection with the first part of the definition recall our convention concerning the similarity type of CA's adopted in the remarks immediately following 1.1.1.

A unary operation O on a Boolean algebra $\mathfrak{A}$ is said to be *normal* if $O0 = 0$. In Jónsson-Tarski [51] a Bo is called normal if all its extra-Boolean operations are normal. In particular, a distinguished element considered as a constant operation of rank one is normal only if it is equal to 0. For this reason we have departed from the Jónsson-Tarski terminology in defining the normality of Bo_α's without reference to the distinguished elements $\mathsf{d}_{\kappa\lambda}$.

REMARK 2.7.2. All of the terminology of cylindric algebras relating to their Boolean parts can be extended to Boolean algebras with operators. Thus, for example, we write $At\mathfrak{A}$ for the set of atoms of $\mathfrak{Bl}\,\mathfrak{A}$, and we say that $\mathfrak{A}$ is atomic if $\mathfrak{Bl}\,\mathfrak{A}$ is atomic. $\sum^{(\mathfrak{A})}$, or simply $\sum$, will denote the infinite generalization of + in $\mathfrak{Bl}\,\mathfrak{A}$. A function $f \in {}^A A$ will be called additive if $f\sum X = \sum f^* X$ for every finite non-empty $X \subseteq A$, and completely additive if $\sum f^* X$ exists and $f\sum X = \sum f^* X$ for every non-empty $X \subseteq A$ such that $\sum X$ exists (cf. the remarks immediately before and after 1.2.6).

Finally, we remark that every additive function is *monotone-increasing*, or simply *increasing*, in the sense that $fx \leq fy$ whenever $x \leq y$ (cf. 1.2.6). Thus in every Bo_α all the operations c_κ are increasing; this fact is used repeatedly throughout this section.

COROLLARY 2.7.3. *Every* CA_α $\mathfrak{A}$ *is a normal* Bo_α; *furthermore, if* $\mathfrak{Bl}\,\mathfrak{A}$ *is complete, then* $\mathfrak{A}$ *is a complete* Bo_α.

PROOF: by 1.2.6 and 2.7.1.

DEFINITION 2.7.4. *Assume* $\mathfrak{A}$ *is an* α*-dimensional Boolean algebra with operators, and let* M *be the set of all maximal proper ideals of* $\mathfrak{Bl}\,\mathfrak{A}$. *For each* $\kappa < \alpha$ *let* $C_\kappa \in {}^{SbM}SbM$ *be defined by the condition that*

$$C_\kappa X = \{J : J \in M, (\mathsf{c}_\kappa^{-1})^* J = 0\} \cup \bigcup_{I \in X} \{J : J \in M, (\mathsf{c}_\kappa^{-1})^* J \subseteq I\}$$

for every $X \subseteq M$, *and for all* $\kappa, \lambda < \alpha$ *let*

$$D_{\kappa\lambda} = \{I : \mathsf{d}_{\kappa\lambda} \notin I \in M\}.$$

Then the algebra

$$\langle SbM, \cup, \cap, {}_M\sim, 0, M, C_\kappa, D_{\kappa\lambda}\rangle_{\kappa,\lambda<\alpha}$$

is called the **canonical embedding algebra** *of* $\mathfrak{A}$ *and is denoted by* $\mathfrak{Em}\,\mathfrak{A}$, *and* SbM *is denoted by* $Em\,\mathfrak{A}$.

The function h *from* A *into* SbM *such that* $hx = \{I : x \notin I \in M\}$ *for every* $x \in A$ *is called the* **canonical embedding function** *and is denoted by* $em^{(\mathfrak{A})}$ *or simply* em.

THEOREM 2.7.5. *Assume* α *is any ordinal and* $\mathfrak{A} \in \mathsf{Bo}_\alpha$, *and let* M *be the set of all maximal proper ideals of* $\mathfrak{Bl}\,\mathfrak{A}$. *We then have*:

(i) $\mathfrak{Em}\,\mathfrak{A}$ *is a complete and atomic* Bo_α *and* $em \in Ism(\mathfrak{A}, \mathfrak{Em}\,\mathfrak{A})$; *furthermore, if* $\mathfrak{A}$ *is normal, then so is* $\mathfrak{Em}\,\mathfrak{A}$;

(ii) *if* K *is an arbitrary subset of* em^*A *such that* $\bigcup K = M$, *then* $\bigcup L = M$ *for some finite* $L \subseteq K$;

(iii) *if* $X, Y \in At\,\mathfrak{Em}\,\mathfrak{A}$ *with* $X \neq Y$, *then there is a* $Z \in em^*A$ *such that* $X \subseteq Z$ *and* $Y \subseteq {}_M\sim Z$;

(iv) *for every* $\kappa < \alpha$ *and* $B \in At\,\mathfrak{Em}\,\mathfrak{A}$ *we have*

$$C_\kappa B = \bigcap_{B \subseteq X \in em^*A} C_\kappa X.$$

PROOF. $\mathfrak{Bl}\,\mathfrak{Em}\,\mathfrak{A}$ is obviously a complete and atomic BA and it follows immediately from 2.7.4 that, for every $\kappa < \alpha$, C_κ is completely additive, i.e., that 2.7.1(ii′) holds when c_κ is taken to be C_κ. Thus $\mathfrak{Em}\,\mathfrak{A}$ is a complete and atomic Bo_α. It is a well known result from the theory of Boolean algebras that em is an isomorphism from $\mathfrak{Bl}\,\mathfrak{A}$ into $\mathfrak{Bl}\,\mathfrak{Em}\,\mathfrak{A}$ and that conditions (ii) and (iii) hold. Thus, since we obviously have $em\,\mathsf{d}_{\kappa\lambda} = D_{\kappa\lambda}$ for all $\kappa, \lambda < \alpha$, to prove that $em \in Is(\mathfrak{A}, \mathfrak{Em}\,\mathfrak{A})$ it only remains to show that

(1) $\quad em\,\mathsf{c}_\kappa x = C_\kappa em\,x$ for all $\kappa < \alpha$ and $x \in A$.

So consider any $\kappa < \alpha$ and $x \in A$, and suppose $J \in C_\kappa em\,x$, i.e., $(\mathsf{c}_\kappa^{-1})^*J = 0$ or $(\mathsf{c}_\kappa^{-1})^*J \subseteq I$ for some $I \in M$ such that $x \notin I$; in either case, $\mathsf{c}_\kappa x \notin J$, i.e., $J \in em\,\mathsf{c}_\kappa x$. Thus we have $em\,\mathsf{c}_\kappa x \supseteq C_\kappa em\,x$. To obtain the inclusion in the opposite direction let J be an arbitrary element of $em\,\mathsf{c}_\kappa x$, i.e., suppose $\mathsf{c}_\kappa x \notin J \in M$. Then clearly

(2) $\quad x \notin (\mathsf{c}_\kappa^{-1})^*J.$

In case $(\mathsf{c}_\kappa^{-1})^*J = 0$, it is clear that $J \in C_\kappa em\,x$; so assume that $(\mathsf{c}_\kappa^{-1})^*J \neq 0$. Then by 2.7.1(ii) (the additivity of c_κ) it follows easily that $(\mathsf{c}_\kappa^{-1})^*J$ is a Boolean ideal of $\mathfrak{A}$, and hence by (2) and the theory of BA's there is an $I \in M$ such that $(\mathsf{c}_\kappa^{-1})^*J \subseteq I$ and $x \notin I$. Thus $J \in C_\kappa em\,x$ also in the case $(\mathsf{c}_\kappa^{-1})^*J \neq 0$, and we have shown that $em\,\mathsf{c}_\kappa x \subseteq C_\kappa em x$. This gives (1) and hence $em \in Ism(\mathfrak{A}, \mathfrak{Em}\,\mathfrak{A})$. Since $\mathfrak{A} \cong|\subseteq \mathfrak{Em}\,\mathfrak{A}$ it follows immediately that $\mathfrak{Em}\,\mathfrak{A}$ is normal whenever $\mathfrak{A}$ is normal; this finishes the proof of (i).

Finally, in order to establish (iv) we observe first of all that from 2.7.4 we have, for every $I \in M$, $J \in C_\kappa\{I\}$ iff $J \in M$ and $\mathsf{c}_\kappa x \notin J$ for every $x \in A$ such that $x \notin I$, i.e.,

$$C_\kappa\{I\} = \bigcap_{x \in A \sim I}\{J : \mathsf{c}_\kappa x \notin J \in M\}.$$

(iv) now follows easily using (1).

It will be seen that conditions (i)–(iv) of 2.7.5 when re-formulated in abstract algebraic terms can be used to characterize up to isomorphism the canonical embedding algebra of a given $\mathsf{Bo}_\alpha\,\mathfrak{A}$. This is essentially the content of our next theorem, 2.7.6. In the whole subsequent discussion it proves possible and almost always convenient to disregard the set-theoretical construction of canonical embedding algebras which was given in 2.7.3, and to employ instead the abstract characterization of these algebras contained in conditions (i)–(iv) of 2.7.5 or 2.7.6.

THEOREM 2.7.6. *Suppose $\mathfrak{A} \in \mathsf{Bo}_\alpha$ and $\mathfrak{B}$ is any algebra for which $\mathfrak{A} \subseteq \mathfrak{B}$. In order that there exist an isomorphism h from $\mathfrak{B}$ onto $\mathfrak{Em}\,\mathfrak{A}$ such that $A \upharpoonright h = em^{(\mathfrak{A})}$, it is necessary and sufficient that the following conditions hold:*

(i) *$\mathfrak{B}$ is a complete and atomic Bo_α;*

(ii) *if X is an arbitrary subset of A such that $\sum^{(\mathfrak{B})}X = 1$, then $\sum^{(\mathfrak{B})}Y = 1$ for some finite $Y \subseteq X$;*

(iii) *if $x, y \in At\,\mathfrak{B}$ with $x \neq y$, then there is a $z \in A$ such that $x \leqq z$ and $y \leqq -z$;*

(iv) *for each $\kappa < \alpha$ and $b \in At\,\mathfrak{B}$ we have*

$$\mathsf{c}_\kappa b = \prod^{(\mathfrak{B})}_{b \leqq x \in A} \mathsf{c}_\kappa x.$$

PROOF. It follows immediately from 2.7.5 that conditions (i)–(iv) are necessary. In order to prove that they are also sufficient we recall first of all the familiar result from the theory of BA's by which conditions (ii) and (iii) along with the completeness and atomicity of $\mathfrak{Bl}\,\mathfrak{B}$ are sufficient for the existence of an isomorphism h from $\mathfrak{Bl}\,\mathfrak{B}$ onto $\mathfrak{Bl}\,\mathfrak{Em}\,\mathfrak{A}$ such that $A \upharpoonright h = em^{(\mathfrak{A})}$. Now assume in addition that $\mathfrak{B} \in \mathsf{Bo}_\alpha$ and that (iv) holds, and recall that $\mathfrak{Em}\,\mathfrak{A} = \langle Sb\,M, \cup, \cap, {}_M\!\sim, 0, M, C_\kappa, D_{\kappa\lambda}\rangle_{\kappa,\lambda<\alpha}$ where M is the set of maximal proper ideals of $\mathfrak{Bl}\,\mathfrak{A}$. By (iv) and 2.7.5(iv) we obtain (using the fact that h is a Boolean isomorphism from $\mathfrak{B}$ onto $\mathfrak{Em}\,\mathfrak{A}$)

(1) $$h\mathsf{c}_\kappa b = C_\kappa hb$$

for every $b \in At\mathfrak{B}$. Thus by the complete additivity of c_κ and C_κ (condition 2.7.6(i) and 2.7.5(i), respectively) we have that (1) holds for every $b \in B$; also, $h\mathsf{d}_{\kappa\lambda} = em\,\mathsf{d}_{\kappa\lambda} = D_{\kappa\lambda}$ for all $\kappa, \lambda < \alpha$. Hence $h \in Is(\mathfrak{B}, \mathfrak{Em}\,\mathfrak{A})$, and the proof is finished.

REMARK 2.7.7. If $\mathfrak{A} \in \mathsf{Bo}_\alpha$ and $\mathfrak{B}$ is any algebra such that $\mathfrak{A} \subseteq \mathfrak{B}$ and conditions (i)–(iv) of 2.7.6 hold, then $\mathfrak{B}$ is said to be a *perfect extension* of $\mathfrak{A}$; cf. Jónsson-Tarski [51], p. 925, Definition 2.14.

Each time we consider two Bo's $\mathfrak{A}$ and $\mathfrak{B}$ such that $\mathfrak{A} \subseteq \mathfrak{B}$, and we use the symbols Σ and Π without explicitly relativizing them to either of the algebras, these symbols should be understood to refer to the operations of $\mathfrak{B}$.

COROLLARY 2.7.8. *Assume* $\mathfrak{A} \in \mathsf{Bo}_\alpha$ *and let* C *and* O *be subsets of* $Em\mathfrak{A}$ *defined by the formulas*

$$C = \{\Pi X : X \subseteq em^*A\} \text{ and } O = \{\Sigma X : X \subseteq em^*A\}.$$

We then have:

(i) $O = \{-x : x \in C\}$, $C = \{-x : x \in O\}$;

(ii) *if* $x \in C$, $X \subseteq O$, *and* $x \leq \Sigma X$, *then there is a finite* $Y \subseteq X$ *such that* $x \leq \Sigma Y$;

(iii) *if* $x \in O$, $X \subseteq C$, *and* $x \geq \Pi X$, *then there is a finite* $Y \subseteq X$ *such that* $x \geq \Pi Y$;

(iv) $At\,\mathfrak{Em}\,\mathfrak{A} \subseteq C$.

PROOF. (i) is obvious. (ii) follows immediately from (i) and 2.7.5(ii), and (iii) is a direct consequence of (i) and (ii). Finally, (iv) follows easily from 2.7.5(iii).

An element x of a canonical embedding algebra $\mathfrak{Em}\,\mathfrak{A}$ is said to be *closed*, or *open*, if $x = \Pi X$, or $x \in \Sigma X$, for some $X \subseteq em^*A$. Thus both 2.7.8(ii) and 2.7.8(iii) express the familiar facts from the theory of Boolean algebras and from general topology that $\mathfrak{Em}\,\mathfrak{A}$, with respect to the topology for which em^*A forms a base, is compact, and that a closed subset of a compact space is compact.

LEMMA 2.7.9. *Assume* $\mathfrak{A} \in \mathsf{Bo}_\alpha$ *and let* a *be any element of* $Em\mathfrak{A}$ *such that* $a = \Pi X$ *for some* $X \subseteq em^*A$. *Then in the algebra* $\mathfrak{Em}\,\mathfrak{A}$ *we have*

$$\mathsf{c}_\kappa a = \Pi_{a \leq x \in em^*A}\, \mathsf{c}_\kappa x$$

for every $\kappa < \alpha$.

PROOF. Since c_κ is increasing (cf. 2.7.2) we have

(1) $$\mathsf{c}_\kappa a \leq \Pi_{a \leq x \in em^*A}\, \mathsf{c}_\kappa x.$$

To obtain the inclusion in the opposite direction let u be any atom of $\mathfrak{Em}\,\mathfrak{A}$ such that $u \cdot \mathsf{c}_\kappa a = 0$. Then for each y such that $a \geqq y \in At\,\mathfrak{Em}\,\mathfrak{A}$ we get $\mathsf{c}_\kappa a \geqq \mathsf{c}_\kappa y$ since c_κ is increasing. Hence $u \cdot \mathsf{c}_\kappa y = 0$ and thus by 2.7.5(iv) there is, for each y such that $a \geqq y \in At\,\mathfrak{Em}\,\mathfrak{A}$, an element x_y with the property that

$$y \leqq x_y \in em^*A \text{ and } u \cdot \mathsf{c}_\kappa x_y = 0.$$

Therefore, setting

$$X = \{x_y : a \geqq y \in At\,\mathfrak{Em}\,\mathfrak{A}\}$$

we have

$$a \leqq \Sigma X \text{ and } u \cdot \Sigma \mathsf{c}_\kappa^* X = 0. \tag{2}$$

By 2.7.8(ii) there is a finite $X' \subseteq X$ such that

$$a \leqq \Sigma X'; \tag{3}$$

but then by (2) and the additivity of c_κ we get

$$u \cdot \mathsf{c}_\kappa \Sigma X' = u \cdot \Sigma \mathsf{c}_\kappa^* X' \leqq u \cdot \Sigma \mathsf{c}_\kappa^* X = 0.$$

Thus, by (3),

$$u \cdot \textstyle\prod_{a \leqq x \in em^*A} \mathsf{c}_\kappa x \leqq u \cdot \mathsf{c}_\kappa \Sigma X' = 0. \tag{4}$$

Since u is an arbitrary atom of $\mathfrak{Em}\,\mathfrak{A}$ such that $u \cdot \mathsf{c}_\kappa a = 0$, (4) implies that $\mathsf{c}_\kappa a \geqq \prod_{a \leqq x \in em^*A} \mathsf{c}_\kappa x$. When combined with (1) this gives the conclusion of the lemma.

In the next portion of this section we will be concerned with showing that the canonical embedding algebra of a CA_α is always again a CA_α. We shall actually obtain a much stronger result in 2.7.13. Indeed, it will be seen from the discussion in 2.7.14 that, as a consequence of 2.7.13, if an equation not involving the fundamental operation of complementation is identically satisfied in an arbitrary $\mathsf{Bo}_\alpha\,\mathfrak{A}$, then it holds also in the canonical embedding algebra $\mathfrak{Em}\,\mathfrak{A}$.

The preceding lemma was the first of three lemmas involved in the proof of 2.7.13. In order to formulate the remaining two lemmas and Theorem 2.7.13 conveniently we introduce some special terminology.

DEFINITION 2.7.10. *Let* $\mathfrak{A} = \langle A, +, \cdot, -, 0, 1, \mathsf{c}_\kappa, \mathsf{d}_{\kappa\lambda} \rangle_{\kappa,\lambda<\alpha}$ *be a* Bo_α. *Then the algebra*

$$\langle A, +, \cdot, 0, 1, \mathsf{c}_\kappa, \mathsf{d}_{\kappa\lambda} \rangle_{\kappa,\lambda<\alpha}$$

is called the ***positive reduct*** *of* $\mathfrak{A}$ *and is denoted by* $\mathfrak{Pr}\,\mathfrak{A}$.

It is well known that the class of all positive reducts of BA's is definitionally equivalent in the sense of 0.1.6 to the class BA itself. Hence, it is easily seen that the classes Bo_α and $\mathfrak{Pr}^*\mathsf{Bo}_\alpha$ are also definitionally equivalent for every α. Frequently in the literature, e.g., in Jónsson-Tarski [51], Bo_α's and, in particular, BA's are defined to be what we call positive reducts of Bo_α's and BA's. However, as compared with BA and Bo_α, the classes $\mathfrak{Pr}^*\mathsf{BA}$ and $\mathfrak{Pr}^*\mathsf{Bo}_\alpha$ have the disadvantage that they are not equational.

In connection with the next lemma recall that $Pl_\omega\mathfrak{Pr}\,\mathfrak{Em}\,\mathfrak{A}$ is the set of all polynomials in ω variables over $\mathfrak{Pr}\,\mathfrak{Em}\,\mathfrak{A}$; cf. 0.4.1. Throughout the proofs of 2.7.11–2.7.13 we shall deal with the Boolean inclusion relations of two distinct but closely related algebras, $\mathfrak{Em}\,\mathfrak{A}$ and ${}^\omega\mathfrak{Em}\,\mathfrak{A}$. To distinguish between them we shall always write $\leqq$ for $\leqq^{(\mathfrak{Em}\mathfrak{A})}$ and $\leqq_\omega$ for $\leqq^{({}^\omega\mathfrak{Em}\mathfrak{A})}$; an analogous convention is adopted for the converse relations.

LEMMA 2.7.11. *Assume* $\mathfrak{A}\in\mathsf{Bo}_\alpha$ *and let* C *be the subset of* $Em\,\mathfrak{A}$ *defined by the formula*

$$C = \{\prod X : X \subseteq em^*A\}.$$

Then for every $P\in Pl_\omega\mathfrak{Pr}\,\mathfrak{Em}\,\mathfrak{A}$ *and every* $a\in{}^\omega Em\,\mathfrak{A}$ *we have*

$$Pa = \sum\{Px:\ x\in{}^\omega C,\ x_\kappa \leqq a_\kappa \text{ for every } \kappa<\omega\}.$$

PROOF. Let $\mathfrak{Em}\,\mathfrak{A} = \langle Em\,\mathfrak{A}, +, \cdot, -, 0, 1, \mathsf{c}_\kappa, \mathsf{d}_{\kappa\lambda}\rangle_{\kappa,\lambda<\alpha}$ and $\mathfrak{Pl}_\omega\mathfrak{Pr}\,\mathfrak{Em}\,\mathfrak{A} = \langle Pl_\omega\mathfrak{Pr}\,\mathfrak{Em}\,\mathfrak{A}, +', \cdot', 0', 1', \mathsf{c}'_\kappa, \mathsf{d}'_{\kappa\lambda}\rangle_{\kappa\lambda<\alpha}$. For brevity we set $Ta = \{x : x\in{}^\omega C,\ x\leqq_\omega a\}$ for any $a\in{}^\omega C$.

Observe first of all that, by 2.7.8(iv),

(1) $$At\,\mathfrak{Em}\,\mathfrak{A}\subseteq C.$$

Notice also that for any $X, Y\subseteq em^*A$ we have $\prod X+\prod Y = \prod\{x+y : x\in X, y\in Y\}$ and hence

(2) $$x+y\in C \text{ whenever } x, y\in C.$$

Set

(3) $$B = \{P:\ P\in Pl_\omega\mathfrak{Pr}\,\mathfrak{Em}\,\mathfrak{A},\ Pa = \textstyle\sum_{x\in Ta} Px \text{ for every } a\in{}^\omega Em\,\mathfrak{A}\}$$

and observe that

(4) $$Px \leqq Py \text{ whenever } P\in B,\ x, y\in{}^\omega Em\,\mathfrak{A}, \text{ and } x\leqq_\omega y.$$

For any $\kappa<\omega$ and $a\in{}^\omega Em\,\mathfrak{A}$ we have, by (1) and the atomicity of $\mathfrak{Em}\,\mathfrak{A}$,

$$({}^\omega Em\,\mathfrak{A}\,1\,\mathsf{pj}_\kappa)a = a_\kappa = \textstyle\sum_{a_\kappa\geqq y\in C} y = \sum_{x\in Ta}({}^\omega Em\,\mathfrak{A}\,1\,\mathsf{pj}_\kappa)x.$$

Thus

(5) $$\{{}^\omega Em\,\mathfrak{A}\,1\,\mathsf{pj}_\kappa : \kappa<\omega\}\subseteq B.$$

Now consider $P, Q \in B$ and $a \in {}^{\omega}Em\,\mathfrak{A}$. Then

$$(6) \qquad (P +' Q)a = Pa + Qa$$
$$= \sum_{x\in Ta} Px + \sum_{x\in Ta} Qx \qquad \text{by (3)}$$
$$= \sum_{x\in Ta} (P +' Q)x.$$

Also

$$\sum_{x,y\in Ta} (Px \cdot Qy) \leqq \sum_{x\in Ta} (P(x+y) \cdot Q(x+y)) \qquad \text{by (4)}$$
$$\leqq \sum_{z\in Ta} (Pz \cdot Qz) \qquad \text{by (2)}$$
$$\leqq \sum_{x,y\in Ta} (Px \cdot Qy);$$

thus we obtain

$$(7) \qquad (P \cdot' Q)a = Pa \cdot Qa$$
$$= \sum_{x\in Ta} Px \cdot \sum_{y\in Ta} Qy \qquad \text{by (3)}$$
$$= \sum_{x,y\in Ta} (Px \cdot Qy)$$
$$= \sum_{z\in Ta} (Pz \cdot Qz)$$
$$= \sum_{z\in Ta} (P \cdot' Q)z.$$

Finally, for every $\kappa < \omega$ we have

$$(8) \qquad (\mathsf{c}'_\kappa P)a = \mathsf{c}_\kappa Pa$$
$$= \mathsf{c}_\kappa \sum_{x\in Ta} Px \qquad \text{by (3)}$$
$$= \sum_{x\in Ta} \mathsf{c}_\kappa Px \qquad \text{by 2.7.5(i)}$$
$$= \sum_{x\in Ta} (\mathsf{c}'_\kappa P)x.$$

From the fact that (6)–(8) hold for all $P, Q \in B$, $a \in {}^{\omega}Em\,\mathfrak{A}$, and $\kappa < \omega$ we conclude that B is a subuniverse of $\mathfrak{Pl}_\omega \mathfrak{Pr}\,\mathfrak{Em}\,\mathfrak{A}$. Therefore, in view of (5), we get $\mathfrak{B} = \mathfrak{Pl}_\omega \mathfrak{Pr}\,\mathfrak{Em}\,\mathfrak{A}$, and this gives the lemma.

LEMMA 2.7.12. *Suppose* $\mathfrak{A} \in \mathsf{Bo}_\alpha$ *and let* C *be the subset of* $Em\,\mathfrak{A}$ *defined by the formula*

$$C = \{\prod X : X \subseteq em^* A\}.$$

Then for every $P \in Pl_\omega \mathfrak{Pr}\,\mathfrak{Em}\,\mathfrak{A}$ *and every* $a \in {}^{\omega}C$ *we have*

$$Pa = \prod\{Px : x \in {}^{\omega}em^*A,\ a_\kappa \leqq x_\kappa \text{ for every } \kappa < \omega\}.$$

PROOF.[1] Let $\mathfrak{Em}\,\mathfrak{A} = \langle Em\,\mathfrak{A}, +, \cdot, -, 0, 1, \mathsf{c}_\kappa, \mathsf{d}_{\kappa\lambda}\rangle_{\kappa,\lambda<\alpha}$ and $\mathfrak{Pl}_\omega \mathfrak{Pr}\,\mathfrak{Em}\,\mathfrak{A} = \langle Pl_\omega \mathfrak{Pr}\,\mathfrak{Em}\,\mathfrak{A}, +', \cdot', 0', 1', \mathsf{c}'_\kappa, \mathsf{d}'_{\kappa\lambda}\rangle_{\kappa,\lambda<\alpha}$. For brevity we set $Ta = \{x : x \in {}^{\omega}em^*\mathfrak{A},\ x \leqq_\omega a\}$ for any $a \in {}^{\omega}C$. Also we put

[1]) In the proof of 2.7.12 presented here we apply, not the original methods of Jónsson-Tarski [51], but an idea suggested in Ribeiro [52*].

(1) $B = \{P : P \in Pl_\omega \mathfrak{Pr}\,\mathfrak{Em}\,\mathfrak{A},\ Pa = \prod_{x\in Ta} Px \text{ for every } a \in {}^\omega C\}$;

observe that

(2) $$Pa \leqq Pb \text{ whenever } P \in B,\quad a, b \in {}^\omega C, \text{ and } a \leqq_\omega b,$$

and

(3) $$Pa \in C \text{ whenever } P \in B \text{ and } a \in {}^\omega C.$$

For any $\kappa < \omega$ and $a \in {}^\omega C$ we have, by the definition of C,

$$({}^\omega Em\,\mathfrak{A}1\ \mathsf{pj}_\kappa)a = a_\kappa = \prod_{a_\kappa \leqq y \in em^*A} y = \prod_{x\in Ta} ({}^\omega Em\,\mathfrak{A}1\ \mathsf{pj}_\kappa)x,$$

thus

(4) $$\{{}^\omega Em\,\mathfrak{A}1\ \mathsf{pj}_\kappa : \kappa < \omega\} \subseteq B.$$

Now consider any $P, Q \in B$ and $a \in {}^\omega C$. Then

$$\begin{aligned}\prod_{x,y\in Ta}(Px+Qy) &\geqq \prod_{x,y\in Ta}(P(x\cdot y) + Q(x\cdot y)) && \text{by (2)}\\ &\geqq \prod_{z\in Ta}(Pz+Qz)\\ &\geqq \prod_{x,y\in Ta}(Px+Qy);\end{aligned}$$

thus we obtain

(5) $$\begin{aligned}(P +' Q)a &= Pa+Qa\\ &= \prod_{x\in Ta} Px + \prod_{y\in Ta} Qy && \text{by (1)}\\ &= \prod_{x,y\in Ta}(Px+Qy)\\ &= \prod_{z\in Ta}(Pz+Qz)\\ &= \prod_{z\in Ta}(P +' Q)z.\end{aligned}$$

Also,

(6) $$\begin{aligned}(P \cdot' Q)a &= Pa\cdot Qa\\ &= \prod_{x\in Ta} Px \cdot \prod_{x\in Ta} Qx && \text{by (1)}\\ &= \prod_{x\in Ta}(P\cdot' Q)x.\end{aligned}$$

Finally, fix an arbitrary $\kappa < \omega$. By (3) and 2.7.9 we have

(7) $$(\mathsf{c}'_\kappa P)a = \mathsf{c}_\kappa Pa = \prod_{Pa \leqq y \in em^*A} \mathsf{c}_\kappa y.$$

If $a \leqq_\omega x \in {}^\omega em^*A$, then $Pa \leqq Px \in em^*A$ by (2), and hence, since c_κ is increasing,

$$(\mathsf{c}'_\kappa P)a = \mathsf{c}_\kappa Pa \leqq \mathsf{c}_\kappa Px = (\mathsf{c}'_\kappa P)x.$$

Thus

(8) $$(\mathsf{c}'_\kappa P)a \leqq \prod_{x\in Ta} (\mathsf{c}'_\kappa P)x.$$

On the other hand, consider any y such that

(9) $$Pa \leqq y \in em^*A.$$

Then by (1) we have $\prod_{z \in Ta} Pz \leqq y$ and hence by 2.7.8(iii) there is a finite set $Z \subseteq {}^{\omega}em^*A$ such that

(10) $$\prod_{a \leqq_{\omega} z \in Z} Pz \leqq y;$$

setting $x = \langle \prod_{a \leqq_{\omega} z \in Z} z_{\lambda} : \lambda < \omega \rangle$ it is obvious that

(11) $$a \leqq_{\omega} x \in {}^{\omega}em^*A,$$

and that $x \leqq_{\omega} z$ for every z such that $a \leqq_{\omega} z \in Z$. Thus from (2) and (10) one obtains $Px \leqq \prod_{a \leqq_{\omega} z \in Z} Pz \leqq y$, and in view of the fact that c_{κ} is increasing this implies that

(12) $$\mathsf{c}_{\kappa} Px \leqq \mathsf{c}_{\kappa} y.$$

For every y such that (9) holds we have shown that there exists an x such that (11) and (12) hold. Therefore by (7) we obtain

(13) $$(\mathsf{c}'_{\kappa} P)a \geqq \prod_{x \in Ta} (\mathsf{c}'_{\kappa} P)x.$$

From the fact that (5), (6), (8), and (13) hold for all $P, Q \in B$, $a \in {}^{\omega}C$, and $\kappa < \omega$ one concludes that B is a subuniverse of $\mathfrak{Pl}_{\omega}\mathfrak{Pr}\mathfrak{Em}\mathfrak{A}$. Therefore in view of (4) we get $\mathfrak{B} = \mathfrak{Pl}_{\omega}\mathfrak{Pr}\mathfrak{Em}\mathfrak{A}$. The proof is complete.

THEOREM 2.7.13. *If* $\mathfrak{A} \in \mathsf{Bo}_{\alpha}$, *then* $\mathfrak{Pr}\mathfrak{Em}\mathfrak{A} \in \mathbf{HSP}\mathfrak{Pr}\mathfrak{A}$ (*and, in fact,* $\mathbf{HSP}\mathfrak{Pr}\mathfrak{Em}\mathfrak{A} = \mathbf{HSP}\mathfrak{Pr}\mathfrak{A}$).

PROOF. Set $F = \langle {}^{\omega}em^*A \upharpoonright P : P \in Pl_{\omega}\mathfrak{Pr}\mathfrak{Em}\mathfrak{A} \rangle$. Then it is easily seen that F is a homomorphism from $\mathfrak{Pl}_{\omega}\mathfrak{Pr}\mathfrak{Em}\mathfrak{A}$ onto $\mathfrak{Pl}_{\omega}\mathfrak{Pr}\,em^*\mathfrak{A}$; compare here the proof of 0.4.11(i). However, by 2.7.11 and 2.7.12 we have, setting $C = \{\prod X : X \subseteq em^*A\}$,

$$Pa = \sum_{a \geqq_{\omega} x \in {}^{\omega}C} \prod_{x \leqq_{\omega} y \in {}^{\omega}em^*A} Py$$

for every $P \in Pl_{\omega}\mathfrak{Pr}\mathfrak{Em}\mathfrak{A}$ and $a \in {}^{\omega}Em\mathfrak{A}$. Thus, if $P, Q \in Pl_{\omega}\mathfrak{Pr}\mathfrak{Em}\mathfrak{A}$ and ${}^{\omega}em^*A \upharpoonright P = {}^{\omega}em^*A \upharpoonright Q$, we conclude that $P = Q$. Hence F is one-one, and we have shown that

$$\mathfrak{Pl}_{\omega}\mathfrak{Pr}\mathfrak{Em}\mathfrak{A} \cong \mathfrak{Pl}_{\omega}\mathfrak{Pr}\,em^*\mathfrak{A}.$$

The conclusion of the theorem now follows from 0.4.43(iii),(vi) and 0.4.50(i).

REMARK 2.7.14. Let us agree to call a term τ of the discourse language of Bo_{α}'s, and hence also of CA_{α}'s, *positive*, if it does not contain any occurrence of the symbol $-$ for complementation; an equation of the same language is called *positive* if both its terms are positive. Obviously, a positive equation is identically satisfied in an arbitrary Bo_α $\mathfrak{A}$ iff it is identically satisfied in the positive reduct of $\mathfrak{A}$. In view of this and of 0.4.64 we conclude from 2.7.13 that, if Γ is any set of positive equations and if K is the class of all Bo_α's in which every equation of Γ holds, then K is closed under the formation of canonical embedding algebras. This observation leads to

THEOREM 2.7.15. *If* $\mathfrak{A} \in \mathsf{CA}_\alpha$, *then* $\mathfrak{Em}\,\mathfrak{A} \in \mathsf{CA}_\alpha$.

PROOF. By the remark immediately after 1.3.6 it follows that for a Bo_α $\mathfrak{B}$ to be a CA_α it is necessary and sufficient that the axioms (C_1)–(C_6) of Definition 1.1.1 hold in $\mathfrak{B}$ and that, in addition, $\mathsf{c}_\kappa(\mathsf{d}_{\kappa\lambda} \cdot x \cdot y) = \mathsf{c}_\kappa(\mathsf{d}_{\kappa\lambda} \cdot x) \cdot \mathsf{c}_\kappa(\mathsf{d}_{\kappa\lambda} \cdot y)$ for all $x, y \in B$ and distinct $\kappa, \lambda < \alpha$. All these conditions are in the form of positive equations and hence the theorem follows by Remark 2.7.14.

REMARK 2.7.16. By 2.7.15 the class CA_α is closed under the formation of canonical embedding algebras; the most important consequence of this result is that every algebra in the class CA_α can be embedded in a complete and atomic algebra belonging to the same class (cf. 2.7.20 below). It is of course immediately seen how 2.7.15 can be generalized: every subclass K of CA_α which can be characterized as the class of all CA_α's in which an arbitrarily given set of positive identities holds is closed under the formation of canonical embedding algebras. Actually, this closure property applies to an even larger category of equational classes of CA_α's and general Bo_α's. We shall call a term τ in the discourse language of Bo_α *positive in the wider sense* if there is no subterm of τ beginning with the symbol $-$ which contains an occurrence of a variable. (We assume here that in the discourse language Λ for Bo_α's the distinguished elements of the algebras are represented by individual constants, say 0, 1, $\mathsf{d}_{\kappa\lambda}$, so that the latter by themselves, without the adjunction of variables, form terms of Λ. The reader may compare here the remarks following 1.1.1.) An equation is called *positive in the wider sense* if both its terms have this property. We then have:

(I) *Let Γ be any set of equations which are positive in the wider sense and let* K *be the class of all* Bo_α*'s in which every equation in Γ is identically satisfied. Then* $\mathfrak{Em}\,\mathfrak{A} \in \mathsf{K}$ *whenever* $\mathfrak{A} \in \mathsf{K}$.

To obtain this result we improve 2.7.11–2.7.13 in the following way. Let T be the set of all terms in the language of Bo_α's in which no variables occur. By the stipulations in the Preliminaries, given a Bo_α $\mathfrak{A}$ and a term $\tau \in T$, $\tilde{\tau}_0^{(\mathfrak{A})}$ is a distinguished element of A. We now correlate with every Bo_α $\mathfrak{A}$ the algebra

$$\mathfrak{Pr}'\mathfrak{A} = \langle A, +, \cdot, 0, 1, c_\kappa, \tilde{\tau}_0^{(\mathfrak{A})} \rangle_{\kappa<\alpha, \tau\in T}.$$

It turns out that, with minimal changes in proofs, 2.7.11–2.7.13 can be shown to hold if $\mathfrak{Pr}$ is replaced everywhere by $\mathfrak{Pr}'$; (I) above is an easy consequence of the re-formulated Theorem 2.7.13.

We have previously come across some important classes of CA's characterized by equations which are positive in the wider sense but not in the ordinary sense. For instance, by 2.4.63 the class of CA's of characteristic κ is such a class for each $\kappa < \omega$. (In opposition to 2.4.61 we assume here that all one-element CA's are included in this class.) Thus it follows easily from (I) that the canonical embedding algebra of a CA of characteristic κ is always of the same characteristic. We do not know whether the property of being of characteristic κ can be expressed by a conjunction of positive equations. However, there are interesting properties which are known to carry over from $\mathfrak{A}$ to $\mathfrak{Em}\,\mathfrak{A}$, but which are not expressible in the form of a conjunction of positive equations or even equations positive in the wider sense. In particular, it will be seen in the next theorem, 2.7.17, that for $\alpha < \omega$ the property of being a simple CA_α is one which is carried over to the canonical embedding algebra; however, this property cannot be characterized by any set of equations, whether positive or not.

The result of 2.7.17 is a special case of a more general result. Like 2.7.16(I) this general result, which we now briefly describe, has a metamathematical character. Let τ, σ, and σ' be fixed but arbitrary terms in the discourse language for Bo_α's which are positive in the wider sense. Let K be the class of all Bo_α's $\mathfrak{A}$ such that every $x \in {}^\omega A$ which satisfies $\tau = \mathbf{0}$ also satisfies $\sigma = \sigma'$, and let L be the class of all Bo_α's $\mathfrak{B}$ such that every $x \in {}^\omega B$ which satisfies $\tau \neq \mathbf{0}$ also satisfies $\sigma = \sigma'$. It can be shown that both K and L are closed under the formation of canonical embedding algebras. We shall not prove this result here; however, the proof of 2.7.17 will clearly illustrate the general method. For a proof of the general result in application to arbitrary Boolean algebras with operators, along with a brief discussion of some related results, see Jónsson-Tarski [51], pp. 919 ff.

In 2.7.24 we will come across a large number of equational classes of CA_α's, which are closed under the formation of embedding algebras, but for which it is not known if they can be characterized relative to CA_α by a set of equations which are positive in the wider sense; cf. 2.7.25.

THEOREM 2.7.17. *Suppose $\mathfrak{A} \in CA_\alpha$ and $\alpha < \omega$. If $\mathfrak{A}$ is simple, then so is $\mathfrak{Em}\,\mathfrak{A}$.*

PROOF. Let $\mathfrak{A} = \langle A, +, \cdot, -, 0, 1, c_\kappa, d_{\kappa\lambda}\rangle_{\kappa,\lambda<\alpha}$. From 2.3.14(i),(ii) we see that

(1) $\mathfrak{A}$ is simple iff $|A| > 1$ and $x \neq 0$ implies $c_{(\alpha)}x = 1$ for every $x \in A$.

We now introduce an auxiliary operation O on A by the formula

(2) $$O = [(A \sim \{0\}) \times \{1\}] \cup \{\langle 0, 0\rangle\}.$$

Then by (1) and (2) we have that $\mathfrak{A}$ is simple iff $|A| > 1$ and $Ox \cdot c_{(\alpha)}x = Ox$ for every $x \in A$, i.e., iff a certain positive equation is identically satisfied in the algebra obtained from $\mathfrak{A}$ by adding O as a fundamental operation. In order to apply 2.7.13, however, we must deal with Boolean algebras with operators as defined in 2.7.1. For this reason we set

(3) $$c_\alpha = O$$

and for each $\kappa \leqq \alpha$ we set $d_{\alpha\kappa}$ and $d_{\kappa\alpha}$ equal to some arbitrarily chosen element of A. Then setting

(4) $$\mathfrak{B} = \langle A, +, \cdot, -, 0, 1, c_\kappa, d_{\kappa\lambda}\rangle_{\kappa,\lambda<\alpha+1}$$

we conclude from (1)–(3) and the assumption that $\mathfrak{A}$ is simple that

(5) $$c_\alpha x \cdot c_{(\alpha)}x = c_\alpha x \text{ for every } x \in A.$$

Since c_α is clearly additive, we see that $\mathfrak{B}$ is a $\mathsf{Bo}_{\alpha+1}$ and thus that $\mathfrak{Pr}\,\mathfrak{Em}\,\mathfrak{B} \in \mathbf{HSP}\,\mathfrak{Pr}\,\mathfrak{B}$ by 2.7.13. Therefore, recalling 0.4.64 and setting $\mathfrak{Em}\,\mathfrak{B} = \langle Em\,\mathfrak{B}, +', \cdot', -', 0', 1', c'_\kappa, d'_{\kappa\lambda}\rangle_{\kappa,\lambda<\alpha+1}$,

(6) $$c'_\alpha x \cdot c'_{(\alpha)}x = c'_\alpha x \text{ for every } x \in Em\,\mathfrak{B}.$$

It is an immediate consequence of (2), (3), and the definition of c'_α (Definition 2.7.4) that

$$c'_\alpha = [(Em\,\mathfrak{B} \sim \{0'\}) \times \{1'\}] \cup \{\langle 0', 0'\rangle\}.$$

Therefore, from (6) one obtains $c'_{(\alpha)}x = 1$ for every non-zero $x \in Em\,\mathfrak{B}$ and thus we see that $\mathfrak{Rd}_\alpha \mathfrak{Em}\,\mathfrak{B}$ is simple by 2.3.14(i),(ii). (Here we extend the notation of 2.6.1 in an obvious way from CA_α's to Bo_α's.) The theorem now follows since it is clear from the definition of the canonical embedding algebra that $\mathfrak{Em}\,\mathfrak{A} = \mathfrak{Rd}_\alpha \mathfrak{Em}\,\mathfrak{B}$.

REMARK 2.7.18. Theorem 2.7.17 does not extend to CA's of infinite dimension. In fact, we now show that for every $\alpha \geqq \omega$ there exists a simple CA_α $\mathfrak{A}$ for which $\mathfrak{Em}\,\mathfrak{A}$ is not directly indecomposable. Assume $\alpha \geqq \omega$ and let $\mathfrak{A}$ be the minimal subalgebra in the full cylindric set algebra in the space ${}^\alpha\omega$. It is easily seen from Remarks 2.3.15 that $\mathfrak{A}$ is simple. Let b be the element of $\mathfrak{Em}\,\mathfrak{A}$ defined by the formula

$$b = \prod_{\Gamma \subseteq \alpha,\ |\Gamma| < \omega} \bar{\mathsf{d}}(\Gamma \times \Gamma).$$

For each finite sequence $\Gamma \in {}^{\kappa}Sb\alpha$ of finite subsets of α we have, setting $\Delta = \bigcup_{\lambda < \kappa} \Gamma_\lambda$,

$$\prod_{\lambda < \kappa} \bar{\mathsf{d}}(\Gamma_\lambda \times \Gamma_\lambda) \geqq \bar{\mathsf{d}}(\Delta \times \Delta) > 0.$$

Thus 2.7.8(iii) implies that $b \neq 0$. The complete additivity of c_κ for all $\kappa < \alpha$ leads easily to

$$\sum_{\Gamma \subseteq \alpha,\ |\Gamma| < \omega} \mathsf{c}_{(\Gamma)} b \in Zd\,\mathfrak{Em}\,\mathfrak{A}.$$

On the other hand, for every finite $\Gamma \subseteq \alpha$ we have, choosing any distinct $\kappa, \lambda \in \alpha \sim \Gamma$,

$$\mathsf{c}_{(\Gamma)} b \cdot \prod_{\Delta \subseteq \alpha,\ |\Delta| < \omega} \mathsf{d}_\Delta \leqq \mathsf{c}_{(\Gamma)} b \cdot \mathsf{d}_{\kappa\lambda} = \mathsf{c}_{(\Gamma)}(b \cdot \mathsf{d}_{\kappa\lambda}) = 0.$$

Thus

$$\sum_{\Gamma \subseteq \alpha,\ |\Gamma| < \omega} \mathsf{c}_{(\Gamma)} b \leqq -\prod_{\Delta \subseteq \alpha, |\Delta| < \omega} \mathsf{d}_\Delta.$$

However, $\prod_{\Delta \subseteq \alpha, |\Delta| < \omega} \mathsf{d}_\Delta \neq 0$ since obviously $\prod^{(\mathfrak{A})}_{\Delta \subseteq \alpha, |\Delta| < \omega} \mathsf{d}^{(\mathfrak{A})}_\Delta \neq 0^{(\mathfrak{A})}$, as one sees by using 2.7.8(iii) again. Therefore, we have

$$\sum_{\Gamma \subseteq \alpha, |\Gamma| < \omega} \mathsf{c}_{(\Gamma)} b \in Zd\,\mathfrak{Em}\,\mathfrak{A} \sim \{0, 1\},$$

which implies that $\mathfrak{Em}\,\mathfrak{A}$ is not directly indecomposable by 2.4.14.

The preceding example shows also that for every $\alpha \geqq \omega$ a CA_α $\mathfrak{A}$ may be in any one of the classes Mn_α, Lf_α, Dc_α, and Ss_α without $\mathfrak{Em}\,\mathfrak{A}$ being in the same class.

REMARK 2.7.19. It is still an open problem whether every equation which holds in a CA $\mathfrak{A}$ continues to hold in its canonical embedding algebra $\mathfrak{Em}\,\mathfrak{A}$. As we shall now see, however, this problem has a negative solution if $\mathfrak{A}$ is allowed to be an arbitrary Boolean algebra with operators.

Consider any $\alpha \geqq 2$ and let A be the set consisting of all finite subsets of ω and of their complements (with respect to ω). Set

$$C_0 = \langle \{2\} \cup \{x+2 : x \in X\} : X \in A \rangle \text{ and } C_1 = [(A \sim \{0\}) \times \{\omega\}] \cup \{\langle 0, 0 \rangle\}.$$

C_0 and C_1 are easily seen to be additive functions from A onto itself. Therefore, choosing any additive functions in ${}^A A$ to be the remaining fundamental operations C_κ with $2 \leqq \kappa < \alpha$, and any members of A for the distinguished elements $D_{\kappa\lambda}$ with $\kappa, \lambda < \alpha$, we have that $\mathfrak{A} = \langle A, \cup, \cap, {}_\omega\sim, 0, \omega, C_\kappa, D_{\kappa\lambda} \rangle_{\kappa,\lambda < \alpha}$ is an α-dimensional Boolean algebra with operators such that $C_1[(C_0X \sim X) \cup (X \sim C_0X)] = \omega$ for every $X \in A$. I.e., consider the equation

(1) $$\mathsf{c}_1[(\mathsf{c}_0 x \cdot - x) + (x \cdot - \mathsf{c}_0 x)] = \mathbf{1}$$

where $+, \cdot, -, \mathbf{1}, \mathsf{c}_0, \mathsf{c}_1$ are the operation symbols of the discourse language for Bo_α's corresponding respectively to the fundamental operations $+, \cdot, -, 1, \mathsf{c}_0, \mathsf{c}_1$. Then (1) is identically satisfied in $\mathfrak{A}$. However, this equation does not hold in the canonical embedding algebra of $\mathfrak{A}$; indeed, it does not hold in any $\mathsf{Bo}_\alpha\, \mathfrak{B} = \langle B, +, \cdot, -, 0, 1, \mathsf{c}_\kappa, \mathsf{d}_{\kappa\lambda}\rangle_{\kappa,\lambda<\alpha}$ for which c_1 is a normal function and the Boolean part of $\mathfrak{B}$ is a complete Boolean algebra with more than one element. In particular, using the fact that c_0 is additive and hence increasing, we obtain an element $b \in B$ such that $\mathsf{c}_0 b = b$ by setting

$$b = \sum\nolimits_{\mathsf{c}_0 x \geqq x \in B} x$$

(cf. Tarski [55a*], p. 286 f.). Then from the normality of c_1 we conclude that $\mathsf{c}_1[(\mathsf{c}_0 b \cdot - b) + (b \cdot - \mathsf{c}_0 b)]$ equals 0 and hence is different from 1 since $\mathfrak{B}$ is non-trivial. Thus (1) is not identically satisfied in $\mathfrak{B}$.

We now give the embedding theorem mentioned in the introductory remarks to this section; the corresponding general result in Jónsson-Tarski [51], p. 226, Theorem 2.15 is called the *extension theorem.*

THEOREM 2.7.20. *For any* $\mathsf{CA}_\alpha\,\mathfrak{A}$ *there exists a complete and atomic* $\mathsf{CA}_\alpha\,\mathfrak{B}$ *such that* $\mathfrak{A} \subseteq \mathfrak{B}$.

PROOF. By 2.7.5(i) and 2.7.15, $\mathfrak{Em}\,\mathfrak{A}$ is a complete and atomic CA_α and we have $\mathfrak{A} \cong\mid\subseteq \mathfrak{Em}\,\mathfrak{A}$. Hence by 0.2.17(ii) we can construct an algebra $\mathfrak{B}$ such that $\mathfrak{A} \subseteq \mathfrak{B} \cong \mathfrak{Em}\,\mathfrak{A}$. From this the conclusion follows immediately.

By analyzing the proof of 2.7.20 we can conclude that $\mathfrak{B}$ is a perfect extension of $\mathfrak{A}$ in the sense of 2.7.7. From this we further conclude that, in case $\mathfrak{A}$ is infinite, there is an infinite set $X \subseteq A$ such that $\sum^{(\mathfrak{A})}X$ exists but is different from $\sum^{(\mathfrak{B})}X$. (Actually, it turns out that $\sum^{(\mathfrak{A})}X = \sum^{(\mathfrak{B})}X$ only if there is a finite set $Y \subseteq X$ such that $\sum^{(\mathfrak{A})}X = \sum^{(\mathfrak{A})}Y$.) Thus we can say that the embedding of $\mathfrak{A}$ in its perfect extension $\mathfrak{B}$ does not preserve sums. On the other hand, it is known from the theory of Boolean algebras that every $\mathsf{BA}\,\mathfrak{A}$ can be embedded as a subalgebra in a complete (but not necessarily atomic) $\mathsf{BA}\,\mathfrak{B}$ in such a way that every element of $\mathfrak{B}$ is a sum of elements of $\mathfrak{A}$; this implies that the embedding preserves all sums. This result can be extended to arbitrary CA's (as shown in Mangani [66b], p. 309). In fact we have:

THEOREM 2.7.21. *For any* $\mathsf{CA}_\alpha\,\mathfrak{A}$ *there is a complete* $\mathsf{CA}_\alpha\,\mathfrak{B}$ *such that* $\mathfrak{A} \subseteq \mathfrak{B}$ *and the following two conditions hold*:

(i) $\sum^{(\mathfrak{A})} X = \sum^{(\mathfrak{B})} X$ *for every* $X \subseteq A$ *for which* $\sum^{(\mathfrak{A})} X$ *exists*;

(ii) $b = \sum_{b \geqq x \in A} x$ *for every* $b \in B$.

PROOF. As was mentioned above, there exists a complete $\mathsf{BA}\,\mathfrak{B}' = \langle B, +, \cdot, -, 0, 1\rangle$ such that $\mathfrak{B}\mathfrak{l}\mathfrak{A} \subseteq \mathfrak{B}'$ and conditions (i), (ii) hold (with $\mathfrak{B}$ replaced by $\mathfrak{B}'$); cf. Sikorski [64*], pp. 152ff. To complete the proof we need only to provide $\mathfrak{B}'$ with a cylindric structure extending that of $\mathfrak{A}$.

For each $\kappa < \alpha$ let c_κ be the function in ${}^B B$ defined by the condition

(1) $$c_\kappa b = \sum_{b \geqq x \in A} \mathsf{c}_\kappa x \text{ for every } b \in B,$$

and set $\mathfrak{B} = \langle B, +, \cdot, -, 0, 1, c_\kappa, \mathsf{d}_{\kappa\lambda}\rangle_{\kappa,\lambda<\alpha}$. It is then obvious that $\mathfrak{A} \subseteq \mathfrak{B}$, so it remains only to check that the postulates of 1.1.1 hold for $\mathfrak{B}$. Postulates (C_0), (C_1), (C_5), (C_6), and (C_7) clearly hold in $\mathfrak{B}$. In establishing the other postulates the following lemma proves convenient

(2) $$c_\kappa \sum X = \sum \mathsf{c}_\kappa^* X \text{ for every } \kappa < \alpha \text{ and } X \subseteq A.$$

Indeed, by (1) we have

(3) $$c_\kappa \sum X = \sum_{\sum X \geqq y \in A} \mathsf{c}_\kappa y.$$

Thus, since $\sum X \geqq x \in A$ for every $x \in X$, we get $\mathsf{c}_\kappa x \leqq c_\kappa \sum X$ for all such x, and hence $\sum \mathsf{c}_\kappa^* X \leqq c_\kappa \sum X$. On the other hand, for any y such that $\sum X \geqq y \in A$ it is easily seen that $y = \sum_{x \in X}(y \cdot x) = \sum^{(\mathfrak{A})}_{x \in X}(y \cdot x)$ by (i), and so, using (i) again,

$$\mathsf{c}_\kappa y = \sum^{(\mathfrak{A})}_{x \in X} \mathsf{c}_\kappa(y \cdot x) = \sum_{x \in X} \mathsf{c}_\kappa(y \cdot x) \leqq \sum \mathsf{c}_\kappa^* X.$$

Hence (3) implies that $c_\kappa \sum X \leqq \sum \mathsf{c}_\kappa^* X$ and thus (2) is established.

To prove (C_2) note that by (1) and (ii) we have, for every $b \in B$,

$$c_\kappa b = \sum_{b \geqq x \in A} \mathsf{c}_\kappa x \geqq \sum_{b \geqq x \in A} x = b.$$

For (C_3) take any $\kappa < \alpha$ and $a, b \in B$. Then

$$\begin{aligned} c_\kappa a \cdot c_\kappa b &= \textstyle\sum_{a \geqq x \in A} \mathsf{c}_\kappa x \cdot \sum_{b \geqq y \in A} \mathsf{c}_\kappa y && \text{by (1)} \\ &= \textstyle\sum_{a \geqq x \in A,\, b \geqq y \in A} (\mathsf{c}_\kappa x \cdot \mathsf{c}_\kappa y) \\ &= \textstyle\sum_{a \geqq x \in A,\, b \geqq y \in A} \mathsf{c}_\kappa (x \cdot \mathsf{c}_\kappa y) \\ &= c_\kappa \textstyle\sum_{a \geqq x \in A,\, b \geqq y \in A} (x \cdot \mathsf{c}_\kappa y) && \text{by (2)} \\ &= c_\kappa [\textstyle\sum_{a \geqq x \in A} x \cdot \sum_{b \geqq y \in A} \mathsf{c}_\kappa y] \\ &= c_\kappa (a \cdot c_\kappa b) && \text{by (1),(ii).} \end{aligned}$$

Finally, consider any $\kappa, \lambda < \alpha$ and $b \in B$. Then by (1) and (2) we get

$$c_\kappa c_\lambda b = c_\kappa \sum_{b \geq x \in A} \mathsf{c}_\lambda x = \sum_{b \geq x \in A} \mathsf{c}_\kappa \mathsf{c}_\lambda x.$$

Similarly, $c_\lambda c_\kappa = \sum_{b \geq x \in A} \mathsf{c}_\lambda \mathsf{c}_\kappa x$ so that $c_\kappa c_\lambda b = c_\lambda c_\kappa b$. Thus ($C_4$) holds and the theorem is proved.

REMARK 2.7.22. The algebra $\mathfrak{B}$ which satisfies the conditions of the preceding theorem is called a *completion* of $\mathfrak{A}$. We see thus that for eveiy CA $\mathfrak{A}$ there are two different kinds of complete extensions of $\mathfrak{A}$, namely, perfect extensions and completions; they coincide only if $\mathfrak{A}$ is finite. Any two completions $\mathfrak{B}$ and $\mathfrak{B}'$ of the same CA_α $\mathfrak{A}$ are isomorphic. In fact, there always exists an isomorphism from $\mathfrak{B}$ onto $\mathfrak{B}'$ which is the identity when restricted to the universe of $\mathfrak{A}$.

The notion of completion will not be involved further in this section. We may mention only that several results established here for canonical embedding algebras hold also for completions. For instance, 2.7.13 when restricted to CA's turns out to hold for completions as well as for canonical embedding algebras, and, as a consequence, every equation positive in the wider sense which holds in a CA $\mathfrak{A}$ continues to hold in each completion of $\mathfrak{A}$. Also, in connection with 2.7.17 we note that by a direct argument, which does not use 2.7.13 (in its formulation for completions), one can easily show that every completion of a simple CA of arbitrary, possibly infinite, dimension is again simple. All these results about completions are due to Monk; the work in this direction is still in progress.

We continue now with the investigation of the canonical embedding algebra. We discussed above in 2.7.14–2.7.19 properties which carry over from a CA $\mathfrak{A}$ to its canonical embedding algebra $\mathfrak{Em}\,\mathfrak{A}$. We shall now concern ourselves, generally speaking, with relations which hold between $\mathfrak{Em}\,\mathfrak{A}$ and $\mathfrak{Em}\,\mathfrak{B}$ whenever they hold between $\mathfrak{A}$ and $\mathfrak{B}$.

Most of the relevant results 2.7.23–2.7.31 were obtained by Henkin and Monk; however 2.7.26–2.7.28 are due to Don Pigozzi.

THEOREM 2.7.23. *Assume* $\mathfrak{B} \in \mathsf{CA}_\beta$.

(i) *If* $\mathfrak{A} \subseteq \mathfrak{B}$, *then* $\mathfrak{Em}\,\mathfrak{A} \cong|\subseteq \mathfrak{Em}\,\mathfrak{B}$.

(ii) *If* $\alpha \leqq \beta$ *and* $\mathfrak{A} \subseteq \mathfrak{Nr}_\alpha \mathfrak{B}$, *then* $\mathfrak{Em}\,\mathfrak{A} \cong|\subseteq \mathfrak{Nr}_\alpha \mathfrak{Em}\,\mathfrak{B}$.

PROOF. In view of 2.6.30(ii) we need only prove part (ii). So assume $\mathfrak{A} \subseteq \mathfrak{Nr}_\alpha \mathfrak{B}$ and set $\mathfrak{A}' = em^{(\mathfrak{B})*}\mathfrak{A}$ and $\mathfrak{B}' = em^{(\mathfrak{B})*}\mathfrak{B}$. We then have

$$\mathfrak{A}' \subseteq \mathfrak{Nr}_\alpha \mathfrak{B}' \subseteq \mathfrak{Nr}_\alpha \mathfrak{Em}\,\mathfrak{B}.$$

Set

(1) $$C = \{\prod X : X \subseteq A'\}$$

and

(2) $$D = \{\sum Y : Y \subseteq C\}.$$

$\mathfrak{Bl}\,\mathfrak{Em}\,\mathfrak{B}$ is the Boolean set algebra of all subsets of a certain set (cf. Definition 2.7.4) and thus the infinitary operations $\sum$ and $\prod$ of $\mathfrak{Em}\,\mathfrak{B}$ coincide respectively with set-theoretic union and intersection. This implies of course that $\mathfrak{Em}\,\mathfrak{B}$ is completely distributive, i.e., that for every system $\langle X_i : i \in I\rangle$ of subsets of $Em\,\mathfrak{B}$ we have

(3) $$\prod_{i\in I}\sum X_i = \sum_{f\in \mathsf{P}_{i\in I}X_i}\prod_{i\in I} f_i.$$

From this fact we easily conclude that D is a subuniverse of $\mathfrak{Bl}\,\mathfrak{Em}\,\mathfrak{B}$. Indeed, consider any $a \in D$. Then there is a system $\langle Y_i : i \in I\rangle$ of subsets of A' such that $a = \sum_{i\in I}\prod Y_i$, and hence

$$-a = \prod_{i\in I}\sum X_i$$

where $X_i = \{-y : y \in Y_i\} \subseteq A'$ for every $i \in I$. Then by (3) one concludes that $-a \in D$. Therefore, D is closed under $-$, and it follows immediately from (2) that D is also closed under $+$. We see then that

(4) $$\langle D, +, \cdot, -, 0, 1\rangle \subseteq \mathfrak{Bl}\,\mathfrak{Em}\,\mathfrak{B}.$$

Consider now an $\kappa < \alpha$ and an arbitrary

(5) $$a \in C.$$

Then by (1), 2.7.3, and 2.7.9 (taking $\mathfrak{B}$ for $\mathfrak{A}$ in 2.7.9) we have

(6) $$\mathsf{c}_\kappa a = \prod_{a \leqq x \in B'} \mathsf{c}_\kappa x.$$

Consider any $u \in At\,\mathfrak{Em}\,\mathfrak{B}$ such that $u \cdot \mathsf{c}_\kappa a = 0$. Then (6) implies the existence of a b such that

(7) $$a \leqq b \in B' \text{ and } u \cdot \mathsf{c}_\kappa b = 0.$$

By (1), (5), the first part of (7), and 2.7.8(iii) there is a finite $Y \subseteq A'$ such that $a \leqq \prod Y \leqq b$. Then the second part of (7) implies that

$$u \cdot \mathsf{c}_\kappa \prod Y = 0.$$

Since $a \leqq \prod Y \in A'$ and u is an arbitrary atom of $\mathfrak{Em}\,\mathfrak{B}$ disjoint from $\mathsf{c}_\kappa a$, we have shown that

(8) $$\mathsf{c}_\kappa a = \prod_{a \leqq x \in A'} \mathsf{c}_\kappa x \text{ for all } a \in C \text{ and } \kappa < \alpha.$$

From (1) and (8) it follows in particular that C is closed under c_κ for each $\kappa < \alpha$, and thus in view of (2) and the complete additivity of c_κ we have that D is closed under c_κ as well; it follows, in addition, that $D \upharpoonright \mathsf{c}_\kappa \subseteq Id$ if $\kappa \in \beta \sim \alpha$.

Thus in view of (4) we see that D is a subuniverse of $\mathfrak{Nr}_\alpha \mathfrak{Em}\,\mathfrak{B}$ which includes A', and it remains only to show that $\mathfrak{D} \cong \mathfrak{Em}\,\mathfrak{A}$ where

$$\mathfrak{D} = \langle D, +, \cdot, -, 0, 1, \mathsf{c}_\kappa, \mathsf{d}_{\kappa\lambda} \rangle_{\kappa,\lambda<\alpha}.$$

In order to do this it suffices to verify that the four conditions of Theorem 2.7.6 hold when $\mathfrak{B}$ and $\mathfrak{A}$ are replaced respectively by $\mathfrak{D}$ and $\mathfrak{A}'$.

Recall that in proving D was a Boolean subuniverse of $\mathfrak{Em}\,\mathfrak{B}$ we used the fact that (2) obviously implied that D is closed under $+$. Actually, (2) just as easily implies that D is closed under the infinitary operation $\sum$, so that we may conclude that $\mathfrak{D}$ is a complete subalgebra of $\mathfrak{Nr}_\alpha \mathfrak{Em}\,\mathfrak{B}$ in the sense that

(9) $\mathfrak{D}$ is a complete CA_α

and

(10) $\sum^{(\mathfrak{D})} X = \sum X$ and $\prod^{(\mathfrak{D})} X = \prod X$ for every $X \subseteq A'$.

From (10) and the fact that (3) holds in particular for every system $\langle X_i : i \in I \rangle$ of subsets of A' we see that $\mathfrak{Bl}\,\mathfrak{D}$ is completely distributive. It follows then from (9) by the theory of BA's that $\mathfrak{D}$ is atomic; cf., e.g., Sikorski [64*], § 25. Therefore we get 2.7.6(i) from (9) and 2.7.3; also (10) and 2.7.5(ii) together immediately give 2.7.6(ii). Finally, from (2) we see that $At\,\mathfrak{D} \subseteq C$, and thus 2.7.6(iii),(iv) follow at once from (1) and (8), respectively. This completes the proof of the theorem.

Notice that the above proof actually establishes more than just $\mathfrak{Em}\,\mathfrak{A} \cong\mid\subseteq \mathfrak{Nr}_\alpha \mathfrak{Em}\,\mathfrak{B}$. Indeed, it shows that, if $\mathfrak{A} \subseteq \mathfrak{Nr}_\alpha \mathfrak{B}$, with $\alpha \leqq \beta$ and $\mathfrak{B} \in \mathsf{CA}_\beta$, then $\mathfrak{Em}\,\mathfrak{A}$ can be embedded in $\mathfrak{Nr}_\alpha \mathfrak{Em}\,\mathfrak{B}$ as a complete subalgebra.

Finally, we remark that, as opposed to what was actually done in the proof of 2.7.23, an isomorphic embedding of $\mathfrak{Em}\,\mathfrak{A}$ into $\mathfrak{Nr}_\alpha \mathfrak{Em}\,\mathfrak{A}$ could be obtained directly. Bearing in mind that elements of $\mathfrak{Em}\,\mathfrak{A}$ and $\mathfrak{Em}\,\mathfrak{B}$ are actually sets of Boolean ideals of $\mathfrak{A}$ and $\mathfrak{B}$, respectively, we take as the image of any $X \in \mathfrak{Em}\,\mathfrak{A}$ the set of all proper maximal Boolean ideals I of $\mathfrak{B}$ such that $I \supseteq J$ for some $J \in X$. The mapping obtained in this manner can be shown to be the desired isomorphic embedding. Several of the other theorems of this section could also be established directly by a similar method.

Corollary 2.7.24. (i) *If* $\beta \geqq \alpha$, *then* $\mathfrak{Em}^*\mathsf{SNr}_\alpha \mathsf{CA}_\beta \subseteq \mathsf{SNr}_\alpha \mathsf{CA}_\beta$;
(ii) *in particular*, $\mathfrak{Em}^*\mathsf{SNr}_\alpha \mathsf{CA}_{\alpha+\omega} \subseteq \mathsf{SNr}_\alpha \mathsf{CA}_{\alpha+\omega}$.

PROOF: by 2.7.15 and 2.7.23(ii).

REMARKS 2.7.25. As was pointed out in 2.6.26, 2.7.24(ii) implies that the canonical embedding algebra of any representable CA is again representable. In this connection it is interesting to note that, while the class of representable CA_α's, and more generally the classes $\mathsf{SNr}_\alpha\mathsf{CA}_\beta$ for every $\beta > \alpha$, are all equational (cf. 2.6.33), we do not know, except in the case $\alpha \leqq 2$, if any of these classes can be characterized by equations which are positive in the wider sense. For this reason we have not been able to derive 2.7.24(ii) as a direct corollary from 2.7.16(I). It will be seen in Part II that the class $\mathsf{SNr}_2\mathsf{CA}_\omega$ can be singled out from among all CA_2's by a pair of equations which are positive in the wider sense; however, the particular equations which are claimed in 2.6.44 to serve the same purpose are obviously not of this kind. The same applies to every class $\mathsf{SNr}_2\mathsf{CA}_\beta$ with $\beta \geqq 3$ since, as was noted in 2.6.44, all these classes coincide.

THEOREM 2.7.26. *Assume $\mathfrak{A} \in \mathsf{CA}_\alpha$ and let O be the subset of $Em\,\mathfrak{A}$ defined by the formula*

$$O = \{\Sigma X : X \subseteq em^*A\}.$$

Then

(i) *the function $\langle \Sigma em^*I : I \in Il\,\mathfrak{A}\rangle$ establishes a one-one correspondence between ideals of $\mathfrak{A}$ and elements of $O \cap Zd\,\mathfrak{Em}\,\mathfrak{A}$;*

(ii) *for any $I \in Il\,\mathfrak{A}$, $\mathfrak{Em}(\mathfrak{A}/I) \cong \mathfrak{Rl}_{-\Sigma em^*I}\,\mathfrak{Em}\,\mathfrak{A}$.*

PROOF. Set

$$\mathfrak{A}' = em^*\mathfrak{A} \text{ and } F = \langle \Sigma I : I \in Il\,\mathfrak{A}'\rangle.$$

Then in view of 2.7.5(i) it obviously suffices to prove that F is a one-one function from $Il\,\mathfrak{A}'$ onto $O \cap Zd\,\mathfrak{Em}\,\mathfrak{A}$ such that

$$\mathfrak{Em}(\mathfrak{A}'/I) \cong \mathfrak{Rl}_{-\Sigma I}\,\mathfrak{Em}\,\mathfrak{A}$$

for every $I \in Il\,\mathfrak{A}'$. If $\kappa < \alpha$ and $I \in Il\,\mathfrak{A}'$, then $c_\kappa^* I \subseteq I$ by 2.3.7(ii), and hence by the complete additivity of c_κ we get $\Sigma I \in Zd\,\mathfrak{Em}\,\mathfrak{A}$; since clearly $\Sigma I \in O$, we conclude that

(1) $$\Sigma I \in O \cap Zd\,\mathfrak{Em}\,\mathfrak{A} \text{ for every } I \in Il\,\mathfrak{A}'.$$

On the other hand, assume $a \in O \cap Zd\,\mathfrak{Em}\,\mathfrak{A}$ and set $I = \{x : x \in A', x \leqq a\}$; then $I \in Il\,\mathfrak{A}'$ by 2.3.10(iii). Obviously $\Sigma I \leqq a$, and, because $a \in O$, there is an $X \subseteq A'$ such that $a = \Sigma X$, which gives $a \leqq \Sigma I$ since clearly $X \subseteq I$. Thus $a = \Sigma I$, and the range of F includes all of $O \cap Zd\,\mathfrak{Em}\,\mathfrak{A}$. Hence in view of (1) we have shown that $Rg F = On Zd \mathfrak{Em}\mathfrak{A}$. To prove that F is one-one it is obviously sufficient to prove that, for any given $I \in Il\mathfrak{A}'$,

$$I = \{x : x \in A', x \leqq \sum I\}. \tag{2}$$

Suppose $x \leqq \sum I$ for some $x \in A'$. Then by 2.7.3 and 2.7.8(ii) there is a finite $X \subseteq I$ such that $x \leqq \sum X$. Clearly $\sum X \in I$ so that $x \in I$, and this shows that $I \supseteq \{x : x \in A', x \leqq \sum I\}$. Since the inclusion in the opposite direction is obvious, we obtain (2) and, consequently, the fact that F is a one-one function from $Il\mathfrak{A}'$ onto $On Zd \mathfrak{Em}\mathfrak{A}$. This proves part (i).

In the remaining portion of the proof assume that I is a fixed but arbitrary ideal of $\mathfrak{A}'$ and let

$$h = \langle x \cdot -\sum I : x \in Em \mathfrak{A} \rangle. \tag{3}$$

Then by (1) and 2.3.26 we have $h \in Ho(\mathfrak{Em}\mathfrak{A}, \mathfrak{Rl}_{-\sum I}\mathfrak{Em}\mathfrak{A})$ and thus $h' \in Ho(\mathfrak{A}', h^*\mathfrak{A}')$, where $h' = A' \upharpoonright h$. Furthermore, from (2) it is easily seen that

$$(h'^{-1})^{\star} 0^{(h^*\mathfrak{A}')} = I$$

and hence by 2.3.6(II) we may conclude that

$$\mathfrak{A}'/I \cong h^*\mathfrak{A}' \subseteq \mathfrak{Rl}_{-\sum I}\mathfrak{Em}\mathfrak{A}. \tag{4}$$

To complete the proof we set

$$\mathfrak{B} = \mathfrak{Rl}_{-\sum I}\mathfrak{Em}\mathfrak{A} \tag{5}$$

and proceed to verify that the four conditions of Theorem 2.7.6 hold when $\mathfrak{A}$ is replaced by $h^*\mathfrak{A}'$. Condition 2.7.6(i) obviously holds. Consider any $X \subseteq h^*\mathfrak{A}'$ such that $\sum^{(\mathfrak{B})} X = 1^{(\mathfrak{B})}$; clearly then $\sum X = -\sum I$ so that, in view of (3)

$$\sum (h'^{-1})^* X \geqq \sum X = -\sum I.$$

Thus by (1) and 2.7.8(i),(ii) there is a finite $Y \subseteq (h'^{-1})^* X$ such that $\sum Y \geqq -\sum I$. Then $h^*Y \subseteq X$, h^*Y is finite, and

$$\sum h^* Y = \sum Y \cdot -\sum I = -\sum I$$

by (3). Consequently, we have $\sum^{(\mathfrak{B})} h^* Y = 1^{(\mathfrak{B})}$, and 2.7.6(ii) has been shown to hold. Turning to 2.7.6(iii), consider any $x, y \in At\,\mathfrak{B}$ with $x \neq y$. From (5) we see that $x, y \in At\,\mathfrak{Em}\mathfrak{A}$ and $x, y \leqq -\sum I$; thus by 2.7.5(iii) there is a $z \in A'$ such that $x \leqq z$ and $y \leqq -z$ and hence, by (3), $x = hx \leqq hz$ and $y = hy \leqq h-z = -^{(\mathfrak{B})} hz$. Finally, in order to obtain 2.7.6(iv), take any

$a \in At\,\mathfrak{B}$. Then also $a \in At\,\mathfrak{Em}\,\mathfrak{A}$ and hence by 2.7.5(iv) we obtain

$$\mathsf{c}_\kappa a = \prod_{a \leqq y \in A'} \mathsf{c}_\kappa y.$$

Therefore, in view of (3) and the fact that $a \leqq -\sum I$, we have

$$\mathsf{c}_\kappa^{(\mathfrak{B})} a = h\mathsf{c}_\kappa a = \prod_{a \leqq y \in A'} \mathsf{c}_\kappa hy = \prod_{a \leqq x \in h^*A'} \mathsf{c}_\kappa^{(\mathfrak{B})} x.$$

Thus 2.7.6(iv) holds, and the proof of (ii) is finished.

COROLLARY 2.7.27. *If* $\mathfrak{A} \in \mathsf{CA}_\alpha$ *and* $\mathfrak{A} \succcurlyeq \mathfrak{B}$, *then* $\mathfrak{Em}\,\mathfrak{B} \mid \mathfrak{Em}\,\mathfrak{A}$.

REMARK 2.7.28. It is of course obvious that, if $\mathfrak{A}, \mathfrak{B} \in \mathsf{CA}_\alpha$ and $\mathfrak{A} \cong \mathfrak{B}$, then $\mathfrak{Em}\,\mathfrak{A} \cong \mathfrak{Em}\,\mathfrak{B}$. On the other hand, making use of Corollary 2.7.27 we can show that the converse does not hold. For this purpose we take any pair $\mathfrak{A}, \mathfrak{B}$ of CA_α's such that $\mathfrak{A} \succcurlyeq \mathfrak{B}$ and $\mathfrak{B} \succcurlyeq \mathfrak{A}$ but not $\mathfrak{A} \cong \mathfrak{B}$. Such algebras exist for any α; for example, we may take $\mathfrak{A} = \mathfrak{Fr}_\omega\mathsf{CA}_\alpha$ and $\mathfrak{B} = \mathfrak{A} \times \mathfrak{C}$ where $\mathfrak{C}$ is a two-element CA_α. Clearly $\mathfrak{A} \succcurlyeq \mathfrak{B}$ and $\mathfrak{B} \succcurlyeq \mathfrak{A}$, but not $\mathfrak{A} \cong \mathfrak{B}$ since $\mathfrak{A}$ is atomless by 2.5.13, while $\mathfrak{B}$ is obviously not atomless. On the other hand, by 2.7.27 we have $\mathfrak{Em}\,\mathfrak{A} \mid \mathfrak{Em}\,\mathfrak{B}$ and $\mathfrak{Em}\,\mathfrak{B} \mid \mathfrak{Em}\,\mathfrak{A}$ and hence $\mathfrak{Em}\,\mathfrak{A} \cong \mathfrak{Em}\,\mathfrak{B}$ by 2.4.24 and 2.7.5(i).

We want to add that it is very easy to find examples of non-isomorphic discrete CA_α's $\mathfrak{A}$ and $\mathfrak{B}$ for which it can be shown directly, without appeal to the results in 2.4.9–2.4.33, that $\mathfrak{Em}\,\mathfrak{A} \cong \mathfrak{Em}\,\mathfrak{B}$; moreover, in view of the next theorem, 2.7.29, this observation extends to non-discrete CA_α's as well; compare in this regard the discussion in 2.4.34. On the other hand, arguing as in the first part of the present remark (and hence using 2.4.24 and 2.7.27 in an essential way), one can also find, for every α, hereditarily non-discrete CA_α's $\mathfrak{A}$ and $\mathfrak{B}$ such that $\mathfrak{Em}\,\mathfrak{A} \cong \mathfrak{Em}\,\mathfrak{B}$ but $\mathfrak{A}$ is not isomorphic to $\mathfrak{B}$. The actual choice of $\mathfrak{A}$ and $\mathfrak{B}$ which serve for this purpose will be left as an exercise; cf. 2.4.36.

THEOREM 2.7.29. *If* $\mathfrak{A} \in {}^I\mathsf{CA}_\alpha$, *then*

$$\mathsf{P}_{i \in I}\,\mathfrak{Em}\,\mathfrak{A}_i \mid \mathfrak{Em}(\mathsf{P}_{i \in I}\,\mathfrak{A}_i);$$

if, in addition, $|I| < \omega$, *then* $\mathsf{P}_{i \in I}\,\mathfrak{Em}\,\mathfrak{A}_i \cong \mathfrak{Em}(\mathsf{P}_{i \in I}\,\mathfrak{A}_i)$.

PROOF. Let $\mathfrak{B} = em^*\mathsf{P}_{i \in I}\,\mathfrak{A}_i$. Then by 2.4.7 and 2.7.5(i) there exists a system $\langle x_i : i \in I\rangle$ of elements of $\mathfrak{B}$ satisfying the following conditions:

(1) $$\Delta x_i = 0 \text{ for all } i \in I;$$

(2) $$x_i \cdot x_j = 0 \text{ if } i, j \in I \text{ and } i \neq j;$$

(3) $$\textstyle\sum_{i \in I}^{(\mathfrak{B})} x_i = 1;$$

(4) $$\mathfrak{Rl}_{x_i}\mathfrak{B} \cong \mathfrak{A}_i \text{ for all } i \in I;$$

(5) $$\text{for all } y \in {}^I\mathfrak{B} \text{ the sum } \Sigma^{(\mathfrak{B})}_{i \in I}(y_i \cdot x_i) \text{ exists.}$$

From (1) and 2.3.10(iii) we obtain $-\Sigma Ig\{-x_i\} = x_i$ for each $i \in I$. Hence by (4), 2.3.27(i), and 2.7.26 we have

(6) $$\mathfrak{Rl}_{x_i}\mathfrak{Em}\mathsf{P}_{i \in I}\mathfrak{A}_i \cong \mathfrak{Em}(\mathfrak{B}/Ig\{-x_i\}) \cong \mathfrak{Em}\mathfrak{A}_i \text{ for all } i \in I.$$

Thus, if

(7) $$\mathfrak{C} = \mathfrak{Rl}_{\Sigma_{i \in I} x_i}\mathfrak{Em}\mathsf{P}_{i \in I}\mathfrak{A}_i,$$

we have by (6) and 2.2.15

(8) $$\mathfrak{Rl}_{x_i}\mathfrak{C} \cong \mathfrak{Em}\mathfrak{A}_i \text{ for each } i \in I;$$

also, it is clear that

(9) $$\Sigma^{(\mathfrak{C})}_{i \in I} x_i = 1^{(\mathfrak{C})}.$$

Using (1), (2), (8), (9), and the obvious fact that $\mathfrak{C}$ is complete, we conclude from (7) and 2.4.7 that

(10) $$\mathsf{P}_{i \in I}\mathfrak{Em}\mathfrak{A}_i \cong \mathfrak{Rl}_{\Sigma_{i \in I} x_i}\mathfrak{Em}\mathsf{P}_{i \in I}\mathfrak{A}_i.$$

The first part of the theorem now follows immediately from (10), and the last part follows from (10) along with the observation that by (3) we have

$$\Sigma_{i \in I} x_i = \Sigma^{(\mathfrak{B})}_{i \in I} x_i = 1^{(\mathfrak{B})} = 1^{(\mathfrak{Em}\mathsf{P}_{i \in I}\mathfrak{A}_i)}$$

whenever I is finite.

REMARK 2.7.30. From 2.4.43 and 2.7.17 we conclude that, if a CA_α $\mathfrak{A}$ is subdirectly, weakly subdirectly, or directly indecomposable, and $\alpha < \omega$, then $\mathfrak{Em}\mathfrak{A}$ has the same property. However, none of these properties is preserved in general for α infinite, as is seen from 0.3.58 and 2.7.18.

REMARK 2.7.31. In accordance with what was said at the beginning of this section we have not been interested here in investigating the canonical embedding algebras of arbitrary Boolean algebras with operators — at least not in their own right. We will only mention that in one form or other almost all of the results obtained exclusively for CA's extend to arbitrary Bo's. To be precise, Theorems 2.7.23(i) and 2.7.29 along with Corollary 2.7.27 continue to hold when "CA" is changed everywhere to "Bo". Theorem 2.7.26 also continues to hold if, in addition to the above changes, $Zd\mathfrak{Em}\mathfrak{A}$ is replaced by

$$\{x : x \in A,\ \mathsf{c}_\kappa x \leqq x + \mathsf{c}_\kappa 0 \text{ for all } \kappa < \alpha\}.$$

Theorem 2.7.23(ii) and its corollary, 2.7.24, also continue to hold when "CA" is changed everywhere to "Bo" and the notion of a neat subreduct is generalized in a suitable way; we will not formulate this generalization, however, since it seems to be of little interest. The proofs of all these results are rather obvious modifications of the proofs of the corresponding results for CA's. Actually, the only results which do not extend in any way to Bo's are 2.7.17 and the positive results of 2.7.30. Indeed, it is not hard to find a simple Bo_α for any given $\alpha > 0$ whose canonical embedding algebra is not even directly indecomposable.

Finally we remark that every Bo can be shown to have a completion in the sense of 2.7.22.

We now turn to our last task in this section, namely, to formulating and establishing a kind of general representation theorem for CA's, which was mentioned at the beginning of the section. To understand the underlying idea, consider first a CA $\mathfrak{A}$ assumed to be complete and atomic. It is well known that $\mathfrak{Bl}\,\mathfrak{A}$, the Boolean part of $\mathfrak{A}$, is determined, up to isomorphism, by the set $At\,\mathfrak{A}$. It is also easily seen that, within the framework of $\mathfrak{Bl}\,\mathfrak{A}$, each element $\mathsf{d}_{\kappa\lambda}$ is uniquely determined by the set $D_{\kappa\lambda}$ of all atoms included in it, while each operation c_κ is so determined by the relation R_κ which holds between x and y iff $x, y \in At\,\mathfrak{A}$ and $x \leq \mathsf{c}_\kappa y$. This suggests the idea of correlating with each such algebra $\mathfrak{A}$ the relational structure formed by the set $At\,\mathfrak{A}$, the binary relations R_κ, and the unary relations $D_{\kappa\lambda}$, and of constructing from this structure an algebra isomorphically representing $\mathfrak{A}$ by purely set-theoretical means. The result thus obtained can then be extended to arbitrary CA's by using the embedding theorem 2.7.20.

We find it convenient to refer the next few definitions and theorems to Bo's and not specifically to CA's.

DEFINITION 2.7.32. *Let $\mathfrak{A}$ be a normal and atomic Bo_α. By the* ***atom structure*** *of $\mathfrak{A}$, $\mathfrak{At}\,\mathfrak{A}$, we mean the relational structure $\langle At\,\mathfrak{A}, R_\kappa, D_{\kappa\lambda}\rangle_{\kappa,\lambda<\alpha}$ where $R_\kappa = \{\langle x, y\rangle : x, y \in At\,\mathfrak{A},\ y \leq \mathsf{c}_\kappa x\}$ and $D_{\kappa\lambda} = \{x : x \in At\,\mathfrak{A},\ x \leq \mathsf{d}_{\kappa\lambda}\}$, for any $\kappa, \lambda < \alpha$.*

DEFINITION 2.7.33. *Let $\mathfrak{B} = \langle B, T_\kappa, E_{\kappa\lambda}\rangle_{\kappa,\lambda<\alpha}$ be any relational structure such that $T_\kappa \subseteq B \times B$ and $E_{\kappa\lambda} \subseteq B$ for all $\kappa, \lambda < \alpha$. We define the* ***complex algebra*** *of $\mathfrak{B}$, $\mathfrak{Cm}\,\mathfrak{B}$, to be the algebra*

$$\langle Sb\,B, \cup, \cap, {}_B\sim, 0, B, T_\kappa^*, E_{\kappa\lambda}\rangle_{\kappa,\lambda<\alpha}$$

(where T_κ^* are unary operations and $E_{\kappa\lambda}$ are distinguished elements of the algebra).

The definition just given automatically extends to arbitrary relational structures and in particular to algebraic structures. If, for instance, we consider a group $\mathfrak{G} = \langle G, \cdot, {}^{-1}\rangle$, then by the complex algebra of $\mathfrak{G}$ we understand the algebra

$$\mathfrak{Cm}\,\mathfrak{G} = \langle SbG, \cup, \cap, {}_G\sim, 0, G, \cdot, {}^{-1}\rangle$$

where the operations $\cdot$ and ${}^{-1}$ are defined for any $X, Y \subseteq G$ by the formulas

$$X \cdot Y = \{x \cdot y : x, y \in G\} \text{ and } X^{-1} = \{x^{-1} : x \in X\}.$$

Thus the notion of complex algebra belongs to the general theory of relational structures. So far complex algebras have rarely been studied in their own right. On the other hand, the fundamental operations of complex algebras correlated with such familiar structures as groups and rings are frequently employed in the discussion of these structures and prove to be useful tools in modern algebraic research.

THEOREM 2.7.34. *If* $\mathfrak{A}$ *is a normal, complete, and atomic* Bo_α, *then* $\mathfrak{A} \cong \mathfrak{Cm}\,\mathfrak{At}\,\mathfrak{A}$.

PROOF. Let $\mathfrak{At}\,\mathfrak{A} = \langle At\mathfrak{A}, R_\kappa, D_{\kappa\lambda}\rangle_{\kappa,\lambda<\alpha}$, and set $h = \langle \sum X : X \subseteq At\mathfrak{A}\rangle$. It is easily seen that h is a Boolean isomorphism from $\mathfrak{Cm}\,\mathfrak{At}\,\mathfrak{A}$ onto $\mathfrak{A}$. Consider any $\kappa < \alpha$ and $X \subseteq At\mathfrak{A}$. If X is not empty, then by the atomicity of $\mathfrak{A}$ we have

$$hR^*_\kappa X = \sum R^*_\kappa X = \sum_{x\in X} \sum R^\star_\kappa x = \sum_{x\in X} \mathsf{c}_\kappa x;$$

thus, as c_κ is completely additive and $X \neq 0$, we get

$$hR^*_\kappa X = \mathsf{c}_\kappa \sum X = \mathsf{c}_\kappa hX.$$

On the other hand, if $X = 0$, then $hR^*_\kappa X = \mathsf{c}_\kappa hX$ by the normality of c_κ. Therefore, since $hD_{\kappa\lambda}$ clearly equals $\mathsf{d}_{\kappa\lambda}$ for all $\kappa, \lambda < \alpha$, we have

$$h \in Is(\mathfrak{Cm}\,\mathfrak{At}\,\mathfrak{A}, \mathfrak{A}).$$

THEOREM 2.7.35. *If* $\mathfrak{B}$ *is any relational structure of the form* $\mathfrak{B} = \langle B, T_\kappa, E_{\kappa\lambda}\rangle_{\kappa,\lambda<\alpha}$ *where* $T_\kappa \subseteq B \times B$ *and* $E_{\kappa\lambda} \subseteq B$ *for all* $\kappa, \lambda < \alpha$, *then* $\mathfrak{Cm}\,\mathfrak{B}$ *is a normal, complete, and atomic* Bo_α, *and* $\mathfrak{B} \cong \mathfrak{At}\,\mathfrak{Cm}\,\mathfrak{B}$.

PROOF. That $\mathfrak{Cm}\,\mathfrak{B}$ is a normal, complete, and atomic Bo_α is immediate from the relevant definitions 2.7.1 and 2.7.33. The atoms of $\mathfrak{Cm}\,\mathfrak{B}$ are the singletons $\{x\}$ of elements x of B. Let B' be the set of all these atoms. Then, by the definition of atom structures, $\mathfrak{At}\,\mathfrak{Cm}\,\mathfrak{B}$ is the relational structure $\langle B', T'_\kappa, E'_{\kappa\lambda}\rangle_{\kappa,\lambda<\alpha}$ where for all $\kappa < \alpha$ and $\{x\}, \{y\} \in B'$ we have $\{x\}T'_\kappa\{y\}$

iff $\{y\} \subseteq T^*_\kappa\{x\}$, i.e., iff $xT_\kappa y$, and $\{x\} \in E'_{\kappa\lambda}$ iff $\{x\} \subseteq E_{\kappa\lambda}$, i.e., iff $x \in E_{\kappa\lambda}$. Thus the function $\langle\{x\}: x \in B\rangle$ is seen to be an isomorphism from $\mathfrak{B}$ onto $\mathfrak{At}\,\mathfrak{Cm}\,\mathfrak{B}$.

COROLLARY 2.7.36. *For every relational structure* $\mathfrak{B} = \langle B, T_\kappa, E_{\kappa\lambda}\rangle_{\kappa,\lambda<\alpha}$ *with* $T_\kappa \subseteq B \times B$ *and* $E_{\kappa\lambda} \subseteq B$ *for all* $\kappa, \lambda < \alpha$ *there is a normal, complete and atomic* $\mathsf{Bo}_\alpha\,\mathfrak{A}$ *such that* $\mathfrak{B} = \mathfrak{At}\,\mathfrak{A}$.

PROOF. By 2.7.35 there is a normal, complete, and atomic $\mathsf{Bo}_\alpha\,\mathfrak{A}'$ for which $\mathfrak{B} \cong \mathfrak{At}\,\mathfrak{A}'$; e.g., $\mathfrak{Cm}\,\mathfrak{B}$ can be taken for $\mathfrak{A}'$. Hence we easily conclude that there is also an algebra $\mathfrak{A} \cong \mathfrak{A}'$ for which $\mathfrak{B} = \mathfrak{At}\,\mathfrak{A}$. (We apply here a familiar method of reasoning from the general theory of algebras, sometimes referred to as exchange method; it was used in the proof of 0.2.15.)

COROLLARY 2.7.37. (i) *For any two given normal, complete, and atomic* Bo's $\mathfrak{A}$ and $\mathfrak{A}'$ *we have* $\mathfrak{A} \cong \mathfrak{A}'$ *iff* $\mathfrak{At}\,\mathfrak{A} \cong \mathfrak{At}\,\mathfrak{A}'$.

(ii) *For any two relational structures* $\mathfrak{B}$ *and* $\mathfrak{B}'$ *of the type described in 2.7.36 we have* $\mathfrak{B} \cong \mathfrak{B}'$ *iff* $\mathfrak{Cm}\,\mathfrak{B} \cong \mathfrak{Cm}\,\mathfrak{B}'$.

PROOF: straightforward, by 2.7.32–2.7.35.

We turn now specifically to atom structures of CA's.

DEFINITION 2.7.38. *The class of all relational structures which are atom structures of complete and atomic* CA's, *or, more specifically, of complete and atomic* CA_α's, *is called the class of* ***cylindric atom structures***, *or* ***cylindric atom structures of dimension*** α, *and is denoted by* Ca, *or* Ca_α, *respectively.*

THEOREM 2.7.39. *For every relational structure* $\mathfrak{B} = \langle B, T_\kappa, E_{\kappa\lambda}\rangle_{\kappa,\lambda<\alpha}$ *with* $T_\kappa \subseteq B \times B$ *and* $E_{\kappa\lambda} \subseteq B$ *for all* $\kappa, \lambda < \alpha$ *we have* $\mathfrak{B} \in \mathsf{Ca}_\alpha$ *iff* $\mathfrak{Cm}\,\mathfrak{B}$ *is a complete and atomic* CA_α.

PROOF. If $\mathfrak{B} \in \mathsf{Ca}_\alpha$, then, by 2.7.38, $\mathfrak{B} = \mathfrak{At}\,\mathfrak{A}$ for some complete and atomic $\mathsf{CA}_\alpha\,\mathfrak{A}$; hence $\mathfrak{Cm}\,\mathfrak{B} \cong \mathfrak{A}$ by 2.7.34 and 2.7.3, so that $\mathfrak{Cm}\,\mathfrak{B}$ is also a complete and atomic CA_α.

Assume, conversely, that $\mathfrak{Cm}\,\mathfrak{B}$ is a complete and atomic CA_α. By 2.7.36 there is a normal, complete, and atomic $\mathsf{Bo}_\alpha\,\mathfrak{A}$ such that $\mathfrak{B} = \mathfrak{At}\,\mathfrak{A}$. Hence $\mathfrak{Cm}\,\mathfrak{B} = \mathfrak{Cm}\,\mathfrak{At}\,\mathfrak{A} \cong \mathfrak{A}$ by 2.7.34. Consequently, $\mathfrak{A}$ is a complete and atomic CA_α, and $\mathfrak{B}$ is a Ca_α by 2.7.38. The proof is complete.

THEOREM 2.7.40. *For every relational structure* $\mathfrak{B} = \langle B, T_\kappa, E_{\kappa\lambda}\rangle_{\kappa,\lambda<\alpha}$ *with* $T_\kappa \subseteq B \times B$ *and* $E_{\kappa\lambda} \subseteq B$ *we have* $\mathfrak{B} \in \mathsf{CA}_\alpha$ *iff the following five conditions hold for all* $\kappa, \lambda, \mu < \alpha$:

(i) T_κ *is an equivalence relation on* B,
(ii) $T_\kappa | T_\lambda = T_\lambda | T_\kappa$,
(iii) $E_{\kappa\kappa} = B$,
(iv) $E_{\kappa\lambda} = T^*_\mu(E_{\kappa\mu} \cap E_{\mu\lambda})$ *provided* $\kappa, \lambda \neq \mu$,
(v) $T_\kappa \cap (E_{\kappa\lambda} \times E_{\kappa\lambda}) \subseteq Id$ *provided* $\kappa \neq \lambda$.

PROOF. Assume first $\mathfrak{B} \in \mathsf{Ca}_\alpha$. Then, by 2.7.39, $\mathfrak{Cm}\mathfrak{B}$ is a (complete and atomic) CA_α, and hence it satisfies postules (C_1)–(C_7) used in 1.1.1 to characterize CA_α's. Therefore, in view of 2.7.33, the following conditions hold for all sets $X, Y \subseteq B$ and all ordinals $\kappa, \lambda, \mu < \alpha$:

(1) $$T^*_\kappa 0 = 0,$$

(2) $$X \subseteq T^*_\kappa X,$$

(3) $$T^*_\kappa(X \cap T^*_\kappa Y) = T^*_\kappa X \cap T^*_\kappa Y,$$

(4) $$T^*_\kappa T^*_\lambda X = T^*_\lambda T^*_\kappa X,$$

(5) $$E_{\kappa\kappa} = B,$$

(6) $$E_{\kappa\lambda} = T^*(E_{\kappa\mu} \cap E_{\mu\lambda}) \text{ provided } \kappa, \lambda \neq \mu,$$

(7) $$T^*_\kappa(E_{\kappa\lambda} \cap X) \cap T^*_\kappa(E_{\kappa\lambda} \cap_B {\sim} X) = 0 \text{ provided } \kappa \neq \lambda.$$

Now (4) clearly implies (ii); (5) and (6) respectively coincide with (iii) and (iv); (v) can easily be derived from (2) and (7). It remains to obtain (i). To this end notice that, letting $X = T^*_\kappa Y$ in (3), we get

$$T^*_\kappa T^*_\kappa Y \subseteq T^*_\kappa Y,$$

and hence T_κ is transitive. Assume $xT_\kappa y$ and let $X = \{y\}$ in (2) as well as $X = \{x\}$ and $Y = \{y\}$ in (3). Then

$$\{y\} \subseteq T^*_\kappa\{x\} \cap T^*_\kappa\{y\} = T^*_\kappa(\{x\} \cap T^*_\kappa\{y\}),$$

whence $\{x\} \cap T^*_\kappa\{y\} \neq 0$ and $yT_\kappa x$. Thus T_κ is symmetric. Finally, the range (and hence the field) of T_κ is B by (2), so that T_κ is an equivalence relation on B, and (i) is shown to hold.

Assume now, conversely, that conditions (i)–(v) hold. We want to show that conditions (1)–(7) hold as well. (1) is satisfied by every relation T_κ. (2) and (4)–(7) either occur among conditions (i)–(v) or can very easily be derived from them (e.g., (7) from (i) and (v)). It remains to derive (3). (i) implies that T_κ is transitive; therefore

(8) $$T^*_\kappa(X \cap T^*_\kappa Y) \subseteq T^*_\kappa X \cap T^*_\kappa T^*_\kappa Y \subseteq T^*_\kappa X \cap T^*_\kappa Y.$$

On the other hand, (i) implies also

$$T^*_\kappa({}_B{\sim}T^*_\kappa Y) = {}_B{\sim}T^*_\kappa Y,$$

whence

$$\begin{aligned} T^*_\kappa X &\subseteq T^*_\kappa[(X \cap T^*_\kappa Y) \cup ({}_B{\sim}T^*_\kappa Y)] \\ &= T^*_\kappa(X \cap T^*_\kappa Y) \cup T^*_\kappa({}_B{\sim}T^*_\kappa Y) \\ &= T^*_\kappa(X \cap T^*_\kappa Y) \cup ({}_B{\sim}T^*_\kappa Y), \end{aligned}$$

and therefore

$$T^*_\kappa X \cap T^*_\kappa Y \subseteq T^*_\kappa(X \cap T^*_\kappa Y). \tag{9}$$

Thus conditions (1)–(7) are satisfied (for all $X, Y \subseteq B$ and $\kappa, \lambda, \mu < \alpha$). Therefore, in view of 2.7.33, $\mathfrak{Cm}\,\mathfrak{B}$ is a CA_α and actually a complete atomic CA_α. Hence $\mathfrak{B} \in \mathsf{Ca}_\alpha$ by 2.7.39, and the proof has been completed in both directions.

REMARK 2.7.41. The characterization of Ca_α's given in Definition 2.7.38 is not intrinsic. An intrinsic characterization of these structures is provided, however, in 2.7.39, since the complex algebra of a relation structure is intrinsically constructed in terms of this structure. The situation becomes even clearer if we notice that 2.7.39 is obviously equivalent to the following statement: $\mathfrak{B} = \langle B, T_\kappa, E_{\kappa\lambda}\rangle_{\kappa,\lambda<\alpha}$ is a Ca_α iff it satisfies identically conditions (1)–(7) stated in the proof of 2.7.40. From this statement it is seen that the characterization in 2.7.39 involves implicitly arbitrary subsets of B and hence can be formally expressed only in the language of second order logic. On the other hand, in 2.7.40 we have succeeded in giving a rather simple first-order characterization of Ca_α's. Several variants of this characterization are known; for instance, condition (v) of 2.7.40 can be replaced by

(v′) $|T^*_\kappa x \cap E_{\kappa\lambda}| = 1$ *provided* $x \in B$ *and* $\kappa \neq \lambda$.

REMARK 2.7.42. From the results obtained it is seen that a "natural" one-one correspondence can be established between the isomorphism types of normal, complete, and atomic Bo_α's and those of arbitrary relational structures $\langle B, T_\kappa, E_{\kappa\lambda}\rangle_{\kappa,\lambda<\alpha}$ with $T_\kappa \subseteq B \times B$ and $E_{\kappa\lambda} \subseteq B$. This is achieved by mapping the isomorphism type of any normal, complete, and atomic Bo_α on that of its atom structure, or — what amounts in the present case to the same — by mapping the isomorphism type of any relational structure $\langle B, T_\kappa, E_{\kappa\lambda}\rangle_{\kappa,\lambda<\alpha}$ on that of its complex algebra. Using the same mappings we obtain a "natural" one-one correspondence between the isomorphism types of complete and atomic CA_α's and those of Ca_α's, i.e. relational structures $\langle B, T_\kappa, E_{\kappa\lambda}\rangle_{\kappa,\lambda<\alpha}$ satisfying conditions 2.7.40(i)–(v).

THEOREM 2.7.43. (i) *Every complete and atomic* CA_α $\mathfrak{A}$ *is isomorphic with the complex algebra of some* Ca_α $\mathfrak{B}$, *i.e. of a relational structure* $\langle B, T_\kappa, E_{\kappa\lambda}\rangle_{\kappa,\lambda<\alpha}$, *with* $T_\kappa \subseteq B \times B$ *and* $E_{\kappa\lambda} \subseteq B$, *satisfying conditions 2.7.40(i)–(v); in fact,* $\mathfrak{At}\,\mathfrak{A}$ *can be taken for* $\mathfrak{B}$.

(ii) *Every* CA_α $\mathfrak{A}$ *is isomorphic with a subalgebra of the complex algebra of some* Ca_α $\mathfrak{B}$; *in fact,* $\mathfrak{At}\,\mathfrak{Em}\,\mathfrak{A}$ *can be taken for* $\mathfrak{B}$.

PROOF: (i) by 2.7.3, 2.7.34, 2.7.39, and 2.7.40; (ii) by (i), 2.7.5(i), and 2.7.15.

REMARK 2.7.44. The notions introduced and the main results established in this section can be extended with the greatest ease to diagonal-free cylindric algebras. Thus, by appropriately defining the notion of a canonical embedding algebra for Df's, we extend automatically the embedding theorem 2.7.20. To carry over Theorem 2.7.43 we modify in the obvious way the notion of an atom structure. It is easily seen that the structures whose role in the discussion of Df's is exactly analogous to that of Ca's in the discussion of CA's are relational structures $\mathfrak{B} = \langle B, T_\kappa\rangle_{\kappa<\alpha}$ with $T_\kappa \subseteq B \times B$, which can be characterized intrinsically by two conditions, (i) and (ii), of 2.7.40; such structures can be called *diagonal-free cylindric atom structure* or else, in view of their intrinsic characterization, *commutative equivalence structures.*

REMARK 2.7.45. Theorem 2.7.43(ii) can, of course, be regarded as a kind of representation theorem. However, the notion of representability underlying this theorem is not the one which was introduced in 1.1.13 and which will be studied thoroughly in Part II. Exclusively for the purposes of the present and the immediately following remarks, we introduce the terms "relational representability" to refer to the notion of 2.7.43(ii) and "geometrical representability" to refer to that of 1.1.13.

By 2.7.43(ii) every CA_α is relationally representable in the sense that it is isomorphic with a subalgebra of the complex algebra of a relational structure and, specifically, of a Ca_α $\mathfrak{B}$ as defined in 2.7.38; if, in addition, $\mathfrak{A}$ is complete and atomic, then by 2.7.43(i) it is isomorphic to the whole complex algebra of $\mathfrak{B}$. On the other hand, according to 1.1.13, $\mathfrak{A}$ is geometrically representable iff it is isomorphic to an α-dimensional generalized cylindric set algebra. It is easily seen that in case $\alpha = 0$ each of the two notions of representability reduces to the one known from the theory of BA's. In general, however, the two notions are defined in entirely different terms. It is of interest, therefore, that a characterization of geometrical representability is known in which this notion appears as a specialized form of relational representability.

In fact, it will be shown in Part II that a CA_α $\mathfrak{A}$ is geometrically representable iff it is isomorphic to a subalgebra of the complex algebra of a Ca_α $\mathfrak{B}$ satisfying a certain additional condition. This additional condition is of the same nature as conditions 2.7.40(i)–(v) characterizing the notion of a Ca_α. In case $\alpha = 2$ the condition is simply

(vi′) $$T_0 \cap T_1 \subseteq Id;$$

if $\alpha \geqq 2$ is finite, it can be expressed by the formula

(vi″) $$\bigcap_{\kappa<\alpha}(T_0|\ldots|T_{\kappa-1}|T_{\kappa+1}|\ldots|T_{\alpha-1}) \subseteq Id;$$

finally, in the general case it assumes the form

(vi) $$(T_\kappa|T_{\tau 0}|\ldots|T_{\tau_{\mu-1}}) \cap (T_\lambda|T_{\tau 0}|\ldots|T_{\tau_{\mu-1}}) = T_{\tau 0}|\ldots|T_{\tau_{\mu-1}}$$
for all $\kappa, \lambda < \alpha$, $\mu < \omega$, *and* $\tau \in {}^\mu\alpha$.

Thus the problem whether a CA_α is geometrically representable reduces to the following question: among the many Ca_α's which provide a relational representation of $\mathfrak{A}$, can we find at least one which satisfies the additional condition (vi)? It is known that, for every $\alpha \geqq 2$, there are CA_α's for which the answer to this question is negative.

The above remarks extend in a natural way from CA_α's to Df_α's. Instead of Ca_α's we consider of course α-dimensional commutative equivalence structures as defined in 2.7.44, and we use the notion of a generalized cylindric set algebra in the form adapted to Df_α's (cf. 1.1.13); however, the additional condition (vi) for passing from relational to geometrical representability remains unchanged. Actually the two notions appear to be even closer to each other in the case of Df's, since both the additional condition and the conditions 2.7.40(i),(ii) characterizing commutative equivalence structures are formulated entirely in terms of the relations T_κ. From what was said in 1.1.13, we know that in case $\alpha \leqq 2$ an appropriate commutative equivalence structure satisfying (vi) can be found for any given Df_α; in case $\alpha > 2$ this is in general not true.

REMARKS 2.7.46. In spite of their formal proximity the two notions of representability discussed in 2.7.44 differ considerably in some important respects, and these differences influence in an essential way the relative value of the corresponding representation theorems.

The results known in modern mathematics as representation theorems have as a rule the following character: in each of them a class K of "abstractly" defined mathematical structures is considered, a subclass L of this class is singled out, and it is shown that every structure in K is isomorphic to some structure in L; the proof frequently consists in effectively correlating, with any given structure $\mathfrak{S}$ in K, its isomorphic image in L — the representative of $\mathfrak{S}$.

The value of a representation theorem depends both on the scope of the class K and on such properties of members of L as simplicity of structure and "concreteness" of notions involved in their construction.

In the case of relational representability the scope of the representation theorem 2.7.43(ii) is wide: the class K consists of all CA's. The class L of representatives is formed by the complex algebras of cylindric atom structures and their subalgebras. The Boolean operations in these representatives are "concrete", well-determined set-theoretical notions used in constructing Boolean set algebras. On the other hand, the extra-Boolean operations are defined in terms of the fundamental relations of cylindric atom structures and have therefore as "abstract" a character as the corresponding notions in arbitrary CA's. Hence 2.7.43(ii) is not what could be regarded as a satisfactory representation theorem for arbitrary cylindric algebras. Nevertheless, it is a rather important result in its own right. It is essentially a combination of two results: 2.7.20, the relational embedding theorem, and 2.7.43(i), the representation theorem for complete and atomic CA's. Due to the first of these results, the discussion of arbitrary CA's reduces in certain situations to that of complete and atomic CA's (or, what amounts almost to the same thing, to the discussion of CA's whose Boolean parts are full Boolean set algebras, i.e., set algebras whose universes consist of all subsets of their unit sets); it will be seen in Part II that such a reduction is very helpful in the study of some problems concerning geometrical representability. The second result leads to a one-one correspondence (up to isomorphism) between complete and atomic CA's and cylindric atom structures. Since these structures are in general simpler and easier to handle than complete and atomic CA's, the result provides us with a convenient method of constructing complete and atomic CA's with various prescribed properties. The method turns out to be particularly efficient when applied to the construction of finite CA's.[1)]

With respect to geometrical representability no general representation theorem equal in scope to 2.7.43(ii) can be hoped for. In fact, as will be seen in Part II, many examples of CA's are known which are not geometrically representable (cf. also 2.6.44). On the other hand, however, the notion of geometrical representability is intuitively much more satisfactory and valuable than that of relational representability. The generalized cylindric set algebras, which serve as representatives under this notion, are in a sense "concrete" algebraic structures (or, at any rate, structures which are much more "concrete"

[1)] This method was first applied in Lyndon [50], though not to cylindric algebras, but to closely related relation algebras. In fact, using this method the first example of a finite non-representable relation algebra was constructed there.

than arbitrary CA's). All the fundamental operations and distinguished elements of these algebras are defined in straightforward set-theoretical terms; the definitions are uniform for all algebras involved, and, as a consequence, each of the algebras is uniquely determined by its universe. The geometrical intuitions inherent in the definitions of fundamental notions provide us with good insight into the structure of the algebras.

Due to these properties of geometrical representatives, the representation theorem of this section should certainly not detract the interest of the reader from the theory of geometrical representation which will be developed in Part II. In this theory, in particular, some interesting necessary and sufficient conditions for geometrical representability and some properties of the class of all representable CA's will be established. Some rather comprehensive and simply defined classes of CA's will be shown to consist exclusively of representable CA's, and certain fairly general methods of constructing non-representable CA's will be developed. The reader may also become interested in two outstanding open problems of the representation theory. One of them is the problem of providing a simple intrinsic characterization for all representable CA's (recall the remarks at the beginning of § 0.2 concerning the sense in which the term "intrinsic" is used in this work). The second problem is to find a notion of representability for which a general representation theorem of the scope of 2.7.43 could be obtained and which at the same time would be close to geometrical representability in its "concrete" character and intuitive simplicity. It is by no means clear that a satisfactory solution of either of these problems will ever be found, or that such a solution is possible.

PROBLEMS

Problem 2.1. *Is the class* Cr_α *with* $\alpha \geqq 3$ *closed under the formation of homomorphic images?*

The definition of Cr_α is given in 2.2.1.

Problem 2.2. *Is the class* Cr_α *with* $\alpha \geqq 3$ *elementary?*

Cf. Remark 2.2.6.

Problem 2.3. *Does there exist a finitely generated simple* CA_α $\mathfrak{A}$ *with* $\alpha \geqq \omega$ *which is not generated by a single element?*

For results related to this problem see 2.3.22, 2.3.23, and 2.6.25.

Problem 2.4. *Is there a denumerable* CA_α $\mathfrak{A}$ *such that* $\mathfrak{A} \cong \mathfrak{A} \times \mathfrak{A} \times \mathfrak{A}$ *holds while* $\mathfrak{A} \cong \mathfrak{A} \times \mathfrak{A}$ *fails?*

The analogous problem for Boolean algebras was stated in Hanf[57*].

Problem 2.5. *Does the sum of all atoms in the algebra* $\mathfrak{Fr}_\beta\mathsf{CA}_\alpha$ *with* $2 < \alpha < \omega$ *and* $0 < \beta < \omega$ *exist?*

Problem 2.6. *Are there any atoms in* $\mathfrak{Fr}_\beta\mathsf{CA}_\alpha$ *with* $\alpha \geqq \omega$ *and* $0 < \beta < \omega$ *which are not zero-dimensional?*

In connection with the last two problems recall various results on atoms in free CA's which are given in Section 2.5 beginning with 2.5.5.

Problem 2.7. *Assume that the algebra* $\mathfrak{Fr}_\beta\mathsf{CA}_\alpha$ *with* $\alpha > 2$ *and* $\beta > \omega$ *is generated by a set* X *of cardinality* β. *Is* $\mathfrak{Fr}_\beta\mathsf{CA}_\alpha$ *necessarily* CA_α*-freely generated by* X*?*

By 2.5.23 the answer to this question is affirmative for $\alpha = 0, 1, 2$.

Problem 2.8. *Assume that* β, β' *are two cardinals* $\neq 0$ *and* Δ, Δ' *are respectively functions from* β, β' *into the set of finite subsets of an ordinal* α. *Is it true that the algebras* $\mathfrak{Fr}_\beta^{(\Delta)}\mathsf{CA}_\alpha$ *and* $\mathfrak{Fr}_{\beta'}^{(\Delta')}\mathsf{CA}_\alpha$ *are isomorphic only if the sets* $\{\xi : \xi < \beta, |\Delta\xi| = \kappa\}$ *and* $\{\xi : \xi < \beta', |\Delta'\xi| = \kappa\}$ *have the same cardinality for* *every $\kappa < \omega$? In particular, is it true that $\mathfrak{Fr}_{\omega}^{(\Delta)}\mathsf{CA}_{\omega}$ and $\mathfrak{Fr}_{\omega}^{(\Delta')}\mathsf{CA}_{\omega}$ are not isomorphic if $\Delta = \omega \times \{1\}$ and $\Delta' = (\Delta \sim \{\langle 0, 1\rangle\}) \cup \{\langle 0, 0\rangle\}$?*

The conjecture formulated in this problem originates with Don Pigozzi. Compare here 2.5.43 and 2.5.54. Notice that, in view of 2.5.40, the second question raised in the problem is equivalent to the question whether for $\Delta = \omega \times \{1\}$ the algebras $\mathfrak{Fr}_{\omega}^{(\Delta)}\mathsf{CA}_{\omega}$ and $\mathfrak{Fr}_{\omega}^{(\Delta)}\mathsf{CA}_{\omega} \times \mathfrak{Fr}_{\omega}^{(\Delta)}\mathsf{CA}_{\omega}$ are not isomorphic.

PROBLEM 2.9. *Let $\mathfrak{A} = \mathfrak{Fr}_{\beta}^{(\Delta)}\mathsf{CA}_{\alpha}$ with $\alpha \geqq \omega$, $0 < \beta < \omega$, $|\Delta\xi| < \omega$ for every $\xi < \beta$, and $|\Delta\eta| \geqq 2$ for some $\eta < \beta$, and let $\mathfrak{B}$ be the zero-dimensional part of $\mathfrak{A}$. Determine whether or not $\boldsymbol{\varepsilon}\mathfrak{B} = \omega^{\omega}$.*

The meaning of $\boldsymbol{\varepsilon}\mathfrak{B}$ for any given BA $\mathfrak{B}$ is explained in 2.5.56. As mentioned in 2.5.46, the specific BA $\mathfrak{B}$ involved in our problem is isomorphic with the Boolean algebra of sentences correlated with a first-order predicate logic whose language contains only finitely many non-logical predicates and not all of them are of rank $\leqq 1$. It was noticed by William Hanf that for this specific $\mathfrak{B}$ we have $\boldsymbol{\varepsilon}\mathfrak{B} \geqq \omega^{\omega}$; compare here 2.5.57 and 2.5.69.

PROBLEM 2.10. *Is it true that, for any ordinal $\alpha \geqq \omega$ and cardinal $\beta \neq 0$, the set of all finite-dimensional elements of $\mathfrak{Fr}_{\beta}\mathsf{CA}_{\alpha}$ is the subuniverse generated by $\{x \cdot \mathsf{c}_{0}^{\partial}\mathsf{d}_{01} : x \in G\}$, where G is the set of standard generators of $\mathfrak{Fr}_{\beta}\mathsf{CA}_{\alpha}$?*

Compare here 2.6.23 and 2.6.24.

PROBLEM 2.11. *Is the class $\mathsf{Nr}_{\alpha}\mathsf{CA}_{\beta}$ with $\beta > \alpha$ closed under the formation of subalgebras and homomorphic images?*

Cf. 2.6.33; also compare 2.6.74(ii).

PROBLEM 2.12. *For $2 < \alpha < \omega$, is there a $\kappa < \omega$ such that $\mathsf{SNr}_{\alpha}\mathsf{CA}_{\alpha+\kappa} = \mathsf{SNr}_{\alpha}\mathsf{CA}_{\alpha+\kappa+1}$?*

For $\alpha \leqq 1$ the equation in Problem 2.13 holds for all $\kappa < \omega$, by 2.6.39. For $\alpha = 2$ the equation fails for $\kappa = 0$ but holds for $0 < \kappa < \omega$ (cf. 2.6.40). Finally, Don Pigozzi has shown that for $\alpha \geqq \omega$ no κ exists such that the equation holds.

PROBLEM 2.13. *Does* (iv) *of 2.6.50 imply* (iii) *of 2.6.50?*

PROBLEM 2.14. *Given $\kappa < \omega$, is $\mathsf{SRd}_{\kappa}\mathsf{CA}_{\kappa+1}$ (the class of κ-dimensional subreducts of $\mathsf{CA}_{\kappa+1}$'s) closed under the formation of homomorphic images?*

Cf. here 2.6.61. In case $\kappa \geqq \omega$ the answer to the question is trivially affirmative in view of 2.6.9.

PROBLEM 2.15. *Does every identity which holds in a CA $\mathfrak{A}$ continue to hold in the canonical embedding algebra $\mathfrak{Em}\,\mathfrak{A}$?*

We know that the answer is affirmative if we restrict ourselves to positive identities (and actually positive identities in the wider sense); cf. 2.7.14 and 2.7.16.

PROBLEM 2.16. *Can the property of being of characteristic* κ *(for any given* $\kappa < \omega$*) be expressed by a set of positive equations? Or the property of being representable?*

Cf. here 2.7.16 and 2.7.25. In formulating the first half of this problem we have assumed (in opposition to 2.4.61) that all one-element CA's are included in the class of CA's of characteristic κ (for every $\kappa < \omega$).

BIBLIOGRAPHY

The bibliography is divided into two parts: the Bibliography of Cylindric Algebras and Related Algebraic Structures, or simply the CA Bibliography, and the Supplementary Bibliography. The CA Bibliography is intended to be a comprehensive bibliography of algebraic theories designed for investigating predicate logic. Thus it will list papers on classes of algebraic structures such as cylindric algebras, polyadic algebras, and relation algebras (but not classes of Boolean algebras and lattices unless enriched with supplementary fundamental operations). On the other hand, works in which algebraic methods are used in studying predicate logic are excluded if they do not deal with such a class, or with a particular member of such a class. These principles explain the absence from the CA Bibliography of several well known works on algebraic logic such as those of Curry and of Rasiowa and Sikorski. Within these limits the CA Bibliography is intended to be complete for the period from 1930 through 1969. However, we are aware that we have not undertaken the necessary effort to secure completeness in the period preceding 1930. The reader interested in this period may find supplementary material listed in the following works:

Lewis, C. I., **A survey of symbolic logic**, University of California Press, Berkeley, 1918, vi + 406 pp.

Church, A., *A bibliography of symbolic logic*. **J. Symbolic Logic,** vol. 1 (1936), pp. 121–218. (See also the indexes in the Journal of Symbolic Logic, vol. 26 (1961).)

We hope that the readers will call our attention to possible omissions so that they can be included in an addition to the CA Bibliography which will appear in the second volume of this work.

The Supplementary Bibliography lists all works referred to in this volume which are outside the scope of the CA Bibliography.

In both bibliographies the items are first arranged alphabetically by the last name of the author. The papers of a given author are then arranged chronologically by year of publication. A work is referred to throughout the volume by the last name of the author and the last two digits of the year of publication, e.g., "Baayen [60]". In a small number of cases involving works published before 1900 the year of publication is given in full, e.g., "De Morgan [1864]".

In the CA Bibliography where a given author has more than one paper included for a given year we distinguish between them by using small Roman letters, for example, "Halmos [56]", "Halmos [56a]", and "Halmos [56b]". A similar convention applies to items in the Supplementary Bibliography except that in addition, an asterisk * is employed to distinguish them from items in the CA Bibliography, for example, "Craig [57*]" and "Craig [57a*]".

An abstract is not listed whenever the information contained in it can be found in some later paper. Journal names have been abbreviated in agreement with the list of abbreviations printed in the index to volume 29 of Mathematical Reviews; journal names not appearing in that list are quoted in full. Finally, the Russian alphabet is transliterated according to the system described in the same volume of Mathematical Reviews.

I. BIBLIOGRAPHY OF CYLINDRIC ALGEBRAS AND RELATED ALGEBRAIC STRUCTURES

Addison, J. W., Henkin, L., and Tarski, A., Editors

[65] **The theory of models, Proceedings of the 1963 International Symposium at Berkeley,** North-Holland Publishing Co., Amsterdam, 1965, xvi + 494 pp.

Baayen, P. C.

[60] *Subdirect oplosbare cylinder-algebra's.* **Math. Centrum Amsterdam Afd. Zuivere Wisk.,** ZW 1960–006 (1960), 11 pp.

[62] *Cylinderalgebra's.* **Math. Centrum Amsterdam Afd. Zuivere Wisk.,** ZW 1962–004 (1962), 12 pp.

[62a] *Partial ordering of quantifiers and of clopen equivalence relations.* **Math. Centrum Amsterdam Afd. Zuivere Wisk.,** ZW 1962–025 (1962), 15 pp.

Banaschewski, B.

[63] *Regular closure operators.* **Arch. Math.,** vol. 14 (1963), pp. 271–274.

Bass, H.

[58] *Finite monadic algebras.* **Proc. Amer. Math. Soc.,** vol. 9 (1958), pp. 258–268.

Behmann, H.

[22] *Beiträge zur Algebra der Logik, insbesondere zum Entscheidungsproblem.* **Math. Ann.,** vol. 86 (1922), pp. 163–229.

Bernays, P.

[59] *Über eine natürliche Erweiterung des Relationenkalküls.* **Constructivity in mathematics, Proceedings of the colloquium held at Amsterdam, 1957,** ed. A. Heyting, North-Holland Publishing Co., Amsterdam, 1959, pp. 1–14.

Bevis, J. H.

[65] *Orthomodular geometries from lattices with quantifiers.* **Notices Amer. Math. Soc.,** vol. 12 (1965), p. 68.

[65a] **Quantifiers and dimension equivalence relations on orthomodular lattices,** Doctoral dissertation, The University of Florida, 1965, 45 pp.

Chin, L. H., and Tarski, A.

[48] *Remarks on projective algebras.* **Bull. Amer. Math. Soc.,** vol. 54 (1948), pp. 80–81.

[51] *Distributive and modular laws in the arithmetic of relation algebras.* **University of California Publications in Mathematics,** new series, vol. 1, no. 9 (1951), pp. 341–384.

Comer, S. D.

[67] **Some representation theorems and the amalgamation property in algebraic logic,** Doctoral dissertation, University of Colorado, Boulder, 1967, vii + 92 pp.

[68] *Galois theory and the amalgamation property in finite dimensional cylindric algebras. Preliminary report.* **Notices Amer. Math. Soc.,** vol. 15 (1968), p. 103.

[69] *Representation of cylindric algebras by sheaves. Preliminary report.* **Notices Amer. Math. Soc.,** vol. 16 (1969), p. 529.

[69a] *Classes without the amalgamation property.* **Pacific J. Math.,** vol. 28 (1969), pp. 309–318.

[69b] *Finite inseparability of some theories of cylindrification algebras.* **J. Symbolic Logic,** vol. 34 (1969), pp. 171–176.

Copeland, A. H., Sr.
[55] *Note on cylindric algebras and polyadic algebras.* **Michigan Math. J.,** vol. 3 (1955–56), pp. 155–157.

Copi, I. M., and Harary, F.
[53] *Some properties of n-adic relations.* **Portugal. Math.,** vol. 12 (1953), pp. 143–152.

Copilowish, I. M.
[48] *Matrix development of the calculus of relations.* **J. Symbolic Logic,** vol. 13 (1948), pp. 193–203.

Craig, W.
[65] *Boolean notions extended to higher dimensions.* Addison-Henkin-Tarski [65], pp. 55–69.
[68] *Two complete algebraic theories of logic.* **Logic, Methodology and Philosophy of Science III, Proceedings of the Third International Congress,** Amsterdam 1967, eds. B. van Rootselaar and J. Staal, North-Holland Publishing Co., 1968, pp. 23–30.

Daigneault, A.
[59] **Products of polyadic algebras and of their representations,** Doctoral dissertation, Princeton University, Princeton, 1959, xviii + 72 pp.
[63] *Tensor products of polyadic algebras.* **J. Symbolic Logic,** vol. 28 (1963), pp. 177–200.
[63a] **Théorie des modèles en logique mathématique, Séminaire de mathématiques supérieures,** no. 6, Les Presses de l'Université de Montréal, Montréal, 1963, xvi + 127 pp.
[64] *Freedom in polyadic algebras and two theorems of Beth and Craig.* **Michigan Math. J.,** vol. 11 (1964), pp. 129–135.
[64a] *On automorphisms of polyadic algebras.* **Trans. Amer. Math. Soc.,** vol. 112 (1964), pp. 84–130.

Daigneault, A., and Monk, J. D.
[63] *Representation theory for polyadic algebras.* **Fund. Math.,** vol. 52 (1963), pp. 151–176.

Davis, A. S.
[66] *An axiomatization of the algebra of transformations over a set.* **Math. Ann.,** vol. 164 (1966), pp. 372–377.

Davis, C.
[50] **Lattices and modal operators,** Doctoral dissertation, Harvard University, Cambridge, Mass., 1950.
[54] *Modal operators, equivalence relations, and projective algebras.* **Amer. J. Math.,** vol. 76 (1954), pp. 747–762.

Demaree, D. B.
[69] *On the Copeland formulation of algebraic logic.* **Notices Amer. Math. Soc.,** vol. 16 (1969), p. 843.

De Morgan, A.
[1864] *On the syllogism, no. IV, and on the logic of relations.* **Transactions of the Cambridge Philosophical Society,** vol. 10 (1864), pp. 331–358.

Driessel, K.
[68] **Significance and invariance in mathematical structures,** Doctoral dissertation, Oregon State University, Corvallis, 1968, vii + 80 pp.

Everett, C. J., and Ulam, S.
[46] *Projective algebra. I.* **Amer. J. Math.,** vol. 68 (1946), pp. 77–88.

Eytan, M.
[66] **Sur une algèbre monadique associée à certaines logiques modales,** Doctoral dissertation, Paris, 1966, 79 pp.

Fenstad, J. E.
[63] *Algebraic logic.* **Seminar Reports, Institute of Mathematics, University of Oslo,** 1963, no. 5, 12 pp.

[64] *Algebraic logic and the foundation of probability.* **Seminar Reports, Institute of Mathematics, University of Oslo,** 1964, no. 6, 29 pp.

[64a] *On representation of polyadic algebras.* **Norske Vid. Selsk. Forh. (Trondheim),** vol. 37 (1964), pp. 36–41.

[66] *A limit theorem in polyadic probabilities.* **J. Symbolic Logic,** vol. 31 (1966), pp. 293–294.

[67] *Representations of probabilities defined on first order languages.* **Sets, models and recursion theory, Proceedings of the Summer School in Mathematical Logic and Tenth Logic Colloquium, Leicester, August–September 1965,** ed. J. N. Crossley, North-Holland Publishing Co., Amsterdam, 1967, pp. 156–172.

Galler, B. A.

[55] **Some results in algebraic logic,** Doctoral dissertation, The University of Chicago, Chicago, 1955, iv + 67 pp.

[57] *Cylindric and polyadic algebras.* **Proc. Amer. Math. Soc.,** vol. 8 (1957), pp. 176–183.

Gao, Héng-shan

[63] *A simple proof of a theorem of H. Bass.* **Shuxue Jinzhan,** vol. 6 (1963), pp. 92–95, 306. [Chinese.]

[63a] *Quantifier operations in relative pseudo-complemented lattices.* **Shuxue Jinzhan,** vol. 6 (1963), pp. 279–285. [Chinese.]

[63b] *Algebraic treatment of the notion of satisfiability in the system* $\mathscr{S}_{\in}$. **Acta Math. Sinica,** vol. 13 (1963), pp. 68–77. [Chinese.] [English translation: **Chinese Math.-Acta,** vol. 4 (1963), pp. 76–87.]

[64] *On a theorem of Bass.* **Sci. Sinica,** vol. 13 (1964), p. 1005.

[64a] *Simple completeness of the predicate calculus* $\mathscr{S}^*_{\in}$ **Acta Math. Sinica,** vol. 14 (1964), pp. 546–548. [Chinese.] [English translation: **Chinese Math.–Acta,** vol. 5 (1964), pp. 588–590.]

Geymonat, L.

[58] *Logica matematica e algebra moderna.* **Confer. Sem. Mat. Univ. Bari,** no. 37 (1958), 20 pp.

Guillaume, M.

[58] *Calculs de conséquences et tableaux d'épreuve pour les classes algébriques générales d'anneaux booléiens à opérateurs.* **C. R. Acad. Sci. Paris,** vol. 247 (1958), pp. 1542–1544.

[64] *Sur les structures hilbertiennes polyadiques.* **C. R. Acad. Sci. Paris,** vol. 258 (1964), group 1, pp. 1957–1960.

[64b] *Quelques remarques sur les "Tableaux de Beth".* **E. W. Beth Memorial Colloquium** (Paris, 1964), pp. 39–45.

[67] *Algèbre monadoïdale.* **An. Acad. Brasil. Ci.,** vol. 39 (1967), pp. 217–220.

[67a] *Sur quelques notions concernant les anneaux hilbertiens de la logique algébrique.* **Annales de la Faculté des Sciences de l'Université de Clermont,** vol. 35 (1967), pp. 37–40.

Halmos, P. R.

[56] *Algebraic logic I. Monadic Boolean algebras.* **Compositio Math.,** vol. 12 (1956), pp. 217–249. [See also Halmos [62], pp. 37–72.]

[56a] *Algebraic logic III. Predicates, terms, and operations in polyadic algebras.* **Trans. Amer. Math. Soc.,** vol. 83 (1956), pp. 430–470. [See also Halmos [62], pp. 169–209.]

[56b] *The basic concepts of algebraic logic.* **Amer. Math. Monthly,** vol. 63 (1956), pp. 363–387. [See also Halmos [62], pp. 9–33.]

[57] *Algebraic logic II. Homogeneous locally finite polyadic Boolean algebras of infinite degree.* **Fund. Math.,** vol. 43 (1957), pp. 255–325. [See also Halmos [62], pp. 97–166.]

[57a] *Algebraic logic IV. Equality in polyadic algebras.* **Trans. Amer. Math. Soc.,** vol. 86 (1957), pp. 1–27. [See also Halmos [62], pp. 213–239.]

[59] *Free monadic algebras.* **Proc. Amer. Math. Soc.**, vol. 10 (1959), pp. 219–227. [See also Halmos [62], pp. 85–93.]

[59a] *The representation of monadic Boolean algebras.* **Duke Math. J.**, vol. 26 (1959), pp. 447–454. [See also Halmos [62], pp. 75–82.]

[60] *Polyadic algebras.* **Summaries of talks presented at the Summer Institute for Symbolic Logic, Cornell University, 1957,** second edition, Communications Research Division, Institute for Defense Analyses, 1960, pp. 252–255.

[62] **Algebraic logic,** Chelsea Publishing Co., New York, 1962, 271 pp.

Harary, F.

[50] *On complete atomic proper relation algebras.* **J. Symbolic Logic,** vol. 15 (1950), pp. 197–198.

See also Copi and Harary.

Henkin, L.

[55] *The algebraic structure of mathematical theories.* **Bull. Soc. Math. Belg.,** vol. 7 (1955), pp. 131–136.

[55a] *The representation theorem for cylindrical algebras.* **Mathematical interpretation of formal systems,** North-Holland Publishing Co., Amsterdam, 1955, pp. 85–97.

[56] **La structure algébrique des théories mathématiques,** Gauthier-Villars, Paris; E. Nauwelaerts, Louvain, 1956, 53 pp.

[57] *Cylindrical algebras of dimension* 2. *Preliminary report.* **Bull. Amer. Math. Soc.,** vol. 63 (1957), p. 26.

[61] **Cylindric algebras, Lectures presented at the 1961 Seminar of the Canadian Mathematical Congress, Université de Montréal, August 14–September 9, 1961,** 54 pp.

[67] **Logical systems containing only a finite number of symbols, Séminaire de mathématiques supérieures,** no. 21, Les Presses de l'Université de Montréal, *Montréal* 1967, 48 pp.

[68] *Relativization with respect to formulas and its use in proofs of independence.* **Compositio Math.,** vol. 20 (1968), pp. 88–106.

See also Addison, Henkin, and Tarski.

Henkin, L., and Tarski, A.

[60] *Cylindrical algebras.* **Summaries of talks presented at the Summer Institute for Symbolic Logic, Cornell University, 1957,** second edition, Communications Research Division, Institute for Defense Analyses, 1960, pp. 332–340.

[61] *Cylindric algebras.* **Lattice theory, Proceedings of symposia in pure mathematics,** vol. 2, ed. R. P. Dilworth, American Mathematical Society, Providence, 1961, pp. 83–113.

Hermes, H.

[63] **La teoria de reticulos y su aplicación a la lógica matemática, Conferencias de Matematica, VI, Consejo Superior de Investigaciónes Cientificas,** Publicaciones del Instituto de Matemáticas "Jorge Juan", Madrid, 1963, 57 pp.

Hiż, H.

[58] *A warning about translating axioms.* **Amer. Math. Monthly,** vol. 65 (1958), pp. 613–614.

Hoehnke, H.-J.

[66] *Zur Strukturgleichheit axiomatischer Klassen.* **Z. Math. Logik Grundlagen Math.,** vol. 12 (1966), pp. 69–83.

Howard, C. M.

[62] *A two dimensional analogue to Boolean algebras.* **Notices Amer. Math. Soc.,** vol. 9 (1962), p. 219.

[65] **An approach to algebraic logic,** Doctoral dissertation, University of California, Berkeley, 1965, 83 pp.

[65a] *Finite-dimensional analogues to Boolean algebras.* Addison-Henkin-Tarski [65], pp. 429–430.

Janowitz, M. F.

[63] *Quantifiers and orthomodular lattices.* **Pacific J. Math.**, vol. 13 (1963), pp. 1241–1249.

[65] *Quantifier theory on quasi-orthomodular lattices.* **Illinois J. Math.**, vol. 9 (1965), pp. 660–676.

[67] *Residuated closure operators.* **Portugal. Math.**, vol. 26 (1967), pp. 221–252.

Jaśkowski, S.

[61] *Cylindric algebras and counterfactual implications.* **Second Hungarian Mathematical Congress, Budapest, 24–31 August 1960 (Magyar Matematikai Kongresszus)**, vol. 2, Akadémiai Kiadó, Budapest, 1961, section 5, p. 24.

Johnson, H. H.

[61] *Realizations of abstract algebras of functions.* **Math. Ann.**, vol. 142 (1961), pp. 317–321.

Johnson, J. S.

[67] *Amalgamation of polyadic algebras. Preliminary report.* **Notices Amer. Math. Soc.**, vol. 14 (1967), p. 361.

[68] **Amalgamation of polyadic algebras and finitizability problems in algebraic logic,** Doctoral dissertation, University of Colorado, Boulder, 1968, vi + 129 pp.

[69] *Nonfinitizability of classes of representable polyadic algebras.* **J. Symbolic Logic**, vol. 34 (1969), pp. 344–352.

Jónsson, B.

[59] *Representation of modular lattices and of relation algebras.* **Trans. Amer. Math. Soc.**, vol. 92 (1959), pp. 449–464.

[62] *Defining relations for full semigroups of finite transformations.* **Michigan Math. J.**, vol. 9 (1962), pp. 77–85.

Jónsson, B., and Tarski, A.

[51] *Boolean algebras with operators. Part I.* **Amer. J. Math.**, vol. 73 (1951), pp. 891–939.

[52] *Boolean algebras with operators. Part II.* **Amer. J. Math.**, vol. 74 (1952), pp. 127–162.

Jurie, P.-F.

[67] *Notion de quasi-somme amalgamée: premières applications à l'algèbre booléienne polyadique.* **C. R. Acad. Sci. Paris Sér. A–B**, ser. A, vol. 264 (1967), pp. 1033–1036.

[67a] *Quasi-sommes amalgamées et constantes transformationnelles.* **C. R. Acad. Sci. Paris Sér. A–B**, ser. A, vol. 264 (1967), pp. 1125–1127.

Kamel, H.

[52] *Relational algebra.* **Bull. Amer. Math. Soc.**, vol. 58 (1952), p. 391.

[54] *Relation algebra and uniform spaces.* **J. London Math. Soc.**, vol. 29 (1954), pp. 342–344.

Keisler, H. J.

[63] *A complete first-order logic with infinitary predicates.* **Fund. Math.**, vol. 52 (1963), pp. 177–203.

Kotas, J., and Pieczkowski, A.

[66] *On a generalized cylindrical algebra and intuitionistic logic.* **Studia Logica**, vol. 18 (1966), pp. 73–81.

[67] *A cylindrical algebra based on the Boolean ring.* **Studia Logica**, vol. 21 (1967), pp. 71–80.

Krasner, M.

[38] *Une généralisation de la notion de corps.* **J. Math. Pures Appl.**, ser. 9, vol. 17 (1938), pp. 367–385.

[45] *Généralisation et analogues de la théorie de Galois.* **Comptes Rendus du Congrès de 1945 de l'Association Française pour l'Avancement des Sciences**, pp. 54–58.

[50] *Généralisation abstraite de la théorie de Galois*. **Algèbre et theorie des nombres, Colloques internationaux du Centre National de la Recherche Scientifique,** no. 24, Centre National de la Recherche Scientifique, Paris, 1950, pp. 163–168.
[58] *Les algèbres cylindriques*. **Bull. Soc. Math. France,** vol. 86 (1958), pp. 315–319.

Kuratowski, C., and Tarski, A.
[31] *Les opérations logiques et les ensembles projectifs*. **Fund. Math.,** vol. 17 (1931), pp. 240–248. [See also Tarski [56a], pp. 143–151.]

L'Abbé, M.
[58] *Structures algébriques suggérées par la logique mathématique*. **Bull. Soc. Math. France,** vol. 86 (1958), pp. 299–314.

LeBlanc, L.
[59] **Non-homogeneous and higher order polyadic algebras,** Doctoral dissertation, University of Chicago, 1959, 81 pp.
[60] *Les algèbres booléiennes topologiques bornées*. **C. R. Acad. Sci. Paris,** vol. 250 (1960), pp. 3766–3768.
[60a] *Représentation des algèbres polyadiques pour anneau*. **C. R. Acad. Sci. Paris,** vol. 250 (1960), pp. 4092–4094.
[61] *Transformation algebras*. **Canad. J. Math.,** vol. 13 (1961), pp. 602–613.
[62] *Duality for Boolean equalities*. **Proc. Amer. Math. Soc.,** vol. 13 (1962), pp. 74–79.
[62a] *Nonhomogeneous polyadic algebras*. **Proc. Amer. Math. Soc.,** vol. 13 (1962), pp. 59–65.
[63] **Introduction à la logique algébrique,** Université de Montréal, Montréal, 1963, vi + 105 pp.
[66] **Représentabilité et définissabilité dans les algèbres transformationnelles et dans les algèbres polyadiques, Séminaire de mathématiques supérieures,** no. 24, Les Presses de l'Université de Montréal, Montréal, 1966, 124 pp.

Löwenheim, L.
[13] *Potenzen im Relativkalkul und Potenzen allgemeiner endlicher Transformationen*. **Sitzungsberichte der Berliner Mathematischen Gesellschaft,** 12th year (1913), pp. 65–71. [Published as appendix to **Archiv der Mathematik und Physik,** ser. 3, vol. 21, no. 1 (1913).]
[15] *Über Möglichkeiten im Relativkalkül*. **Math. Ann.,** vol. 76 (1915), pp. 447–470.
[40] *Einkleidung der Mathematik in Schröderschen Relativkalkül*. **J. Symbolic Logic,** vol. 5 (1940), pp. 1–15.

Lucas, Th.
[67] **Algèbres cylindriques et polyadiques,** Doctoral dissertation, Université Catholique de Louvain, Louvain, 1967, vii + 139 pp.
[68] *Sur l'équivalence des algèbres cylindriques et polyadiques*. **Bull. Soc. Math. Belg.,** vol. 20 (1968), pp. 236–263.

Lüroth, J.
[04] *Aus der Algebra der Relative* (*nach dem dritten Bande von E. Schröders Vorlesungen über die Algebra der Logik*). **Jber. Deutsch. Math.-Verein.,** vol. 13 (1904), pp. 73–111.

Lyndon, R. C.
[50] *The representation of relational algebras*. **Ann. of Math.,** ser. 2, vol. 51 (1950), pp. 707–729.
[56] *The representation of relation algebras, II*. **Ann. of Math.,** ser. 2, vol. 63 (1956), pp. 294–307.
[61] *Relation algebras and projective geometries*. **Michigan Math. J.,** vol. 8 (1961), pp. 21–28.

Macfarlane, A.
[1880] *On a calculus of relationship*. **Proceedings of the Royal Society of Edinburgh,** vol. 10 (1880), pp. 224–232.

[1882] *Algebra of relationship – Part II.* **Proceedings of the Royal Society of Edinburgh,** vol. 11 (1882), pp. 5–13.

[1882a] *Algebra of relationship – Part III.* **Proceedings of the Royal Society of Edinburgh,** vol. 11 (1882), pp. 162–163.

Magari, R.

[65] *Sulle connessioni fra i vari modi di rappresentare un'algebra monadiča.* **Matematiche (Catania),** vol. 20 (1965), pp. 22–40.

[65a] *Sulle topologie associate a un'algebra monadica.* **Atti Accad. Sci. Torino Cl. Sci. Fis. Mat. Natur.,** vol. 99 (1965), pp. 49–69.

Magari, R., and Mangani, P.

[64] *Alcune osservazioni sugli assiomi delle "Algebre monadiche" di Halmos.* **Atti Accad. Sci. Torino Cl. Sci. Fis. Mat. Natur.,** vol. 98 (1964), pp. 265–278.

[64a] *Sulle topologie "compatibili" con una data algebra monadica.* **Atti Accad. Sci. Torino Cl. Sci. Fis. Mat. Natur.,** vol. 98 (1964), pp. 299–314.

Mangani, P.

[64] *Su alcune questioni riguardanti le "algebre monadiche" e le "algebre monadiche concrete".* **Atti Accad. Sci. Torino Cl. Sci. Fis. Mat. Natur.,** vol. 98 (1964), pp. 1073–1084.

[65] *Alcune applicazioni del concetto di "stabilizzante" al problema dell'estensione di un quantore.* **Atti Accad. Sci. Torino Cl. Sci. Fis. Mat. Natur.,** vol. 99 (1965), pp. 689–699.

[65a] *Alcune osservazioni sulle algebre monadiche.* **Boll. Un. Mat. Ital.,** ser. 3, vol. 20 (1965), pp. 439–445.

[65b] *Sul problema dell'estensione di un quantore.* **Atti Accad. Sci. Torino Cl. Sci. Fis. Mat. Natur.,** vol. 99 (1965), pp. 31–47.

[66] *Su certe algebre connesse con sistemi di logica elementare dotati dell'operatore τ di Hilbert.* **Matematiche (Catania),** vol. 21 (1966), pp. 65–82.

[66a] *Sulla quantificazione simultanea nelle algebre cilindriche.* **Boll. Un. Mat. Ital.,** ser. 3, vol. 21 (1966), pp. 302–311.

See also Magari and Mangani.

Mangione, C.

[64] *Ricerche sulle algebre cilindriche.* **Atti Accad. Sci. Torino Cl. Sci. Fis. Mat. Natur.,** vol. 98 (1964), pp. 366–374.

McKenzie, R.

[65] *On representing relation algebras in groups.* **Notices Amer. Math. Soc.,** vol. 12 (1965), p. 821.

[66] **The representation of relation algebras,** Doctoral dissertation, University of Colorado, Boulder, 1966, vi + 128 pp.

McKinsey, J. C. C.

[40] *Postulates for the calculus of binary relations.* **J. Symbolic Logic,** vol. 5 (1940), pp. 85–97.

[48] *On the representation of projective algebras.* **Amer. J. Math.,** vol. 70 (1948), pp. 375–384.

Monk, J. D.

[60] *Nonrepresentable polyadic algebras of finite degree.* **Notices Amer. Math. Soc.,** vol. 7 (1960), p. 735.

[60a] *Polyadic Heyting algebras.* **Notices Amer. Math. Soc.,** vol. 7 (1960), p. 735.

[61] *On the representation theory for cylindric algebras.* **Pacific J. Math.,** vol. 11 (1961), pp. 1447–1457.

[61a] *Relation algebras and cylindric algebras.* **Notices Amer. Math. Soc.,** vol. 8 (1961), p. 358.

[61b] **Studies in cylindric algebra,** Doctoral dissertation, University of California, Berkeley, 1961, vi + 83 pp.

[64] *On representable relation algebras.* **Michigan Math. J.,** vol. 11 (1964), pp. 207–210.

[64a] *Singulary cylindric and polyadic equality algebras.* **Trans. Amer. Math. Soc.,** vol. 112 (1964), pp. 185–205.

[65] *Model-theoretical methods and results in the theory of cylindric algebras.* Addison-Henkin-Tarski [65], pp. 238–250.

[65a] *Substitutionless predicate logic with identity.* **Arch. Math. Logik Grundlagenforsch.,** vol. 7 (1965), pp. 102–121.

[69] *Nonfinitizability of classes of representable cylindric algebras.* **J. Symbolic Logic,** vol. 34 (1969), pp. 331–343.

[69a] *On the lattice of equational classes of one- and two-dimensional polyadic algebras.* **Notices Amer. Math. Soc.,** vol. 16 (1969), p. 183.

[70] *On an algebra of sets of finite sequences.* **J. Symbolic Logic,** vol. 35 (1970), pp. 19–28.

See also Daigneault and Monk.

Monteiro, A.

[57] *Normalidad en las álgebras de Heyting monádicas.* **Unión Matemática Argentina, Actas de las X Jornadas,** Instituto de Matemáticas, Universidad Nacional del Sur, Bahía Blanca, 1957, pp. 50–51.

[60] *Álgebras monádicas.* **Atas do Segundo Colóquio Brasileiro de Matemática, Poços de Caldas, 5 a 18 de julho de 1959,** Conselho Nacional de Pesquisas, São Paulo, 1960, pp. 33–52.

[67] *Construction des algèbres de Lukasiewicz trivalentes dans les algèbres de Boole monadiques I.* **Math. Japan.,** vol. 12 (1967), pp. 1–23.

Monteiro, A., and Varsavsky, O.

[57] *Álgebras de Heyting monádicas.* **Unión Matemática Argentina, Actas de las X Jornadas,** Instituto de Matemáticas, Universidad Nacional del Sur, Bahía Blanca, 1957, pp. 52–62.

Murphy, J. J.

[1882] *On the addition and multiplication of logical relatives.* **Memoirs of the Manchester Literary and Philosophical Society,** ser. 3, vol. 7 (27) (1882), pp. 201–224.

Nadiu, Gh. S.

[67] *Asupra unei metode de construcţie v algebrelor Łukasiewicz trivalente.* **Stud. Cerc. Mat.,** vol. 19 (1967), pp. 1063–1070.

Nolin, L.

[58] *Sur l'algèbre des prédicats.* **Le raisonnement en mathématiques et en sciences expérimentales, Colloques internationaux du Centre National de la Recherche Scientifique,** no. 70, Centre National de la Recherche Scientifique, Paris, 1958, pp. 33–37.

Peirce, C. S.

[33] **Collected papers of Charles Sanders Peirce,** eds. C. Hartshorne and P. Weiss, vol. III, **Exact logic (Published papers),** Harvard University Press, Cambridge, 1933, xiv + 433 pp. [Reprinted by The Belknap Press of Harvard University Press, Cambridge, 1960 (vol. III bound with vol. IV).]

Penzov, Ju. E.

[61] *K arifmetike n-otnošeniĭ.* **Izv. Vysš. Učebn. Zaved. Matematika,** 1961, no. 4 (23), pp. 78–92.

Pieczkowski, A.

[64] *On the equivalence of the calculus of dependent sentential variables and the cylindrical algebra without diagonal elements.* **Bull. Acad. Polon. Sci. Sér. Sci. Math. Astronom. Phys.,** vol. 12 (1964), pp. 143–146.

See also Kotas and Pieczkowski.

Pinter, C. C.
[69] *On π-polyadic algebras.* **Notices Amer. Math. Soc.,** vol. 16 (1969), p. 179.
Ponasse, D.
[62] *Problèmes d'universalité s'introduisant dans l'algébrisation de la logique mathématique I.* **Nagoya Math. J.,** vol. 20 (1962), pp. 29–73.
[62a] *Problèmes d'universalité s'introduisant dans l'algébrisation de la logique mathématique II.* **Nagoya Math. J.,** vol. 21 (1962), pp. 61–110.
Preller, A.
[67] *La catégorie des algèbres quantifiées.* **Publications du Département de Mathématiques. Faculté des Sciences de Lyon.** (Lyon), vol. 4 (1967), no. 1, pp. 91–135.
[68] *Quantified algebras.* **Syntax and semantics of infinite languages,** ed. J. Barwise, Springer-Verlag, Berlin, 1968, pp. 182–203.
[68a] **Logique algèbrique infinitaire,** Thése Doct. Sci. Math. Lyon, 1968, 218 pp.
[69] *Logique algèbrique infinitaire. Complétude des calculs $L_{\alpha\beta}$,* **C. R. Acad. Sci. Paris,** vol. 268 (1969), pp. 1509–1511; 1589–1592.
[69a] *Interpolation et amalgamation.* **Publications du Département de Mathématiques. Faculté des Sciences de Lyon.** (Lyon), vol. 6 (1969), no. 1, pp. 49–65.
[69b] *On the weak representability of σ-complete dimension complemented cylindric algebras.* **Algebra i Logika Sem.,** vol. 8 (1969), pp. 695–711.
[70] *Substitution algebras in their relation to cylindric algebras.* **Arch. Math. Logik Grundlagenforsch.,** vol. 30 (1970), pp. 91–96.
Purdea, I.
[69] *Relations généralisées.* **Rev. Roumaine Math. Pures Appl.,** vol. 14 (1969), pp. 533–556.
Quine, W. V.
[36] *Toward a calculus of concepts.* **J. Symbolic Logic,** vol. 1 (1936), pp. 2–25.
Ribeiro, H.
[55] *Topological groups and Boolean algebras with operators.* **Univ. Lisboa Revista Fac. Ci. A,** vol. 4 (1955), pp. 195–200.
Rieger, L.
[64] *Zu den Strukturen der klassischen Prädikatenlogik.* **Z. Math. Logik Grundlagen Math.,** vol. 10 (1964), pp. 121–138.
[67] **Algebraic Methods of Mathematical Logic,** Academic Press, New York, 1967, 210 pp.
Riguet, J.
[48] *Relations binaires, fermetures, correspondences de Galois.* **Bull. Soc. Math. France,** vol. 76 (1948), pp. 114–155.
[50] *Produit tensoriel de treillis et théorie de Galois généralisée.* **Algèbre et théorie des nombres, Colloques internationaux du Centre National de la Recherche Scientifique,** no. 24, Centre National de la Recherche Scientifique, Paris, 1950, pp. 173–178.
[54] *Sur l'extension du calcul des relations binaires au calcul des matrices à éléments dans une algèbre de Boole complète.* **C. R. Acad. Sci. Paris,** vol. 238 (1954), pp. 2382–2385.
Rosenblatt, D.
[63] *On the graphs of finite idempotent Boolean relation matrices.* **J. Res. Nat. Bur. Standards Sect. B,** vol. 67B (1963), pp. 249–256.
Rozen, V. V.
[68] *Predstavlenija algebr n-otnošeniĭ v algebry binarnyh otnošeniĭ.* **Izv. Vysš. Učebn. Zaved. Matematika,** 1968, no. 12 (79), pp. 82–91.
Rubin, J. E.
[56] *Remarks about a closure algebra in which closed elements are open.* **Proc. Amer. Math. Soc.,** vol. 7 (1956), pp. 30–34.
Rutledge, J. D.
[60] *On the definition of an infinitely-many-valued predicate calculus.* **J. Symbolic Logic,** vol. 25 (1960), pp. 212–216.

Šaĭn, B. M.

[65] *Relation algebras.* **Bull. Acad. Polon. Sci. Sér. Sci. Math. Astronom. Phys.**, vol. 13 (1965), pp. 1–5.

[66] *Algebry otnošeniĭ.* **Tartuskiĭ Gosudarstvennyĭ Universitet, Mežvuzovskiĭ Naučnyĭ Simpozium po Obščeĭ Algebre, Doklady, coobščenija, rezjume,** Tartu, 1966, pp. 130–168.

Schröder, E.

[1890] **Vorlesungen über die Algebra der Logik (exakte Logik),** vols. 1, 2 (parts 1 and 2), 3 (part 1). **Algebra und Logik der Relative,** edited in part by E. Müller and B. G, Teubner, Leipzig, 1890–1905 (published in 4 vols.). [second edition, in 3 vols. published by Chelsea Publishing Co., Bronx, N.Y., 1966.]

[1895] *Note über die Algebra der binären Relative.* **Math. Ann.**, vol. 46 (1895), pp. 144–158.

Schweitzer, A. R.

[09] *A theory of geometrical relations.* **Amer. J. Math.**, vol. 31 (1909), pp. 365–410.

Schweizer, B., and Sklar, A.

[60] *The algebra of functions.* **Math. Ann.**, vol. 139 (1960), pp. 366–382.

[61] *The algebra of functions. II.* **Math. Ann.**, vol. 143 (1961), pp. 440–447.

Scognamiglio, G.

[62] *Estensione della nozione di algebra: algebra di secondo ordine su un'algebra di Boole.* **Giorn. Mat. Battaglini** ser. 5, vol. 10 (90) (1962), pp. 93–108.

Sebastião e Silva, J.

[45] *Sugli automorfismi di un sistema matematico qualunque.* **Pontificia Academia Scientiarum. Commentationes,** vol. 9 (1945), pp. 327–357.

Seidenberg, A.

[69] *On k-constructible sets, k-elementary formulae, and elimination theory,* **J. Reine Angew. Math.**, vols. 239–240 (1969), pp. 256–267.

Sklar, A. See Schweizer and Sklar.

Špaček, A.

[60] *Statistical estimation of provability in Boolean logic.* **Transactions of the Second Prague Conference on Information Theory, Statistical Decision Functions, Random Processes,** Publishing House of the Czechoslovak Academy of Sciences, Prague, 1960, pp. 609–626.

[61] *Statistical estimation of semantic provability.* **Proceedings of the Fourth Berkeley Symposium on Mathematical Statistics and Probability,** vol. 1, **Contributions to the theory of statistics,** ed. J. Neyman, University of California Press, Berkeley, 1961, pp. 655–668.

Stamm, E.

[31] *Über Relativfunktionen und Relativgleichungen.* **Monatshefte für Mathematik und Physik** vol. 38 (1931), pp. 147–166.

Svenonius, L.

[60] **Some problems in logical model-theory,** CWK Gleerup, Lund; Ejnar Munksgaard, Copenhagen, 1960, 43 pp.

Tarski, A.

[31] *Sur les ensembles définissables de nombres réels. I.* **Fund. Math.**, vol. 17 (1931), pp. 210–239. [See also Tarski [56a], pp. 110–142.]

[41] *On the calculus of relations.* **J. Symbolic Logic,** vol. 6 (1941), pp. 73–89.

[52] *A representation theorem for cylindric algebras. Preliminary report.* **Bull. Amer. Math. Soc.**, vol. 58 (1952), pp. 65–66.

[52a] *Some notions and methods on the borderline of algebra and metamathematics.* **Proceedings of the International Congress of Mathematicians, Cambridge, Massachusetts, U.S.A., August 30–September 6, 1950,** vol. 1, American Mathematical Society, Providence, 1952, pp. 705–720.

[53] *A formalization of set theory without variables.* **J. Symbolic Logic,** vol. 18 (1953), p. 189.

[53a] *Some metalogical results concerning the calculus of relations.* **J. Symbolic Logic,** vol. 18 (1953), pp. 188–189.

[54] *On the reduction of the number of generators in relation rings.* **J. Symbolic Logic,** vol. 19 (1954), pp. 158–159.

[56] *Equationally complete rings and relation algebras.* **Nederl. Akad. Wetensch. Proc. Ser. A,** vol. 59 (**Indag. Math.,** vol. 18) (1956), pp. 39–46.

[56a] **Logic, semantics, metamathematics, papers from 1923 to 1938,** translated by J. H. Woodger, Clarendon Press, Oxford, 1956, xiv + 471 pp.

[66] *On direct products of Boolean algebras with additional operations.* **Notices Amer. Math. Soc.,** vol. 13 (1966), pp. 728–729.

Tarski, A., and Thompson, F. B.

[52] *Some general properties of cylindric algebras. Preliminary report.* **Bull. Amer. Math. Soc.,** vol. 58 (1952), p. 65.

See also Addison, Henkin, and Tarski; Chin and Tarski; Henkin and Tarski; Jónsson and Tarski; Kuratowski and Tarski.

Thompson, F. B.

[52] **Some contributions to abstract algebra and metamathematics,** Doctoral dissertation, University of California, Berkeley, 1952, v + 78 pp.

See also Tarski and Thompson.

Ulam, S. M.

[64] **Problems in modern mathematics,** John Wiley & Sons, Inc., New York, 1964, xviii + 150 pp. [First published in 1960 under the title **A collection of mathematical problems.**]

See also Everett and Ulam.

Vagner, V. V.

[65] *Teoriya otnošeniĭ i algebra čactičnyh otobraženiĭ.* **Teoriya polugrupp i ee priloženiya, Sbornik ctateĭ,** Vypusk 1, Izdatel′stvo Saratovskogo Universiteta (1965), pp. 3–178.

Varsavsky, O.

[56] *Quantifiers and equivalence relations.* **Revista Matemática Cuyana** (San Luis), vol. 2 (1956), pp. 29–51.

[57] *Individuos despreciables en álgebras de Boole monádicas.* **Unión Matemática Argentina, Actas de las X Jornadas,** Instituto de Matemáticas, Universidad Nacional del Sur, Bahía Blanca, 1957, pp. 67–68.

See also Monteiro and Varsavsky.

Venne, M.

[65] **Langues, théories et algèbres polyadiques d'ordres supérieurs,** Doctoral dissertation, Université de Montréal, Montréal, 1965, viii + 110 pp.

[66] *Langues et théories d'ordres supérieurs.* **C. R. Acad. Sci. Paris Sér. A–B,** ser. A, vol. 262 (1966), pp. 1142–1143.

[66a] *Algèbres polyadiques d'ordres supérieurs.* **C. R. Acad. Sci. Paris Sér. A–B,** ser. A, vol. 262 (1966), pp. 1236–1238.

[66b] *Représentation des algèbres polyadiques d'ordres supérieurs.* **C. R. Acad. Sci. Paris Sér. A–B,** ser. A, vol. 262 (1966), pp. 1293–1294.

Whitlock, H. I.

[63] *Abstract characterization of an algebra of multi-place functions. I.* **Notices Amer. Math. Soc.,** vol. 10 (1963), p. 664.

Wooyenaka, Y.

[59] *On postulate-sets for relation algebras.* **Notices Amer. Math. Soc.,** vol. 6 (1959), p. 534.

Wright, F. B.

[57] *Ideals in polyadic algebra.* **Proc. Amer. Math. Soc.,** vol. 8 (1957), pp 544–546.

[57a] *Some remarks on Boolean duality.* **Portugal. Math.,** vol. 16 (1957), pp. 109–117.

[61] *Generalized means.* **Trans. Amer. Math. Soc.,** vol. 98 (1961), pp. 187–203.

[61a] *Recurrence theorems and operators on Boolean algebras.* **Proc. London Math. Soc.,** ser. 3, vol. 11 (1961), pp. 385–401.

[62] *Convergence of quantifiers and martingales.* **Illinois J. Math.,** vol. 6 (1962), pp. 296–307.

[63] *Boolean averages.* **Canad. J. Math.,** vol. 15 (1963), pp. 440–455.

II. SUPPLEMENTARY BIBLIOGRAPHY

Adjan, S. I.

[66*] *Opredeljajuščie cootnošenija i algoritmičeskie problemy dlja grupp i polugrupp.* **Trudy Mat. Inst. Steklov**, vol. 85 (1966), 122 pp. [English translation: *Defining relations and algorithmic problems for groups and semigroups.* **Proceedings of the Steklov Institute of Mathematics** (Providence, R. I.), no. 85 (1966), 152 pp.]

Bernays, P., and Fraenkel, A. A.

[58*] **Axiomatic set theory**, North-Holland Publishing Co., Amsterdam, 1958, viii + 226 pp.

See also Hilbert and Bernays.

Bernstein, B. A.

[50*] *A dual-symmetric definition of Boolean algebra free from postulated special elements.* **Scripta Math.**, vol. 16 (1950), pp. 157–160.

Beth, E. W.

[53*] *On Padoa's method in the theory of definition.* **Nederl. Akad. Wetensch. Proc. Ser. A**, vol. 56 (**Indag. Math.**, vol. 15) (1953), pp. 330–339.

Birkhoff, G.

[35*] *On the structure of abstract algebras.* **Proc. Cambridge Philos. Soc.**, vol. 31 (1935), pp. 433–454.

[44*] *Subdirect unions in universal algebras.* **Bull. Amer. Math. Soc.**, vol. 50 (1944), pp. 764–768.

[46*] *Sobre los grupos de automorfismos.* **Rev. Un. Mat. Argentina**, vol. 11 (1946), pp. 155–157.

[67*] **Lattice theory**, third edition, **American Mathematical Society colloquium publications**, vol. 25, American Mathematical Society, Providence, 1967, vi + 418 pp.

Birkhoff, G., and Frink, O.

[48*] *Representations of lattices by sets.* **Trans. Amer. Math. Soc.**, vol. 64 (1948), pp. 299–316.

Bradford, R. E.

[65*] **Cardinal addition and the axiom of choice**, Doctoral dissertation, University of California, Berkeley, 1965, iv + 97 pp.

Bruck, R. H.

[66*] **A survey of binary systems**, second printing, Springer-Verlag New York Inc., 1966, viii + 85 pp.

Carnap, R.

[46*] *Modalities and quantification.* **J. Symbolic Logic**, vol. 11 (1946), pp. 33–64.

Chang, C. C.

[59*] *On unions of chains of models.* **Proc. Amer. Math. Soc.**, vol. 10 (1959), pp. 120–127.

Chang, C. C., Jónsson, B., and Tarski, A.

[64*] *Refinement properties for relational structures.* **Fund. Math.**, vol. 55 (1964), pp. 249–281.

Church, A.
[56*] **Introduction to mathematical logic,** vol. 1, Princeton University Press, Princeton, 1956, x + 378 pp.
Cohen, P. J.
[63*] *The independence of the continuum hypothesis I.* **Proc. Nat. Acad. Sci. U.S.A.,** vol. 50 (1963), pp. 1143–1148.
[64*] *The independence of the continuum hypothesis II.* **Proc. Nat. Acad. Sci. U.S.A.,** vol. 51 (1964), pp. 105–110.
Cohn, P. M.
[65*] **Universal algebra,** Harper and Row Publishers, New York, 1965, xviii + 333 pp.
Corner, A. L. S.
[64*] *On a conjecture of Pierce concerning direct decompositions of Abelian groups.* **Proceedings of the Colloquium on Abelian Groups, Tihany (Hungary), September 1963,** eds. L. Fuchs and E. T. Schmidt, Akadémiai Kiadó, Budapest, 1964, pp. 43–48.
Craig, W.
[57*] *Linear reasoning. A new form of the Herbrand-Gentzen theorem.* **J. Symbolic Logic,** vol. 22 (1957), pp. 250–268.
[57a*] *Three uses of the Herbrand-Gentzen theorem in relating model theory and proof theory.* **J. Symbolic Logic,** vol. 22 (1957), pp. 269–285.
Crawley, P., and Jónsson, B.
[64*] *Refinements for infinite direct decompositions of algebraic systems.* **Pacific J. Math.,** vol. 14 (1964), pp. 797–855.
Day, G.
[67*] *Superatomic Boolean algebras.* **Pacific J. Math.,** vol. 23 (1967), pp. 479–489.
Ehrenfeucht, A.
[61*] *An application of games to the completeness problem for formalized theories.* **Fund. Math.,** vol. 49 (1961), pp. 129–141.
Feferman, S., and Vaught, R. L.
[59*] *The first order properties of products of algebraic systems.* **Fund. Math.,** vol. 47 (1959), pp. 57–103.
Felscher, W.
[65*] *Zur Algebra unendlich langer Zeichenreihen.* **Z. Math. Logik Grundlagen Math.,** vol. 11 (1965), pp. 5–16.
[68*] *Equational maps.* **Contributions to mathematical logic,** ed. K. Schütte, North-Holland Publishing Co., Amsterdam, 1968, pp. 121–161.
Fraenkel, A. See Bernays and Fraenkel.
Fraïssé, R. J.
[54*] *Sur quelques classifications des systèmes de relations.* **Publ. Sci. Univ. Alger Sér. A,** vol. 1 (1954), pp. 35–182. [English summary.] [Thesis (Paris, 1953).]
[55*] *Sur quelques classifications des relations, basées sur des isomorphismes restreints. I. Étude générale. II. Application aux relations d'ordre, et construction d'exemples montrant que ces classifications sont distinctes.* **Publ. Sci. Univ. Alger Sér. A,** vol. 2 (1955), pp. 15–60, 273–295.
Frayne, T., Morel, A. C., and Scott, D. S.
[62*] *Reduced direct products.* **Fund. Math.,** vol. 51 (1962), pp. 194–228.
Frink, O. See Birkhoff and Frink.
Fujiwara, T.
[55*] *Note on the isomorphism problem for free algebraic systems.* **Proc. Japan Acad.,** vol. 31 (1955), pp. 135–136.
Galvin, F.
[67*] *Reduced products, Horn sentences, and decision problems.* **Bull. Amer. Math. Soc.,** vol. 73 (1967), pp. 59–64.

Goodman, N.
[43*] *On the simplicity of ideas.* **J. Symbolic Logic,** vol. 8 (1943), pp. 107–121.
Grätzer, G.
[68*] **Universal algebra,** D. Van Nostrand Co., Inc., Princeton, 1968, xvi + 368 pp.
Grätzer, G., and Schmidt, E. T.
[63*] *Characterizations of congruence lattices of abstract algebras.* **Acta Sci. Math. (Szeged),** vol. 24 (1963), pp. 34–59.
Halmos, P. R.
[63*] **Lectures on Boolean algebras,** D. Van Nostrand Co., Inc., Princeton, 1963, iv + 147 pp.
Hanf, W.
[56*] *Representations of lattices by subalgebras. Preliminary report.* **Bull. Amer. Math. Soc.,** vol. 62 (1956), p. 402.
[57*] *On some fundamental problems concerning isomorphism of Boolean algebras.* **Math. Scand.,** vol. 5 (1957), pp. 205–217.
[62*] *Isomorphism in elementary logic. Preliminary report.* **Notices Amer. Math. Soc.,** vol. 9 (1962), pp. 146–147.
[62a*] **Some fundamental problems concerning languages with infinitely long expressions,** Doctoral dissertation, University of California, Berkeley, 1962, 95 pp.
[64*] *On a problem of Erdös and Tarski.* **Fund. Math.,** vol. 53 (1964), pp. 325–334.
Hashimoto, J.
[57*] *Direct, subdirect decompositions and congruence relations.* **Osaka Math. J.,** vol. 9 (1957), pp. 87–112.
Hilbert, D., and Bernays, P.
[34*] **Grundlagen der Mathematik,** vol. 1, J. Springer, Berlin, 1934, xii + 471 pp.
Horn, A., and Tarski, A.
[48*] *Measures in Boolean algebras.* **Trans. Amer. Math. Soc.,** vol. 64 (1948), pp. 467–497.
Huntington, E. V.
[04*] *Sets of independent postulates for the algebra of logic.* **Trans. Amer. Math. Soc.,** vol. 5 (1904), pp. 288–309.
[33*] *New sets of independent postulates for the algebra of logic, with special reference to Whitehead and Russell's Principia Mathematica.* **Trans. Amer. Math. Soc.,** vol. 35 (1933), pp. 274–304.
[33a*] *Boolean algebra. A correction.* **Trans. Amer. Math. Soc.,** vol. 35 (1933), pp. 557–558.
Jakubík, J.
[54*] *O otnošenijah kongruèntnosti na abstraktnyh algebrah.* **Czechoslovak Math. J.,** vol. 4 (1954), pp. 314–317. [English summary.]
Jónsson, B.
[57*] *On isomorphism types of groups and other algebraic systems.* **Math. Scand.,** vol. 5 (1957), pp. 224–229.
[57a*] *On direct decompositions of torsion-free Abelian groups.* **Math. Scand.,** vol. 5 (1957), pp. 230–235.
[65*] *Extensions of relational structures.* Addison-Henkin-Tarski [65], pp. 146–157.
[66*] *The unique factorization problem for finite relational structures.* **Colloq. Math.,** vol. 14 (1966), pp. 1–32.
Jónsson, B., and Tarski, A.
[47*] **Direct decompositions of finite algebraic systems, Notre Dame mathematical lectures,** no. 5, University of Notre Dame, Notre Dame, 1947, vi + 64 pp.
[61*] *On two properties of free algebras.* **Math. Scand.,** vol. 9 (1961), pp. 95–101.
See also Chang, Jónsson, and Tarski; Crawley and Jónsson; Tarski [49*].
Karrass, A. See Magnus, Karrass and Solitar.

Keisler, H. J.

[61*] *On some results of Jónsson and Tarski concerning free algebras.* **Math. Scand.**, vol. 9 (1961), pp. 102–106.

[61a*] *Ultraproducts and elementary classes.* **Nederl. Akad. Wetensch. Proc. Ser. A,** vol. 64 (**Indag. Math.**, vol. 23) (1961), pp. 477–495.

[63*] *Limit ultrapowers.* **Trans. Amer. Math. Soc.**, vol. 107 (1963), pp. 382–408.

[65*] *Reduced products and Horn classes.* **Trans. Amer. Math. Soc.**, vol. 107 (1965), pp. 307–328.

Keisler, H. J., and Tarski, A.

[64*] *From accessible to inaccessible cardinals.* **Fund. Math.**, vol. 53 (1964), pp. 225–308.

Kelley, J. L.

[55*] **General topology,** D. Van Nostrand Co., Inc., Princeton, 1955, xiv + 298 pp.

Kerkhoff, R.

[65*] *Eine Konstruktion absolut freier Algebren.* **Math. Ann.**, vol. 158 (1965), pp. 109–112.

Kinoshita, S.

[53*] *A solution of a problem of R. Sikorski.* **Fund. Math.**, vol. 40 (1953), pp. 39–41.

Kochen, S.

[61*] *Ultraproducts in the theory of models.* **Ann. of Math.**, ser. 2, vol. 74 (1961), pp. 221–261.

Łoś, J.

[55*] *On the extending of models (I).* **Fund. Math.**, vol. 42 (1955), pp. 38–54.

[55a*] *Quelques remarques, théorèmes et problèmes sur les classes définissables d'algèbres.* **Mathematical interpretation of formal systems,** North-Holland Publishing Co., Amsterdam, 1955, pp. 98–113.

[65*] *Free product in general algebras.* Addison-Henkin-Tarski [65], pp. 229–237.

Łoś, J., and Suszko, R.

[57*] *On the extending of models (IV).* **Fund. Math.**, vol. 44 (1957), pp. 52–60.

Lovász, L.

[67*] *Operations with structures.* **Acta Math. Acad. Sci. Hungar.**, vol. 18 (1967), pp. 321–328.

Łukasiewicz, J.

[63*] **Elements of mathematical logic,** The Macmillan Co., New York, 1963, xii + 124 pp.

Lyndon, R. C.

[59*] *Properties preserved under homomorphism.* **Pacific J. Math.**, vol. 9 (1959), pp. 143–154.

[59a*] *Properties preserved in subdirect products.* **Pacific J. Math.**, vol. 9 (1959), pp. 155–164.

[62*] *Metamathematics and algebra: an example.* **Logic, methodology and philosophy of science, Proceedings of the 1960 International Congress,** eds. E. Nagel. P. Suppes, and A. Tarski, Stanford University Press, Stanford, 1962, pp. 143–150.

MacLane, S.

[63*] **Homology,** Springer-Verlag, Berlin, 1963, x + 422 pp.

Magnus, W., Karrass, A., and Solitar, D.

[66*] **Combinatorial group theory: presentations of groups in terms of generators and relations,** Interscience Publishers, New York, 1966, xii + 444 pp.

Mal'cev, A. I.

[54*] *K obščeĭ teorii algebraičeskih sistem.* **Mat. Sb.**, n.s., vol. 35 (77) (1954), pp. 3–20. [English translation: *On the general theory of algebraic systems.* **Amer. Math. Soc. Transl.**, ser. 2, vol. 27 (1963), pp. 125–142.]

[58*] *Strukturnaja harakteristika nekotoryh klassov algebr.* **Dokl. Acad. Nauk SSSR,** n.s., vol. 120 (1958), pp. 29–32.

Marczewski, E.

[66*] *Independence in abstract algebras, results and problems.* **Colloq. Math.**, vol. 14 (1966), pp. 169–188.

Mayer, R. D., and Pierce, R. S.
[60*] *Boolean algebras with ordered bases.* **Pacific J. Math.**, vol. 10 (1960), pp. 925–942.
Mazurkiewicz, S., and Sierpiński, W.
[20*] *Contribution à la topologie des ensembles dénombrables.* **Fund. Math.**, vol. 1 (1920), pp. 17–27.
McKenzie, R.
[70*] *Cardinal multiplication of structures with a reflexive relation.* **Fund. Math.**, vol. 69 (1970), to appear.
McKinsey, J. C. C., and Tarski, A.
[44*] *The algebra of topology.* **Ann. of Math.**, ser. 2, vol. 45 (1944), pp. 141–191.
[46*] *On closed elements in closure algebras.* **Ann. of Math.**, ser. 2, vol. 47 (1946), pp. 122–162.
Monk, J. D.
[62*] *On pseudo-simple universal algebras.* **Proc. Amer. Math. Soc.**, vol. 13 (1962), pp. 543–546.
[69*] **Introduction to set theory,** McGraw-Hill Book Co., New York, 1969, xii + 193 pp.
Montague, R. M., Scott, D. S., and Tarski, A.
[72*] **An axiomatic approach to set theory,** North-Holland Publishing Co., Amsterdam, to appear.
Morel, A. See Frayne, Morel and Scott.
Morse, A. P.
[65*] **A theory of sets,** Academic Press, New York, 1965, xxxi + 130 pp.
Mostowski, A.
[52*] *On direct products of theories.* **J. Symbolic Logic,** vol. 17 (1952), pp. 1–31.
Mostowski, A., and Tarski, A.
[39*] *Boolesche Ringe mit geordneter Basis.* **Fund. Math.**, vol. 32 (1939), pp. 69–86.
[49*] *Arithmetical classes and types of well ordered systems. Preliminary report.* **Bull. Amer. Math. Soc.**, vol. 55 (1949), p. 65.
See also Tarski, Mostowski, and Robinson.
Neumann, B. H.
[62*] **Special topics in algebra: universal algebra,** Courant Institute of Mathematical Sciences, New York University, 1962, ii + 78 pp.
Pierce, R. S. See Mayer and Pierce.
Pigozzi, D.
[66*] *On some operations on classes of algebras.* **Notices Amer. Math. Soc.**, vol. 13 (1966), p. 829.
Preller, A.
[68*] *On the relationship between the classical and the categorical direct products of algebras.* **Nederl. Akad. Wetensch. Proc. Ser. A,** vol. 71 (**Indag. Math.**, vol. 30) (1968), pp. 512–516.
Presburger, M.
[30*] *Über die Vollständigkeit eines gewissen Systems der Arithmetik ganzer Zahlen, in welchem die Addition als einzige Operation hervortritt.* **Sprawozdanie z I Kongresu Matematyków Krajów Słowiańskich, (Comptes-rendus du I Congrès des Mathématiciens des Pays Slaves), Warszawa 1929,** Warsaw, 1930, pp. 92–101, 395.
Quine, W. V. O.
[55*] **Mathematical logic,** revised edition, Harvard University Press, Cambridge, 1955, xii + 346 pp.
Rasiowa, H., and Sikorski, R.
[63*] **The mathematics of metamathematics,** Państwowe Wydawnictwo Naukowe, Warszawa, 1963, 519 pp.

Resnikoff, I.

[65*] *Tout ensemble de formules de la logique classique est équivalent à un ensemble indépendant.* **C. R. Acad. Sci. Paris,** vol. 260 (1965), pp. 2385–2388.

Ribeiro, H.

[52*] *A remark on Boolean algebras with operators.* **Amer. J. Math.,** vol. 74 (1952), pp. 163–167.

Robinson, R. M.

[45*] *Finite sequences of classes.* **J. Symbolic Logic,** vol. 10 (1945), pp. 125–126.

See also Tarski, Mostowski, and Robinson.

Šaĭn, B. M.

[65*] *K teoreme Birkgofa–Kogalovskogo.* **Uspehi Mat. Nauk,** n.s., vol. 20 (1965), no. 6, pp. 173–174.

Samuel, P.

[48*] *On universal mappings and free topological groups.* **Bull. Amer. Math. Soc.,** vol. 54 (1948), pp. 591–598.

Sąsiada, E.

[61*] *Negative solution of I. Kaplansky's first test problem for Abelian groups and a problem of K. Borsuk concerning cohomology groups.* **Bull. Acad. Polon. Sci. Sér. Sci. Math. Astronom. Phys.,** vol. 9 (1961), pp. 331–334.

Schmidt, E. T. See Grätzer and Schmidt.

Schmidt, J.

[52*] *Über die Rolle der transfiniten Schlussweisen in einer allgemeinen Idealtheorie.* **Math. Nachr.,** vol. 7 (1952), pp. 165–182.

[53*] *Einige grundlegende Begriffe und Sätze aus der Theorie der Hüllenoperatoren.* **Bericht über die Mathematiker-Tagung in Berlin, January,** 1953, Deutscher Verlag der Wissenschaften, Berlin, 1953, pp. 21–48.

Scholz, H.

[52*] *Ein ungelöstes Problem in der symbolischen Logik.* **J. Symbolic Logic,** vol. 17 (1952), p. 160.

Scott, D. See Frayne, Morel and Scott; Montague, Scott, and Tarski.

Sierpiński, W.

[64*] **Elementary theory of numbers,** translated by A. Hulanicki, Państwowe Wydawnictwo Naukowe, Warszawa, 1964, 480 pp.

[65*] **Cardinal and ordinal numbers,** second edition revised, translated by J. Smólska, Państwowe Wydawnictwo Naukowe, Warszawa, 1965, 491 pp.

See also Mazurkiewicz and Sierpinski.

Sikorski, R.

[64*] **Boolean algebras,** second edition, Springer-Verlag, Berlin, 1964, x + 237 pp.

See also Rasiowa and Sikorski.

Słomiński, J.

[59*] *The theory of abstract algebras with infinitary operations.* **Rozprawy Mat.,** vol. 18 (1959), pp. 1–67.

Smith, E. C., and Tarski, A.

[57*] *Higher degrees of distributivity and completeness in Boolean algebras.* **Trans. Amer. Math. Soc.,** vol. 84 (1957), pp. 230–257.

Solitar, D. See Magnus, Karrass, and Solitar.

Solovay, R.

[65*] $2^{\aleph_0}$ *can be anything it ought to be.* Addison-Henkin-Tarski [65], p. 435.

Stone, M. H.

[34*] *Boolean algebras and their application to topology.* **Proc. Nat. Acad. Sci., U.S.A.,** vol. 20 (1934), pp. 197–202.

Suppes, P.
[60*] **Axiomatic set theory,** D. Van Nostrand Co., Inc., Princeton, 1960, xii + 265 pp.
Suszko, R. See Łoś and Suszko.
Szélpál, I.
[49*] *Die abelschen Gruppen ohne eigentliche Homomorphismen.* **Acta Sci. Math. (Szeged),** vol. 13 (1949), pp. 51–53.
Szmielew, W.
[55*] *Elementary properties of Abelian groups.* **Fund. Math.,** vol. 41 (1955), pp. 203–271.
Taĭmanov, A. D.
[61*] *Harakteristika aksiomatiziruemyh klassov modeleĭ. I, II.* **Izv. Akad. Nauk SSSR Ser. Math.,** vol. 25 (1961), pp. 601–620, 755–764.
Tarski, A.
[30*] *Fundamentale Begriffe der Methodologie der deduktiven Wissenschaften. I.* **Monatshefte für Mathematik und Physik,** vol. 37 (1930), pp. 361–404. [See also Tarski [56a], pp. 60–109.]
[35*] *Grundzüge des Systemenkalküls. Erster Teil.* **Fund. Math.,** vol. 25 (1935), pp. 503–526. [See also Tarski [56a], pp. 342–383.]
[37*] *Ideale in Mengenkörpern.* **Rocznik Polskiego Towarzystwa Matematycznego (Annales de la Société Polonaise de Mathématique),** vol. 15 (1937), pp. 186–189.
[49*] **Cardinal algebras,** with an appendix, **Cardinal products of isomorphism types,** by B. Jónsson and A. Tarski, Oxford University Press, New York, 1949, xii + 326 pp.
[51*] *Remarks on the formalization of the predicate calculus. Preliminary report.* **Bull. Amer. Math. Soc.,** vol. 57 (1951), pp. 81–82.
[53*] *Universal arithmetical classes of mathematical systems. Preliminary report.* **Bull. Amer. Math. Soc.,** vol. 59 (1953), pp. 390–391.
[54*] *Contributions to the theory of models. I, II.* **Nederl. Akad. Wetensch. Proc. Ser. A,** vol. 57 (**Indag. Math.,** vol. 16) (1954), pp. 572–588.
[55*] *Contributions to the theory of models. III.* **Nederl. Akad. Wetensch. Proc. Ser. A,** vol. 58 (**Indag. Math.,** vol. 17) (1955), pp. 56–64.
[55a*] *A lattice-theoretical fixpoint theorem and its applications.* **Pacific J. Math.,** vol. 5 (1955), pp. 285–309.
[57*] *Remarks on direct products of commutative semigroups.* **Math. Scand.,** vol. 5 (1957), pp. 218–223.
[58*] *Remarks on predicate logic with infinitely long expressions.* **Colloq. Math.,** vol. 6 (1958), pp. 171–176.
[65*] *A simplified formulation of predicate logic with identity.* **Arch. Math. Logik Grundlagenforsch.,** vol. 7 (1965), pp. 61–79.
[68*] *Equational logic and equational theories of algebras.* **Contributions to mathematical logic,** ed. K. Schütte, North-Holland Publishing Co., Amsterdam, 1968, pp. 275–288.
Tarski, A., Mostowski, A., and Robinson, R. M.
[53*] **Undecidable theories,** North-Holland Publishing Co., Amsterdam, 1953, xii + 98 pp.
Tarski, A., and Vaught, R. L.
[57*] *Arithmetical extensions of relational systems.* **Compositio Math.,** vol. 13 (1957), pp. 81–102.
See also Chang, Jónsson, and Tarski; Horn and Tarski; Jónsson and Tarski; Keisler and Tarski; McKinsey and Tarski; Montague, Scott, and Tarski; Mostowski and Tarski; Smith and Tarski.
Telgársky, R.
[68*] *Derivatives of Cartesian product and dispersed spaces.* **Colloq. Math.,** vol. 19 (1968), pp. 59–66.

Valucè, I. I.

[63*] *Universal'nye algebry s pravil'nymi, no ne perestanovočnymi kongruèncijami.* **Uspehi Mat. Nauk,** n.s., vol. 18 (1963), no. 3, pp. 145–148.

Vaught, R. L.

[54*] *Remarks on universal classes of relational systems.* **Nederl. Akad. Wetensch. Proc. Ser. A,** vol. 57 (**Indag. Math.,** vol. 16) (1954), pp. 589–591

[66*] *The elementary character of two notions from general algebra.* **Essays on the foundations of mathematics,** second edition, eds. Y. Bar-Hillel, E. I. J. Poznanski, M. O. Rabin, and A. Robinson, The Magnes Press, The Hebrew University, Jerusalem, 1966, pp. 226–233.

See also Feferman and Vaught; Tarski and Vaught.

INDEX OF SYMBOLS

SET THEORY

$x \in A$	x is a member of A, 26
$\{x: \varphi(x)\}$	class of elements x such that $\varphi(x)$, 26
$\{x: \varphi(x), \psi(x)\}$	class of elements x such that $\varphi(x)$ and $\psi(x)$, 26
$\{x \in A: \varphi(x)\}$	class of $x \in A$ such that $\varphi(x)$, 26
${}_x\{\tau(x): \varphi(x)\}$	class of elements $\tau(x)$ such that $\varphi(x)$, 26
${}_{x,y}\{\tau(x, y): \varphi(x, y)\}$	class of elements $\tau(x, y)$ such that $\varphi(x, y)$, 26
0	empty set, 27
$\{x\}$	set whose only member is x, 27
$\{x, y\}$	set whose only members are x and y, 27
$\subseteq$	inclusion, 27
$\subset$	proper inclusion, 27
$\nsubseteq$, $\not\subset$, $\supseteq$, $\supset$	negations and converses of above, 27
$Sb\, A$	class of subsets of A, 27
$A \cup B$	union of A and B, 27
$A \cap B$	intersection of A and B, 27
$\bigcup C$	union of members of C, 27
$\bigcap C$	intersection of members of C, 27
${}_x\bigcup_{\varphi(x)}\tau(x)$	union of all $\tau(x)$ such that $\varphi(x)$, 27
${}_x\bigcap_{\varphi(x)}\tau(x)$	intersection of all $\tau(x)$ such that $\varphi(x)$, 27
$A \sim B$, ${}_A{\sim} B$	difference of A and B, 27
$\langle x, y \rangle$	ordered pair, 27
$\langle x, y, z \rangle$	ordered triple, 27
Id	identity relation, 27
xRy	$\langle x, y \rangle \in R$, 28
$x\varphi y\psi z$	$x\varphi y$ and $y\psi z$, 28
$A \times B$	product of A and B, 28
$A \times B \times C$	product of A, B, and C, 28
$R \mid S$	relative product of R and S, 28

R^{-1}	converse of R, 28
$Do\ R$	domain of R, 28
$Rg\ R$	range of R, 28
$Fd\ R$	field of R, 28
$A \upharpoonleft R$	R domain-restricted to A, 28
R^*A	R-image of A, 28
$R^\star x$	$R^*\{x\}$, 28
$fx, f(x), f_x, f^x, f^{(x)}$	x^{th} value of f, 28
${}_x\langle\tau(x):\varphi(x)\rangle$	$\{\langle x, \tau(x)\rangle:\varphi(x)\}$, 28
${}_x\langle\tau(x)\rangle_{\varphi(x)}$	${}_x\langle\tau(x):\varphi(x)\rangle$, 28
$f\circ g$	composition of f and g, 29
${}^AB$	power of B to the exponent A, 29
$\mathsf{P}F$	product of the function F, 29
${}_x\mathsf{P}_{\varphi(x)}\tau(x)$	product of sets $\tau(x)$ such that $\varphi(x)$, 29
pj_a	a^{th} projection, 30
x/R	R-equivalence class of x, 30
X/R	$R^{\star*}X$, 30
S/R	$\{\langle x/R, y/R\rangle:xSy\}$, 30
$\tau_R x$	R-type of x, 30
$\leqq$	symbol for partial ordering, 31
$\alpha, \beta, \gamma, \ldots$	ordinal numbers, 31
$\Gamma, \Delta, \Theta, \ldots$	sets of ordinals, 31
$\xi < \eta$	ξ is less than η, 32
$\eta > \xi$	η is greater than ξ, 32
$\xi \leqq \eta, \eta \leqq \xi$	$\xi < \eta$, or $\xi = \eta$ and ξ is an ordinal, 32
$\xi + \eta$	ordinal sum, 32
0, 1, 2, etc.	least ordinals, 32
ω	least limit ordinal, 32
$\|A\|$	number of elements of A, 32
ω_ξ	ξ^{th} infinite cardinal, 32
$\alpha\cdot\beta$	ordinal product, 32
α^β	power of ordinals, 32
$\alpha \# \beta$	natural sum of α and β, 366
α^+	cardinal successor of α, 33
$\mathcal{GCH}$	generalized continuum hypothesis, 33
$\langle x_0, \ldots, x_\xi, \ldots\rangle_{\xi<\alpha}$	α-termed sequence, 33
$\langle x_0, \ldots, x_{\alpha-1}\rangle$	α-termed sequence, $0 < \alpha < \omega$, 33
$\{x_0, \ldots, x_\xi, \ldots\}_{\xi<\alpha}$	range of an α-termed sequence x, 33
$\{x_0, \ldots, x_{\alpha-1}\}$	range of an α-termed sequence x, $0 < \alpha < \omega$, 33

$\langle a\rangle$	one-termed sequence, 33
$\langle a, b\rangle$	two-termed sequence, 33, 34
$\overline{X}$	X or $\{X\}$, 34
$O(x_0, \ldots, x_\xi, \ldots)_{\xi<\alpha}$	Ox, 35
$O(x_\xi : \xi < \alpha)$	Ox, 35
$O(x_0, \ldots, x_{\alpha-1})$	Ox, 35
ρO	rank of O, 35
$x_0 O \ldots O x_{\kappa-1}$	extended operation, 35
$x_0 O x_1 O \ldots O x_{\kappa-1}$	extended operation, 35
$S^{[\kappa]}$	κ^{th} relative power of S, 36
$[x_\kappa / y_\kappa : \kappa < \nu]_A$	finite transformation, 36
$[x_0/y_0, \ldots, x_{\nu-1}/y_{\nu-1}]_A$	finite transformation, 36
$[x/y]$	replacement, 36
$[x/y, y/x]$	transposition, 36
$\langle A, R_i, O_j\rangle_{i\in I, j\in J}$	relational structure, 36

METALOGIC

$\Lambda_{\Sigma,\upsilon,\Gamma,\Delta}$	language of predicate logic, 39
$\langle \boldsymbol{v}_0, \ldots, \boldsymbol{v}_\xi, \ldots\rangle_{\xi<\alpha}$	sequence of variables, 40
$\boldsymbol{x}, \boldsymbol{y}$, etc.	special variables, 40
$\boldsymbol{A}$	disjunction symbol, 40
$\boldsymbol{K}$	conjunction symbol, 40
$\boldsymbol{N}$	negation symbol, 40
$\boldsymbol{Q}$	existential quantifier, 40
$\boldsymbol{E}$	equality symbol, 40
$\boldsymbol{T}$	truth symbol, 40
$\boldsymbol{F}$	falsehood symbol, 40
$\varphi \vee \psi$	disjunction of φ and ψ, 40
$\varphi \wedge \psi$	conjunction of φ and ψ, 40
$\neg\varphi$	negation of φ, 40
$\exists_\upsilon \varphi$	existential quantification of φ, 40
$\rightarrow$	implication, 41
$\leftrightarrow$	equivalence, 41
$\forall_\upsilon \varphi$	universal quantification of φ, 41
$\exists_{\upsilon_0 \ldots \upsilon_{\kappa-1}}$	composition of existential quantifications, 41
$\forall_{\upsilon_0 \ldots \upsilon_{\kappa-1}}$	composition of universal quantifications, 41
$\boldsymbol{C}$	implication symbol, 41

$T\mu^{(\Lambda)}, T\mu, T\mu^{(\alpha)}$	set of terms, 41, 42
$\sigma = \tau$	equation, 42
$\sigma \neq \tau$	$\neg(\sigma = \tau)$, 42
$\Phi\mu^{(\Lambda)}, \Phi\mu$	set of all formulas, 42
$\Sigma\nu^{(\Lambda)}, \Sigma\nu$	set of all sentences, 42
$[\varphi]$	closure of φ, 42
$\tilde{\tau}^{(\mathfrak{A},\alpha)}, \tilde{\tau}^{(\mathfrak{A})}$	operation in $\mathfrak{A}$ represented by τ, 43
$\tilde{\tau}_\beta^{(\mathfrak{A},\alpha)}, \tilde{\tau}_\beta^{(\mathfrak{A})}$	β-ary operation in $\mathfrak{A}$ represented by τ, 43
$\tilde{\varphi}^{(\mathfrak{A},\alpha)}, \tilde{\varphi}^{(\mathfrak{A})}$	relation in $\mathfrak{A}$ defined by φ, 43
$\tilde{\varphi}_\beta^{(\mathfrak{A},\alpha)}, \tilde{\varphi}_\beta^{(\mathfrak{A})}$	β-ary relation in $\mathfrak{A}$ defined by φ, 44
$\mathfrak{A} \models \varphi[x]$	x satisfies φ in $\mathfrak{A}$, 44
$\mathfrak{A} \models \varphi$	φ holds in $\mathfrak{A}$, 44
$\mathsf{K} \models \varphi$	φ holds in K, 44
$\theta\rho\mathfrak{A}$	theory of $\mathfrak{A}$, 44
$\mathfrak{A} \equiv \mathfrak{B}$	$\mathfrak{A}$ is elementarily equivalent to $\mathfrak{B}$, 113, 44
$\theta\rho\mathsf{K}$	theory of K, 44
$\mathsf{Md}\varphi$, $\mathsf{Md}\Phi$	class of models of φ, Φ, 44
$\mathscr{E}\mathscr{C}$	elementary class in the narrower sense, 44
$\mathscr{E}\mathscr{C}_\Delta$	elementary class, 44
$\mathscr{P}\mathscr{C}$	pseudo-elementary class in the narrower sense, 154
$\mathscr{P}\mathscr{C}_\Delta$	pseudo-elementary class, 154
$\mathscr{U}\mathscr{C}$	universal class in the narrower sense, 44
$\mathscr{U}\mathscr{C}_\Delta$	universal class, 44
$\mathscr{E}\mathscr{Q}\mathscr{C}$	equational class in the narrower sense, 44
$\mathscr{E}\mathscr{Q}\mathscr{C}_\Delta$	equational class, 44
$\Sigma \vdash \varphi$	φ is a consequence of Σ, 45
$A\xi$	set of axioms, 45
$\sigma \equiv_\Sigma \tau$	σ is equivalent to τ with respect to Σ, 45
$\varphi \equiv_\Sigma \psi$	φ is equivalent to ψ with respect to Σ, 45

GENERAL ALGEBRA

$\mathfrak{A} = \langle A, Q_i \rangle_{i \in I}$	algebra, 50
$Uv\mathfrak{A}$	universe of $\mathfrak{A}$, 50
$In\mathfrak{A}$	index set of $\mathfrak{A}$, 50
$Op_i^{(\mathfrak{A})}$	i^{th} fundamental operation of $\mathfrak{A}$, 50
$\mathfrak{A}, \mathfrak{B}, \ldots$	algebras, 50

$+, \cdot, \circ$	binary operations, 50
0	zero element, 53
1	unit element, 53
∞	infinity element, 54
$\sum X$	least upper bound of X, 55
$\prod X$	greatest lower bound of X, 55
$Su\mathfrak{A}$	set of subuniverses of $\mathfrak{A}$, 58
$\mathfrak{B} \subseteq \mathfrak{A}, \mathfrak{A} \supseteq \mathfrak{B}$	$\mathfrak{A}$ is a subalgebra of $\mathfrak{B}$, 58
$\mathbf{S}\mathfrak{A}$	set of subalgebras of $\mathfrak{A}$, 58
$\mathbf{S}\mathsf{K}$	class of subalgebras of members of K, 58
$Sg^{(\mathfrak{A})}X, SgX$	subuniverse of $\mathfrak{A}$ generated by X, 61
$\mathfrak{Sg}^{(\mathfrak{A})}X, \mathfrak{Sg}X$	subalgebra of $\mathfrak{A}$ generated by X, 61
$\mathbf{S}_\alpha\mathfrak{A}$	set of subuniverses of $\mathfrak{A}$ generated by $<\alpha$ elements, 61
$\mathbf{S}_\alpha\mathsf{K}$	$\bigcup\{\mathbf{S}_\alpha\mathfrak{C} : \mathfrak{C} \in \mathsf{K}\}$, 61
$\mathbf{U}\mathsf{K}$	union of K, 63
$Ho\mathfrak{A}$	class of homomorphisms on $\mathfrak{A}$, 67
$Is\mathfrak{A}$	class of isomorphisms on $\mathfrak{A}$, 67
$h^*\mathfrak{A}$	h-image of $\mathfrak{A}$, 68
$Ho(\mathfrak{A}, \mathfrak{B})$	set of homomorphisms from $\mathfrak{A}$ onto $\mathfrak{B}$, 68
$Is(\mathfrak{A}, \mathfrak{B})$	set of isomorphisms from $\mathfrak{A}$ onto $\mathfrak{B}$, 68
$Hom(\mathfrak{A}, \mathfrak{B})$	set of homomorphisms from $\mathfrak{A}$ into $\mathfrak{B}$, 68
$Ism(\mathfrak{A}, \mathfrak{B})$	set of isomorphisms from $\mathfrak{A}$ into $\mathfrak{B}$, 68
$h: \mathfrak{A} \rightarrow \mathfrak{B}$	$h \in Hom(\mathfrak{A}, \mathfrak{B})$, 68
$\mathfrak{A} \succcurlyeq \mathfrak{B}, \mathfrak{B} \preccurlyeq \mathfrak{A}$	$\mathfrak{A}$ is homomorphic to $\mathfrak{B}$, 68
$\mathfrak{A} \cong \mathfrak{B}$	$\mathfrak{A}$ is isomorphic to $\mathfrak{B}$, 68
$\mathbf{H}\mathfrak{A}$	class of homomorphic images of $\mathfrak{A}$, 68
$\mathbf{H}\mathsf{K}$	class of homomorphic images of members of K, 68
$\mathbf{I}\mathfrak{A}$	class of isomorphic images of $\mathfrak{A}$, 68
$\mathbf{I}\mathsf{K}$	class of isomorphic images of members of K, 68
$\succcurlyeq \mid \subseteq$	relative product of $\succcurlyeq$ and $\subseteq$, 68
$\tau\mathfrak{A}$	isomorphism type of $\mathfrak{A}$, 71
$Co\mathfrak{A}$	set of congruence relations on $\mathfrak{A}$, 73
$\mathfrak{A}/R$	quotient algebra of $\mathfrak{A}$ over R, 74
$Cg^{(\mathfrak{A})}X, CgX$	congruence relation generated by X, 75
$Il_z\mathfrak{A}$	set of z-ideals of $\mathfrak{A}$, 78
$Ig_z{}^{(\mathfrak{A})}, Ig_zX$	z-ideal generated by X, 79

$\mathfrak{A}/(I, z)$	quotient algebra of $\mathfrak{A}$ over I, 80
$x/(I, z)$, $Y/(I, z)$, $S/(I, z)$, $(I, z)^\bigstar$	notions for z-ideals corresponding to notions for congruence relations, 80
$\mathsf{P}\mathfrak{A}$, $\mathsf{P}_{i\in I}\mathfrak{A}_i$	direct product of $\mathfrak{A}$, 83
$\mathfrak{C}\times\mathfrak{D}$	direct product of $\mathfrak{C}$ with $\mathfrak{D}$, 83
${}^I\mathfrak{C}$	I^{th} direct power of $\mathfrak{C}$, 83
$\mathfrak{C}\mid\mathfrak{B}$	$\mathfrak{C}$ divides $\mathfrak{B}$ directly, 83
$\mathbf{P}\mathfrak{C}$	class of isomorphic images of direct powers of $\mathfrak{C}$, 83
$\mathbf{P}\mathsf{K}$	class of isomorphic images of direct products of systems of algebras in K, 83
$\mathsf{P}\alpha$, $\mathsf{P}_{i\in I}\alpha_i$	direct product of α, 86
$\alpha\times\beta$	direct product of α, with β, 86
K_α	direct power of α, 86
$\mathfrak{B}\subseteq_{\mathsf{d}}\mathsf{P}_{i\in I}\mathfrak{A}_i$, $\mathsf{P}_{i\in I}\mathfrak{A}_i\supseteq_{\mathsf{d}}\mathfrak{B}$	$\mathfrak{B}$ is a subdirect product of $\mathfrak{A}$, 99
$\overline{F}^{(\mathfrak{A})}$	congruence relation associated with F, 106
$\mathsf{Up}\,\mathfrak{A}$	class of isomorphic images of ultrapowers of $\mathfrak{A}$, 106
$\mathsf{Up}\,\mathsf{K}$	class of isomorphic images of ultraproducts of systems of algebras in K, 106
$\mathsf{P}_{i\in I}A_i/\overline{F}$	ultraproduct of sets, 110
$\mathfrak{A}\triangleq\mathfrak{B}$	$\mathsf{Up}\,\mathfrak{A}\cap\mathsf{Up}\,\mathfrak{B}\neq 0$, 113
$\mathsf{Up}'\mathsf{K}$	class of isomorphic images of ultrapowers of members of K, 115
UfK	$\{\mathfrak{A}:\mathsf{Up}\,\mathfrak{A}\cap\mathsf{K}\neq 0\}$, 115
$\mathsf{Ps}\,\mathfrak{A}$	set of partial subalgebras of $\mathfrak{A}$, 117
$\mathsf{Ps}_\omega\,\mathfrak{A}$	set of finite partial subalgebras of $\mathfrak{A}$, 117
$\mathsf{Ps}\,\mathsf{K}$	class of partial subalgebras of members of K, 117
$\mathsf{Ps}_\omega\,\mathsf{K}$	class of finite partial subalgebras of members of K, 117
$Pl_\alpha\mathfrak{A}$	set of polynomials of rank α over $\mathfrak{A}$, 119
$\mathfrak{Pl}_\alpha\mathfrak{A}$	polynomial algebra of rank α over $\mathfrak{A}$, 119
$Pl\,\mathfrak{A}$	class of all polynomials over $\mathfrak{A}$, 119
Fr_α	set of words of $\mathfrak{Fr}_\alpha$, 130
$\mathfrak{Fr}_\alpha$	absolutely free algebra with α generators, 130
$Cr_\alpha\mathsf{K}$	congruence relation defining $\mathfrak{Fr}_\alpha\mathsf{K}$, 130
$Fr_\alpha\mathsf{K}$	universe of $\mathfrak{Fr}_\alpha\mathsf{K}$, 131
$\mathfrak{Fr}_\alpha\mathsf{K}$	free algebra over K with α generators, 131

$Cr_\alpha\mathfrak{A}$, $Fr_\alpha\mathfrak{A}$, $\mathrm{Fr}_\alpha\mathfrak{A}$	special case of above with $\mathsf{K} = \{\mathfrak{A}\}$, 131
$Pd_\alpha^{(\mathfrak{A})}$	function assigning polynomials over $\mathfrak{A}$ to words, 138
$Pd_\alpha^{(\mathfrak{A},\mathsf{K})}$	generalization of $Pd_\alpha^{(\mathfrak{A})}$, 138
$\mathfrak{Tm}_\alpha$	free term algebra, 143
$\mathfrak{Tm}_\alpha\mathsf{K}$	term algebra of K, 143
$\mathfrak{Fr}_\alpha^{(S)}\mathsf{K}$	free algebra over K with α generators and with defining relations S, 147
$Cr_\alpha^{(S)}\mathsf{K}$	congruence relation defining $\mathfrak{Fr}_\alpha^{(S)}\mathsf{K}$, 147
$\mathfrak{Rd}_J^{(r)}\mathfrak{A}$, $\mathfrak{Rd}^{(r)}\mathfrak{A}$	r-reduct of $\mathfrak{A}$, 149
$\mathfrak{Rd}_J\mathfrak{A}$	J-reduct of $\mathfrak{A}$, 149
$\mathsf{Rd}_J^{(r)}\mathsf{K}$	class of r-reducts of members of K, 149
$\mathsf{Rd}_J\mathsf{K}$	class of J-reducts of members of K, 149
$\mathfrak{B} \subseteq^{\mathrm{r}} \mathfrak{A}$, $\mathfrak{A} \supseteq^{\mathrm{r}} \mathfrak{B}$	$\mathfrak{B}$ is a subreduct of $\mathfrak{A}$, 151
$\mathsf{U}^{\mathrm{r}}\mathsf{K}$	reduct union of K, 151
$\mathfrak{Rd}^{(r,s)}\mathfrak{A}$, $\mathfrak{Rd}_{I',J'}\mathfrak{A}$	generalization of above, 155
$\mathsf{Rd}^{(r,s)}\mathsf{L}$, $\mathsf{Rd}_{I',J'}\mathsf{L}$	generalization of above, 155

CYLINDRIC ALGEBRAS

$\langle A, +, \cdot, -, 0, 1\rangle$	Boolean algebra, 161
$x - y$	difference of x and y, 161
$x \oplus y$	symmetric difference of x and y, 161
BA	class of all Boolean algebras, 161
$\sum^{(\mathfrak{A})}$, $\sum$	infinitary generalization of $+$, 162
$\prod^{(\mathfrak{A})}$, $\prod$	infinitary generalization of $\cdot$, 162
$\langle A, +, \cdot, -, 0, 1, \mathsf{c}_\kappa, \mathsf{d}_{\kappa\lambda}\rangle_{\kappa,\lambda<\alpha}$	cylindric algebra, 162
c_κ	κ^{th} cylindrification, 162
$\mathsf{d}_{\kappa\lambda}$	κ, λ-diagonal element, 162
$(C_0) - (C_7)$	postulates for cylindric algebras, 162
CA	class of all cylindric algebras, 163
CA_α	class of all cylindric algebras of dimension α, 163
$\mathfrak{Bl}\mathfrak{A}$	Boolean part of $\mathfrak{A}$, 163, 341
Df_α	class of diagonal-free cylindric algebras of dimension α, 164
$\mathfrak{Df}\mathfrak{A}$	diagonal-free part of $\mathfrak{A}$, 164
$\boldsymbol{C}_\kappa^{(U,\alpha)}$, $\boldsymbol{C}_\kappa^{(U)}$, $\boldsymbol{C}_\kappa$	set-theoretical cylindrification, 166

$\mathbf{D}_{\kappa\lambda}^{(U,\alpha)}$, $\mathbf{D}_{\kappa\lambda}^{(U)}$, $\mathbf{D}_{\kappa\lambda}$	set-theoretical diagonal element, 166
$\mathfrak{Fm}^{(\Lambda)}$	free algebra of formulas in Λ, 168
$\mathbf{C}_{\kappa}^{[V]}$	generalized set-theoretical cylindrification, 171
$\mathbf{D}_{\kappa\lambda}^{[V]}$	generalized set-theoretical diagonal element, 171
$\mathfrak{A}^{\partial}$, $\mathsf{c}_{\kappa}^{\partial}$	duals of usual notions, 185ff.
$\mathsf{s}_{\lambda}^{\kappa}$	substitution operation, λ for κ, 189
${}_{\mu}\mathsf{s}(\kappa,\lambda)$	substitution operation, interchanging κ and λ, 191
$\Delta^{(\mathfrak{A})}x$, Δx	dimension set of x, 199
$Cl_{\Gamma}\mathfrak{A}$	set of (Γ)-closed elements of $\mathfrak{A}$, 203
$\mathfrak{Cl}_{\Gamma}\mathfrak{A}$	Boolean algebra of (Γ)-closed elements of $\mathfrak{A}$, 203
$Zd\mathfrak{A}$	set of zero-dimensional elements of $\mathfrak{A}$, 203
$\mathfrak{Zd}\mathfrak{A}$	Boolean algebra of zero-dimensional elements of $\mathfrak{A}$, 203
$\mathsf{c}_{(\Gamma)}$	generalized cylindrification, 205
$\mathbf{C}_{(\Gamma)}$	generalized set-theoretical cylindrification, 205
d_{Γ}	generalized diagonal element, 209
$\mathbf{D}_{\Gamma}$	generalized set-theoretical diagonal element, 209
$\breve{\mathsf{d}}R$	generalized co-diagonal element, 215
Lf_{α}	class of all locally finite CA's of dimension α, 231
Dc_{α}	class of all dimension-complemented CA's of dimension α, 231
s, $\mathsf{s}^{(\mathfrak{A})}$	substitution operator, 236
s^{+}	generalized substitution operator, 241
$\mathsf{c}_{(\Gamma)}^{+}$	generalized cylindrification, 241
Mn_{α}	class of all minimal CA's of dimension α, 254
$Rl_{b}\mathfrak{A}$	universe of $\mathfrak{Rl}_{b}\mathfrak{A}$, 261
$\mathfrak{Rl}_{b}\mathfrak{A}$	cylindric-relativized algebra, 261
Cr_{α}	class of all cylindric-relativized algebras of dimension α, 261
${}^{\alpha}U^{(p)}$	weak Cartesian space, 267
L_{α}	class of all weakly separably countably complete Lf_{α}'s, 311
${}_{\mathsf{l}}\mathsf{P}\mathfrak{A}$, ${}_{\mathsf{l}}\mathsf{P}_{i\in I}\mathfrak{A}_i$	Lf-direct product, 311

C_α	class of all separably countably complete CA_α's, 313
Ss_α	class of all semisimple CA_α's, 326
H_α	class of all hereditarily non-discrete CA_α's, 343
$Cr_\beta^{(\Delta)}\mathsf{K}$	congruence relation to define $\mathfrak{Fr}_\beta^{(\Delta)}\mathsf{K}$, 347
$\mathfrak{Fr}_\beta^{(\Delta)}\mathsf{K}$	dimension-restricted free algebra over K with β generators, 348
$\langle\gamma\mathfrak{B}, \boldsymbol{\kappa}\mathfrak{B}\rangle$	characteristic pair of $\mathfrak{B}$, 365
$\boldsymbol{\delta}\mathfrak{B}$	union of all order-types of well-ordered sets $\subseteq B$, 368
$\boldsymbol{\varepsilon}\mathfrak{B}$	union of all $\boldsymbol{\delta}\mathfrak{Rl}_x\mathfrak{B}$ with $\mathfrak{Rl}_x\mathfrak{B}$ of class 1, 368
$\mathfrak{Rd}_\alpha^{(\rho)}\mathfrak{B}$, $\mathfrak{Rd}^{(\rho)}\mathfrak{B}$	ρ-reduct of $\mathfrak{B}$, 381
$\mathfrak{Rd}_\alpha\mathfrak{B}$	α-reduct of $\mathfrak{B}$, 381
$\mathsf{Rd}_\alpha^{(\rho)}\mathsf{K}$	class of all ρ-reducts of members of K, 381
$\mathsf{Rd}_\alpha\mathsf{K}$	class of all α-reducts of members of K, 382
$\mathfrak{C} \subseteq^{\mathrm{r}} \mathfrak{B}$	$\mathfrak{C}$ is a subreduct of $\mathfrak{B}$, 382
$Nr_\alpha\mathfrak{B}$	$Cl_{\beta\sim\alpha}\mathfrak{B}$, 401
$\mathfrak{Nr}_\alpha\mathfrak{B}$	neat α-reduct of $\mathfrak{B}$, 401
$\mathsf{Nr}_\alpha\mathsf{K}$	class of all neat α-reducts of members of K, 401
Bo	class of all Boolean algebras with operators, 430
Bo_α	class of all α-dimensional Boolean algebras with operators, 430
$Em\mathfrak{A}$	universe of $\mathfrak{Em}\mathfrak{A}$, 432
$\mathfrak{Em}\mathfrak{A}$	canonical embedding algebra of $\mathfrak{A}$, 432
$em^{(\mathfrak{A})}$, em	canonical embedding function, 432
$\mathfrak{Pr}\mathfrak{A}$	positive reduct of $\mathfrak{A}$, 435
$\mathfrak{At}\mathfrak{A}$	atom structure of $\mathfrak{A}$, 453
$\mathfrak{Cm}\mathfrak{B}$	complex algebra of $\mathfrak{B}$, 453
Ca	class of all cylindric atom structures, 455
Ca_α	class of all α-dimensional cylindric atom structures, 455

INDEX OF NAMES AND SUBJECTS

The main reference to a subject is given in boldface type.

D

E

T

U

Lightning Source UK Ltd.
Milton Keynes UK
UKOW06n1303110516

274039UK00001B/12/P

9 780720 420432